T0180082

Communications
in Computer and Information Science

763

Commenced Publication in 2007
Founding and Former Series Editors:
Alfredo Cuzzocrea, Orhun Kara, Dominik Ślęzak, and Xiaokang Yang

More information about this series at http://www.springer.com/series/7899

Kang Li · Yusheng Xue · Shumei Cui
Qun Niu · Zhile Yang · Patrick Luk (Eds.)

Advanced Computational Methods in Energy, Power, Electric Vehicles, and Their Integration

International Conference on Life System Modeling
and Simulation, LSMS 2017
and International Conference on Intelligent Computing
for Sustainable Energy and Environment, ICSEE 2017
Nanjing, China, September 22–24, 2017
Proceedings, Part III

 Springer

Editors
Kang Li
Queen's University Belfast
Belfast
UK

Qun Niu
Shanghai University
Shanghai
China

Yusheng Xue
Nanjing Automation Research Institute
Nanjing
China

Zhile Yang
Queen's University Belfast
Belfast
UK

Shumei Cui
Harbin Institute of Technology
Harbin
China

Patrick Luk
Cranfield University
Bedford
UK

ISSN 1865-0929   ISSN 1865-0937 (electronic)
Communications in Computer and Information Science
ISBN 978-981-10-6363-3   ISBN 978-981-10-6364-0 (eBook)
DOI 10.1007/978-981-10-6364-0

Library of Congress Control Number: 2017952379

Printed on acid-free paper

This Springer imprint is published by Springer Nature
The registered company is Springer Nature Singapore Pte Ltd.
The registered company address is: 152 Beach Road, #21-01/04 Gateway East, Singapore 189721, Singapore

Preface

This book constitutes the proceedings of the 2017 International Conference on Life System Modeling and Simulation (LSMS 2017) and the 2017 International Conference on Intelligent Computing for Sustainable Energy and Environment (ICSEE 2017), which were held during September 22–24, in Nanjing, China. These two international conference series aim to bring together international researchers and practitioners in the fields of advanced methods for life system modeling and simulation as well as advanced intelligent computing theory and methodologies and engineering applications for sustainable energy and environment. The two conferences held this year were built on the success of previous LSMS and ICSEE conferences held in Shanghai and Wuxi, respectively. The success of the LSMS and ICSEE conference series were also based on several large-scale RCUK/NSFC funded UK–China collaborative projects on sustainable energy and environment, as well as a recent government funded project on the establishment of the UK-China University Consortium in Engineering Education and Research, with an initial focus on sustainable energy and intelligent manufacturing.

At LSMS 2017 and ICSEE 2017, technical exchanges within the research community took the form of keynote speeches, panel discussions, as well as oral and poster presentations. In particular, two workshops, namely, the Workshop on Smart Grid and Electric Vehicles and the Workshop on Communication and Control for Distributed Networked Systems, were held in parallel with LSMS 2017 and ICSEE 2017, focusing on the two recent hot topics on green and sustainable energy systems and electric vehicles and distributed networked systems for the Internet of Things.

The LSMS 2017 and ICSEE 2017 conferences received over 625 submissions from 14 countries and regions. All papers went through a rigorous peer review procedure and each paper received at least three review reports. Based on the review reports, the Program Committee finally selected 208 high-quality papers for presentation at LSMS 2017 and ICSEE 2017. These papers cover 22 topics, and are included in three volumes of CCIS proceedings published by Springer. This volume of CCIS includes 79 papers covering 7 relevant topics.

Located at the heartland of the wealthy lower Yangtze River region in China and being the capital of several dynasties, kingdoms, and republican governments dating back to the 3rd century, Nanjing has long been a major center of culture, education, research, politics, economy, transport networks, and tourism. In addition to academic exchanges, participants were treated to a series of social events, including receptions and networking sessions, which served to build new connections, foster friendships, and forge collaborations. The organizers of LSMS 2017 and ICSEE 2017 would like to acknowledge the enormous contribution of the Advisory Committee, who provided guidance and advice, the Program Committee and the numerous referees for their efforts in reviewing and soliciting the papers, and the Publication Committee for their editorial work. We would also like to thank the editorial team from Springer for their support and guidance. Particular thanks are of course due to all the authors, as without their high-quality submissions and presentations the conferences would not have been successful.

Finally, we would like to express our gratitude to our sponsors and organizers, listed on the following pages.

September 2017

Bo Hu Li
Sarah Spurgeon
Mitsuo Umezu
Minrui Fei
Kang Li
Dong Yue
Qinglong Han
Shiwei Ma
Luonan Chen
Sean McLoone

Organization

Sponsors

China Simulation Federation (CSF), China
Chinese Association for Artificial Intelligence (CAAI), China
IEEE Systems, Man & Cybernetics Society Technical Committee on Systems Biology,
  USA
IEEE CC Ireland Chapter, Ireland

Technical Support Organization

National Natural Science Foundation of China (NSFC), China

Organizers

Shanghai University, China
Queen's University Belfast, UK
Nanjing University of Posts and Telecommunications, China
Southeast University, China
Life System Modeling and Simulation Technical Committee of CSF, China
Embedded Instrument and System Technical Committee of China Instrument
  and Control Society, China
Intelligent Control and Intelligent Management Technical Committee of CAAI, China

Co-sponsors

Shanghai Association for System Simulation, China
Shanghai Association of Automation, China
Shanghai Instrument and Control Society, China
Jiangsu Association of Automation, China

Co-organizers

Swinburne University of Technology, Australia
Queensland University of Technology, Australia
Tsinghua University, China
Harbin Institute of Technology, China
China State Grid Electric Power Research Institute, China
Chongqing University, China
University of Essex, UK
Cranfield University, UK
Peking University, China

Nantong University, China
Shanghai Dianji University, China
Jiangsu Engineering Laboratory of Big Data Analysis and Control for Active
  Distribution Network, China
Shanghai Key Laboratory of Power Station Automation Technology, China

Honorary Chairs

Li, Bo Hu, China
Spurgeon, Sarah, UK
Umezu, Mitsuo, Japan

Advisory Committee Members

Bai, Erwei, USA
Ge, Shuzhi, Singapore
He, Haibo, USA
Hu, Huosheng, UK
Huang, Biao, Canada
Hussain, Amir, UK
Liu, Derong, USA
Mi, Chris, USA

Nikolopoulos,
  Dimitrios S., UK
Pardalos, Panos M., USA
Pedrycz, Witold, Canada
Polycarpou, Marios M.,
  Cyprus
Qin, Joe, HK
Scott, Stan, UK

Tan, KC, Singapore
Tassou, Savvas, UK
Thompson, Stephen, UK
Wang, Jun, HK
Wang, Zidong, UK
Wu, Qinghua, China
Xue, Yusheng, China
Zhang, Lin, China

General Chairs

Fei, Minrui, China
Li, Kang, UK
Yue, Dong, China

International Program Committee

Chairs

Chen, Luonan, Japan
Han, Qinglong, Australia
Ma, Shiwei, China
McLoone, Sean, UK

Local Chairs

Chiu, Min-Sen, Singapore
Cui, Shumei, China
Deng, Mingcong, Japan
Ding, Yongsheng, China
Ding, Zhengtao, UK
Fang, Qing, Japan

Fridman, Emilia, Israel
Gao, Furong, HK
Gu, Xingsheng, China
Guerrero, Josep M.,
  Demark
Gupta, Madan M., Canada

Hunger, Axel, Germany
Lam, Hak-Keung, UK
Liu, Wanquan, Australia
Luk, Patrick, UK
Maione, Guido, Italy
Park, Jessie, Korea

Peng, Chen, China
Su, Zhou, China
Tian, Yuchu, Australia
Xu, Peter, New Zealand

Yang, Taicheng, UK
Yu, Wen, Mexico
Zeng, Xiaojun, UK
Zhang, Huaguang, China

Zhang, Jianhua, China
Zhang, Wenjun, Canada
Zhao, Dongbin, China

Members

Andreasson, Stefan, UK
Adamatzky, Andy, UK
Altrock, Philipp, USA
Asirvadam, Vijay S.,
    Malaysia
Baig, Hasan, UK
Baker, Lucy, UK
Barry, John, UK
Best, Robert, UK
Bu, Xiongzhu, China
Cao, Jun, UK
Cao, Yi, UK
Chang, Xiaoming, China
Chen, Jing, China
Chen, Ling, China
Chen, Qigong, China
Chen, Rongbao, China
Chen, Wenhua, UK
Cotton, Matthew, UK
Deng, Jing, UK
Deng, Li, China
Deng, Shuai, China
Deng, Song, China
Deng, Weihua, China
Ding, Yate, UK
Ding, Zhigang, China
Du, Dajun, China
Du, Xiangyang, China
Ellis, Geraint, UK
Fang, Dongfeng, USA
Feng, Dongqing, China
Feng, Zhiguo, China
Foley, Aoife, UK
Fu, Jingqi, China
Gao, Shouwei, China
Gu, Dongbin, UK
Gu, Juping, China
Gu, Zhou, China
Guo, Lingzhong, UK

Han, Bo, China
Han, Xuezheng, China
Heiland, Jan, Germany
Hong, Xia, UK
Hou, Weiyan, China
Hu, Liangjian, China
Hu, Qingxi, China
Hu, Sideng, China
Huang, Sunan, Singapore
Huang, Wenjun, China
Hwang, Tan Teng,
    Malaysia
Jia, Dongyao, UK
Jiang, Lin, UK
Jiang, Ming, China
Jiang, Ping, China
Jiang, Yucheng, China
Kuo, Youngwook, UK
Laverty, David, UK
Li, Chuanfeng, China
Li, Chuanjiang, China
Li, Dewei, China
Li, Donghai, China
Li, Guozheng, China
Li, Jingzhao, China
Li, Ning, China
Li, Tao, China
Li, Tongtao, China
Li, Weixing, China
Li, Xin, China
Li, Xinghua, China
Li, Yunze, China
Li, Zhengping, China
Lin, Zhihao, China
Lino, Paolo, Italy
Liu, Chao, France
Liu, Guoqiang, China
Liu, Mandan, China
Liu, Shirong, China

Liu, Shujun, China
Liu, Tingzhang, China
Liu, Xianzhong, China
Liu, Yang, China
Liu, Yunhuai, China
Liu, Zhen, China
Ljubo, Vlacic, Australia
Lu, Ning, Canada
Luan, Tom, Australia
Luo, Jianfei, China
Ma, Hongjun, China
McAfee, Marion, Ireland
Menary, Gary, UK
Meng, Xianhai, UK
Menhas, Muhammad
    Ilyas, Pakistan
Menzies, Gillian, UK
Naeem, Wasif, UK
Nie, Shengdong, China
Niu, Yuguang, China
Nyugen, Bao Kha, UK
Ouyang, Mingsan, China
Oyinlola, Muyiwa, UK
Pan, Hui, China
Pan, Ying, China
Phan, Anh, UK
Qadrdan, Meysam, UK
Qian, Hua, China
Qu, Yanbin, China
Raszewski, Slawomir, UK
Ren, Wei, China
Rivotti, Pedro, UK
Rong, Qiguo, China
Shao, Chenxi, China
Shi, Yuntao, China
Smyth, Beatrice, UK
Song, Shiji, China
Song, Yang, China
Su, Hongye, China

Sun, Guangming, China
Sun, Xin, China
Sun, Zhiqiang, China
Tang, Xiaoqing, UK
Teng, Fei, UK
Teng, Huaqiang, China
Trung, Dong, UK
Tu, Xiaowei, China
Vlacic, Ljubo, UK
Wang, Gang, China
Wang, Jianzhong, China
Wang, Jihong, UK
Wang, Ling, China
Wang, Mingshun, China
Wang, Shuangxin, China
Wang, Songyan, China
Wang, Yaonan, China
Wei, Kaixia, China
Wei, Lisheng, China
Wei, Mingshan, China
Wen, Guihua, China
Wu, Jianguo, China
Wu, Jianzhong, UK

Wu, Lingyun, China
Wu, Zhongcheng, China
Xie, Hui, China
Xu, Sheng, China
Xu, Wei, China
Xu, Xiandong, UK
Yan, Huaicheng, China
Yan, Jin, UK
Yang, Aolei, China
Yang, Kan, USA
Yang, Shuanghua, UK
Yang, Wankou, China
Yang, Wenqiang, China
Yang, Zhile, UK
Yang, Zhixin, Macau
Ye, Dan, China
You, Keyou, China
Yu, Ansheng, China
Yu, Dingli, UK
Yu, Hongnian, UK
Yu, Xin, China
Yuan, Jin, China
Yuan, Jingqi, China

Yue, Hong, UK
Zeng, Xiaojun, UK
Zhang, Dengfeng, China
Zhang, Hongguang, China
Zhang, Jian, China
Zhang, Jingjing, UK
Zhang, Lidong, China
Zhang, Long, UK
Zhang, Qianfan, China
Zhang, Xiaolei, UK
Zhang, Yunong, China
Zhao, Dongya, China
Zhao, jun, China
Zhao, Wanqing, UK
Zhao, Xiaodong, UK
Zhao, Xingang, China
Zheng, Xiaojun, UK
Zhou, Huiyu, UK
Zhou, Wenju, China
Zhou, Yu, China
Zhu, Yunpu, China
Zong, Yi, Demark
Zuo, Kaizhong, China

Organization Committee

Chairs

Li, Xin, China
Wu, Yunjie, China
Naeem, Wasif, UK
Zhang, Tengfei, China
Cao, Xianghui, China

Members

Chen, Ling, China
Deng, Li, China
Du, Dajun, China
Jia, Li, China
Song, Yang, China
Sun, Xin, China
Xu, Xiandong, China
Yang, Aolei, China
Yang, Banghua, China
Zheng, Min, China
Zhou, Peng, China

Special Session Chairs

Wang, Ling, China
Meng, Fanlin, UK

Publication Chairs

Zhou, Huiyu, UK
Niu, Qun, China

Publicity Chairs

Jia, Li, China
Yang, Erfu, UK

Registration Chairs

Song, Yang, China
Deng, Li, China

Secretary-General

Sun, Xin, China
Wu, Songsong, China
Yang, Zhile, UK

Contents

Intelligent Methods for Energy Saving and Pollution Reduction

Intelligent Methods in Developing Electric Vehicles, Engines and Equipment

Optimization Methods

Computational Methods for Sustainable Environment

Computational Intelligence in Utilization of Clean and Renewable Energy Resources

Research on Wind Speed Vertical Extrapolation Based on Extreme Learning Machine

Hui Lv and Guochu Chen[✉]

Electric Engineering School,
Shanghai DianJi University, Shanghai 201306, China
770440679@qq.com, chengc@sdju.edu.cn

Abstract. In engineering, the method of wind speed vertical extrapolation is based on the actual wind data of the wind tower, and the wind shear index is used to calculate the wind speed at any height in the near ground. The wind shear index is only considered in the neutral state of the atmosphere, without considering the impact of atmospheric stability on the wind shear index, which has some limitations. At the same time, the calculation of the wind shear index is a rather complicated task when considering the atmospheric stability. In order to solve these problems, this paper puts forward to use extreme learning machine for fitting the relationship between wind speed at different heights. Extreme learning machine has the advantages of fast learning speed, good generalization ability and so on. In this paper, the results obtained by the extreme learning machine and traditional methods are compared with the measured values. The results show that the extreme learning machine has a better application prospect in the vertical wind speed extrapolation.

Keywords: Extreme learning machine · Wind speed · Vertical extrapolation · Wind shear index

1 Introduction

The accurate calculation of the wind resource data at pre assembled hub height is a prerequisite for the evaluation of wind farm power generation. In the feasibility study of wind farm, wind turbine power generation should be calculated according to the wind speed and wind directions at the height of the wind turbine hub. But in the actual wind measurement, the height of the wind measuring instrument can not completely meet the installation height of the wind turbine. Therefore, when estimating the generating capacity of a wind turbine, it is necessary to use wind shear index to extrapolate the wind speed at any height of the near ground according to the actual wind measurement data of the wind tower [1, 7]. The measured wind speed is as close as possible to the wind speed at the hub height. For the existing practice, the wind shear index is calculated according to the measured wind speed at 2 different heights. However, this method may result in large errors in the calculation results. The main reason is that the wind shear index is only considered in the neutral state of the atmosphere, without considering the impact of atmospheric stability on the wind shear index.

© Springer Nature Singapore Pte Ltd. 2017
K. Li et al. (Eds.): LSMS/ICSEE 2017, Part III, CCIS 763, pp. 3–11, 2017.
DOI: 10.1007/978-981-10-6364-0_1

The wind shear index is affected by the atmospheric stability, which has been proved by more and more domestic and foreign scholars. It is pointed out in the document [2] that when the atmosphere is stable, the speed of wind speed increasing with the increase of height is faster than logarithmic relationship in the neutral atmosphere. Based on the M-O similarity theory and a series of studies, Panofsky and Dutton proposed an empirical formula for evaluating the wind shear index using the atmospheric stability function and the roughness. The wind shear index calculated by the empirical formula is more in line with the actual situation. The atmosphere is from stable to neutral to unstable, and the wind shear index decreases in turn [3]. However, it is a very complex task to obtain the wind shear index when considering the atmospheric stability. Therefore, this method has not been applied in engineering practice.

Aiming at these problems, this paper puts forward the use of Extreme Learning Machine (ELM) to fit the relationship between wind speed at each height. In this paper, the wind speed extrapolated by ELM and traditional methods is compared with the measured wind speed and the vertical extrapolation of wind speed is studied.

2 Wind Shear Index

The wind shear index is a comprehensive parameter to characterize the variation of wind speed with height, atmospheric stability, surface roughness and so on. In practical engineering applications, the wind shear index is obtained by only considering the atmosphere in a neutral state. At this time, the turbulence will depend entirely on the dynamic factors, and the variation of wind speed with height follows Prandtl's empirical formula:

$$u(z) = \frac{u^*}{k} \ln \frac{z}{z_0} \tag{1}$$

In the formula, u^* is the friction velocity; k is the von Karman constant, $k = 0.4$. Friction velocity is defined as: $u^* = (\tau/\rho)^{1/2}$, τ is the surface shear stress, ρ is the air density.

The power exponent formula is derived as follows:

$$u_2 = u_1 \left(\frac{z_2}{z_1}\right)^{\alpha} \tag{2}$$

$$\alpha = \ln(u_1/u_2)/\ln(z_2/z_1) \tag{3}$$

In the formula, u_2 and u_1 are wind speeds at height z_2 and z_1, α is the wind shear index.

In the flat interior, the increase of wind speed with height is submitted to exponential law. But in some areas affected by terrain factors, the wind speed will reach the peak value at a certain height, and the wind shear index may even be negative. The main reason for this phenomenon is the terrain effect. When the airflow through the mountains, due to the obstruction of the terrain, the airflow changes, some of the airflow flows over the top of the mountain, some of the air flows around both sides of the mountain.

The more unstable the atmosphere, the more the air flows over, the less the air flows around. There is a updraft in the windward side of the mountain. At the top of the hill and on both sides, as the streamline is dense, the wind speeds up. On the leeward side of the mountain, the wind speed is weakening due to streamline divergence [4].

Based on the M-O similarity theory and a series of studies, Panofsky and Dutton proposed an empirical formula for evaluating the wind shear index using the atmospheric stability function and the roughness [6, 8]:

$$\alpha = \phi_m(z/L)/[\ln(z/z_0) - \psi_m(z/L)] \tag{4}$$

In the formula, z is height, L is M-O length, ϕ_m and ψ_m are M-O functions.

(Stable atmosphere: $\phi_m = 1 + 5\,(z/L)$; Unstable atmosphere: $\phi_m = [1 - 16\,(z/L)]^{-1/4}$; Neutral atmosphere: $\phi_m = 1$).

3 Extreme Learning Machine

3.1 ELM Principle

ELM is developed from a single hidden layer feed forward neural network (SLFN) [5]. ELM has many advantages: fast learning speed; less intervention. The weights between the input layer and the hidden layer of the ELM and the bias of the hidden layer neuron are given randomly, and do not need to adjust during the training process, which greatly improves the training time of the algorithm. Set N different random samples $(\vec{x}_i,\ \vec{t}_i)$, $\vec{x}_i = [x_{i1},\ x_{i2}, \ldots x_{in}]^T \in R^n$, $\vec{t}_i = [t_{i1},\ t_{i2}, \ldots t_{im}]^T \in R^n$. Single hidden layer node number is $\tilde{N}$ and the excitation function is $g(x)$. Standard SLFN model:

$$\sum_{i=1}^{\tilde{N}} \beta_i g_i(x_j) = \sum_{i=1}^{\tilde{N}} \beta_i g(a_i \bullet x_j + b_i) \quad j = 1, \ldots, N \tag{5}$$

In the formula, $\vec{a}_i = [a_{i1},\ a_{i2}, \ldots a_{in}]^T$ is the input weight vector that connects the first i hidden layer node; b_i is the threshold of the first i hidden layer node; $\vec{\beta}_i = [\beta_{i1},\ \beta_{i2}, \ldots \beta_{im}]^T$ is the output weight vector that connects the first i hidden layer node; $a_i \bullet x_j$ represents the inner product of $\vec{a}_i$ and $\vec{x}_j$.

Standard SLFN can zero error approximate N training samples, and the presence of $\vec{\beta}_i, \vec{a}_i, \vec{x}_j$ can make the following formula true:

$$\sum_{i=1}^{\tilde{N}} \beta_i g(a_i \bullet x_j + b_i) = t_j, j = 1, \ldots, N \tag{6}$$

Formula (6) written in matrix form:

$$\mathbf{H\beta} = \mathbf{T} \tag{7}$$

In the formula: $\mathbf{H}$ is the hidden layer output matrix. Its row i represents the hidden layer output of the first i training sample, column j represents the hidden layer output of the first j hidden layer node of the relative input $x_1, x_2, \cdots, x_N$.

$$H = \begin{bmatrix} G(a_1,b_1,x_1) & \cdots & G(a_{\tilde{N}},b_{\tilde{N}},x_1) \\ G(a_1,b_1,x_2) & \cdots & G(a_{\tilde{N}},b_{\tilde{N}},x_2) \\ \vdots & \vdots & \vdots \\ G(a_1,b_1,x_N) & \cdots & G(a_{\tilde{N}},b_{\tilde{N}},x_N) \end{bmatrix}_{N \times \tilde{N}} = \begin{bmatrix} g(a_1 \cdot x_1 + b_1) & \cdots & g(a_{\tilde{N}} \cdot x_1 + b_{\tilde{N}}) \\ g(a_1 \cdot x_2 + b_1) & \cdots & g(a_{\tilde{N}} \cdot x_2 + b_{\tilde{N}}) \\ \vdots & \vdots & \vdots \\ g(a_1 \cdot x_N + b_1) & \cdots & g(a_{\tilde{N}} \cdot x_N + b_{\tilde{N}}) \end{bmatrix}_{N \times \tilde{N}}$$

$$\beta = \begin{bmatrix} \beta_{11} & \beta_{12} & \cdots & \beta_{1m} \\ \beta_{21} & \beta_{22} & \cdots & \beta_{2m} \\ \vdots & \vdots & \vdots & \vdots \\ \beta_{\tilde{N}1} & \beta_{\tilde{N}2} & \cdots & \beta_{\tilde{N}m} \end{bmatrix}_{\tilde{N} \times m}, \quad T = \begin{bmatrix} t_{11} & t_{12} & \cdots & t_{1m} \\ t_{21} & t_{22} & \cdots & t_{2m} \\ \vdots & \vdots & \vdots & \vdots \\ t_{N1} & t_{N2} & \cdots & t_{Nm} \end{bmatrix}_{N \times m}$$

$\mathbf{H}$ and $\mathbf{T}$ are determined by the matrix, so the learning process of ELM is the process of solving the output weights $\boldsymbol{\beta}$. According to the document [5], the formula (7) can be expressed as:

$$\hat{\boldsymbol{\beta}} = \mathbf{H}^+ \mathbf{T} \tag{8}$$

In the formula: $\mathbf{H}^+$ is a generalized inverse matrix of the matrix $\mathbf{H}$ of the hidden layer.

3.2 Establishment of ELM Model

The purpose of this paper is to solve the problem of vertical extrapolation of wind speed. The ELM model creation process can be described as: First, two different height wind speed data at the same time are used as training sets for ELM training. Then, the wind speed data at a certain height during a certain period are used as testing sets for ELM testing, and the results are the fitting wind speed of the required height at the same time. Finally, the wind speed is compared with the actual wind speed, and the performance of ELM is evaluated. Specific steps are shown in Fig. 1:

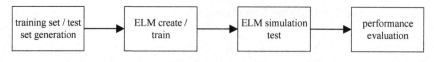

Fig. 1. Modeling steps

The training and testing procedures for ELM are shown in Figs. 2 and 3 respectively:

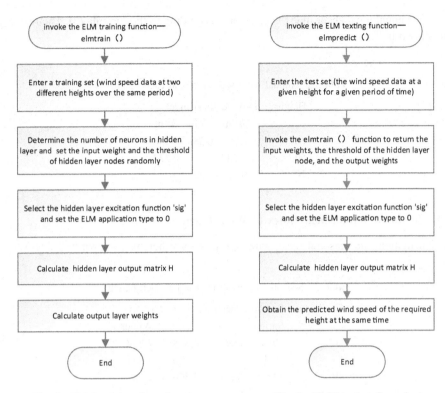

Fig. 2. ELM training flow chart   **Fig. 3.** ELM testing flow chart

4   Case Study and Analysis

4.1   Data Sources

The object of this case is a wind farm in Kaifeng Weishi of Henan, which belongs to the mountainous terrain. According to the wind data of the wind farm provided by a certain wind power company, some measured wind speed data of No.1 wind tower are selected as the original data. The chosen time is from March 1, 2013 to February 28, 2014. The height options are 70 m, 50 m and 10 m. The anemometer outputs data every 10 min. The complete rate of wind data is 99.7% and the integrity is good.

4.2   Annual Wind Shear Index of Wind Farm

By calculation, the annual mean wind speed of each height is shown in Table 1:

Table 1.   Annual mean wind speed at each height

Height (m)	Annual mean wind speed (m/s)
70	5.012
50	4.397
10	2.609

The annual wind shear index is calculated according to the formula (3). The results are shown in Table 2:

Table 2. Annual wind shear index at each height

Height (m)	Wind shear index		
	70 m	50 m	10 m
70	–	0.389	0.336
50	0.389	–	0.324
10	0.336	0.324	–

4.3   Seasonal Wind Shear Index of Wind Farm

By calculation, the seasonal mean wind speed of each height is shown in Table 3:

Table 3. Seasonal mean wind speed at each height

Height (m)	Seasonal mean wind speed (m/s)			
	Spring	Summer	Autumn	Winter
70	5.638	4.692	4.835	4.899
50	4.948	4.067	4.236	4.354
10	3.069	2.254	2.360	2.769

The seasonal wind shear index is calculated according to the formula (3). The results are shown in Table 4:

Table 4. Seasonal wind shear index at each height

Height (m)	Wind shear index (spring)			Wind shear index (summer)			Wind shear index (autumn)			Wind shear index (winter)		
	70	50	10	70	50	10	70	50	10	70	50	10
70	–	0.388	0.313	–	0.425	0.377	–	0.393	0.369	–	0.351	0.293
50	0.388	–	0.297	0.425	–	0.367	0.393	–	0.363	0.351	–	0.281
10	0.313	0.297	–	0.377	0.367	–	0.369	0.363	–	0.293	0.281	–

4.4   Wind Speed Vertical Extrapolation

Using the following three methods for wind speed vertical extrapolation, and the results are compared with the measured values. Height is from 50 m to 70 m. The time is: March 5th 14:50–23:50 (time period one), June 5th 14:50–23:50 (time period two), September 14th 14:50–23:50 (time period three), December 5th 14:50–23:50 (time period four).

(1) Method one: Using the annual mean wind speed data to extrapolate wind speed of the required height according to formula (2).
(2) Method two: Using the seasonal mean wind speed data to extrapolate wind speed of the required height according to formula (2).
(3) Method three: Using ELM algorithm to fit wind speed of each height.

4.4.1 Wind Speed Comparison Chart

Two charts of comparison between the extrapolated wind speed and the measured value are shown as follows:

It can be seen from Figs. 4 and 5 that the wind speed extrapolated by these three methods can well reflect the measured value. The results of method one and method two are almost the same, while the wind speed extrapolated by method three (ELM) is closer to the measured value.

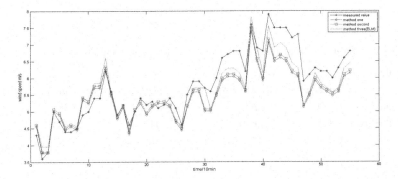

Fig. 4. Comparison of extrapolated wind speed and measured value in the time period two

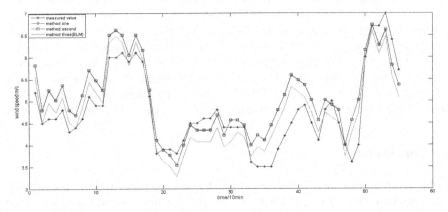

Fig. 5. Comparison of extrapolated wind speed and measured value in the time period three

4.4.2 Performance Evaluation

(1) Evaluation Standard

The mean relative error (MRE) is the average of ratio of the absolute error to the actual wind speed. The formula is as follows:

$$MRE = \frac{1}{n}\sum_{i=1}^{n}\left|\frac{Y_i - \widehat{Y}_i}{Y_i}\right| \tag{9}$$

In the formula, Y_i is the actual wind speed of the moment i, $\widehat{Y}_i$ is the predicted wind speed of the moment i.

(2) The mean relative error is shown in Table 5:

As can be seen from Table 5, the method one and the method two have the same fitting effect, while the method three (ELM) has a better fitting effect than the other two methods.

Table 5. Mean relative error of each method in each time period

Method	Average relative error (%)			
	Time period one	Time period two	Time period three	Time period four
Method one	6.184	6.997	8.852	7.587
Method two	6.185	6.483	9.047	7.324
ELM	6.232	5.572	7.313	6.994

4.5 Case Analysis

As can be seen from Table 5, the average relative error of the three methods is almost the same in the time period one. This may be in the time period one, the atmosphere is relatively close to the neutral state. Therefore, the wind speed extrapolated by formulas (2) and (3) is close to the measured wind speed. In the period of two or three and four, the average relative error of method three (ELM) is less than method one and method two. It is possible that during these periods, the atmosphere is in a state of instability or stability. If the formulas (2) or (3) are used to extrapolate the wind speed, a large deviation will occur. Regardless of the state of the atmosphere, method three (ELM) can be more rapid and accurate extrapolate the required height of wind speed.

5 Conclusions

In this paper, the method of using the extreme learning machine for wind speed vertical extrapolation is presented. The results are compared with the measured values by the extreme learning machine and the traditional method. The conclusions are as follows:

(1) Because of the influence of surface roughness and topography, the variation of wind shear index with different height gradient is larger.
(2) The average relative error obtained by using the seasonal wind shear index to extrapolate wind speed is smaller than that of the annual mean wind shear index. That is to say, the wind shear index of the nearest time interval is usually better.
(3) It is too idealistic to use the formulas (2), (3) to extrapolate wind speed, without considering the influence of atmospheric stability on wind shear index, there are some limitations.
(4) Method three (ELM) can be used to directly, quickly and accurately extrapolate the required wind speed without the influence of atmospheric stability. Therefore, this method can be extended to the practical application of engineering.

Acknowledgements. This work was supported by the Scientific Research Innovation Project of Shanghai Municipal Education Committee (Grant No. 13YZ140) and the Project Sponsored by the Scientific Research Foundation for the Returned Overseas Chinese Scholars, Education Ministry. (Grant No. [2014]1685).

References

1. Yanjun, D., Changqing, F.: Application of wind shear index in wind resource assessment of wind farm. Power Syst. Clean Energy **26**(5), 68–72 (2010)
2. Weiguo, S.: Ren bodybuilding. The surface temperature and wind profiles stratification correction. Shanxi Meteorol. (1), 23–29 (1995)
3. Gualtieri, G., Secci, S.: Methods to extrapolate wind resource to the turbine hub height based on power law: a 1-h wind speed vs. Weibull distribution extrapolation comparison. Renew. Energy **43**, 183–200 (2012)
4. Xiaomei, M.: Study on the effect of atmospheric stability on wind resource characteristics. North China Electric Power University (Beijing) (2016)
5. Huang, G.B., Zhu, Q.Y., Siew, C.K.: Extreme learning machine: theory and applications. Neurocomputing **70**(1–3), 489–501 (2006)
6. Feng, Y., Yan, H., Qihao, Z., et al.: Coastal and offshore wind resources spatio-temporal characteristics. J. Tsinghua Univ. (Nat. Sci. Edn.) **5**, 522–529 (2016)
7. Baoqing, X., Tingting, W., et al.: Study on wind shear index of wind resource assessment. Electr. Power Sci. Eng. **30**, 99–104 (2014)
8. Lange, B., Larsen, S., Højstrup, J., et al.: Importance of thermal effects and sea surface roughness for offshore wind resource assessment. J. Wind Eng. Ind. Aerodyn. **92**(11), 959–988 (2004)

Optimal Scheduling of Wind Turbine Generator Units Based on the Amount of Damage of Impeller

Kai Lin and Guochu Chen[✉]

Department of Electrical Engineering, Shanghai Dianji University,
Shanghai 201306, China
361928561@qq.com, chengc@sdju.edu.cn

Abstract. Impeller (blade and wheel) is one of the key components of wind turbine. According to different degrees of leaf and root damage and hub damage amount, a multi-objective scheduling model of wind farm with wind turbine impeller damage, wind turbine startup rate, and the uncertainty of the output of generator is established to improve the model output allocation strategy. Then optimize the model with the adaptive discrete particle swarm (ADPSO) and artificial bee colony algorithm (ABC), then obtaining the target power value and start-stop group. In combination with the practical example, the simulation results show that the proposed method optimizes the start-up and shut-down times of the wind turbines and improves the operating life of the wind turbines.

Keywords: The damage of the impeller · ADPSO · ABC · The mixed integer nonlinear programming

1 Introduction

The main components of wind turbine are blades, wheels, gearboxes, generators, control systems and other parts, because the wind turbine output influenced by the wind speed, wind direction uncertainty and variable speed constant frequency power generation control constraints, the operating state need to be frequently switched in different conditions, and the impeller is the most complex and the highest reliability requirements key component with high annual failure rate, long fault downtime.

What's more, Impeller accounts for about 23% of the total cost of wind turbines [1–3]. The operation and maintenance costs of wind farms can be reduced by extending the impeller life. In paper [4], the genetic algorithm is used to optimize the scheduling model considering the health status of the wind turbines. In paper [5], only the damage amount of the blade as the single target is considered to optimize the wind turbine working condition, and the unit output of the unit is not optimized.

In this paper, one multi-objective optimal scheduling is carried out by combining the times of start-stop and the relative fatigue damage value of the impeller. And this is a mixed integer nonlinear multi-objective optimization problem, in which the integer variable is the wind turbine start-up and shut-down, the continuous variable is the output power. This paper uses the improved ADPSO algorithm to optimize the

© Springer Nature Singapore Pte Ltd. 2017
K. Li et al. (Eds.): LSMS/ICSEE 2017, Part III, CCIS 763, pp. 12–21, 2017.
DOI: 10.1007/978-981-10-6364-0_2

combination of the generator start-up and shut-down, uses ABC algorithm to optimize the output, and combing the improved power output allocation strategy to solve the optimal scheduling problem.

2 Optimization Model of Wind Turbine Considering Impeller Damage Quantity and Start-Up and Shut-Down Times

2.1 Objective Function

Paper [6] takes 1.5 MW wind turbine as the research object, use the method of rain flow cycle counting to get the relative fatigue damage value of wind turbine blade and hub under different working conditions. Damage values of wind turbine under different working conditions are listed in Table 1.

Table 1. Damage values of wind turbine under different working conditions

Working condition	Damage coefficient	Code	Power output range (MW)	Blade damage values	Wheel hub damage values
Stop	A	000	0	0	0
Normal operation		001	0–0.093	3.80E−08/min	1.11E−10/min
		010	0.093–0.326	4.32E−08/min	2.24E−09/min
		011	0.326–0.753	6.02E−08/min	2.13E−08/min
		100	0.753–1.5	9.51E−08/min	1.84E−08/min
		101	≥1.5	1.53E−07/min	4.55E−09/min
Start-up	B	000-100	0–1.5	1.12E−09/time	2.95E−11/time
		000-101	≥1.5	1.18E−09/time	5.17E−11/time
Shutdown	C	100-000	0–1.5	2.77E−09/time	9.54E−09/time
		101-000	≥1.5	1.35E−09/time	4.85E−09/time

In the Table 1, 'time' means frequency, indicates the number of changes in the operating state of the wind turbine, 'min' means minute. And in the Table 1, A indicates the damage value of the impeller in normal operation, like the codes: 001, 010, 011, 100, 101. B indicates the damage value of the impeller in the process of start-up the wind turbine, like the codes: 000-100, 000-101. C indicates the damage value of the impeller in the process of shut-down the wind turbine, like the codes: 100-000, 101-000.

Under the different conditions, the impeller damage quantity of each condition has the corresponding numerical value, and the impeller damage value function can be defined as [7]:

$$h(u) = \sum_{j=1}^{T} \sum_{i=1}^{n} [A_i^j u_i^j t + B_i^j u_i^j (1 - u_i^{j-1}) + C_i^j u_i^{j-1}(1 - u_i^j)] \tag{1}$$

In the Eq. (1), $h(u)$ indicates the total amount of damage of the impeller wind farm, T indicates wind farms scheduling period, n indicates the total number of wind turbines, and A_i^j, B_i^j, C_i^j are respectively corresponding to damage values of wind turbine i in the wind farm during the normal running process, start-up process and shut-down process, at time j. u_i^j is the working condition of wind turbine i at time j. And the start-up and shut-down states are respectively represented by 0 and 1.

In order to avoid unnecessary start-up and shut-down of wind turbine, the times of start-up and shut-down is optimized in the scheduling period, and it can be defined as:

$$f(u) = \sum_{j=1}^{T}\sum_{i=1}^{n}\left|u_i^j - u_i^{j-1}\right| \tag{2}$$

According to Eq. (2), it is necessary to balance the proportion between the damage quantity and the start-up and shut-down times:

$$F(u,p) = \min[ah(u) \times 10^5 + bf(u)] \tag{3}$$

In the Eq. (3), a and b respectively corresponding to the weight coefficient of wind turbine impeller damage and wind turbine start-up and shut-down times.

2.2   Constraint Condition

Wind Turbine Predictive Power Constraints

$$0 \le p_{yuce}^{i,j} \le p_{max}^{i,j} \tag{4}$$

In the Eq. (4), $p_{yuce}^{i,j}$ is power prediction value of wind turbine i at time j. $p_{max}^{i,j}$ is the output power limit of the wind turbine i at time j.

Wind Turbine Output Range Constraint

$$p_{i,min} \le p_{i,j} \le p_{yuce}^{i,j} \tag{5}$$

In the Eq. (5), $p_{i,min}$ indicates the lower limit of the output power of wind turbine i. In this paper, it is set 20% of the rated power of the wind turbine. $P_{i,j}$ represents the actual output power of wind turbine i at time j

Load Scheduling Constraints

$$\sum_{j=1}^{T}\sum_{i=1}^{n}u_i^j p_i^j = p_w^j \tag{6}$$

In the Eq. (6), p_w^j represents dispatching command of grid to wind farm, that is, the total power required for the wind farm, at the number j.

Spinning Reserve Constraints

$$\sum_{j=1}^{T}\sum_{i=1}^{n} u_i^j (p_{max}^{i,j} - p_i^j) \geq p_{by}^j \tag{7}$$

In the Eq. (7), p_{by}^j represents the requirements of spinning reserve, at time j. This page set it to 5% of the total output power.

2.3 Improvement of Wind Turbine Output Allocation Strategy

Because of the wind power mainly depends on the wind, rather than the installed capacity, so this paper improved the traditional strategy for the output distribution, so that the wind turbine can be controlled according to the wind power prediction value and working state adjust distribution, when the output power is greater than the total scheme generation scheduling instructions, according to the Eq. (8) of each unit of proportional output:

$$p_{ij}^* = p_{ij} - \frac{u_{ij}(p_{ij} - p_{i,min})}{u_j(p_j - p_{i,min})} (p_j \times u_j - p_w^j) \tag{8}$$

When the total power output is less than the dispatching instruction, according to Eq. (9) the output of each turbine and the ratio of the reverse will be increased:

$$p_{ij}^* = p_{ij} + \frac{u_{ij}(P_{yuce}^{i,j} - p_{ij})}{u_j(P_{yuce}^j - p_j)} (p_w^j - p_j \times u_j) \tag{9}$$

3 Adaptive Discrete Particle Swarm - Artificial Bee Colony Algorithm Optimization Scheduling

The mathematical model is a mixed integer nonlinear programming problem contains integer discrete variables and continuous variables. In order to avoid the combinatorial explosion caused by large-scale scheduling, this paper uses adaptive discrete particle swarm algorithm (ADPSO) [8, 9] combined with the artificial bee colony algorithm (ABC) [10, 11] to solve the model within the outer and inner optimization way [12].

Step 1: Set the particle population M of the adaptive discrete particle swarm algorithm, set the inertia weight parameter ω_{min} and ω_{max} (in this paper, take 0.01 and 1 respectively), the learning factor C_1 and C_2 (both are set to 2) and the number of iterations, set the number of artificial Bee colony population NP and other basic parameters.

Step 2: According to load scheduling constraints and spare constraints to produce the start-up and shut-down combinations:

$$U_m = \begin{vmatrix} u_{1,1} & \cdots & u_{1,j-1} & u_{1,j} \\ \vdots & \vdots & \vdots & \vdots \\ u_{i-1,1} & \cdots & u_{i-1,j-1} & u_{i-1,j} \\ u_{i,1} & \cdots & u_{i,j-1} & u_{i,j} \end{vmatrix} \tag{10}$$

In the Eq. (10), U_m represents start-up and shut-down combination matrix. $u_{i,j}$ represents the start-up and shut-down state of wind turbine i at time j.

Step 3: After selecting the start-up and shut-down combinations which satisfies the load schedule constraint and the reserve constraint, the ABC algorithm randomly generates the power allocation of each wind turbine in the scheduling cycle that satisfies the output range constraint and the predicted power constraint:

$$P_k = \begin{vmatrix} p_{1,1} & \cdots & p_{1,j-1} & p_{1,j} \\ \vdots & \vdots & \vdots & \vdots \\ p_{i-1,1} & \cdots & p_{i-1,j-1} & p_{i-1,j} \\ p_{i,1} & \cdots & p_{i,j-1} & p_{i,j} \end{vmatrix} \tag{11}$$

In the Eq. (11), P_k represents power output condition matrix. p_{ij} represents the power output condition of wind turbine i at time j.

Step 4: Calculate fitness value. According to the Eq. (3) calculates objective value, then calculates the fitness value by the Eq. (12):

$$fit_{i,k} = \begin{cases} 1/1 + F_{i,k}, & F_{i,k} \geq 0 \\ 1 + abs(F_{i,k}), & F_{i,k} < 0 \end{cases} \tag{12}$$

In the Eq. (12), $F_{i,k}$ refers to the objective function value of the kth honey source under the ith particle of the particle swarm. $fit_{i,k}$ refers to the fitness function value of the kth honey source under the ith particle of the particle swarm. And this is the minimum optimization problem, so the greater the fitness value, the corresponding solution of the better.

Step 5: Search for new solutions and make greedy choices. The colony produces a new solution by the following search method:

$$V_{ij} = p_{ij} + \varphi_{ij}(p_{ij} - p_{kj}) \tag{13}$$

In the Eq. (13), p_{kj} is different from p_{ij}, and $k \in \{1, 2, ..., SN\}$, SN is the number of honey source. ϕ_{ij} is the random number between -1 and 1. After searching process, calculate the fitness value and choose the optimal solution through greed selection.

Step 6: Selection. According to the following bee roulette selection method with the probability $r_{i,k}$ of honey source, Calculate $r_{i,k}$ [11] according to Eq. (14):

$$r_{i,k} = \frac{fit_{i,k}}{\sum_{n=1}^{SN} fit_n} \tag{14}$$

Step 7: When the honey source always doesn't not change, honey source is abandoned, at the same time, bees for Scout bees, and by Eq. (15) random search to generate a new alternative raw honey source. And through Eqs. (3) and (12) to calculate the fitness value of new honey source.

$$p_{i,j} = p_{\min}^{i,j} + rand(0,1)(p_{\max}^{i,j} - p_{\min}^{i,j}) \qquad (15)$$

Step 8: After the ending of the artificial bee colony algorithm, ADPSO algorithm choose global optimal value and individual optimal values, and then update the particle velocity and position, and generate new population processes are as follows:

(1) Generating inertia weight parameter

$$fit_{av} = \left(\sum_{m=1}^{M} fit_m\right)/M \qquad (16)$$

$$w_i = w_{\min} + \frac{w_{\max} - w_{\min} * (fit_i - fit_{av})}{fit_{\max} - fit_{\min}} \qquad (17)$$

In the Eqs. (16) and (17), fit_{av} is fitness value.
(2) Update particle velocity and position

$$v_i^d = w_i v_i^{d-1} + c_1(p_{best,i}^d - U_i^{d-1}) + c_2(g_{best,i}^d - U_i^{d-1}) \qquad (18)$$

$$U_i^d = U_i^{d-1} + v_i^d \qquad (19)$$

In the Eqs. (18) and (19), v_i^d is the velocity of ith particle at dth iteration. $p_{best,i}^d$ is the individual optimal values at dth iteration. g_{best}^d is the global optimal value at dth iteration.
(3) Discretization

$$u_{ij} = \begin{cases} 1 & if \ rand(0,1) < 1/1 + e^{-v_i^d} \\ 0 & else \end{cases} \qquad (20)$$

Step 9: Loop iteration until the number of iterations reaches the maximum iterations.

4 Example Analysis

4.1 Example Introduction

By random selecting 10 sets of 1.5 MW wind turbine from a 49.5 MW wind farm in North China as the analysis objects, because the one hour in the short-term wind power prediction has higher credibility, this paper divided one hour to 4 scheduling cycle, each scheduling period is 15 min. The initial state and the predicted power value of the 10 wind turbines set are shown in Table 2, 0 indicates the shutdown, and 1 indicates the working. For adaptive discrete particle swarm optimization algorithm, the number of population is set to 20, the number of iterations is set to 30. For artificial bee colony algorithm, the number of population is set to 40, the number of iterations is 10, limit is 4.

Table 2. The initial state and the predictive power of the 10 wind turbines in a wind farm

Generator code	Initial state	Upper limit of predicted power (MW)			
		T1	T2	T3	T4
1	1	1.04	1.19	1.20	1.28
2	1	1.09	1.24	1.28	1.34
3	0	1.17	1.16	1.29	1.40
4	1	0.84	1.27	1.24	1.30
5	1	0.75	1.26	1.17	1.31
6	1	0.64	1.13	1.10	1.11
7	0	0.96	1.16	1.14	1.20
8	1	0.70	1.27	1.26	1.29
9	1	1.12	1.10	1.22	1.41
10	1	0.82	1.35	1.29	1.40
Total power		9.13	12.13	12.19	13.04

Under the condition of limited wind power, the sum of dispatching and the system reserve should be less than the upper limit of the wind farm. Therefore, the dispatching orders of the wind farm in the 4 scheduling periods are 7 MW, 9 MW, 9 MW, 9.5 MW, and the corresponding reserve power are 0.35 MW, 0.45 MW, 0.45 MW, 0.475 MW.

4.2 Objective Function Weight

It can be seen that only the damage values of the impeller and the times of start-up and shut-down of the wind turbine have effect on the optimal scheduling results. Therefore, the comparison of the two extreme examples is shown in Table 3.

As can be seen from Table 3, in the first case, the spinning reserve of the 4 scheduling cycles are respectively 0.74, 0.9, 0.92, 1.03 MW, the times of start-up and shut-down is 10, and the damage value is 4.1043e−05. Therefore, when the weight coefficient is relatively large, the damage value of the wind turbine plays an important role on the optimization of the scheduling trend. In another case, the spinning reserve of the 4 scheduling cycles are respectively 2.13, 3.13, 3.19, 3.54 MW, start-up and shut-down times is 2, the damage value is 5.13e−05, so when the weight coefficient is larger, wind turbine's start-up and shut-down times plays a leading role, leading to the failure to utilize the advantages of the wind turbine starting and stopping convenient, resulting in spare capacity is much greater, resulting in waste, and the damage value is bigger.

Table 3. Results of optimal dispatching of 10 wind turbines under different extreme weight

Generator code	a = 1000, b = 1				a = 1, b = 1000			
	T_1	T_2	T_3	T_4	T_1	T_2	T_3	T_4
1	0.84	1.04	1.18	1.1	0.83	1.03	1.17	1.08
2	0.96	1.03	1.01	1.21	0.63	1.06	1.26	0.94
3	1.15	1.13	1.23	1.35	0.74	0.93	0.91	0.89
4	0.74	1.21	1.22	1.12	0.54	0.81	1.2	0.8
5	0	1.18	0	0	0.47	0.83	0.75	1.2
6	0	0	0	0.97	0.45	0.74	0.84	0.78
7	0.96	1.11	1.02	0	0.91	0.77	0.75	0.78
8	0.64	1.09	1.08	1.11	0.68	0.84	0.77	0.91
9	0.94	0	1.21	1.35	1.06	0.97	0.68	1.18
10	0.77	1.21	1.05	1.29	0.69	1.02	0.67	0.94
Spinning reserve	0.74	09	0.92	1.03	2.13	3.13	3.19	3.54
Damage amount	4.1043e−05				5.1300e−05			
Start stop times	10				2			

Two kinds of extreme circumstances results are explained in this paper. In order to take into account the two indexes in engineering, scheduling planner can adjust the weight coefficients. It requires a large number of optimization tests to roughly determine a reasonable proportion. In this paper, set "8:1", "7:1", "6:1", "5:1", "4:1", "3:1", "2:1", "1:1", "1:2", "1:3" 10 groups, the results can be seen in Fig. 1.

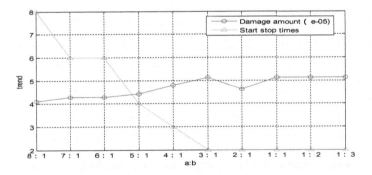

Fig. 1. Effect of a:b on optimization results

As can be seen from Fig. 1, in the process of a:b from 8:1 to 5:1, start-up and shut-down times decrease from 8 to 4, a larger decline, while the damage from 4.1e−05 to 4.45e−05, a smaller increase. When a:b from 5:1 to 1:3, the amount of damage increased from 4.45e−05 to 5.13e−05, the start-up and shut-down up to the minimum times. In order to take into account the comprehensive performance of the two indicators, therefore, take a:b as 5:1, the corresponding optimal scheduling results can be seen in Table 4.

Table 4. a = 5, b = 1, 10 wind turbines scheduling results

Wind turbine code	Wind turbine output power (MW)			
	T1	T2	T3	T4
1	0.86	1.05	1.10	1.18
2	0.73	0.95	1.09	0.99
3	0.99	0.99	1.16	1.28
4	0.71	0.93	1.02	1.26
5	0.74	1.13	1.13	1.22
6	0.00	0.00	0.00	0.00
7	0.67	0.70	0.00	0.00
8	0.68	1.20	1.23	0.97
9	0.93	1.00	1.06	1.36
10	0.69	1.05	1.22	1.24
Spinning reserve	1.49	2	0.95	1.23
Damage amount	4.2750e−05			
Start stop times	4			

Table 4 shows the column of the spinning reserve system relative to each cycle instruction scheduling, respectively is 21.28%, 22.22%, 10.5%, 12.95%, are more than 5%, and the wind turbine start-up and shut-down times is 4, scheduling iteration diagram shown in Fig. 2.

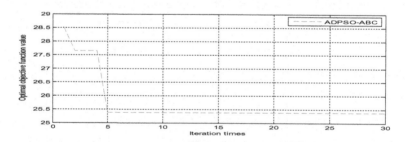

Fig. 2. Iterative process of optimal dispatching of 10 wind turbines.

5 Conclusion

In the limit of power generation, based on the premise of meet the power dispatching command, a mathematical model of multi-objective including wind turbine impeller damage quantity and start-up and shut-down times of wind turbine is established. An example is given to validate, the dispatch of 10 wind turbines in a wind farm in North China. The ADPSO-ABC algorithm is used to solve model. ADPSO algorithm has better searching ability in discrete optimization problem, and has the characteristics of simplicity and generality. ABC algorithm has better searching ability in multidimensional optimization problem, and has the characteristics of simple, practical and less parameter. The calculation results show that the reasonable use of the wind turbine's start-up and shut-down advantages, can ensure the wind power output reliability, reduce the damage of impeller, reduces the wind turbine operation and maintenance costs, improve the market competitiveness of wind power.

Acknowledgments. This work was financially supported by the Innovation Program of Shanghai Municipal Education Commission (Grant No. 13YZ140), Scientific Research Foundation for the Returned Overseas Chinese Scholars Education Ministry of China (Grant No. [2014]1685).

References

1. Li, H., Hu, Y.G., Li, Y., et al.: High power grid connected wind turbine condition monitoring and fault diagnosis. Electr. Power Autom. Equip. **1**, 6–16 (2016)
2. Carroll, J., Mcdonald, A., Mcmillan, D.: Failure rate, repair time and unscheduled O&M cost analysis of offshore wind turbines. Wind Energy **19**(6), 1107–1119 (2016)
3. Zhang, J.H., Liu, Y.Q., Tian, D., et al.: Optimal power dispatch in wind farm based on reduced blade damage and generator losses. Renew. Sustain. Energy Rev. **44**(44), 64–77 (2015)
4. Tong Ji, X., Wang, L., Feng, H.: Optimal scheduling method for wind farm considering wind power generator. Int. J. Hydroelectr. Energy **1**, 198–202 (2016)
5. Lin, Z.Y.: Optimal Scheduling of Wind Farm for the Minimum Blades Damage. North China Electric Power University (2013)
6. Zhang, J.H.: Research on Unit Optimal Dispatch in Wind Farm. North China Electric Power University (2014)
7. Chai, L.J., Feng, H.: Optimal scheduling of wind farms based on reducing the damage of wind turbine blades. Yangtze River **02**, 95–100 (2016)
8. Liu, S.C., Zhang, J.H.: Application of parallel adaptive particle swarm optimization in reactive power optimization of power system. Power Syst. Technol. **1**, 108–112 (2012)
9. Sun, R.: A modified adaptive particle swarm optimization algorithm. In: International Conference on Computational Intelligence and Security, pp. 209–214. IEEE (2016)
10. Karaboga, D.: Artificial bee colony algorithm. Scholarpedia **5**(3), 6915 (2010)
11. Qin, Q.D., Cheng, Y., Li, L., et al.: Overview of artificial bee colony algorithm. CAAI Trans. Intell. Syst. **2**, 127–135 (2014)
12. Xiao, F., Chen, G.: Scheduling of power system economic optimization with wind farms based on power prediction. Electr. Power Sci. Eng. (2016)

A Short Term Wind Speed Forecasting Method Using Signal Decomposition and Extreme Learning Machine

Sizhou Sun[1,2], Jingqi Fu[1(✉)], and Feng Zhu[1]

[1] School of Mechatronics Engineering and Automation,
Shanghai University, Shanghai 200072, China
jqfu@staff.shu.edu.cn
[2] School of Electrical Engineering, Anhui Polytechnic University,
Wuhu 241000, China

Abstract. In this study, a novel hybrid model using signal decomposition technique and extreme learning machine (ELM) is developed for wind speed forecasting. In the proposed model, signal decomposition technique, namely wavelet packet decomposition (WPD), is utilized to decompose the raw non-stationary wind speed data into relatively stable sub-series; then, ELMs are employed to predict wind speed using these stable sub-series, eventually, the final wind speed forecasting results are calculated through combination of each sub-subseries prediction. To evaluate the forecasting performance, real historical wind speed data from a wind farm in China are employed to make short term wind speed forecasting. Compared with other forecasting method mentioned in the paper, the proposed hybrid model WPD-ELM can improve the wind speed forecasting accuracy.

Keywords: Wind speed forecasting · ELM · WPD · Multi-step forecasting

1 Introduction

Wind power has been developed at the fastest rate among renewable low-carbon resources globally. Duic and Rosen [1] stated that the reduction of greenhouse gas emissions will reach to 85–95% by 2050 with the fast development of wind energy harvesting techniques. However, integration of amounts of wind power in the power system has great impacts on the operation scheduling, planning, safety and stability. Accurate wind power prediction can mitigate these problems effectively.

Many researchers and scientists have developed different reliable methods for higher forecasting accuracy. Autoregressive moving average (ARMA) [2], fractional autoregressive integrated moving average (f-ARIMA) [3] and Gaussian regressions [4] are typical statistical methods which build relationship among the historical wind power/speed for wind speed forecasting. In recent years, artificial intelligent (AI) algorithms combined with multi-scale decomposition techniques have been developed to improve forecasting accuracy [5–19]. It is obviously seen from the results in the literatures that artificial intelligent methods with multi-scale decomposition can yield

© Springer Nature Singapore Pte Ltd. 2017
K. Li et al. (Eds.): LSMS/ICSEE 2017, Part III, CCIS 763, pp. 22–31, 2017.
DOI: 10.1007/978-981-10-6364-0_3

better forecasting performance as compared with the traditional statistical methods and the single artificial intelligent methods. For example, Meng et al. [6] utilized the wavelet packet decomposition (WPD) to make the original wind speed decomposed into components. For each sub-series, the back-propagation neural network (BPNN) was applied to predict wind speed. The forecasting results of the model had a significant improvement over previously reported methodologies. Afshin et al. [8] employed wavelet transform (WT) to filter wind power series, made use of radial basis function neural network (RBFNN) as primary prediction and three multi-layer perceptron neural networks (MLPNN) as main prediction. Wang et al. [9] developed a new wind speed forecasting method combined signal decomposition EEMD with BP neural network (BPNN) model tuned by genetic algorithm (GA). In the combinational model, EEMD decomposed the original data into more stationary signals with different frequencies, GA was applied to optimize the ANN's initial weights, then, each signal was employed as an input data for the GA-BPNN model. Ali [10] applied ELM with variational mode decomposition (VMD) technique to 10-min wind power time series forecasting and aimed to enhance forecasting accuracy, in this study, VMD technique decomposed wind power time series into different modes, Gram-Schmidt orthogonalization (GSO) used as the feature selection method was applied to eliminate redundant properties, the well-trained ELM algorithm made the wind speed forecasting, the final forecasting results justified the superiority of the proposed method.

WPD is an effective nonlinear stochastic signals analyzing method [6] and ELM is an efficient and fast regression tool [10]. Compared with Fourier transformation and Wavelet Transform (WT) technique, WPD has some advantages, such as good multi-resolution. ELM has very rapid learning process because it requires only one single-pass training process for weights adjustment. In addition, the quantity of the hidden neurons of ELM is easily determined only through the dimensions of the input and output samples. Therefore, this paper develops a novel wind speed forecasting approach using WPD and ELM. The original wind speed data is decomposed into relatively stable sub-series components by WPD, then, each component is taken as inputs of ELM forecasting model. The final forecast value is calculated by aggregating the predicted value of individual signals.

The remains of the paper are organized as follows: the proposed methodology is described in the Sect. 2. In Sect. 3, the modeling steps of the WPD-ELM model are given; the wind speed forecasting results, comparison and analysis are presented in Sect. 4. In the end, the conclusions are drawn.

2 Methodology

The proposed hybrid method for short term wind speed prediction makes multi-step forecasting by usage of the historical wind data. The main forecasting procedure is realized as follows:

Step 1: WPD is applied to decompose the non-stationary wind speed data; several different wind speed sub-series with different frequency can be obtained after decomposition;

Step 2: Partial autocorrelation function (PACF) method is used to determine the input variables of ELM model after normalization of each subseries;

Step 3: Train ELM with every decomposed subseries, the 1st–600th sampling wind speed is utilized as the training data set. The ELM is adopted as the regression core for wind speed forecasting and the number of ELM is equal to the number of subseries;

Step 4: Apply the trained ELM to forecast the multi-step ahead wind speed, the 601th–700th sampling wind speed is used as the test data set;

Step 5: Obtain final wind speed forecasting results by aggregation of the de-normalized values of each ELM outputs;

Step 6: Compare the performance between the proposed method and other models.

3   Modeling Process

3.1   Evaluation Criteria

To evaluate the proposed model and reduce the statistic errors, all simulations experiments are carried out 30 times independently by Matlab 2012a. In this study, three statistical indices including mean absolute error (MAE), root mean square error (RMSE) and mean absolute percent error (MAPE) are utilized to evaluate forecasting performance, and they are expressed as Eqs. (1)–(3).

$$RMSE = \sqrt{\frac{1}{N} \sum_{i=1}^{N} |W_i^{act} - W_i^{fore}|^2}, \tag{1}$$

$$MAE = \frac{1}{N} \sum_{i=1}^{N} |W_i^{act} - W_i^{fore}|, \tag{2}$$

$$MAPE = \frac{1}{N} \sum_{i=1}^{N} \frac{|W_i^{act} - W_i^{fore}|}{W_i^{act}} \times 100\%, \tag{3}$$

where W_{act}^i and W_{fore}^i are the real measured and the forecasting wind speed data at time period i, respectively. N is the total number of wind speed samples used for test and comparison.

3.2   Experimental Wind Speed

The wind speed data at 30-min intervals are collected from a wind farm in Jiuquan of China (Fig. 1). In this study, we randomly select 700 continuous data sampling points as the training and test data set. The Fig. 2 shows the original selected wind speed and Table 1 depicts statistical properties of wind speed samples. From Fig. 2 and Table 1, it is obviously seen that there are large fluctuation and no apparent regularity in the original wind speed data.

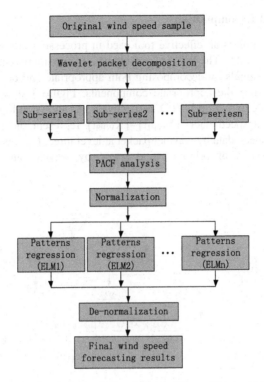

Fig. 1. Framework of the wind speed forecasting method

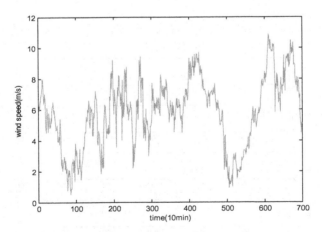

Fig. 2. Original wind speed samples

Table 1. Statistical properties of wind speed (m/s)

Location	Max	Min	Std.	Mean
Jiuquan	10.8609	0.5019	2.2601	5.8062

3.3   Wind Speed Decomposition

Wavelet decomposition is an effective tool used in processing non-stationary or non-linear sets widely [21, 22]. The WPD is considered as some improvement of WT. WPD is used to analyse signals by decomposing both appropriate and detailed components while WD decompose the appropriate components. Figure 3 shows the three-layer binary trees structure of WT and WPD, respectively. As shown in Fig. 3(b), suppose S represents the wind speed data, $\{X_{WP1t}\}$–$\{X_{WP8t}\}$ represent the final decomposed subseries of wind speed data by wavelet packet at level three. Figures 4 and 5 show the decomposed sub-series of original wind speed by wavelet and wavelet packet, respectively.

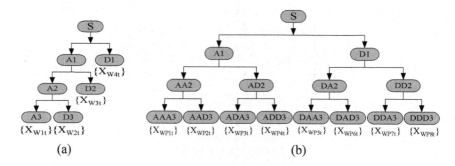

(a)   (b)

Fig. 3. Signal method. (a) Wavelet decomposition at level three; (b) wavelet pocket decomposition at level three

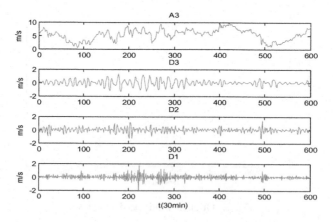

Fig. 4. Results of WD for the 1st–600th wind speed

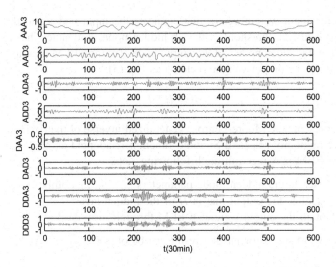

Fig. 5. Decomposed sub-series of wind speed by WPD

In the model, the WPD technique is utilized to decompose wind speed data into individual components which are data inputs of ELM model. Morlet, Haar, Mexican, Hat, Meyer and Daubechies can be used as a mother wavelet for wavelet packet. Among these wavelet basis functions, Daubechies have better performance than other ones [21]. Thus, Daubechies of order 4 (db4) is utilized as mother wavelet in this study. To reduce the inappropriate influence of different range of wind speed subseries on the regression procedure, each decomposed subseries are normalized into interval [0, 1]. After normalization of the subseries, the PACF technique is used to analyse the correlation between the candidate variables and the historical wind speed data for determination of the inputs of ELM model. The PACF values of the original wind speed subseries are two, five, four, four, three, six, four and three, respectively.

3.4   ELM Modelling

ELM is a novel tool of learning algorithm with a single-hidden-layer-feed-forward neural network (SLFN) architecture [20]. The main characteristic of the ELM algorithm is that it selects the parameters of hidden node randomly without adjustment and determines the output weights through simple mathematical calculation during the whole learning process.

For a training set $(x_i, y_i) \in R_n \times R_m$, where $x_i = [x_{i1}, x_{i2}, \ldots, x_{in}]^T$ and $y_i = [y_{i1}, y_{i2}, \ldots, y_{im}]^T$. Supposed that SLFN with L hidden neurons and ELM can be mathematically expressed as Eq. (4):

$$\sum_{i=1}^{N} \beta_i G(a_i, b_i, x_i) = y_i, \tag{4}$$

where a_i is the input weight vector that connects the input nodes and the ith hidden node, b_i represents the threshold of the ith hidden node, β_i denotes the weight vector connecting the ith hidden node and the output node. Training ELM is simply equivalent to calculation of the output weight vector β as Eq. (5):

$$\beta = H^+ T, \tag{5}$$

where H^+ represents the Moore-Penrose generalized inverse of matrix H. In this study, a_i and b_i are randomly selected; Sigmoid function is adopted as $g(x)$. The regression procedure of ELM is realized through following steps:

Step 1: Generate the hidden mode parameters $\omega_{i,j}$, b_j, randomly;
Step 2: Calculate the output matrix H of the hidden layer;
Step 3: Obtain the output weight of ELM through formula $\beta = H^+ T$.

The final wind speed forecasting result can be obtained as:

$$\hat{s}(t) = \hat{X}_{PW1t}(t) + \hat{X}_{PW2t}(t) + \cdots \hat{X}_{PW8t}(t), \tag{6}$$

where t is the sampling time.

4 Numerical Results

The 601st–700th sample points of wind speed in Fig. 2 are utilized to make the wind speed forecasting. All the experiments are implemented in Matlab 2012a environment on windows 7 PC. All the experiments are implemented 30 times independently to reduce the statistical errors.

4.1 WPD-ELM vs. WT-ELM vs. ELM

The original wind speed samples shown in Fig. 2 are selected randomly from a wind farm in Jiuquan of China. Firstly, the wind speed forecasting model, including WPD-ELM, WT-ELM and ELM, are used to do wind speed forecasting. Table 2 shows the wind speed forecasting errors. The one-step MAPE errors of WPD-ELM, WT-ELM and ELM are 6.013%, 7.8049% and 10.4903%, the MAE errors are 0.485 m/s, 0.6401 m/s and 0.8411 m/s, the RMSE errors are 0.5197 m/s, 0.7045 m/s and 0.9557 m/s, respectively. It is obviously obtained that both WPD-ELM model and WT-ELM model have better forecasting results than ELM model, the reasons of the forecasting results are that ELM model does wind speed forecasting by direct usage of the original wind speed without decomposition, while WPD and WT decompose the non-stationary wind speed time series into relatively stable components which can improve the forecasting performance of the ELM model. From the statistical indexes in Table 2, it can be seen that WPD-ELM yields smaller statistical errors than WT-ELM; the reason is that WPD can offer more decomposed wind speed sub-series signal than WT. In the two-step and three-step ahead forecasting, the proposed hybrid model also outperforms WT-ELM and ELM.

Table 2. Statistical errors obtained by different forecasting model

Index	One-step	Two-step	Three-step
WPD-ELM			
MAPE (%)	6.0130	7.2843	8.5266
MAE (m/s)	0.4850	0.5933	0.6865
RMSE (m/s)	0.5197	0.6308	0.7062
WT-ELM			
MAPE (%)	7.8049	8.5142	9.6255
MAE (m/s)	0.6401	0.6877	0.7797
RMSE (m/s)	0.7045	0.7149	0.8243
ELM			
MAPE (%)	10.4903	12.4889	13.2816
MAE (m/s)	0.8411	0.9927	1.0841
RMSE (m/s)	0.9557	1.0531	1.1250

4.2 WPD-ELM vs. WT-SVM vs. ARMA

To further evaluate the proposed method, the WPD-ELM is compared with WT-SVM [12] and ARMA. WT-SVM is recently developed hybrid model by Liu et al. forecast short-term wind speed. The parameters of the WT-SVM model are set according to the corresponding reference. Compared with WT-SVM and ARMA in one-step ahead forecasting, the MAPE errors of the proposed model are cut by 2.5652%, 7.2012%, respectively. As seen from Table 3, the proposed model also outperforms the WT-SVM model and ARMA model in the other multi-step forecasting. The forecasting performances of WPD-ELM and WT-SVM are much better than the single ARMA model. The forecasting results further substantiate that the WPD-ELM model can forecast short-term wind speed effectively.

Table 3. Statistical errors obtained by other model

Index	One-step	Two-step	Three-step
WPD-ELM			
MAPE (%)	6.0130	7.2843	8.5266
MAE (m/s)	0.4850	0.5933	0.6865
RMSE (m/s)	0.5197	0.6308	0.7062
WT-SVM			
MAPE (%)	8.5782	9.7934	10.9279
MAE (m/s)	0.6962	0.7939	0.8994
RMSE (m/s)	0.7166	0.8482	0.9483
ARMA			
MAPE (%)	13.2142	14.4175	15.6667
MAE (m/s)	1.0618	1.1550	1.2579
RMSE (m/s)	1.2013	1.2426	1.3617

5 Conclusions

In the paper, a novel hybrid method namely Wavelet Packet Decomposition-ELM is proposed for multi-step ahead wind speed forecasting. From the analysis and comparisons, conclusion can be drawn: (a) the single ELM approach has higher wind forecasting accuracy than the single statistical method ARMA, this reason is that ELM algorithm is an intelligent algorithm which can tackle the nonlinear issues better than the ARMA method; (b) all the hybrid forecasting models mentioned in the paper, namely WPD-ELM, WT-ELM and WT-SVM, can obtain higher forecasting accuracy than ELM and ARMA models considerably in that the raw wind speed data is characterized by high fluctuations; (c) the proposed WPD-ELM model outperforms the WT-ELM and WT-SVM methods because the WPD algorithm can get more information from the raw wind speed data than WT. Therefore, it is obtained that the hybrid wind speed forecasting model WPD-ELM is an effective wind speed prediction method.

Acknowledgments. This work was supported by the Open Research Fund of Anhui Key Laboratory of Detection Technology and Energy Saving Devices, Anhui Polytechnic University; the Projects of Science and Technology Commission of Shanghai Municipality of China under Grant (No. 17511107002 and No. 15JC1401900); Natural Capital Project of Anhui Province under Grant Nos. 1408085ME105 and 1501021015.

References

1. Duic, N., Rosen, M.A.: Sustainable development of energy systems. Energy Convers. Manag. **87**, 1057–1062 (2014)
2. Torres, J., Garca, A., Deblas, M.: Forecast of hourly average wind speed with ARMA models in Navarre (Spain). Sol. Energy **79**, 65–77 (2005)
3. Rajesh, G., Kavasseri, K.S.: Day-ahead wind speed forecasting using f-ARIMA models. Renew. Energy **34**, 1388–1393 (2009)
4. Niya, C., Zheng, Q., Nabney, I.T., Xiaofeng, M.: Wind power forecasts using Gaussian processes and numerical weather prediction. IEEE Trans. Power Syst. **29**, 656–665 (2014)
5. Chao, R., Ning, A., Jianzhou, W., et al.: Optimal parameters selection for BP neural network based on particle swarm optimization: a case study of wind speed forecasting. Knowl. Based Syst. **91**, 226–239 (2014)
6. Anbo, M., Jiafei, G., Hao, Y., et al.: Wind speed forecasting based on wavelet packet decomposition and artificial neural networks trained by crisscross optimization algorithm. Energy Convers. Manag. **114**, 75–88 (2016)
7. Osamah, B., Muhammad, S.: Daily wind speed forecasting through hybrid KF-ANN model based on ARIMA. Renew. Energy **76**, 637–647 (2015)
8. Afshin, A., Rasool, K., Afshin, E.: A novel hybrid approach for predicting wind farm power production based on wavelet transform, hybrid neural networks and imperialist competitive algorithm. Energy Convers. Manag. **121**, 232–240 (2016)
9. Shouxiang, W., Na, Z., Leix, W., et al.: Wind speed forecasting based on the hybrid ensemble empirical mode decomposition and GA-BP neural network method. Renew. Energy **94**, 629–636 (2016)

10. Abdoos, A.: A new intelligent method based on combination of VMD and ELM for short term wind power forecasting. Neurocomputing **203**, 111–120 (2016)
11. Mladenovi, C., Markovi, C., Milovan, C.: Extreme learning approach with wavelet transform function for forecasting wind turbine wake effect to improve wind farm efficiency. Adv. Eng. Softw. **96**, 91–95 (2016)
12. Si, W., Youyi, W., Shijie, C.: Extreme learning machine based wind speed estimation and sensorless control for wind turbine power generation system. Neurocomputing **102**, 163–175 (2013)
13. Salcedo, S., Sánchez, A., Prieto, L., et al.: Feature selection in wind speed prediction systems based on a hybrid coral reefs optimization-extreme learning machine approach. Energy Convers. Manag. **87**, 10–18 (2014)
14. Hui, L., Hongqi, T., Difu, P., et al.: Forecasting models for wind speed using wavelet, wavelet packet, time series and artificial neural networks. Renew. Energy **107**, 191–208 (2013)
15. Jianming, H., Jianzhou, W., Kailiang, M.: A hybrid technique for short-term wind speed prediction. Energy **81**, 563–574 (2015)
16. Da, L., Dongxiao, N., Hui, W.: Short-term wind speed forecasting using wavelet transform and support vector machines optimized by genetic algorithm. Renew. Energy **62**, 592–597 (2014)
17. Tascikaraoglu, A., Sanandaji, B.M., Poolla, K., et al.: Exploiting sparse of interconnections in spatio-temporal wind speed forecasting using wavelet transform. Appl. Energy **165**, 735–747 (2016)
18. Yun, W., Jianzhou, W., Xiang, W.: A hybrid wind speed forecasting model based on phase space reconstruction theory and Markov model: a case study of wind farms in Northwest China. Energy **91**, 556–572 (2015)
19. Guoyong, Z., Yonggang, W., Yuqi, L.: An advanced wind speed multi-step ahead forecasting approach with characteristic component analysis. J. Renew. Sustain. Energy **6**(053139), 1–14 (2014)
20. Huang, G.B., Zhu, Q.Y., Siew, C.K.: Extreme learning machine: a new learning scheme of feed forward neural networks. In: Proceedings of International Joint Conference on Neural Networks, vol. 2, pp. 985–990 (2004)
21. Yan, W., Zhi, L.: Study on feature extraction method in border monitoring system using optimum wavelet packet decomposition. Int. J. Electron. Commun. **66**, 575–580 (2012)
22. Wenyu, Z., Jujie, W., Jianzhou, W., Zeng, Z., et al.: Short-term wind speed forecasting based on a hybrid model. Appl. Soft Comput. **13**, 3225–3233 (2013)
23. Osório, G.J., Matias, J.C.O., Catal, J.P.S.: Short-term wind power forecasting using adaptive neuro-fuzzy inference system combined with evolutionary particle swarm optimization, wavelet transform and mutual information. Renew. Energy **75**, 301–307 (2014)
24. Jianzhou, W., Jianming, H., Kailiang, M., et al.: A self-adaptive hybrid approach for wind speed forecasting. Renew. Energy **78**, 374–385 (2015)

A Novel Method for Short-Term Wind Speed Forecasting Based on UPQPSO-LSSVM

Wangxue Nie[1], Jingqi Fu[1(✉)], and Sizhou Sun[1,2]

[1] School of Mechatronic Engineering and Automation, Shanghai University,
No. 149 Yanchang Road, Jing'an District, Shanghai 200072, China
jqfu@staff.shu.edu.cn
[2] School of Electrical Engineering, Anhui Polytechnic University,
Wuhu 241000, China

Abstract. In order to improve the accuracy of the short-term wind speed forecasting, this paper presents a novel wind speed forecasting model based on least square support vector machine (LSSVM) optimized by an improved Quantum-behaved Particle Swarm Optimization algorithm called up-weighted-QPSO (UPQPSO), which uses a non-linearly decreasing weight parameter to render the importance of particles in population in order to have a better balance between the global and local searching. The developed method is examined by a set of wind speeds measured at mean half an hour of two windmill farms located in Shandong province and Hebei province, simulation results indicate UPQPSO-LSSVM model yields better predictions compared with QPSO-LSSVM and ARIMA model both in prediction accuracy and computing speed.

Keywords: LSSVM · QPSO · UPQPSO · Wind speed forecasting

1 Introduction

Nowadays, wind energy has become one of the most popular and fastest growing renewable energy resources around the world. According to the published data from the world wind energy association (WWEA), by the end of June 2016, the worldwide wind power capacity had reached 456.48 GW, where 24 GW equivalent to about 50 nuclear power generation capacity were new added in the first 6 months of 2016. However, the intermittency and uncertainty related to wind power bring austere challenge to the power quality and system reliability. Wind power is greatly affected by the wind speed and strong positive correlation exits between them. Forecasting the wind speed accurately improves performance and reliability of wind turbines as well as power systems [1]. Recently, various methods and approaches have been put forward to predict power and speed of wind in the literature, such as time series model [2], neural network model [3], support vector machines model (SVM) [4] and LSSVM [5]. Compared with SVM, LSSVM has equality constraints rather than inequality constraints with slack variables and performs well in tackling nonlinear problems. However, learning parameters, kernel function types and kernel parameters have direct effects on the learning accuracy and generalization ability of LSSVM model. In literature [6], LSSVM parameters are optimized by chaotic particle swarm optimization (CPSO) algorithm. In paper [7],

© Springer Nature Singapore Pte Ltd. 2017
K. Li et al. (Eds.): LSMS/ICSEE 2017, Part III, CCIS 763, pp. 32–42, 2017.
DOI: 10.1007/978-981-10-6364-0_4

parameters of LSSVM are optimized and short-term wind power prediction accuracy is improved by genetic algorithm. In literature [8], Gorjaei RG et al. use Particle Swarm Optimization algorithm to optimize LSSVM parameters.

Compared with PSO, QPSO has even better ability to be global convergent. However, it also tends to trap into such local optima as PSO. In this paper, an improved QPSO algorithm called up-weighted-QPSO, using a non-linearly decreasing weight parameter to render the importance of particles in population to have a better balance between the global and local searching, is applied to optimize the learning parameters of LSSVM and build a wind speed forecasting model. Simulation results demonstrate the LSSVM optimized by UPQPSO algorithm perform better than models based on QPSO-LSSVM and ARIMA both in prediction accuracy and computing speed.

2 Methodology

2.1 LSSVM

Based on statistical learning, SVM modeling converts a nonlinear separable problem in the sample space to a linear separable problem in Hilbert space, which avoids the drawback of over fitting existing in traditional machine learning and curse of dimensionality. However, for large-scale problems, the optimization process of SVM has high computational complexity which demands more computing time. LSSVM, possessing equality constraints rather than the inequality constraints with slack variables, is a variant of the standard SVM. On the other hand, a squared loss function is performed in the objective function of LSSVM model, whereas the standard SVM model has a linear combination of slack variables in its objective function.

The principle of LS-SVM is introduced as follows.

Suppose a set of training samples $\{(\mathbf{x}_i, y_i)\}$. $\mathbf{x}_i \in R^d$, $y_i \in R$, y_i is the output variable from $\varphi(\mathbf{x})$, a high dimensional feature space of the input $\mathbf{x}_i$. The regression model is

$$f(x) = \mathbf{w} \cdot \varphi(\mathbf{x}) + b \tag{1}$$

where w and b are the weight vector and bias term respectively.

LSSVM determines the optimal weight vector and bias term by minimizing the following cost function J,

$$\min J(\mathbf{w}, \mathbf{e}) = \frac{1}{2}\|\mathbf{w}\|^2 + \frac{1}{2}C\sum_{i=1}^{n} e_i^2 \tag{2}$$

Subject to the equality constraint:

$$y_i = \mathbf{w} \cdot \varphi(\mathbf{x}) + b + e_i \tag{3}$$

where C is the regularization parameter. e_i is the slack variables.

The Lagrangian of Eq. (2) is

$$L(\mathbf{w}, b, \mathbf{e}, \lambda) = J(\mathbf{w}, \mathbf{e}) - \sum_{i=1}^{l} \alpha_i(\mathbf{w} \cdot \varphi(\mathbf{x}_i) + b + e_i - y_i) = 0 \qquad (4)$$

where α_i is the Lagrange multiplier. By the KKT Theorem, the conditions of optimality are:

$$\frac{\partial L}{\partial \mathbf{w}} = 0 \rightarrow \mathbf{w} = \sum_{i=1}^{l} \alpha_i \varphi(\mathbf{x}_i) \qquad (5)$$

$$\frac{\partial L}{\partial b} = 0 \rightarrow \sum_{i=1}^{l} \alpha_i = 0 \qquad (6)$$

$$\frac{\partial L}{\partial e_i} = 0 \rightarrow \alpha_i = Ce_i \qquad (7)$$

$$\frac{\partial L}{\partial \alpha_i} = 0 \rightarrow \mathbf{w} \cdot \varphi(\mathbf{x}_i) + b + e_i - y_i = 0 \qquad (8)$$

Thus, b and α can be solved from the following set of linear equations after eliminating w and e.

$$\begin{bmatrix} 0 \\ \mathbf{Y} \end{bmatrix} = \begin{bmatrix} 0 & \mathbf{Q}^T \\ \mathbf{Q} & \mathbf{K} + C^{-1}\mathbf{I} \end{bmatrix} \begin{bmatrix} b \\ \alpha \end{bmatrix} \qquad (9)$$

where I is the identity matrix, $\mathbf{Y} = (y_1, y_2 \ldots y_l)^T$, $\mathbf{K} = (k(\mathbf{x}_i, \mathbf{x}_j))_{i,j=1}^{n}$ is the kernel matrix and $k(\mathbf{x}_i, \mathbf{x}_j) = \varphi^T(\mathbf{x}_i)\varphi(\mathbf{x}_j)$ is the kernel function. The Gaussian kernel function is applied in this paper, which is described as follows:

$$k_G(\mathbf{x}, \mathbf{z}) = \exp(-\|\mathbf{x} - \mathbf{z}\|^2)/\sigma^2 \qquad (10)$$

where σ is a constant determining the width of Gaussian kernel.

As a result, given vectors and x_i, the LSSVM regression model for estimating y becomes:

$$f(x) = \sum_{i=1}^{l} \alpha_i k(\mathbf{x}, \mathbf{x}_i) + b \qquad (11)$$

At this time, LSSVM parameter optimization is the optimal combination of kernel function parameter σ and regularization parameter C. An improved quantum particle swarm optimization algorithm is used to obtain the optimal σ and C values of LSSVM.

2.2   UPQPSO

PSO originates from the social behavior simulation of bird flocking or fish schooling, using a set of particles, representing potential solutions to the global optimization problems. Where M particles move through a n-dimensional searching space, each particle i has a position $X_i = (x_{i1}, x_{i2}, \ldots, x_{id})$ and a flight velocity $V_i = (v_{i1}, v_{i2}, \ldots, v_{id})$. The updating of velocity and particle position using the current velocity and the distance from personal best and global best demonstrate as follows:

$$v_{ij}^{k+1} = wv_{ij}^{k} + c_1 r_1 \left(p_{ij} - x_{ij}^{k} \right) + c_2 r_2 \left(p_{gj} - x_{ij}^{k} \right) \tag{12}$$

$$x_{ij}^{k+1} = x_{ij}^{k} + v_{ij}^{k+1} \tag{13}$$

For $j = 1, 2, \ldots, d$. w is inertia weight, c_1 is cognition learning factor, c_2 is social learning factor, r_1 and r_2 are independent random numbers uniformly distributed between 0 and 1. p_{id} represents the best ever position of particle i and p_{gd} is the best position in the swarm.

Clerc and Kennedy express in [9] that convergence of whole particle swarm can be achieved if each particle converges to its local attractor $p_{id} = (p_{i1}, p_{i2}, \ldots, p_{id})$ with coordinates:

$$p_{ij}(t) = \left(c_1 r_1 P_{ij}(t) + c_2 r_2 P_{gj}(t) \right) / (c_1 r_1 + c_2 r_2) \tag{14}$$

We obtain the fundamental iterative equation of QPSO by Monte Carlo method.

$$X_{i,j}(t+1) = p_{i,j}(t) \pm \frac{L_{i,j}(t)}{2} \ln(1/u) \quad u = rand(0, 1) \tag{15}$$

where L is standard deviation. Paper [10] introduces a global attractor point called Mean Best Position (M_{best}) of the population to evaluate the value of $L_{i,j}(t)$. Where:

$$m(t) = (m_1(t), m_2(t) \ldots m_d(t))$$
$$= \left(\frac{1}{M} \sum_{i=1}^{M} P_{i,1}(t), \frac{1}{M} \sum_{i=1}^{M} P_{i,2}(t) \ldots \frac{1}{M} \sum_{i=1}^{M} P_{i,d}(t) \right) \tag{16}$$

Then the value of L is obtained by

$$L_{i,j}(t) = 2\beta \cdot \left| m_j(t) - X_{i,j}(t) \right| \tag{17}$$

And the positions are updated by

$$X_{i,j}(t+1) = p_{i,j}(t) \pm \beta \cdot \left| m_j(t) - X_{i,j}(t) \right| \cdot \ln(1/u) \tag{18}$$

where parameter β is Contraction-Expansion Coefficient and must be set as $\beta < 1.782$ to guarantee the convergence of the particle. We let β decrease from 1.0 to 0.5 linearly which can lead to a generally good performance of the algorithm [11].

In order to balance the convergence speed and diversity, M_{best} is taken place by the global best position P_g, and the new equation is as follows:

$$L_{i,j}(t) = 2\beta \cdot |g_j(t) - X_{i,j}(t)| \tag{19}$$

Information exchanging between particles is of great importance for global searching ability of algorithm, therefore, mean value of global best position and mean best position of swarm instead of single M_{best} position is employed, and then proposed equation is given.

$$L_{i,j}(t) = \beta \cdot \left(|g_j(t) - X_{i,j}(t)| + |m_j(t) - X_{i,j}(t)| \right) \tag{20}$$

In [12], Xi et al. suggested that the determination of the mean best position is not so matched with decision making in real society and thus introduced a linear weighted coefficient into QPSO to calculate M_{best} position.

$$\begin{aligned} m(t) &= (m_1(t), m_2(t) \ldots m_d(t)) \\ &= \left(\frac{1}{M} \sum_{i=1}^{M} a_{i,1} P_{i,1}(t), \frac{1}{M} \sum_{i=1}^{M} a_{i,2} P_{i,2}(t) \ldots \frac{1}{M} \sum_{i=1}^{M} a_{i,d} P_{i,n}(t) \right) \end{aligned} \tag{21}$$

However, "The Golden Mean" is always recognized by most of particles and only the coefficient with non-linear can demonstrate the reasonable decision process. Then up parabolic curve is used to compute the coefficient of particles in this paper and its equation is:

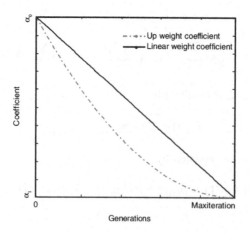

Fig. 1. Different variants of weighted coefficient

$$\alpha = (\alpha_0 - \alpha_1) \times (t/MaxIter)^2 + \alpha_1 (\alpha_1 - \alpha_0) \times (2t/MaxIter) + \alpha_1 \qquad (22)$$

where $\alpha_0 \geq \alpha_1$, α_0 is initial value and α_1 is terminal value, which are set as 1.5 and 0.5. *MaxIter* is the maximum number of iterations. The two different variants of weighted coefficient are given in Fig. 1. In this work, the QPSO with open side up weighted coefficient is called Up-Weighted-QPSO (UPQPSO).

2.3   UPQPSO-LSSVM Modeling

1. Normalize the input data to [0, 1] by Eq. (24) and initialize the least squares support vector machine model.

$$f_i(x) = \frac{f(x) - f_{\min}}{f_{\max} - f_{\min}} \qquad (23)$$

2. Particle initialization and UPQPSO parameters setting: number of particles is 25, particle dimension is 2, number of maximal iterations is 200, nonlinear weighted coefficient α_0 and α_1 are 1.5 and 0.5, Contraction-Expansion Coefficient is decreased from 1.0 to 0.5 linearly.
3. Calculate and compare the fitness function values of each particle by Eq. (24)

$$MAPE = \frac{1}{N} \sum_{i=1}^{N} \frac{|W_{Ai} - W_{Fi}|}{W_{Ai}} \times 100\% \qquad (24)$$

4. Update the particle position by UPQPSO algorithm.
5. Update termination condition: The fitness value is less than the set value 0.08 or the evolutionary algebra is equal to the maximum evolutionary algebra 200, then the parameters σ and c are obtained.
6. The LSSVM regression model of least squares support vector machine is trained by the optimized parameters σ and c.
7. The normalized test data is brought into the trained LSSVM model, and the prediction results are obtained.

The flow chart of short-term wind speed prediction based on LSSVM optimized by UPQPSO algorithm is shown in Fig. 2.

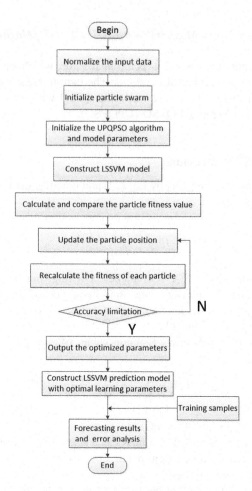

Fig. 2. Flow chart of wind speed forecasting

3 Application

3.1 Data Processing

In order to verify the validity and reliability of the proposed model, we conduct the wind speed forecasting based on the wind speed observations at half an hour intervals measured from two wind power factories in Shandong province (Case 1) and Hebei province (Case 2). The training samples, as shown in Fig. 3, in which the wind speed of Case 2 is more volatile than Case 1, consist of 240 data respectively measured from August 1, 2016 to August 5 in order to forecast wind speed of August 6. And the original 48 data of August 6 is used as testing data in this study. All simulation experiments are carried out 25 times independently by Matlab 2014a in order to evaluate the proposed model and reduce the statistic errors.

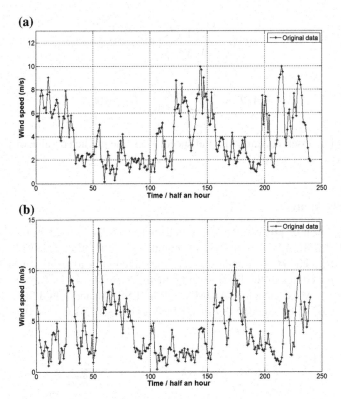

Fig. 3. (a) Original wind speed time series in Shandong. (b) Original wind speed time series in Hebei

For Case 1, the UPQPSO algorithm get the optimal fitness value in generation 10, and the QPSO algorithm obtains the optimal fitness value in generation 25. For Case 2, the UPQPSO algorithm get the optimal fitness value in generation 16, and the QPSO algorithm obtains the optimal fitness value in generation 28, which can be concluded that the UPQPSO algorithm is better than QPSO algorithm in computing speed.

3.2 Evaluation Indices for Wind Power Forecasting Performance

To evaluate the forecasting capability of UPQPSO-LSSVM, three statistical indices are utilized to measure the forecasting accuracy. These indices are the mean absolute error (MAE), root mean square error (RMSE) and mean absolute percent error (MAPE), where small values indicate high forecast performance. These indices are defined as follows:

$$MAE = \frac{1}{N} \sum_{i=1}^{N} |W_{Ai} - W_{Fi}| \tag{25}$$

$$RMSE = \sqrt{\frac{1}{N} \sum_{i=1}^{N} (W_{Ai} - W_{Fi})^2} \qquad (26)$$

$$MAPE = \frac{1}{N} \sum_{i=1}^{N} \frac{|W_{Ai} - W_{Fi}|}{W_{Ai}} \times 100\% \qquad (27)$$

where W_A is the observed value of wind speed, W_F is the predicted value of wind speed, N is the size of data sample.

3.3   Results Analysis

After the optimization process, the 48 normalized testing data is used for prediction. The results of UPQPSO-LSSVM model are compared with QPSO-LSSVM model and ARIMA model both in Case 1 and Case 2, as shown in Fig. 4(a) and (b), meanwhile, the prediction error curves of the three learning methods are shown in Table 1(a) and (b).

For Case 1, MAE errors of ARIMA, QPSO-LSSVM and UPQPSO-LSSVM are 1.2542 m/s, 0.6234 m/s and 0.5489 m/s, RMSE errors are 1.3135 m/s, 0.7134 m/s and

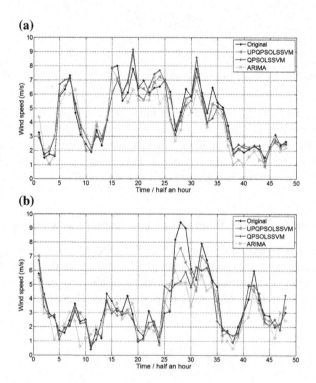

Fig. 4. (a) Comparison between three models in Case 1. (b) Comparison between three models in Case 2

Table 1. Error statistics of two models in (a) Case 1 and (b) Case 2

Methods	MAE (m/s)	RMSE (m/s)	MAPE (%)
(a)			
ARIMA	1.2542	1.3135	14.8323
QPSO-LSSVM	0.6234	0.7134	7.3218
UPQPSO-LSSVM	0.5489	0.6592	6.9849
(b)			
ARIMA	1.3021	1.4213	15.2132
QPSO-LSSVM	0.7234	0.8167	9.3967
UPQPSO-LSSVM	0.6247	0.7436	7.7942

0.6952 m/s, MAPE errors are 14.8323%, 7.3218% and 6.9849% respectively. For Case 2, the forecasting result of all the three models for Case 2 doesn't perform well compared with Case 1 due to volatile wind in Hebei. However, both the UPQPSO-LSSVM and QPSO-LSSVM model outperform ARIMA model for two cases, which suggests the effectiveness of LSSVM model in wind forecasting. It is also obvious that the LSSVM model optimized by UPQPSO algorithm has better forecasting results than model optimized by QPSO, for the reason that UPQPSO algorithm uses a non-linearly decreasing weight parameter to render the importance of particles in population in order to better balance between the global and local searching.

However, it can be found from Fig. 4(b) that the output wind speed of the 28th to 32nd sample point are fluctuating due to the strong volatility and randomness of wind energy. Although the output of UPQPSO-LSSVM prediction model is a little well fitted than QPSO-LSSVM, it is also not ideal for prediction to some extent.

4   Conclusions

In this paper, the improved algorithm UPQPSO, which has stronger global search capability than QPSO algorithm, is proposed to select parameters of LSSVM. The real data set is used to investigate its feasibility in forecasting wind speed of two windmill farms, and forecasting results indicate the proposed model can achieve greater esti-mating accuracy than QPSO-LSSVM model and ARIMA model. However, as shown in Sect. 3.3, it is not ideal for singular wind speed prediction, therefore, the study for data smoothing is sure to fall into the scope of our research.

Acknowledgment. This work was financially supported by the Science and Technology Commission of Shanghai Municipality of China under Grant (No. 17511107002).

References

1. Wei, L., Lin, X.: A new modeling and ultra-short term forecasting method for wind speed time series of wind farm. Power Syst. Clean Energy **31**(9), 78–84 (2015)

2. Zhang, S., Zeng, J., et al.: Application of time series model to prediction of wind speed in wind field. Water Resour. Hydropower Eng. **47**(12), 32–44 (2016)
3. Federico, C., Massimiliano, B.: Short wind speed prediction based on neural network and wavelet analysis. Renew. Energy Resour. **34**(5), 705–711 (2016)
4. Yan, X., Gong, R., Zhang, Q.: Application of optimization SVM based on improved genetic algorithm in short-term wind speed prediction. Power Syst. Prot. Control **44**(9), 38–42 (2016)
5. Fang, B., Liu, D., et al.: Short-term wind speed forecasting based on WD-CFA-LSSVM model. Power Syst. Prot. Control **44**(8), 88–93 (2016)
6. Zhao, C., et al.: Soft sensor modeling for wastewater treatment process based on adaptive weighted least squares support vector machines. Chin. J. Sci. Instrum. **36**(8), 1972–1980 (2015)
7. Ling, W.N., Hang, N.S.H., Li, R.Q.: Short-term wind power forecasting based on cloud SVM model. Electr. Power Autom. Equip. **33**(7), 34–38 (2013)
8. Gorjaei, R.G., et al.: A novel PSO-LSSVM model for predicting liquid rate of two phase flow. J. Nat. Gas Sci. Eng. **24**(5), 228–237 (2015)
9. Clerc, M., Kennedy, J.: The particle swarm: explosion, stability, and convergence in a multi-dimensional complex space. IEEE Trans. Evol. Comput. **1**(6), 58–73 (2002)
10. Sun, J., et al.: A global search strategy of quantum-behaved particle swarm optimization. In: Proceedings on 2004 IEEE Conference on Cybernetics Systems, Singapore, pp. 111–115 (2004)
11. Sun, J., Xu, W.-B., Feng, B.: Adaptive parameter control for quantum-behaved particle swarm optimization on individual level. In: Proceedings on 2005 IEEE International Conference on Systems, Man and Cybernetics. Piscataway, NJ, pp. 3049–3050 (2005)
12. Xi, M., Sun, J., Xu, W.: An improved quantum-behaved particle swarm optimization algorithm with weighted mean best position. Appl. Math. Comput. **205**(2), 751–759 (2008)

Structure Design and Parameter Computation of a Seawater Desalination System with Vertical Axis Wind Turbine

Yihuai Hu[✉], Kai Li, and Hao Jin

Marine Engineering Department,
Shanghai Maritime University, Shanghai 201306, China
yhhu@shmtu.edu.cn

Abstract. This paper proposes a method to firstly convert wind energy into thermal energy by a vertical axis wind turbine and then use thermal energy to evaporate seawater for fresh water generation. The working principle and structure characteristics of the S type vertical axis wind turbine, liquid-stirring heater and seawater evaporation chamber are described. Mathematical calculation of bucket diameter and other parameters of the liquid-stirring heater are carried out according to the driving torque of wind turbine. Evaporating chamber capacity for seawater desalination is determined according to the liquid-stirring heater. These calculation models are introduced including power, torque, tip speed ratio, height and diameter of wind turbine; power, torque, rotating speed, blades diameter and other structural parameters of liquid-stirring heater; diameter, height of evaporator chamber. Generated fresh water from the seawater desalination system under rated wind speed is estimated at 239.1 g/h, which verify the feasibility of this kind of wind-powered seawater desalination method.

Keywords: Vertical axis wind turbine · Liquid-stirring heater · Wind-powered heater · Seawater desalination · Wind energy application · Thermal energy application

1 Introduction

At present, the way of wind-powered seawater desalination is to firstly convert wind energy into electric energy and then use electric energy to drive desalination plant for fresh water generation. Due to several processes of energy conversion this method has low conversion efficiency around 40% [1–3]. Because of the strict requirements of energy conversion from wind power to electricity in the respect of high cutting in and cutting out wind velocity, the wind turbine investment is usually high. The proportion of wind turbine investment accounts for about 75% of the whole wind-powered system, resulting in high cost of wind-powered seawater desalination [4].

China is one of the countries with the lowest fresh water resources. The per capita fresh water resources is only 1/4 of the world average, and the water resources are unevenly distributed [5, 6]. By now, there have been 617 cities in China, but 300 cities are in the shortage of water and 110 cities are in seriously water shortage [7]. Fourteen

© Springer Nature Singapore Pte Ltd. 2017
K. Li et al. (Eds.): LSMS/ICSEE 2017, Part III, CCIS 763, pp. 43–51, 2017.
DOI: 10.1007/978-981-10-6364-0_5

coastal cities amongst eighteen Chinese coastal cities are in water shortage, of which nine cities are in serious water shortage. This situation has become the bottleneck of economic development in these regions [8]. On the other hand, wind energy resources in China are quite rich. The available wind energy resources are about one billion kW [3], which could be used for fresh water generation from seawater by wind power.

A kind of wind-powered seawater desalination device is introduced in this paper, including its structure, working principle and the characteristic analysis of main parts. Mathematical models are established in order to provide a theoretical basis for the design of vertical axis wind turbine, liquid-stirring heater and seawater desalination system.

2   The Structure Design of Wind-Powered Seawater Desalination

The wind-powered seawater desalination device is mainly composed of S type vertical axis wind turbine, liquid-stirring heater, seawater evaporation chamber and bracket, etc. (Fig. 1). The S type vertical axis wind turbine is located on the top of liquid-stirring heater through a bracket, and the seawater evaporation chamber is fixed on the side of liquid-stirring heater through a bracket. The S type vertical axis wind turbine captures wind energy and converts it into mechanical energy of stirring blades. By stirring liquid the stirring blades convert mechanical energy into thermal energy, which is then induced into a vacuum water evaporation chamber to evaporate seawater into fresh water. The structure of wind-powered seawater desalinations is as shown in Fig. 1.

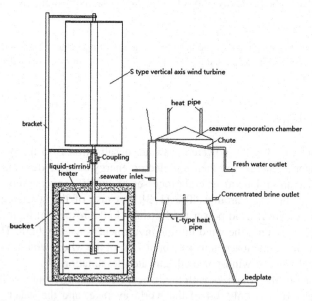

Fig. 1. Structural schematic of wind-powered seawater desalination

The S type vertical axis wind turbine, rotated by wind power, drives the stirring shaft and stirring blades in the mixing bucket through a coupling. Under the action of stirring blades and damping plate in the bucket, running water frictions between water and water, water and bucket wall, water and blades could generate heat. An L-type heat pipe transmits the heat into an evaporation chamber to evaporate the sea water inside. The working principle of wind-powered seawater desalination system is described in Fig. 2.

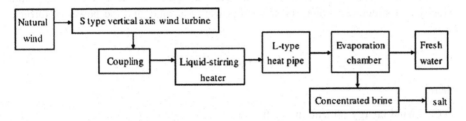

Fig. 2. Working principle of wind-powered seawater desalination

The liquid-stirring heater requires lower wind velocity and could be used more widely than wind-powered electrical heater. The wind-powered seawater desalination has advantages of small size, simple structure, low investment, high flexibility with just natural wind energy. This system could be used for rescue and disaster relief where electrical power is cut off or there are not enough fresh water resources in island, offshore platform or some crowded coastal cities.

3   S Type Vertical Axis Wind Turbine

The two semi-cylindrical blades of S type vertical axis wind turbine are symmetrically installed on two sides of rotating shaft. The cylinders are opposite, and the two blades are staggered as shown in Fig. 3. The diameter of the blades is d_T, the staggered

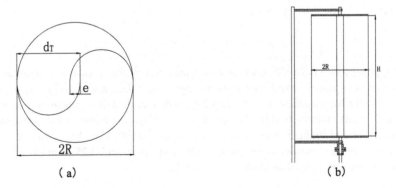

Fig. 3. Structure of type vertical-axis wind turbine

distance is e. When wind blows to the turbine, resistance difference could rotate the turbine. The concave part of air will flow through the staggered space into convex surface behind the surface, and the air flow could be offset by air convex resistance, which improves the working efficiency of wind turbine. When the ratio of staggered distance e and blade diameter d_T is 0.17, the effect will be the best. At this time, the wind energy utilization coefficient will be 0.3, and the corresponding tip speed ratio λ_E will be 0.9 [9]. If there is a rotating shaft in the gap, it should be as small as possible with larger distance.

Mechanical energy generated by the wind-powered turbine is related to wind velocity and structure parameters of wind-powered turbine, and the power of turbine P_T is

$$P_T = C_p \rho_A A V^3 / 2 \tag{1}$$

where

C_p—wind energy utilization coefficient;
ρ_A—air density [kg/m^3];
A—wind turbine sweeping area [m^2];
V—wind velocity [m/s].

The S type wind turbine sweeping area A could be calculated as

$$A = 2HR \tag{2}$$

where

H—the height of wind turbine [m];
R—the radius of wind turbine [m].

The torque of S type wind turbine T_T could be calculated as

$$T_T = \frac{P_T}{\omega_T} = \frac{\rho_A A C_P V^3}{2\omega_T} \tag{3}$$

where ω_T refers to the angular velocity of wind turbine [rad/s]. The tip speed ratio of wind turbine λ could be calculated as

$$\lambda = \omega_T R / V \tag{4}$$

Compared to traditional horizontal axis wind turbine, the vertical axis wind turbine has several advantages as wider utilization range with different blade type, low running noise, suitable for installation near people's residence, beautiful appearance, compatible with surrounding buildings, simple structure, automatic wind orientation, ground-set heat exchanger, long working duration and broad market prospect.

As for a vertical axis wind turbine of 1.08 m in height and 0.25 m in diameter, the parameters are determined as shown in Table 1.

Table 1. Parameters of a vertical axis wind turbine

Parameter	Value
Air density ρ_A	1.225 kg m^{-3}
The maximum wind-powered coefficient C_p^{max}	0.3
Corresponding tip speed ratio λ_E	0.9
Rated wind velocity V	12 m s^{-1}
Wind turbine height H	1.08 m
Rotation diameter of wind turbine R	0.25 m
Blade overlap ratio e/d_T	0.17
Wind turbine sweeping area A	0.54 m^2
Wind turbine power P	170 W

4   Liquid-Stirring Heater

4.1   Characteristic Analysis

The liquid-stirring heater is mainly composed of a stirring bucket, a stirring shaft, several stirring blades and a damping plate as shown in Fig. 4. The stirring bucket is a double-layer structure. The stirring shaft drives the stirring blades, which is connected with the central shaft of wind turbine through a coupling. There are four damping plates attached to the inner wall of stirring bucket. When wind drives the wind turbine to rotate, the stirring shaft drives the stirring blades to rotate and stir the water. Then the water is mixed, impacted and heated between stirring blades, damping plate and inner wall of stirring bucket.

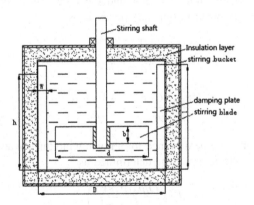

Fig. 4. Structure schematic of liquid-stirring heater

As wind turbine directly drives the stirring blades, the liquid-stirring heater could absorb the energy at any wind velocity and convert most of energy into thermal energy. The liquid-stirring heater could match wind turbine perfectly at any condition, suitable for different wind velocity and different working liquid. The liquid-stirring heater has simple structure, low cost, high reliability and automatic thermal circling. But the heater could be easily corroded by cavitations.

4.2   The Parameter of Liquid-Stirring Heater

When liquid-stirring heater runs steadily, the heating power is equal to the absorbed power of stirring liquid P_L [10].

$$P_L = N_P \rho_L n^3 d^5 \tag{5}$$

$$N_P = f(R_e) \tag{6}$$

where

N_P—stirring power coefficient;
ρ_L—density of the stirred liquid [kg/m^3];
n—stirring speed of blades [r/s];
d—stirring blade diameter [m];
R_e—Reynolds number.

The rotating speed of blades n could be calculated as

$$n = \omega G / 2\pi \tag{7}$$

The rotating torque of stirring blades could be calculated as

$$T_G = \frac{P_L}{\omega_G} = \frac{N_P \rho_L n^3 d^5}{\omega_G} \tag{8}$$

where ω_G is the angular velocity of stirring blades [rad/s].

The transmission between wind turbine and liquid-stirring heater is connected by the coupling. The transmission ratio is 1 and the transmission efficiency is η. The relationship between the wind turbine and heater is as

$$\omega_T = \omega_G \tag{9}$$

$$T_G = \eta T_T \tag{10}$$

$$\frac{\eta \rho_A A C_P V^3}{2\omega_T} = \frac{N_P \rho_L n^3 d^5}{\omega_G} \tag{11}$$

A matched heater could be designed for a given wind turbine according to formula (11). The diameter of heater could be determined as

$$d = \left(\frac{\eta \rho_A A C_P V^3}{2 N_P \rho_L n^3} \right)^{0.2} \tag{12}$$

From formula (4) the diameter could be as

$$d = \left(\frac{4\pi^3 \eta \rho_A A C_P R^3}{N_P \rho_L \lambda^3} \right)^{0.2} \tag{13}$$

For the liquid-stirring heater, the blade tip end velocity $\Lambda = nd$ is often used as its calculated velocity, so the stirring Reynolds number is defined as:

$$Re = \frac{d\Lambda\rho_L}{\mu} = \frac{\rho_L nd^2}{\mu} \tag{14}$$

where μ is the viscosity of liquid [kg/m s].

The Reynolds number R_e not only determines the flowing condition of liquid in the heater bucket, but also plays a decisive role in the heater characteristics. The flowing of liquid in the bucket could be classified as laminar flow when $R_e < 10$, turbulent flow when $R_e > 10^4$, transitional flow when R_e is between 10 and 10^4 [10]. The direct relationship between power coefficient N_P and Reynolds number R_e is as [11]

$$N_P = BRe^z \tag{15}$$

where B and z constants are related to the stirred liquid Reynolds number.

If stirred liquid is water, its Reynolds number could be larger than 10^4 even under low wind velocity. For a S type vertical axis wind turbines running at rated wind velocity with 3 to 6 blades, B equals 5 and z equals 0. Then the liquid absorption power coefficient will be [10, 11]:

$$\frac{P}{\rho_L n^3 d^5} = 5 \tag{16}$$

In this case, the rotating torque of stirring blades is

$$T_G = \frac{5\rho_L d^5}{8\pi^3} \omega_G^2 \tag{17}$$

The optimum diameter of the stirring blades d is

$$d = \left(\frac{0.8\pi^3 \eta\rho_A AC_P^{MAX} R^3}{\rho_L \lambda_E^3}\right)^{0.2} \tag{18}$$

As for the vertical axis wind turbine mentioned above if η is 0.99 and ρ_L is 1000 kg/m^3, the diameter of matched heater stirring blades d will be about 0.16 m.

Other parameters of the matched liquid-stirring heater could be determined as [12, 13]:

(1) For the flat type stirring blade, the stirring bucket inner diameter D could be calculated as $D = 1.25d$.
(2) Stirring blade width b could be calculated as $b = 0.25d$.
(3) Depth of the liquid h could be calculated as $1.11d \leq h \leq 1.38d$.
(4) Damping plate width W could be calculated as $W = 0.1D$.
(5) Damping plate length l has little influence on the liquid-stirring heater and could be calculated as $l = 1.2 h$.

The parameters of matched liquid-stirring heater are determined as shown in Table 2.

Table 2. Parameters of matched liquid-stirring heater

Parameter	Value
Number of stirring blades	4
Number of damping plates	4
Diameter of stirring blades d	0.160 m
Width of stirring blade b	0.04 m
Inner diameter of stirring bucket D	0.40 m
Length of damping plate l	0.265 m

5 Seawater Evaporation Chamber

The seawater evaporation chamber is a bucket-shaped structure with a tapered hood at the top. The seawater evaporation chamber and the liquid-stirring heater are connected with five 30 W L-type heat pipes to transfer the heat generated by the liquid-stirring heater to the seawater evaporation chamber. In the bottom of conical cover of the evaporation chamber, there is a chute along inner wall of cylindrical bucket. Check valve and vacuum tube are connected at the highest point of chute to form air pressure regulating mechanism. The lowest point of chute is connected to fresh water outlet pipe. There are water inlet pipe and a concentrated brine outlet pipe at the bottom of seawater evaporation chamber. There are two 57 W cylindrical heat pipes under inner wall of conical cover, which could speed up the condensation process of water vapor. Sea water enters the evaporation chamber through water inlet pipe, heated and evaporated under vacuum pressure. The water vapor then condenses into water droplets on the conical cover, flows into the chute along conical cover, and flows out of water outlet pipe.

As the inner diameter of heater stirring bucket is 0.4 m, the evaporator chamber is designed as 0.30 m in diameter, 0.20 m in height with a conical cover of 0.15 m in radius. An L type heat pipe of 0.05 m in diameter is inserted into the evaporating chamber and 0.10 m deep into sea water.

The heat transfer between air and conical cover is simplified as air sweeping plate model. The conical cover thickness is about 0.002 m with 49.8 W/(m K) carbon steel heat transfer coefficient and its thermal resistance R is 4.02×10^{-5} m² K/W. So the heat transfer coefficient K between the conical cover and outside air is 35.371 W/(m² K). If ambient air temperature is 20 °C, and the condensation temperature on the inner wall of the conical cover is 60 °C, the condensing power of evaporation chamber φ will be 99.84 W.

If two 57 W heat pipe are installed outside conical cover, the condensing power will be 213.84 W larger than the wind turbine power of 170 W. This will not only ensure the greater condensing power than the heating power, but also accelerate the evaporation rate of seawater. If latent heat of seawater vaporization at atmospheric pressure is 2258 kJ/kg, the working capacity of seawater evaporation chamber will be 150 W driven by the 170 W wind turbine, and the effluent rate of fresh water generator will be 239.1 g/h. If the rated power of wind turbine is designed to be 2 kW, the corresponding effluent rate will be 2.812 kg/h.

6 Conclusion

For the design of a wind-powered seawater desalination system the rated power of vertical axis wind turbine should be firstly determined according to the required amount of fresh water, and then the parameters of wind turbine, liquid-stirring heater and seawater evaporation chamber could be determined according to the mathematical models introduced in this paper. The amount of generated fresh water under 12 m/s wind velocity is estimated to be about 239.1 g/h. If vacuum pressure and heat recovery technologies are used in the sea water evaporation chamber, the amount of generated fresh water will be increased. This kind of wind-powered seawater desalination method could be used with different heat exchangers and in different ways such as water heating, air conditioning and so on.

References

1. Li, C., Li, L.: Research and application of wind power seawater desalination. Electr. Power Surv. Des. **2**, 72–75 (2014)
2. Wang, S., Liu, Y.: A review of desalination system for island development. J. Eng. Stud. **5** (1), 100–114 (2013)
3. Hu, Y.: New Energy and Ship Energy Saving Technology. Science Press, Beijing (2015)
4. Wang, Z., Lu, Z.: The economic analysis on the investment and operation of the wind power project. Renew. Energy Resour. **26**(6), 21–24 (2008)
5. Qi, Q.: Present situation of water resources and the problems and Countermeasures in water resources management in China. China Water Transp. **8**(2), 180–181 (2008)
6. Guoling, R.: Design of Seawater Desalination Project. China Electric Power Press, Beijing (2013)
7. Hao, X., Li, H.: Desalination wind electricity salt production—a proposal for a trinity technology of cleaner production. Energy Conserv. Environ. Prot. **10**, 25–28 (2006)
8. Li, Q.: Suggestions on urban water shortage in China. Resour. Dev. Prot. **1**, 12 (1991)
9. Menet, J.L.: A double-step savonius rotor for local production of electricity: a design study. Renew. Energy **29**(11), 1843–1862 (2004)
10. Hemrajani, R.R., Tatterson, G.B.: Handbook of Industrial Mixing: Science and Practice. Wiley Interscience, New Jersey (2004)
11. Chakirov, R., Vagapov, Y.: Direct conversion of wind energy into heat using joule machine. In: 2011 International Conference on Environmental and Computer Science, Singapore (2011)
12. Wang, K., Yu, J.: Mixing Equipment. Chemical Industry Press, Beijing (2003)
13. Liu, Y., Hu, Y.: Parameter design of stirring wind heating device. Acta Energiae Sol. Sin. **35** (10), 1977–1980 (2014)

Inertial Response Control Strategy of Wind Turbine Based on Variable Universe Fuzzy Control

Le Gao, Guoxing Yu, Lan Liu, and Huihui Song[(⊠)]

School of Information and Electrical Engineering,
Harbin Institute of Technology at Weihai, Weihai 264209, China
songhh@hitwh.edu.cn

Abstract. Wind turbines connect to the power grid through power converters, which makes the lack of effective synchronization relationship between generator speed and grid frequency. Existing strategies usually add an additional controller to utilize the hidden kinetic energy in wind turbines for frequency modulation, but this controller heavily dependents on the droop coefficient and the inertia constant. To avoid the influence of the two factors on the frequency modulation, this paper proposes an adaptive fuzzy control strategy for the inertial response of wind turbine. Its universe can be changed with frequency deviation and rate of change of frequency (ROCOF). By comparison of the synthesize inertia control and common fuzzy control, results of our adaptive fuzzy controller show the advantages of accuracy and applicability.

Keywords: Wind turbine · Inertial response · Synthesize inertia control · Adaptive fuzzy control

1 Introduction

Normally, synchronous generators have inertial response inherently, their rotors respond to the change of frequency automatically, hence they can participate in system frequency modulation. While wind turbines connected to the grid have almost no contribution to the inertial response of power system because of the output power and the grid frequency are decoupled by their power converters [1–3]. As the growing penetration of wind power in the grid, the fluctuation of grid frequency is increasingly notable.

To harness the hidden kinetic energy of wind turbines, extensive study has been carried out and many control strategies have been proposed, which can be generally classified in two types. One type is to allocate a certain capacity of energy storage equipment in wind turbines to augment their participation in frequency modulation [4]. Whereas the energy storage equipment will raise the cost and impair the economic efficiency. The other type is to add additional frequency control, such as inertia control [3], over-speed control [5], and pitch control [6]. As a typical additional frequency control, synthesize inertia control is the most common used method to achieve the

© Springer Nature Singapore Pte Ltd. 2017
K. Li et al. (Eds.): LSMS/ICSEE 2017, Part III, CCIS 763, pp. 52–62, 2017.
DOI: 10.1007/978-981-10-6364-0_6

inertia response of wind turbines [7–9], which consists of inertial control and droop control. It varies according to the rate of change of frequency df/dt and frequency deviation, and offers rotational inertia promptly like traditional synchronous generators. This method depends on the inertia constant and droop coefficient, which have crucial effects on system stability and response time. However, these two parameters are hard to be calculated, thus the promotion of synthesize inertia control is constrained.

Fuzzy control is a rule-based control method with greater robustness and less dependence on the model, thus the effects of disturbance and changing parameters are reduced. In [10], a fuzzy control for inertia response is proposed and its feasibility is validated. Nevertheless, as a conventional fuzzy control, its control performance varies largely with different universes. To solve this problem, adaptive fuzzy control for variable universe introduces a stretch factor that can vary with the control or the output, which is successively applied in the area of robot control [11, 12]. It has an obvious advantage in accuracy and response speed.

In order to solve the problem that the traditional integrated inertial control is too dependent on the parameter setting, improve the adaptability of the fuzzy control, and Reconstruct the coupling between speed and frequency. In this paper, an adaptive fuzzy controller for a variable universe will be designed for a wind turbine with "hidden" kinetic energy, and the reminder of this paper is organized as follows: In Sect. 2, synthesize inertia control is presented briefly; in Sect. 3, The variable universe fuzzy controller of wind turbine is designed; and in Sect. 4, the influence of different parameters of the synthesize inertia control and the performance of the conventional fuzzy control and the adaptive fuzzy control are testified and analyzed.

2 Synthesize Inertial Control

For synchronous generator, the kinetic energy stored in the rotor is:

$$E_k = \frac{1}{2}J\omega^2 \tag{1}$$

When the system frequency changes, the kinetic energy released from the synchronous generator rotor can be converted into additional active power to support frequency:

$$P = \frac{dE_k}{dt} = J\omega\frac{d\omega}{dt} \tag{2}$$

Then, the inertia time constant H can be described as:

$$H = \frac{E}{S} = \frac{J\omega_s^2}{2S} \tag{3}$$

where S and ω_s are base values of the active power and the rotor speed respectively. Combining (2) and (3), we can have:

$$\bar{P} = 2H\bar{\omega}\frac{d\bar{\omega}}{dt} \tag{4}$$

where $\bar{P} = P/S$ and $\bar{\omega} = \omega/\omega_s$ are per-unit variable.

The synthesize inertial controller is designed basing on the consideration that making wind turbine simulates the response characteristics of the synchronous motor. The control scheme is shown in Fig. 1.

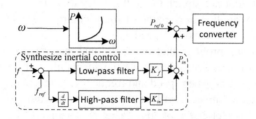

Fig. 1. Synthesize inertial controller

In this controller, the existence of the high pass filter is to avoid the influence of low frequency deviation thus keep clear differential signal. There, the control coefficients K_{in} and K_f are generally taken as:

$$K_{in} = -2H, \ K_f = -\frac{1}{R} \tag{5}$$

Inertial output is:

$$P_{in} = K_{in}\frac{d\Delta f}{dt} + K_f\Delta f \tag{6}$$

It can be see that the coefficients K_{in} and K_f have great influence on the compensation power and the system frequency.

3 Fuzzy Control

3.1 Conventional Fuzzy Control

Fuzzy Control doesn't depend on droop coefficient and inertia constant and it has strong robustness against parameter changes. The control scheme of fuzzy control is illustrated as Fig. 2. The fuzzy controller takes the frequency deviation and the ROCOF as the input signals, and handles them by three steps: fuzzification, fuzzy reasoning, defuzzification, then obtains the compensation power P_{in} as output.

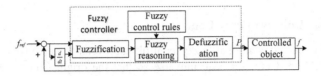

Fig. 2. Control scheme of fuzzy control

The three steps are described as follows.

In order to map the actual values to the fuzzy sets, a fuzzification process is needed. We set the universe of the frequency deviation, the ROCOF and the output variable as $[-E,E]$, $[-E_c,E_c]$ and $[-U,U]$ respectively. The fuzzy set of each variable above contains seven fuzzy linguistic variables: "negative big (NB)", "negative middle (NM)", "negative small (NM)", "zero (ZO)", "positive small (PS)", "positive middle (PM)", "positive big (PB)". And the membership functions are illustrated as Fig. 3.

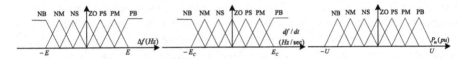

Fig. 3. Membership functions of the inputs and the output variable

The next step is to realize fuzzy reasoning. Fuzzy rules are the core of controller. And the rules of our fuzzy controller are designed as Table 1.

Table 1. Fuzzy rules

E_C	E						
	NB	NM	NS	ZO	PS	PM	PB
NB	PB	PB	PM	PM	PS	ZO	ZO
NM	PB	PM	PS	PS	PS	ZO	NS
NS	PM	PM	PS	PS	ZO	NS	NS
ZO	PS	PS	PS	ZO	NS	NS	NM
PS	PS	PS	ZO	NS	NS	NM	NM
PM	PS	ZO	NS	NM	NM	NM	NB
PB	ZO	ZO	NM	NM	NM	NB	NB

Defuzzification is the third step to design fuzzy controller. Here we use the centroid method for defuzzification. If the membership function of the set P on the universe U is $P(u)$, and the corresponding abscissa of the area center is u_o, then:

$$u_o = \frac{\int_U P(u)u\,du}{\int_U P(u)\,du} \tag{7}$$

3.2   Variable Universe Fuzzy Control

Different from the conventional fuzzy control, the variable universe fuzzy control introduces a stretch factor for each universe so that the original universe of the frequency deviation, the ROCOF and the output variable turns into:

$$E_1 = [-E\alpha_1(r_1), E\alpha_1(r_1)] \tag{8}$$

$$E_{C1} = [-E_c\alpha_2(r_2), E_c\alpha_2(r_2)] \tag{9}$$

$$U_1 = [-U\beta(u), U\beta(u)] \tag{10}$$

We set the stretch factors of input variable universe as:

$$\alpha_i = 1 - \lambda_i e^{-k_i r_i^2} \tag{11}$$

where k is relevant to the sensibility of the control. The larger the values of k is, the bigger stretch factor α_i is; and λ determines the range of the universe. The bigger the values of λ is, the smaller stretch factor α_i is; r_i represents the input variable. When r_i increases, α_i also increases and enlarges the range of input variable universe to accelerate response speed. Otherwise, α_i will decreases so that the range of input variable universe becomes smaller and improve control accuracy.

The stretch factor of output variable universe is:

$$\beta(u) = \sum_{i=1}^{n} \int_0^u K_i r_i(\tau)\,d\tau + \beta_0 \tag{12}$$

where n represents the number of input variables and K_i is a parameter can be tuned. β_0 is the initial value that we set into constant 1.

In sum, the diagram of the variable universe fuzzy controller we designed for wind turbine is described as Fig. 4.

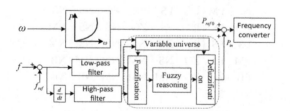

Fig. 4. Variable universe fuzzy controller

4 Results and Discussion

A typical model is established in MATLAB/Simulink as shown in Fig. 5. It contains two DFIG wind turbines (300 MW each), two diesels (700 MW each), a load (1500 MW) and a resectable load (150 or 300 MW). These devices are connected via a 30 kV bus.

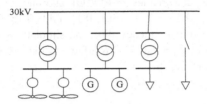

Fig. 5. Models and parameters

4.1 The Influence of Parameters on Synthesize Inertia Response

In order to study the effect of parameters K_{in} and K_f in the synthesize inertial inertia control, a 150 MW load is added to produce a frequency dip at 20 s. For the one hand, K_{in} is set at 0 and K_f is variable. Then we get the frequency with different value of K_f whose base value is $K_1 = 0.4$ shown in Fig. 6.

It can be seen from Fig. 6. That when K_f is less than K_1, the effect of frequency modulation is not obvious and the response is slow. With the increase of K_f, the frequency dip will be alleviate gradually and the response speed becomes faster. When K_f is more than $4K_1$, the transient response becomes better however the steady response becomes worse thus the stability is influenced. When K_f is more than $6K_1$, the stability of the system is broken.

For the other hand, K_f is set at K_1 and K_{in} is variable, we can get different output frequency with different value of K_{in} whose base value $K_0 = 0.216$ as shown in Fig. 7.

When K_{in} is less than K_0, the effect of frequency modulation is not obvious. With the increasing in K_{in}, its dynamic effect in ROCOF is getting better, but the transition time is gradually increasing. when K_{in} reaches $4K_0$, the transition time has been more than 15 s.

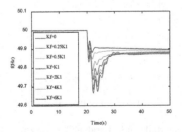

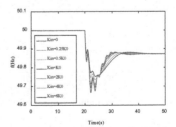

Fig. 6. The frequency with different K_f **Fig. 7.** The frequency with different K_{in}

In brief, the parameters K_f and K_{in} in inertia control have significant effects on the frequency modulation. The setting of K_{in} contributes to a good dynamic response about the frequency changing rate, but the transient time will be relatively prolonged. The setting of K_f provides more energy to participate in the frequency modulation, but the system will be unstable because a large K_f.

4.2   Fuzzy Control with Different Universe

Setting universe of the input variables Δf ROCOF and output inertia power P_{in} in the fuzzy control which are $[-0.25, 0.25]$, $[-0.1, 0.1]$ and $[-0.15, 0.15]$. A 150 MW load is added at 20 s and causing frequency dip. The simulations are carried out by the synthesize inertial control ($K_{in} = K_0$, $K_f = K_1$) the and fuzzy control respectively. The Comparisons are shown in Fig. 8.

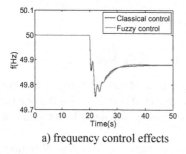
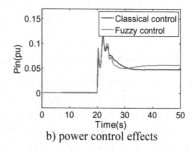

a) frequency control effects   b) power control effects

Fig. 8. Comparison of the effects between synthesize inertial control and fuzzy control when 150 MW load is added

In Fig. 8, the dynamic of the frequency and compensation power of the synthesize inertial control and the fuzzy control are similar. It means that the fuzzy control can realize the function of the synthesize inertial control.

Without changing the fuzzy control universes, we alter a 150 MW load into 300 MW which is added at 20 s. Then the synthesize inertial control and fuzzy control comparison results shown in Fig. 9.

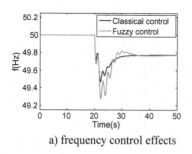
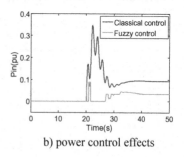

a) frequency control effects   b) power control effects

Fig. 9. Comparison of the effects between synthesize inertial control and fuzzy control when 300 MW load is added

When the load becomes larger, the frequency dip increases, the maximum frequency deviation is about 0.7 Hz and the output power is more than 0.3 pu. The little range of the input and output universe lead to almost no contribution to the inertial response of the fuzzy control.

Changing the value of the input variables Δf, df/dt and output inertia power to $[-1, 1]$, $[-0.5, 0.5]$ and $[-0.5, 0.5]$. The 150 MW and 300 MW loads are added in 20 s respectively, and the results are shown in Figs. 10 and 11.

In Fig. 10 although the fuzzy control can modulate the frequency by utilizing the inertia of wind turbine, its accuracy and response speed are lower than that of the synthesize inertial control. In Fig. 11, the frequency dip becomes larger, the frequency control effects of two kinds of control are similar. The changed universes are more suitable for the 300 MW load dip.

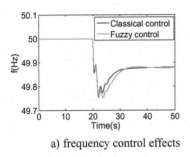
a) frequency control effects

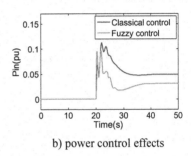
b) power control effects

Fig. 10. Comparison of the effects between synthesize inertial control and fuzzy control after changing the universes when 150 MW is added

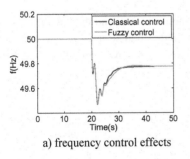
a) frequency control effects

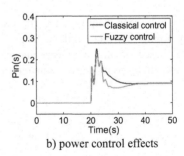
b) power control effects

Fig. 11. Comparison of the effects between synthesize inertial control and fuzzy control after changing the universes when 150 MW is added

Figures 8, 9, 10 and 11 simulation results show that:

(1) The fuzzy control can utilize the kinetic energy of wind turbine to realize the inertial response. The conventional fuzzy control can achieve similar control effects as the synthesize inertial controller if the fuzzy control universe is suitable.

(2) The fuzzy control avoid the existing of K_{in} and K_f in the fuzzy control, thus it is not affected by the change of parameters. However, the universe of inputs and outputs affect it's performance heavily. When the universe is too large, the accuracy and the response speed are not ideal; when the universes are too small, the fuzzy control has little contribution to the inertial response

4.3   Variable Universe Fuzzy Control

To avoid the dependence of the conventional fuzzy controller on universe, we designed the variable universe fuzzy controller, whose initial input variable Δf is set to [−1.2, 1.2], initial universe of df/dt to [−0.5, 0.5], initial universe of output variable to [−0.5, 0.5]. 150 MW load is added at 20 s. The simulation results based on the synthesize inertia control ($K_{in} = K_0$, $K_f = K_1$) and the variable universe fuzzy control are compared, as shown in Fig. 12.

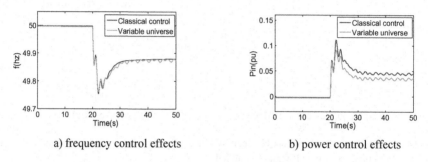

a) frequency control effects   b) power control effects

Fig. 12. Variable universe fuzzy control when 150 MW load is added

It can be seen from Fig. 12. The synthesize inertia control and the variable universe fuzzy control can get similar results. 300 MW load is added at 20 s. The simulation results based on the synthesize inertia control ($K_{in} = K_0$, $K_f = K_1$) and the variable universe fuzzy control are compared, as shown in Fig. 13.

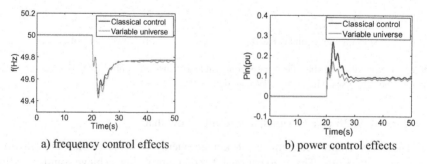

a) frequency control effects   b) power control effects

Fig. 13. Variable universe fuzzy control when 300 MW load is added

The variable universe fuzzy controller is able to vary with stretch factor, and it can still achieve the similar control effect as synthesize inertia controller when load is changed. Figures 12 and 13 indicate that variable universe fuzzy controller is able to regulate frequency via wind turbine inertia, and can achieve the same control effect as the synthesize inertia controller does. Meanwhile it also can solve the problem that fuzzy control relies too heavily on universe by modulating universe automatically.

5   Conclusions

In this paper, the fuzzy control strategy is applied to the inertial response control of wind turbine, and the main conclusions are as follows:

(1) When using synthesize inertial controller, K_{in} and K_f have a significant effect on the control effects in terms of the transition time and stability.
(2) The fuzzy control can achieve the same control effects as the synthesize inertial controller if the fuzzy control universe is suitable, but the universe of the input and the output variables have a great influence on the effects of the fuzzy control.
(3) The universe of the adaptive fuzzy controller we designed can automatically change according to the stretch factor, and no longer depends on the parameter settings. The controller can achieve the same control effects as the synthesize inertial controller with good parameter setting.

Acknowledgment. The work was supported by the NNSF of China (nos. 61403099 and 61511140293), the NNSF of Shandong Province (nos. 2014BSA10007 and 2014J14LN92), and Foundation in Harbin Institute of Technology (No. HIT.NSRIF.2014138).

References

1. Lias, R., Damian, F.: Emulated inertial response from wind turbines: gain scheduling and resource coordination. IEEE Trans. Power Syst. **31**(5), 3737–3755 (2016)
2. Zhang, Z.S., Sun, Y.Z., Lin, J., Li, G.J.: Coordinated frequency regulation by doubly fed induction generator-based wind power plants. IET Renew. Power Gener. **6**, 38–47 (2012)
3. Johan, M., Sjoerd, W.H., Wil, L.K., Ferreira, J.A.: Wind turbines emulatinng inertia and supporting primary frequency control. IEEE Trans. Power Syst. **21**(1), 433–434 (2006)
4. Muyeen, S., Hasanien, H., Tamura, J.: Reduction of frequency fluctuation for wind farm connected power systems by an adaptive artificial neural network controlled energy capactior systems. IET Renew. Power Gener. **6**(4), 226–235 (2012)
5. Teninge, A., Jecu, C., Roye, D.: Contribution to frequency control through wind turbine inertial energy storage. IET Renew. Power Gener. **3**(3), 358–370 (2009)
6. Wu, Z.P., Gao, W.Z., Wang, J.H.: A coordinated primary frequency regulation from permanent magnet synchronous wind turbine generation. In: IEEE in Power Electronics and Machines in Wind Applications, pp. 1–6. IEEE Press, Denver (2012)
7. Xue, Y.C., Tai, N.L.: Review of contribution to frequency control through variable speed wind turbine. Renew. Energy **36**(6), 1671–1677 (2011)

8. Hwang, M., Muljadi, E., Park, J.W., Sorensen, P., Kang, Y.C.: Dynamic droop based inertial control of a doubly fed induction generator. IEEE Trans. Sustain. Energy **7**(3), 924–933 (2016)
9. Itani, S.E., Annakkage, U.D., Joos, G.: Short-term frequency support utilizing inertial response of DFIG wind turbine. In: IEEE Power and Energy Society General Meeting, pp. 1–8 (2011)
10. Konstantina, M., Raquel G., Monica, A., Evangelos, R.: Implementation of fuzzy logic controller for virtual inertia emulation. In: 2015 International Symposium on Smart Electric Distribution Systems and Technologies, pp. 606–611 (2015)
11. Huang, W.X., Sun, P., Long, H.Y.: Design and realization of a four-wheeled robot based on fuzzy variable universe control. In: 35th Chinese Control Conference, pp. 3761–3765. CCC Press, Chengdu (2016)
12. Lu, W.M., Gao, Y.: Variable universe fuzzy adaptive PID control in the Digital servo system. In: 8th International conference on Intelligent Human-Machine Systems and Cybernetics, pp. 204–207. IHMSC Press, Beijing (2016)

System Frequency Control of Variable Speed Wind Turbines with Variable Controller Parameters

Guoyi Xu[1(✉)], Chen Zhu[1], Libin Yang[2], Chunlai Li[2], Jun Yang[2], and Tianshu Bi[1]

[1] State Key Laboratory of Alternate Electrical Power System with Renewable Energy Sources, North China Electric Power University, Beijing, China
xu_gy@ncepu.edu.cn

[2] QingHai Province Key Laboratory of Photovoltaic Gird Connected Power Generation Technology, Xining, China

Abstract. System operators are require wind power plants to provide system frequency control to secure safe operation of power systems. This paper first discussed the available amount of kinetic energy from wind turbines which could be released to provide extra power, minimum rotor speed of wind turbines operate at different condition to provide system frequency control is defined. The strategy to determine the wind turbine frequency controller parameter values is proposed, which will release all the available kinetic energy to provide system frequency support, this strategy also make sure the wind turbine rotor speed drop during frequency support is within the limited range, which ensure the stable operation of the turbine. The proposed strategy is tested by simulations carried out with Matlab/Simulink, which demonstrated the improvements on wind turbine operation and system frequency control effect.

Keywords: Wind power · Kinetic energy · Minimum rotor speed · Variable parameters

1 Introduction

With wide installation of wind power, the penetration of wind power in power system keep increasing. As the wind speeds are intermittent, in order to maximize the power generation, power electronic interface is adapted to connect wind turbines (WT) to the power system, the power system has become more complicated than ever. Normally, the variable speed WTs are not capable to participate system frequency response, since the WT is decoupled from the gird by the power electronic interface, which means decrease of inertia of the system. The system will have a large frequency nadir and large rate of change of frequency (ROCOF) when there is a power imbalance in the system.

In order to secure safe operation of the power system, some system operators are require WTs to provide ancillary services, like frequency control [1], especially when the renewable generation is high in the system. The additional control strategies of the WTs to participate grid frequency regulation have been proposed in many papers. The de-load energy [2, 3] and kinetic energy [4–7] from the variable speed WTs are the two

© Springer Nature Singapore Pte Ltd. 2017
K. Li et al. (Eds.): LSMS/ICSEE 2017, Part III, CCIS 763, pp. 63–73, 2017.
DOI: 10.1007/978-981-10-6364-0_7

extra energy sources to provide addition power for WTs. The variable speed WTs are operate at the optimal speed to harvest the maximum power from the wind under normal operation, the WTs are not able to provide power reserve like the conventional generators. De-load operation of the WT can operate the WT with a power reserve by rotor speed control (low wind speeds) or pitch control (high wind speeds), the reserved power can be released by a droop controller when system frequency drops. For de-load control of WT, some power generation is lost, it is not economic.

Due to the large rotating mass of the WTs in normal operation, although there is no power reserve to provide extra power from the turbine, the large kinetic energy can be utilized to provide extra power for a short time from the WT. Different strategies have been reported [8–10]. One method uses $K_1 d\Delta f/dt + K_2 \Delta f$ to generate the extra power from the WT's kinetic energy, when using this method, the parameter needs to be carefully selected, if the parameters K_1 and K_2 is too large, the WT will have a large speed drop, which will cause unstable of the WT, the other method is to generate a constant power output during the WT frequency control process, it has been implemented in industrial [10]. However, after the WT release the kinetic energy in the recovery period, a second frequency drop might occur due to the large power imbalance. Some research propose to use energy storage system with WT to provide system frequency support [11], however, it requires extra investment.

The available kinetic energy of the WT is different due to different operation status of the WT, it is important to consider the operation status of the WT when design the frequency controller parameters. Frequency control strategies of variable speed WTs with variable controller parameters is discussed in this paper. The control strategy is based on the method of constant output power utilizing WT kinetic energy. The available kinetic energy of the WT under different operation points is quantified, then controller parameters are determined according to the available kinetic energy. This control strategy aims to fully utilize WT's available kinetic energy under different operation status and ensure the stable operation of the WT. The rest of the paper is organized as follows. The WT model used in this paper is presented in Sect. 2. Quantification analysis of the kinetic energy of the WT and controller parameter selection method is studied in Sect. 3. Simulation analysis is carried out in Sect. 4. Finally, Sect. 5 draws conclusions.

2   WT Model

There are different types of WTs, in the power system, the most installed types of WTs are the variables speed WTs. According to the aerodynamic characteristics, the power captured by the turbine and the conversion efficiency can be described by a set of mathematics expressions [12] as

$$P_m = \frac{1}{2}\rho A V_w^3 c_p(\lambda, \beta) \tag{1}$$

$$c_p(\lambda, \beta) = c_1\left(\frac{c_2}{\lambda_i} - c_3\beta - c_4\right)e^{\frac{-c_5}{\lambda_i}} + c_6\lambda \tag{2}$$

$$\frac{1}{\lambda_i} = \frac{1}{\lambda + 0.08\beta} - \frac{0.035}{\beta^3 + 1} \tag{3}$$

where P_m is the mechanical power, ρ is air density, A is the blades swept area, V_w is the wind speed, λ is the tip speed ratio, β is the pitch angle, c_1, c_2, c_3, c_4, c_5, c_6 are the coefficients.

The WT with the power curve shown in Fig. 1 is used for analysis, due to the variable speed nature of the WT, the WT rotor speed can vary between 0.7 pu and 1.2 pu according to different wind speeds. During operation of the variable speed WTs, when wind speeds are low, WT control its rotor speed at the optimal values and operate on the maximum power curve (curve AB in Fig. 1). When the rated rotor speed increase to 1.2 pu, the WT will increase active power production with the speed controller limit the rotor speed to 1.2 pu (curve BC in Fig. 1). At high wind speeds, the WT generated power will be limited by the pitch control to 1.0 pu (point C in Fig. 1).

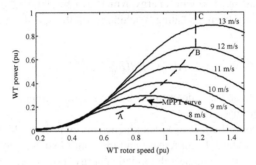

Fig. 1. WT operation power curve.

In this paper we chose the fully rated converter based WT for analysis, the generator is a permanent magnet machine and driven by the turbine directly. The stator of the generator is connected by an AC/DC converter to the DC link, and the DC side is connected by a DC/AC converter to the power system. Compare to the DFIGs, this type of WT has a better performance during fault, and the use of gear box can be avoid, which has attracted much interest of the developers. The proposed frequency control strategy in this paper also can be used with DFIGs. Detailed description and control of the WTs can be found in paper [12].

3   Frequency Control of WT with Variable Controller Parameters

Normally the WT is operate in maximum power production mode, the WT generate the maximum active power, thus there is no extra power which could be generated from the WT to provide system frequency control. Due to variable operation characteristics of the variable speed WTs, kinetic energy from the rotating mass of the WT can be released to generate a higher output power for a short period to provide system frequency support, although it is temporary available, as the power electronics has a fast power control capability, the WT could play an important part in system frequency control. Furthermore, unlike the conventional generators, the amount of extra energy from WT for system frequency support is dependent on the operation status of the WT before frequency event, when design the frequency controller and select the controller parameter values, it is important to consider the operation status of the WT, otherwise, excessive power supply from the WT will cause unstable operation of the WT. In this section, the available kinetic energy of the WT under different operation status is discussed, and then the controller parameter is selected according to the operation status of the WT before system frequency event.

3.1   Kinetic Energy of WT

When the WTs are controlled in normal optimal rotor speed operation, the WTs have large inertia and kinetic energy [9]. The capacity and size of WTs are still increasing, which means increasing of kinetic energy which can be obtained from the WT. The released kinetic energy from a rotating machine when the speed is drop from ω_0 to ω_1 is calculated by

$$\Delta E = 2HS(\omega_0^2 - \omega_1^2) \tag{4}$$

where H is the inertia and it is determined by the specification of the machine, S is the rated capacity of the machine.

The frequency control effect of WT's kinetic energy is dependent on the amount of kinetic energy which the WT can release and the control strategy to release them. WT rotor speed will decrease to release the kinetic energy, how much energy the WT can release is dependent on the rotor speed variation. Normally, when the kinetic energy is adopted to provide frequency control, the low rotor speed limit is 0.7 pu for most of the variable speed WTs. However, as the power production of the WT is highly related with the rotor speed, the mechanical power captured by the turbine will drop to a very low value when the rotor speed varies to 0.7 pu, thus it reduces the total power from the WT. By calculating the mechanical power drop rate with the variation of rotor speed, we could get the mechanical power decrease speed at different operation point, which is shown in Fig. 2.

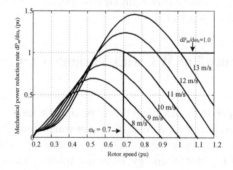

Fig. 2. Mechanical power drop rate dpm/ωr.

From Fig. 2 we can observe the mechanical power drop rate for different wind speeds in the WT operation range is similar to the mechanical power curve of the WT. As observed for different wind speeds, $dP_m/d\omega_r$ is larger for high wind speeds, which indicates larger mechanical power drop for high wind speeds for the same rotor speed variation. For a particular wind speed, if the rotor speed gets lower, the mechanical power of the WT drops faster. For example, in normal condition, with wind speed of 12 m/s, the WT is controlled at optimal speed to produce maximum power, which is 1.2 pu and the WT generate 0.7 pu active power, with the rotor speed varies down to 0.7 pu, power generation from the WT will change from 0.7 pu to 0.33 pu, although the speed drop will release some kinetic energy, it is not worth to operate the WT at such speeds to sacrifice the mechanical power abstract from wind. In order for the WT to harvest more mechanical power during frequency control, rather than to operate the WT to 0.7 pu, we define a minimum rotor speed for different wind speeds. By increase the low rotor speed limit, although the released kinetic energy is reduced due to a narrow rotor speed variation, the turbine can have a higher mechanical production which is beneficial for the total output power and the speed recovery after the WT release kinetic energy. The pre-defined low rotor speed limitation for WT during frequency control is shown in Fig. 3. For wind speeds <11 m/s, rotor speed variation range is relatively narrow, in order to provide enough kinetic energy during WT's frequency control, the low speed limit is set to 0.7 pu which is the limit of the WT in normal operation. For the wind speeds between 11 m/s and 12 m/s, the WT can be operate from the optimal speed to 0.7 pu which has a larger operation range, in order to reduce the large mechanical drop, the low rotor speed limit is set to ensure that $dP_m/d\omega_r < 1.0$, and the corresponding rotor speed is illustrated in Fig. 3. For wind speeds >12 m/s, the WT operate at 1.2 pu, $dP_m/d\omega_r$ is very large, thus the low rotor speed limit is set to be the value of wind speed of 12 m/s to reduce mechanical power drop.

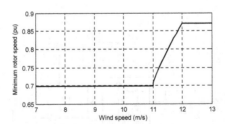

Fig. 3. Pre-defined rotor speed low limit.

3.2 Frequency Control Strategy

To carry out the frequency control strategy, extra controller shown in Fig. 4 is designed. This extra controller is added in the WT generator side controller, which generate the power reference P_{ref} for the WT. In order to activate the controller when system frequency drops, a phase lock loop is used to measure system frequency and then the deviation of the frequency is filter before enter the hysteresis controller, the hysteresis controller is used to enable WT's frequency control when the frequency deviation is below certain limit, to avoid activate WT's frequency control too frequent for small frequency drop, the value of f_m can be set to a lower value. P_{ref} can be set to have different shapes during frequency support. a constant power during frequency support for the WT is adopted in this paper, which is set as

$$P_{ref} = P_0 + \Delta P_{dec} \tag{5}$$

where P_0 is the output power of the WT when system frequency drops, ΔP_{dec} is the increased output power from the kinetic energy, ΔP_{dec} is the controller parameter, which is selected according to the operation status and the frequency support duration of the WT, the selection of ΔP_{dec} will be investigated in the next section. The value of ΔP_{dec} for different operation status is stored in a table, when system frequency drop occurs, it is read from the table according to wind speed and duration to release kinetic energy T_{dec}. When P_{ref} is higher than the mechanical power, the imbalance torque will drop the WT rotor speed from the initial value, in this strategy, after the WT's rotor speed drops to the minimum value which is discussed in Sect. 3.1. The WT is controlled to increase its speed in order to return to optimal speed to capture the maximum power. In the rotor speed recover stage, power reference is set as $P_{ref} = P_m - \Delta P_{acc}$. In order to reduce the power imbalance during the speed transit, ΔP_{acc} is selected to be a small value. Which is less than the mechanical power, thus the rotor will increase its speed. At some point in the speed recovery stage, P_{ref} will be smaller than the maximum power on the MPPT curve, then the WT will change to operate on the MPPT curve and back to normal operation. To secure safe operation of the WT, a logic gate is added in the frequency controller as shown in Fig. 4, during WT frequency control process, if wind speed drops, the rotor speeds drops below the allowed minimum value, then the frequency control should stop.

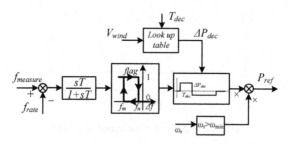

Fig. 4. Frequency controller of wind turbine.

3.3 Controller Parameter Calculation

In order to effectively utilize the available kinetic energy and make sure the WT rotor speed does not drop below its limitation, the controller parameter ΔP_{dec} is determined according to the operation status of the WT when it enters the frequency control. The motion Eq. (6) is used to describe the speed variation of a machine

$$2H\frac{d\omega}{dt} = P_m - P_e \tag{6}$$

where P_m and P_e are mechanical power and electromagnetic power of the WT. Based on (6), we can get the diagram shown in Fig. 5 to calculate the rotor speed variation of the WT which is controlled to carry out frequency support using the control strategy of (5). By setting different values of ΔP_{dec} in (5), rotor speed variation can be obtained. The duration of the WT from the initial operation rotor speed to reach the pre-defined minimum speed can be obtained from the simulation. Take a 2 MW variable speed WT as an example, assume the inertia constant H of the WT is 6 s, the result of the relationship between ΔP_{dec} and T_{dec} for 11.5 m/s wind speed is plotted in Fig. 6. After calculate the results of different wind speeds for the WT, the results can be stored in a table, system operators can determine T_{dec}, when WT's frequency control is activated, the value for the frequency controller ΔP_{dec} can be get from the table as shown in Fig. 4.

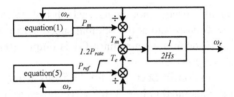

Fig. 5. Calculation of WT rotor speed.

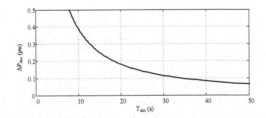

Fig. 6. ΔPdec and Tdec with wind speed of 11.5 m/s.

4 Simulation Analysis

To test the proposed control strategy and verify the correction of the controller parameters, simulation analysis has been carried out in this part. The power system model shown in Fig. 7 is developed with Matlab/Simulink. There are two synchronous generators in the power system, G1 rated at 150 MW represent local generation,

G2 rated at 350 MW represent remote generation. The total load of the simulated system is 420 MW, all the loads are constant power loads. In this power system, it has 50 variable speed WTs each rated at 2 MW. In the simulation, an equivalent fully rated converter base WT model is used to represent the wind power plant. The parameter of the transmission lines are illustrated in the figure.

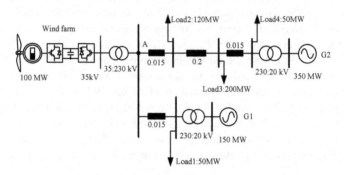

Fig. 7. Simulated power system.

During simulation, wind speed is not change, which is 11.5 m/s, and the WT is in unit power factor control and operate at the optimal speed 1.15 pu generating 60 WM active power to the grid. Synchronous generator G1 is output 100 MW active power, G2 output 260 MW active power.

According to the discussion in Sect. 3.1, for 11.5 m/s wind speed, the low rotor speed limit is at 0.8 pu as illustrated in Fig. 3. If T_{dec} is chosen to be 20 s, the frequency controller parameter ΔP_{dec} is 0.18 according to the calculation in Sect. 3.3. In order to compare the results, WT minimum rotor speed of 0.7 pu during frequency control is simulated. The duration for the WT to release kinetic energy is also 20 s, the corresponding value for ΔP_{dec} is 0.2.

In the simulation, a load increase of 30 MW at 5 s is applied, the power imbalance will decrease the system frequency. The frequency drop is detected by the frequency controller at around 5.7 s, and then the WT is controlled to release kinetic energy, the simulation results are illustrated in Fig. 8. In order to show the improvement of the proposed methods with the existing emulated inertial response of the WT, the emulated inertial control strategy of the WT is simulated, trial and error is used to determine the optimal value for the controller parameter. It can be observed from Fig. 8(a), without WT's frequency support, system frequency drops to 49.78 Hz, WT operation is not disturbed, it produces 60 MW active power, and the rotor keeps running at 1.15 pu. With the WT's frequency control, the ROCOF and frequency nadir is significantly reduced as illustrated in Fig. 8(a), the proposed method also has a reduced ROCOF and frequency nadir than the emulated inertial response. As observed from the figure, for the case of $\omega_{min} = 0.7$ pu, because it has larger kinetic energy, the WT provide a higher power output than the case of $\omega_{min} = 0.8$ during frequency control, thus, it has a lower frequency variation. However, due to the large drop of rotor speed, the mechanical

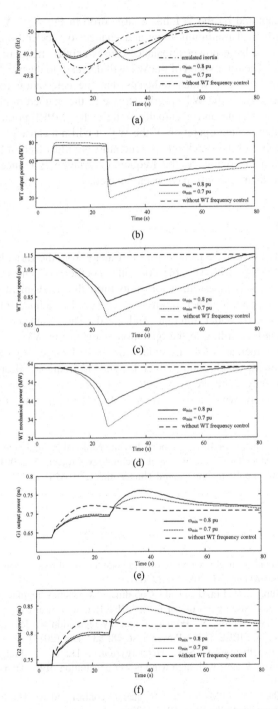

Fig. 8. Simulation results. (a) Frequency of the system. (b) WT generated power. (c) Rotor speed variation of WT. (d) Mechanical power captured by WT. (e) Output power of G1. (f) Output power of G2.

power which is captured by the WT has a large drop in Fig. 8(d). When rotor speed drops to the pre-defined minimum value and change to acceleration, it has a larger power imbalance, the large power imbalance causes another frequency drop which is shown in Fig. 8(a). The rotor speed of the WT will decrease to the pre-defined values in 20 s as shown in Fig. 8(b), which indicates that the calculation for the controller parameter is correct. After the WT speed decrease to the defined minimum value, the WT will start to accelerate and eventually settle at the MPPT point. The controller parameter ensures the stable operation of the WT, it will not cause over speed drop of the WT. The response of the synchronous generators are illustrated in Fig. 8(e) and (f), it is observed from the figure, as the frequency variation is reduced by the WT's frequency control, their response have been delayed.

5   Conclusion

The paper presented system frequency control strategy from the variable speed WTs by releasing kinetic energy with variable controller coefficients. The releasable kinetic energy from the WTs is assessed, in order to limit the mechanical power decrease during frequency support, a higher minimum rotor speed is defined for WT to provide system frequency control under different wind speeds. The parameters of the frequency controller is selected according to the operation status of the WTs. Simulation results show that by proper increase of the minimum rotor speed, large drop of the mechanical power could be avoid, which could reduce the second frequency drop. The proposed strategy for frequency controller parameters selection will ensure fully release of available kinetic energy, which also ensure the rotor speed variation of the WT is in the stable range.

Acknowledgments. The research was supported by the National Natural Science Foundation of China (51507062), State Grid Scientific and Technology Project (5228001600DT) and the Fundamental Research Funds for the Central Universities (2016MS16).

References

1. Tsili, M., Papathanassiou, S.: A review of grid code technical requirements for wind farms. IET Renew. Power Gener. **3**, 308–332 (2009)
2. Zhang, Z.S., Sun, Y.Z., Lin, J., et al.: Coordinated frequency regulation by doubly fed induction generator based wind power plants'. IET Renew. Power Gener. **6**, 38–47 (2012)
3. Vidyanandan, K.V., Senroy, N.: Primary frequency regulation by deloaded wind turbines using variable droop. IEEE Trans. Power Syst. **28**, 837–846 (2013)
4. Margaris, D., Papathanassiou, S.A., Hatziargyriou, N.D., et al.: Frequency control in autonomous power systems with high wind power penetration. IEEE Trans. Sustain. Energy **3**, 189–199 (2012)
5. Lalor, G., Mullane, A., O'Malley, M.: Frequency control and wind turbine technologies. IEEE Trans. Power Syst. **20**, 1905–1913 (2005)

6. Morren, J., de Haan, S.W.H., Kling, W.L., et al.: Wind turbines emulating inertia and supporting primary frequency control. IEEE Trans. Power Syst. **21**, 433–434 (2006)
7. Ullah, N.R., Thiringer, T., Karlsson, D.: Temporary primary frequency control support by variable speed wind turbines- potential and applications. IEEE Trans. Power Syst. **23**, 601–612 (2008)
8. Attya, B.T., Hartkopf, T.: Control and quantification of kinetic energy released by wind farms during power system frequency drop. IET Renew. Power Gener. **7**, 210–224 (2013)
9. Diaz, G., Casielles, P.G., Viescas, C.: Proposal for optimizing the provision of inertial response reserve of variable-speed wind generators. IET Renew. Power Gener. **7**, 225–234 (2013)
10. Shao, M., Miller, N.W.: GE wind power control design. In: 2nd Workshop on Active Power Control from Wind Power (2015)
11. Zhang, S., Mishra, Y., Shahidehpour, M.: Fuzzy-logic based frequency controller for wind farms augmented with energy storage systems. IEEE Trans. Power Syst. **31**, 1595–1603 (2015)
12. Chinchilla, M., Arnaltes, S., Burgos, J.C.: Control of permanent-magnet generators applied to variable-speed wind-energy systems connected to the grid. IEEE Trans. Energy Convers. **21**, 130–135 (2006)

Base-Load Cycling Capacity Adequacy Evaluation in Power Systems with Wind Power

Jingjie Ma$^{(\boxtimes)}$, Shaohua Zhang, and Liuhui Wang

Key Laboratory of Power Station Automation Technology,
Department of Automation, Shanghai University, Shanghai 200072, China
mjj_staff@163.com

Abstract. Large scale penetration of intermittent wind power may result in base-load cycling capacity (BLCC) shortage problem, which poses an adverse impact on secure operation of power systems. The integration scale of wind power is heavily relevant to the BLCC adequacy. Therefore, it is important to evaluate the BLCC adequacy of power systems. Using probabilistic production simulation technology, a BLCC adequacy evaluation method considering the forced outage of conventional generation units is developed in this paper. In this method, several BLCC adequacy indexes are defined, namely the probability of BLCC shortage index, the expectation of BLCC shortage index, and the expectation of BLCC margin index. A scenario reduction technique is employed to tackle the uncertainty of wind speed. Numerical examples are presented to verify the reasonableness and effectiveness of the proposed method. This work is helpful to determine the appropriate wind power integration scale in power systems.

Keywords: Wind power integration · Base-load cycling capacity adequacy · Probabilistic production simulation · Scenario reduction

1 Introduction

The increasing daily peak-valley load difference poses great cycling pressure on power systems and base-load cycling pressure in particular [1]. With the increasing shortage of fossil energy resources, intermittent renewable energy, notably wind power, is expected to play an increasing role to the sustainable development in the near future. Because of the inverse peak-regulation characteristics in wind power output, large scale penetration of wind power will increase the daily peak-valley difference of net loads, which will result in base-load cycling capacity (BLCC) shortage problem. Compared with other countries, the BLCC problem is much more severe in China. The BLCC shortage problem will pose an adverse impact on secure operation of power systems. In addition, the integration scale of wind power is heavily relevant to the BLCC adequacy. As such, how to evaluate the BLCC adequacy is a crucial task all around the world, especially in China.

Some related works have been published. Reference [2] shows the serious impact increasing levels of wind power will have on the operation of base-load units. The importance of base-load cycling and a new model to calculate the limit of capability of

© Springer Nature Singapore Pte Ltd. 2017
K. Li et al. (Eds.): LSMS/ICSEE 2017, Part III, CCIS 763, pp. 74–83, 2017.
DOI: 10.1007/978-981-10-6364-0_8

base-load cycling is proposed in Reference [3]. Two indexes are proposed based on Monte-Carlo simulation method to evaluate the peak-load regulating adequacy in [4]. A multitude of other cycling related issues have been documented in the literature [5–7]. Most of related works focus on the evaluation of peak-load cycling adequacy, and the forced outage of conventional generation units is not considered. To date, the evaluation of BLCC adequacy has not been investigated.

A BLCC adequacy evaluation method considering the forced outage of conventional generation units is proposed in this paper. Using the probabilistic production simulation technology, several BLCC adequacy indexes are introduced, namely the probability of BLCC shortage index, the expectation of BLCC shortage index, and the expectation of BLCC margin index. A scenario reduction technique is employed to deal with the uncertainty of wind speed. Numerical examples are presented to verify the effectiveness of the method. The BLCC adequacy indexes for power systems with and without wind power are compared. The impacts of wind power integration scale on the BLCC adequacy are examined. Some sensitivity analyses are also conducted to show the impacts of the minimum output of conventional units and the daily peak-valley load difference.

2 Estimation of BLCC Neglecting Forced Outage of Conventional Generation Units

The base-load cycling capability of conventional units is determined by the minimum output limit. The minimum output limit of large scale thermal power units is generally to be 50% of their installed capacity [8]. When load demand declines, there are two commitment approaches to balance the power: shutting down some power plants or running some power plants at minimum stable level. Assume that the ramp-down constraints and the forced outage of the conventional units are not considered, a power system's BLCC for a time period t can be estimated as follows.

$$BLCC(t) = L(t) - P_{Gmin}(t) \qquad (1)$$

where $BLCC(t)$ is the BLCC at time t, $L(t)$ is the load level at time t, $P_{Gmin}(t)$ is the sum of minimum output of all available thermal units at time t. If $BLCC(t) < 0$, the system is lack of cycling capability at time t, the BLCC shortage is $|BLCC(t)|$. Otherwise, the system has sufficient cycling capability at time t, the BLCC margin is $BLCC(t)$.

Figure 1 depicts the impact of wind power integration on the BLCC. The inverse peak-regulation characteristics in wind power output is considered. Curve A is the chronological initial load curve. The net load is defined as the initial load minus the wind generation, and the chronological net load curve is shown in the Curve B of Fig. 1. The chronological net load curve is the primary determinant of thermal units commitment. The BLCC with wind power integration can be estimated by:

$$BLCC'(t) = L(t) - P_W(t) - P_{Gmin} \qquad (2)$$

where $P_W(t)$ is the wind power output of time t.

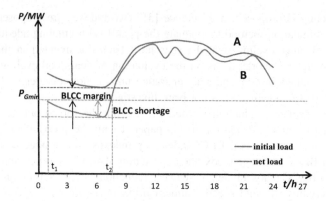

Fig. 1. Impact of wind power integration on the BLCC

It can be found from Fig. 1 that after the wind power is integrated, the BLCC decreases. When the net base-load level is less than the sum of minimum output of all thermal units, the system is lack of base-load cycling capability. As Fig. 1 shows, before wind power integration, the system has a certain BLCC margin. However, after wind power integration, the system has a BLCC shortage problem during time t_1 to t_2.

3 Estimation of BLCC Adequacy Considering Forced Outage of Conventional Generation Units

Probabilistic production simulation is widely used to predict production cost and reliability level of a power system at a certain time in the future, in which the uncertainties inherent in both the system load demand and the forced outages of generating units can be taken into account. In the same fashion, the probabilistic production simulation method originally developed for reliability evaluation could be applied to evaluate the BLCC adequacy.

3.1 Probabilistic Distributions of Available Minimum Output

Assume that a power system consists of I units with $\{i = 1, 2, \ldots, I\}$ being the merit loading order. The ith generating unit is represented by a two-state available capacity model with a forced outage rate of FOR_i. During the base-load time period, the unit either generates with the minimum output or has no output because of the forced outage. Let x_i denote the available minimum output of the ith generating unit. x_i is a random variable and can be expressed as follows:

$$x_i = \begin{cases} P_{Gmin,i} & \text{with probability: } 1 - FOR_i \\ 0 & \text{with probability: } FOR_i \end{cases} \tag{3}$$

where, $P_{Gmin,i}$ denotes the minimum output limit of unit i.

Let A_k denote the sum of available minimum output after the first k units are loaded, i.e.

$$A_k = \sum_{i=1}^{k} x_i \tag{4}$$

Assume that the available minimum output of each generating units are independent, the classic recursive convolution method or Z transform method [9, 10] can be used to obtain the probability distribution function of the system's available minimum output after the first k units are loaded. The convolution method and Z transform method are accurate methods, for they can strictly deal with the discrete characteristic of the available minimum output distribution. Therefore using the Z transform method of probabilistic production simulation technique, a discrete cumulative probability distribution function of the system's available minimum output after the first k units are loaded, can be obtained as follows:

$$F_{A_k}(x) = \sum_{i=1}^{N_k} p_k(i)\, u[x - X_k(i)], \quad k = 1, 2, \ldots, I \tag{5}$$

where, $X_k(i)$ and $p_k(i)$ denote the ith available minimum output state in MW and the state probability after the first k units are loaded, respectively. N_k is the number of available minimum output states after the first k units are loaded. The available minimum output states are arranged in strictly ascending order, i.e. $X_k(i) < X_k(i + 1)$. $u(x)$ is the unit step function.

3.2 Estimation of BLCC Adequacy Indexes

In this paper, three indexes, namely the probability of BLCC shortage, the expectation of BLCC shortage, and the expectation of BLCC margin, are proposed to evaluate the BLCC adequacy. The estimation method of these indexes is presented below.

The notation $L(t)$ is used to represent the system load during base-load time period t. After all units are loaded, the probability of BLCC shortage at time t, $p_S(t)$, the expectation of BLCC shortage, $E_S(t)$, the expectation of BLCC margin, $E_M(t)$, can be, respectively, given by:

$$p_S(t) = \sum_{i=M}^{N} p(i) \tag{6}$$

$$E_S(t) = \sum_{i=M}^{N} [X(i) - L(t)]p(i) \tag{7}$$

$$E_M(t) = \begin{cases} \sum_{i=1}^{M-1} [L(t) - X(i)]p(i), & p_S(t) > 0 \\ \sum_{i=1}^{N} [L(t) - X(i)]p(i), & p_S(t) = 0 \end{cases} \tag{8}$$

where M is defined such that $X_i(M)$ is the least available minimum output state that would cause a BLCC shortage for the given load $L(t)$ after all units are loaded; more precisely:

$$X(M-1) \leq L(t) < X(M) \tag{9}$$

Let T_G denote the number of base-load time periods. The average p_S, E_S and E_M during base-load periods, can be, respectively, calculated as follows:

$$p_S = \frac{\sum_{t=1}^{T_G} p_S(t)}{T_G} \tag{10}$$

$$E_S = \frac{\sum_{t=1}^{T_G} E_S(t)}{T_G} \tag{11}$$

$$E_M = \frac{\sum_{t=1}^{T_G} E_M(t)}{T_G} \tag{12}$$

3.3 Estimation of BLCC Indexes Considering Wind Power Integration

2-parameter Weibull distribution [11] is used to describe the probability distribution of wind speeds. N samples of wind speed are selected randomly, and the probability of each sample is $1/N$. A recursive backward scenario reduction method [12] is employed to get N' wind speed scenarios. According to the relationship between wind speed and power output described in [11], the N' wind speed scenarios at time t are computed to get N' wind power output scenarios. Net load forecast scenarios are generated in a similar manner.

After wind power integration, $p_S(t)$, $E_S(t)$, $E_M(t)$, at the load level $L(t)$, can be, respectively, expressed as:

$$p_S(t) = \sum_{k=1}^{N'} p_{S,k}(t) \cdot p_{wk} \tag{13}$$

$$E_S(t) = \sum_{k=1}^{N'} E_{S,k}(t) \cdot p_{wk} \tag{14}$$

$$E_M(t) = \sum_{k=1}^{N'} E_{M,k}(t) \cdot p_{wk} \tag{15}$$

where p_{wk} is the probability of occurrence of the kth scenario, $p_{S,k}(t)$ is the probability of BLCC shortage of the kth scenario at time t, $E_{S,k}(t)$ is the expectation of BLCC shortage of the kth scenario at time t, $E_{M,k}(t)$ is the expectation of BLCC margin of the kth scenario at time t.

4 Numerical Analysis

The proposed method is applied to the generation system in [9] that is patterned after the IEEE reliability test system. Table 1 shows the generating units in their loading order, with assumptions that each unit be represented by a two-state model. Figure 2 depicts the initial load data of 24 h and the historical wind speed of Gansu province. According to the load characteristics, the base-load period is set to 0: 00-6: 00, 23: 00-24: 00. The cut-in wind speed is 2 m/s, the rated wind speed is 11 m/s, and the cut-out wind speed is 20 m/s, the standard deviation of wind speed SD = 2. The wind energy turbine's power output satisfies $P_w = 0.7646v^3$ [13]. The number of scenarios $N' = 10$. It is assumed that the wind turbine generators in the wind farm are the same.

Table 1. Generating units' data

Unit	Capacity/MW	FOR_i	Generating cost ($/MWh)	Minimum output/ installed capacity
1	1000	0.1000	4.50	50%
2	900	0.1034	5.00	50%
3	700	0.0977	5.50	50%
4–5	600	0.0909	5.75	50%
6–8	500	0.0873	6.00	55%
9–13	400	0.0756	8.50	55%
14	300	0.0654	10.00	55%
15–19	200	0.0535	14.50	55%
20–26	100	0.0741	22.50	0%
27–32	100	0.0331	44.00	0%

In traditional power system reliability evaluation, the loss-of-load probability (*LOLP*) and the expected unserved energy (*EUE*) are employed to assess the generation capacity adequacy. In this paper, three indexes, namely the probability of BLCC shortage, the expectation of BLCC shortage, and the expectation of BLCC margin, are proposed to evaluate the BLCC adequacy.

The generation capacity adequacy indexes *LOLP*, *EUE*, and the BLCC adequacy indexes p_S, E_S, E_M before and after 100 wind power units integration are calculated and listed in Table 2. It can be seen that before wind power integration, *LOLP* are very small, p_S and E_S are all zero, which indicates that the system's generation capacity and BLCC are both adequate. There are BLCC margins in base-load time periods.

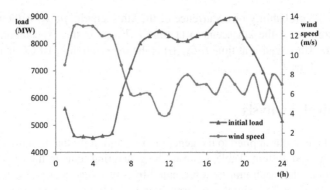

Fig. 2. The initial load data and annual average wind speed data of Gansu

It can be observed that after wind power integration, *LOLP* and *EUE* are reduced, which means that the generation capacity adequacy is enhanced. However, p_S increases, E_M decreases, E_S for part of the periods is greater than zero, which means the BLCC adequacy is lowered.

Table 2. Capacity adequacy indexes and BLCC adequacy indexes before and after wind power integration

	Time	*LOLP*	*EUE*/MWh	p_S	E_S/MW	E_M/MW
Before wind power integration	0–1	3.4E−5	6.3E−3	0	0	1454
	1–2	2.6E−7	4.3E−5	0	0	474
	2–3	1.5E−7	3.3E−5	0	0	429
	3–4	1.5E−7	2.6E−5	0	0	385
	4–5	2.6E−7	3.8E−5	0	0	456
	5–6	4.6E−7	7.1E−5	0	0	563
	23–24	3.4E−6	7.5E−4	0	0	1008
After wind power integration	0–1	2.2E−5	4.9E−3	0	0	1394
	1–2	1.5E−7	2.5E−5	0.18	1.55	379
	2–3	9.7E−8	1.9E−5	0.22	10.62	343
	3–4	8.7E−8	1.5E−5	0.22	20.32	308
	4–5	1.5E−7	2.3E−5	0.19	4.66	368
	5–6	2.6E−7	4.2E−5	0	0	470
	23–24	3.3E−6	6.5E−4	0	0	976

After wind power integration, during base-load period *LOLP* are very small, p_S and E_S are very large, e.g. during 3:00–4:00, *LOLP* is just $8.7 * 10^{-8}$, while p_S reaches 0.2192. It can be seen that the generation capacity adequacy indexes *LOLP* and *EUE* cannot be used to evaluate the BLCC adequacy. Hence it is necessary to introduce the BLCC adequacy indexes. Figure 3 depicts the impact of wind power integration scale on p_S and E_S. As the scale of wind power integration increases, the average p_S and E_S

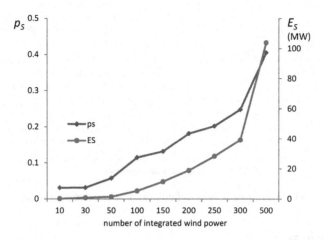

Fig. 3. Impact of wind power integration scale on p_S and E_S

increase. As such, with the increase of wind power integration, power system capacity adequacy evaluation should consider both generation capacity adequacy indexes and the BLCC adequacy indexes.

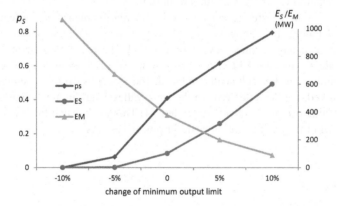

Fig. 4. Impact of units' minimum output on p_S, E_S and E_M

The conventional generators' minimum output plays an important impact on the BLCC adequacy indexes. Impact of conventional generators' minimum output on p_S, E_S and E_M after 500 wind power units integration is depicted in Fig. 4. As can be observed, with the decrease of conventional generators' minimum output, p_S and E_S decrease, while E_M increases. If conventional generators' minimum output can be reduced technically, the system's BLCC adequacy can be greatly improved.

Figure 5 shows the impact of peak-valley load difference on p_S and E_S before wind power integration. It can be found that as the peak-valley load difference increases, the p_S and E_S increase, which leads to a heavier pressure of base-load cycling.

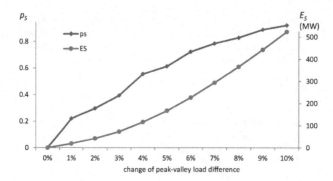

Fig. 5. Impact of peak-valley load difference on p_S and E_S

5 Conclusions

Increasing wind penetration in power system will lead to base-load cycling capacity (BLCC) shortage problem. A BLCC adequacy evaluation method is proposed, in which a scenario reduction technique is employed to tackle the uncertainty of wind speed. The probabilistic production simulation technology is used. Several BLCC adequacy indexes, namely the probability of BLCC shortage, the expectation of BLCC shortage and the expectation of BLCC margin, are introduced.

Numerical examples are presented to verify the reasonableness and effectiveness of the proposed model. It is shown that: (1) with increase of wind power integration, power system capacity adequacy evaluation should consider both generation capacity adequacy indexes and the BLCC adequacy indexes. (2) If conventional generators' minimum output can be technically reduced, the system's BLCC adequacy can be greatly improved. (3) With decrease in the peak-valley load difference, the pressure of base-load cycling can be effectively mitigated. (4) This work is helpful to determine the appropriate wind power integration scale in power systems.

References

1. Batlle, C., Rodilla, P.: An enhanced screening curves method for considering thermal cycling operation costs in generation expansion planning. IEEE Trans. Power Syst. **28**(4), 3683–3691 (2013)
2. Troy, N., Denny, E., O'Malley, M.: Base-load cycling on a system with significant wind penetration. IEEE Trans. Power Syst. **25**(2), 1088–1097 (2010)
3. Yang, H., Liu, J.X., Yuan, J.S.: Research of peak load regulation of conventional generators in wind power grid. Proc. CSEE **30**(16), 26–31 (2010)
4. Zhang, H.Y., Yin, Y.H., Shen, H., et al.: Peak-load regulating adequacy evaluation associated with large-scale wind power integration. Proc. CSEE **31**(22), 26–31 (2011)
5. Ummels, B.C., Gibescu, M., Pelgrum, E., et al.: Impacts of wind power on thermal generation unit commitment and dispatch. IEEE Trans. Energy Convers. **22**(1), 44–51 (2007)

6. Bakdick, R.: Wind and energy markets: a case study of Texas. IEEE Syst. J. **6**(1), 27–34 (2012)
7. Babrowski, S., Jochem, P., Fichtner, W.: How to model the cycling ability of thermal units in power systems. Energy **103**, 397–409 (2016)
8. Cao, F., Zhang, L.Z.: Determination of pumped storage plant capacity with peak regulation proportion. Electr. Power Autom. Equip. **27**(6), 47–50 (2007)
9. Zhang, S.H., Li, Y.Z.: Concise method for evaluating the probability distribution of the marginal cost of power generation. IEE Proc. Gener. Transm. Distrib. **147**(3), 137–142 (2000)
10. Sutanto, D., Outhred, H.R., Lee, Y.B.: Probabilistic power system production cost and reliability calculation by the Z-transform method. IEEE Trans. Energy Convers. **4**(4), 559–565 (1989)
11. Wang, L.H., Wang, X., Zhang, S.H.: Electricity market equilibrium analysis for strategic bidding of wind power producer with demand response resource. In: IEEE PES Asia-Pacific Power and Energy Conference, pp. 181–185, Xi'an (2016)
12. Morales, J.M., Pineda, S., Conejo, A.J., et al.: Scenario reduction for futures market trading in electricity markets. IEEE Trans. Power Syst. **24**(2), 878–888 (2009)
13. Wang, S.X., Xu, Q., Zhang, G.L., et al.: Modeling of wind speed uncertainty and interval power flow analysis for wind farms. Autom. Electr. Power Syst. **21**, 82–86 (2009)

MFAC-PID Control for Variable-Speed Constant Frequency Wind Turbine

Qingye Meng[1], Shuangxin Wang[1(✉)], Jianhua Zhang[2],
and Tingting Guo[1]

[1] School of Mechanical, Electronic and Control Engineering,
Beijing Jiaotong University, Beijing 100044, China
{15121251, shxwang1, 16121261}@bjtu.edu.cn
[2] State Key Laboratory of Alternate Electrical Power System with Renewable
Energy Sources, North China Electric Power University, Beijing 102206, China
zjh@ncepu.edu.cn

Abstract. Due to the randomness and fluctuation characteristics of wind power, those model-based systems having intrinsically nonlinear are harder to be controlled. Based on the variable-speed constant frequency wind power generator, this paper presents a MFAC-PID control strategy to realize model-free, I/O data based dynamic control. Firstly, a control input criterion is established for optimal design, which realizes the targets of maximum wind energy capture and smoothing power point tracking. Then, by the usage of model free adaptive control (MFAC), a series of equivalent local linearization models are built using time-varying pseudo-partial derivative (PPD), which could be estimated only by I/O measurement data. Finally, considering that both MFAC and PID will generate incremental output, a constrained MFAC-PID algorithm is proposed in order to obtain the optimal input. The proposed strategy is verified with comparison to PID and MFAC methods. Results prove that MFAC-PID algorithm guarantees the convergence of tracking error at full wind speed.

Keywords: Wind power · Variable pitch control · Data-driven · MFAC · PID self-tuning

1 Introduction

Wind power is one of the most promising renewable energies owing to many merits they have, such as zero or low emission and inexhaustibility [1]. However, due to the randomness and fluctuation characteristics of wind power, it consequently brings challenges to grid-connected operation and dynamic planning of wind farm. Especially for the AGC-related control which inhibits active power and frequency fluctuations caused by wind power, variable pitch control plays an important role in wind energy conversion system (WECS) in order to maximize the capture of wind energy. Till now, some acceptable control methods have been developed, such as PI regulator [2], optimal control in LQ [3], and LQG form [4].

© Springer Nature Singapore Pte Ltd. 2017
K. Li et al. (Eds.): LSMS/ICSEE 2017, Part III, CCIS 763, pp. 84–93, 2017.
DOI: 10.1007/978-981-10-6364-0_9

Previous studies have examined the random uncertainty and time-varying characteristics of wind power on different model-based methods. In [5], sliding mode control is used to cope with system uncertainty and reduce mechanical efforts and chattering. Wang et al. [6] established a super short-term prediction model of wind speed and power in order to reduce mechanical stresses. Zhang et al. [7] proposed a model-reference adaptive control system for controlling nonlinear blade pitch with robustness and servo-performance. Wei et al. [8] presented a variable-speed pitch control based on fuzzy control strategy which overcomes nonlinearity influence, but it lacks perfect fuzzy rule and membership concerning on the random uncertainty. Hamidreza et al. [9] put forward an adaptive control based on RBF neural network for different operation modes of wind turbines. However, there needed a large number of data to establish RBF neural network model, which brought about tedious operation. According to the front analyses, these methods have not fully considered possible random noises involved in WECS caused by load disturbance and frequent action of pitch switch. The mathematical model is formidable to be accurately established because of the time-varying and nonlinear transmission. Therefore, the model-based control strategy is not desirable for wind energy with uncertainty.

Actually, wind turbines in operation produce large amounts of real-time data stored in the Supervisory Control and Data Acquisition (SCADA) system at any moment, which contains all the useful information related to their operation and equipment status. In cases where mathematical models cannot be obtained accurately, the data-driven control may be used to achieve satisfactory performance, in which the design of controller depends only on the I/O measurement data, without explicitly or implicitly using the plant structure or dynamics information of controlled plant, and whatever the plant is linear or nonlinear [10, 11]. Liu et al. [12] developed individual pitch controller to mitigate the rotor unbalance load for variable speed wind turbine. Xu et al. [13] adopted a data-based adaptive control, in which gradient-like vector can be obtained by using input and output data of the wind energy conversion system. Kusiak and Zhang [14] developed two models using data collected from a large wind farm, and then introduced an anticipatory control scheme for optimizing power and vibration of wind turbines. The data-driven strategy has been shown to be effective in accommodating uncertainties and random disturbance in a systematic and straightforward way in WECS.

Model free adaptive control (MFAC) is a kind of typical data-driven control algorithm, which is accomplished by effectively building a series of CFDL (compacted form dynamic linearization) dynamic linear data model using time-varying pseudo-partial derivative (PPD) and calculating PPD to adjust controller parameters on-line. In fact, MFAC has been widely applied to solve the key problems encountered in industrial and real-life fields [15, 16].

Although the theoretical study of MFAC is much active so far, the research and application is not deep enough in the field of wind energy utilization and control. Meanwhile, PID control is most widely used and its parameters need to be identified exactly. Considering that both MFAC and PID generate incremental output, the MFAC-PID control is proposed in this paper for variable-speed constant frequency wind turbines, which focuses on MFAC method, and uses I/O data of the system to realize the on-line self-tuning of PID parameters.

In this paper, Sect. 2 describes optimal power control strategy of wind turbine for different operating modes and builds WECS model in Matlab/Simulink to obtain on-line I/O data used in the design of MFAC-PID control. Control input criterion function is also put forward to be adapted in control process. Section 3 firstly reviews the control algorithm of MFAC, and subsequently proposes the constrained MFAC-PID strategy that could simultaneously satisfy the increment constrain of MFAC and PID methods. In Sect. 4, dynamic possesses under several operational conditions are analyzed with comparison of PID, MFAC, and constrained MFAC-PID methods. Simulation results show that MFAC-PID algorithm not only guarantees the stability and converges of power at full wind speed, but also realizes the optimal control of blade angle and rotor speed, and further serves the actual wind field and grid, which will enjoy great realistic significance and academic value. Section 5 ends the paper by high lighting the main achievements of the work.

2   Optimal Control of Random-Determining WECS

2.1   Modeling of WECS Used for Generating On-line Data

In this section, the mathematical model of WECS is carried out in Matlab/Simulink, with which the working condition is Pg = 600 kw. The model is just used to serve as I/O data generator and demonstrate the applicability and efficiency of the control system, with no any information is included in the design of MFAC-PID controller.

WECS is made of wind turbine, drive train, generator, pitch blade servo system and AC-DC-AC converter. Firstly the natural wind is applied to wind turbine blades, and subsequently the mechanical torque transferred to the hub is converted to generator, which drives the generator rotor to produce electric energy.

During system operation, the input power of wind turbine is expressed as

$$p_0 = \frac{\pi}{2}\rho R^2 V^3 \tag{1}$$

where, ρ is the air density, R is the rotor radius, and V is wind speed.

Nevertheless, the input power cannot be fully absorbed by wind turbine, and the theoretical wind energy captured by wind turbine rotor is given by

$$P_m = \frac{1}{2}C_p(\beta, \lambda)\rho\pi R^2 V^3 \tag{2}$$

where, P_m is wind energy captured by the rotor, and C_p is called rotor power coefficient, which depends on blade pitch angle β and tip-speed-ratio λ determined from

$$\lambda = \frac{\omega R}{V} \tag{3}$$

where, ω is the rotational speed of rotor.

2.2 Description of Optimization Control Target

The main control objective is to capture maximum wind energy and obtain smoothing power point tracking, which can be achieved by manipulating the desired β and the speed of generator rotor ω.

Figure 1 gives the relationship between λ and the theoretical C_p for a quintessential wind turbine. It is widely accepted that C_p has a unique maximum C_{pmax} with the corresponding values of β_{opt} and λ_{opt}. In addition, the value of C_p changes with λ when β takes different values. Thus, in the full wind speed, to regulate the power at its rated value, the power coefficient should be reduced by changing β, λ, or both variables.

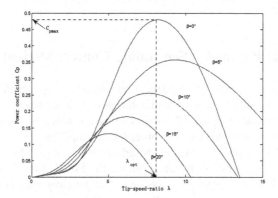

Fig. 1. Relation curve between C_p and λ.

In accordance with the variation of wind speed, under the circumstance that wind speed is lower than rated speed, pitch angle is fixed and the rotor speed is tracked to capture maximum available power. In condition that wind speed above rated speed, wind turbine should work at constant rated power due to the limitation of mechanical strength and physical performance of generator and inverter. Therefore, the generator torque is constant at its rated value and the pitch control should limit the output power.

2.3 Optimal Control Strategy at Full Wind Speed

Since wind turbine is a complicated random-determining coupled power system, a control input cost function is accordingly adopted and given by

$$J(u(k)) = |y^*(k+1) - y(k+1)|^2 + \lambda |u(k) - u(k-1)|^2 \tag{4}$$

where, $y(k)$ and $u(k)$ are the system output and input signals, $y^*(k+1)$ is the expected system output signal, and k is a positive weighted constant. $\lambda > 0$ is a weight factor used to limit the control input changes.

For wind speed below rated speed, the pitch angle is fixed while the rotor speed is tracked to capture maximum available power. Calculate the actual power coefficient C_p, then take the difference $C_{pmax} - C_p$ as the control system input and the wind turbine rotor speed as output. Therefore, the control input objective function is described as

$$J(\omega(k)) = |C_{\text{pmax}}(k+1) - C_{\text{p}}(k+1)|^2 + \lambda|\omega(k) - \omega(k-1)|^2 \qquad (5)$$

For wind speed above rated speed, a fixed power output is obtained by adjusting the pitch angle. Then take the difference between rated power and the actual output power $P_e - P(t)$ as control system input, while the pitch angle β as output. In that way the control input objective function is described as

$$J(\beta(k)) = |P_e(k+1) - P(k+1)|^2 + \lambda|\beta(k) - \beta(k-1)|^2 \qquad (6)$$

3 MFAC-PID Parameter Self-tuning Control Method

Incremental PID control has been widely used because of its simple and applicable features, while the control accuracy and robustness may decrease when the control system is nonlinear and complicated. Motivated by this idea, the constrained MFAC-PID method is proposed in this paper in order for accomplishing the self-tuning of PID parameters to realize model-free, I/O data based dynamic control. On the basis that both MFAC and PID will generate incremental output, the MFAC-PID algorithm is used to obtain the optimal input in order to guarantee the output close to set-value of the plant and satisfy the increment constrain of MFAC and PID methods simultaneously.

3.1 Model Free Adaptive Control

The following is a SISO nonlinear discrete-time system

$$y(k+1) = f(y(k), \ldots, y(k-n_y), u(k), \ldots, u(k-n_u)) \qquad (7)$$

where, n_y, n_u are the unknown orders of output $y(k)$ and input $u(k)$, $f(\cdots)$ is an unknown nonlinear function.

When the nonlinear system (7) satisfies some assumptions, it can be described as the following CFDL model

$$y(k+1) = y(k) + \phi_c(k)\Delta u(k) \qquad (8)$$

By substituting (8) into (4) and then differentiating (4) with respect to $u(k)$, we get

$$u(k) = u(k-1) + \frac{\rho\phi_c(k)}{\lambda + |\phi_c(k)|^2}(y^*(k+1) - y(k)) \qquad (9)$$

where, ρ is a step-size constant, which is added to make (9) general. Therefore, the design parameters of the system will be obtained only when PPD is obtained.

As the unknown $\varphi_c(k)$ is time-varying, the conventional projection or least squares algorithm cannot track it well. Consequently, a slice of time-varying algorithms is used to estimate $\varphi_c(k)$. The projection algorithm is give by

$$J(\phi_c(k)) = |y(k) - y(k-1) - \phi_c(k)\Delta u(k-1)|^2 + \mu \left|\phi_c(k) - \hat{\phi}_c(k-1)\right|^2 \quad (10)$$

Thus, the parameter estimation algorithm can be expressed as

$$\hat{\phi}_c(k) = \hat{\phi}_c(k-1) + \frac{\eta \Delta u(k-1)}{\mu + \Delta u(k-1)^2}(\Delta y(k) - \hat{\phi}_c(k-1)\Delta u(k-1)) \quad (11)$$

where, η is a step-size constant, and μ is a weighting factor.

3.2   Constrained MFAC-PID Self-tuning Algorithm

As one of the most widely used controllers in industrial process control, the expression of incremental PID control is given by:

$$\Delta u(k) = K_p(k)[e(k) - e(k-1)] + K_i(k)e(k) + K_d(k)[e(k) - 2e(k-1) + e(k-2)] \quad (12)$$

where, $K_p(k)$, $K_i(k)$, and $K_d(k)$ are three parameters at the moment of k.

By substituting $\beta = \frac{\rho \phi_c(k)}{\lambda + |\phi_c(k)|^2}$ into Eq. (9), we get

$$\Delta u(k) = \beta(y^*(k+1) - y(k)) \quad (13)$$

Taking Eqs. (12) and (13), we have

$$K_p(k)[e(k) - e(k-1)] + K_i(k)e(k) + K_d(k)[e(k) - 2e(k-1) + e(k-2)] \\ = \beta[y^*(k+1) - y(k)] \quad (14)$$

Equation (14) is the algorithm of constrained MFAC-PID self-tuning proposed in the paper. Assuming that $K_p(k)$, $K_i(k)$ and $K_d(k)$ are unchanged at any three consecutive sampling time of $k-2$, $k-1$ and k. Then a three-variable linear equation is obtained as

$$AP = B \quad (15)$$

where,

$$A = \begin{bmatrix} e(k) - e(k-1) & e(k) & e(k) - 2e(k-1) + e(k-2) \\ e(k-1) - e(k-2) & e(k-1) & e(k-1) - 2e(k-2) + e(k-3) \\ e(k-2) - e(k-3) & e(k-2) & e(k-2) - 2e(k-3) + e(k-4) \end{bmatrix}$$

$$P = [K_p \quad K_i \quad K_d]^T$$

$$B = \beta[y^*(k+1) - y(k) \quad y^*(k) - y(k-1) \quad y^*(k-1) - y(k-2)]^T$$

The rank of matrix A is calculated in order to realize the self-tuning of PID parameters. If $rank\ (A) = 3$, the equation has one solution $P = (A)^{-1}B$, from which the new PID parameters are obtained. But if $rank\ (A) < 3$, the equation has either no solution or infinitely a multitude of solutions, the control system still uses last value.

4 Simulation Results

The model technical parameters of considered WECS in this paper are given in Table 1.

Table 1. Model technical parameters of WECS.

Parameter	Unit	Value
Rated power	kw	600
Impeller diameter	m	45
Air density	kg/m^3	1.225
Cut-in wind speed	m/s	3
Cut-out wind speed	m/s	25
Rated wind speed	m/s	12
Rated rotation speed of rotor	rad/s	3.6

Based on the technical data in Table 1, it is obtained that wind turbine starts up in wind speed at 3 m/s and rises up to rated power at 12 m/s. In following analysis of simulation, MFAC-PID controller performance is firstly verified as wind speed is higher than normal speed by inputting step wind speed. Subsequently, the traditional PID and MFAC strategies will be used to comparatively analyze and validate the performance of MFAC-PID, in which the PID parameter has been determined by the trial and error method.

4.1 Simulation Analysis Above Rated Wind Speed

Step wind speed is firstly inputted to validate the availability of the MFAC-PID controller. Figure 2 gives the variety of pitch angle under high wind speed. As it can be seen, when the wind speed has a step change at some moment such as t = 0.5, the pitch angle modification has commendable tracking performance to achieve system control requirement as wind speed changes. At the same time, it has shown favorable stability and small variations, which is of great benefit to reduce the mechanical stress and decrease the failure rate of variable pitch system.

The output power curve of wind turbine system is shown in Fig. 3. It is observed that the output power could be maintained near rated power of 600 kw when the blade pitch angle is adjusted in real time. Beyond that, we also obtain that there is small overshoot and steady state error with short regulating time of 0.2 s or so, which indicates that the proposed strategy possesses virtues of high control precision, strong robust and against disturbance.

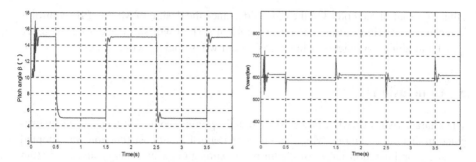

Fig. 2. Pitch angle at high wind speeds.   **Fig. 3.** Output power at high wind speeds.

4.2 Simulation Analysis Under Rated Wind Speed

In order to get comparisons evidently of three different control methods of PID, MFAC, and MFAC-PID, the slope wind speed is input to observe output signal in system response as wind speed changes.

The output rotation rate is shown in Fig. 4. The experimental result is suggestive that as wind speed changes during startup process, the MFAC-PID control accurately exports stable and real-time rotating speed, and therefore tracks reference speed ω_{ref} best. What's more, it also could be concluded that the generator speed quickly stabilized at 3.2 rad/s after a very small jump, so the MFAC-PID controller has both rapid compliant ability in changing process and pleasurable robust performance in practical applications, which meet the needs of solving problems of traditional control methods, such as time lag, frequent start-up of wind turbine, and so on.

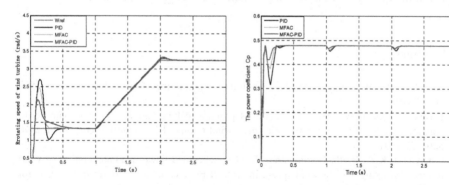

Fig. 4. Rotating speed of wind turbine   **Fig. 5.** The power coefficient C_p.

Figure 5 gives the value of C_p along with the continuous fluctuation of wind speed. It's qualitatively indicated that by using MFAC-PID control method, C_p has a certain jump and recovers to about 0.48 in 0.2 s. Wind turbine not only runs to follow the mechanism of maximum wind energy in different wind speed, but also quickly enters steady running state subsequently, getting the best control effect compared with the other two control methods. On the other side, since the design of MFAC-PID control is independent of wind turbine parameters, control system will show better convergence speed and ability to quickly achieve maximum wind tracking control target.

5 Conclusions

The characteristics of wind power system and optimize control strategy were investigated in this paper, among which an effective control input cost function was adopted to satisfy optimization target of capturing maximization wind energy and outputting stable wind power.

The MFAC-PID control algorithm was proposed for variable-speed variable-pitch system. Firstly, the MFAC method was introduced by building the CFDL model to establish a series of dynamic linear data model and calculating PPD effectively to adjust controller parameters on-line. Then, the MFAC-PID control strategy was put forward in which the self-tuning of PID parameters was realized just based on I/O data. This scheme is model-free, which makes it possible to adapt to various working conditions in solving AGC and frequency stability problems under the background of high wind power penetration.

The simulation at full wind speed was set up by the established mathematical model. Meanwhile, MFAC-PID control strategy was tested and verified that it eliminated the complex and tedious work of system identification and modeling, as well as met the system dynamics and steady-state requirements. By comparison with traditional PID and MFAC control strategy, MFAC-PID proved superior robust and dynamic characteristic. It not only realized rapidly tracking of rotor speed and pitch angle, but also eliminated disturbance triggered by wind speed variation, and further improved the quality of wind power in whole work range of wind turbine.

References

1. Kusiak, A., Li, W., Song, Z.: Dynamic control of wind turbines. Renew. Energy **35**(2), 456–463 (2010)
2. Iyasere, E., Salah, M., Dawson, D., et al.: Optimum seeking-based non-linear controller to maximise energy capture in a variable speed wind turbine. IET Control Theory Appl. **6**(4), 526–532 (2012)
3. Ostergaard, K.Z., Brath, P., Stoustrup, J.: Gain-scheduled linear quadratic control of wind turbines operating at high wind speed. In: 16th IEEE International Conference on Control Applications, pp. 276–281. IEEE Xplore, Singapore (2007)
4. Munteanu, I., Cutululis, N.A., Bratcu, A.I., et al.: Optimization of variable speed wind power systems based on a LQG approach. Control Eng. Pract. **13**(7), 903–912 (2005)
5. Beltran, B., Benbouzid, M.E.H., Ahmed-Ali, T.: Second-order sliding mode control of a doubly fed induction generator driven wind turbine. IEEE Trans. Energy Convers. **27**(2), 261–269 (2012)

6. Wang, X.L., Li, J.L., Ma, C.X.: Optimization control of variable-speed variable-pitch wind power generation system based on power prediction. Power Syst. Prot. Control **13**, 88–92 (2013)
7. Zhang, C.M., Yao, X.J., Zhang, Z.C., et al.: A model-reference adaptive blade-pitch control for a wind generator system. Control Theory Appl. **25**(1), 148–150 (2008)
8. Wei, Z., Chen, R., Chen, J., et al.: Wind turbine-generator unit variable-speed pitch control based on judgment of power changes and fuzzy control. Proc. CSEE **31**(17), 121–126 (2011)
9. Hamidreza, J., Jeff, P., Julian, E.: Adaptive control of a variable-speed variable-pitch wind turbine using RBF neural network. IEEE Trans. Control Syst. Technol. **21**(6), 2264–2272 (2013)
10. Pan, T.L., Sun, C.Q., Ji, Z.C., et al.: Data-driven constant power control of variable speed variable pitch wind energy conversion system. J. Nanjing Univ. Sci. Technol. **39**(1), 115–121 (2015)
11. Ji, Z.C., Feng, H.Y., Shen, Y.X.: Data-driven predictive control for wind turbine pitch angle. Control Eng. China **20**(2), 327–330 (2013)
12. Liu, Y.M., Zhu, J.S., Yao, X.J., et al.: Individual pitch control of wind turbine based on model free adaptive control. Acta Energiae Sol. Sin. **36**(1), 1–5 (2015)
13. Xu, L.L., Shen, Y.X., Ji, Z.C.: The data-based adaptive control for wind energy conversion system. Small Spec. Electr. Mach. **39**(9), 62–65 (2011)
14. Kusiak, A., Zhang, Z.J.: Control of wind turbine power and vibration with a data-driven approach. Renew. Energy **43**, 73–82 (2012)
15. Li, Z.H., Xia, Y.J., Qu, Z.W.: Data-driven background representation method to video surveillance. J. Opt. Soc. Am. A **34**(2), 193–202 (2017)
16. Ran, X., Ting, S., Peng, S., et al.: Model-free adaptive control for spacecraft attitude. J. Harbin Inst. Technol. (New Ser.) **23**(6), 61–66 (2016)

A Multivariate Wind Power Fitting Model Based on Cluster Wavelet Neural Network

Ruiwen Zheng[1], Qing Fang[2], Zhiyuan Liu[3], Binghong Li[1],
and Xiao-Yu Zhang[1(✉)]

[1] Department of Mathematics, Beijing Forestry University,
Beijing 100083, People's Republic of China
xyzhang@bjfu.edu.cn
[2] Faculty of Science, Yamagata University, Yamagata 990-8560, Japan
[3] Posts and Telecommunications, Chongqing University,
Chongqing 400065, People's Republic of China

Abstract. In this paper, we select the hierarchical cluster method to classify the wind energy level with the meteorological data, and then apply the 0–1 output method to quantify the wind energy level. Next, we utilize wavelet neural network to fit multivariate wind power data, which solves the problem of randomness, intermittency and volatility of wind power data. Finally, a wind-power numerical experiment shows the ideal fitting results with an error precision of 1.71% and demonstrates the effectiveness of our model.

Keywords: Wind power · Multivariate data · Fitting model · Wavelet neural network

1 Introduction

As a kind of emerging energy, wind energy has the most mature technology and the relatively low cost. It means the wind energy has the potential for large-scale development. However, because of the changing nature of wind resources, the output power of wind turbines has strong randomness, intermittency and volatility. The controllability and predictability of wind power units are far lower than those of conventional energy units.

The fitting wind power data is the basis for forecasting wind power and wind energy level. There are lots of wind power fitting models in machine learning. In 2006, Potter and Negnevitsky [1] applied the adaptive neural-fuzzy inference system to fit the wind time series. On this basis, the wind forecasting system developed wind direction forecasting by 2.5 min, resulting in an average absolute percentage error of less than 4%. In 2008, Fan et al. [2] constructed the fitting

X.-Y. Zhang—The research is supported by the Fundamental Research Funds for the Central Universities (Grant No. 2017ZY30) and the second author is supported by Scientific Research Grant-in-Aid from JSPS under grant 15K04987.

K. Li et al. (Eds.): LSMS/ICSEE 2017, Part III, CCIS 763, pp. 94–102, 2017.
DOI: 10.1007/978-981-10-6364-0_10

model based on artificial neural network (ANN) according to the influence factors of wind power, and the results show that the fitting model can help increase forecasting accuracy in 30 min ahead of time. In 2011, Lin and Liu [3] combined empirical mode decomposition (EMD) and support vector machine (SVM) to reduce the influence between different feature information and improve the accuracy of forecasting with the result that forecasting error decreased by 5% to 10%. In 2014, De Giorgi et al. [4] take advantage of Least-Squares Support Vector Machine (LS-SVM) with Wavelet Decomposition (WD) at different time horizons to forecast the power production of a wind farm located in complex terrain, which fully embodies the advantages of wavelet theory in deal with multivariate data.

In fact, for different types of wind turbines, the classification of wind energy levels is different [5]. Therefore, we take advantage of the hierarchical clustering method to classify the wind energy level. Then, the wavelet neural network is trained to fit the classification result and meteorological data. The experiment shows that this model has a good fitting effect.

2 Hierarchical Cluster Analysis

2.1 Statistics of Clustering—Euclidean Distance

Let x_{ij} $(i = 1, 2, \cdots, n, \ j = 1, 2, \cdots, p)$ be the observation data of the jth index of the ith sample. That is, each sample can be seen as a point in the p-dimensional space and n samples are n points of the p-dimensional space. Define $d_{i_1 i_2}$ as the distance between sample x_{i_1} and x_{i_2}. Then get a $n \times n$-dimensional distance matrix $D = (d_{i_1 i_2})_{n \times n}$.

$$D = (d_{i_1 i_2})_{n \times n} = \begin{pmatrix} d_{11} & d_{12} & \cdots & d_{1n} \\ d_{21} & d_{22} & \cdots & d_{2n} \\ \vdots & \vdots & \cdots & \vdots \\ d_{n1} & d_{n2} & \cdots & d_{nn} \end{pmatrix}.$$

The Euclidean Distance is chosen as distance calculation formula.

$$d_{i_1 i_2} = \sqrt{\sum_{j=1}^{m}(x_{i_1 j} - x_{i_2 j})^2} \qquad (i_1, i_2 = 1, 2, \cdots, n).$$

2.2 The Fundamental of Ward Method

This method was first proposed by Ward [6] and this basic idea of the method is from the analysis of variance. The method is to first separate the samples into n clusters, after that to choose and aggregate the two clusters which make the deviation squared sum of the aggregation increases least as a new cluster until all the samples are aggregated as one cluster.

Considering the deviation squared sum

$$S_t = \sum_{i=1}^{n_t} X_{ti}^2 - \frac{1}{n_t} \left(\sum_{i=1}^{n_t} X_{ti} \right)^2, \tag{1}$$

where on condition that n samples were divided into k clusters $G_1, G_2,$ $\cdots, G_k$, X_{ti} is the ith sample of the tth cluster G_t, n_t is the capacity of G_t.

Let the G_p and G_q can be aggregated as a new cluster $G_r (G_r = G_p \cup G_q)$, we have the deviation squared sum of G_r from the Eq. (1)

$$S_r = \sum_{i=1}^{n_r} X_{ri}^2 - \frac{1}{n_r} \left(\sum_{i=1}^{n_r} X_{ri} \right)^2. \tag{2}$$

Thus, the square distance between G_p and G_q is denoted as

$$D_{pq}^2 = S_r - S_p - S_q. \tag{3}$$

It can be proved that the recursive formula of D^2 is

$$D_{kr}^2 = \frac{n_k + n_p}{n_r + n_k} D_{kp}^2 + \frac{n_k + n_q}{n_r + n_k} D_{kq}^2 - \frac{n_k}{n_r + n_k} D_{pq}^2,$$

where n_k is the capacity of the cluster G_k that can be aggregated as a new cluster with G_r.

3   Wavelet Neural Network

Wavelet neural network (WNN) is a neural network with BP algorithm whose transfer function of the hidden layer is a wavelet basis function. Proposed by Zhang [7] in 1995, wavelet neural network is widely used in recent years as an efficient non-linear data processing model. It combines the multiscale decomposition of wavelet with feed-forward neural network, which can easily determine the parameters and structure of the network and is able to deal with the multivariate data effectively.

The structure of WNN is shown as Fig. 1.

Here $X = [X_1, X_2, \cdots, X_M]$ is the input sample of the wavelet neural network, $Y = [Y_1, Y_2, \cdots, Y_L]$ is the output of the wavelet neural network, w_{jk} is the strength of the coupling between the kth unit in input layer and the jth unit in hidden layer, w_{ij} is the strength of the coupling between the jth unit in hidden layer and ith unit in output layer.

After inputting the sample sequence, the output of hidden layer is calculated from the following equation.

$$h(j) = \psi \left(\frac{\sum_{j=1}^m w_{jk} x_k - b_j}{a_j} \right), \quad j = 1, 2, \cdots, H, \tag{4}$$

where ψ is a wavelet basis function, and the ψ chosen in this paper is morlet function [8].

$$\psi(x) = \cos(1.75x)e^{-\frac{x^2}{2}}.$$

The graph of ψ is shown as Fig. 2. b_j is the translation factor of the ψ meanwhile a_j is the telescopic factor of that, H is the total number of units in the hidden layer.

The output of output layer is calculated by the following equation

$$y(k) = \sum_{j=1}^{H} w_{jk}h(j), \quad i = 1, 2, \cdots, L,\tag{5}$$

where L is the total number of units in the output layer. The total error of this performance of the WNN is denoted as

$$E = \frac{1}{2}\sum_{i=1}^{L}\left(\hat{y}(i) - y(i)\right)^2,\tag{6}$$

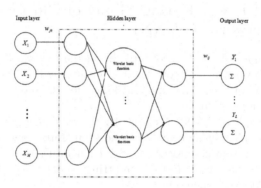

Fig. 1. The network structure of WNN

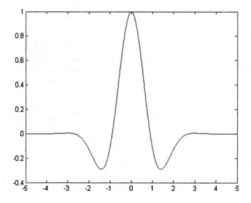

Fig. 2. The graph of morlet function

where $\hat{y}$ is the actual value of the output unit i, and $y(i)$ is the target value of unit i. Let all the parameters of Wavelet neural network form the parameter space $\Theta = \{w_{ij}, w_{jk}, a_j, b_j\}$, and this Θ are updated according to the following equation [8].

$$\Theta(T+1) = \Theta(T) - \alpha \frac{\partial E}{\partial \Theta(T)} + \beta(\Theta(T+1) - \Theta(T)),$$

where α is the learning rate and β is the adaptive momentum factor. T represents the sweep number. The value of β is updated according to the following equation.

$$\beta = \begin{cases} \beta, & E(T+1) < E(T), \\ 0, & E(T+1) > E(T). \end{cases}$$

We choose the mean absolute percentage error (MAPE) to measure the fitting effect

$$\delta = \frac{1}{N} \sum_{i=1}^{N} \frac{|\hat{y}(i) - y(i)|}{y_i} \times 100\%. \tag{7}$$

4   The Wind Power Fitting Model of Cluster-Wavelet Neural Network (CWNN)

The training samples of the CWNN are the meteorological data, wind power and wind energy level which are obtained by clustering the meteorological data for a certain period of time. Meteorological data for the input, wind power and wind energy level for the output.

The basic principle of CWNN is to classify the wind energy level according to the result of hierarchical cluster analysis, and then train the wavelet neural network by the training samples. After the wavelet neural network has completing training, we can evaluate the effect of the fitting model by specific index.

The flow of CWNN is shown in Fig. 3.

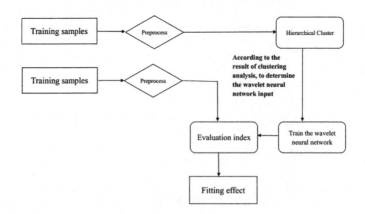

Fig. 3. The flow of CWNN

5   A Numerical Example

In this experiment, we take wind-power data and the meteorological data (70 m average wind speed, 70 m average wind direction, 30 m average wind speed and 10 m average wind direction) per ten minutes (from 0 to 24 o'clock, 144 groups in all) from some wind farm in Zhejiang Province on December 1, 2008 as the training samples.

5.1   The Data Normalization

In order to avoid the effect of the input data magnitude on the fitting results, it is necessary to normalize the training samples [9].

For wind speed data

$$v^* = \frac{v(t) - \min(v(t))}{\max(v(t)) - \min(v(t))}, \tag{8}$$

where $v(t)$ is the value of the original wind speed data at time t, $\max(v(t))$ is the maximum in the original wind speed data sequence, $\min(v(t))$ is the minimum value in the original wind speed data sequence, v^* is the result of the data normalization.

For wind direction data, because the wind direction is the vector, in order to effectively distinguish all the wind direction, we take the sin and cos of the wind direction as input.

$$\sin^* \theta(t) = \frac{\sin \theta(t) - \min(\sin \theta(t))}{\max(\sin \theta(t)) - \min(\sin \theta(t))}, \tag{9}$$

$$\cos^* \theta(t) = \frac{\cos \theta(t) - \min(\cos \theta(t))}{\max(\cos \theta(t)) - \min(\cos \theta(t))}. \tag{10}$$

The definitions of symbols in Eqs. (9) and (10) are similar to the Eq. (8).

5.2   Classification of Wind Energy Level Based on Hierarchical Clustering

According to the impact of wind resources on the grid load and economic benefits, the wind energy level in a short period of time can be divided into poor, basically available, available, relatively rich and rich in five grades [5].

If we regard the classification of the various wind energy levels as a number of different clusters, then classifying the wind energy level of a certain period of time is attributed to clustering based on meteorological data. The result of hierarchical cluster is shown as Fig. 4.

The graph of wind power based on the result of cluster is shown as Fig. 5.

It is obvious that the wind power of the black part is the lowest in the day, and the duration is short, so it is classified into poor wind energy level; the wind power of the yellow part which changes a large range, is classified into

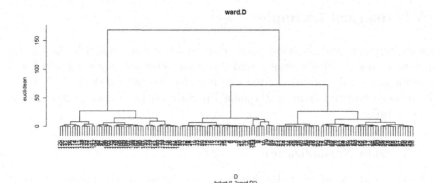

Fig. 4. The result of hierarchical cluster

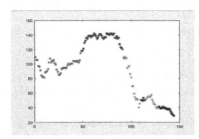

Wind energy level	Network output
Poor (Black)	[1,0,0,0,0]
Basically available (Yellow)	[0,1,0,0,0]
Available (Red)	[0,0,1,0,0]
Relatively rich (Green)	[0,0,0,1,0]
Rich (Blue)	[0,0,0,0,1]

Fig. 5. The graph of wind power (Color figure online)

Fig. 6. The result of cluster with 0–1 output method

basically available; red part is more stable, and the wind power is higher, which is classified into available; the wind power of green part of the wind power is just lower than the blue part, which is classified into relatively rich; the wind power of the blue part is the highest, and very stable, which is classified into rich. Using 0–1 output method to quantify results of hierarchical cluster as one of the input samples of WNN, the relationship is shown in Fig. 6.

5.3   Initialization of Wavelet Neural Network

In this paper, the maximum of sweep number is set to 5000 times, the average error of the target network is set to 0.005, the operating environment is MATLAB R2015a, and refer to a number of tests to select the appropriate parameters [8]. The specific values are shown as Fig. 7.

5.4   The Fitting Effect of Wind Power and Wind Energy Level

To demonstrate how the network's outputs are approaching the target value in the training process, we define the error in the Tth sweep as follow. It's unit is same as that of powerdata.

parameter setting	Wavelet neural network
Mexican hat wavelet basis	$\psi(x) = \cos(1.75x)e^{-\frac{x^2}{2}}$
a	[-5,5]
b	[-5,5]
w_{ij}, w_{jk}	[-1,1]
α	0.4
β	0.9
the number of hidden layer unit	24

Fig. 7. The parameters of WNN

$$error(T) = \sum_{i=1}^{L} \hat{y}(i) - y(i)$$

Then, we plot the graph of final fitting effect and the curve of error shown as the Figs. 8 and 9.

The MAPE of the fitting result is shown as the Fig. 10.

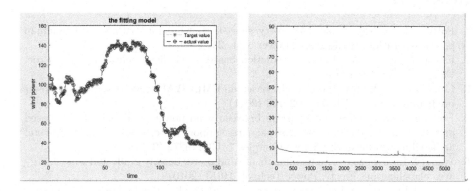

Fig. 8. The graph of fitting effect   **Fig. 9.** The curve of error

Fitting method	MAPR
WNN	1.71%

Fig. 10. The MAPR of fitting result

We also choose the ANN to fit the same data sample. The parameters of ANN is initialized according to the FAN [2]. Compared with the effect of ANN (15.62%), it is obvious that the fitting effect of wind power and wind energy level by WNN is more accurate.

6   Conclusion

(1) Using the hierarchical clustering can get the clusters which reflect the features of the wind farm's meteorological conditions at that time.

(2) The wavelet neural network can primely fit the wind power data with the MAPR of 1.71%, and the changing trend of network's output is nearly the same as that of the target value.

(3) The classification of the five wind energy levels is not clear enough which may be caused by abnormal data. Next we need to seek a kind of model to search and eliminate the abnormal data.

(4) It is time-consuming to use the clustering wavelet neural network for fitting the samples, because the gradient descent method is currently used. The next step is to adopt the conjugate gradient method to improve the efficiency of algorithm and obtain higher accuracy

Acknowledgments. The authors thank Beijing Forestry University and Yamagata University. The research is supported by the Fundamental Research Funds for the Central Universities (Grant No. 2017ZY30) and the second author is supported by Scientific Research Grant-in-Aid from JSPS under grant 15K04987.

References

1. Potter, C.W., Negnevitsky, M.: Very short-term wind forecasting for Tasmanian power generation. IEEE Trans. Power Syst. **21**, 2 (2006)
2. Fan, G.F., et al.: Wind power prediction based on artificial neural network. Proc. CSEE **34**, 118–123 (2008)
3. Lin, Y.E., Liu, P.: Combined model based on EMD-SVM for short-term wind power prediction. Proc. CSEE **31**, 102–108 (2011)
4. De Giorgi, M.G., et al.: Comparison between wind power prediction models based on wavelet decomposition with least-squares support vector machine (LS-SVM) and artificial neural network (ANN). Energies **7**, 5251–5272 (2014)
5. Wang, X.L.: Proposals of amendments of wind power density grade and wind energy area classification. Electr. Power Technol. **19**(8) (2010)
6. Ward, J.J.H.: Hierarchical grouping to optimize an objective function. J. Am. Stat. Assoc. **58**(301), 236–244 (1963)
7. Zhang, J., et al.: Wavelet neural networks for function learning. IEEE Trans. Signal Process. **43**(6), 1485–1497 (1995)
8. Chen, W.G., Ling, Y., Gan, D.G., Wei, C., Yue, Y.F.: Method to identify developing stages of air-gap discharge in oil-paper insulation based on cluster-wavelet neural network. Power Syst. Technol. **36**(7), 126–132 (2012)
9. Yang, Q., Zhang, J.H., Wang, X.F., Li, W.G.: Wind speed and generated wind power forecast based on wavelet-neural network. Power Syst. Technol. **33**(17), 44–48 (2009)

Control Strategy for Isolated
Wind-Solar-Diesel Micro Grid System
Considering Constant Load

Xuejian Yang$^{(\boxtimes)}$, Dong Yue, and Tengfei Zhang

Institute of Advanced Technolgy, Automatic College,
Nanjing University of Posts and Telecommunications, Nanjing, China
cloudy1231@yeah.net

Abstract. With the technologies of renewable energy maturing, the micro grid will become more competitive. Based on constant load, the paper builds an experiment of the isolated wind-solar-diesel micro grid system. Furthermore, the experiment achieves the goal that it can rationally regulate the diesel generators group, allocate the output power of the renewable energy according to the changes of the weather conditions and the power scheduling strategy. The feasibility and the effectiveness of the proposed approach are proved by the result of an isolated micro-grid experiment.

Keywords: Constant load · Wind-Solar-Diesel · Power scheduling strategy · Experiment

1 Introduction

Fossil energy accounts for the major component of world energy consumption, especially the total consumption of oil, gas and coal, which has reached 80% of the global energy consumption. However, the increasingly serious environmental problems and energy crisis have become the chief problem to be solved in this century with the rapid development of the world economy. Therefore, because of the advantage of the renewable energy, such as non-pollution, abundant resource, the research on the application of renewable energy is the focus topic of domestic and foreign experts in the current energy development.

In order to make full use of the renewable energy, the concept of micro grid emerges. The micro-grid system is an important form for the increasing energy supply of the renewable energy and distributed energy permeability. It consists of different distributed energy, all kinds of load and relevant monitoring, protection facilities. It contains two major forms—grid-connected mode and grid-off mode. Grid-connected mode can be run in parallel with the large power network through the distribution network to form a joint operation system of large power network and small power grid. Grid-off mode can provide electricity demand for local load independently by distributed energy without the large power grid network.

Grid-off micro grid is electrical isolation to large power grid, so it has become an important alternative as power provider in islands, remote area or rural power supply.

© Springer Nature Singapore Pte Ltd. 2017
K. Li et al. (Eds.): LSMS/ICSEE 2017, Part III, CCIS 763, pp. 103–110, 2017.
DOI: 10.1007/978-981-10-6364-0_11

The traditional grid-off micro-grid always contains battery as the storage to store the excess energy, maintain system stability and improving power quality. But its charging circuit is quite complex because of its limited charging voltage and charging current. Furthermore, it is limited by the service life with high cost of maintenance due to the long charging time and confined charging and discharging times. Another problem is that the wind turbine and the photovoltaic turbine need different control strategy for their own controllers. It cannot use the different kinds of renewable energy as efficiently as possible with a unified and efficient strategy, due to there is not a practicable parallel operation for it. So, it needs to establish a new isolated Wind-PV-Diesel micro grid to solve the above problem.

This paper built an isolated Wind-Solar-Diesel micro grid system. The system includes diesel generators, photovoltaic generators, wind generators, grid-connected inverter control integrated machine, energy management system, load control device and all kinds of controllers. Based on an experiment modeling a constant load in a changing climate day, the isolated Wind-Solar-Diesel micro grid system is verified feasible and effective.

2   The Isolated Wind-Solar-Diesel Micro Grid System

This paper established a simulation model for the isolated Wind-Solar-Diesel micro-grid system by a simulation experiment. The generation configuration of grid is set as following: diesel generation capacity is 270 + 108 kW, photovoltaic power system capacity is 240 kWp, each 120 kW array meet grid-connected configuration and access to the load; the wind turbine capacity is 2 * 150 kWp, each array connects to the grid-connected inverter control integrated machine though its controller. The grid-connected inverter control integrated machine accesses to the energy management system and the load. The structure of the grid system is shown in Fig. 1.

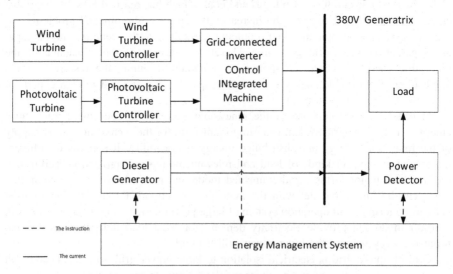

Fig. 1. The isolated Wind-Solar-Diesel micro-grid system structure diagram

3 Isolated Micro-grid Control Mode and Strategy

The control strategy of the micro-grid is using master slave control mode. The master control generation is V/f (constant Voltage and constant frequency), and the slave control generation is PQ control. The system has only one main control power generation, and the rest ate slave power supply. Diesel generator is the best choice of master control power for its control characteristics. If the diesel generator is running, then wind and PV are the slave supply.

3.1 Diesel Generators Mode

The renewable energy generators are running at constant PQ control mode with the diesel generator as the master power supply. Another role of the diesel generator is supplying the rest of the rated system power. Traditional diesel generator group is in a same power type. From the economic aspect, we could optimize the generators mode. As we know, the best coincidence rate of the diesel generator is 75%, with a certain power margin and a high economy. The least load always keeps 30% for its operation life and its unit power consumption, so we could save 70% diesel mostly. According to this, we designed a new diesel generator mode. There were two power types, the 75% power of the little power diesel generator is equal with the 30% power of the big power one. When the renewable energy power changes, we could switch the access of the diesel units flexibly. We could use the diesel generator group in an economic way when the weather is not well, and it could save 18% more than the traditional ways at the highest renewable energy usage rate by calculating.

3.2 Grid-Connected Inverter Control Integrated Machine Mode

Traditional method is that a distributed energy needs a grid-off inverter with a specific strategy. The system will increase the complexity with more distributed energy connecting. The result is that it cannot combine the different kinds of distributed energy with a unified and efficient strategy and improve the cost of the system. So, we use another means to solve the problem.

The grid-connected inverter control integrated machine is the actuator of the power scheduling strategy. It collects the data from the renewable energy generators and sends it to the energy management system. The energy management system gives the control order back according to the data and its power scheduling strategy. Then, the grid-connected inverter control integrated machine follows the order to adjust the output power of the renewable energy by turning on or turning off the renewable energy generators.

3.3 Energy Management System

The energy management system is the most important unit of this system. It undertakes the following functions mainly. Firstly, it collects the data of the renewable energy from the grid-connected inverter control integrated machine and monitors the stability of the system operation. Then, it schedules the running state of the renewable generations to achieve the ideal renewable energy utilization and the most economical power generation scheduling. Finally, it adjusts the power of the diesel generators due to the change of the renewable energy power.

3.4   The Power Scheduling Strategy

The power scheduling strategy is based on the change of the renewable energy power. Its strategy process is shown in Fig. 2. Here is the interpretation of the symbols in Fig. 2.

P_{ref} the output power of the all renewable energy generators;
P_1 the rated power of the load;
P_2 45% of P_1;
P_3 70% of P_1;
P_{net} the net load of the system;
P_w the power of an array of the wind turbines;
P_{PV} the power of an array of the PV turbines;
P_{th} the threshold power for allowing the renewable energy connecting to the micro grid;

$$P_{net} = P_1 - P_{ref}.$$

The detailed description of the strategy steps:

(1) The group of the diesel generators starts the system.
(2) Start the renewable energy generators, calculate the output power of the all renewable energy generators (P_{ref}). If P_{ref} is less than the threshold power for allowing the renewable energy connecting to the micro grid (P_{th}), the diesel generators keep working for supplying the system power;
(3) If P_{ref} is more than P_{th}, an array of wind turbines and PV turbines connects to the micro grid;
(4) If the sum of an array of both renewable energy generators cannot afford 45% of the rated power of the load (P_2), start another array of the renewable generators. If P_{ref} is less than 70% of the rated power of the load, use the big power diesel generator for the net load of the system (P_{net}). Otherwise, use the big power diesel generator for P_{net};
(5) If the sum of an array of both renewable energy generators is more than P_2, compare the power of each renewable energy (P_{PV}, P_w) with P_2. If P_{PV} or P_w cannot afford P_2, start another array of the renewable generators, use unloading device to consume extra electricity for keeping P_{net} at 88% of P_1 and use the little power diesel generator for the lest P_{net}.
(6) If one of P_{PV} and P_w can afford P_2, turn off the little power renewable energy generators, use the single strategy to make the full use of the big power one,

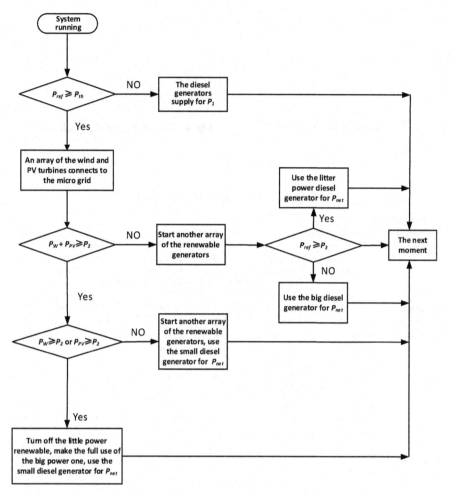

Fig. 2. The power scheduling strategy

use unloading device to consume extra electricity for keeping P_{net} at 88% of P_1 and use the little power diesel generator for P_{net}, use unloading device to consume extra electricity for keeping P_{net} at 88% of P_1.

4   Experiment Example

In order to verify the validity of the system, this paper designs an experiment. We design the threshold power of PV and wind turbines are 27 kW and 30 kW. The load is fixed at 270 kW. The maximum fixed renewable energy usage power is 238 kW. The environment of the experiment is shown in the Figs. 3 and 4. From the figures, we can find that the weather was so volatile. The result of the experiment is in the Table 1.

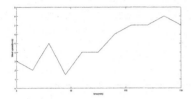

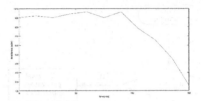

Fig. 3. The change of the wind speed   **Fig. 4.** The change of the irradiance

Table 1. The result of the experiment

The record of the experiment									
The system of the generators	Time	PV turbines			Wind turbines		Diesel generators		Load
		Irradiance (w/m)	The output power (The PV controller/integrated machine) (kw)		Wind speed (m/s)	The output power (The wind controller/integrated machine) (kw)	Power (kw)	The total fuel (ml)	Power (kw)
The isolated wind-solar-diesel micro-grid system	11:00	903	163.7/160		3	6.4/0	110	46442	270
	11:15	921	168.6/167		2	3.2/0	103		270
	11:30	901	167.5/162		5	62.6/61	47		270
	11:45	942	173.7/171		1.5	1.8/0	99		270
	12:00	964	177.8/175		4	36.8/35	60		270
	12:15	899	166.1/163		4	34.8/32	75		270
	12:30	964	176.3/174		6	94.9/64	32		270
	12:45	783	110.2/87		7	152.9/151	32		270
	13:00	658	92.8/90		7	145.6/143	37		270
	13:15	455	66.2/23		8	218.6/215	32		270
	13:30	158	21.8/0		7	164.1/162	108		270
The single diesel generators group	14:00						270	180000	270
	14:30						270		270
	15:00						270		270
	15:30						270		270
	16:00						270		270

Obviously, from Figs. 3, 4, 5 and Table 1, we can find that because of the irradiance keeping in a steady interval in the beginning one and a half hours, the output power of the PV turbines was retained at about 168 kW. The output power of the wind turbines fluctuated wildly, sometimes below the threshold power. So, the PV power was the main renewable supply, the wind power was the reserve energy. After 12:30, the wind speed kept growing and the irradiance was reducing. According to the power scheduling strategy, the wind power became the main renewable power supply for the system. When the sum of the both renewable power was more than the maximum fixed value, the system made full use of the big power one and adjusted the litter one by the strategy. Furthermore, it smoothly switched to the litter power diesel generator as the

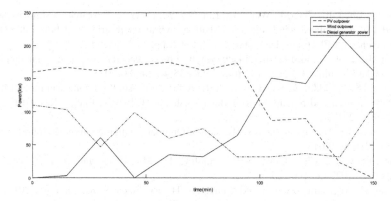

Fig. 5. The line chart of every energy generators output power

renewable energy rate more than 70%. According to the calculating result, it saved 74.8% diesel than the single use of the diesel. It also proves that it could save more diesel than the traditional grid-off micro gird strategy.

5   Conclusion

This paper built an isolated Wind-Solar-Diesel isolated micro-grid system. The result of the simulation experiment showed the feasibility of the system and the availability of the control strategy. To achieve the biggest use of the renewable energy, we took forward the smooth switch control strategy according to the renewable energy change. The result analysis proved the usability of the system and the effectiveness of the control strategy.

References

1. Ram, J.P., Rajasekar, N., Miyatake, M.: Design and overview of maximum power point tracking techniques in wind and solar photovoltaic systems: a review. Renew. Sustain. Energy Rev. **73**, 1138–1159 (2017)
2. Rashid, S., et al.: Optimized design of a hybrid PV-wind-diesel energy system for sustainable development at coastal areas in Bangladesh. Environ. Prog. (2016)
3. Barzola, J., Espinoza, M., Cabrera, F.: Analysis of hybrid solar/wind/diesel renewable energy system for off-grid rural electrification. **6**(3), 1146–1152 (2016). www.researchgate. net
4. Ying, T., Zhilin, L., Jie, L.I.: Multi-objective optimal sizing method for distributed power of wind-solar-diesel-battery independent microgrid based on improved electromagnetism-like mechanism. Power Syst. Prot. Control (2016)
5. Shezan, S.K.A., et al.: Feasibility analysis of a hybrid off-grid wind–DG-battery energy system for the eco-tourism remote areas. Clean Technol. Environ. Policy **17**(8), 1–14 (2015)

6. Cai, G., Chen, R., Chen, J., Wang, K., Lin, Y., Wu, L., et al.: Research on simulations and model of the isolated wind-solar-diesel-battery hybrid micro-grid. In: China International Conference on Electricity Distribution, vol. 8562, pp. 1–4, Shanghai (2012)
7. Chen, B., et al.: Wind solar diesel battery hybrid power generation system simulation based on matlab/simulink. J. Electr. Electron. Educ. **35**(4), 84–88 (2013)
8. Lipu, M.S.H., Uddin, M.S., Miah, M.A.R.: A feasibility study of solar-wind-diesel hybrid system in rural and remote areas of Bangladesh. Int. J. Renew. Energy Res.-IJRER **3**(4), 892–900 (2013)
9. Zhao, B.: Key Technology and Application of Optimal Configuration of Microgrid. Science Press, Beijing (2015)
10. Chowdhury, S., Chowdhury, S.P., Crossley, P.: Microgrids and Active Distribution Networks. IET, London (2009)
11. Wang, C.: Microgrid Analysis and Simulation Theory. Science Press, Beijing (2013)
12. Arthur, R., Johannes, J.: Optimization of PV-wind-hydro-diesel hybrid system by minimizing excess capacity. Eur. J. Sci. Res. **22**, 1331–1333 (2009)

Equilibrium Analysis of Electricity Market with Wind Power Bidding and Demand Response Bidding

Kai Zhang, Xian Wang, and Shaohua Zhang$^{(\boxtimes)}$

Key Laboratory of Power Station Automation Technology,
School of Mechatronic Engineering and Automation,
Shanghai University, Shanghai, China
zjkhzk@gmail.com, xianwang@shu.edu.cn,
eeshzhan@126.com

Abstract. In electricity markets with strategic bidding of wind power, it is important to handle wind power's output deviation. In this paper, the scenario where customers in Demand Response (DR) program matches the wind power's output deviation through strategic bidding is studied, and a stochastic equilibrium model of the electricity market with wind power bidding and demand response bidding is proposed. In this model, linear supply function bidding is applied by both wind power producers and traditional power producers to match power demand in wholesale market. In order to compensate for the wind power's output deviation, two market models in balancing market for demand response are proposed where supply function bidding and demand function bidding are applied by DR customers to match supply deficit and surplus respectively. Furthermore, the penalty cost for output deviation of the wind power producer is determined by the equilibrium price in balancing market. The equilibrium problems are solved by being reverse-engineered into convex optimization problems and the existence and uniqueness of the Nash equilibrium is theoretically proved. A distributed dual gradient algorithm is further proposed to achieve the equilibrium. Numerical examples are presented to verify the validity of the proposed model and effectiveness of the algorithms.

Keywords: Electricity market · Wind power bidding · Demand response bidding · Distributed algorithm · Equilibrium analysis

1 Introduction

In recent years, as a kind of clean energy, wind energy generation has developed rapidly around the world and it has been a trend for wind power producers to bid into the wholesale electricity markets like conventional producers [1]. Because of the uncertainty in wind power output, there may be a difference between the bids and actual outputs [2]. Thus, it's an important issue to deal with the deviation in electricity markets with wind power bidding.

In the literature, there are several studies about how to reduce the deviation, such as improving the prediction accuracy of wind power's output [3, 4], using penalty

© Springer Nature Singapore Pte Ltd. 2017
K. Li et al. (Eds.): LSMS/ICSEE 2017, Part III, CCIS 763, pp. 111–125, 2017.
DOI: 10.1007/978-981-10-6364-0_12

mechanism for wind power bidding deviation [1], proposing a joint planning and operation strategy of wind power producers and other clean energy or energy storage facilities [5–8]. In smart grid, the customers are active parts of the grid. Thus, demand response (DR) programs can be used as an effective way to accomplish different load shaping objectives through altering the electricity consumption patterns [9] offset the output deviation for improving the system reliability.

With respect to the impact of DR on the wholesale electricity market where wind power is involved, relevant works have provided many optimization models [10–12]. In addition to the optimization models, equilibrium models are proposed in some investigations. Reference [13] proposed a new wind offering strategy that a wind power producer employs demand response to cope with the power production uncertainty and market violations. The interaction between the wind power producer and conventional generators is modeled as a Stackelberg game in which the wind power producer as the leader of the game, and the conventional generators compete for maximizing their profits as followers. But it assumes that only the wind power producer has market power and the conventional generators bid in the electricity market in perfect competition, which is impractical. Reference [14] proposed a multi-period stochastic Cournot equilibrium model of electricity market. In this model, both the conventional power producers and wind power producers with DR resource can bid in the market and it's assumed that the wind power producer makes contracts with the DR aggregator for the usage of the flexible DR resource. The downside of the model is, the demand response price is jointly determined by the wind power producer and the DR aggregator by contractual approach in advance and the price is assumed to be a fixed parameter rather than a variable. Reference [15] proposed an abstract market model for demand response where supply function bidding is applied to match power supply deficit and it studies how a cost function affects a customer's demand response and the price in balancing market. The DR prices being determined by the customers' bids in balancing market can give full play to the competitive edge of the market and it has good prospects for the researches on the DR. Thus, apart from the bids in wholesale market with power producers, the bids in balancing market with DR customers to cover the wind power's output deviation are studied in this paper.

In most of the aforementioned studies, the equilibrium analysis of electricity market with wind power producer and DR resource has not yet considered the impacts of both the producers' bids in wholesale market and the DR customers' bids in balancing market simultaneously. In this paper, a stochastic equilibrium model of the electricity market with wind power bidding and demand response bidding is proposed. In this model, supply function bidding is applied by power producers in the wholesale market. In order to compensate for the wind power's output deviation, two market models in balancing market for demand response are proposed where supply function bidding and demand function bidding are applied by DR customers to match supply deficit and surplus respectively. Furthermore, the penalty cost of the wind power producer is determined by the equilibrium price in balancing market, which is caused by wind power's output deviation. Besides, the equilibrium problems are solved by being reverse-engineered into convex optimization problems and the existence and uniqueness of the Nash equilibrium is theoretically proved. We further propose a distributed algorithm to achieve the equilibrium. Numerical examples are presented to verify reasonableness of the proposed model and investigate the impacts of the competition models of the balancing market on the equilibrium results of the electricity market.

2 Equilibrium Model of Electricity Market with Wind Power Bidding and Demand Response Bidding

2.1 Equilibrium Model of Wholesale Market

Consider a wholesale market with n conventional power producers and one wind power producer that are served by one utility company. Here, supply function bidding is applied by the producers to match load demand in the wholesale market. Let $D > 0$ denote the aggregate load demand of the customers at a given time slot. Associated with each conventional producer $i(i = 1, 2, ..., n)$ is a power output Q_i that it is willing to supply in the wholesale market and let Q_{n+1} denote the power output that the wind power producer is willing to supply. It's assumed that the total output needs to meet the aggregate load demand, that is

$$D = \sum_{i=1}^{n} Q_i + Q_{n+1} \tag{1}$$

Assume that conventional power producer i incurs a cost $C_{1i}(Q_i)$ when it supplies a generation of Q_i, that is

$$C_{1i}(Q_i) = a_{1i}Q_i + h_{1i}Q_i^2, \quad i = 1, \ldots, n \tag{2}$$

where a_{1i} and h_{1i} are constants with $a_{1i} \geq 0$, $h_{1i} \geq 0$.

A market mechanism based on supply function bidding [15] for the power generation allocation is considered and takes the form of

$$Q_i(B_i, p_1) = B_i p_1, \quad B_i \geq 0, \quad i = 1, \ldots, n+1 \tag{3}$$

And the price p_1 which clears the market is determined by the utility company based on the bids of the producers, that is

$$p_1(B) = \frac{D}{\sum_{i=1}^{n+1} B_i} \tag{4}$$

(1) Optimization problems of conventional producers:
   Each conventional producer $i(i = 1,2,...,n)$ chooses B_i to maximize its own net benefit $\pi_i(B_i, B_{-i})$ given others' bidding strategy

$$\max_{B_i \geq 0} \pi_i = p_1 Q_i - C_{1i}(Q_i) \tag{5}$$

$$s.t. \sum_{i}^{n} Q_i + Q_{n+1} = D \tag{6}$$

(2) The optimization problem of wind power producer:
It's assumed that Q_w is the actual output of wind power in real time which is a random variable decided by the weather condition and the difference between Q_{n+1} and Q_w is defined as wind power's output deviation, that is

$$\Delta_w = Q_w - Q_{n+1} \tag{7}$$

The supply of wind power is surplus when $\Delta_w > 0$ and deficit when $\Delta_w < 0$. And the wind power producer is supposed to be obliged to sell(or buy) corresponding quantity of electricity in balancing market to clear supply surplus (or deficit) so as to achieve the balance between supply and demand of the electricity system in real time.

In this paper, Weibull distribution is used to describe the fluctuations in the wind speed and Monte Carlo simulation is adopted to obtain M wind speed values. Then a recursive backward scenario reduction method [16] is employed to get N' wind speed scenarios based on the M wind speed values. And the wind speed scenarios can be converted into wind power output scenarios represented by $Q_{wm}(m = 1, 2, ..., N')$ through the relationship between wind speed and wind power output. We refer readers to [14] and [16] for more details about the scenario reduction technique.

The wind power producer also chooses B_{n+1} to maximize its own net benefit $\pi_{n+1}(B_{n+1}, B_{-(n+1)})$ given others' bidding strategy $B_{-(n+1)}$

$$\max_{B_{n+1} \geq 0} \pi_{n+1} = p_1 Q_{n+1} + \sum_{m=1}^{N'} [s_m p_{2m}(Q_{wm} - Q_{n+1})] \tag{8}$$

$$s.t. \sum_{i}^{n} Q_i + Q_{n+1} = D \tag{9}$$

Here, p_{2m} is the equilibrium price in the balancing market of wind power output scenario $m(m = 1, 2..., N')$, Q_{wm} is the actual output of scenario m, s_m is the occurring probability of scenario m with

$$\sum_{m=1}^{N'} s_m = 1 \tag{10}$$

The wholesale market equilibrium model is composed by the optimization models of all conventional producers and the wind producer's decision model. And the conditions of the equilibrium of the wholesale market can be reverse-engineered into optimality conditions of the optimization problem as

$$\min_{0 \le Q_i < D/2} \sum_{i=1}^{n} D_{1,i}(Q_i) + D_{1,n+1}(Q_{n+1}) \tag{11}$$

$$s.t. \sum_{i=1}^{n+1} Q_i = D \tag{12}$$

with

$$D_{1,i}(Q_i) = \left(1 + \frac{Q_i}{D - 2Q_i}\right) C_{1i}(Q_i) + \int_0^{Q_i} \frac{D}{(D - 2x_i)^2} C_{1i}(x_i) dx_i, \quad i = 1,\ldots,n \tag{13}$$

$$D_{1,n+1}(Q_{n+1}) = \sum_{m=1}^{N'} s_m p_{2m} \int_0^{Q_{n+1}} \left(1 + \frac{x_j}{D - 2x_j}\right) dx_j \tag{14}$$

It's proved that the Nash equilibrium of the game satisfies the optimality conditions of (11)–(14), and solves the optimization problems (11)–(14). Hence, the existence and uniqueness of the Nash equilibrium is a result of the existence and uniqueness of the optimal solutions of (11)–(14). It is easy to reach the conclusion that the optimization problem (11)–(14) is a strictly convex problem and has a unique optimal solution.

2.2   Equilibrium Model of Balancing Market

In order to compensate for the deviation, two different bidding schemes for allocating load adjustments among customers in the DR program to match the supply surplus and deficit are proposed respectively in this section.

Associated with each customer $i(i = 1, 2, \ldots, N)$ is a load $q_i(q_i \ge 0)$ that it is willing to shed (or increase) in the DR program. It's assumed that the total load shed (or increase) needs to meet the wind power's output deviation, that is

$$\sum_{i=1}^{N} q_i = \begin{cases} -\Delta_w, & \text{if } \Delta_w \le 0 \\ \Delta_w, & \text{else} \end{cases} \tag{15}$$

(1) Supply function bidding for load shedding. When the supply of wind power is deficit ($\Delta_w < 0$), the customers need to shed their loads to clear supply deficit with incentive payments. Similarly, it's assumed that each customer's "supply" function (for load shedding) is parameterized by a single parameter $b_i \ge 0$ as

$$q_i(b_i, p_2) = b_i p_2, \quad b_i \ge 0, \quad i = 1,\ldots,N \tag{16}$$

The utility company will decide the market-clearing price p_2 in the balancing market, that is

$$p_2(b) = \frac{-\Delta_w}{\sum_{i=1}^{N} b_i} \tag{17}$$

When customer i sheds a load of q_i, it will incur a cost (or disutility) $C_{2i}(q_i)$ as

$$C_{2i}(q_i) = a_{21i}q_i + h_{21i}q_i^2, \quad i = 1, \ldots, N \tag{18}$$

where a_{21i} and h_{21} are constants with $a_{21i} \geq 0$, $h_{21} \geq 0$.

(2) Demand function bidding for load increasing. When the supply of wind power is surplus ($\Delta_w > 0$), each customer bids a "demand" function [17] (for load increasing) which takes the form of

$$q_i(\beta_i, p_2) = \alpha_i - \beta_i p_2, \quad q_i \geq 0 \tag{19}$$

Here, α_i is a constant with $\alpha_i \geq 0$.
The market-clearing price p_2 in the balancing market can be written as

$$p_2(\beta) = \frac{\sum_{i=1}^{N} \alpha_i - \Delta_w}{\sum_{i=1}^{N} \beta_i} = \frac{K}{\sum_{i=1}^{N} \beta_i} \tag{20}$$

And the customers will increase their loads (buy more electricity) to clear supply surplus and each customer i has a utility function for energy consumption as

$$U_{2i}(q_i) = a_{22i}q_i - h_{22i}q_i^2, \quad i = 1, \ldots, N \tag{21}$$

where a_{22i} and h_{22i} are constants with $a_{22i} \geq 0$, $h_{22i} \geq 0$.

When $Q_{wm} - Q_{n+1} < 0$ in the scenario m, let q_{im} and θ_{im} denote the positive load adjustment and benefit at the Nash equilibrium of customer i respectively. Let b_{im} and β_{im} denote the supply and demand function profile of customer i respectively. Each DR customer $i(i = 1, 2, \ldots, N)$ chooses b_i to maximize its own net benefit $\theta_{im}(b_{im}, b_{-im})$ given others' bidding strategy.

$$\max_{b_{im} \geq 0} \theta_{im} = p_{2m}q_{im} - C_{2i}(q_{im}) \tag{22}$$

$$s.t. \sum_{i=1}^{N} q_{im} = -\Delta_{wm} \tag{23}$$

$$q_{im} = b_{im}p_{2m}, \quad b_{im} \geq 0, \quad p_{2m} > 0 \tag{24}$$

$$C_{2i}(q_{im}) = a_{21i}q_{im} + h_{21i}q_{im}^2, \quad i = 1, \ldots, N \tag{25}$$

Similarly, the conditions of the equilibrium with N customers can be reverse-engineered into optimality conditions of the optimization problem as

$$\min_{0 \le q_{im} < -\Delta_{wm}/2} \sum_{i=1}^{N} d_{1,i}(q_{im}) \tag{26}$$

$$s.t. \sum_{i=1}^{N} q_{im} = -\Delta_{wm} \tag{27}$$

with

$$d'_{1,i}(q_{im}) = \left(1 + \frac{q_{im}}{d_m - 2q_{im}}\right) C'_{2i}(q_{im}) \tag{28}$$

When $Q_{wm} - Q_{n+1} > 0$ in the scenario m, each DR customer i chooses β_{im} to maximize its own net benefit $\theta_{im}(\beta_{im}, \beta_{-im})$ given others' bidding strategy β_{-im}

$$\max_{\beta_{im} \ge 0} \theta_{im} = -p_{2m}q_{im} + U_{2i}(q_{im}) \tag{29}$$

$$s.t. \sum_{i=1}^{N} q_{im} = \Delta_{wm} \tag{30}$$

$$q_{im} = \alpha_{im} - \beta_{im}p_{2m}, \quad \beta_{im} \ge 0, \quad p_{2m} > 0, \quad q_{im} \ge 0 \tag{31}$$

$$U_{2i}(q_{im}) = a_{22i}q_{im} - h_{22i}q_{im}^2, \quad i = 1, \ldots, N \tag{32}$$

where $K_m = \sum_i \alpha_{im} - \Delta_{wm}$, $\alpha_{im} = d_m$.

The conditions of the equilibrium with N customers can be reverse-engineered into optimality conditions of the optimization problem as

$$\min_{0 \le q_{im} < \frac{a_{22i}}{2h_{22i}}} - \sum_{i=1}^{N} d_{2,i}(q_{im}) \tag{33}$$

$$s.t. \sum_{i=1}^{N} q_{im} = \Delta_{wm} \tag{34}$$

with

$$d'_{2,i}(q_{im}) = \left(1 - \frac{q_{im}}{K_m - \alpha_{im} + 2q_{im}}\right) U'_{2i}(q_{im}) \tag{35}$$

It's proved that the Nash equilibrium of the DR game satisfies the optimality conditions of (22)–(25) and (29)–(32), and solves the convex optimization problems respectively. The existence and uniqueness of the Nash equilibrium is a result of the existence and uniqueness of the optimal solutions of (26)–(28) and (33)–(35).

2.3 Distributed Algorithm

One way to find the equilibrium is to solve the convex optimization problems, requiring to know the cost (or utility) functions of all customers and producers. However, the electricity market typically involves a very large number of participants who are usually unwilling to report their bidding functions, motivating the needs of a distributed algorithm where the utility company sets the price while the customers and producers submit a bid based on the price. Such bidding schemes require only light communication and computation and can be easily scaled to large systems.

Note that, the convex optimization problems (11)–(14), (26)–(28) and (33)–(35) can be easily solved in a distributed way by the dual gradient algorithm in [18]. Initially, the utility company randomly picks a price $p1(0)$ in the wholesale market and $p2(0)$ in the balancing market respectively and then announces the prices to each customer and producer over the communication network. Set the precision value $\varepsilon = 10^{-6}$.

At kth iteration.

(1) Solution of the wholesale market

At k_1th iteration.

(1.1) Upon receiving price $p_1(k_1)$ announced by the utility company, each conventional producer i updates its supply function $B_i(k_1)$, according to

$$B_i(k_1) = \left[\frac{\left(D'_{1,i}\right)^{-1}(p_1(k_1))}{p_1(k_1)} \right]^+ , \quad i = 1, \ldots n \tag{36}$$

The wind power producer $i = n + 1$ updates its supply function $B_{n+1}(k_1)$, according to

$$B_{n+1}(k_1) = \left[\frac{\left(D'_{1,n+1}\right)^{-1}(p_1(k_1))}{p_1(k_1)} \right]^+ \tag{37}$$

(1.2) Upon gathering bids $B_i(k_1)$ from the producers, the utility company updates the price in the wholesale market according to

$$p_1(k_1 + 1) = \left[p_1(k_1) - \gamma_1 \left(\sum_i^{n+1} B_i(k_1)p_1(k_1) - D \right) \right]^+ \tag{38}$$

where parameter γ_1 denotes the step size. $[\cdot]^+$ is the projection onto the feasible set.

(2) Upon reaching the equilibrium of the wholesale market, the utility company updates the wind power's output deviation of each scenario according to

$$\Delta_{wm} = Q_{wm} - B_{n+1}p_1, \quad m = 1, \ldots, N' \tag{39}$$

Based on the wind power's output deviation, the utility company announces the electricity supply deficit (or surplus) in the balancing market as

$$d_m = \begin{cases} -\Delta_{wm}, & \text{if } \Delta_{wm} \leq 0 \\ \Delta_{wm}, & \text{else} \end{cases} \tag{40}$$

(3) Solution of the balancing market

At k_2th iteration in the scenario m ($m = 1, 2, ..., N'$).

(3.1) Upon receiving the supply deficit (or surplus) announced by the utility company, when the supply is deficit ($\Delta_w < 0$), each customer i updates its demand function $b_{im}(k_2)$, according to

$$b_{im}(k_2) = \left[\frac{\left(d'_{1,i}\right)^{-1}(p_{2m}(k_2))}{p_{2m}(k_2)} \right]^+, \quad i = 1, ...N \tag{41}$$

And when the supply is surplus ($\Delta_w > 0$), each customer i updates its supply function $b_{im}(k_2)$, according to

$$b_{im}(k_2) = \left[\frac{\left(d'_{2,i}\right)^{-1}(p_{2m}(k_2))}{p_{2m}(k_2)} \right]^+, \quad i = 1, ...N \tag{42}$$

(3.2) Upon gathering bids $b_{im}(k_2)$ from customers, the utility company updates the price in the balancing market according to

$$p_{2m}(k_2 + 1) = \left[p_{2m}(k_2) - \gamma_2 \left(\sum_i^N b_{im}(k_2)p_{2m}(k_2) - d_m \right) \right]^+ \tag{43}$$

where parameters γ_2 denote the step size.

(4) Upon reaching the equilibrium of the balancing market for all N' scenarios, the utility company updates the expected value of the price in the balancing market according to

$$p_2(k + 1) = \sum_{m=1}^{N'} s_m p^*_{2m} \tag{44}$$

(5) Upon receiving the expected value of price $p_2(k + 1)$ which determines the penalty cost, each power producer updates its supply function again. Set $k = k + 1$ and repeat until both of the wholesale market and balancing market reach the equilibrium.

It is straightforward to verify that the bidding scheme is equivalent to the dual gradient algorithm of (11)–(14), (26)–(28) and (33)–(35). Therefore, the bidding scheme possess all the convergence properties of the dual gradient algorithm. For example, when γ_1 and γ_2 are small enough, the above algorithm converges exponentially to the equilibrium. Besides, it does not need the profile of the other customers or producers since (36)–(37) and (41)–(42) only depend on its own bidding profile and the information announced by the utility company.

3   Numerical Studies

Consider a wholesale market with two conventional power producers (G1 and G2) and one wind power producer (G3). It's assumed that each conventional producer has a cost function $C_{1i}(Q_i) = a_{1i}Q_i + h_{1i}Q_i^2$, with $a_{11} = 10$ \$/MWh, $h_{11} = 1.0$ \$/(MW)^{2}h and $a_{12} = 10$ \$/MWh, $h_{12} = 1.1$ \$/(MW)^{2}h. Both the installed capacity of G1 and G2 are 20 MW. The wind power producer has 10 wind turbines. The power output of the wind turbine satisfies $Q_w = 0.7646v^3$. The cut-in wind speed $V_{in} = 2$ m/s, the rated wind speed $V_N = 11$ m/s, and the cut-out wind speed $V_{out} = 20$ m/s. The aggregate load demand $D = 12$ MW with step size $\gamma_1 = 0.5$ for strategic supply function bidding.

Assume there are 30 customers in the balancing market who participate in the DR program. When the electricity supply is deficit, each customer i has a cost function $C_{2i}(q_i) = a_{21i}q_i + h_{21i}q_i^2$. And a_{21i} and h_{21i} are randomly drawn from [60, 61] and [12, 15] respectively with step size $\gamma_2 = 0.5$. And when the electricity supply is surplus, each customer i has a utility function $C_{2i}(q_i) = -a_{22i}q_i + h_{22i}q_i^2$. And a_{22i} and h_{22i} are randomly drawn from [15, 16] and [0.1, 0.2] respectively with step size $\gamma_3 = 0.05$. The number of scenarios is $N' = 10$.

3.1   Convergence of Distributed Algorithms

Consider the weather condition of the average wind speed being 7 m/s with the standard deviation being 3, and the balancing market being in oligarch competition. Figure 1 shows the evolution of the price p_1 and the producers' supply functions for strategic supply function bidding in the wholesale market, respectively. Figures 2 and 3 shows the evolution of the price p_{2m} and five customers' supply (or demand) functions in the balancing market of scenario 1 ($m = 1$) and scenario 3 ($m = 3$) in which the wind

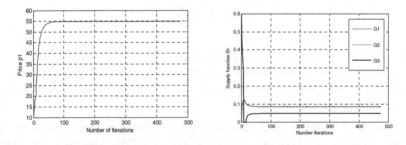

Fig. 1. Price and supply function evolution of strategic bidding in the wholesale market

power supply is deficit and surplus respectively. We can see that the prices and bidding functions approach the market equilibrium within less than 100 iterations. When the iteration time $t = 12$ s or so, the equilibrium of the wholesale market and the balancing market are reached simultaneously.

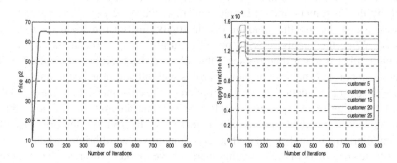

Fig. 2. Price and supply function evolution of strategic supply function bidding in the balancing market of scenario 1 ($m = 1$) where the wind power supply is deficit

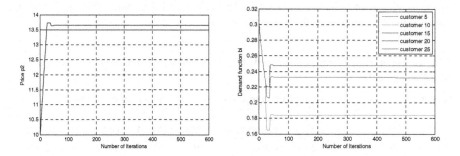

Fig. 3. Price and demand function evolution of strategic demand function bidding in the balancing market of scenario 3 ($m = 3$) where the wind power supply is surplus

3.2 The Impacts of Competition Models of DR Bidding on Electricity Market Equilibrium

Consider the weather condition of the average wind speed being 7 m/s with the standard deviation being 3, and study participants' behavior in two different balancing markets, oligopolistic and competitive [15]. Table 1 shows the equilibrium results in the balancing market of the demand response in two markets with different competition models of DR bidding. Table 2 shows the equilibrium results in the wholesale market of the demand response in the two different balancing markets.

In Table 1, Q_{wm} is the actual output of wind power output scenario m ($m = 1, 2, \ldots, 10$), s_m is the occurring probability of scenario m, Δ_{wm} is the wind power's output deviation and p_{2m} is the equilibrium price in the balancing market, φ_m is the total net revenue of DR customers. It can be seen from Table 1 that when the supply of wind power is deficit ($\Delta_{wm} < 0$), the price p_{2m} in the balancing market is higher than p_1 in the wholesale market (see Table 2). It implies that the price of the quantity of the electricity ($-\Delta_{wm}$) the wind power producer buys in the balancing market is higher than that in the wholesale market so the wind power producer can not get any profits and will be subject to additional economic penalties. Because the supply function bidding is applied by the customers, the customers will raise the price p_{2m} if the value of $|\Delta_{wm}|$ increases which means the amount of electricity the customers need to shed is increased. When the supply of wind power is surplus ($\Delta_{wm} > 0$), the price p_{2m} in the balancing market is lower than p_1 (see Table 2). And the wind power producer will be subject to economic penalties with lower profits without additional economic penalties. Because the demand function bidding is applied by the customers, the price p_{2m} will get lower if the value of $|\Delta_{wm}|$ increases which means the amount of electricity that the customers need to purchase additionally is increased. In addition, the price p_{2m} at the Nash equilibrium is higher than that at competitive equilibrium when $\Delta_{wm} < 0$. Oligopolistic customers in the balancing market have market power, so they will report a higher "cost" in order to obtain higher returns in the DR program, resulting in a rise in p_{2m}. When $\Delta_{wm} > 0$, the price p_{2m} at the Nash equilibrium is lower than that at competitive equilibrium. Oligopolistic customers will report a lower "benefit" in order to obtain higher returns in the DR program, resulting in a decrease in p_{2m}. Therefore, the customers in the oligopolistic balancing market will get more benefits due to the market power regardless of the value of Δ_{wm}.

Table 1. The impacts of competition models of the balancing market on equilibrium results in the balancing market

Scenario	s_m	Q_{wm} (MW)	In oligarch competition			In perfect competition		
			Δ_{wm} (MW)	p_{2m} (USD/MWh)	φ_m (USD/h)	Δ_{wm} (MW)	p_{2m} (USD/MWh)	φ_m (USD/h)
1	0.07	0.23	−2.38	64.9	7.93	−2.42	62.7	2.72
2	0.07	8.23	5.63	13.0	10.58	5.58	13.4	7.92
3	0.07	1.54	−1.07	63.7	2.91	−1.11	61.5	0.65
4	0.10	2.28	−0.33	62.7	0.78	−0.37	60.8	0.12
5	0.04	0	−2.60	65.1	8.97	−2.65	62.9	3.24
6	0.16	6.00	3.40	13.6	4.22	3.35	14.0	2.73
7	0.10	10.2	7.57	12.4	18.5	7.52	12.8	15.2
8	0.14	0.86	−1.75	64.4	5.31	−1.79	62.1	1.53
9	0.14	3.16	0.55	14.4	0.32	0.51	14.9	0.06
10	0.11	0.58	−2.03	64.6	6.42	−2.07	62.4	2.01
Expected value		3.54	0.93	40.2	5.94	0.88	39.3	3.32

Table 2 shows that if the balancing market is in oligarch competition, each power producer tends to bid a lower bidding value B_i compared to that in perfect competition, resulting in a higher wholesale market price p_1. Loosely speaking, the customers in the oligopolistic balancing market will get more benefits due to a higher market power, leading to a rise in the penalty cost of the wind power producer. Therefore, the wind power producer will lower its bidding value B_{n+1} to reduce the penalty cost and other producers will change their bidding strategies to lower bidding values so that they can obtain more profits. Besides, a lower aggregate bidding value leads to a higher price in the wholesale market. Above all, when the balancing market is in oligarch competition, the profits of conventional producers are higher than that in competitive balancing market, but the profits of wind power is lower because of the rise in the penalty cost.

Table 2. The impacts of competition models of the balancing market on equilibrium results in the wholesale market

Competition models of the balancing market		In oligarch competition	In perfect competition
G1	Optimal bidding value (MW/USD)	0.0850	0.0859
	Bid output (MW)	4.7290	4.7077
	Profit (USD/h)	193.5507	188.6511
G2	Optimal bidding value (MW/USD)	0.0838	0.0847
	Bid output (MW)	4.6648	4.6426
	Profit (USD/h)	189.0451	184.1876
G3	Optimal bidding value (MW/USD)	0.0468	0.0484
	Bid output (MW)	2.6062	2.6497
	Profit (USD/h)	113.68	114.6417
p_l (USD/MWh)		55.6573	54.7804

4 Conclusion

In this paper, a stochastic equilibrium model of the electricity market considering both wind power bidding and demand response bidding is proposed. The contribution of this work can be summarized as follows:

(1) The scenario where customers in the DR program match the wind power's output deviation through strategic bidding is studied due to the deviation between the bids and actual outputs when the wind power producer bids in the wholesale electricity market like conventional producers.
(2) In order to compensate for the deviation, two market models in the balancing market for demand response are proposed where supply function bidding and demand function bidding are applied by DR customers to match supply deficit and surplus respectively. Besides, the penalty cost for the output deviation of the wind power producer is determined by the equilibrium price in the balancing market, which can give full play to the competitive edge of the market.
(3) The equilibrium problems are reverse-engineered into convex optimization problems in this paper and it's proved that the existence and uniqueness of the Nash equilibrium is a result of the existence and uniqueness of the optimal solutions.

(4) Considering the circumstances where the participants in the electricity market can only get the information announced by the utility company except their own information, a distributed algorithm is applied to achieve the equilibrium.

(5) Numerical examples are presented to verify the reasonableness of the proposed model and investigate the impacts of the competition models of DR bidding on the equilibrium results of the electricity market. It shows that compared to the balancing market in perfect competition, the DR customers and conventional producers will get more profits while the profit of the wind power producer gets lower when it is in oligarch competition.

References

1. Huang, M., Wang, X., Zhang, S.: Analysis of an electricity market equilibrium model with penalties for wind power's bidding deviation. In: Proceedings of the 2015 5th International Conference on Electric Utility Deregulation and Restructuring and Power Technologies (DRPT), pp. 35–40 (2015)
2. Rodríguez, O., Del Río, J.A., Jaramillo, O.A., Martínez, M.: Wind power error estimation in resource assessments. PLoS ONE **10**, e0124830 (2015)
3. Liu, L., Meng, S., Junji, W.U.: Dynamic economic dispatch based on wind power forecast error interval. Electr. Power Autom. Equip. **9**, 013 (2016)
4. Hu, Y.: A forecasting accuracy improvement method for wind power based on phase and level errors translating and interpolating correction. Power Syst. Technol. **39**, 2758–2765 (2015)
5. Huang, Y., Hu, W., Min, Y., et al.: Risk-constrained coordinative dispatching for large-scale wind-storage system. Autom. Electr. Power Syst. **38**(9), 41–47 (2014)
6. Ju, L., Li, H., Chen, Z., et al.: A benefit contrastive analysis model of multi grid-connected modes for wind power and plug-in hybrid electric vehicles based on two-step adaptive solving algorithm. Power Syst. Technol. **38**(6), 1492–1498 (2014)
7. Wang, J., Lu, J.: Research on optimal operation mode of power generation system doubly driven by wind power and hydraulic power based on equal incremental rate criterion. Power Syst. Technol. **32**(9), 80–83 (2008)
8. Zhang, H., Gao, F., Wu, J., et al.: Dynamic economic dispatching model for power grid containing wind power generation system. Power Syst. Technol. **37**(5), 1298–1303 (2013)
9. Yang, X., Zhou, M., Li, G.: Survey on demand response mechanism and modeling in smart grid. Power Syst. Technol. **40**(1), 220–226 (2016)
10. Zeng, D., Yao, J., Yang, S., et al.: Optimization dispatch modeling for price-based demand response considering security constraints to accommodate the wind power. Proc. CSEE **34**(31), 5571–5578 (2014)
11. Ju, L., Yu, C., Tan, Z.: A two-stage scheduling optimization model and corresponding solving algorithm for power grid containing wind farm and energy storage system considering demand response. Power Syst. Technol. **39**(5), 1287–1293 (2015)
12. Ju, L., Qin, C., Wu, H., et al.: Wind power accommodation stochastic optimization model with multi-type demand response. Power Syst. Technol. **39**(7), 1839–1846 (2015)
13. Mahmoudi, N., Saha, T.K., Eghbal, M.: Modeling demand response aggregator behavior in wind power offering strategies. Appl. Energy **133**, 347–355 (2014)

14. Liuhui, W., Wang, X., Zhang, S.: Electricity market equilibrium analysis for strategic bidding of wind power producer with demand response resource. In: Proceedings of the 2016 IEEE PES Asia-Pacific Power and Energy Engineering Conference (APPEEC), pp. 181–185 (2016)
15. Li, N., Chen, L., Dahleh, M.A.: Demand response using linear supply function bidding. IEEE Trans. Smart Grid **6**, 1827–1838 (2015)
16. Zhang, X., Yan, K., Lu, Z., Zhong, J.: Scenario probability based multi-objective optimized low-carbon economic dispatching for power grid integrated with wind farms. Power Syst. Technol. **38**(7), 1835–1841 (2014)
17. Chen, X., Yu, Y., Xu, L.: Linear supply function equilibrium with demand side bidding and transmission constrain. Proc. CSEE **24**(8), 17–23 (2004)
18. Cheng, Y.C.: Dual gradient method for linearly constrained, strongly convex, separable mathematical programming problems. J. Optim. Theory Appl. **53**(2), 237–246 (1987)

Stability Analysis of Wind Turbines Combined with Rechargeable Batteries Based on Markov Jump Linear Systems

Xiao-kun Dai[1], Yang Song[1,2(✉)], Mira Schüller[3],
and Dieter Schramm[3]

[1] School of Mechatronic Engineering and Automation,
Shanghai University, Shanghai, China
xiaokundai@126.com, y_song@shu.edu.cn
[2] Shanghai Key Laboratory of Power Station Automation Technology,
Shanghai, China
[3] Department Mechanical Engineering, University of Duisburg-Essen,
Duisburg, Germany
mira.schueller@uni-due.de,
schramm@mechatronik.uni-duisburg.de

Abstract. To maximize the output power in low wind speed and to maintain the demanded power of the turbine in high wind speed, switch control strategy is applied to wind turbines combined with rechargeable batteries. A mathematical model of a Markov jump linear system is established for such wind turbine systems. The method for determining the transition probability of the Markov chain is also presented. Then sufficient conditions for almost sure ability are proposed for this combined wind turbine system.

Keywords: Wind turbine · Battery storage system · Markov jump system · Almost-sure stability

1 Introduction

Wind energy is widely regarded as one of the key solution for sustainable energy supply. The volatility of the wind turbines output power has a significant impact on power grid. Many researchers have investigated effective control strategies to overcome this problem. The basic control object for wind turbines is to maximize wind energy conversion efficiency in low wind speed [1, 2], and to maintain the output power to a demanded value in high wind speed [3, 4]. Therefore, switch control strategy is often used in wind turbine control. By using the theory of switch system, the stability of wind turbines is discussed [5, 6].

Similarly, due to the randomness and intermittence of the wind, the output power of the wind turbine generally fluctuates significantly. Combining Battery storage systems are an effective approach to minimize the fluctuation of the out power of wind turbine system. The battery can complement the output power of the turbine when the wind

© Springer Nature Singapore Pte Ltd. 2017
K. Li et al. (Eds.): LSMS/ICSEE 2017, Part III, CCIS 763, pp. 126–136, 2017.
DOI: 10.1007/978-981-10-6364-0_13

speed is low. Meanwhile, the battery can be charged for the operation conditions high the case of wind speed. For wind power systems with combined battery storage, the investigation becomes more complicated. The hybrid system of a wind turbine combined with a rechargeable battery presents significant challenges for control.

Considering the hybrid system of combined wind turbine and rechargeable battery, many experts and scholars have put effort to create better models to describe its operation and control. For example, the switch control system was established in [5]. However, only the change of load of the combined system was discussed. Actually, the battery cannot always be in a state of charge due to capacity constraints. Moreover, the battery's SOC (State Of Charge), whose value denoting the percentage of total energy capacity of the battery in the range of [0, 1] was established by making use of Markov Chain method in [7], but only considered the battery model.

The contribution of this paper is that the model of Markov jump linear system (MJLS) is established for wind turbines connected with an energy storage system. The hybrid system is divided into three parts. When the required wind speed is not reached, the control strategy for the wind turbine is that the maximum aerodynamic efficiency is tracked and the influence of the battery is not needed to be considered. In case that the required wind speed is exceeded and the battery is not full of charge, the load change of the wind turbine has to be considered due to the influence of the battery charge power. In case of a fully charged battery, the battery has to be curtailed. In addition, the switching sequence is a Markov chain. Then, considering the model of MJLS established for wind turbines connected with an energy storage system, the almost sure stability (AS stability) can be used to illustrate the stability.

The paper is structured as follows: Considering the operating point of the wind turbine and SOC change of battery, the wind turbine connected with an energy storage system is modeled as a Markovian jump system in Sect. 2. Section 3 describes the Markov process model of the switching sequence. Then, the almost sure stability is discussed for the linear Markovian jump model of the wind turbine and an example is given in Sect. 4. Section 5 concludes this paper.

2 Wind Turbine Model

In general, the instantaneous power P_w captured by a wind turbine from wind energy can be expressed as

$$P_w = \frac{1}{2}\rho\pi R^2 v^3 C_p, \tag{1}$$

is the instantaneous power available of the wind through a turbine. In (1), ρ is the air density, R is the rotor radius, the instantaneous wind speed v is time varying. And the power coefficient C_p is a nonlinear function of the turbine's blade pitch angle β and tip-speed ratio λ, which can be expressed as [8]

$$C_p = 0.4654 \left(\frac{116}{\lambda'} - 0.4\beta - 5 \right) e^{\frac{-20.24}{\lambda'}},$$

where $\frac{1}{\lambda'} = \frac{1}{\lambda + 0.08\beta} - \frac{0.035}{\beta^3 + 1}$. Figure 1 illustrates $Cp - \lambda$ characteristic curves for different pitch angles. When the pitch angle increases, as the Fig. 1 shows, the Cp value is reduced.

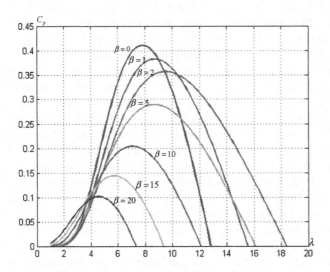

Fig. 1. $C_p - \lambda$ curves for different pitch angle

The tip-speed ratio is the ratio of the tangential speed of the blade tip to the wind speed

$$\lambda = \frac{\omega R}{v}$$

where the rotor angular velocity ω, is time varying for a variable-speed turbine.

When only the rotor speed is considered, the dynamic model of the wind turbine can be expressed as

$$\dot{\omega} = \frac{1}{J}(\tau_{aero} - \tau_{load}), \tag{2}$$

where J is the rotational inertia of the turbine, τ_{aero} is the aerodynamic torque, and τ_{load} is the torque of load. The τ_{aero} is given by

$$\tau_{aero} = \frac{1}{2}\rho\pi R^2 v^3 \frac{C_p(\lambda, \beta)}{\omega}, \tag{3}$$

According to [9], linearizing τ_{aero} applying the Taylor expansion around the operating point (ω_s, β_s, v_s) yields

$$\tau_{aero} \approx \tau_{aero_s} + \Gamma_\omega(\omega - \omega_s) + \Gamma_\beta(\beta - \beta_s) + \Gamma_v(v - v_s), \tag{4}$$

where $\Gamma_\omega = \frac{\partial \tau_{aero}}{\partial \omega}|_{\omega=\omega_s}, \Gamma_\beta = \frac{\partial \tau_{aero}}{\partial \beta}|_{\beta=\beta_s}, \text{ and } \Gamma_v = \frac{\partial \tau_{aero}}{\partial v}|_{v=v_s}.$ The generator load torque τ_{load} in (2) is $\tau_{load} = K\omega^2$, where $K = \frac{1}{2}\rho\pi R^5 \frac{C_{p_{max}}}{\lambda_*^3}$, and λ_* is the tip-speed ratio at the maximum power coefficient $C_{p_{max}}$.

The linearized dynamic model of the wind turbine is then given by

$$\dot{\omega} \approx \frac{1}{J}\left(\tau_{aero_s} + \Gamma_\omega(\omega - \omega_s) + \Gamma_\beta(\beta - \beta_s) + \Gamma_v(v - v_s) - \tau_{load_s} - 2K\omega_s(\omega - \omega_s)\right).$$

It can be considered that the air torque is equal to the load torque around the operating point (ω_s, β_s, v_s)

$$\dot{\omega} \approx \frac{1}{J}\left(\Gamma_\omega(\omega - \omega_s) + \Gamma_\beta(\beta - \beta_s) + \Gamma_v(v - v_s) - 2K\omega_s(\omega - \omega_s)\right) \tag{5}$$

Assume $0 < P_b \leq P_{b-limit}$ is the chargeable power range of the rechargeable battery. When the required wind speed is not reached, the control strategy for the wind turbine is that the maximum aerodynamic efficiency is tracked. Therefore when the wind speed is below the required value, as the Fig. 1 shows, we set to $\beta = 0$, which is the installed angle in order to reach the maximum C_p. The blade pitch control can be expressed as

$$\frac{d\beta}{dt} = K_\beta(\beta_s - \beta)$$

Then the linearized system model for the region below the required wind speed can be expressed as

$$\begin{bmatrix} \dot{\beta} \\ \dot{\omega} \end{bmatrix} = \begin{bmatrix} -K_\beta & 0 \\ \frac{\Gamma_\beta}{J} & \frac{\Gamma_\omega - 2K\omega_s}{J} \end{bmatrix} \times \begin{bmatrix} \beta - \beta_s \\ \omega - \omega_s \end{bmatrix} + \begin{bmatrix} 0 \\ \frac{\Gamma_v}{J} \end{bmatrix} (v - v_s). \tag{6}$$

When exceeds the demand power, the generator power is limited. The pitch control in Region 3 is performed using PI controller

$$\beta(t) = K_P\omega_e(t) + K_I \int_0^t \omega_e(\tau)d\tau, \tag{7}$$

where $\omega_e = \omega_s - \omega$ is the rotor speed error. Therefore

$$\dot{\beta} = -K_P\dot{\omega} + K_I(\omega_d - \omega). \tag{8}$$

When the battery is not fully charged and the output power of the turbine exceeds $P_{load} + P_{b-limit}$, the output power is limited to $P_{load} + P_{b-limit}$ and

$$\tau_{load} = \frac{P_{load} + P_{b-limit}}{\omega}.$$

Substituting this term into (2) and (8), we arrive at

$$\begin{bmatrix} \dot{\beta} \\ \dot{\omega} \end{bmatrix} = \begin{bmatrix} \frac{-K_P\Gamma_\beta}{J} & \frac{-K_P\Gamma_\omega}{J} - K_I \\ \frac{\Gamma_\beta}{J} & \frac{\Gamma_\omega + (P_{load} + P_{b-\lim it})/\omega_s^2}{J} \end{bmatrix} \times \begin{bmatrix} \beta - \beta_s \\ \omega - \omega_s \end{bmatrix} + \begin{bmatrix} \frac{-K_P\Gamma_v}{J} \\ \frac{\Gamma_v}{J} \end{bmatrix}(v - v_s). \quad (9)$$

When the battery is fully charged and the output power of the turbine exceeds P_{load}, the output power is limited to the P_{load}.

$$\begin{bmatrix} \dot{\beta} \\ \dot{\omega} \end{bmatrix} = \begin{bmatrix} \frac{-K_P\Gamma_\beta}{J} & \frac{-K_P\Gamma_\omega}{J} - K_I \\ \frac{\Gamma_\beta}{J} & \frac{\Gamma_\omega + P_{load}/\omega_s^2}{J} \end{bmatrix} \times \begin{bmatrix} \beta - \beta_s \\ \omega - \omega_s \end{bmatrix} + \begin{bmatrix} \frac{-K_P\Gamma_v}{J} \\ \frac{\Gamma_v}{J} \end{bmatrix}(v - v_s). \quad (10)$$

It can be considered $v - v_s$ is the disturbance variable. Therefore, the switch control has been obtained

$$\dot{x}(t) = \tilde{A}_{\sigma(t)}x(t) \quad (11)$$

where $x = [\beta - \beta_s, \omega - \omega_s]$, and the form process $\sigma(t)$ taking values in a finite set $S = \{1, 2, 3\}$, and

$$\tilde{A}_1 = \begin{bmatrix} -K_\beta & 0 \\ \frac{\Gamma_\beta}{J} & \frac{\Gamma_\omega + 2K\omega_s}{J} \end{bmatrix}, \tilde{A}_2 = \begin{bmatrix} \frac{-K_P\Gamma_\beta}{J} & \frac{-K_P\Gamma_\omega}{J} - K_I \\ \frac{\Gamma_\beta}{J} & \frac{\Gamma_\omega + (P_{load} + P_{b-limit})/\omega_s^2}{J} \end{bmatrix},$$

$$\tilde{A}_3 = \begin{bmatrix} \frac{-K_P\Gamma_\beta}{J} & \frac{-K_P\Gamma_\omega}{J} - K_I \\ \frac{\Gamma_\beta}{J} & \frac{\Gamma_\omega + P_{load}/\omega_s^2}{J} \end{bmatrix}$$

According to the sampling period T, the above continuous-time system can be discretized as

$$x(k+1) = A_{\sigma(k)}x(k) \quad (12)$$

And the switching sequence $\sigma(k)$ is a Markov chain taking values on the finite set $\{1, 2, 3\}$. The transition probability of the Markov chain $\sigma(k)$ is given by $p_{ij} = \Pr\{\sigma(k+1) = j | \sigma(k) = i\}$. The state space transition diagram of a wind turbine connected with a rechargeable battery using a Markov jump linear system can be described as shown in Fig. 2.

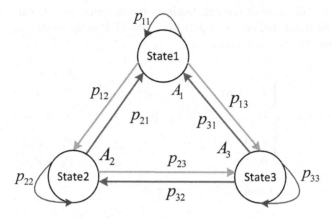

Fig. 2. The state space diagram of wind turbine connected with a battery. State1 represents that the required wind speed is not reached. State2 represents that the wind speed exceeds the required wind speed and the battery is not fully charged. State3 represents that the wind speed exceeds the required wind speed and the battery is fully charged.

3 Determination of Transition Probability

As the energy comes mainly from wind energy, the state of the wind turbine is determined by the change of the wind speed. In addition, the battery status will be also affected by wind speed. Motivated by [10, 11], the wind speed can be represented as a Markov process. Assume v_d is the wind speed when the output power of the turbine is equal to the demanded power P_d. The wind speed is divided into 2-regions between the curt-in wind speed and the cut-out wind speed in this paper. Hence, there will be 2×2 transitions between Region 2 ($v_{in} \leq V_{wind} < v_d$) and Region 3 ($v_d \leq V_{wind} < v_{out}$). The transition probabilities p'_{ij} from a state at time k to another state at time $k+1$, i.e. can be represented as [12]

$$p'_{ij} = P(r_{k+1} = j | r_k = i).$$

Accordingly, the transition probability matrix $p'_{k,k+1}$ can be obtained from the wind speed sample data:

$$p'_{k,k+1} = \begin{bmatrix} p'_{11} & p'_{12} \\ p'_{21} & p'_{22} \end{bmatrix},$$

with $p'_{ij} \geq 0$ for $i,j \in \{1,2\}$, and $\sum_{j=1}^{n} p'_{ij} = 1$.

By regarding the wind speed as a random variable, its probabilistic law at a certain location is generally represented by a Weibull distribution, which is given by

$$f(v) = \frac{k}{c} \left(\frac{v}{c}\right)^{k-1} e^{-\left(\frac{v}{c}\right)^k} \tag{13}$$

where c is the scale factor of Weibull distribution with unit of speed, and k is the shape factor of Weibull distribution. According to [6], the PDF of the output power of a wind turbine can be obtained as follows:

$$f_P(P_w) = \begin{cases} 0, & P_w < 0, \\ 1 - \exp\left(-\left(\frac{v_{in}}{c}\right)^k\right) + \exp\left(-\left(\frac{v_{out}}{c}\right)^k\right), & P_w = 0, \\ \left(\frac{k'}{c'}\right) \cdot \left(\frac{P_w + \gamma}{c'}\right)^{k'-1} \cdot \exp\left(\left(\frac{P_w + \gamma}{c'}\right)^{k'}\right), & 0 < P_w < P_d, \\ \exp\left(-\left(\frac{v_d}{c}\right)^k\right) - \exp\left(-\left(\frac{v_{out}}{c}\right)^k\right), & P_w = P_d, \\ 0, & P_w > P_d, \end{cases}$$

where c', k', and γ are given as aforementioned.

$$c' = \frac{P_d \cdot c^2}{v_d^2 - V_{in}^2}, \quad k' = \frac{k}{2}, \quad \gamma = -\frac{P_d \cdot v_{in}^2}{v_d^2 - v_{in}^2}.$$

The expression for the CDF of generator power of is given by

$$F_p(P'_w) = Pr\{P_w \le P'_w\} = \int_{\infty}^{P'_w} f_P(P_w) d(P_w) \tag{14}$$

The battery behavior is mainly characterized by the battery's SOC, whose value denoting the percentage of total energy capacity of battery is in the range of [0, 1]. As shown in [6], at any time kT, the power supply is equivalent to the output power demanded, namely,

$$SOC(k+1) = SOC(k) + \frac{T}{V_b C_b}(P_w(k) - P_o), \tag{15}$$

where V_b is the terminal voltage of battery, C_b is the energy capacity in kAh.

The SOC value of battery is divided into $[a_1, a_2]$ and $[a_2, a_3]$ between SOC_{min} and SOC_{max}, while $[a_1, a_2]$ represents the state of not fully charged and $[a_2, a_3]$ represents the state of fully charged. Given $SOC(k) = x$ and $b_i \le V_{wind}(k) < b_{i+1}$, define $f_j(x)$ as the probability of $d(SOC(k+1)) = j$. It can be obtained

$$Pr\{d(Sa_{[a,b]}(SOC(k+1) \cap V_{wind}(k+1)) = j | SOC(k) = x \cap b_i \le V_{wind}(k) < b_{i+1}\},$$
$$= Pr\{a_j \le x + \alpha(P_w(k) - P_{load}) < a_{j+1} \cap b_j \le V_{wind}(k+1) < b_{j+1} | b_i \le V_{wind}(k) < b_{i+1}\},$$
$$= f_j(x) \times Pr\{b_j \le V_{wind}(k+1) < b_{j+1} | b_i \le V_{wind}(k) < b_{i+1}\} \, i,j = 1,2,3$$

and by letting $\gamma_j = \frac{a_j - x}{\alpha} + P_{load}$, $\alpha = \frac{T}{V_b C_b}$,

$$f_j(x) = F_P(\gamma_{j+1}) - F_P(\gamma_j) j = 1, 2, \ldots, N, \tag{16}$$

Assume that $d(Sa_{[a,b]}(SOC(k) \cap V_{wind}(k))) = 1$ represents $v_{in} \leq V_{wind}(k) < v_d$, $d(Sa_{[a,b]}$ $(SOC(k) \cap V_{wind}(k))) = 2$ represents $a_1 \leq SOC(k) = x \leq a_2$ and $v_d \leq V_{wind}(k) < v_{out}$, and $d(Sa_{[a,b]}(SOC(k) \cap V_{wind}(k))) = 3$ represents $a_2 \leq SOC(k) = x \leq a_3$ and $v_d \leq V_{wind}(k) < v_{out}$. It can be obtained

$$p_{11} = p'_{11} = \Pr\{v_{in} \leq V_{wind}(k+1) < v_d | v_{in} \leq V_{wind}(k) < v_d\}.$$

$$p_{12} = \int_{a_1}^{a_3} \frac{f_1(x)}{\Pr\{v_{in} \leq V_{wind}(k) < v_d\}} dx \times \Pr\{v_d \leq V_{wind}(k+1) < v_{out} | v_{in} \leq V_{wind}(k) < v_d\}$$

$$= p'_{12} \times \int_{a_1}^{a_3} \frac{f_1(x)}{\Pr\{v_{in} \leq V_{wind}(k) < v_d\}} dx.$$

$$p_{13} = \int_{a_1}^{a_3} \frac{f_2(x)}{\Pr\{v_{in} \leq V_{wind}(k) < v_d\}} dx \times \Pr\{v_d \leq V_{wind}(k+1) < v_{out} | v_{in} \leq V_{wind}(k) < v_d\}$$

$$= p'_{12} \times \int_{a_1}^{a_3} \frac{f_2(x)}{\Pr\{v_{in} \leq V_{wind}(k) < v_d\}} dx.$$

$$p_{21} = p'_{21} = \Pr\{v_{in} \leq V_{wind}(k+1) < v_d | v_d \leq V_{wind}(k) < v_{out}\},$$

$$p_{22} = \int_{a_1}^{a_2} \frac{f_1(x)}{\Pr\{v_{in} \leq V_{wind}(k) < v_d\}} dx \times \Pr\{v_d \leq V_{wind}(k+1) < v_{out} | v_d \leq V_{wind}(k) < v_{out}\}$$

$$= p'_{22} \times \int_{a_1}^{a_2} \frac{f_1(x)}{\Pr\{v_{in} \leq V_{wind}(k) < v_d\}} dx,$$

$$p_{23} = \int_{a_1}^{a_2} \frac{f_2(x)}{\Pr\{v_{in} \leq V_{wind}(k) < v_d\}} dx \times \Pr\{v_d \leq V_{wind}(k+1) < v_{out} | v_d \leq V_{wind}(k) < v_{out}\}$$

$$= p'_{22} \times \int_{a_1}^{a_3} \frac{f_2(x)}{\Pr\{v_{in} \leq V_{wind}(k) < v_d\}} dx.$$

$$p_{31} = p'_{31} = \Pr\{v_{in} \leq V_{wind}(k+1) < v_d | v_d \leq V_{wind}(k) < v_{out}\},$$

$$p_{32} = 0.$$

$$p_{33} = p'_{33} = \Pr\{v_d \leq V_{wind}(k+1) < v_{out} | v_d \leq V_{wind}(k) < v_{out}\},$$

Therefore, the transition probability matrix $p_{k,k+1}$ is

$$p_{k,k+1} = \begin{bmatrix} p_{11} & p_{12} & p_{13} \\ p_{21} & p_{22} & p_{23} \\ p_{31} & p_{32} & p_{33} \end{bmatrix}.$$

4   Almost Sure Stability

Since the MJLS established for wind turbines connected with an energy storage system is a random system, the stability analysis is of fundamental importance. The stability of MJLS is analyzed by AS stability in this paper. Consider a discrete-time MJLS of wind turbines combined with rechargeable batteries

$$x(k+1) = A_{\sigma(k)}x(k), k \in Z, \tag{17}$$

where Z is the non-negative integer set, state $x \in R^n$, the Markov chain $\sigma(k)$ taking values on the finite set $\{1, 2, 3\}$ is irreducible and aperiodic. Therefore it is ergodic and has a unique invariant distribution $\pi = [\pi_1, \pi_2, \ldots, \pi_N]$ which can be calculated by

$$\begin{cases} \pi = \pi P \\ \sum_{j=1}^{N} \pi_j = 1 \end{cases}.$$

According to *Definition 1* [13]: If there exists $\rho > 0$ such that for any $x_0 \in R^n$ and any initial distribution F, the MJLS (17) is said to be exponentially almost surely stable,

$$\Pr\left\{ \limsup_{k \to \infty} \frac{1}{k} \ln\|x_k\| \leq -\rho \right\} = 1$$

According *Theorem 2* in [14], consider MJLS (17), if there exist a set of $Q_i > 0$ and scalars λ_i, μ_{ij}, such that (18)–(20) hold, then the system is exponentially almost surely stable

$$A_i^T Q_i A_i - \lambda_i Q_i < 0 \tag{18}$$

$$Q_j \leq \mu_{ij} Q_i \tag{19}$$

$$\sum_{\substack{i=1, j=1 \\ i \neq j}}^{N} \pi_i p_{ij} \ln \mu_{ij} + \sum_{i=1}^{N} \pi_i \ln \lambda_i < 0 \tag{20}$$

where $i, j = 1, 2, 3, i \neq j$.

For example, considering a practical system of a wind turbine connected with an energy storage system, where rated wind power is 100 kW, rotor radius $R = 9$ m, rotor inertia $J = 26000$ kg $\cdot$ m^2, $C_{pmax} = 0.4101$, battery charging–discharging power threshold $P_{b-limit} = 5$ kW, and nominal load demand $P_{b-limit} = 25$ kW. The controller parameters are designed as $K_\beta = 0.5$, $K_p = 2$, and $K_i = 1$. The discrete-time MJLS is described as

$$x(k+1) = A_{\sigma(k)}x(k)$$

where $A_1 = \begin{bmatrix} 0.7788 & 0 \\ -0.0026 & 0.2188 \end{bmatrix}$, $A_2 = \begin{bmatrix} 0.8855 & 0.4880 \\ -0.0056 & 0.9728 \end{bmatrix}$, $A_3 = \begin{bmatrix} 0.8855 & 0.4883 \\ -0.0056 & 0.9743 \end{bmatrix}$,

and the transition probability $P = \begin{bmatrix} 0.6 & 0.3 & 0.1 \\ 0.4 & 0.1 & 0.5 \\ 0.4 & 0 & 0.6 \end{bmatrix}$. The unique invariant distribution π is

$\left[\frac{1}{3}, \frac{1}{3}, \frac{1}{3}\right]$. By using Theorem 2 in [14], we choose $\lambda_1 = 0.8$, $\lambda_2 = 0.2$, $\lambda_3 = 0.3$; $\mu_{12} = 10.3$, $\mu_{13} = 5.4$, $\mu_{21} = 8.9$, $\mu_{23} = 4.7$, $\mu_{31} = 1.2$, $\mu_{32} = 4.1$ and

$$Q_1 = \begin{bmatrix} 0.4732 & -0.2716 \\ -0.2716 & 0.6453 \end{bmatrix}, Q_2 = \begin{bmatrix} 0.2809 & -0.1931 \\ -0.1931 & 0.3559 \end{bmatrix}, Q_3 = \begin{bmatrix} 0.3660 & -0.2975 \\ -0.2975 & 0.5432 \end{bmatrix}.$$

Then, all the conditions in Theorem 2 are satisfied. Therefore this MJLS of wind turbine connected with energy storage system is AS stable. Figure 3 illustrates the simulation of $x(k)$ starting from the initial state $[1.5, -0.5]^T$, which shows that the MJLS is exponentially almost surely stable.

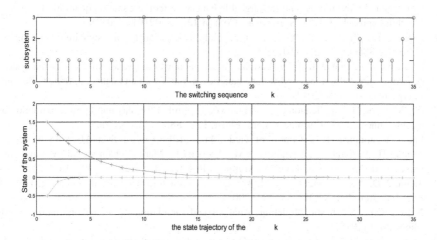

Fig. 3. The switching sequence and the state trajectory of the system

5   Conclusions

This paper proposes a Markovian jump linear system model for the wind turbine connected with storage system by using the technique of linearization. The approach for determining the transition probability of discrete-time Markov chain mode is presented. Then sufficient conditions of almost sure stability for the MJLS of wind turbine are proposed. A numerical example is given to demonstrate the effectiveness of the proposed results.

Acknowledgements. This work was supported by the National Natural Science Funds of China (61573237, 61533010) and Shanghai Natural Science Foundation (13ZR1416300).

References

1. Balas, M.J., Pao, L.Y.: Methods for increasing region 2 power capture on a variable-speed wind turbine (2004)
2. Johnson, K.E., Pao, L.Y., Balas, M.J., et al.: Control of variable-speed wind turbines: standard and adaptive techniques for maximizing energy capture. IEEE Control Syst. **26**(3), 70–81 (2006)
3. Yilmaz, A.S., Özer, Z.: Pitch angle control in wind turbines above the rated wind speed by multi-layer perceptron and radial basis function neural networks. Expert Syst. Appl. **36**(6), 9767–9775 (2009)
4. El-Tous, Y.: Pitch angle control of variable speed wind turbine. Am. J. Eng. Appl. Sci. **1**(2), 118–120 (2008)
5. Palejiya, D., Hall, J., Mecklenborg, C., et al.: Stability of wind turbine switching control in an integrated wind turbine and rechargeable battery system: a common quadratic Lyapunov function approach. J. Dyn. Syst. Meas. Control **135**(2), 021018 (2013)
6. Palejiya, D., Shaltout, M., Yan, Z., et al.: Stability of wind turbine switching control. Int. J. Control **88**(1), 193–203 (2015)
7. Li, J., Wei, W.: Probabilistic evaluation of available power of a renewable generation system consisting of wind turbines and storage batteries: a Markov chain method. J. Renew. Sustain. Energy **6**(1), 013139 (2014)
8. Ro, K., Choi, H.: Application of neural network controller for maximum power extraction of a grid-connected wind turbine system. Electr. Eng. **88**(1), 45–53 (2005)
9. Pao, L.Y., Johnson, K.E.: Control of wind turbines. IEEE Control Syst. **31**(2), 44–62 (2011)
10. Nfaoui, H., Essiarab, H., Sayigh, A.A.M.: A stochastic Markov chain model for simulating wind speed time series at Tangiers, Morocco. Renew. Energy **29**(8), 1407–1418 (2004)
11. Sahin, A.D., Sen, Z.: First-order Markov chain approach to wind speed modelling. J. Wind Eng. Ind. Aerodyn. **89**(3), 263–269 (2001)
12. Nelson, B.L.: Stochastic Modeling: Analysis and Simulation. Courier Corporation, North Chelmsford (2012)
13. Li, C., Chen, M.Z.Q., Lam, J., et al.: On exponential almost sure stability of random jump systems. IEEE Trans. Autom. Control **57**(12), 3064–3077 (2012)
14. Song, Y., Dong, H., Yang, T., et al.: Almost sure stability of discrete-time Markov jump linear systems. IET Control Theory Appl. **8**(11), 901–906 (2014)

Modeling and Simulation Study of Photovoltaic DC Arc Faults

Zhihua Li, Zhiqun Ye$^{(\boxtimes)}$, Chunhua Wu, and Wenxin Xu

Shanghai Key Laboratory of Power Automation Technology, Shanghai University,
Jing'an District, Shanghai 200072, China
zhiqunye@163.com

Abstract. The DC arc fault is a major threat to the safety of photovoltaic systems, a large amount of heat from sustained arcs leads to fire accidents. Therefore, detecting the arc faults for PV systems is receiving considerable concern. In order to develop accurate and rapid detection and location methods for arc faults, it is important to establish an arc model to characterize and predict arc characteristics and transient response. In this paper, a new DC arc model is developed from a hyperbolic approximation by observing the arc current and voltage waveforms. Based on the derived model, pink noise is superimposed to obtain better frequency domain characteristics. After comparing the simulation and the experimental results, the model is proved to be suitable for transient simulations. Furthermore, developing detection algorithm and location strategies will also be based on the DC arc model.

Keywords: Photovoltaic system · DC arc fault · Model

1 Introduction

With the increase of PV capacity and the service time of photovoltaic power generation system, the arc faults often occur in the PV system. The arc faults are discharge of electricity in a conductive ionized gas [1]. It may cause great hazards such as electrical shocks and fires and produce significant damage and loss to both PV systems and people [2]. Since the arc faults may not be sensed and stopped by conventional over current protection devices such as fuses or circuit breakers, due to their complex characteristics, detecting and locating arc faults occurring in power systems are important. Arc faults can be difficult to detect and locate due to their randomness, intermittent and chaotic nature [3]. According to the types of power distribution systems, arc faults can be classified as AC and DC. Recently, there are more studies of AC arc faults study than DC arc faults. However, the DC arc will be more sustainable and may cause greater danger than AC arc faults because there is no zero-crossings [4]. In order to prevent hazards caused by electric arc, National Electrical Code (NEC) 2011 added the requirements of DC arc-fault circuit protection in the section of 690.11, that PV system with maximum systems voltage of 80 V or above shall be protected

© Springer Nature Singapore Pte Ltd. 2017
K. Li et al. (Eds.): LSMS/ICSEE 2017, Part III, CCIS 763, pp. 137–146, 2017.
DOI: 10.1007/978-981-10-6364-0_14

by DC arc-fault circuit interrupter (AFCI) [5]. In 2011, UL (Underwriter Laboratories Inc.) promulgated new standards to evaluate and certify PV DC AFCI equipment. In UL subject 1699B, the test platform and methods for PV AFCI products are introduced in detail [6]. The establishment and the research of electric arc model contribute to a more comprehensive understanding of PV system, providing an important basis for the detection and location [7–9]. However, the parameters of these models are too difficult to get and limited by voltage and current. In addition, the characteristics of PV system are different from the other power system, the direct use of these model will lead to inaccurate simulation results. So it is very necessary to develop the arc model suitable for PV system. This paper contributes to the description of the arc model by referring to the heuristic approach. The arc behavior is strongly dependent on surrounding environmental conditions after analyzing the typical experimental voltage-current signatures of the arc faults. According to the analysis results, a DC arc model is developed to be used for the study of the detection method. The experiments reveal that the measured and simulated arc voltage and current have a good agreement.

2   Characteristic Analysis of DC Arc Fault

2.1   Arc Fault Experiment Platform

In this paper, according to the UL1699B standard, the arc fault experiment platform is built, and its configuration is shown in Fig. 1. In order to simulate the arc fault to the DC side of the photovoltaic system, experimental platform directly uses the DC output voltage of the PV arrays as power supply and the parameters of PV arrays are shown in Table 1. Data acquisition is carried out by a Tektronix oscilloscope, a voltage probe and a current probe. MATLAB is used for processing the data.

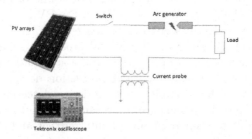

Fig. 1. Configuration of the arc fault experiment platform.

The configuration of the arc generator is shown in Fig. 2. In order to eliminate the influence of random factors on the frequency domain characteristics of the arc, the 42 step motor is used to control the velocity of the electrode.

Table 1. Parameters of the PV arrays

Name	Abbreviation	Value
Maximum power	Pm	200/W
Maximum-power-point voltage	Vm	36.27/V
Maximum-power-point current	Im	5.37/A
Open-circuit voltage	Voc	45.58/V
Short-circuit current	Isc	5.63/A

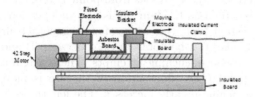

Fig. 2. Configuration of the arc generator

2.2 Arc Fault Signatures

During the ongoing investigation, the source and load networks were frequently changed to assess their effect on the arc faults. Although the test conditions are constantly changing, measurements from numerous faults show consistent voltage and current envelope behaviors. One experimental data after processing in MATLAB is shown in Fig. 3.

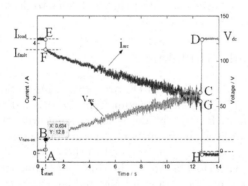

Fig. 3. Processed experiment data (open-circuit voltage: 124.8 V short circuit current: 4.24 A)

The following describes the key characteristics of DC arc faults labeled as points A through H in Fig. 3.

A: This point indicates that the arc fault generates (at time t_{start}). Before the point A, v_{arc} is equal to the electrode voltage which is about 0 V and i_{arc} is equal to the load current I_{load}. When the electrode starts to separate, the gap distance d_{gap} increases from 0 to 0^+ (i.e., an infinitesimal gap distance greater than 0);

B: The arc forms and a turn-on voltage appears across the electrodes. The voltage $v_{turn-on}$,which is independent on the gap distance d_{gap} and V_{dc} depends on the material composition of the electrodes. Referring to [10], the minimum turn-on voltage is around 12 V and $v_{turn-on}$ is 12.8 V in Fig. 3. As d_{gap} continues to increase, v_{arc} increases from point B to C. The gradient of this voltage rise is recorded as KV/mm. K is also independent of d_{gap} and V_{dc}, and is nearly constant for a wide range of experimental source voltage levels;

C: Several voltage spikes can be noted in v_{arc}, between B and C. These spikes represent unsuccessful quenching attempts. These attempts also appear in i_{arc}. Near the point C,the fluctuation of the arc is more severe and the possibility of quenching greatly increases. The length of the arc reaches the limit length $d_{\lim}$. The near-instantaneous relationship suggests that the arc impedance is resistive.

D: When d_{gap} is greater than $d_{\lim}$, the arc is extinguished. This extinction is noticed by v_{arc} rising abruptly from points C to D where v_{arc} is equal to the system voltage;

E: Before the fault, the load current is I_{load}. When a arc begins, the load current slightly reduces to I_{fault}, there are negligible changes in load current which increase the difficulty of detection;

F: It is indicated that i_{arc} decreases from I_{load} to I_{fault} is related to $v_{turn-on}$. As v_{arc} increases with a gradient of K, i_{arc} decreases with a related gradient $f(K)$ from point F to G;

G: When i_{arc} reaches G,the current of the arc begins to decrease abruptly;

H: The arc extinguishes and $i_{arc} = 0$.

3   DC Arc Fault Modeling

Section 2 described common traits of DC arc faults. This section presents an arc model according to a hyperbolic approximation of the arcs dynamic component of voltage and current. But the model is inconsistent with the real arc in the frequency domain. In fact, the frequency spectrum of the arc is similar to that of pink noise, so pink noise is superimposed to the model and the resulting arc model is consistent with the real arc in the time and frequency domain and can be used in transient simulations, and is validated later in this paper.

3.1   Function Fitting

After the current and voltage waveform of the arc in the experiment are averaged, the arc model curves shown in Fig. 4 are obtained. The current and voltage values are normalized. The abscissa ($n = d_{gap}/d_{\lim}$) is the ratio of the distance between the electrodes to the arc limit distance.

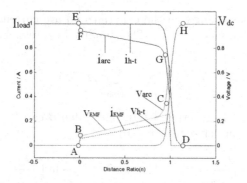

Fig. 4. The arc model curves

Referring to Fig. 4, v_{arc} can be decomposed into a nonlinear hyperbolic-tangent function voltage v_{h-t} and an electromotive force (EMF) pulse term v_{EMF}. This decomposition is given by $v_{arc} = v_{h-t} + v_{EMF}$.

According to the previous section, v_{EMF} is 0, before point A, is $v_{turn-on}$, at point A, linearly increases with the arc length between point B to C and immediately drops to 0 after point C. The curve of v_{EMF} is shown in Fig. 3. and is approximated with the hyperbolic-tangent function as

$$v_{EMF} = \frac{1}{2}\left(v_{turn-on} + K \cdot d_{gap}\right)\left(\tanh\left(\beta n\right) - \tanh\left(\beta\left(n-1\right)\right)\right). \quad (1)$$

v_{h-t} is used to describe the arc voltage between point C to D. It can be found that the non-linearity can be described by using the appropriate parameters and the hyperbolic-tangent function as

$$v_{h-t} = V_{dc}\left(0.5 + 0.5\tanh\left(\beta\left(n-1\right)\right)\right) = V_{dc}\frac{e^{2n\beta}}{e^{2\beta} + e^{2n\beta}}. \quad (2)$$

In (2), V_{dc} is the average DC voltage of the PV system without the arc and β is a variable that controls the slope of v_{h-t}.

Similar to v_{arc}, the arc current i_{arc} also has two compositions including a nonlinear hyperbolic-tangent function i_{h-t} and the systems response v_{EMF}. So the formula is $i_{arc} = i_{h-t} - i_{EMF}$.

The term i_{EMF} represents the response to v_{EMF} of the PV system and can be given by

$$i_{EMF} = \frac{V_{EMF}}{R_{arc} + R_{load}} \approx \frac{V_{dc} \cdot V_{EMF}}{I_{load}}. \quad (3)$$

In (3), R_{arc} is the arc resistance which is negligible in the region $0 < n < 1$ when compared to R_{load}.

The trace for i_{h-t} noted in Fig. 3 is approximated with the hyperbolic-tangent function

$$i_{h-t} = I_{load}\left(0.5 - 0.5\tanh\left(\beta\left(n-1\right)\right)\right) = I_{load}\left(\frac{1}{1 + e^{2\beta(n-1)}}\right). \quad (4)$$

The simultaneous changes in the experimental arc current voltage indicate that the arc impedance is resistive. The simplified expression for R_{arc} is given as follows

$$R_{arc} = \frac{v_{arc}}{i_{arc}} \approx \frac{V_{dc}}{I_{load}} e^{2\beta(n-1)}. \tag{5}$$

The value R_{closed} is generally 0.001. The value β can be calculated by (6) when the electrodes are closed (at $n = 0$)

$$R_{arc}|_{n=0} = \frac{V_{dc}}{I_{load}} e^{-2\beta} = R_{closed} \Rightarrow \beta = \frac{-1}{2} \ln \left(\frac{R_{closed} I_{load}}{V_{dc}} \right). \tag{6}$$

During the arc combustion process, the voltage and current have a strong fluctuation, especially when the arc length approaches the limit. The fluctuation can be seen as an attempt to quench the arc. In the arc model, this fluctuation is mainly reflected in the limit length of the arc $d_{\lim}$. The experimental data show that the surrounding environment and electrode materials have a great impact on $d_{\lim}$. So the limit length of the arc can be described by a random function. When the arc length is closer to the limit length of the arc, the more intense the arc fluctuates, indicating that the limit distance of the arc is a parameter that fluctuates near a fixed constant as

$$d_{\lim} = L + (random() - 0.5)^3. \tag{7}$$

The term L is the average value of the limit arc length. The time of the arc fault t_{fault} is obtained in Fig. 3. The 42 step motor rotates a circle with 1600 pulses and the moving electrode moves 8 mm. So L can be calculated by as follows

$$L = \frac{t_{fault} \cdot f_{pulse}}{1600} \times 0.8. \tag{8}$$

The term f_{pulse} is the frequency of the pulse. Random() is a function that can generates a random number between 0–1.

The arc model is developed in the PSIM simulation software. Since the arc is a nonlinear resistive load, The programmable non-linear resistor element is applied. The output characteristics of the programmable non-linear resistor are controlled by the C module and the part of the C code is as follows:

```
ran=rand()/(RANDMAX+1.0) −0.5;
  ran3=ran*ran*ran;
   if( g_nStepCount >800){
    dgap=dgap+0.005;
        dlim=5+ran3;
        n=dgap/dlim;
Vemf=(a+b*dgap+(1+8*dgap)*ran3)*(0.5 −0.5*tanh(beta*(n−1)));
        Varc=Vemf+Vdc*(0.5+0.5*tanh(beta*(n−1)));
        if(iarc >1.2)
    iarc=Iload*(0.5 −0.5*tanh(beta*(n−1)))−0.4*Vemf*Iload/Vdc;
```

```
if(iarc <1.2)
  iarc =0.01;}
else{
  Varc=1*ran ;
  iarc=Iload ;}
Rarc=Varc/iarc ;
out[0]= Rarc ;
```

The simulation result after processing in MATLAB is shown in Fig. 5 and show that a good agreement between simulated and measured current and voltage (Fig. 3) are obtained.

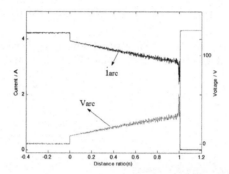

Fig. 5. Processed simulation data (Open-circuit voltage: 124.8 V Short circuit current: 4.24 A)

3.2   Pink Noise Superimposition

Figure 6(b) shows that there are a lot of difference between the simulation results and the experiment results in the frequency-domain. Johnson have obtained that the frequency of the arc is declined from 0 Hz to 100 KHz through many arc experiment in Sandia National Laboratories [13]. The current of the system without arc and with arc from our experiment platform after spectrum analysis is shown in Fig. 6(a).

The pink noises occur widely in nature and are a source of considerable interest in many fields [14]. In terms of constant bandwidth power, the pink noise is reduced by 3 dB per octave. At a sufficiently high frequency, pink noise is never dominant. White noise has equal energy at each frequency interval, so pink noise shown in Fig. 7(a) is usually generated by filtering white noise. So arc noise and pink noise have similar frequency domain characteristics, a better arc model can be developed by superimposing pink noise on the model. Figure 7(b) shows that a better agreement between the simulation results and the experiment results is obtained compared with Fig. 6(b).

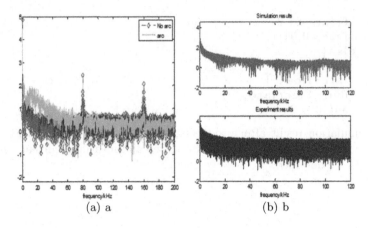

Fig. 6. (a) Comparison of the current spectrum between the normal system and system with arc; (b) Arc current spectrum of the simulation results and the experiment results

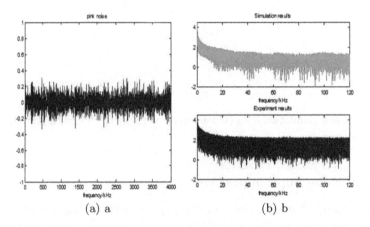

Fig. 7. (a) Spectrum of the pink noise; (b) Arc current spectrum of the simulation results superimposed the pink noise and the experiment results

4   Model Parameter Validation

The model is suitable for PV system of 124.8 V open-circuit voltage with resistive load. In this section, another 5 experiments are applied to verify the model is also suitable for other PV systems. The PV system configuration of all cases is shown in Fig. 1. The experiment conditions are shown in Table 2. The parameters of the proposed model calculated by the corresponding experimental data are also shown in Table 2. The parameters of the developed model, $v_{turn-on}$, K, and β can be considered as constant that is independent of open-circuit voltage, short-circuit current and load in PV systems.

Table 2. Experiment conditions and parameters of the model

Experiment number	Open-circuit voltage	Short-circuit current	Load	$v_{turn-on}$	K	β
1–1	124.8 V	4.24 A	Resistor	9.33	5.68	5.14
1–2	120.8 V	6.36 A	Inverter	9.6	6.11	5.23
2–1	113.2 V	2.32 A	Resistor	9.4	6.49	5.4
2–2	120.8 V	3.38 A	Inverter	9.6	6.37	5.38
3–1	161.6 V	3.04 A	Resistor	10.4	6.58	5.66
3–2	160.8 V	3.14 A	Inverter	9.6	6.33	5.57

5   Conclusion

An arc model for the arc faults study in the PV system is proposed. The model does not require physics (difficult to obtain) the time constant knowledge that can be implemented using the programmable non-linear resistor in PSIM. Since the instantaneous current and voltage characteristics of the arc model are consistent with the experimental results, the model parameters are proved to be independent of the source and load. The proposed model can be used in the transient simulation to study arc faults. But many improvements can be made. In this model, the environment and the electrode material are not taken into account. Studying the relationship between the environment and the electrode material can obtain that the model is more consistent with what occurs in practice.

References

1. Li, J., Thomas, D.W.P., Sumner, M., et al.: DC series arc generation, characteristic and modelling with arc demonstrator and shaking table. In: IET International Conference on Developments in Power System Protection, pp. 1–5 (2014)
2. Terzija, V.V., Ciric, R., Nouri, H.: Improved fault analysis method based on a new arc resistance formula. IEEE Trans. Power Deliv. **26**, 120–126 (2011)
3. Strobl, C., Meckler, P.: Arc faults in photovoltaic systems. In: Proceedings of the IEEE Holm Conference on Electrical Contacts, pp. 1–7. IEEE (2010)
4. Liu, Y., Ji, S., Wang, J., et al.: Study on characteristics and detection of DC arc fault in power electronics system. In: International Conference on Condition Monitoring and Diagnosis, pp. 1043–1046. IEEE (2012)
5. National Electrical Code 2014 Handbook. 13th edn. National Fire Protection Associations, Quincy (2013)
6. Underwriters Laboratories, Photovoltaic (PV) DC Arc-fault Circuit Protection, UL 1699B (2011)
7. Ziani, A., Moulai, H.: Extinction properties of electric arcs in high voltage circuit breakers. J. Phys. D Appl. Phys. **42**, 105205 (2009)
8. Ammerman, R.F., Gammon, T., Sen, P.K., et al.: DC-arc models and incident-energy calculations. IEEE Trans. Ind. Appl. 1810–1819 (2010)

9. Andrea, J., Schweitzer, P., Tisserand, E.: A new DC and AC arc fault electrical model. In: Electrical Contacts, pp. 1–6. IEEE (2010)
10. Mu, L., Wang, Y., Jiang, W., Zhang, F.: Study on characteristics and detection method of DC arc fault for photovoltaic system. In: Proceedings of the CSEE, pp. 5236–5244 (2016)
11. Gao, Y., Zhang, J., Lin, Y., et al.: An innovative photovoltaic DC arc fault detection method through multiple criteria algorithm based on a new arc initiation method. In: Photovoltaic Specialist Conference, pp. 3188–3192. IEEE (2014)
12. Chen, W., Shifeng, O.U., Wang, L.: Simulation and experimental study of arc characteristics of dc contactor. Low Volt. Appar. (2013)
13. Johnson, J., Pahl, B., Luebke, C., et al.: Photovoltaic DC arc fault detector testing at Sandia National Laboratories. In: Photovoltaic Specialists Conference, pp. 3614–3619. IEEE (2011)
14. Milonni, P.W.: Electronic Noise and Fluctuations in Solids, by Kogan, S. Contemporary Physics, p. 555 (2010)

Data Management of Water Flow Standard Device Based on LabVIEW

Shaoshao Qin, Bin Li[✉], and Chao Cheng

School of Mechatronic Engineering and Automation,
Shanghai University, Shanghai 200072, China
sulibin@shu.edu.cn

Abstract. It is very important to strengthen the research on water flow standard device. The software LabVIEW is used as the development platform of the software control system for the flow standard device. A large amount of data is involved in the calibration process. A method of combination of the database system and file system is adopted to manage all data involved in the device on the basis of the actual needs. It can simplify data operations, process the calibration data automatically. In this way, a high accuracy for the calibration will be ensured and the automation level of the device will be improved.

Keywords: Flow standard device · LabVIEW · Database · Data management

1 Introduction

With the rapid development of modern industrial production, flow detection is widely used. At the same time, the requirements for flow measurement are getting higher and higher. As the standard of the unity and transmission of the flow unit value, the flow standard device can ensure the unity value of all regions and departments of our country on a unified standard and provide the basis for economic accounting and the arbitration work [1, 2]. It is of great significance to strengthen the research on various flow devices and establish a thorough flow standard measurement system for the development of flow measurement technology and even the national economy [3].

Manual or semi-automatic operation is adopted in traditional verification systems which complicates the operation process. Verification data has to be recorded and calculated manually, which results in slow speed, low efficiency and high error rate and hinders improvement of labor productivity [4]. Although in recent years control systems of flow standard device based on different development platform emerge in an endless stream, there are some shortcomings such as insufficient accuracy, high cost, poor versatility and low use efficiency. With powerful analysis and data processing capabilities of modern computer, the virtual instrument technology can complete a variety of test and analysis functions by the software with great flexibility. It can improve the intelligent level of instruments, simplify the connection and debug and greatly reduce costs with high versatility [5]. Therefore the virtual instrument software LabVIEW is used as the platform in this control system.

© Springer Nature Singapore Pte Ltd. 2017
K. Li et al. (Eds.): LSMS/ICSEE 2017, Part III, CCIS 763, pp. 147–156, 2017.
DOI: 10.1007/978-981-10-6364-0_15

In the calibration process, a large amount of data will be collected and some information needs to be entered in the flow standard device. How to manage data effectively is one of the most basic and important issues [6]. Methods of data management mainly include manual management, file system management and database system management. However, the workload for manual data management is too large to suit for large quantities of data processing and convenient and flexible management can't be achieved when the data saved in the form of file [7, 8]. To solve the problem, a method of combination of the database system and file system is adopted to manage the data on the basis of functional requirements.

2 Field-Service Flow Standard Device Based on Master Meter Method

2.1 Design of Hardware Structure

The flow standard device based on master meter method has been paid more and more attention in the field of flow measurement because of its short construction period, low cost, high verification efficiency, wide flow range and high accuracy [9–11]. Therefore, the master meter method is used to study the field-service flow standard device in this paper.

The basic principle of flow standard device based on master meter method is the continuity equation of fluid mechanics [12]. From the equation, it can be seen that the cumulative volume of the fluid passing through the flowmeters in series on the same pipe is equal for a certain period of time. When the flow gets stable, the output flow values of the tested flowmeter and the master meter are respectively measured and then compared to determine the measurement performance of the tested flowmeter (Fig. 1).

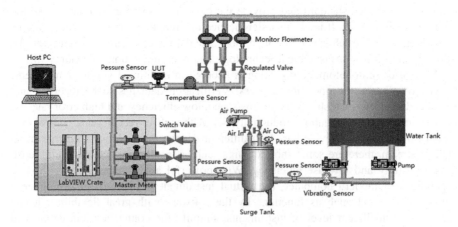

Fig. 1. The hardware structure design.

The hardware structure design of the whole set of equipment is shown in the figure including a water tank, variable frequency water pumps, a surge tank, an air pump, switch valves, a group of master meters, a tested flowmeter, regulated valves, monitor flowmeters, control cabinets, a PXI Chassis, a PC and some sensors [13].

The hardware part mainly executes the operation command from the software system. It adjusts the flow rate with regulating the flow to keep stable. Calibration data collected and the monitor equipment status by the hardware part are sent to the software system [14]. In the whole device, the surge tank and variable frequency pumps work to realize the stability control in the process of flow regulation. One master meter is selected by opening the corresponding switch valve while fine adjustment of the flow is realized by the regulated valves. The monitor flowmeters measure the flow to monitor the flow regulation. The signals of temperature sensors and pressure sensors are used to compensate for the calibration results. The vibration sensor and level gauges monitor the operating environment. Obtained by the data acquisition module in the hardware, the flow values of the standard flowmeter and the tested flowmeter and the signal of each sensor are sent to the software system for processing through communication.

2.2   Introduction to the Software Development Platform

Virtual instrument technology includes efficient software, modular I/O hardware and the hardware and software platform for integration. The flexible and efficient software can create a fully customized user interface which is the most important part of virtual instrument technology [15]. In many development platform softwares for instrument virtual, LabVIEW developed by NI is the oldest and most influential. LabVIEW is a graphical programming language and development environment. It is easy to learn and use, which can greatly improve the programming efficiency and interaction experience of users for the software system. Now it has become the most widely used, the fastest growing and most powerful and popular virtual instrument development platform.

Applications developed in LabVIEW are called VIs (virtual instruments) whose expanded-name is VI by default. These VIs are similar to the program modules in the conventional language, through which the software system of the flow standard device is built. All VIs include three parts: front panel, block diagram, and icon and connector panel [16]. The front panel is a graphical user interface that is equivalent to the standard instrument panel with interactive input and output controls. The block diagram is graphical source code for functional components inside the standard instrument box. The icon is used to identify sub VI called in the master VI and the connector is equivalent to the graphical subroutine parameter.

The data connectivity kit of LabVIEW is used to access databases and has been included in the LabVIEW Pro. A series of advanced functional modules that encapsulate most of the database operations and some advanced database access functions is integrated in the kit [17]. Users can complete the operations of query, inserting, update and deleting for database records without learning SQL syntax. And the toolkit has a high degree of portability and supports for all ODBC-compatible database drivers, which simplifies the connection to databases.

3 Data Management Method

3.1 Database Structure Design

The database used in this system is MySQL, which is a small open source relational database management system (DBMS), and it has become the most popular open source database in the world because of small size, high speed and low cost [18].

It is necessary to avoid space waste and ensure a high query efficiency when designing the database structure. At this time the three normal forms can provide the necessary guidance. The first normal form (1NF) is the basic requirement for the relational schema. Each column of the data table is an indivisible basic data item, and there are no more values in the same column. The second normal form (2NF) is established on the basis of 1NF which requires that each row in the data table must be uniquely distinguished. And 2NF should be met before satisfying the third normal form (3NF). The third normal form (3NF) requires that in a data table there can't be non-primary keyword information that has been included in other tables.

Table 1. Database structure design.

Tables	Components of the table
user_information	user, password, authority, last login time
entrust_info	entrusting party, sample number, entrusting party address, manufacturer, verification officer, contacts, telephone, calibration date
standard_info	model, number, measuring range, accuracy, certificate no., calibration medium, medium temperature
testedmeter_info	sample no., sample name, model/specification, measuring range, caliber, accuracy grade, full scale value, pulse equivalent, output type

According to the structure and function requirements of the system [19], the data that needs to be saved is stored in a database and the data tables are designed separately as shown in the following table (Table 1).

All data that needs to be stored in the database is divided into four parts: user information, entrusting order information, the information of master meters and information of tested flowmeters. The user information is stored in the user_information table, which is used for the user login and user management in the control system. The business information of the entrusting order is stored in the entrust_info table and the standard_info table is used to store details of the master flowmeters in the device, and the testedmeter_info table is used to store detailed parameters for the table to be checked. The last three tables are mainly used to implement the management of late information. Database structure is designed as above, which can not only avoid repeated storage of data in the system but also improve the efficiency of data query and is convenient for late business information query.

3.2   File System Design

In the flow standard device only a type of file needs to be used which is the calibration original record file of the flowmeter to be calibrated (Fig. 2).

(a)   **Original Record of Flowmeters**

No. : J16434100

Sample name			entrusting party			
Model / Specification			Address of entrusting party			/
Sample No.			Date			
Accuracy grade		Caliber DN	Temperature	℃	Relative humidity	%
Flow range			Manufacturer			
Instrument status	□normal □abnormal	location	Site production workshop of the entrusting party			
Basis of Calibration /verification						
Standard device and matching equipment						
Name / Model		No.	Measuring range / Accuracy		Certificate No. / Period of validity	
Signal outputting method: /		Calibration medium:	Medium temperature:		Medium pressure: /	

(b)

No.	Calibration point (m³/h)	Indication of standard meter (L)	Indication of tested meter (L)	Single error E_{ij} (%)	Error E_i (%)	Repeatability $(E_r)_i$ (%)
1						
2						
3						

(c)

Expanded uncertainty : U_{rel} =	(k=2)	
Remarks :		
Conclusion :	Verification officer :	Check personnel :

Fig. 2. (a) The upper part of original record template, (b) The middle part of original record template, (c) The lower part of original record template.

According to the requirements of the national metrological verification regulation [20], it's necessary to generate the calibration original record file with a given template. At last a corresponding verification certificate or verification results notice will be provided based on the record results.

As shown in the figure, the original record includes the examiner, some order information, some information of tested flowmeters, master meter information selected and all calibration data. After obtained with the data acquisition module, calibration data is directly filled in the original record file through the software system. The rest data is filled in the file during running of software system in turn.

The report generation tool in LabVIEW is used to generate an original record document with the above original record as a template. A word document is created with the "new report", and the parameters and the corresponding text information that need to be saved are provided for the document through the "add report text" respectively. After the calibration work is completed, the original log file named with the sample number is saved in the given path with "Save Report to File". After the calibration work is completed, the original record file named with the sample number is saved in the given path with the "Save Report to File".

4 Implementation of Data Management in the Flow Standard Device

4.1 User Login Management

Before starting to calibrate with the flow standard device, the user need to log in to his own account in the software system at first. The user login authentication can ensure the safety of calibration work, automatically identify the examiner and judge user permissions to achieve different operations management according to different permissions (Fig. 3).

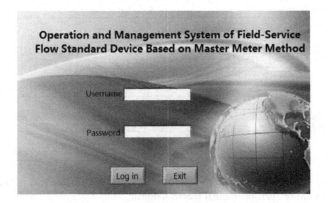

Fig. 3. User login

When a user logs in, the user name and password entered in the front panel are retrieved in the user_information table. If the query is successful, the user can successfully login to the account and the user permission is judged. The user permission is divided into two levels including administrator and ordinary staff. The "user management" button is only visible to administrator. Users with different permissions can modify their password while the user with administrator authority enjoy a higher right to operate, such as modifying information of users, adding new users and deleting users. The user information table in the MySQL is managed with these operations.

In LabVIEW, some events are created for different operations with a while loop and an event structure, and then these operations are completed by binding the operation controls to events. The shortcut menu in the event structure is used to define the right click functions for the contents of the table control. Sub VI programs are established respectively for the operations of query, inserting, update, deleting, which can be called in corresponding places (Figs. 4 and 5).

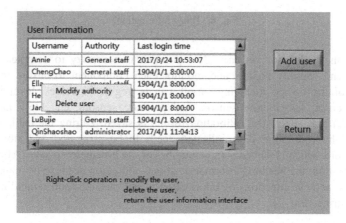

Fig. 4. User management

Fig. 5. The management interface of administrator

4.2 Business Management

After the user logins successfully, the customer information in the entrusting order and the details of the tested flowmeter need to be entered into the system and respectively stored in the entrust_info table and the testedmeter_info table in the database in order to facilitate other modules to call and late management.

When clicking the "Save" button, the "Save" event will be triggered and two sub VI programs that respectively inserting the customer information into the entrust_info table and inserting the tested meter information into the testmeter_info table are called to realize the function. When clicking the "Load History" button, the "Load History" event is triggered. Then the corresponding sub VI is called to carry out connection condition query of the entrust_info table and testedmeter_info table and then the query results are assigned to text frames of the front panel in the form of property control. The "set" and "save" functions are both implemented by binding the function buttons to events in the event structure (Fig. 6).

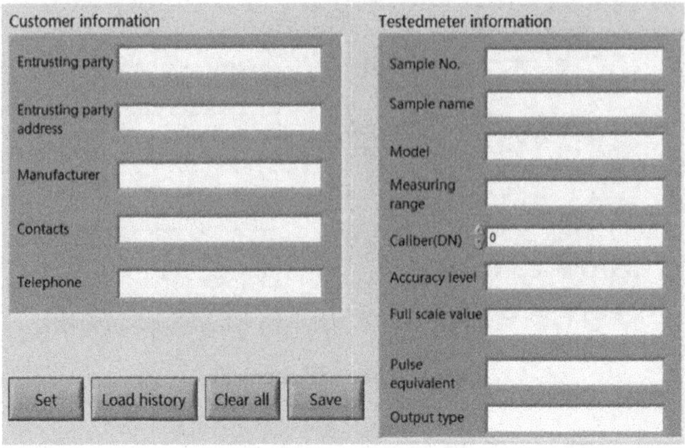

Fig. 6. Entry information.

Before the calibration on each flow point, it is necessary to complete the flow regulation and stability control and monitor each component to ensure the normal operation of the device including on-off valves, regulated valves, level gauges, a vibration sensor, temperature sensors and pressure sensors, etc. Only part of the monitor data that affects the calibration results is saved. During the calibration of a flowmeter, the calibration data collected needs to be processed and stored in the original record which is named sample number of the flowmeter to save. After the end of the calibration, the business information about calibration needs to be managed. It is necessary to achieve functions of information query, statistics and reminding of calibration expiring (Fig. 7).

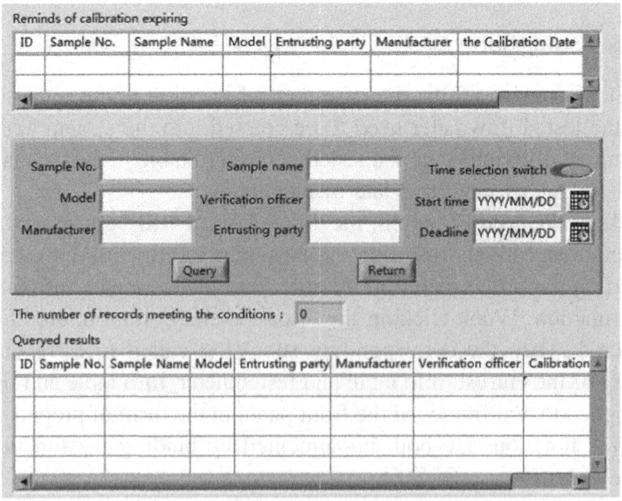

Fig. 7. Calibration information query.

In LabVIEW, the tab control is used to switch the operation interface and the three functions mentioned above are achieved by using the event structure to call the sub VI with the corresponding function. According to the demands, the sub VI for business information query is established which completes single query or multiple fuzzy query of some conditions including sample number, mode/specification, the time of verification, calibrater, entrusting party and manufacturer with connection condition query of the entrust_info table and testedmeter_info table. The results are shown in the table below. When clicking at a grid in the column of sample number, the original record named the sample number would be opened to query the detailed calibration information. The business statistics can be achieved by single or multiple connection query of some conditions including sample name, entrusting party, verification officer, and the time of verification with the two tables. The flowmeter needs to be calibrated regularly according to the verification regulations. Based on the specified period of calibration, the information of instruments that is about to expire in a month is automatically retrieved and displayed in a table in order of time so that users can know these information of instruments in time.

5   Conclusions

5.1   A Subsection Sample

On the basis of the functional requirements a method of combination of the database system and file system is applied to data management with powerful function of LabVIEW and connection with MySQL in the paper. All data involved in the flow standard device before calibration, during calibration and after calibration is stored in the database and files in the appropriate form. Customized data management interface and functional design are completed in the way of graphical programming which can simplify data operations and maintain consistency and integrity of the data throughout the calibration process of the device so that processing the data automatically is realized. In a word, the effective management of these data information can not only make the flow standard device work more smoothly to complete the calibration also ensure the accuracy and reliability of calibration results and improve the automation degree of the device.

References

1. Su, Y., Liang, G., Sheng, J.: Flow Measurement and Testing, 2nd edn. China Metrology Publishing House, Beijing (2007)
2. Guo, L.: Research on the Key Questions of Influence on the Performance of a Water Flowrate Calibration Facility. Shandong University (2014)
3. Ma, K.: Research and Design on Efficient Combination Type Water Flow Standard Facility. Tianjin University. Master Degree Thesis (2004)
4. Ji, H.: A Study on the Computer Control System for Water Flow Calibration Facilities Based on MCGS. Tianjin University. Master Degree Thesis (2005)

5. Wu, G., Wang, G., Guo, Y.: Modern Monitoring and Control Technology and Application. Electronic Industry Press, Beijing (2007)
6. Li, J., Wang, R., Wang, R., Chen, W.: The labview based portable data management for fault diagnosis system of large rotating machines. Process. Autom. Instrum. **22**(8), 12–14 (2001)
7. Yang, H.: Research on Calibration of Flowmeter System with LabVIEW. Northwest A&F University, Master Degree Thesis (2009)
8. Huang, J.: The Design of High Speed Data Acquisition and Management System Based on LabVIEW. Beijing Institute of Technology (2016)
9. Liu, D.: Research on Gas Flow Standard Facility by Parallel Master Meter Method. China Jiliang University (2014)
10. Li, J., Zhang, C., Wang, Y.: Design and implement of flow calibration facilities based on a master meter. China Instrum. **17**(12), 78–81 (2006)
11. Liu, Y., Sun, L., Qi, L., Li, S., Wei, Y.: Development of a gas flow and velocity calibration facility. In: International Conference on Consumer Electronics, Communications and Networks (pp. 276–279). IEEE (2012)
12. Tan, J.: The Development of the Gas Flow Standard Facility by Master Meter. Southwest Jiaotong University, Master Degree Thesis (2015)
13. Duan, H.: Fluid Flow Calibration Facilities and Flow Calibration Facilities with Master Meter Method. China Metrology Publishing House, Beijing (2004)
14. Baker, R.C., Gautrey, D.P., Mahadeva, D.V., Sennitt, S.D., Thorne, A.J.: Case study of the electrical hardware and software for a flowmeter; calibration facility. Flow Meas. Instrum. **29** (1), 9–18 (2013)
15. Xu, G.: The Application of the Technology of Virtual Instrument in the Flowmeter Test. Xinjiang University. Master Degree Thesis (2009)
16. Huang, S., Wu, J.: Basic Tutorial of Virtual Instrument Design. Tsinghua University Press, Beijing (2008)
17. Li, W., Cao, Y., Bu, X.: Implementation of database accessing technique of LabVIEW and its application. J. Ind. Mine Autom. **38**(3), 69–72 (2012)
18. Liu, Z., Li, K.: MySQL 5.6 Learn from Scratch. Tsinghua University Press, Beijing (2013)
19. Baker, R.C.: Flow Measurement Handbook. Cambridge University Press, Cambridge (2016)
20. JJG643-2003: Verification Regulation of Flow Standard Facilities by Master Meter Method. China Metrology Publishing House, Beijing (2003)

Design and Research of Water Flow Standard Facilities Based on Field Service

Chao Cheng, Bin Li[✉], and Shaoshao Qin

School of Mechatronical Engineering and Automation,
Shanghai University, Shanghai 200072, China
sulibin@shu.edu.cn

Abstract. A flow standard facility based on field service is designed to solve the problem of on-site high-precision parameter metering for water flow standard facilities, in which the water is take as the medium, the mass flowmeters are used as the transfer standard of liquid flow, the variable frequency pump and the surge tank are acted as the secondary regulation system, and the hardware and software platform of LabVIEW are conducted as the development system. The entire facility is small and light weight so that it can be carried to the scene for rapid calibration. Because it changes from sent by customers to sending calibration to customers and saves the standby time caused by disassembling and sending flowmeters, it improves the production efficiency of enterprises greatly. Ultimately, the facility can be produced and it will fill the domestic technical gap of water flow calibration based on field service.

Keywords: Flow standard facility · Field service · Mass flowmeter · LabVIEW

1 Introduction

Metering is the eye of industrial production, and flow metering is one of the components of the metering science and technology [1]. It has a close relationship with the national economy, the national defense construction and the scientific research. Doing flow metering well has an important role for guaranteeing the product quality, improving the production efficiency and promoting the development of science and technology [2]. In the guidance of current social policy named "energy saving and environmental protection", flow standard facilities as the basis of the flow value transfer get more research and development [3].

With the adjustment of national policy, the competition of flow calibration market has been intensified and the customer-oriented concept has been rooted deeply [4]. The traditional flow standard facilities are fixed and very large, the calibration of flowmeters needs to remove all the flowmeters to the metering institution, after the calibration is finished and then the workers need take them back. Sending back and forth usually takes one or two days. Count the calibration time on, a week is spent to reinstall them on the production line. Thus traditional flow standard facilities reduce the production efficiency greatly [5]. In addition, due to the characteristic of low efficiency, long measurement time, maintenance difficulties, pipeline size and high cost, the traditional calibration methods such as volumetric method and weighing method, are difficult to

© Springer Nature Singapore Pte Ltd. 2017
K. Li et al. (Eds.): LSMS/ICSEE 2017, Part III, CCIS 763, pp. 157–166, 2017.
DOI: 10.1007/978-981-10-6364-0_16

meet the need of on-site calibration and can only be widely used in enterprises and laboratories [6]. Therefore, the production of a set of water flow standard facility by master meter method based on field service is necessary for the grass-roots service [7]. The set of facility is of great importance to promote the development of water flow metering, improve the production efficiency and promote the sustainable development of water resources.

2 System Structure

According to the verification regulation of flow standard facility by master meter method, a set of flow standard facility consists of five parts: fluid source, test piping, master meters, a timer and a control system [8]. Base on national standards, the design of this facility makes some improved innovations for adapting the need of on-site calibration. The system structure is shown in Fig. 1. Three parallel high-precision mass flowmeters are acted as master meter in this facility. Before starting calibration, it is necessary to adjust and stabilize the flow point through the variable frequency pump and the surge tank. When the flow point is stable, the flow signal of master meters and UUT (Unit Under Test) is collected, then the calibration results are calculated by comparative method. Meanwhile, the paper collects the signal of some sensors for the purpose of compensating the calibration results. And the results display on HMI (Human Machine Interface). Finally, according to the needs of customers, the corresponding original records and certificates can be printed by the printer.

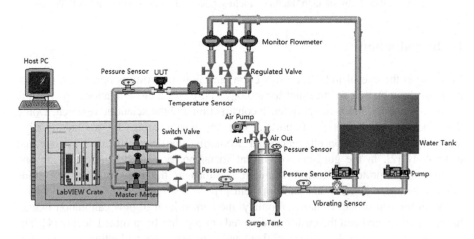

Fig. 1. System components

3 Operating Principle

The mass flow through master meters and UUT is equal at the same time interval, therefore, the basic (or indication) error value of the facility is determined by a comparative method [9].

According to the principle of liquid incompressibility, the mass flow through different sections in the same pipe is equal at any time.

$$q_m = q_{v_s} \cdot \rho_s = q_{v_t} \cdot \rho_t \tag{1}$$

q_{v_s} is the volume flow of master meter and q_{v_t} is the volume flow of UUT, m^3/h.
ρ_s is the medium density of master meter and ρ_t is the medium density of UUT, kg/m^3.

At the beginning of the calibration, the flow is adjusted to the specified value, after the flow rate is stable, reading the count of master meter and UUT. After a time, stop counting of the two flowmeters and record data. For bus meters, analog meters and pulse meters, data conversion process will be different. Bus meters (HART, PROFIBUS, ModBus) can be read the cumulative flow and instantaneous flow directly, which can be transfer to the host computer through the serial port or Ethernet. And the value can be compared with the value of the master meters directly, then get the accuracy of the UUT; Pulse meters can be collect the number of pulses during the calibration period and convert to cumulative flow by multiplying pulse equivalent, then the value can be compared with the value of the master meters directly; Analog meters (4–20 mA) need convert the instantaneous flow to the accumulated flow according to the integral principle and calculate the calibration results [10].

Taking pulse meters as an example:

$$q_{v_s} = \frac{N_s}{K_s.t}; \quad q_{v_s} = \frac{N_t}{K_t.t} \tag{2}$$

N_s is the pulses of master meter during calibration time, N_t is the pulses of UUT during calibration time;

K_s is the meter factor of master meter, K_t is the meter factor of UUT;

t is the calibration time.

Bring Formula (2) into Formula (1):

$$\frac{N_t}{K_t \cdot t} \cdot \rho_t = \frac{N_s}{K_s \cdot t} \cdot \rho_s \tag{3}$$

$$K_t = \frac{N_t}{N_s} \cdot \frac{\rho_t}{\rho_s} K_s \tag{4}$$

If the density of the media passing through the master meter and the UUT is close or equal, then:

$$K_t = K_s \frac{N_t}{N_s} \tag{5}$$

4 Hardware Design

4.1 Mechanical Structure of the Facility

As is shown in Fig. 2, the facility consists of detection unit A, the regulator unit B and the water supply unit C. Three parts are equipped with locking sub, which can be transported to the scene by vehicle.

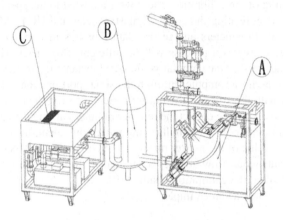

Fig. 2. Mechanical structure of the facility

In the water supply unit, two centrifugal pumps are installed below the water tank, and the flow range of the facility can be covered through regulating the frequency of transducer; The unit body is equipped with power supply cabinet, which can supply power for electrical equipments of the entire facility; The environment temperature and humidity sensor is placed in its outlet pipe to monitor whether the calibration environment is in the limits of the national verification procedure.

The regulator unit is connected between the detection unit and the water supply unit via hoses; The inlet and outlet pipes of the surge tank are equipped with pressure sensors to detect the water pressure; There is a drain valve at the bottom of the surge tank; And the top is equipped with a intake valve, a outlet valve, a water intake and a safety valve; At the same time, there is a radar level gauge to monitor the surge tank level, according to the change of the level, the outlet valve or the intake valve is adjusted to ensure the gas pressure inside the tank constant.

In the detection unit, the master meter group (three mass flowmeters) is installed in the unit body by the tilt of the column, each master meter line is equipped with a three-valve valve to strobe which one is used; There is a control cabinet in the unit body to complete the data acquisition and control tasks; In the water inlet of the master meter group, a vibration sensor is installed to ensure the vibration demand of the mass flowmeters [11]. The water outlet of the master meter group is connected to UUT by the upper connection pipe; There are three control valves in the water outlet of UUT, each control valve is equipped with a monitoring flowmeter to observe the change of flow when we are adjusting a flow point; A temperature sensor is installed in the front of UUT and a pressure sensor is installed behind UUT, which can make a compensate for the calibration results.

4.2 Selection of Electrical Equipments

The technical data of the facility is shown in Table 1, all the selection of electrical equipments must be corresponding to those parameters.

Table 1. The technical data of the facility

Name	Data
Dimension	2.2 m × 1.5 m × 1.8 m
Weight	500 kg
Environment temperature and humidity	5–45 °C, 35%–95%
Flow range	0.5–38 m^3/h
Apply caliber	DN8–80
Uncertainty	0.16%

Firstly, in order to solve the problem that calibration accuracy is not high when using master meter method, the facility is equipped with three mass flowmeters made by Endress+Hauser for high-precision. The series of mass flowmeters requires small installation space and has no requirements for straight pipe, which are good for the need of field service. Secondly, in order to reduce the space of water tank and transport conveniently, the paper select two horizontal centrifugal pumps that can be installed under the tank. And the two pumps can cover the flow range of facility. Thirdly, in order to compensate or reduce the calibration error, we install a environment temperature and humidity sensor, five pressure sensor, a temperature sensor, a radar level gauge, a differential pressure level gauge and a vibration sensor. The selection of electrical equipments is shown in Table 2.

Table 2. The selection of electrical equipments

Name	Version
Environment sensor	Honeywell SCTHWA43SNS
Temperature sensor	EMERMON Model 248/0086
Pressure sensor	EMERMON Model 2088
Radar level gauge	Endress+Hauser FMP50
Differential pressure level gauge	Endress+Hauser FUN2051
Vibration sensor	BENTLY 177230-001-01
Mass flowmeter	Endress+Hauser Promass 83F
Pump	WILO MVI 3202, MVI 1602/6
Transducer	ABB ACS510-01-09A4-4

4.3 Regulator System

The verification regulation of flow standard facility by master meter method regulates that the test fluid should be stable. Only when the fluid is stable, can we use the average flow value to approximate its instantaneous flow value [12]. Therefore, the water flow standard facility must ensure that the flow can maintain a high degree of stability during the calibration. But the stability is difficult to control, it is related to many factors, such as pumps, pipes, valves, automatic control system and so on. After analyzing the commonly pressure regulation method (water tower regulator, container regulator, variable frequency regulator, variable frequency plus container regulator) of flow standard facilities [13], the paper select the method of variable frequency plus container regulator. We make full use of its advantages (low construction costs, small footprint, big pressure range and wide flow range) to complete the water source stabilization system of the facility. The process diagram is shown in Fig. 3.

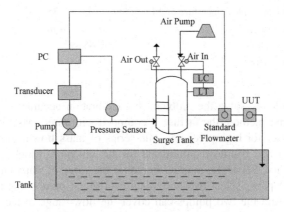

Fig. 3. Variable frequency plus container regulator

(1) First Regulator: Adjusting the frequency of transducer realizes first regulator, the control process is shown in Fig. 4. Cascade PID control technology is used in this paper. The flow before master meter is used as the master controlled variable to achieve the final stability of export flow. The pressure of pump outlet is used as the secondary controlled variable to eliminate the pressure changes of pump outlet caused by the instability of power voltage and frequency [14].

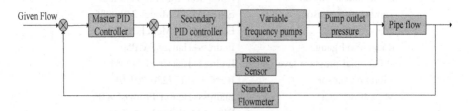

Fig. 4. Cascade PID control

(2) Second Regulator: In order to eliminate the high frequency pulsation of the pump outlet, the surge tank is used as secondary regulator. The radar level gauge measures the height of the tank, once the height is change, we use controller to control the solenoid valve for air in or air out, then the air pressure of surge tank maintains stable. It makes a buffer for fluid fluctuations and precipitates the air bubbles in the fluid. As is shown in Fig. 5, the tank volume is designed to meet the maximum flow rate of 45 m³/h, the maximum pressure of 3×10^5 Pa and the pressure fluctuation value of outlet flow within 0.1% [15].

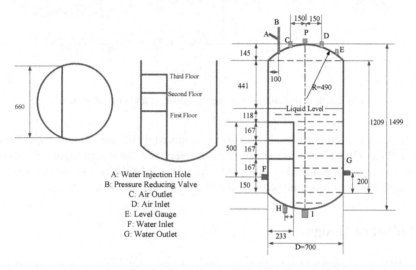

Fig. 5. The structure of surge tank

4.4   Data Acquisition and Control System

As is shown in Fig. 6, the LabVIEW platform is used as the software development system of the facility. The paper chooses PXI chassis equipped with different board to complete the data acquisition and control work. At the same time, in order to collect the real-time data properly and react the current state truly, the paper introduces LabVIEW RT (Real Time) technology. Real-time data acquisition tasks are segmented from the numerous man-machine interface functions, which run on RT processor independently. So even if the host program collapses, LabVIEW RT program still will continue to run, greatly enhance the reliability of the flow calibration system [16].

The embedded processor PXIe-8840 acts as the RT Target, which is running the LabVIEW RT operating system and the real-time program. The laptop acts as the Host PC, which is running the windows operating system and the monitor program. The RT Target transmits the real-time data to the HOST PC via Ethernet. The HOST PC receives and displays the parameters, and transfers the control command parameters that need to be changed to the RT target.

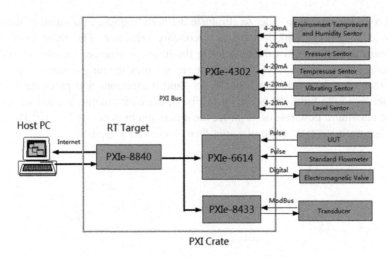

Fig. 6. Structure diagram of data acquisition and control system

The 4–20 mA current signal of sensors is acquired by the current acquisition board named PXIe-4302. The pause signal of flowmeters and the digital signal of electromagnetic valves are acquired by the counter board named PXIe-6614. The transducer is controlled by serial board named PXI-843 based on the ModBus protocol.

5   Software Design

LabVIEW is a graphical development environment, providing a large number of functions and controls. It does not require us to have more software programming foundation, but you can develop data acquisition program and control program conveniently [17]. Meanwhile, we can design a professional and beautiful HMI (Human Machine Interface). According to the calibration process shown in Fig. 7, the software development system of the facility mainly completes the functions of user login, information entry, steady regulation of water supply, calibration monitoring, data management, printing report.

Due to the introduction of LabVIEW RT technology, the software system is divided into two parts: real-time program and non real-time program. Data acquisition and transducer control are as the real-time program, when the development of real-time program is completed in the LabVIEW environment, it will be downloaded to the RT target. User login, information entry, monitoring display, data management and print report are as the non real-time program, the non real-time program will run on the Windows host when it is developed. The real-time program exchange information with the non real-time program through TCP and RT FIFO [18].

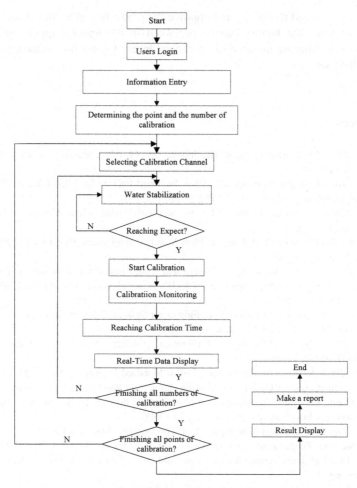

Fig. 7. Calibration process

6 Conclusions

The accuracy of the flow standard facility by master meter depends on the accuracy of the master meter. Improving the accuracy of the master meter improves the accuracy of the facility. The paper takes the mass flowmeters for 0.05 grades as the master meter group and develops two-stage regulator system (variable frequency plus container regulator). The relative expansion uncertainty of the facility is less than 0.2% (k = 2), meeting the need of high-precision calibration. At the same time, the facility can calibrate the flowmeters whose caliber is DN8–DN80, and the flow range of calibration is 0.5–38 m^3/h.

The biggest innovation point of the paper is that the entire facility is used for field service, which means that it can be carried to the scene for rapid calibration. And the HMI developed by virtual instrument is flexible and efficient. Thus the facility will save a lot of manpower and resources, and improve the production efficiency of enterprises. At the same time, the facility can be promoted in the national quality inspection industry so as to improve the domestic technical level of water flow standard facilities based on field service.

References

1. Baker, R.C.: Flow Measurement Handbook, pp. 61–64. Cambridge University Press, Cambridge (2010)
2. Tan, J.: The Development of the Gas Flow Standard Facility by Master Meter. Southwest Jiaotong University, Master Degree Thesis, pp. 6–7 (2015)
3. Su, Y., Liang, G., Sheng, J.: Flow Measurement and Testing. China Metrology Publishing House, Beijing (2007)
4. Zhao, N.: The Research of Mobile Gas Flow Standard Equipment. Hebei University, Master Degree Thesis, p. 9 (2010)
5. Engel, R., Beyer, K., Baade, H.-J.: Design and realization of the high-precision weighing systems as the gravimetric references in PTB's national water flow standard. Meas. Sci. Technol. **23**(7), 45 (2012)
6. Huancai, Y.: Research on Calibration of Flowmeter System with LabVIEW. Northwest A&F University, Master Degree Thesis, p. 18 (2009)
7. Jiang, M.: Mobile Flowmeter Calibration Facility Based on Master Meter. CN204831485UP, China (2015)
8. JJG 643-2003: Verification Regulation of Flow Standard Facility by Master Meter Method. State Administration of Quality Supervision, Inspection and Quarantine, Beijing
9. Li, J., Zhang, C., Wang, Y.: Design and implement of flow calibration facilities based on a master meter. China Instrum. Meters (12), 78 (2006)
10. Tu, B.: The Research and Development of The Gasflow Standard Facilities Using Mater Meter Method. Zhejiang Sci-Tech University, Master Degree Thesis, p. 20 (2015)
11. Mobile FlowLab User Manual for the Flow Calibration Systems Solution Package. Endress +Hauser, pp. 4–7 (2013)
12. Li, J., Zhang, Z., Su, Y.: Stability test of liquid flow standard facility with a flowmeter. In: International Conference on Future Energy, Environment and Materials (2012)
13. Li, B.: Research on Pressure Stabilization System of Water Flow Standard Facility Based on Frequency Conversion. Tian Jin University, Master Degree Thesis, p. 7 (2009)
14. He, Z.: Research on Pressure Stabilization System of Water Flow Standard Facility Based on Frequency Converter and Container. Tian Jin University, Master Degree Thesis, p. 3 (2010)
15. Li, Z.: Research on Uncertainty and Flow Stability of Water Flow Standard Facility. Tian Jin University, Master Degree Thesis, p. 7 (2009)
16. Giannone, L., Eich, T., Fuchs, J.C., Ravindran, M.: Data acquisition and real-time bolometer tomography using LabVIEW RT. Fusion Eng. Des. **86**(6/8), 12 (2011)
17. Chen, S., Liu, X.: A Valuable Book of LabVIEW. Electronic Industry Press, Beijing (2012)
18. Gadzhanov, S.D., Nafalski, A., Nedic, Z.: An application of NI SoftMotion RT system in a motion control workbench. In: International Conference on Remote Engineering and Virtual Instrumentation 10th (2013)

An Improved Multi-objective Bare-Bones PSO for Optimal Design of Solar Dish Stirling Engine Systems

Qun Niu$^{(\boxtimes)}$, Ziyuan Sun, and Dandan Hua

Shanghai Key Laboratory of Power Station Automation Technology,
School of Mechatronic Engineering and Automation, Shanghai University,
Shanghai 200071, China
comelycc@hotmail.com

Abstract. An improved bare-bones multi-objective particle swarm optimization, namely IMOBBPSO is proposed to optimize the solar-dish Stirling engine systems. A new simple strategy for updating particle's velocity is developed based on the conventional bare-bones PSO, aiming to enhance the diversity of the solutions and accelerate the convergence rate. In order to test the effectiveness of IMOBBPSO, four benchmarks are used. Compared with the non-dominated sorting genetic algorithm-II (NSGAII) and multi-objective particle swarm optimization algorithm (MOPSO), it is revealed that IMOBBPSO can quickly converge to the true Pareto front and efficiently solve practical problems. IMOBBPSO is then used to solve the design of the solar-dish Stirling engine. It is shown that IMOBBPSO obtains the best optimization results than NSGAII and MOPSO. It further achieves significant improvements 25.6102% to 29.2926% in terms of the output power and entropy generation rate when it is compared with existing results in the literature.

Keywords: Stirling engine · Solar dish · Multi-objective PSO · Bare bones

1 Introduction

Solar energy is a clean and renewable energy source, and its large-scale deployment can help to reduce the heavy reliance on fossil fuels and benefits the environment enormously. Solar-powered Stirling engine are one of the most promising solutions due to the advantages such as high efficiency, modularity, hardness against deflection and wind load, versatility, durability against moisture and temperature changes, long lifetime and low construction cost [1]. In the last two decades, it has received increasing attentions to research the modeling, optimization and application issue [2].

Some most recent efforts have been focused on the optimal parametric design of solar-powered Stirling engine, which is a complex nonlinear multi-objective problem. Ahmadi et al. [3] investigated the optimization performance of solar dish Stirling engine using multi-objective evolutionary algorithms. Arora [4] simultaneously optimized the power output, overall thermal efficiency and thermo-economic function objectives for solar driven Stirling engines based on NSGA-II, considering both the regenerative heat losses and conductive thermal bridging losses. Punnathanam [5, 6]

© Springer Nature Singapore Pte Ltd. 2017
K. Li et al. (Eds.): LSMS/ICSEE 2017, Part III, CCIS 763, pp. 167–177, 2017.
DOI: 10.1007/978-981-10-6364-0_17

considered a solar dish Stirling engine system with seven design parameters and also acquired satisfactory optimization performance by NSGA-II. Other related researches on multi-objective optimization of solar dish Stirling engine systems were carried out in [7, 8]. Some of these works focused on how to establish the numerical models for solar-dish Stirling systems, and often employed the conventional MOP methods such as NSGAII and MOEAs. Given the significance of the solar-powered Stirling engine where a number of parameter with conflicting objectives need to be optimized, it is vital to explore some alternative effective MOP approaches to achieve better solutions exist.

Over the past decades, many multi-objective algorithms have been proposed, such as SPEA2 [9], NSGAII [10] and MOEA/D [11], which have attracted substantial interests. In recent years, many particle swarm-based multi-objective methods have been developed due to their computational advantage. Parsopoulos [12] proposed a vector evaluated PSOin 2004 which used a ring migration topology. Coello and Pulido [13] developed a MOPSO method, incorporating the Pareto dominance and a special mutation operation. They further proposed an improving OMPSO by using crowding, mutation and ε-dominance for multi-objective optimization [14]. Nebro [15] proposed SMPSO, which designed a strategy to limit the velocity of the individuals. Reddy and Kumar [16] proposed EM-MOPSO for reservoir operation problems. Cabrera and Coello [17] developed Micro-MOPSO to handle small population sizes, where an auxiliary archive is employed to store non-dominated solutions. Although these algorithms have achieved satisfactory results, some of them have introduced a complex algorithmic structure such as introducing other operators, or have several parameters, which may cause algorithm becoming more sensitive to the settings of the algorithmic parameters. Therefore, it is crucial to explore a more concise and efficient algorithm for complex engineering optimization problems such as the solar-dish Stirling engine design.

In this paper, an improvedmulti-objective bare-bones PSO algorithm, namely IMOBBPSO, is proposed to optimize the design of solar dish Stirling engine, which considers the maximization of the power output and thermo-economic objectives. The bare-bones PSO was first designed in 2003 [18]. Compared to the traditional PSO, it is simpler and has only a few control parameters to be tuned. Zhang [19] proposed a bare-bones multi-objective particle swarm optimization (BB-MOPSO) for solving the economic dispatch problems and achieved good performance. In IMOBBPSO, an improved update operation is proposed to make the algorithm simpler. Each particle has the same probability to select different operation learning from the best individual. The introduction of *round* (*1 + rand*) not only increases the diversity of the solutions but also accelerates the convergence rate. Further, it combines the external archive with non-dominated sorting and crowding distance technology, resulting in significant improvement of the performance. In summary, the proposed IMOBBPSO has the advantages of fewer parameters, simpler structure and faster convergence speed.

2 Thermodynamic Modeling of the Solar-Dish Stirling Engine

Nomenclature			
C	heat capacitance rate (W K^{-1})	t	time (s)
C_v	specific heat capacity (J mol^{-1}K^{-1})	**Subscripts**	
M	proportionality constant	1	inlet
n	number of moles	2	outlet
Q	heat (J)	ave	average
R	the gas constant (J mol^{-1}K^{-1})	c	cold side
T	temperature	H	heat source
ε	effectiveness and emissivity (W m^{-2}K^{-1})	h	hot side
λ	ratio of volume during the regenerative processes	L	heat sink

The solar-dish Stirling system consists of a Stirling engine and a parabolic mirror of a parabolic shaped concentrator. When the solar dish Stirling is working, the mirror focuses solar energy into a cavity absorber located at the Stirling engine where the solar energy is absorbed and transferred to heat energy. Figure 1 presents a schematic for a solar dish-Stirling engine and shows an ideal Stirling cycle which includes two isothermal processes (1–2 and 3–4) and two isochoric processes (2–3 and 4–1). In a real cycle, the heat energy absorbed by the regenerator during the 4–1 process is transferred to the working fluid during process 2–3.

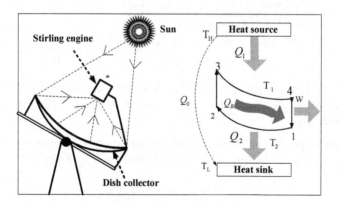

Fig. 1. Schematic of a solar dish Stirling engine

The modeling of the solar dish Stirling engines briefly introduced first, which has two conflicting objectives, namely maximization of the output power and minimization of the rate of entropy generation. Similar models are widely used in [5, 6]. The seven design variables of the problem are considered in this paper, namely the effectiveness of effectiveness of regenerator-side heat exchanger (ε_R), effectiveness of sink-side heat exchanger (ε_L), source-side heat exchanger (ε_H), source side temperature of working fluid (T_h), heat source temperature (C_H), and sink side temperature of the working fluid (T_c). Some related parameters in the model, as well as variables boundaries can be found in [5, 6].

(1) The output power is defined as the ratio between output work and time,

$$P = \frac{Work}{Time} = \frac{Q_H - Q_L}{t_{cycle}} \tag{1}$$

where Q_H and Q_L are the net heat released from the heat source and the net heat absorbed by the heat sink respectively [5, 6], t_{cycle} is the cyclic time period.

$$\begin{aligned} Q_H &= Q_h + Q_0 \\ Q_L &= Q_c + Q_0 \end{aligned} \tag{2}$$

The heat released by the heat source to the working fluid Q_h and the heat absorbed by the cold sink from the working fluid Q_c are given below:

$$\begin{aligned} Q_h &= nRT_h \ln(\lambda) + nC_v(1 - \varepsilon_R)(T_h - T_c) \\ Q_c &= nRT_c \ln(\lambda) + nC_v(1 - \varepsilon_R)(T_h - T_c) \end{aligned} \tag{3}$$

The equation for the cyclic time period is given as follows [5, 6].

$$\begin{aligned} t_{cycle} &= \frac{nRT_h \ln(\lambda) + nC_v(1 - \varepsilon_R)(T_h - T_c)}{C_H \varepsilon_H (T_{H1} - T_h) + \xi C_H \varepsilon_H (T_{H1}^4 - T_h^4)} \\ &+ \frac{nRT_c \ln(\lambda) + nC_v(1 - \varepsilon_R)(T_h - T_c)}{C_L \varepsilon_L (T_c - T_{L1})} + \left(\frac{1}{M_1} + \frac{1}{M_2}\right)(T_h - T_c) \end{aligned} \tag{4}$$

where C_L is the heat capacitance rate of the heat sink, M_1 and M_2 are proportionality constants known as the regenerative time constants for the heating and cooling processes respectively.

(2) The rate of production of entropy of the engine (r) is determined as the power output per unit investment cost, which is shown as follows [5, 6].

$$\sigma = \frac{1}{t_{cycle}} \left(\frac{Q_L}{T_{Lave}} - \frac{Q_H}{T_{Have}}\right) \tag{5}$$

The average temperatures of the heat source and heat sink are given below.

$$T_{Have} = \frac{T_{H1} + T_{H2}}{2}, \quad T_{H2} = (1 - \varepsilon_H)T_{H1} + \varepsilon_H T_h$$

$$T_{Lave} = \frac{T_{L1} + T_{L2}}{2}, \quad T_{L2} = (1 - \varepsilon_L)T_{L1} + \varepsilon_L T_c \tag{6}$$

3 The Improved Multi-objective Bare-Bones Particle Swarm Optimization Algorithm

The particle swarm optimization (PSO) is a population-based optimization algorithm proposed by Eberhart and Kennedy [20]. The simplicity and efficiency of the PSO led to the extension to the multi objective problem domain. There have been several recent attempts to use PSO for MOPs [17, 21]. Among them, the bare-bones multi-objective particle swarm optimization algorithm, is a powerful, almost parameter-free multi-objective optimization algorithm [21] and is highly competitive in terms of convergence, diversity, and distribution. In this paper, a new simple formula for updating particle's velocity is proposed based on the bare-bones PSO to increase the diversity of the solutions and accelerate the convergence rate. In addition, the external archive technique combined with non-dominated sorting and crowding distance technique can also greatly improve the performance of the proposed IMOBBPSO. The structure and steps of IMOBBPSO is presented as follows.

(1) Initialization

The particle swarms with size of N is randomly generated and evaluated and then store the non-dominated solutions in the archive. The initial local best position of each particle is set to be itself.

(2) Update the local best position

The local best position (*Pbest*) is searched by them, if the new position is dominated by the current position, we will keep the local best position in the memory; otherwise, the current local best position will be replaced by the new position. The update equation of *Pbest* is given as follows:

$$Pbest_i^{t+1} = \begin{cases} Pbest_i^t, & if \quad F\left(Pbest_i^t\right) \prec F\left(x_i^{t+1}\right) \\ x_i^{t+1}, & otherwise \end{cases} \tag{7}$$

(3) Update the global best position

The global best position (*Gbest*) in multi-objective optimization problems is different from the global best one in single-objective problems. It is difficult to choose a best solution for the conflicting nature of multiple objectives. In this paper, we use the crowding distance to estimate the diversity of the non-dominated solutions stored in the archive and chose a solution as the *Gbest*.

(4) Update each particle position

In the algorithm, we use the follow equation to update each particle's position:

$$x_{i,j}(t+1) = \begin{cases} Pbest_{i,j}(t) + round(1 + rand) * \left[Gbest_{i,j}(t) - Pbest_{i,j}(t)\right], & U(0,1) < 0.5 \\ Gbest_{i,j}(t), & otherwise \end{cases}$$

(8)

where $U(0,1)$ is a random number between 0 and 1. The introduction of *round* (*1 + rand*) widens the searching range of a particle, the random selection of different update strategies for particle not only keep the diversity of the particles, but also accelerate the convergence of the resultant solutions.

(5) Mutation operator

Mutation operator is used to avoid getting stuck into local best solution and to efficiently explore the search space. In this paper, the mutation operator in [19] is employed to reduce the harmful influence on the performance of algorithm for fast convergence speed of IMOBBPSO.

(6) Update the external archive

To provide each particle with a good *Gbest*, the non-dominated solutions should be stored in the archive during the entire search process. All the non-dominated solutions from both the solutions stored in the external archive and the ones reached after each iteration will be stored in the archive, but the capacity of the external archive is often limited, therefore, if the number of stored solutions reached to the archive's maximal capacity N_a, an approach based upon the crowing distance will be adopted to maintain the achieve size.

4   Tests on Benchmark Functions

In this paper, four well-known benchmarks including ZDT1, ZDT2, ZDT3 and DTLZ1 are used to test the performance of the proposed IMOBBPSO. To ensure a fair comparison, IMOBBPSO are compared with other two conventional algorithms, NSGA-II [10] and MOPSO [13]. All of the methods are coded in MatlabR2010b and run on an Intel Core i3-2100 CPU @3.10 GHz. A number of tests were carried out using the above platform to acquire the statistical results.

All algorithms are run for a maximum of 15000 function evaluations (FES) for ZDT1, ZDT2 and ZDT3, and 30000 FES for DTLZ1. The population size for all algorithms is set to 100 on the four benchmarks.

Three performance metrics are used to compare the performance of IMOBBPSO and the other approaches. The first metric generation distance [22] is defined as the distance between the obtained Pareto front and the exact Pareto front, which is to be minimized. The max spread metric [23] is the distance between the extreme solutions in the non-dominated set and the closer the metric is to 1, the better the searched solutions are. The hype volume metric [23] is the volume of the space dominated by the obtained solution set, which considers both convergence and diversity.

The approximated Pareto fronts obtained by each algorithm on four benchmarks are illustrated in Fig. 2. It is clear that IMOBBPSO produces better converged and distributed Pareto fronts on all functions than NSGAII and MOPSO. MOPSO can cover the whole PFs well for ZDT1 and ZDT3 compared with NSGAII, but is not as promising as IMOBBPSO on ZDT2 and DTLZ1.

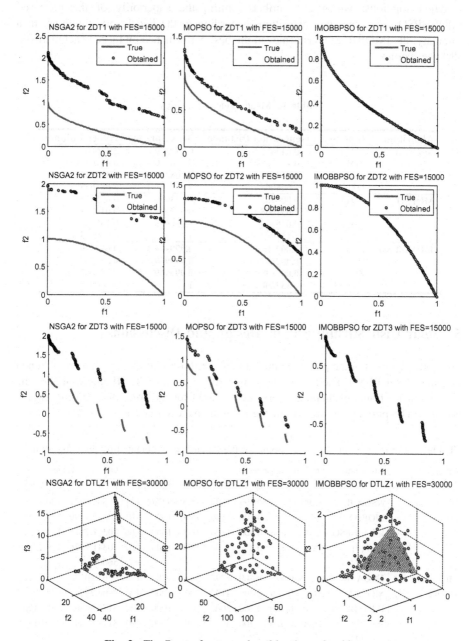

Fig. 2. The Pareto fronts produced by three algorithms

The metrics comparison results on four benchmarks are summarized in Table 1. The GD results show that IMOBBPSO performs competitively well on all the selected benchmarks. It is revealed that for the MS metric, IMOBBPSO is the best on ZDT1, ZDT2 and ZDT3 while the MOPSO is the second best and is slightly worse than IMOBBPSO. The HV metric is more effective to measure convergence and diversity of searched solutions, which can further confirm the superiority of the proposed IMOBBPSO over other two algorithms according to the benchmark solutions. In a summary, the IMOBBPSO has achieved the best performance on these benchmarks in all three metrics.

Table 1. Mean values of metrics

Algorithm	function	Generation Distance	Max Spread	Hyper Volume
NSGA-II	ZDT1	0.047671	0.770619	0.466400
	ZDT2	0.109885	0.450750	0.067033
	ZDT3	0.045424	0.768264	0.479033
	DTLZ1	1.608008	1	0.681627
MOPSO	ZDT1	0.006963	0.973479	0.625133
	ZDT2	1.051668	0.967528	0.019533
	ZDT3	0.016529	0.936633	0.735467
	DTLZ1	5.111094	1	0.998433
IMOBBPSO	ZDT1	**0.000106**	**0.999106**	**0.654800**
	ZDT2	**0.000102**	**0.999949**	**0.322367**
	ZDT3	**0.000145**	**0.995969**	**0.789067**
	DTLZ1	**0.132000**	1	0.766900

5   Solar Dish Stirling Engine Design and Discussions

The optimal values of the power output and the thermo-economic function are obtained by using IMOBBPSO. The results are compared with both NSGAII and MOPSO. The maximum functional evaluations were set at 10000 for the solar dish Stirling engine problem. The population sizes of all the methods were set at 100.

Table 2. Comparison of the corner points of the Pareto fronts obtained by three algorithms

Criteria	Algorithm	Decision variables							Objective values	
		ε_R	ε_H	ε_L	C_H	C_L	T_h	T_c	P (W)	σ (W/K)
Maximum power	NSGA-II	0.899	0.800	0.799	1800	1800	977.6	433	65506.6227	120.6070
	MOPSO	0.899	0.799	0.800	1800	1800	1000	510	70319.9399	139.3743
	IMOBBPSO	0.900	0.800	0.800	1800	1800	1000	510	**70341.7216**	139.2852
	Ref. [5]	0.9	0.8	0.8	1800	1800	1000	510	70341.7026	139.2852
	Ref. [24]	–	–	–	–	–	–	–	56000	–
Minimum entropy	NSGA-II	0.899	0.570	0.400	1292	300	978.8	400	9863.5022	25.1722
	MOPSO	0.88	0.705	0.403	419.5	300	1000	400	9004.7906	24.0833
	IMOBBPSO	0.900	0.400	0.400	300	300	1000	400	8177.8449	**21.9193**
	Ref. [5]	0.900	0.400	0.400	300	300	1000	400	8177.8449	21.9193
	Ref. [24]	–	–	–	–	–	–	–	–	32

Figure 2 presents the Pareto fronts obtained by IMOBBPSO, NSGAII and MOPSO for the maximization of the output power along with the minimization of the entropy generation rate when FES is set to 10000. In this case, it is clear that IMOBBPSO obtains better Pareto fronts than the others comparators. The NSGA-II Pareto front terminates prematurely at the region of the maximum power (right side end of the Pareto front) compared with other algorithms. While, the front obtained by IMOBBPSO is observed to extend further than the others towards the minimum extreme of entropy generation rate, which can also be verified in Table 2, which provides the corner points obtained by the three algorithms. In Table 2, IMOBBPSO obtains the best solutions in terms of both maximizing the power and minimizing the entropy generation rate. In terms of the values of output power and entropy generation rate, IMOBBPSO shows great improvements of about 25.6102% to 29.2926% respectively over literature values [24] and achieves the same literature results in [5] but with smaller FES.

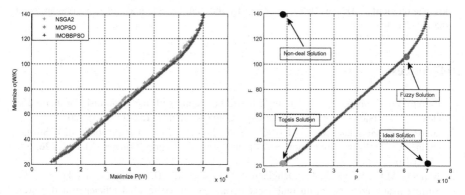

Fig. 3. Comparison of solutions obtained by three algorithms

Fig. 4. The Pareto solutions obtained by IMOBBPSO

Table 3. Comparison of obtained Pareto solutions with two decision-making methods

Algorithms	Decision variables							Objective values	
	ε_R	ε_H	ε_L	C_H	C_L	T_h	T_c	P (W)	σ (W/K)
Fuzzy	0.900	0.800	0.800	1757.2	1800	1000	400	60991.2157	106.0212
Topsis	0.900	0.400	0.400	300	300	1000	400	8177.8449	21.9193

In this paper, Fuzzy algorithm and Topsis method, which are well-known and commonly used decision-making methods are used to make decision for this multi objective problem. The ideal and non-ideal solutions in coordination (70342, 21.9193) and (8177.8, 139.2852) respectively, which are shown in Fig. 3. The results in Table 3 and Fig. 4 indicate that the chosen result through Fuzzy algorithm is the close to the ideal solution.

6 Conclusions

A new multi-objective particle swarm algorithm namely IMOBBPSO is proposed and successfully applied to the optimal design of the solar-dish Stirling engine for maximizing power output and minimizing thermos-economic function. IMOBBPSO extends the framework of BB-MOPSO to improve the performance using a simple update strategy. The proposed IMOBBPSO has the advantages of fewer parameters, simpler structure and faster convergence speed. As for the design of the solar-dish Stirling engine, IMOBBPSO can obtain the best optimization results than NSGAII and MOPSO. IMOBBPSO shows great improvement of about 25.6102% to 29.2926% respectively over literature values by the evolution algorithm in terms of the values of output power and entropy generation rate, and achieves the same results compared with the results by NSGAII but with smaller FES. Future works will focus on extending IMOBBPSO to other real world complex multi-objective optimization problems.

Acknowledgements. This work is supported by the National Natural Science Foundation of China (61273040).

References

1. Barreto, G., Canhoto, P.: Modelling of a Stirling engine with parabolic dish for thermal to electric conversion of solar energy. Energy Convers. Manag. **132**, 119–135 (2017)
2. Hafez, A.Z., Soliman, A., El-Metwally, K.A., Ismail, I.M.: Solar parabolic dish Stirling engine system design, simulation, and thermal analysis. Energy Convers. Manag. **126**, 60–75 (2016)
3. Ahmadi, M.H., Sayyaadi, H., Dehghani, S., Hosseinzade, H.: Designing a solar powered Stirling heat engine based on multiple criteria: maximized thermal efficiency and power. Energy Convers. Manag. **75**, 282–291 (2013)
4. Arora, R., Kaushik, S.C., Kumar, R., Arora, R.: Multi-objective thermo-economic optimization of solar parabolic dish Stirling heat engine with regenerative losses using NSGA-II and decision making. Int. J. Electr. Power Energy Syst. **74**, 25–35 (2016)
5. Punnathanam, V., Kotecha, P.: Effective multi-objective optimization of Stirling engine systems. Appl. Therm. Eng. **108**(5), 261–276 (2016)
6. Punnathanam, V., Kotecha, P.: Multi-objective optimization of Stirling engine systems using front-based Yin-Yang-pair optimization. Energy Convers. Manag. **133**(1), 332–348 (2017)
7. Ahmadi, M.H., Sayyaadi, H., Mohammadi, A.H., Barranco-Jimenez, M.A.: Thermo-economic multi-objective optimization of solar dish-Stirling engine by implementing evolutionary algorithm. Energy Convers. Manag. **73**, 370–380 (2013)
8. Ferreira, A.C., Nunes, M.L., Teixeira, J.C.F., Martins, L.A.S.B., Teixeira, S.F.C.F.: Thermodynamic and economic optimization of a solar-powered Stirling engine for micro-cogeneration purposes. Energy. **111**, 1–17 (2016)
9. Zitzler, E., Laumanns, M., Thiele, L.: SPEA2: Improving the strength pareto evolutionary algorithm. Technical report Computer Engineering and Networks Laboratory, Department of Electrical Engineering, Swiss Federal Institute of Technology (ETH) Zurich, Switzerland (2001)

10. Deb, K., Pratap, A., Agarwal, S., Meyarivan, T.: A fast and elitist multi objective genetic algorithm: NSGA-II. IEEE Trans. Evol. Comput. **6**(2), 182–197 (2002)
11. Zhang, Q.F., Liu, W., Li, H.: The performance of a new version of MOEA/D on CEC09 unconstrained MOP instances. In: IEEE Congress on Evolutionary Computing (CEC), Trondheim, pp. 18–21 (2009)
12. Parsopoulos, K.E., Tasoulis, D.K., Vrahatis, M.N.: Multi-objective optimization using parallel vector evaluated particle swarm optimization. In: International Conference on Artificial Intelligence and Applications (AIA 2004), vol. 2, pp. 823–828 (2004)
13. Coello, C.A.C., Pulido, G.T., Lechuga, M.S.: Handling multiple objectives with particle swarm optimization. IEEE Trans. Evol. Comput. **8**(3), 256–279 (2004)
14. Sierra, M.R., Coello Coello, C.A.: Improving PSO-Based Multi-objective Optimization Using Crowding, Mutation and ∈-Dominance. In: Coello Coello, C.A., Hernández Aguirre, A., Zitzler, E. (eds.) EMO 2005. LNCS, vol. 3410, pp. 505–519. Springer, Heidelberg (2005). doi:10.1007/978-3-540-31880-4_35
15. Nebro, A.J., Durillo, J., Garcia-Nieto, J., Coello, C.A., Luna, F., Alba, E.: SMPSO: a new pso-based metaheuristic for multi-objective optimization. In: IEEE Symposium on Computational Intelligence in Multi-criteria Decision-Making, pp. 66–73 (2009)
16. Reddy, M.J., Kumar, D.N.: Multi-objective particle swarm optimization for generating optimal trade-offs in reservoir operation. Hydrol. Process. **21**, 2897–2909 (2007)
17. Cabrera, J.C.F., Coello, C.A.C.: Micro-MOPSO: a multi-objective particle swarm optimizer that uses a very small population size. In: Nedjah, N., dos Santos Coelho, L., de Macedo Mourelle, L. (eds.) Multi-Objective Swarm Intelligent Systems, vol. 261, pp. 83–104. Springer, Heidelberg (2010). doi:10.1007/978-3-642-05165-4_4
18. Kennedy, J.: Bare bones particle swarms. In: Proceedings of the 2003 IEEE Swarm Intelligence Symposium, pp. 80–87 (2003)
19. Zhong, Y., Gong, D.W., Ding, Z.H.: A bare-bones multi-objective particle swarm optimization algorithm for environmental/economic dispatch. Inf. Sci. **192**(1), 213–227 (2012)
20. Kennedy, J., Eberhart, R.: Particle swarm optimization. In: Proceedings of the IEEE International Conference on Neural Networks, vol. 4, pp. 1942–1948 (1995)
21. Zhong, Y., Gong, D.W., Ding, Z.H.: A bare-bones multi-objective particle swarm optimization algorithm for environmental/economic dispatch. Inf. Sci. **192**(1), 213–227 (2012)
22. Van Veldhuizen, D.A., Lamont, G.B.: Multi Objective Evolutionary Algorithm Research: A History and Analysis (1998)
23. Zitzler, E., Thiele, L.: Multi objective evolutionary algorithms: a comparative case study and the strength Pareto approach. IEEE Trans. Evol. Comput. **3**(4), 257–271 (1999)
24. Nedjah, N., Mourelle, L.D.M.: Evolutionary multi-objective optimization: a survey. Int. J. Bio Inspired Comput. **7**(1), 1–25 (2015)

Fault Diagnosis Method of Ningxia Photovoltaic Inverter Based on Wavelet Neural Network

Guohua Yang[1,2(✉)], Pengzhen Wang[1], Bingxuan Li[1], Bo Lei[1],
Hao Tang[1], and Rui Li[1]

[1] Department of Electrical Engineering and Automation,
Ningxia University, Yinchuan 750021, China
ghyangchina@126.com
[2] Ningxia Key Laboratory of Intelligent Sensing & Intelligent Desert,
Yinchuan 750021, China

Abstract. Accurate fault diagnosis is the premise to ensure the safe and reliable operation of photovoltaic three-level inverter. A fault diagnosis method based on wavelet neural network is researched in the paper. First of all, the topology and the fault characteristics of three-level inverter are analyzed, the fault features are analyzed for three-level inverter when single and double IGBTs fault, the eigenvectors of phase voltage, the upper bridge arm and the lower bridge arm voltage are extracted by three-layer Wavelet Package Transform, the BP neural network is designed for training data and testing. The simulation model is built by Matlab/Simulink, the simulation results show that the method can accurately diagnose for various fault circumstances.

Keywords: Three level inverter · Neural network · Wavelet transform · Fault feature · Space vector modulation

1 Introduction

With the gradual development of photovoltaic power generation system in Ningxia, the photovoltaic inverter is used as network interface of photovoltaic power generation unit [1, 2], and it is particularly important to ensure its operation safely. Currently, the three-level inverter is widely used in photovoltaic power generation system. Compared with the two-level inverter, the switching device number of the three-level inverter is doubled, which increases the possibility of faults. When a seriously fault occurs, it will cause the entire inverter to shut down. Therefore, it is necessary to study the fault diagnosis method of the three-level inverter [3, 4]. So as to solve the above problem, a swarm of scholars have put forward different control strategies. For the single pipe fault, they proposed the fault diagnosis method based on real-time waveform analysis [5]. Whereas, the above method will fail if a double pipe failure occurs. With regards to it, we need to study the effective strategies. So far, there are two modes for double pipe failures. We use the basic information such as voltage, current and power to diagnose the fault [6, 7]. Another commonly use intelligent optimization algorithm, including

© Springer Nature Singapore Pte Ltd. 2017
K. Li et al. (Eds.): LSMS/ICSEE 2017, Part III, CCIS 763, pp. 178–184, 2017.
DOI: 10.1007/978-981-10-6364-0_18

cuckoo optimization algorithm [8], FFT transform and neural network combination of diagnostic methods [9] and C3C3 fault diagnosis algorithm [10], etc. The above intelligent methods have high diagnostic precision but they are difficult to construct the model, and the program design is more complicated. For the environmental impact of sand and other relatively volatile inverter PV power plant in Ningxia, more is needed is a low cost, simple and easy algorithm of fault diagnosis. In view of this, a kind of the three-level inverter fault diagnosis method is adopted by using wavelet transform and neural network unit. A simulation model was built by Matlab/Simulink simulation software. The wavelet transform and neural network program are written through M file, and the method is verified by simulation.

2 Fault Characteristic Analysis of Three-Level Inverter

2.1 Topological Structure

Three-phase three-level inverter topology is shown in Fig. 1.

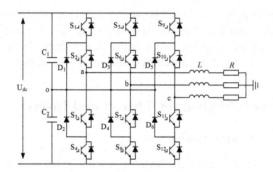

Fig. 1. The topology of three-phase three-level inverter

2.2 Fault Characteristics of Three-Level Inverter

A three-level inverter model is built by Matlab/Simulink simulation software. The DC bus voltage is set to 720 V, and load resistance is set to 10 Ω. The LC filter that the inductance is 5 mH and the capacitance is 12 μF is selected for filtering. The three-level inverter's open circuit faults with single IGBT are studied. These open circuit faults include no fault, S_{a1} open circuit failure, S_{a2} open circuit failure, S_{a3} open circuit failure and S_{a4} open circuit failure. The waveforms of a phase voltage are shown in Fig. 2.

The same simulation model is used to the open circuit faults with double IGBTs. These open circuit faults include $S_{a1} S_{a2}$ fault, $S_{a1} S_{a3}$ fault, $S_{a1} S_{a4}$ fault, $S_{a2} S_{a3}$ fault, $S_{a2} S_{a4}$ fault and $S_{a3} S_{a4}$ fault. By analyzing the simulation results of single IGBT and two IGBTs open faults, we find that $S_{a1} S_{a2}$ open circuit failure waveform is same to S_{a2} single failure waveform and $S_{a3} S_{a4}$ open circuit failure is same to S_{a3} single failure waveform. The fault information cannot be effectively distinguished by phase voltage waveform.

In order to distinguish, the upper bridge arm voltage and the lower bridge arm voltage are introduced. We can learn that $S_{a1} S_{a2}$ fault waveform of the upper bridge arm

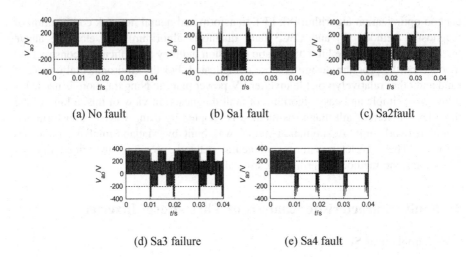

(a) No fault   (b) Sa1 fault   (c) Sa2fault

(d) Sa3 failure   (e) Sa4 fault

Fig. 2. The phase voltage waveforms of three level inverter under single IGBT open fault

voltage is different from S_{a2} single failure waveform of the upper bridge arm voltage, and similarly S_{a3} S_{a4} fault waveform of the lower bridge arm voltage is different from S_{a3} single failure waveform of the lower bridge arm voltage. In summary, the fault information can be effectively separated by introducing new voltage information.

3 Fault Diagnosis Method for Three-Level Inverter

3.1 Extraction of Feature Components

At present, the commonly used methods of feature component extraction include Fourier Transform, Concordia Transform and Wavelet Transform. As the number of three-level inverter switch tube, fault mode is complex. Compared with other transformations, Wavelet Transform has a great advantage. Therefore, this paper uses the wavelet transform to extract the characteristics of the inverter [11].

In order to extract the fault feature component of the three-level inverter, a three-tier wavelet packet decomposition method is adopted, and its structure is shown in Fig. 3.

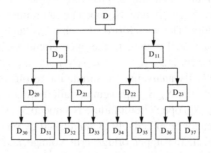

Fig. 3. The structure of three layer wavelet packet decomposition

First of all, the fault waveform data is read and the wavelet packet is decomposed to obtain the wavelet packet decomposition coefficients. And then calculate the wavelet packet reconstruction coefficient, and finally the concept of wavelet packet energy. The energy values of each band are calculated by the wavelet coefficients [12], and the energy can be expressed as:

$$E_j = \sum_k |C_i(k)^2|\qquad(1)$$

In the formula: $C_i(K)$ is wavelet packet coefficient.

The total energy of the wavelet packet can be expressed as:

$$E = \sum_j \sum_k |C_j(k)^2| = \sum_j E_j\qquad(2)$$

Then each band energy value can be normalized according to the total energy value, and the processed data is used as the training data and test data for the neural network.

3.2   Fault Diagnosis Method

In this paper, wavelet neural network is used to diagnose open faults and its detection structure is shown in Fig. 4.

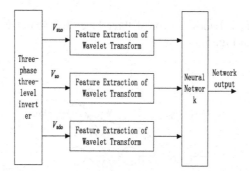

Fig. 4. The fault detection structure of three-level inverter

According to the Fig. 4, the phase voltage V_{ao}, the upper bridge arm voltage V_{auo} and the lower bridge arm voltage V_{ado} of the different failure modes are collected by simulation model.

The eigenvectors of phase voltage, the upper bridge arm and the lower bridge arm voltage are extracted by three-layer wavelet package transform, and the feature vector is delivered to the neural network. Finally, network output is obtained.

After determining the input data of 11 kinds of fault conditions, and a desired target output data is coded. As the three-phase three-level inverter with the same arm there are four IGBT switching devices, but also because the previous feature extraction using three-tier wavelet packet decomposition method, there are 8 different band energy values in each set of data feature vectors. So there are 11 kinds of fault codes defined, and each code is 8 bits, of which the first four bits correspond to the working state of four IGBT switching devices. When the device is open-circuit failure, the code is 1, while the code of the normal device is 0, and back four corresponding codes are 0. According to the above analysis, the 11 open-circuit fault codes are shown in Table 1.

Table 1. Three level inverter open circuit fault code

The fault type (open circuit)	Fault code
Trouble-free	00000000
S_{a1} Fault	10000000
S_{a2} Fault	01000000
S_{a3} Fault	00100000
S_{a4} Fault	00010000
S_{a1} S_{a2} Fault	11000000
S_{a1} S_{a3} Fault	10100000
S_{a1} S_{a4} Fault	10010000
S_{a2} S_{a3} Fault	01100000
S_{a2} S_{a4} Fault	01010000
S_{a3} S_{a4} Fault	00110000

In the BP neural network, a typical three-layer BP neural network structure is selected, as shown in Fig. 5.

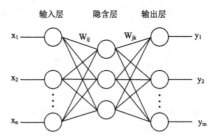

Fig. 5. The structure of a typical three layer neural network

The basic principle of BP neural network is the steepest descent method. By adjusting the weights, the total network error is minimized. By gradient search method, error in the actual output and the target output is minimized. The learning process is a process of modifying the weights of the error edges [13].

According to Fig. 5, the three-layer neural network includes the input layer, the output layer and the hidden layer, which x_1, x_2, x_n for input data, y_1, y_2, y_n for output data, W_{ij} W_{jk} respectively for the input layer to hidden, hidden layer to output layer weights.

The three-layer neural network includes input layer, output layer and hidden layer. According to previous analysis, the number of input layer is set to 11, and the number of output layer is set to 8. According to empirical formula, the hidden layer number can be expressed as:

$$q = 2 \times M + 1 \tag{3}$$

In the formula: M is the number for the input layer.

The activation function of the input layer to the hidden layer is set to tan-sigmod function, and the activation function of the hidden layer to the output layer is set to log-sigmod function.

4 Fault Diagnosis Result

In order to verify the validity and feasibility of fault diagnosis method based on wavelet and neural network, the DC-side voltage is set to 720 V, 700 V and 680 V, and the modulation ratio is chosen to be 0.2 to 0.9. The interval between each modulation ratio is 0.1, and a total of 24 sets of sample data can be obtained. The data of 1/3 were selected for training data and the data of 2/3 were selected for testing data. The eigenvector extraction is realized according to the previous wavelet transform. The input sample data are trained and tested by BP neural network. The simulation results are shown in Table 2.

Table 2. Fault diagnosis results based on wavelet neural network

Fault type	Actual output								Fault coding	Diagnosis results
Trouble-free	0.0000	0.0000	0.0000	0.0000	0.0000	0.0002	0.0001	0.0001	00000000	√
S_{a1}	0.9943	0.0000	0.0029	0.0001	0.0000	0.0000	0.0001	0.0001	10000000	√
S_{a2}	0.0019	0.9957	0.0000	0.0028	0.0000	0.0000	0.0000	0.0000	01000000	√
S_{a3}	0.0000	0.0000	0.9853	0.0011	0.0000	0.0000	0.0116	0.0045	00100000	√
S_{a4}	0.0000	0.0078	0.0074	0.9987	0.0000	0.0000	0.0000	0.0000	00010000	√
$S_{a1}\&S_{a2}$	1.0000	0.9956	0.0000	0.0032	0.0000	0.0000	0.0000	0.0000	11000000	√
$S_{a1}\&S_{a3}$	0.9978	0.0000	0.9860	0.0031	0.0000	0.0000	0.0113	0.0043	10100000	√
$S_{a1}\&S_{a4}$	0.0000	0.0086	0.0067	0.9987	0.0000	0.0000	0.0000	0.0000	10010000	√
$S_{a2}\&S_{a3}$	0.0021	0.9979	0.9710	0.0155	0.0000	0.0000	0.0164	0.0055	01100000	√
$S_{a2}\&S_{a4}$	0.0027	1.0000	0.0018	0.9981	0.0000	0.0000	0.0000	0.0000	01010000	√
$S_{a3}\&S_{a4}$	0.0001	0.0000	0.9903	0.9993	0.0000	0.0000	0.0004	0.0002	00110000	√

As can be seen from Table 2, the IGBT open circuit faults of the three-level inverter can be diagnosed accurately by using wavelet neural network algorithm.

5 Conclusion

In this paper, a three-level converter fault diagnosis method based on wavelet neural network is proposed. Based on the single and two IGBTs fault characteristics of the same bridge arm, the fault diagnosis method of wavelet neural network is given. The fault information is extracted by wavelet transform, and the fault diagnosis is carried out by BP neural network. The simulation results show that the proposed method can diagnose the open circuit fault quickly. The method has high diagnostic accuracy, easy implementation and certain engineering application value.

Acknowledgments. The work described in this paper is fully supported by a grant from the National Natural Science Foundation (No. 71263043).

References

1. Ding, M., Wang, W., Wang, X., et al.: A review on the effect of large-scale PV generation on power systems. Proc. CSEE **34**(1), 1–14 (2014)
2. Li, N., Wang, Y., Lei, W., et al.: Research on equivalent relations between two kinds of SVPWM strategies and SPWM strategy for three-level neutral point clamped inverter. Power Syst. Technol. **38**(5), 1283–1290 (2014)
3. Bendre, A., Cuzner, R., Krstic, S.: Three-level inverter system. IEEE Ind. Appl. Mag. **15**(2), 12–23 (2009)
4. Quntao, A., Li, S., Lizhi, S., et al.: Recent developments of fault diagnosis methods for switches in three-phase inverters. Trans. China Electrotech. Soc. **26**(4), 135–144 (2011)
5. Chen, D., Ye, Y., Hua, R.: Fault diagnosis for three-level inverter of CRH based on real-time waveform analysis. Trans. China Electrotech. Soc. **29**(6), 106–113 (2014)
6. Wan, X., Hu, H., Yu, Y., et al.: Survey of fault detection and diagnosis technology for three-level inverter of photovoltaic. J. Electr. Meas. Instrum. **29**(12), 1727–1738 (2015)
7. Shang, W., He, Z., Hu, H., et al.: An IGBT output power-based diagnosis of open-circuit fault in inverter. Power Syst. Technol. **37**(4), 1140–1145 (2013)
8. Junbo, L., Mahemuti, P., Chan, Z., et al.: Study on open-circuit fault diagnosis of the IGBT in three-level inverter. Electr. Meas. Instrum. **52**(20), 35–40 (2015)
9. Chen, C., Chen, D., Ye, Y.: The neural network-based diagnostic method for atypical faults in NPC three-level inverter. In: Chinese Control and Decision Conference (CCDC), pp. 4740–4745(2013)
10. Fan, J., Yi, Y.: Fault diagnosis of photovoltaic grid-connected inverter based on wavelet analysis. High Power Invert. Technol. **5**, 12–16 (2014)
11. Chen, D., Ye, Y., Hua, R.: Fault diagnosis of three-level inverter based on wavelet analysis and bayesian classifier. In: Chinese Control and Decision Conference (CCDC), pp. 4777–4779 (2013)
12. Jiang, Y., Wang, Y., Youren, T., et al.: Online multiple fault diagnosis for PV inverter based on wavelet packet energy spectrum and extreme learning machine. Chin. J. Sci. Instrum. **36**(9), 2145–2152 (2015)
13. Guoyong, L.: Neural Fuzzy Predictive Control MATLAB Implementation. Publishing House of Electronics Industry, Beijing (2013)

Research on Expert Knowledge Base of Intelligent Diagnosis Based on Tubing Leakage of High-Pressure Heater in Nuclear Power Plant

Miao Zheng[1,2(✉)], Hong Qian[1,2], Siyun Lin[1], Bole Xiao[3], and Xiaoping Chu[1]

[1] School of Automation Engineering, Shanghai University of Electric Power, Shanghai 200090, China
zhengmiao000207@163.com
[2] Shanghai Power Station Automation Technology Key Laboratory, Shanghai 200072, China
[3] Shanghai Power Equipment Research Institute, Shanghai 200240, China

Abstract. In order to improve the accuracy and timeliness of the tubing leakage of the high-pressure heater in the nuclear power plant, the fault diagnosis is carried out with the terminal difference triggered by the heat economy of the unit. Through the mechanism modeling of the tubing leakage of the high-pressure heater, the set of the symptom parameters related to the fault are obtained. The expert knowledge base of the tubing leakage in fault diagnosis system of the high-pressure heater is analyzed by using the mathematical statistics and the experience of the on-site experts. Through the insert of man-made fault in the 1000 MW nuclear power model, the intelligent diagnosis expert system is used to fault diagnosis. The results show that the method can accurately diagnose the tubing leakage of the high-pressure heater by analyzing the monitoring parameters at the beginning of the fault, and prove the validity and feasibility of the knowledge base.

Keywords: The tubing leakage of high-pressure heater · Model of failure mechanism · Mathematical statistics · Fault diagnosis · Knowledge base

1 Introduction

High-pressure heater is the important heat transfer equipment in a nuclear power station. The operation of the high pressure heater is influenced by high temperature and pressure of the fluid erosion, vibration and other factors, which lead to the possibility of the tubing leakage and affect the normal operation of the unit. Therefore, it is necessary to carry out timely diagnosis which can not only improve the efficiency of maintenance, but also ensure the reliability, safety and economy.

At present, there are many fault diagnosis systems of high pressure heater at home and abroad. In [1, 2], the neural network is used to realize fault. In [3], it is using fuzzy-cluster analysis theory and BP network for the fault diagnosis in high pressure heater. In [4], the improved particle swarm optimization algorithm and Elman neural

© Springer Nature Singapore Pte Ltd. 2017
K. Li et al. (Eds.): LSMS/ICSEE 2017, Part III, CCIS 763, pp. 185–194, 2017.
DOI: 10.1007/978-981-10-6364-0_19

network are used to diagnose system. However, most of the above methods are based on a large number of historical operation data in order to ensure the accuracy and effectiveness of the method, but the actual history running data is often not adequate.

In this paper, the failure mechanism model of high pressure heater is used as the criterion. Based on the mathematical statistics theory and the experience of the on-site experts, the diagnosis system of the high-pressure heater is able to ensure accuracy and timeliness of fault diagnosis of the leakage of high-pressure heater.

2   Construction of Expert Knowledge Base Based on Event Triggering Intelligent Diagnosis

When equipment of the nuclear power plant has failed, we can select some of the larger change parameters or related to the secondary parameters as the alarm triggering parameters. The alarm status of the alarm parameters are as trigger event of the fault diagnosis system. In nuclear power plant the construction of knowledge base of intelligent diagnosis expert system shown in Fig. 1.

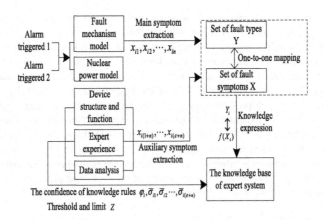

Fig. 1. Construction of knowledge base of intelligent diagnosis expert system

As the different faults may cause the alarm of the same status of the same parameters, and different failures may correspond to the same symptoms. So it formats the three logical structure of the "alarm - failure - symptoms". In order to improve the accuracy of fault diagnosis, it is necessary to realize the mapping relationship between the set of fault types and the set of fault symptoms under the same alarm condition. In other words, this is $Y_i \leftrightarrow f(X_i)$.

2.1   Construction of Fault Model Based on Event Triggered of Intelligent Diagnostic of the Expert Knowledge Base

In this knowledge base, set the fault alarm parameter to G. Based on the alarm trigger parameters to establish the mechanism model associated with it. Set the model set to $\mathbf{F} = [f_1, f_2, \ldots f_k, \ldots f_l]$, and it can be described as follows:

$$\begin{cases} f_1\left(G, v_1^1, v_2^1, \ldots v_{p_1}^1, \ldots v_{q_1}^1\right) = 0 \\ f_2\left(G, v_1^2, v_2^2, \ldots v_{p_2}^2, \ldots v_{q_2}^2\right) = 0 \\ \vdots \\ f_k\left(G, v_1^k, v_2^k, \ldots v_{p_k}^k, \ldots v_{q_k}^k\right) = 0 \\ \vdots \\ f_l\left(G, v_1^l, v_2^l, \ldots v_{p_l}^l, \ldots v_{q_l}^l\right) = 0 \end{cases} \quad (1)$$

In the formula (1): G—the alarm trigger parameter; $v_{p_k}^k$—The No. p operating parameter in the No. k subsystem of mechanism model; q_k—The number of operating parameters in the No. k subsystem of mechanism model. Then the formula (1) to sort out:

$$G = g\left(v_1, v_2, \ldots v_p \ldots, v_{q'}\right) \quad (2)$$

The self-independent derivative in formula (2) is used to derive from the fault mechanism model under the alarm triggered by this parameter, which can be described as:

$$\Delta G = \sum_{i=1}^{q'} (\varphi_i \Delta v_i) = \sum_{i=1}^{q'} \left(\frac{\partial g\left(v_1, v_2, \ldots v_p, \ldots v_{q'}\right)}{\partial v_i} \Delta v_i \right) \quad (3)$$

In formula (2)–(3): v_p—under the alarm trigger parameter is G, the fault corresponds to the No. p symptom characteristic parameter in this system; q'—the number of symptom characteristic parameters; $\Delta v_i (i = 1, 2 \ldots q')$—the increment of independent variable v_i; ΔG G—increment of the alarm parameter G; φ_i—a constant whose size is related to the capacity of the equipment and the capacity of the nuclear power plant and can be obtained by solving the nuclear power model.

The elements of the main symptom set $X_i' = [x_{i1}, x_{i2}, \ldots, x_{in}]$ of the No. i fault in the fault type are derived from the larger change elements which are in the set of independent variables of the formula (2). When the main symptoms of equipment failure can not fully achieve the one-to-one mapping relationship between the fault types set and fault symptoms set under the alarm state, we need to extract other auxiliary symptoms. The auxiliary symptom set $X_i'' = [x_{i(1+n)}, \ldots, x_{i(\varepsilon+n)}]$ are selected by combining the change parameters when the nuclear power plant equipment is in fault condition.

2.2 Intelligent Diagnosis of Expert Knowledge Base Based on Event Triggered

2.2.1 Representation of Rules of the Knowledge Base

The representation of the rules in the expert knowledge base of intelligent diagnostic is based on the IF-THEN rule. Assuming the No. i fault set is $[x_{i1}, x_{i2}, \ldots x_{ij}, \ldots x_{i(\varepsilon+n)}]$ in the knowledge base, the output of the rule is the result of the fault diagnosis. The No. i rule in the knowledge base can be expressed as follows:

$$R_i : If \; x_{i1} \; is \; A_1^i \wedge x_{i2} \; is \; A_2^i \wedge \cdots \wedge x_{ij} \; is \; A_j^i \wedge \cdots \wedge x_{i(\varepsilon+n)} \; is \; A_{(\varepsilon+n)}^k$$

$$Then \;\; Y_i(\gamma_i) \; with \; attribute \; weight \; \bar{\sigma}_1, \bar{\sigma}_2, \ldots \bar{\sigma}_{(\varepsilon+n)}$$

The $x_{ij}(j = 1, 2, \ldots, (\varepsilon+n))$ represents the No. j premise condition of the No. i rule in the knowledge base; Y_i represents the conclusion of the No. i rule; γ_i represents the confidence of the conclusion of the corresponding No. i rule; $\bar{\sigma}_i(i = 1, 2, \ldots, (\varepsilon+n))$ indicates the confidence of each premise condition; Each fault symptom corresponds to the states of the characteristic parameters which state can be divided into unusually low and unusually high, $A_j^i = \{L, \quad H\}$, $(j = 1, 2, \ldots, (\varepsilon+n))$. The above rule describes the mapping between the input and the output evaluation results.

2.2.2   To Determine the Value of the Rule Variables of Knowledge Base

The value of initial rule variable is determined by the experimental method. We can obtain the basic data from the fault model. Then the initial value of the limit is obtained by statistically analyzing the normal distribution of the large number of basic data. In the practical application, the adjustment of the threshold and limit combines the real-time data in the actual operation by the self-learning function of knowledge base of the intelligent diagnosis expert system.

The confidence level of conclusion is $\gamma_{mi}(m = 1, 2 \cdots N)$, which set according to the probability of occurrence of the corresponding fault in the course of the actual operation of the high pressure heater. The $\bar{\sigma}_j(j = 1, 2, \cdots, (\varepsilon+n))$ is the fault corresponds to the preference confidence vector of the symptoms. The threshold of the fault symptom is selected according to the change point of the possible failure with a certain margin. The selection of limit of fault symptom need to take multiple groups of the symptom parameters to do the analysis of mathematical statistics at fault alarm trigger time.

Assume that the symptom parameters conform to the normal distribution, and it is $X \sim N(\mu, \delta^2)$.

$$f(x) = \frac{1}{\sqrt{2\pi}\delta} \exp\left(-\frac{(x-\mu)^2}{2\delta^2}\right) \tag{4}$$

μ for the mean, δ for the standard deviation. The limit of the symptoms are set by the mean and standard deviation which are calculated for a large number of parameters of the symptom, that is $(\mu + p\delta)$. The p is the real number, which is set by combining with the real-time operating parameters.

3 Establish the Expert Knowledge Base of Intelligent Diagnosis on Tubing Leakage of High-Pressure Heater Based on Terminal Error Triggering Alarm

From the outage reasons of high pressure heater in the actual operation, the tubing leakage in the failures of high pressure heater has a larger proportion. The tubing leakage of high pressure heater is more likely to occur in the tubing section of feed water entrance [6]. In order to provide accurate information to the field operators in practical applications. And it can improve the inspection and maintenance efficiency and the operating economy of high-pressure heater.

3.1 Construction of Tubing Leakage Failure Model of High-Pressure Heater Based on Terminal Difference Trigger

Through the analysis of the operating mechanism of the high pressure heater, the mathematical model is established with the influencing factors of terminal temperature difference:

$$F_{f1}c_{p1}t_1 - F_{f2}c_{p1}t_2 + F_{di}c_{p2}t_{di} - F_{do}c_{p2}t_{do} + F_{cq}c_g t_s$$
$$= (V_t\rho_1 c_{p1} + M_t c_t)\frac{dt_2}{d\tau} + (V_l\rho_2 c_{p2} + V_g\rho_3 c_g)\frac{dt_s}{d\tau} \quad (5)$$

$$\frac{dK_1}{d\tau} = \frac{\alpha_2}{\alpha_1} * \frac{dt_s}{d\tau} - \frac{\alpha_1}{\alpha_2} * \frac{dt_2}{d\tau} + \left(\frac{1}{\alpha_2} - \frac{1}{\alpha_1}\right)$$
$$* \left(F_{f1}c_{p1}t_1 - F_{f2}c_{p1}t_2 + F_{di}c_{p2}t_{di} - F_{do}c_{p2}t_{do} + F_{cq}c_g t_s\right) \quad (6)$$

Equation (6) can be derived from Eq. (5), in the sixth equation:

$$\alpha_1 = V_t\rho_1 c_{p1} + M_t c_t \quad \alpha_2 = V_l\rho_2 c_{p2} + V_g\rho_3 c_g$$

$$F_{do} = l_1\varepsilon\sqrt{l_2(P_s - P_{snext}) + \Delta H_1} \quad (7)$$

Comments in the above formula: K_1—terminal temperature difference; K_2—the hydrophobic end error; F_{f1}—feed water inlet flow; F_{f2}—feed water inlet flow; F_{di}—hydrophobic inlet flow; F_{do}—hydrophobic outlet flow; t_1—feed water inlet temperature; t_2—feed water outlet temperature; t_{di}—hydrophobic inlet temperature; t_{do}—hydrophobic outlet temperature; t_s—saturated steam temperature; c_{p1}—the specific heat capacity of the tubing side water; c_{p2}—the specific heat capacity of shell side water; c_g—the specific heat capacity of shell side vapor; c_t—the specific heat capacity of metal; V_t—the volume of the tubing; V_l—volume of shell side water; V_g— volume of shell side vapor; ρ_1—water density in tubing; ρ_2—water density in shell side; ρ_3—vapor density in shell side; l_1, l_2—coefficient; ε—the opening of hydrophobic control valve; P_s—pressure of heater internal; P_{snext}—pressure of the lower heater internal; M_t—quality of heater metal tubing; ΔH_l—water level variation.

Through the analysis of the mechanism model of the high pressure heater and combine with formula (5)–(7), the mechanism model based on the terminal temperature difference can be described as:

$$K_1 = f(t_1, t_2, t_{di}, t_{do}, \varepsilon) \tag{8}$$

In the formula, t_1, t_2, t_{di}, t_{do}, ε represent the influence factors of the terminal temperature difference of the high-pressure heater mechanism model, and then the fault model based on the alarm terminal temperature difference of high-pressure heater is obtained by deriving the derivative from the independent variable of the above equation.

$$\Delta K_1 = \lambda_1 \Delta t_1 + \lambda_2 \Delta t_2 + \lambda_3 \Delta t_{di} + \lambda_4 \Delta t_{do} + \lambda_5 \Delta \varepsilon \tag{9}$$

$$\lambda_1 = \frac{\partial f(t_1, t_2, t_{di}, t_{do}, \varepsilon)}{\partial t_1} \quad \lambda_2 = \frac{\partial f(t_1, t_2, t_{di}, t_{do}, \varepsilon)}{\partial t_2} \quad \lambda_3 = \frac{\partial f(t_1, t_2, t_{di}, t_{do}, \varepsilon)}{\partial t_{di}}$$

$$\lambda_4 = \frac{\partial f(t_1, t_2, t_{di}, t_{do}, \varepsilon)}{\partial t_{do}} \quad \lambda_5 = \frac{\partial f(t_1, t_2, t_{di}, t_{do}, \varepsilon)}{\partial \varepsilon}$$

In the formula ΔK_1 indicates the increments of terminal temperature difference, Δt_1, Δt_2, Δt_{di}, Δt_{do}, $\Delta \varepsilon$ are respectively the differential of the independent variable t_1, t_2, t_{di}, t_{do}, ε. In the formula, λ_1, λ_2, λ_3, λ_4, λ_5 is constant. Through the analysis of mechanism model and combine with changing a large degree of parameters of the leakage fault in the actual operation, then extract ε, t_2, t_{do} in the failure model as the main symptoms of the fault. The extraction steam inlet pressure P_c is selected as the auxiliary symptom.

By the mechanism model, select K_1 as one of the alarm parameters, and take the K_2 which has closely related to change symptom parameters as another alarm parameter. When the two alarm parameters at the same time exceed the alarm limits, it triggers the operation of the fault diagnosis program (Figs. 2 and 3).

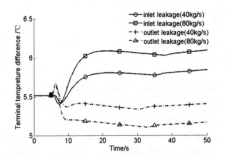

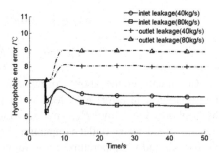

Fig. 2. Terminal temperature difference of the tubing leakage in high-pressure heater

Fig. 3. Hydrophobic end error of the tubing leakage in high-pressure heater

From the graphs (2)–(3), the change of terminal temperature difference is Δk_{1i} and the hydrophobic end error is Δk_{2i}. After adding the fault, by the data analysis of the alarm parameters and the expert experience of the nuclear power, when the requirement is $\Delta k_{1i} \geq 0.301$ and $\Delta k_{2i} \leq -0.98$, the fault diagnosis of inlet tubing leakage program will be alarm triggered in high-pressure heater. when the requirement is $\Delta k_{1i} \leq -0.1345$ and $\Delta k_{2i} \geq 0.8132$, the fault diagnosis of outlet tubing leakage program will be alarm triggered in high pressure heater.

3.2 Determination of Rule Variables of Leakage Fault for High Pressure Heater Based on Differential Alarm

Figure 4 shows the change of symptoms with different levels of tubing leakage which are near the feed water entrance of high-pressure heater and near the feed water outlet of high-pressure heater.

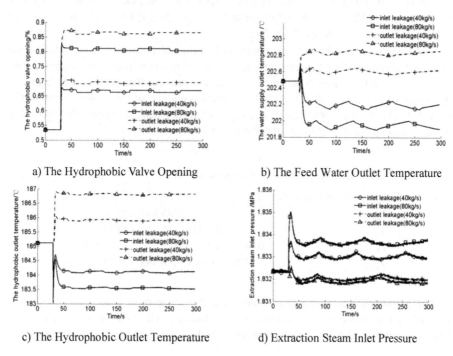

a) The Hydrophobic Valve Opening

b) The Feed Water Outlet Temperature

c) The Hydrophobic Outlet Temperature

d) Extraction Steam Inlet Pressure

Fig. 4. The varying curves of symptom characteristic parameters on the tubing leakage of high-pressure heater

When terminal temperature difference and hydrophobic end error to reach the alarm conditions, the moment corresponding to parameters of the symptoms which include ε, t_2, t_{do}, P_c as the sample extraction parameters in the same leakage fault. Then respectively select 30 sets of data of a fault as sample parameters.

The sample of hydrophobic valve opening of the inlet tubing leakage of high-pressure heater is $x_{\varepsilon 1} = [0.64457, 0.65324 \ldots 0.74953, 0.88684]$; The sample of feed water outlet temperature of the inlet tubing leakage of high-pressure heater is $x_{t1} = [202.1835, 202.1834 \ldots 202.1747, 202.1729]$; The sample of hydrophobic outlet temperature of the inlet tubing leakage of high-pressure heater is $x_{d1} = [184.1636, 184.1268 \ldots 183.9825, 184.0331]$; The sample of extraction steam inlet pressure of the inlet tubing leakage of high-pressure heater is $x_{p1} = [1833.036, 18333.037 \ldots 1833.612, 1835.181]$. The sample of hydrophobic control valve opening of the outlet tubing leakage of high-pressure heater is $x_{\varepsilon 2} = [0.6656, 0.6910 \ldots 0.8491, 0.9302]$; The sample of feed water outlet temperature of the outlet tubing leakage of high-pressure heater is $x_{t2} = [202.6194,$ $202.62 \ldots 202.6564, 202.6581]$; The sample of hydrophobic outlet temperature of the outlet tubing leakage of high-pressure heater is $x_{d2} = [185.8046, 185.854 \ldots$ $186.3649, 186.5985]$; The sample of extraction steam inlet pressure of the inlet tubing leakage of high-pressure heater is $x_{p2} = [1832.013, 1832.006 \ldots 1832.242, 1832.165]$. Suppose that the sample obeys normal distribution and obtain the parameters: $x_{\varepsilon 1} \sim N(0.7339, 0.0563)$, $x_{t1} \sim N(202.178, 0.026)$, $x_{d1} \sim N(184.056, 0.0458)$, $x_{p1} \sim N$ $(1833.5879, 0.4192)$, $x_{\varepsilon 2} \sim N(0.7394, 0.0639)$, $x_{t2} \sim N(202.6313, 0.0146)$, $x_{d2} \sim N$ $(186.13, 0.2236)$, $x_{p2} \sim N(1832.14, 0.199)$.

The limit value of the rule parameter is determined according to calculating the mean and variance of each parameter, that is $(\mu + p\delta)$ where p is a real number. The value of the p of the different parameters is set according to the speed of the change of the symptom parameter after the fault and the experience of operation expert in the nuclear power. In the Table 1, Z_H, Z_L, Z_{max}, Z_{min} represent the upper threshold, the lower threshold, the upper limit, and the lower limit respectively.

Table 1. Reference values for thresholds and limits of symptoms

Symptom characteristics	$\varepsilon/\%$	$t_2/°C$	$t_{do}/°C$	P_c/KPa
Z_{min}	–	202.1	183.9639	1831.941
Z_L	–	202.475	185.012	1832.33
Z_H	0.554	202.5	185.2	1832.387
Z_{max}	0.6776	202.661	186.356	1833.177

The graphs (4) shows the change degree of the symptoms, you can set the corresponding the precondition confidence of the various symptoms of each fault. The diagnostic rules for the inlet tubing leakage of high-pressure heater can be described as:

R_1: *If* ε *is* $H \wedge t_2$ *is* $L \wedge t_{do}$ *is* $L \wedge P_c$ *is* H

Then $Y_1(0.9)$

With attribute degree $0.9, 0.8, 0.8, 0.8$

The diagnostic rules for the outlet tubing leakage of high-pressure heater can be described as:

R_2: *If* ε *is* $H \wedge t_2$ *is* $H \wedge t_{do}$ *is* $H \wedge P_c$ *is* L

Then $Y_1(0.9)$

With attribute degree $0.9, 0.8, 0.9, 0.8$

From the thresholds and limits in Table 1, the evidence confidence of current symptom can be calculated using the trapezoidal unilateral membership function [5]. From the confidence of the confidence and the premise of confidence, then according to a certain matching algorithm [7] and rules in rule base perform a reasoning match calculation, where the matching index is set to 0.3. If the matching conditions in certain matching algorithm [7] are met, then the fault confidence is calculated based on calculation formula the of the conclusion confidence [7] to obtain the diagnosis result.

4   The Case of Tubing Leakage of High-Pressure Heater of Intelligent Diagnosis Expert System Based on the Trigger of Terminal Difference in Nuclear Power Plant

Based on the nuclear power model of 1000 MW, this system will connect the diagnostic software of the expert knowledge base about the high pressure heater in the nuclear power plant based on the terminal difference with the intelligent diagnosis platform, and collect the diagnostic information from the nuclear power model through the OPC access. The nuclear power model is personally introduced into the faults such as the leakage rate of 5% of the total flow of the inlet tubing leakage fault of high-pressure heater. As shown in Fig. 5:

Fig. 5.   Fault diagnosis interface of the tubing leakage of 6 high-pressure heater in nuclear power conventional island

Diagnostic screen shows that the diagnostic fault is the inlet tubing leakage of high-pressure heater, and the fault confidence degree is 0.7614. At the same time, there are the curves of the main parameters corresponding to the fault, the reasons for analysis and operational guidance and other content on the screen. This can improve the human reliability of the operation in the failure of high pressure heater.

5   Conclusion

Based on the combination of nuclear power model, mechanism model and experience, the fault symptom set is extracted on the condition of terminal difference alarm and the rule variable in the knowledge base is set. The results show that the diagnosis system based on this knowledge base can accurately and timely diagnose the leakage of the high pressure heater in the nuclear power plant.

In the practical application, the initial knowledge base is used, and the adaptive knowledge base is used to carry out continuous depth of learning by the actual operation in the nuclear power plant, in order to meet the requirements on the time and accuracy of the on-site diagnostic system.

Acknowledgments. This work was partially supported by National Natural Science Foundation of China, Grant No. 61503237; Shanghai Natural Science Foundation (No. 15ZR1418300); Shanghai Key Laboratory of Power Station Automation Technology (No. 13DZ2273800).

References

1. Ma, L., Ma, X., Feng, Z., et al.: Fault diagnosis of high pressure heater based on radial basis probabilistic neural network. J. North China Electr. Power Univ. **34**, 81–84 (2007)
2. Liang, N., Ding, C., Ding, Z.: Fault diagnosis and simulation of high pressure heater based on probabilistic neural network. Power Gener. Equip. **24**, 97–100 (2010)
3. Ding, C., Dai, Z., Tian, S., et al.: Fault diagnosis of high pressure heater combined with fuzzy clustering analysis and BP network. Power Gener. Equip. **21**, 60–63 (2007)
4. Wang, X., Ma, L., Qi, Z.: Based on particle swarm and nearest neighbor of the variable condition of the thermal system fault diagnosis method of dynamic process. J. Power Eng. **6**, 469–476 (2014)
5. Qian, H., Luo, J., Jin, W.: The research on alarm trigger type of steam generator of the pipe burst accident diagnosis system. Nuclear power Eng. **4**, 98–103 (2015)
6. Niu, X., Gong, C.: Talking about the causes and countermeasures of erosion and leakage of high pressure heater tube system. Power Station Syst. Eng. **19**, 35–36 (2003)
7. Luo, J.: In nuclear power plant intelligent diagnosis expert system base on alarm triggered. Shanghai Institute of Electric Power (2015)

Research on Intelligent Early-Warning System of Main Pipeline in Nuclear Power Plants Based on Hierarchical and Multidimensional Fault Identification Method

Hong Qian[1,2], Siyun Lin[1,2(✉)], Miao Zheng[1,2], and Qiang Zhang[3]

[1] School of Automation Engineering, Shanghai University of Electric Power, Shanghai 200090, China
1002361520@qq.com
[2] Shanghai Power Station Automation Technology Key Laboratory, Shanghai 200072, China
[3] Shanghai Power Equipment Research Institute, Shanghai 200240, China

Abstract. In order to improve the timeliness and accuracy of the fault identification for SB-LOCA (small break-loss of coolant accident), a hierarchical and multidimensional fault identification method is proposed, and a intelligent early-warning system is established to locate and evaluate the degree of the fault in the early stage, which can improve the operating safety of nuclear power plants. The faults in different kinds of locations and degrees are artificially inserted into the nuclear power simulator and are recognized by the early-warning system based on the method researched above. The results show that it can accurately locate and evaluate the tiny degree of fault, which verifies the validity and feasibility of the intelligent early-warning system.

Keywords: SB-LOCA · Intelligent early-warning system · Hierarchical and multi-dimensional · Fault location · Fault degree evaluation

1 Introduction

The nuclear safety affects the environment and the future of nuclear development. Actually, SB-LOCA is the design basis accident in nuclear power plants. The position and size of a break will lead to the variation of operating parameters, and the extent of the break usually increases with time. Presently, there are no corresponding measures for a small break. Since it is not easy to be found, and may cause reactor to shutdown if letting it go. Therefore, it is urgent to establish an effective intelligent early-warning system to detect the small break through monitoring and diagnosis of the main pipeline.

Currently, researchers are committed to improving the correctness of location and the accuracy of diagnosis for SB-LOCA. Some identify and locate this fault successfully, but did not assess the fault degree [1, 2]. Some use the method of SVM to identify fault degree, but the recognition between different fault degrees is not adequate, which can not meet the requirements of diagnostic accuracy [3]. Actually, the neural network algorithm and optimal scaling technology have been successfully applied for the identification of

© Springer Nature Singapore Pte Ltd. 2017
K. Li et al. (Eds.): LSMS/ICSEE 2017, Part III, CCIS 763, pp. 195–205, 2017.
DOI: 10.1007/978-981-10-6364-0_20

the similar faults [4]. However, the neural network algorithm needs to be based on a large amount of experimental data, which requires high quality of data samples, and the rules of identification are complex. The optimization technique is limited to a certain system structure of topology when applied for estimating fault degree. These are not conducive to improving the timeliness and accuracy of fault diagnosis for SB-LOCA.

Thus, at the aim of improving the reliability and safety of nuclear operation, a intelligent early-warning system is established based on a method of hierarchical and multi-dimensional fault identification.

Firstly, determine the state of system initially through warning signals, then use the limit type of symptoms to locate the fault, and use the trend type of symptoms to estimate the extent of fault in order to detect leakages in the early stage of the accident, optimizing overhaul strategies and maintenance of the main pipeline and related equipment. In addition, the method has been certificated through the nuclear simulation platform.

2 Construction of Intelligent Early-Warning System

2.1 Construction on the Early-Warning Set of Multi-parameter

The pressurizer is an important equipment in nuclear power plants. When SB-LOCA occurs, the operating state of the pressurizer will be abnormal. Thus, make analysis of the operation mechanism of the pressurizer, whose mechanism model can be described in expression 1.

$$\mathbf{F} = [f_1, f_2] \tag{1}$$

$$f_1 : \frac{dH_Z}{dt} = \frac{1}{A}[W_{SU}v_{SU} + (W_{SP} + W_{SC} + W_{RO} + W_{WC} - W_{FL})v_f] + \frac{M_M}{A}\left(\frac{\partial v_M}{\partial h_M}\frac{dh_M}{dt} + \frac{\partial v_M}{\partial P_Z}\frac{dP_Z}{dt}\right)$$

$$f_2 : k_1 \frac{dW_{SU}}{dt} = P_Z - P_{SG} + \rho_M g \Delta H_Z - k_2 W_{SU}$$

where H_z is the water level of pressurizer, P_Z is the pressure of pressurizer, P_{SG} is the pressure of steam generator, W_{SU} is the fluctuating flow of pressurizer, W_{SP} is the structure parameters of pressurizer.

Actually, H_Z and P_Z are important parameters in nuclear operation. Take these two parameters as independent variables, the mechanism model of SB-LOCA can be described in expression 2.

$$F = g(P_Z, H_Z) \tag{2}$$

Then make derivation of these independent variables in Eq. 2, the fault model of SB-LOCA is shown in Eq. 3.

$$F = c_1 \Delta P_Z + c_2 \Delta H_Z \tag{3}$$

$$c_1 = \frac{\partial g(H_Z, P_Z)}{\partial P_Z}, c_2 = \frac{\partial g(H_Z, P_Z)}{\partial H_Z} \tag{4}$$

Where c_1, c_2 are constants whose values are related to the structure of the equipment and the capacity of nuclear power unit, ΔH_Z is the increment of H_Z, ΔP_Z is the increment of P_Z.

In order to improve the reliability of the diagnosis triggering, it is necessary to select several parameters to establish a multi-parameter early-warning signal set, which is shown in Eq. 5.

$$[x_{j1}, x_{j2}] = [\Delta P_Z, \Delta H_Z] \tag{5}$$

Through the summary of data, the change of pressure and water level of pressurizer in cold or hot leg at different fault degrees are shown in Figs. 1 and 2.

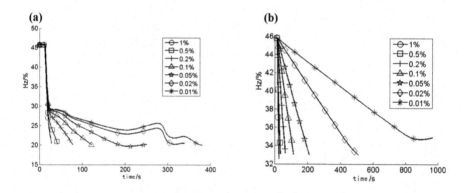

Fig. 1. (a) The curves of Hz of cold leg, (b) the curves of Hz in hot leg

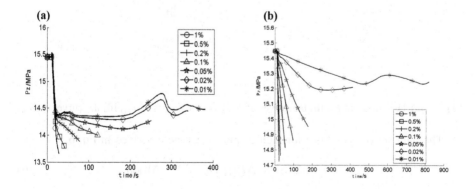

Fig. 2. (a) The curves of Pz in cold leg, (b) the curves of Pz in hot leg

Therefore, there are determination of the limits of these two early-warning parameters. If the pressure value of pressurizer is less than 15.3 MPA, in the meantime, the water level value of pressurizer is less than 35%, this intelligent system will produce the early-warning signal, which triggers the fault diagnosis section.

2.2   Extraction of Symptoms for Fault Diagnosis

SB-LOCA directly leads to the great change of the water level of pressurizer, thus it is reasonable to select the fault symptoms on the condition that the water level of pressurizer is in the same change process. Combined with the operation mechanism model of nuclear power simulator, 4 main operating parameters are identified as the symptoms for fault diagnosis. Figure 3 shows the change of these symptoms on the altering of the water level.

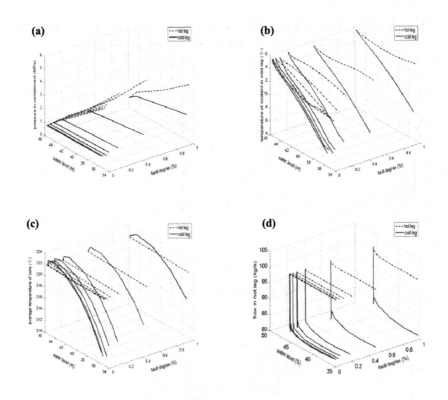

Fig. 3. (a) The curves of P_c, (b) the curves of T_{cold}, (c) the curves of T_{av}, (d) the curves of W_{hot}

The fault set of symptoms for fault diagnosis can be described in Eq. 6.

$$[x_{i1}, x_{i2}, x_{i3}, x_{i4}] = [\Delta T_{cold}, \Delta P_c, \Delta T_{av}, \Delta W_{hot}] \tag{6}$$

Where ΔP_c is the increment of the pressure of containment, ΔT_{cold} is the increment of the temperature of cold leg, ΔT_{av} is the increment of the average temperature of core coolant, ΔW_{hot} is the increment of the flow of hot leg.

And the state of these symptoms in different fault positions are summarized in Table 1.

Table 1. The symptoms of limit type in different fault locations

Serial number	Symptom characteristic parameter	In cold leg	In hot leg
1	Pressure in containment (P_c)	N	H
2	Temperature in cold leg (T_{cold})	L	L
3	Average temperature of coolant in core reactor (T_{av})	L	N
4	Flow of hot leg (W_{hot})	L	N

Where H means the symptom is abnormally high; N means the symptom is normal; L means the symptom is abnormally low.

2.3 Construction of Hierarchical and Multidimensional Fault Recognition Expert Database in the Early-Warning System

Firstly we need to determine the fault location Y and fault degree d to construct the hierarchical and multidimensional fault recognition expert database, as shown in Eq. 7.

$$Y = \left[Y_1, \ldots, Y_j, \ldots, Y_n\right]^T, \quad j = 1, 2, \ldots, n; \ F = (Y, d) \tag{7}$$

$$d = \begin{bmatrix} d_1 \\ \vdots \\ d_j \\ \vdots \\ d_n \end{bmatrix} = \begin{bmatrix} d_{11} & \cdots & d_{1k} & \cdots & d_{1l} \\ \vdots & \vdots & \vdots & \vdots & \vdots \\ d_{j1} & \cdots & d_{jk} & \cdots & d_{jl} \\ \vdots & \vdots & \vdots & \vdots & \vdots \\ d_{n1} & & d_{nk} & & d_{nl} \end{bmatrix} \quad F = \begin{bmatrix} (Y_1, d_{11}) & \cdots & (Y_1, d_{1k}) & \cdots & (Y_1, d_{1l}) \\ \vdots & \vdots & \vdots & \vdots & \vdots \\ (Y_j, d_{j1}) & \cdots & (Y_j, d_{jk}) & \cdots & (Y_j, d_{jl}) \\ \vdots & \vdots & \vdots & \vdots & \vdots \\ (Y_n, d_{n1}) & \cdots & (Y_n, d_{nk}) & \vdots & (Y_n, d_{nl}) \end{bmatrix} \tag{8}$$

In the above equations, Y is the set of fault location, Y_j is the No.j position of fault. n is the number of the position of fault in the set of fault positions. d is the set of fault degree. d_j is the sub set of fault degree in the No.j position of fault. d_{jk} is the No.k evaluation grade of fault degree in the No.j fault position, and $k = 1, 2, \ldots, l$; l is the number of fault degree evaluation grade in the No.j sub set of fault degree. Therefore, the multidimensional fault set F can be described in Eq. 8.

In the above equations, F is the fault set of this diagnosis system. (Y_j, d_{jk}) is the No.j fault position in the No.k fault degree, which can be denoted as F_{jk}.

The fault location and fault degree needed to be identified in this system are revealed in Tables 2 and 3.

Table 2. The location of fault location and its description

The name of fault location	Description of fault location
A small break in cold leg (LOCA$_{cold}$)	Y_1
A small break in hot leg (LOCA$_{hot}$)	Y_2

Table 3. The description of fault location

Fault degrees	The name of fault degree in rating level	The description of rating level
(0–0.01%)	Leakage (VS)	d_{11}, d_{21}
[0.01–0.1%)	Minimal break (S)	d_{12}, d_{22}
[0.1–1%)	Tiny break (M)	d_{13}, d_{23}
[1–2%]	Small break (L)	d_{14}, d_{24}

The set of fault location Y and the set of fault degree d can be described in Eqs. 9 and 10.

$$Y = [Y_1, Y_2]^T = [\text{LOCA}_{cold}, \text{LOCA}_{hot}]^T \tag{9}$$

$$d = \begin{bmatrix} d_{11} & d_{12} & d_{13} & d_{14} \\ d_{21} & d_{22} & d_{23} & d_{24} \end{bmatrix} = \begin{bmatrix} \text{VS} & \text{S} & \text{M} & \text{L} \\ \text{VS} & \text{S} & \text{M} & \text{L} \end{bmatrix} \tag{10}$$

Thus, the set of fault F can be described in Eq. 11.

$$F = \begin{bmatrix} (\text{LOCA}_{cold}, \text{VS}) & (\text{LOCA}_{cold}, \text{S}) & (\text{LOCA}_{cold}, \text{M}) & (\text{LOCA}_{cold}, \text{L}) \\ (\text{LOCA}_{hot}, \text{VS}) & (\text{LOCA}_{hot}, \text{S}) & (\text{LOCA}_{hot}, \text{M}) & (\text{LOCA}_{hot}, \text{L}) \end{bmatrix} \tag{11}$$

To ensure the accuracy of fault recognition, it is necessary to agree one-to-one mapping relationship between the symptom set S and fault set F. This paper use a method of combining limit symptoms (X) and tendency symptoms (c) to recognize the location (Y) and degree of faults (d) respectively, which are shown in Eq. 12.

$$F \leftrightarrow S; \ Y \leftrightarrow X; \ d \leftrightarrow c \tag{12}$$

In the above equations, S is the set of symptoms for fault recognizing, X is the set of limit symptoms, c is the set of trend symptoms.

And the set of symptoms (S) can be described in Eqs. 13, 14 and 15.

$$S = (X, c) \tag{13}$$

$$X = \begin{bmatrix} X_1 \\ \vdots \\ X_j \\ \vdots \\ X_n \end{bmatrix} = \begin{bmatrix} x_{11} & \cdots & x_{1i} & \cdots & x_{1m} \\ \vdots & \vdots & \vdots & \vdots & \vdots \\ x_{j1} & \cdots & x_{ji} & \cdots & x_{jm} \\ \vdots & \vdots & \vdots & \vdots & \vdots \\ x_{n1} & \cdots & x_{ni} & \cdots & x_{nm} \end{bmatrix} ; \ c = \begin{bmatrix} c_1 \\ \vdots \\ c_j \\ \vdots \\ c_n \end{bmatrix} = \begin{bmatrix} c_{11} & \cdots & c_{1k} & \cdots & c_{1l} \\ \vdots & \vdots & \vdots & \vdots & \vdots \\ c_{j1} & \cdots & c_{jk} & \cdots & c_{jl} \\ \vdots & \vdots & \vdots & \vdots & \vdots \\ c_{n1} & \cdots & c_{nk} & \cdots & c_{nl} \end{bmatrix} \tag{14}$$

$$c_{jk} = [c_1^{jk}, \ldots, c_p^{jk}, \ldots, c_q^{jk}] \quad p = 1, 2, \ldots, q;\ j = 1, 2, \ldots, n;\ k = 1, 2, \ldots, l \quad (15)$$

In the above equations, X_j is the sub set of limit symptoms in accordance with the No.j position of fault. x_{ji} is the No.i symptom in the sub set of limit symptoms corresponding to the No.j position of fault. m is the number of symptoms in the sub set of limit symptoms. c_j is the sub set of trend symptoms corresponding to all the fault degree evaluation grades in the No.j fault position. c_{jk} is the sub set of trend symptoms corresponding to the No.k fault degree evaluation grades in the No.j fault position. c_p^{jk} is the No.p symptom in the sub set of trend symptoms corresponding to the No.k fault degree evaluation grades in the No.j fault position. q is the number of symptoms in the sub set of trend symptoms.

Limit symptoms $(x_{ji},\ j = 1, \ldots, n;\ i = 1, \ldots, m)$ are applied for fault location. The method to determine limit symptoms is to select nuclear parameters whose tendency are the same, and they are of different fault degrees in the same fault location as symptom characteristic parameters. Then compare current value of parameters with normal setting value, and take different threshold of deviation into consideration to determine the changing degree of parameters. If the trend of this symptom characteristic parameter with time is $v_{ji}(t)$, make its instantaneous value $v_{ji}(t_0)$ at certain time t_0 as its input when regarded as limit symptom.

Trend symptoms $(c_p^{jk}, j = 1, \ldots, n; k = 1, \ldots, l; p = 1, \ldots, q)$ are applied for fault degree evaluation. The method to determine trend symptoms is to select parameters whose changing slope is changing monotonically with the growing of fault degree as symptom characteristic parameters, and they are of different fault degrees in the same fault location. Then according to the slope of parameters in per unit time, determine the change degree of parameters. If the trend of this symptom characteristic parameter with time is $v_p^{jk}(t)$, make its instantaneous variation value at certain time t_0 as its input when regarded as trend symptom, which is shown in Eq. 16.

$$v_p^{jk\prime}(t_0) = \left.\frac{dv_p^{jk}(t)}{dt}\right|_{t=t_0} = \left.\frac{v_p^{jk}(t) - v_p^{jk}(t - \Delta t)}{\Delta t}\right|_{t=t_0} \quad (16)$$

The system use four symptoms for fault diagnosis as the limit characteristic parameters. Take these factors into consideration: the consistency of the symptoms from the same fault location, the diversity of the symptoms at the same fault location, and the difference of the symptoms in the different fault location. The final set of limit symptoms for the small break in the cold or hot leg are described in Eqs. 17 and 18.

$$X_1 = [x_{11}, x_{12}, x_{13}, x_{14}] = [0, T_{\text{cold}}, T_{\text{av}}, W_{\text{hot}}] \quad (17)$$

$$X_2 = [x_{21}, x_{22}, x_{23}, x_{24}] = [P_c, T_{\text{cold}}, 0, 0] \quad (18)$$

Then, take the derivative of parameters in Table 1 with respect of time, and the curves of operating parameters in different fault location and degrees can be obtained.

Finally, we select the variation rates of the pressure and water level of pressurizer as trend characteristic symptoms. Figure 4 reveals the varying curves of these main operating parameters.

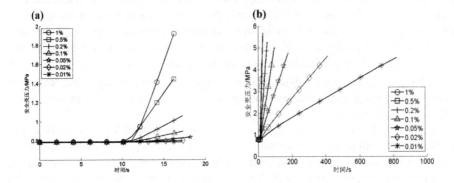

Fig. 4. (a) The curves of Pc in cold leg, (b) the curves of Pc in hot leg

According to Figs. 2 and 4, they reveal the change of trend symptom characteristic parameters with fault degree. Combined with Table 3, the reference value of each trend symptoms can be described in Eq. 19.

$$A_p^{jk} \in \{N_4, N_3, N_2, N_1, Z_0, P_1, P_2, P_3, P_4\} \quad j = 1, 2; \ k = 1, 2, 3, 4; \ p = 1, 2 \quad (19)$$

Therefore, the states of trend symptoms characteristic parameters are revealed in Table 4.

Table 4. The symptoms of trend type in different fault degrees

Serial number	Symptom characteristic parameters	Fault location	Fault degree			
			Leakage (VS)	Minimal break (S)	Tiny break (M)	Small break (L)
1	The variation of Pc (P'_c)	Cold leg	P_1	P_2	P_3	P_4
		Hot leg	P_1	P_2	P_3	P_4
2	The variation of Hz (H'_Z)	Cold leg	N_4	N_4	N_4	N_4
		Hot leg	N_1	N_2	N_3	N_4

According to Table 4, the sub set of trend symptoms can be described in Eq. 20.

$$c_{11} = c_{12} = c_{13} = c_{14} = [P'_c, 0]; \quad c_{21} = c_{22} = c_{23} = c_{24} = [P'_c, H'_Z] \quad (20)$$

According to the mentioned method [5], the limit and threshold of symptoms can be determined, which are shown in Table 5. Following the calculating principle of trapezoidal membership function, the confidence degree of limit symptoms can be determined.

Table 5. The thresholds and limits of signs of limit type

Serial number	Symptom characteristic parameter	Lower limit value	Lower threshold	Upper threshold	Upper limit value	Unit
1	P_c	–	–	1.30	3.81	MPa
2	T_{cold}	279.54	279.58	–	–	°C
3	T_{av}	297.5	300	–	–	°C
4	W_{hot}	83	94	–	–	Kg/s

In addition, according to nuclear operating principle, expert experience and relative fault model, combined with the improved AHP (analytic hierarchical process) [2], there is the diagnosis order that cold leg ranks before hot leg.

Therefore, these rules can be described in expression 21.

R^Y: if P_Z is $L(1) \wedge T_{cold}$ is $L(1) \wedge T_{av}$ is $L(0.9) \wedge W_{hot}$ is $L(0.9)$, then $(\text{LOCA}_{cold}, 0.95)$

else if P_Z is $L(0.8) \wedge P_c$ is $H(0.9) \wedge T_{cold}$ is $L(0.65)$, then $(\text{LOCA}_{hot}, 0.9)$

$$(21)$$

Therefore, rules can be described in expression 22 and 23.

R_1^d: if $\text{LOCA}_{cold}(0.6) \wedge P_c'$ is $P_4(0.8)$, then $(L, 0.95)$

else if $\text{LOCA}_{cold}(0.6) \wedge P_c'$ is $P_3(0.65)$, then $(M, 0.9)$

else if $\text{LOCA}_{cold}(0.6) \wedge P_c'$ is $P_2(0.9)$, then $(S, 0.95)$

else $\text{LOCA}_{cold}(0.6) \wedge P_c'$ is $P_1(0.65)$, then $(VS, 0.95)$

$$(22)$$

R_2^d: if $\text{LOCA}_{hot}(0.6) \wedge P_c'$ is $P_4(0.8) \wedge H_Z'$ is $N_4(0.8)$, then $(L, 0.95)$

else if $\text{LOCA}_{hot}(0.6) \wedge P_c'$ is $P_3(0.6) \wedge H_Z'$ is $N_3(0.6)$, then $(M, 0.95)$

else if $\text{LOCA}_{hot}(0.6) \wedge P_c'$ is $P_2(0.9) \wedge H_Z'$ is $N_2(0.9)$, then $(S, 0.95)$

else $\text{LOCA}_{hot}(0.6) \wedge P_c'$ is $P_1(0.6) \wedge H_Z'$ is $N_1(0.6)$, then $(VS, 0.95)$

$$(23)$$

3 Application of Early-Warning System Based on the Method of Hierarchical and Multidimensional Fault Recognition

In order to verify the correctness and validity of this system, lots of fault experiments have been carried out. Presently, 14 groups of experimental data are listed, which contains 2 kinds of different fault location and 7 kinds of different fault degree.

As shown in Table 6, the results are consistent with the fault condition inserted into the nuclear model, namely, the accuracy rate of results is 100%. Furthermore, it is proved that the validity of this early-warning system for fault diagnosis.

Table 6. The examples of hierarchical and multidimensional fault identification of SB-LOCA in primary loop

Number	Fault in nuclear model	Real time value of symptoms							Results of fault recognition and its confidence degree
		$P_Z/$ MPa	$P_c/$ MPa	$T_{cold}/°$ C	$T_{av}/°$ C	$W_{hot}/$ Kg/s	$H'_Z/\%/$ s	$P'_c/$ 10^{-3} MPa/s	
1	(cold, 0.01%)	14.78	–	278.58	294.2	82.2	–	0.008	{LOCA$_{cold}$(0.95), S(0.834)}
2	(cold, 0.02%)	14.7	–	278.6	293.5	83	–	2.347	{LOCA$_{cold}$(0.95), S(0.9)}
3	(cold, 0.05%)	14.62	–	278.7	293.2	83.1	–	8.084	{LOCA$_{cold}$(0.95), S(0.9)}
4	(cold, 0.1%)	14.7	–	278.54	293.6	82.4	–	19.821	{LOCA$_{cold}$(0.95), M(0.9)}
5	(cold, 0.2%)	14.66	–	278.62	294.3	82.2	–	46.783	{LOCA$_{cold}$(0.95), M(0.9)}
6	(cold, 0.5%)	14.7	–	278.65	295.3	82.1	–	123.88	{LOCA$_{cold}$(0.95), M(0.9)}
7	(cold, 1%)	14.71	–	278.82	297.8	81.4	–	238.374	{LOCA$_{hot}$(0.931), L(0.949)}
8	(hot, 0.01%)	15.25	4.5	279.15	–	–	−1.709	0.051	{LOCA$_{hot}$(0.765), S(0.95)}
9	(hot, 0.02%)	14.9	4.9	279.2	–	–	−1.81	2.277	{LOCA$_{hot}$(0.9), S(0.95)}
10	(hot, 0.05%)	14.96	4.7	279.08	–	–	−1.667	8.462	{LOCA$_{hot}$(0.9), S(0.903)}
11	(hot, 0.1%)	14.88	4.95	279.21	–	–	−1.921	19.576	{LOCA$_{hot}$(0.9), M(0.927)}
12	(hot, 0.2%)	14.82	5.28	279.38	–	–	−2.079	47.207	{LOCA$_{hot}$(0.9), M(0.95)}
13	(hot, 0.5%)	14.75	5.6	279.51	–	–	−2.049	121.780	{LOCA$_{hot}$(0.9), M(0.95)}
14	(hot, 1%)	14.73	4.26	279.56	–	–	−2.444	246.944	{LOCA$_{hot}$(0.765), L(0.95)}

4   Conclusion

A new method of hierarchical and multidimensional fault recognition for SB-LOCA is proposed, and its relative intelligent early-warning system has been constructed. The application shows that this system is able to rapidly obtain the correct diagnosis result of fault location and degree, namely, it can recognize the faults of tiny degree timely and effectively. Actually, in the process of fault recognition in nuclear power plants, this expert database of the intelligent early-warning system can be considered as original rule database. On this basis, according to the operating condition and data collection, it is able to learn and modify constantly, and furthermore improve the accuracy of fault recognition in the actual situation.

Acknowledgements. This work was partially supported by National Natural Science Foundation of China, Grant No. 61503237; Shanghai Natural Science Foundation (No. 15ZR1418300); Shanghai Key Laboratory of Power Station Automation Technology (No. 13DZ2273800).

References

1. Gang, Z., Li, Y.: Advance in study of intelligent diagnostic method for nuclear power plant. J. At. Energy Sci. Technol. **S1**, 92–99 (2008)
2. Jianbo, L.: Research on Alarm Triggered Intelligent Fault Diagnosis Expert System of Nuclear Power Plant. Shanghai University of Electric Power, Shanghai (2015)

3. Xiaolong, W., Qi, C., Xiaoqi, Z.: Study on diagnostic technique for nuclear power plant primary coolant circuit system break characteristics based on multi-classification SVM. J. At. Energy Sci. Technol. **03**, 462–468 (2014)
4. Liangyu, M., Yongguang, M., Bingshu, W.: A new approach to diagnose variable-degree faults under different operating points for high-pressure feedwater heater system. J. Proc. CSEE **02**, 115–121 (2010)
5. Zhijie, Z., Jianbo, Y., Changhua, H.: Confidence Rule Base Expert System and Complex System Modeling. Science Press, Beijing (2011)
6. Hong, Q., Jianbo, L., Weixiao, J.: Research on alarm triggered fault-diagnosis expert system for U-shaped tube breaking accident of steam generators. J. Nucl. Power Eng. **36**(1), 98–103 (2015)

The Early Warning System of Nuclear Power Station Oriented to Human Reliability

Shuai Ren[1(✉)] and Hong Qian[1,2]

[1] Shanghai University of Electric Power, Shanghai, China
15151827597@163.com
[2] Shanghai Key Laboratory of Power Station Automation Technology,
Shanghai, China

Abstract. In order to improve the reliability of nuclear power plant operators in the face of the abnormal operation of nuclear island, this paper studies the early-warning system of nuclear power plant through the abnormal operation parameters of nuclear island. In this paper, the object studied about is the fault of the passive equipment of the reactor. After the reaction shutdown, operators can take the emergency measures in an accurate and timely manner. The early warning system will show how the fault is expected in the operational measures, so that operators can do respond to prepare. The early warning system studied in this paper is applied to nuclear power simulation system. Through the research on abnormal operation simulated by nuclear power simulation system, the results show that the early-warning system can improve human reliability of nuclear power plant in the face of abnormal operation.

Keywords: Early warning system · Human reliability · Belief rule base

1 Preface

The passive safety system is designed to improve the safety of nuclear power plant. However, abnormality of the equipment of the passive safety system can have an impact on normal operation condition, and even lead to reactor abnormal shutdown. Therefore, it is necessary to design the early warning system of the passive system to avoid the normal maintenance of the passive system in order to avoid the unnecessary consequence of the abnormality of the passive equipment.

Paper [1] describes the traditional nuclear power early warning technology from the characteristics of conditions of early warning system, and puts forward the realization of the early warning system combined with simulation model. In paper [2], a diagnostic system with alarm condition as the trigger condition is proposed, and the belief rule base of the expert system is studied. The mathematical model between the threshold level of the fault degree and the belief rule is established. Paper [3] studied about human error event through HCR + THERP method, and disadvantage and advantage of both two methods is described.

In this paper, the combination of early warning set and fault identification is proposed. When the warning system is triggered, the system will locate the fault quickly through the expert system belief rule base. The early warning system enhances the

© Springer Nature Singapore Pte Ltd. 2017
K. Li et al. (Eds.): LSMS/ICSEE 2017, Part III, CCIS 763, pp. 206–216, 2017.
DOI: 10.1007/978-981-10-6364-0_21

reliability of human factors from the knowledge-based level, as when the fault occur, early warning system will locate and identify the fault, then show how the fault is expected in the operational measures on the interface. Through the test platform the faults were artificially inserted. In this way, timeliness and reliability of passive equipment early warning system is verified.

2 Design of Early Warning System

For achieving the timely position and operational guidance information when there is a failure occur, the early warning system is designed to be time periodically triggered. The early warning parameter values are obtained from the real-time database firstly, then they are compared with the warning threshold set, and the warning value is output. Only when the early warning system is triggered, the early warning system will diagnose the current state of nuclear power operation based on the belief rule base. The abnormal information is used to achieve fault location and identification.

2.1 Build Belief Rule Base

Extract Fault Symptoms
In the implementation of the belief rule base, the first step should be determining the set of fault, assuming that the system may occur n types of faults, then all the faults of the collection called the fault set Y, can be described as:

$$Y = \{y_1, y_2, y_3, \ldots y_i, \ldots y_n\} \tag{1}$$

The fault of the nuclear power plant equipment will lead to the relevant system parameters exceed a certain limit and raise an alarm, so according to fault signs at the alarm moment to build the belief rule base. If y_i represents one of the fault types, the set of fault symptoms corresponding to the y_i fault type is:

$$X_j = \{X_{j1}, X_{j2}, X_{j3}, \ldots X_{ji} \ldots X_{jm}\} \tag{2}$$

Representation of Rules in the Confidence Rule Base
In the belief rule base, the set of faults and the set of symptoms exist in the form of rules. The rule base consists of prerequisites and conclusions. The prerequisites are composed of logical concatenations. When the prerequisite is true, the conclusion is obtained. The basic form is generally: "$P \rightarrow Q$" or "*IF P THEN Q*". However, considering the complexity of the nuclear power system and the fuzzy uncertainty of the fault symptom itself, a belief rule based on distributed confidence is used in the system, which can be described as follows:

$$R : IFA_1(\varepsilon_1) \wedge A_2(\varepsilon_2) \wedge A_3(\varepsilon_3) \wedge \cdots A_i(\varepsilon_i) \cdots A_m(\varepsilon_m)$$
$$THEN\{(D_1, \beta_1), (D_2, \beta_2), (D_3, \beta_3) \cdots (D_j, \beta_j), \cdots (D_n, \beta_n)\} \tag{3}$$

where A_i is the prerequisite numbered with i; εi is the confidence of A_i; m means a total number of m prerequisites; D_i is the conclusion numbered with i; βi is the confidence of D_i; n is the total number of conclusions.

Set the Variables of Belief Rule Base

In the process of building the belief rule base, the setting of the belief rule variable includes the setting of the rule confidence and the setting of the threshold value and limit value of the symptom, in which the rule confidence is divided into symptom confidence and fault confidence. The threshold and limit value determines the accuracy of the diagnose results.

2.2   Build Warning Set

The early warning system takes the form of time periodic triggered. Elements in warning parameter set show the characteristic that it can reflect the fault information before the alarm system is alarmed in the event of a fault. When the parameter crosses the warning threshold value in the set of early warning threshold, it is assumed that there is a risk of a corresponding fault in the operation of the nuclear power.

Construct Set of Early Warning Parameters

The establishment of the early warning system should correspond to a certain set of faults, that is, the ability to identify and locate faults within the scope of the fault set, the fault set is as follow:

$$Y = \{y_1, y_2, y_3, \ldots y_i, \ldots y_n\} \tag{4}$$

Fault corresponding to the set of symptom:

$$X_j = \{X_{j1}, X_{j2}, X_{j3}, \ldots X_{ji} \ldots X_{jm}\} \tag{5}$$

Warning parameter set:

$$Z = \{z_1, z_2, \ldots z_i \ldots z_k\} = X_1 \cup X_2 \cup \ldots \cup X_n \tag{6}$$

Construct Set of Warning Threshold Value

The setting of the threshold set refer to the threshold value setting method of belief rule base. For each symptom value, there are corresponding lower limit value, lower threshold value, upper threshold value and upper limit value, respectively for:

$$v_{\min i}, v_{Li}, v_{Hi} \text{ and } v_{\max i}$$

Corresponding to the early warning parameters $z_i \in Z$, the threshold is expressed as the corresponding set of forms.

The upper threshold value for early warning is:

$$\{\delta_1, \delta_2, \ldots \delta_i \ldots \delta_m\}$$

The lower threshold value for early warning is:

$$\{\delta'_1, \delta'_2, \ldots \delta'_i \ldots \delta'_m\}$$

According to the characteristics of threshold and limit value, in order to be able to obtain early fault information, to achieve early warning, set the threshold value as follow:

$$\delta_i = \frac{v_{Hi} + \theta * v_{\max i}}{1 + \theta} \tag{7}$$

$$\delta'_i = \frac{v_{Li} + \theta * v_{\min i}}{1 + \theta} \tag{8}$$

where θ is the adjusting parameters of early warning threshold value.

Warning Triggering Rules

During the operation of the nuclear power plant, the early warning system operates in a fixed time period triggered form, corresponding to the warning parameter $z_i \in Z$, there is a unique warning value α_i.

$$\alpha_i = \begin{cases} 0 & \delta'_i < Z_i < \delta_i \\ 1 & else \end{cases} \tag{9}$$

The system warning value takes the OR operation of the warning value corresponding to each warning parameter, that is, the true value of any one of the early warning values would make the system early warning value "true", that is, the warning information is issued.

$$\alpha = \bigvee_{i=1}^{m} \alpha_i = \alpha_1 \vee \alpha_2 \vee \alpha_3 \ldots \vee \alpha_m \tag{10}$$

3 Construct Passive Equipment Early Warning System

The passive reactor core cooling system consists of two parts, the passive core waste heat exhaust system and the passive safety injection system. Two core makeup tanks (CMT) can provide a larger injection stream over a longer period of time. The box of core makeup tanks is filled with low temperature boron water. In the main steam pipe rupture accident, the water tank within the concentrated boron water can provide adequate shutdown margin [4].

3.1 Construct Belief Rule Base for Passive Equipment Early Warning System

For the research and realization of the passive safety system early warning system, three typical faults are selected as the fault types of the early warning system.

1. Passive core makeup tank (CMT) outlet isolation valve misplaced
2. Passive core makeup tank inlet pipeline rupture
3. Pressure container Directly Vessel Injection (DVI) pipeline rupture

Corresponding to y_1, y_2, y_3 respectively.

Extract the Main Symptoms of Fault
Firstly, the mathematical model [5, 6] is established by P_Z, and the operation mechanism of the regulator is analyzed, a simple model of the regulator pressure is obtained.

$$P_Z = k_1 \frac{dW_{su}}{dt} + k_2 W_{su} + P_{SG} - \rho_M g \Delta H_Z \tag{11}$$

where k_1, k_2 is constants; W_{su} is the flow rate of Pressurizer Surge Line (*kg/s*); ρ_M is density of W_{su}; ΔH_Z is the change of water level in Pressurizer.

The model can be written as a simple transformation:

$$W_{su} = f_1 \left(P_Z, \frac{dW_{su}}{dt}, P_{SG}, \Delta H_Z \right) \tag{12}$$

Secondly, the fault model is analyzed by mechanism, and the model object is a primary loop reactor coolant system which contains a passive safety injection system. The simple diagram is as follows (Fig. 1):

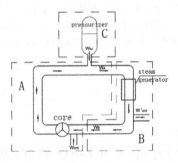

Fig. 1. Diagram of primary loop and CMT

The study system is divided into three subsystems: system C including the Pressurizer part, system A including the pressure vessel and DVI pipeline, including the coolant system hot leg, and B system containing the steam generator, CMT inlet pipeline and coolant system cool leg [7].

The mass conservation equation of the systems:

$$\begin{cases} \frac{dM_C}{dt} = W_{su} \\ \frac{dM_A}{dt} = W_A - W_B + W_{CMT} - W_{su} \\ \frac{dM_B}{dt} = W_B - W_A - W'_{CMT}A \end{cases} \quad (13)$$

The energy conservation equation of the systems:

$$\begin{cases} \frac{d(M_C h_C)}{dt} = W_{su} h_{su} + q_C + \sum_C V \frac{dp}{dt} \\ \frac{d(M_A h_A)}{dt} = -W_{su} h_{su} + q_A + W_A h_A + \sum_A V \frac{dp}{dt} + W_{CMT} h_{CMT} \\ \frac{d(M_B h_B)}{dt} = -W_B h_B - q_B + \sum_B V \frac{dp}{dt} - W'_{CMT} h'_{CMT} \end{cases} \quad (14)$$

W_{su} is the flow rate of the Pressurizer Surge Line; W_{CMT} is the flow rate of CMT outlet pipeline; W'_{CMT} is the CMT flow rate of inlet pipeline; W_A is the flow rate of coolant flow from the system A into the system B; W_B is the flow rate of coolant flow from the system B into the system A; M_B, M_C for the coolant quality of each system; h_A, h_B, h_C is the enthalpy of each system's coolant; q_A, q_B, q_C for the system inner heat source.

The transformation of above Eqs. (13) and (14) are:

$$W'_{CMT} h'_{CMT} - W_{CMT} h_{CMT} = V \frac{dP_A}{dt} + W_A h_A - W_B h_B - W_{su} h_{su} - h_{avg} M$$
$$+ \frac{M_A(dh_A)}{dt} + \frac{M_B(dh_B)}{dt} + q_A - q_B \quad (15)$$

where P_A is the pressure of coolant system; h_{avg} is the average enthalpy of coolant system.

We can describe Eq. (15) as

$$W'_{CMT} h'_{CMT} - W_{CMT} h_{CMT} = f_2 \left(\frac{dP_A}{dt}, W_{su}, \frac{dh_A}{dt}, \frac{dh_B}{dt} \right) \quad (16)$$

Infer form Eqs. (12) and (15), equation can be described as:

$$W'_{CMT} h'_{CMT} - W_{CMT} h_{CMT} = f \left(P_Z, P_{SG}, \Delta H_Z, \frac{dW_{su}}{dt}, \frac{dP_A}{dt}, \frac{dh_A}{dt}, \frac{dh_B}{dt} \right) \quad (17)$$

Combined with the fault test of nuclear power simulation model, the main symptoms of each fault are identified as below:

Fault y_1: passive core makeup tank isolation valve misplaced
The main symptom: steam generator pressure P_{SG}.
Failure y_2: passive reactor core makeup tank inlet pipeline rupture
The main symptom: Pressurizer pressure P_Z.
Failure y_3: pressure container direct vessel injection pipeline rupture
The main symptom: Pressurizer pressure P_Z.

Extract Associated Symptom of Fault

With the fault inserted into the nuclear power model, a series of experiments were made. Associated signs of fault is identified respectively:

Fault y_1: passive core makeup tank isolation valve misplaced

The associated symptoms are: direct vessel injection pipeline temperature T_{CMT}, passive core makeup tank internal temperature T_{in}.

$$X_1 = \{P_{SG}, T_{CMT}, T_{in}\}$$

Fault y_2: passive reactor core makeup tank inlet pipeline rupture

The associated symptoms are: passive reactor core makeup tank inlet pipeline temperature T'_{CMT}, container pressure P_C.

$$X_2 = \{P_Z, T'_{CMT}, P_C\}$$

Fault y_3: pressure container direct vessel injection pipeline rupture

The associated symptoms are: direct vessel injection pipeline temperature T_{CMT}, passive reactor core makeup tank inlet pipeline temperature T'_{CMT}.

That is

$$X_3 = \{P_Z, T_{CMT}, T'_{CMT}\}$$

Threshold and Limit Value of Belief Rule Base

According to the method of probabilistic analysis of data in paper [8], the value of the symptom parameter is obtained through multiple experiment feedback of nuclear power model and combined with experts' experience, the final set of the threshold and the limit value are as follows (Table 1).

Rules in Belief Rule Base

(1) passive core makeup tank isolation valve misplaced rule:
   IF steam generator pressure drop (0.85)
   AND direct vessel injection pipeline temperature drop (0.9)
   AND passive core makeup tank internal temperature rise (0.85)
   THEN passive core makeup tank isolation valve misplaced (0.9)

Table 1. Threshold and limit value of symptoms

	v_{mini}	v_{Li}	v_{Hi}	v_{maxi}
P_{SG} (MPa)	5.23	5.395	–	–
P_Z (MPa)	14.69	15.185	–	–
P_C (MPa)	0.766	0.78	1.3	3.81
T_{CMT} (°C)	30.3	32.31	60.448	93.915
T'_{CMT} (°C)	146.3	153.45	197.2	265.8
T_{in} (°C)	–	–	37.2	43.7

(2) passive reactor core makeup tank inlet pipeline rupture rule:
IF pressurizer pressure drop (0.85)
AND passive reactor core makeup tank inlet pipeline temperature rise (0.9)
AND container pressure rise (0.8)
THEN passive reactor core makeup tank inlet pipeline rupture (0.9)
(3) pressure container direct vessel injection pipeline rupture rule:
IF pressurizer pressure drop (0.85)
AND direct vessel injection pipeline temperature drop (0.85)
AND passive reactor core makeup tank inlet pipeline temperature rise (1)
THEN pressure container direct vessel injection pipeline rupture (0.9).

3.2 Construct Early Warning Set of Passive Equipment Early Earning System

Set of Early Warning Parameters

Through the above research, the set of symptoms corresponding to the fault of three passive safety devices has been determined. According to the establishment rules of the warning parameter set, the set of symptoms of each fault is taken together to get the set of warning parameters:

$$Z = \{Z_1, Z_2, \ldots Z_i \ldots Z_m\} = X_1 \cup X_2 \cup X_3$$
$$= \{P_{SG}, P_Z, P_C, T_{CMT}, T'_{CMT}, T_{in}\} \tag{18}$$

Set of Early Warning Threshold Value

The threshold value of the warning parameter is determined by the corresponding threshold and limit value of the symptoms, which is determined by the Eqs. (7) and (8).

Considering the condition of some other factors and experiments of inserting fault into the simulation model, select θ equal with 4.

Calculate the threshold value (Table 2):

Table 2. Threshold value of early warning parameters

	P_{SG}	P_Z	P_C	T_{CMT}	T'_{CMT}	T_{in}
δ'	5.263	14.789	0.768	30.942	147.73	–
δ	–	–	3.308	87.221	252.8	42.4

3.3 System Case Verify

In the nuclear power model, the passive core makeup tank outlet isolation valve misplaced and the real-time value of each warning parameter is obtained. After the fault is inserted, at $t = 240\ s$, the warning parameter crosses the warning threshold and the warning system sends an early warning signal (Table 3).

Table 3. Real time value of simulation system at t = 240 s

P_{SG}	P_Z	P_C	T_{CMT}	T'_{CMT}	T_{in}
5.2545	15.5994	0.7693	31.05	157.27	41.72

According to the threshold value in the belief rule base, the confidence degree of each evidence is obtained according to the trapezoidal membership formula [9] (Table 4).

$$\alpha_i = \max\left(0, \frac{Z_i - v_{L(H)i}}{\left|v_{L(H)i} - v_{\min(\max)i}\right|}\right) \tag{19}$$

Table 4. Evidence believe value of parameters

α_1	α_2	α_3	α_4	α_5	α_6
0.85125	0	0.75774	0.73685	0	0.60280

Finally, the confidence of the three fault types is calculated by the rule confidence [1] in the belief rule base (Table 5).

$$\gamma = \beta \prod_{i=1}^{m} \left(1 - \max(0, \varepsilon_i - \alpha_i)\right) \tag{20}$$

Table 5. Confidence of conclusions

Fault	Passive core makeup tank isolation valve misplaced	Passive reactor core makeup tank inlet pipeline rupture	Pressure container direct vessel injection pipeline rupture
Confidence	0.60335	0.02930	0.17110

It can be seen that the final fault identification is consistent with the type of fault inserted, and the difference in confidence is significant in the fault distinction, which verifies the effectiveness of the early warning system.

4   Analysis and Evaluation of Human Reliability

The early warning system enhances the reliability of human factors from the knowledge-based level, and transform human action into a rule-based level.

The human reliability analysis is based on Human Cognitive Reliability (HCR) method. HCR is mainly used to analysis the cognitive reliability of human after an initiator event. HCR is based on Rasmussen cognitive process SRK theory. It divided human cognitive process into three levels: skill-based level, rule-based level and knowledge-based level.

There are two forms of fitting function in HCR model, one (also the fitting function used in this paper) is described as below:

$$P(t) = \exp\left\{-\left[\frac{\left(\frac{t}{T_{1/2}}\right) - C_{ri}}{C_{\eta i}}\right]^{\beta i}\right\}, \ \frac{t}{T_{1/2}} \geq C_{ri} \tag{21}$$

Where $T_{1/2}$ is the median of time the operator needed to finish a diagnosis or decision action; t is the time window of operator finish some actions; $P(t)$ is the non-response probability in given time t. C_{ri}, $C_{\eta i}$, β_i can be obtained from the table provided by IAEA.

For parameter t and $T_{1/2}$, each of t and $T_{1/2}$ is different from events, and also based on simulators and expert decisions. Several typical condition of $(t/T_{1/2})$ is chosen to analysis the Human Reliability of early warning system.

When faced with event, some actions operator should take may be a knowledge-based level, which can lead to a human error in a big probability. Since the early warning system can obtain fault information and identify the fault in an early time, then operational measures will be shown to help operator makes action. In this way, early warning system can enhance the human reliability, transform the human cognitive process and action from knowledge-based level to a rule-based level (Table 6).

Table 6. P(t) calculated in IAEA data source

Level	$t/T_{1/2}$					
	P(t)					
	1.5	1.8	2.1	2.4	2.7	3.0
Knowledge-based level	0.2993	0.2258	0.1726	0.1332	0.1037	0.0812
Rule-based level	0.2373	0.1552	0.1025	0.0683	0.0458	0.0309
Human reliability promotion (%)	20.71	31.26	40.61	48.72	55.83	61.94

We can see from the tables above, early warning system decrease $P(t)$ (the non-response probability in given time t). The early warning system enhance the human reliability in a way that changed the human action level when faced with initiator event.

5   Conclusion

The realization of the passive equipment early warning system is obtained through the combination of the nuclear power simulation model and several times of experiment. As a system to improve the safety of nuclear power, the early warning system is not only applicable to passive safety equipment. For the realization of early warning system on different equipment, study and research should be taken in a further way.

Therefore, the study of passive equipment early warning system, for the improvement of nuclear human reliability and power security, has a positive meaning.

Acknowledgements. This work was partially supported by National Natural Science Foundation of China, Grant No. 61503237; Shanghai Natural Science Foundation (No. 15ZR1418300); Shanghai Key Laboratory of Power Station Automation Technology (No. 13DZ2273800).

References

1. Wang, L., Cheng, J., Zhou, Z.: Design of early warning system for passive safety nuclear power plant. Power Equip. **29**(5), 377–380 (2015)
2. Luo, J.: Research on Alarm Triggered Intelligent Fault Diagnosis Expert System of Nuclear Power Plant. Shanghai University of Electric Power, Shanghai (2015)
3. Zhang, L., Huang, S., Huang, X.: Application and analysis model of human event based on THERP + HCR. Nucl. Power Eng. **24**(3), 273–275 (2003)
4. Lin, C., Yu, Z.: Passive Safety Advanced Nuclear Power Plant AP1000. Atomic Energy Press, Peking (2008)
5. Tsai, C.-W., Shih, C., Wang, J.-R., et al.: Parametric analysis of pressure control based on feedback dissipation and backstepping. IEEE Trans. Nucl. Sci. **57**(03), 1577–1588 (2010)
6. Chen, Y., Qiu, H., Fu, J.: Fault diagnosis for nuclear power plant regulator pressure transient fluctuations. Nucl. Power Eng. **35**(3), 92–95 (2014)
7. Chen, S.: Simulation Research on Passive Safety Injection System of Nuclear Power Plant. Harbin Engineering University, Harbin (2008)
8. Ma, C.: Research on Intelligent Fault Diagnosis Expert System of Passive PWR based on Pressurizer Operating State. Shanghai University of Electric Power, Shanghai (2017)
9. Qian, H., Luo, J., Jin, W., Wang, D., Zhou, J.: Research on alarm triggered fault-diagnosis expert system for U-shaped tube breaking accident of steam generators. Nucl. Power Eng. **36**(1), 98–103 (2015)

Research on Energy Interconnection Oriented Big Data Sharing Platform Reference Architecture

Wei Rao[1(⊠)], Jing Jiang[1], Ming Yang[2], Wei Peng[2], and Aihua Zhou[1]

[1] Advanced Computing and Big Data Laboratory,
Global Energy Interconnection Research Institute,
State Grid Corporation of China, Beijing 102209, China
{raowei,jiangjing,zhouaihua}@geiri.sgcc.com.cn
[2] Shanghai Electric Power Corporation Information Communications Branch,
Shanghai 200122, China
{yangm,pengw}@sh.sgcc.com.cn

Abstract. In order to provide a unified data sharing service support for energy interconnection business, big data application development and operation, the large power data sharing platform for energy interconnection will integrate data storage, data calculation, data analysis and data service functions. This platform will not only invigorate the power data assets, bringing enormous economic benefits, but also promote economic restructuring and energy saving and emission reduction. This paper analyzes the development requirements of energy interconnection, and then designs the general framework, functional framework, technical framework and deployment framework of the power big data platform which is suitable for the energy interconnection. Finally, this paper lists the application flow and strategy of the big data platform, which is under the typical scenarios of transmission monitoring and status assessment real-time analysis and distribution network planning off-line analysis.

Keywords: Big data · Energy interconnection · Reference architecture · Sharing platform

1 Introduction

Since the concept of big data has been proposed, it is attracted considerable attention by the information service providers, the industry, the government and international organizations. They have lain out of the big data strategy through internal research and development, purchase, technology integration and other means, in order to achieve a unified, efficient processing of structured, semi-structured and unstructured data. China's State Council issued the "action to promote the development of big data platform" [1] on September 5, 2015, pointing out that "big data is a kind of data set taking the large capacity, multi type, fast access speed, high application value as main

© Springer Nature Singapore Pte Ltd. 2017
K. Li et al. (Eds.): LSMS/ICSEE 2017, Part III, CCIS 763, pp. 217–225, 2017.
DOI: 10.1007/978-981-10-6364-0_22

features. It is rapidly developing into a new generation of information technology and service formats by discovering new knowledge, creating new values, and enhancing new abilities from data acquisition, storage and correlation analysis".

In order to adapt to the big data and Internet technology development opportunities, State Grid Corporation of China proposed the development strategy of "Global Energy Interconnection, GEI" in early 2015. The GEI uses ultra high voltage power grid as a channel to transport clean energy, and it aims at building a strong global grid of smart grids. By applying new technologies, such as big data, cloud computing, Internet of things, mobile internet, intelligent wear, computer vision and other technologies, the GEI is able to enhance those abilities, such as intelligent perception, real-time evaluation of power transmission, transformation equipment, equipment precision positioning and automatic failure warning, and is able to optimize control mode of large grid distribution and centralized coordination [2]. The GEI combines the Internet with renewable energy sources such as wind and solar energy to achieve easy energy sharing. It also connects hundreds of millions of equipment, machines and systems from energy production side, energy consumption side and energy transmission side. Through the integration of operational data, weather data, meteorological data, grid data, electricity market data, etc., it can open up and optimize the operational efficiency of energy production and energy consumption by big data analysis, load forecasting, power generation forecasting, and machine learning [3].

Power big data in the Global Energy Interconnection environment emphasizes the diversity, complexity and real-time of multi-source data processing [4]. In order to achieve above goal, it needs to design a set of complete and adaptive power big data architecture based on large scale distributed storage and processing, and to build the storage structure and computing engine which is suitable for real time data analysis and processing [5].

2 Basic Architecture System

2.1 Power Big Data General Structure for Global Energy Interconnection

In order to analyze the general structure of power big data, it is needed to comprehend the main modules of the big data platform from the life cycle of data. The overall process of power big data analysis contains four stages: data acquisition, data management, data computing and business application. As the data computing is the most expensive one of the four phases, a hybrid computing architecture is introduced, which is composed of real-time computing, batch computing and flow computing [6]. Computing tasks are decomposed according to the structure characteristics of multi-source data and application requirements, which fully reflects the flexible characteristics of big data computing system. A general structure covering electric power industry of the Global Energy Interconnection business requirement and scenarios is shown in Fig. 1.

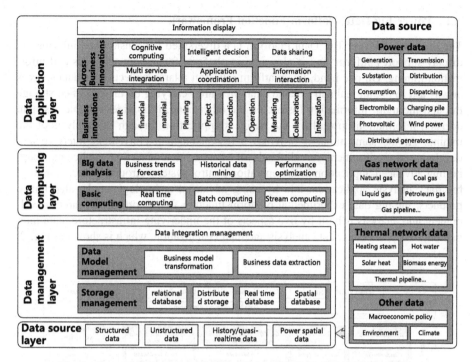

Fig. 1. Power big data general structure for Global Energy Interconnection

(a) **Data source layer:** By integrating the multi energy data, it exchanges with the data collected from acquisition points and the data transmitted by system, including power grid data, gas grid data, heat grid data, environmental climate data, etc. Those multi data source forms the ocean of data classified into structured data, videos, pictures, texts, logs, and other unstructured data, hypo-real-time data and GIS data, which is the data base and source of big data technology application.

(b) **Data management layer:** It carries on the expansion to the relational database of structured data, and uses distributed storage to improve unstructured database, and uses real-time database to upgrade hypo-real-time database, and uses spatial database to inherit and expand the GIS data of power grid [7]. Based on the SG-CIM model, the data of different scenarios is unified, and finally the unified management of data integration is realized.

(c) **Data computing layer:** Through real-time calculation for real-time data, batch calculation for mass data, and stream processing for streaming data, it provides integrated services, intelligent decision, information display and other middleware services to meet the needs of the upper application [8].

(d) **Data application layer:** By providing interactive analysis, decision reference, trend prediction, value mining, and other typical application of big data, it serves the construction of smart grid and Global Energy Interconnection.

2.2   Power Big Data Platform Function Framework

The position of Power Big Data Platform is a diversified data fusion and information display support platform, and is also a basic platform for big data analysis and information mining. On one hand, this platform relies on a large collection of data sources downward to obtain real and reliable data, then after carrying on the classified storage through the multi information fusion; on the other hand, it provides reliable assistant decision information upward for the business application through the analysis of the mining. At the same time, it provides a reasonable and convenient interface and development environment for all kinds of energy interconnection application, supporting advanced application in the operation control system. Around the ecosystem of big data analysis application, and by using advanced technology, tools, algorithms, products, the big data experimental platform function framework for research, development and analysis is built by five aspects of the underlying infrastructure, data integration, data processing, data analysis and data visualization, which is shown in Fig. 2.

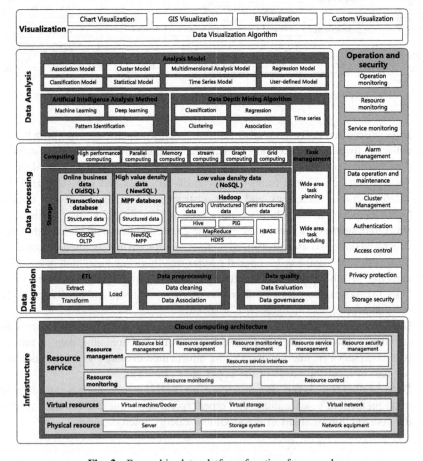

Fig. 2. Power big data platform function framework

(a) **Data Integration:** Through improving existing ETL, file adapter and other acquisition techniques, combining the distributed high speed and high reliable data crawling or acquisition and other new acquisition technology, it completes data analysis, conversion and reprint of the existing data center's data, business data, terminal data and other massive data.

(b) **Data Storage:** By improving the existing relational database, data warehouse storage technologies, fusing reliable distributed file system technology and big data storage technology with high efficiency and low-cost, it completes the storage of big data, then after establishing the correlation index, management and calling service. Usually unstructured data stores in the distributed file system; semi structured data stores column database or key database; structured data stores line storage database; data of high real-time performance and high computational performance stores in memory database and real-time database [9].

(c) **Data Computing:** By improving the existing query calculation, batch calculation, memory calculation technologies, integrating real-time stream computing, parallel computing and other new computing technologies, it supports big data analysis and mining applications.

(d) **Data Analysis:** By upgrading distributed ability of existing analysis platform, integrating open source analysis mining tools and distributed algorithms library, it achieves that application of real-time and off-line analysis of business system is supported by big data analysis modeling, digging and display.

(e) **Data Visualization:** Through the data graph, image and animation of display technologies, it provides services for big data applications by report, query, analysis, forewarning, search, data opening, service interface and other forms.

(f) **Security:** By improving the data destruction, transparent encryption and decryption, it solves a number of security issues such as authentication, authorization, and input validation, which is produced from data collection, storage, analysis, application and other processes in big data environment [10].

(g) **Maintenance:** By centralized monitoring and management of big data platform service cluster, using configuration expansion technology, it solves the management problems of large-scale clusters of hardware and software, and can dynamically configure and adjust the system function of big data platform.

2.3   Power Big Data Platform Technical Framework

Based on Hadoop, Spark, Stream framework of a high degree of integration, depth optimization, Power big data core platform can achieve high performance computing with high availability. It integrates Tableau, Pluto, R language environment to achieve statistical data analysis and data mining capabilities in data analysis. It uses Ganglia to achieve cluster monitoring, service monitoring, node monitoring, performance monitoring, alarm monitoring and other monitoring in monitoring management [11]. It builds visual display module based on GIS, Flash, ECHART, HTML5 and so on in visualization. The technical framework is shown in Fig. 3.

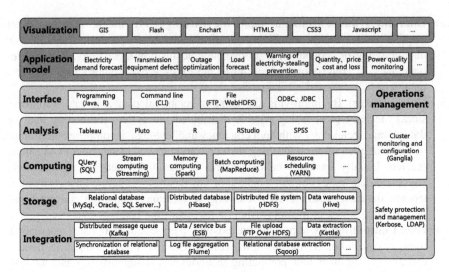

Fig. 3. Power big data platform technical framework

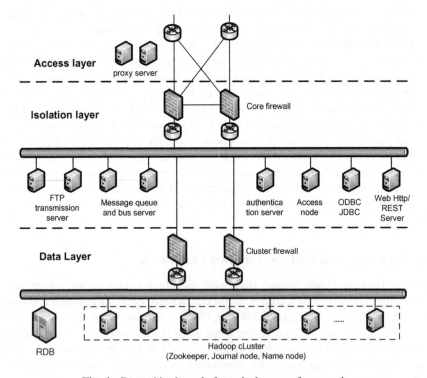

Fig. 4. Power big data platform deployment framework

2.4   Power Big Data Platform Deployment Framework

In infrastructure and capacity planning, big data platform cluster mainly consists of data storage server, interface server, cluster management server and application server. Using X86 servers to build distributed two level deployment, headquarters as the core data collection point, while the provincial companies as regional data collection points, we suggest a total of twenty-nine servers cluster shown in Fig. 4, which consist of sixteen Hadoop cluster servers, four stream computing servers, four interface servers, three application servers, two relational databases.

3   Application Scenarios

3.1   Transmission Monitoring and Status Assessment Real-Time Analysis

The development of transmission monitoring and state evaluation business, involves line ledger, on-line monitoring, testing, daily patrol, helicopter or unmanned aerial vehicle inspection, and satellite remote sensing data, as shown in Fig. 5. According to high throughput in distributed storage systems, the transmission monitoring and state evaluation business which obtains real-time flow data, can realize synchronous storage of massive data [12]. According to predefined business rules and data processing logic, the business can support quick data retrieval; according to the technology for flow computing, it can not only evaluate the health status of electric transmission line in real time, identifying accurately lines' fault to alarm in the abnormal states, but also forecast the trend of the line under poor natural conditions, providing theoretical and practical guideline for the maintenance decision of electric transmission lines [13].

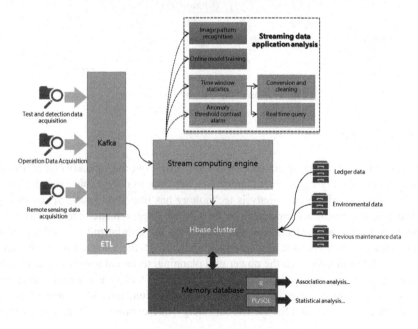

Fig. 5. Power big data real-time processing

3.2   Distribution Network Planning Off-Line Analysis

Massive data produced by distribution network in the planning and running process have many characteristics, such as multi temporal and spatial, multiple source, hybrid and uncertainty. Figure 6 provides the brief process of the off-line application analysis. Firstly, we should obtain types and formats of the massive data, and build a unified big data storage interface to bring about the integrated distributed fast storage for off-line data. Then based on the unified data storage interface, we design the data analysis interface which supports the data statistics processing task. From this process, we have better to meet the needs of data analysis, risk assessment and early warning and other advanced applications, and provide scientific and rational basis for the management of load adjustment, operation mode, formulation of electricity price policy and so on [14].

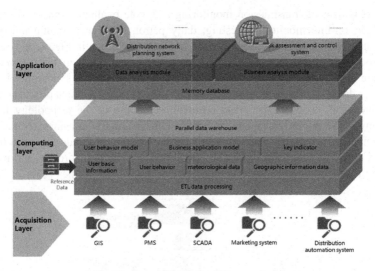

Fig. 6. Power big data off-line analysis

4   Conclusion

The Global Energy Interconnection can not only bring out the fundamental revolution of the energy industry by internet conception, but also produce explosive information and data. The use of big data analysis technology has important practical significance for better management and operation to the Global Energy Interconnection. Big data technology is not only a technical part needed by the specialized technology of the energy industry, but also the technology foundation of the Global Energy Interconnection, which will impact on the power grid planning, technical revolution, equipment upgrade, power grid renovation, design specifications, technical standards, operation rules and marketing policies. This paper presents the fundamental system of the big data which can be extended for the Global Energy Interconnection. This system builds up the model of functional, technical and deployment architecture, and then forms into real-time and offline analysis framework of power big data, which can strongly support the refined energy management for future energy interconnection.

References

1. The State Council of the People's Republic of China. Action to promote the development of big data platform. Group Technol. Prod. Mod. **32**(3), 51–58 (2015)
2. Wang, Y.: Energy internet promoting energy revolution. Autom. Panor. 68–69 (2015)
3. Li, D., Geng, S., Zheng, J.: The development trend of power big data under the situation of energy internet. Mod. Electr. Power **32**(5), 10–14 (2015)
4. Cao, Y.: The road of China's energy Internet white paper. Electr. Ind. (2015)
5. Zhang, P.: Present situation and prospect of power big data application. Electr. Age 24–27 (2014)
6. Yan, H., Di, F., Yuan, R., et al.: Research on big data and application scenario in intelligent dispatching of power network. Power Inf. Commun. Technol. 7–12 (2014)
7. Zhang, D., Miao, X., Liu, L., et al.: Research on the development of smart grid technology. J. Chin. Electr. Eng. Sci. **1**, 2–12 (2015)
8. Yan, M.: Research on big data technology and application in electric power industry. J. Nanjing Inst. Ind. Technol. 1–5 (2015)
9. Zhang, J., Dong, N., Peng, W., et al.: Application of big data platform in electric power enterprise. Hebei Electr. Power (2016)
10. Lin, W., Yu, Y., Liang, Y., et al.: Research on information communication technology supporting Global Energy Interconnection. Smart Grid (2015)
11. Peng, X., Deng, D., Cheng, S., et al.: Key technologies of large power data for smart grid applications. J. Chin. Electr. Eng. Sci. 503–511 (2015)
12. Huang, Z., Shu, N., Liu, H., et al.: Unified data platform for smart grid and its application. Hubei Electr. Power **39**(2), 56–58 (2015)
13. Shi, M., Han, X., Cheng, Z., et al.: Research on application of big data for power demand side. Distrib. Util. 20–23 (2014)
14. Lu, R., Fan, H., Zhou, X.: Optimization method of distribution network repair point based on big data technology. Distrib. Util. **32**(8), 31–36 (2015)

Intelligent Methods for Energy Saving and Pollution Reduction

Study on Lightweight Design and Connection of Dissimilar Metals of Titanium Alloy TC4/T2 Copper/304 Stainless Steel

Shun Guo[1], Qi Zhou[1(✉)], Peng Xu[2], Qiong Gao[1], Tianyuan Luo[1], Yong Peng[1(✉)], Jian Kong[1], KeHong Wang[1], and Jun Zhu[3]

[1] School of Materials Science and Engineering,
Nanjing University of Science and Technology, Nanjing 210094, China
{cheezhou,ypeng}@njust.edu.cn
[2] School of Mechanical Engineering and Automation, Shanghai University,
Shanghai 200072, China
[3] School of Materials Science and Engineering, Nanjing Institute of Technology,
Nanjing 211167, China

Abstract. Under the background of lightweight design, manufacturing and improvement of comprehensive performance, dissimilar metals connection has been becoming a research focus recently and will have a broad application prospect. Titanium alloy TC4 and 304 stainless steel have many excellent properties, achievement of effective connection between these two materials has a significant promoting effect on Industrial Technology. However, the bonding connection of TC4/304 is very poor, so it is necessary to redesign the joint and further study the strengthening mechanism. In the paper, connection experiment of TC4/304 was carried out using two methods: Electron Beam Welding and Friction Stir Welding. Optical microscopy, SEM, EDS were applied for the analysis of microstructure and phase structure. The results state that EBW and FSW are effective and the maximum strength are 196 Mpa and 178 Mpa respectively. Both failure mode are brittle fracture.

Keywords: Lightweight design · TC4 · 304 stainless steel · Dissimilar metals · Microstructure

1 Introduction

Materials with the properties of lightweight, energy saving, high strength and multi-function are increasingly gaining ground, especially in the areas of automobile industry, aviation and aerospace [1]. However, it has been a knotty problem to further improve performance of materials today. Therefore, it is necessary to redesign and put forward a new proposal for the material manufacturing. Titanium alloy [2–4] is one kind of important structural metals and first developed in the 1950s. Now, it has been widely used in various fields because of its high strength, good corrosion resistance and high heat resistance. Meanwhile, stainless steel [5–7] is the most commonly used in the field of metal materials due to its good corrosion resistance, high temperature and low temperature performance and weldability, and low price. Effective connection of

© Springer Nature Singapore Pte Ltd. 2017
K. Li et al. (Eds.): LSMS/ICSEE 2017, Part III, CCIS 763, pp. 229–237, 2017.
DOI: 10.1007/978-981-10-6364-0_23

dissimilar materials of titanium alloy and stainless steel not only meets the lightweight requirements but also can combine both performance advantages. Accordingly, it will have broad application prospects.

But, there are many problems [8, 9] in the connection between titanium alloy and stainless steel due to differences in physical and chemical properties. Brittle inter-metallic compounds and microcracks are very easy to gather at the front of connection interface. Shanmugarajan [10] studied on the welding of titanium alloy and 304 stainless steel using 3.5 kW CO_2 laser. High cooling rate of laser welding could reduce the formation of brittle intermetallic compounds, thereby increase the ductility of joints and improve the strength. But the experimental results show that even if a higher welding speed or manually adjust the laser position relative to the docking surface, a large number of microcracks will still be produced. Thus, improved design of TC4/304 joint need to be further studied.

Kumar et al. [11] studied the friction welding of TC4 titanium alloy and 304 stainless steel adding copper as middle layer. Its results show that the strength of TC4/304 joint formed by directly welding is very low. The reason lies in the formation of brittle intermetallic compounds, which results in microcracks. By adding copper as middle layer, the formation of intermetallic compounds and microcracks are restrained. At the same time, the thickness of intermetallic layer decreases. It can be obtained from EDS that there is no direct contact between the two base metal and intermetallic compounds of Ti/Cu are formed at the interface of Ti substrate. Compared to the direct friction welding of TC4/304, the joint strength has been significantly improved due to the reduce of formation of Ti/Fe brittle phase. Although there are some Cu_3Ti generated at the interface, it has little effect on the strength of joints.

Wang [12–14] reported electron beam welding of TA15 and 304 stainless steel by adding Ni, V, Cu and Ag filler metal. The results show that the formation of Ti/Fe would be restrained by these filler metals. The weld obtained by using different filler metals is composed of solid solution and intermetallic compound, and the type of solid solution and intermetallic compounds depend on the metallurgical reaction between filler metal and base metal. $Fe_2Ti + Ni_3Ti + NiTi_2$, $TiFe$, Ti_2Ag and $Cu_2Ti + CuTi + CuTi_2$ are obtained respectively by adding Ni, V, Ag. The tensile strength of the joint is mainly determined by the hardness of the intermetallic compounds at the interface. Ag is used as the filling layer, the strength of the joint can reach the maximum value of 310 MPa.

To summarize, combination of both properties of titanium alloy TC4 and 304 stainless steel, and also lightweight design are the common goal of researchers. However, there are lots of problems needed to be further studied, the aim of this paper is to realize high-strength connection of TC4/304 by joint design. Microstructural and mechanical properties, distribution, and evolution of IMCs layer in the weld were investigated.

2 Experimental Procedure

Titanium alloy Ti-6Al-4 and 304 stainless steel (GB/T 2059-2008) plates were applied with the dimension of $100 \times 50 \times 4$ mm and no groove. In addition, a thin copper sheet with the dimension of $100 \times 5 \times 4$ mm was used as the middle layer. Chemical compositions and mechanical properties of these three kinds of metals are listed in Table 1. All specimens were prepared carefully by fine polishing with sandpapers of grit sizes (240# and 400#) and surface cleaning with absolute alcohol before welding. Specimens were firmly fixed in the form of butt joint and copper sheet was sandwiched at the middle of plates of TC4 and 304 stainless steel. Thermo-physical properties of the metals are shown in Table 2. It indicates that there are some significant differences between titanium and ferrum. Thereby, electron beam welding and friction stir welding were used to realize the connection of TC4/304 due to the advantages in dissimilar metals welding.

Table 1. Chemical compositions (wt.%) and mechanical properties of the materials.

Materials	Elements (wt.%)					Hv	δ	Rm/MPa
Steel 304	Cr	Ni	Mn	Si	C	200	≥ 40	520
	20	8.0	2.0	0.8	0.08			
Copper (T2)	Cu	O	Fe	S	Ni	70–90	≥ 20	236
	99.9	0.06	0.005	/	/			
Ti-6Al-4V	Ti	Al	V	Fe	C	310–330	≥ 10	895
	89.12	6.42	4.30	0.05	0.03			

Table 2. Thermo-physical properties of Ti, Fe and Cu

Ele.	Crystal lattice	Density $kg \cdot m^{-3}$	Specific heat $J \cdot kg^{-1} \cdot K^{-1}$	Melting point K	Fusion heat $kJ \cdot mol^{-1}$	Thermal conductivity $W \cdot m^{-1} \cdot K^{-1}$	Expansion coefficient $1E\text{-}6 \cdot K^{-1}$
Ti	hcp	4540	544.2	1943	15.45	15.7	7.14
Fe	bcc(α) fcc(γ)	7900	502.0	1535	13.78	77.5	11.5
Cu	fcc	8960	380.9	1357	13.26	393.6	16.4

A Zeiss optical microscope and a Quant 250 FEG scanning electron microscope (SEM) equipped with an energy-dispersive X-ray spectrometry (EDS) system were applied for the observation of microstructure and chemical composition of samples. All samples cut off from weld with the dimension of $10 \times 5 \times 4$ mm were prepared with standard grinding, polishing, and corrosion before observation. Fractional corrosion method was used. To investigate the mechanical properties of the joints, a tensile testing machine (maximum tensile strength 10 kN) with the speed of 3 mm/min at room temperature was used to test strength and average value was adopted. Tensile standard was according to GB/T228-2002.

3 Results and Discussion

3.1 Electron Beam Welding of 304 Stainless Steel/Titanium Alloy TC4

Connection of titanium alloy and 304 stainless steel was first carried out by the electron beam welding without any middle layer, the parameters of EBW are shown in Table 3. Due to the characteristics of deep penetration, small heat affected zone and cutting off the adverse effects of air under vacuum [15], EBW has lots of good advantages for dissimilar metals welding.

Table 3. Electron beam welding parameters

Parameters	Accelerating voltage U/kV	Electron beam I_b /mA	Welding speed v/mm·s^{-1}	Focus current I_f/mA	Working distance H/mm
Value	60	25	10	678	295

Figure 1 shows the schematic diagram of electron beam welding with offset of 0.3 mm on titanium side. But the actual picture of EBW shows that EBW of TC4/304 stainless steel is very poor and cannot achieve the connection of the joint. The macro morphology of EBW seam indicates that crack is very serious. After unloading fixture, the sample was broken directly and the joint strength is 0 Mpa. The reason for the results is that a large number of intermetallic compounds are formed in the joints, at the same time, because of the rapid cooling speed, a great deal of residual stress are retained in the weld. These two unfavorable factors eventually lead to direct fracture of the joint.

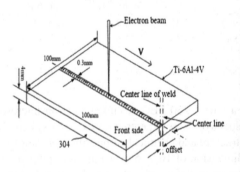

Fig. 1. Schematic diagram of EBW with offset of 0.3 mm on titanium side and actual picture of 304/TC4

Based on the previous experiments, a thin copper layer was added in the middle of TC4/304 stainless steel plates. Lots of former studies show that the connection between copper and steel is practical and the strength is enough, moreover, the welding of titanium and copper is also possible and the strength coefficient could reach almost 90% of copper substrate. Thus, copper was presented to be the middle layer to weld TC4/304 stainless steel.

Figure 2 shows the actual picture of EBW of 304 stainless steel/T2 copper/titanium alloy TC4, indicating that there are no obvious cracks in the surface and the connection is good. Figure 3 presents the microstructures and tensile fracture of the joints, from which it can be known that the joint is composed of kinds of areas including TC4 substrate, IMCs layer, T2 copper interlayer, fusion zone and 304 stainless steel substrate. There are some microcracks in the seam after welding. Due to the formation of the intermetallic compounds of Ti/Cu, there is a IMCs layer occurred between TC4 and T2 copper. In addition, because of low mutual solubility between Fe and Cu, there are some ferric ball embedded in the copper matrix. Fracture occurred in the IMCs layer and the fracture surface was shown in Fig. 3(d), revealing that the joint is brittle fracture. The reason is IMCs layer destroyed the toughness of joint resulting in the decrease of strength, maximum tensile strength is 196 Mpa.

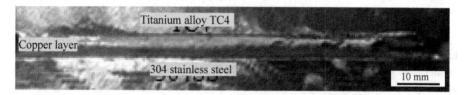

Fig. 2. Actual picture of EBW of 304 stainless steel/T2 copper/titanium alloy TC4

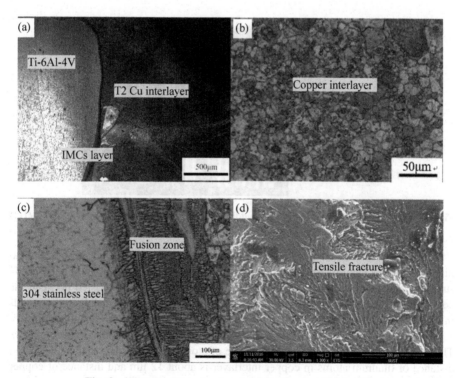

Fig. 3. Microstructures (a–c) and tensile fracture of joint (d)

3.2   Friction Stir Welding of 304 Stainless Steel/Titanium Alloy TC4

As another effective method for the dissimilar metals welding, friction stir welding has been applied in various fields such as aerospace, automotive industries and other areas recently. It is one kind of solid state joining process and is able to eliminate most of the side-effects associated with the remelting of conventional fusion welding. The friction stir welding of 304 stainless steel and titanium alloy TC4 with the middle layer of copper was carried out. The parameters of FSW are listed in the Table 4.

Table 4. Friction stir welding parameters

Parameters	Frictional pressure	Rotational speed r/min	Frictional time t/s	Upsetting force MPa	Upsetting time t/s
Value	1	1800	4–14	2	3

Figure 4 shows the schematic diagram of FSW with middle layer of copper (two kinds of thicknesses of copper sheet: 0.5 and 1 mm) and actual picture of FSW. The results state that FSW is effective for the 304/TC4 welding and the macroscopic joints are in good connection. With the increase of friction time, the copper is gradually squeezed out of the gap. A ring structure is formed by the copper flash and joint are slightly uplifted at the titanium side. Maximum tensile strength is 178 Mpa.

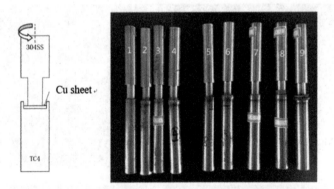

Fig. 4. Schematic diagram of FSW with middle layer of copper and actual picture of 304/TC4

Figure 5(a) shows the microstructural diagram of FSW joint indicating that Titanium - Copper - Steel interface could be divided into four regions, including TC4, 304 substrate, reaction layer and copper interlayer. The reaction layer existed between the copper interlayer and TC4 substrate. There are lots of Ti-Cu intermetallic compounds in the reaction layer, which is similar to the IMCs layer of electron beam welding.

Figure 5(c) presents the EDS line scanning results of yellow line mark in Fig. 5(a). The results shows that the diffusion distance of atoms in solid phase is short, diffusion distance of titanium atoms in copper interlayer is about 12 μm and distance of copper atoms in titanium substrate is about 7 μm. Width of reaction layer is about 10 μm. The addition of copper middle layer has a good effect to block the interdiffusion of Ti and Fe atoms, thus avoiding the formation of intermetallic compounds and improving the joint strength. In addition, high ductility of copper reduces the residual stress of joint, which is beneficial to strength. Fracture measurements show that the mode attributed to cleavage fracture.

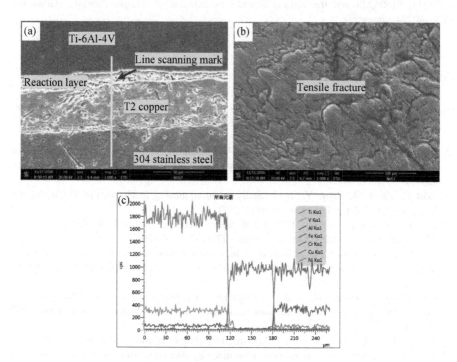

Fig. 5. (a) Microstructural diagram of FSW joint, (b) tensile fracture, and (c) EDS results of line scanning of the yellow lines shown in (a) (Color figure online)

4 Conclusions

1. Based on the former studies of titanium alloy and 304 stainless steel welding, a new joint design of adding copper as middle layer was studied by electron beam welding and friction stir welding. Both joints are in a good connection. The study has important sense to the lightweight manufacturing.
2. Electron beam welding of titanium alloy and 304 stainless steel is effective. But the existence of IMCs destroyed the toughness of joint resulting in the decrease of strength, maximum tensile strength is 196 Mpa, thus it need to be further studied.
3. Friction stir welding of titanium alloy and 304 stainless steel is also a good attempt. But bonding strength is not enough due to the insufficient diffusion. Fracture measurements show that the mode attributed to cleavage fracture, and the maximum tensile strength is 178 Mpa.

4. Dissimilar metals connection has a broad application prospect due to combination of both properties and lightweight design. Although there are lots of problems, the core of this is embrittlement of joints caused by intermetallic compounds. Thus, IMCs layer is necessary to be further studied.

Acknowledgements. Thanks for the National Natural Science Foundations of China, Grant nos. 51375243, 51505226 and the Natural Science Foundation of Jiangsu Province, Grant no. BK20140784 supporting.

References

1. Isaev, V.I., Cherepanov, A.N., Shapeev, V.P.: Numerical study of heat modes of laser welding of dissimilar metals with an intermediate insert. Int. J. Heat Mass Transf. **99**, 711–720 (2016)
2. Junaid, M., Baig, M.N., Shamir, M., et al.: A comparative study of pulsed laser and pulsed TIG welding of Ti-5Al-2.5 Sn titanium alloy sheet. J. Mater. Process. Technol. **242**, 24–38 (2017)
3. Guo, S., Zhou, Q., Peng, Y., et al.: Study on strengthening mechanism of Ti/Cu electron beam welding. Mater. Des. **121**, 51–60 (2017)
4. Cui, L.I., Bin, L.I., Ze-Feng, W.U., et al.: Stitch welding of Ti-6Al-4V titanium alloy by fiber laser. Trans. Nonferr. Met. Soc. China **27**(1), 91–101 (2017)
5. Lu, Z., Shi, L., Zhu, S., et al.: Effect of high energy shot peening pressure on the stress corrosion cracking of the weld joint of 304 austenitic stainless steel. Mater. Sci. Eng. A **637**, 170–174 (2015)
6. Ma, N., Cai, Z., Huang, H., et al.: Investigation of welding residual stress in flash-butt joint of U71Mn rail steel by numerical simulation and experiment. Mater. Des. **88**, 1296–1309 (2015)
7. Ma, H., Qin, G., Geng, P., et al.: Microstructure characterization and properties of carbon steel to stainless steel dissimilar metal joint made by friction welding. Mater. Des. **86**, 587–597 (2015)
8. Zhang, Y., Sun, D.Q., Gu, X.Y., et al.: A hybrid joint based on two kinds of bonding mechanisms for titanium alloy and stainless steel by pulsed laser welding. Mater. Lett. **185**, 152–155 (2016)
9. Chen, S., Zhang, M., Huang, J., et al.: Microstructures and mechanical property of laser butt welding of titanium alloy to stainless steel. Mater. Des. **53**(1), 504–511 (2014)
10. Shanmugarajan, B., Padmanabham, G.: Fusion welding studies using laser on Ti-SS dissimilar combination. Opt. Lasers Eng. **50**(11), 1621–1627 (2012)
11. Kumar, R., Balasubramanian, M.: Experimental investigation of Ti-6Al-4V titanium alloy and 304L stainless steel friction welded with copper interlayer. Def. Technol. **11**, 65–75 (2015)
12. Wang, T., Zhang, B., Wang, H., et al.: Microstructures and mechanical properties of electron beam-welded titanium-steel joints with vanadium, nickel, copper and silver filler metals. J. Mater. Eng. Perform. **23**(4), 1498–1504 (2014)
13. Wang, T., Zhang, B., Feng, J.C.: Influences of different filler metals on electron beam welding of titanium alloy to stainless steel. Trans. Nonferr. Met. Soc. China **24**(1), 108–114 (2014)

14. Wang, T., Zhang, B., Feng, J., et al.: Effect of a copper filler metal on the microstructure and mechanical properties of electron beam welded titanium-stainless steel joint. Mater. Charact. **73**(7), 104–113 (2012)
15. Guo, S., Zhou, Q., Kong, J., et al.: Effect of beam offset on the characteristics of copper/304 stainless steel electron beam welding. Vacuum **128**, 205–212 (2016)

Research on Warehouse Scheduling Optimization Problem for Broiler Breeding

Wenqiang Yang$^{(\boxtimes)}$ and Yongfeng Li

Henan Institute of Science and Technology, Xinxiang, China
yangwqjsj@163.com

Abstract. Feeding on time, which is a key factor for the healthy growth of broilers. To minimize the feeding delay, a mathematical model considering the time spent on transferring feed is proposed. To solve the model above, a fruit fly algorithm (FFA) is adopted. Considering its disadvantages of trapping into local optima and low convergence accuracy, mutation operator and adaptive step-length is imposed to form an improved fruit fly algorithm (IFFA), which not only enhanced the convergence efficiency, but also ensured the global optimization. Finally, to verify the performance of the proposed algorithm, it is compared with FFA and genetic algorithm (GA). Simulation results prove that the feasibility and superiority of the proposed algorithm.

Keywords: Broiler breeding · Warehouse scheduling · Fruit fly algorithm · Mutation operator · Adaptive step-length

1 Introduction

Agricultural modernization is an effective way to promote the development of rural economy and solve the three agricultural problems [1]. However, livestock breeding is an important part of modern agriculture. This paper takes warehouse-style breeding of broiler as research subject. Not only is feeding on time a key factor for the healthy growth of broilers, but also for the benefits of breeders. Therefore, how to optimize the time spent on transferring feed to minimize the feeding delay which has become important in warehouse-style breeding of livestock. This basically belongs to warehouse scheduling problem. Up to now, scholars have did much work and gained their research results. For example, Hsu et al. [2] develops an order batching approach based on genetic algorithm to deal with order batching problem. Petersen [3] evaluates various routing polices and the impact of warehouse shape and pick-up/drop off locations. Litvak et al. [4] give an overview of recent research on the performance evaluation and design of carousel systems. Boysen et al. [5] survey the crane scheduling problems with crane interference, and present a classification scheme. Erdogan et al. [6] model the problem of scheduling twin robots on a line, and solve it with exact algorithms and heuristic algorithms. Briskorn et al. [7] propose a novel gantry scheduling algorithm used for solving scheduling co-operating stacking cranes. Carlo and Vis [8] research the scheduling problem of the automated material handling

K. Li et al. (Eds.): LSMS/ICSEE 2017, Part III, CCIS 763, pp. 238–246, 2017.
DOI: 10.1007/978-981-10-6364-0_24

system and introduce two functions to characterize the system and help identify and resolve situations where the lifts would interfere with each other. Guan et al. [9] study the quay crane scheduling problem, and develop different solution approaches for different scale problems. Diabat and Theodorou [10] propose a novel genetic algorithm to deal with the integration of quay crane assignment and scheduling problem. Peterson et al. [11] develop a heuristic algorithm, which is used for scheduling multiple factory cranes to avoid interference. Xie et al. [12] address multi-crane scheduling problem with shuffling decisions in steel coil warehouse based on mixed integer linear programming model and heuristic algorithm. Rath and Gutjahr [13] develop a math-heuristic for a three-objective warehouse location-problem in disaster relief. Wang et al. [14] adopt an elitist non-dominated sorting genetic algorithm to solve task scheduling of multi-tier shuttle warehousing systems. As can be seen above, few scholars pay close attention to warehouse style breeding of broiler. In effect, with the acceleration of agriculture modernization, warehouse-style breeding is becoming more and more important. Thus, it is necessary to study warehouse scheduling optimization problems related to warehouse-style breeding.

Fruit fly algorithm (FFA) is a kind of swarm intelligence algorithm, inspired by the knowledge from the foraging behavior of fruit flies [15]. Compared with conventional evolutionary algorithms, its principle is simple, easy to understand and implement, with good converging speed and local searching ability. FFA has been widely used to solve continuous function optimization issue [16–18], but its application in discrete combination optimization problems is still relatively few. To address the scheduling problem of warehouse-style breeding, an improved discrete fruit fly algorithm (IFFA) is presented. Mutation operator and adaptive step-length is introduced to improve the convergence rate and accuracy of IFFA, a numerical simulation is utilized to verify the effectiveness of the algorithm in the end.

2 Problem Description and Modeling

Suppose one farm adopts warehouse to breed broiler, and has several breeds of broilers without loss of generality. Sketches of the warehouse used for breeding broilers is shown in Fig. 1.

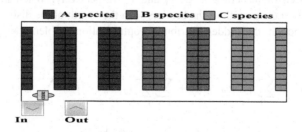

Fig. 1. Layout of the warehouse used for breeding broilers

As you can see in Fig. 1, there are three species broilers to be bred in the warehouse, and the crane that transports fodder to locations specified on prior schedule.

2.1 Problem Statement

Suppose fodder distribution task that has the following features.

(I) The number of the distribution destinations is m, the assigned locations are represented respectively as $(p_1, p_2, \cdots, p_m)$. Similarly, p_0 corresponding to input buffer or output buffer, then the distance between them can be expressed as d_{ij} $(i, j \in \{0, 1, 2, \cdots, m\})$.

(II) For each distribution location, q_{iA}, q_{iB} and q_{iC} $(i \in \{1, 2, \cdots, m\})$ are represented by orders of three species broilers.

Definition 1: If the crane consecutively accesses p_i and p_j during distribution, $e_{ij} = 1$; Otherwise, $e_{ij} = 0$.

Definition 2: If the sub-route r belongs to one of the routes which the crane complete fodder distribution tasks and p_i belongs to sub-route r, $g_{ir} = 1$; Otherwise, $g_{ir} = 0$.

Definition 3: If the task of one distribution destination is completed, $l_i = 0$; Otherwise, $l_i = 1$.

Definition 4: For each distribution destination p_i, if p_i belongs to the region where one species broilers is located, $o_{ij} = 1$ $(i \in \{1, 2, \cdots, m\}, j \in \{A, B, C\})$; Otherwise, $o_{ij} = 0$.

Definition 5: To quantify the degree of effect on the growth of the broilers with the length of the distribution timeout, the character of three species broilers are denoted respectively with w_A, w_B and w_C, which indicate how much of the growth of each species broilers is affected in unit time. Based on prior experience, w_A, w_B and w_C are assigned to 5, 1 and 3, respectively.

2.2 Mathematical Modeling

To deal with fodder distribution optimization problem of multi-species broilers, need to establish a mathematical model. Suppose that maximum load, the horizontal speed and the vertical speed of the crane is Q, v_x and v_y, respectively. When minimizing the degree of effect on the growth of the broilers with the length of the distribution timeout as objective function, the fodder distribution optimization problem can be formulated as follows:

$$\min f(e) = \sum_{j=1}^{m} t_{0j} \cdot e_{0j} \cdot \phi(i) + \sum_{i=1}^{m}\sum_{j=1}^{m} t_{ij} \cdot e_{ij} \cdot \phi(i) + \sum_{i=1}^{m} t_{i0} \cdot e_{i0} \cdot \phi(i) \quad (1)$$

where, $t_{ij} = \max\{(d_{ij})_x / v_x, (d_{ij})_y / v_y\}$, $\phi(i) = (\sum_{i=1}^{m}\sum_{j=A}^{C} o_{ij} \cdot w_j) \cdot l_i$.

s.t.

$$\sum_{j \in \{A,B,C\}} q_{ij} \leq Q \quad \forall i \in \{1, 2, \cdots, m\} \tag{2}$$

$$\sum_{i=1}^{m} (\sum_{j \in \{A,B,C\}} q_{ij}) \cdot g_{ir} \leq Q \tag{3}$$

$$\sum_{i=1}^{m} (\sum_{j \in \{A,B,C\}} q_{ij}) \cdot g_{ir} \leq Q \quad \&\& \quad \sum_{i=1}^{m} (\sum_{j \in \{A,B,C\}} q_{ij}) \cdot g_{ir} + \sum_{j \in \{A,B,C\} \&\& p_i \notin r} q_{ij} > Q \tag{4}$$

$$\sum_{r} g_{ir} = 1 \quad \forall i \in \{1, 2, \cdots, m\} \tag{5}$$

Equation (1) is the objective function. Constraints that are from Eqs. (2) to (5). Equation (2) requires the order of each distribution destination must not exceed maximum load of the crane. Equation (3) prevents the overload of the crane. Equation (4) makes sure that the crane is almost fully loading. Equation (5) ensures that the order of each distribution destination is completed by the crane at a time.

3 Improved Fruit Fly Algorithm (IFFA)

Standard fruit fly algorithm is a new method for finding global optimization based on the food finding behavior of the fruit fly, which is originally used to solve optimization problems in continuous domain. First, for the sake of solving discrete combination optimization problems, a new kind of encoding schema needs to be adopted. Second, the fixed foraging step-size and the homogenization of foraging behavior can't coordinate the exploration and exploitation ability of the algorithm. Based on the conditions mentioned above, this paper makes some improvements to standard fruit fly algorithm, thereinafter.

3.1 Solution Encoding and Decoding

To use fruit fly algorithm conveniently, random numbers-encoded is adopted. However, fresh agricultural products distribution is discrete combination optimization problem, need to create a one-to-one mapping between solutions and fruit fly located in continuous domain. In this case, all dimensions of the fruit fly position is to be sorted in ascending order. Thus to each dimension corresponds to the distribution task number. Meanwhile, one dimension at which the order totals approach or reach fully loading of the crane, insert here number 0. This avoids the generation of infeasible solutions. As Fig. 2 indicates the detailed encoding process.

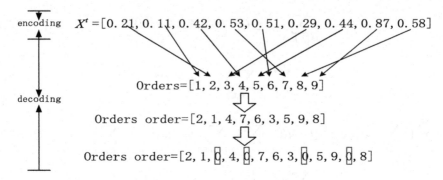

Fig. 2. The schematic for solution encoding and decoding

3.2 Adaptive Foraging Step-Size

The fixed foraging step-size reduces the flexibility of searching optimum. In other words, too large foraging step-size makes it easier to execute global search, but leads to the blindness of search, severely degrades the accuracy of convergence. And yet, too small foraging step-size makes fruit fly trap into local optima easily and have low convergence accuracy. For the purpose of reaching balance between accuracy and efficiency, a mechanism of foraging step-size self-adaptive adjustment is designed.

$$S_i^{(t+1)} = (1-\lambda) \cdot (X_i^{(t)} - X_{pbest}) + \lambda \cdot (X_i^{(t)} - X_{gbest}) \tag{6}$$

where $\lambda = (N_a - N_c)/N_a$.

Furthermore, where L_i^{t+1}, $X_i^{(t)}$, X_{pbest}, X_{gbest}, N_a and N_c refers to the step of the $(t + 1)$th iteration, the position of the ith fruit fly in the tth iteration, the best position of all the fruit flies until the tth iteration, the best position of the ith fruit fly until the tth iteration, the maximum number of iteration and the number of current iteration.

This mechanism ensures that early optimization is prone to global search, and late optimization is prone to global search, and improves the accuracy and efficiency for optimizing.

3.3 Mutation Operator

The homogenization of foraging behavior lowers population diversity, which makes fruit fly trap into local optima easily. Thus, the mutation operator is built into the fruit fly algorithm, which can be used to maintain population diversity and improve the ability of global exploration. The mutation probability is devised in terms of the distribution characteristics of population.

$$p_m^i = (1/\eta_i) \Big/ \sum_{k=1}^{n} (1/\eta_k) \tag{7}$$

where $\eta_i = |X_i - X_j| \Big/ \sum_{k \in (1,2,\cdots,n) \&\& k \neq j} |X_k - X_j|$;

where p_m^i, X_i, X_j, X_k, η_i refers to the mutation probability of the ith fruit fly, the position of the ith fruit fly, the position of the jth fruit fly, the best position of all the fruit flies and the ith fruit fly whose contribution to population diversity. This not only keeps the diversity of the population but good individuals.

3.4   Representation of IFFA

Step 1. Initialize parameters: the generation counter t, the number of fruit flies N_n, N_a, and generate N_n fruit flies;

Step 2. Calculate the food concentration of N_n fruit flies based on the reciprocal of Eq. (1) and find the best fruit fly X_j, and record its food concentration $Con(X_j)$;

Step 3. Other fruit flies except for X_j forage as follows;

$$X_i = X_j + S_i \tag{8}$$

Step 4. If fruit fly X_j is better than the current best fruit fly X_{best}, update the current best fruit fly as follows;

$$X_{best} = X_j \tag{9}$$

$$con\_large = Con(X_j) \tag{10}$$

Step 5. Implement mutation operation based on Eq. (7);

Step 6. Evaluate all the fruit flies, and find the best fruit fly X_j;

Step 7. If $t \leq N_a$, and then go to step 3; Otherwise, output the optimal solution.

4   Numerical Example and Analysis

To analyze the performance of algorithms mentioned above objectively, the examples of warehouse scheduling optimization problem for broiler breeding that are given, where some comparisons are made with FFA and GA. In addition, simulations are made under the same condition, such as windows10 operating system, 3.7 GHz processor, 4 GB of RAM and development environment Matlab R2014b. Population size and the number of generations of all algorithms are 60 and 600, respectively. For GA, crossover probability and mutation probability are set 0.8, 0.2, respectively. For the crane, the maximum load $Q = 5$ kg, the length, the width and the height of the location $L = 30$ cm, $W = 50$ cm, $H = 40$ cm, respectively, the width of the lane $D = 180$ cm, the horizontal speed and the vertical speed of the crane $v_x = 1$ m/s, $v_y = 0.5$ m/s, respectively, the number of the location for each row $C = 80$. 60 feeding orders are randomly generated for three species broilers.

Figure 3 gives the optimization results with IFFA, FFA and GA, respectively. In the meantime, experiments of five different-scaled populations are conducted to analyze the sensitivity of variation of populations, which is illustrated in Fig. 4. Moreover, the optimized sequence of fodder distribution for each species is shown in Fig. 5.

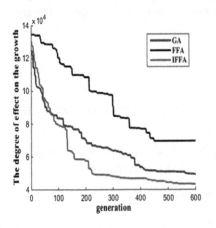

Fig. 3. Evolution comparison   **Fig. 4.** The sensitivity analysis of population size

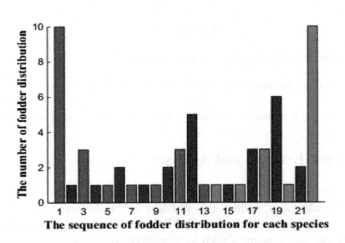

Fig. 5. The sequence of fodder distribution for each species

To further test stability and accuracy of IFFA which is relative to other two algorithms, different-sized fodder distribution for each species is tested. Each algorithm runs thirty times, results are presented in Table 1, in which n, opt, and t indicate fodder distribution size, optimal value, the times of finding optimal, respectively.

Table 1. Performance comparison among algorithms

n	IFFA			FFA			GA		
	opt	ave	t(s)	opt	ave	t(s)	opt	ave	t(s)
10	1.35	1.35	1.21	1.35	1.35	1.89	1.35	1.35	2.04
20	17.72	17.72	1.14	18.97	21.10	2.04	19.73	20.54	1.88
30	18.88	18.88	3.01	20.87	21.91	2.19	18.88	20.08	2.49
40	22.25	22.94	2.86	27.33	30.52	4.04	24.28	25.83	3.79
50	34.59	35.74	3.34	43.68	47.16	3.75	36.59	39.92	4.22
60	43.59	44.19	3.94	60.34	62.99	4.13	49.30	52.04	4.59

As shown in Fig. 4 and Table 1, compared with FFA and GA, IFFA shows a better performance in the accuracy and efficiency. Furthermore, GA, by using mutation, can reduce the chance of falling into local minimum to some extent. Nevertheless, the fixed mutation probability makes them be blindness or randomness. But for FFA, few tunable parameters easily lead to the premature problem. For IFFA, firstly, step-length is adapted by the global optimum and the local optimum, and hence increases both the efficiency of the algorithm; Secondly, the mutation operator increases the multiplicity of individuals, which helps IFFA to get the best optimum result of the problem to the most extent. In other words, it further improves the solving precision and convergent rate of optimization procedure. In addition, as seen in Table 1, the performance difference among algorithms is a little when the size of problems is smaller; however, the performance of IFFA is much better than that of the other two when the size of problems is larger. It further demonstrates that IFFA has better global optimization ability. Meanwhile, the results of these experiments once again indicate that the improvement strategy proposed is valid in this paper. Furthermore, Table 1 also shows that initial solutions affect the quality of the optimum. Besides, in Fig. 4, in spite of the volatility in the optimum, it generally reflects that population size is related to the scale of the problem. Therefore, how to choose the right population is a problem worth researching. In the end, Fig. 5 gives the optimized sequence of fodder distribution for each species, which generally gives us enlightenments as: the broiler having larger degree of effect on the growth of itself with delay should be distributed firstly when distribution cost is considered together.

5   Conclusion

This paper proposes an improved fruit fly algorithm, and then applied to solve warehouse scheduling optimization problem for broiler breeding. Adaptive step-length and mutation operator are introduced to the fruit fly algorithm, which speed up the convergence and improve the accuracy of optimal solution, respectively. Therefore, global and local search ability of the IFFA is promoted in general. In the end, the simulation results demonstrate the effectiveness and superiority of the IFFA, which could have some enlightening that the IFFA is used on discrete combination optimization problem. In view of the impacts of initial solutions and population size on the quality of optimum, on which deep research will be carried out in future.

Acknowledgments. This work is supported by the scientific and technological project of Henan province under Grant No. 172102110031, and by the high-tech people project of Henan institute of science and technology under Grant No. 203010616001.

References

1. Zhu, N., Qin, F.: Influence of mechanization on technical efficiency of large-scale layer breeding. Trans. Chin. Soc. Agric. Eng. **31**(22), 63–69 (2015)
2. Hsu, C.M., Chen, K.Y., Chen, M.C.: Batching orders in warehouses by minimizing travel distance with genetic algorithms. Comput. Ind. **56**(2), 169–178 (2005)
3. Petersen, C.G.: An evaluation of order picking routing policies. Int. J. Oper. Prod. Manag. **17**(11), 1098–1111 (1997)
4. Litvak, N., Vlasiou, M.: An survey on performance analysis of warehouse carousel systems. Stat. Neerl. **64**(4), 401–447 (2010)
5. Boysen, N., Briskorn, D., Meisel, F.: A generalized classification scheme for crane scheduling with interference. Eur. J. Oper. Res. **258**(1), 343–357 (2017)
6. Erdogan, G., Battarra, M., Laporte, G.: Scheduling twin robots on a line. Naval Res. Logist. **61**(2), 119–130 (2014)
7. Briskorn, D., Angeloudis, P.: Scheduling co-operating stacking cranes with predetermined container sequences. Discrete Appl. Math. **201**, 70–85 (2016)
8. Carlo, H.G., Vis, I.F.A.: Sequencing dynamic storage systems with multiple lifts and shuttles. Int. J. Prod. Econ. **140**(2), 844–853 (2012)
9. Guan, Y., Yang, K.H., Zhou, Z.: The crane scheduling problem: models and solution approaches. Ann. Oper. Res. **203**(1), 119–139 (2013)
10. Diabat, A., Theodorou, E.: An integrated quay crane assignment and scheduling problem. Comput. Ind. Eng. **73**, 115–123 (2014)
11. Peterson, B., Harjunkoski, L., Hoda, S., et al.: Scheduling multiple factory cranes on a common track. Comput. Oper. Res. **48**, 102–112 (2014)
12. Xie, X., Zheng, Y.Y., Li, Y.P.: Multi-crane scheduling in steel coil warehouse. Expert Syst. Appl. **41**(6), 2874–2885 (2014)
13. Rath, S., Gutjahr, W.J.: A math-heuristic for the warehouse location–routing problem in disaster relief. Comput. Oper. Res. **42**, 25–39 (2014)
14. Wang, Y.Y., Mou, S.D., Wu, Y.H.: Task scheduling for multi-tier shuttle warehousing systems. Int. J. Prod. Res. **53**(19), 5884–5895 (2015)
15. Pan, W.T.: A new fruit fly optimization algorithm: taking the financial distress model as an example. Knowl. Based Syst. **26**(2), 69–74 (2012)
16. Sheng, W., Bao, Y.: Fruit fly optimization algorithm based fractional order fuzzy-PID controller for electronic throttle. Nonlinear Dyn. **73**(1), 611–619 (2013)
17. Lin, S.M.: Analysis of service satisfaction in web auction logistics service using a combination of fruit fly optimization algorithm and general regression neural network. Neural Comput. Appl. **22**(3), 783–791 (2013)
18. Li, H.Z., Guo, S., Li, C.J., et al.: A hybrid annual power load forecasting model based on generalized regression neural network with fruit fly optimization algorithm. Knowl. Based Syst. **37**(1), 378–387 (2013)

A Discrete Fourier Transform Based Compensation Task Sharing Method for Power Quality Improvement

Jianbo Chen[✉], Dong Yue, Chunxia Dou, and Chongxin Huang

Institute of Advanced Technology and the Jiangsu Engineering Laboratory
of Big Data Analysis and Control for Active Distribution Network,
Nanjing University of Posts and Telecommunications, Nanjing 210046, China
jianbo686@aliyun.com

Abstract. In this paper, a discrete Fourier transform (DFT) based compensation task sharing method is proposed for the improvement of power quality of main grid. Power quality problem induced by typical nonlinear loads is tackled by the cooperation of multi-functional grid-tied inverters (MFGTIs) with the compensation instruction as part of its reference. Unlink the ordinary method where communication is avoided, Low-bandwidth channel is used to transmit the compensation reference after the current data are calculated by DFT. Simulation results are presented to demonstrate the effectiveness of the proposed method.

1 Introduction

Due to the growing concerns regarding traditional fossil fuel shortages and environmental issues, renewable energy has drawn considerable attention in recent years and a large number of power electronics interfaced Distributed Generation Systems (DGSs) have been installed in the low voltage power distribution systems [1–3]. In DGSs, harmonic, reactive and unbalanced current generated by nonlinear loads may not only degrade the power quality at the point of common coupling (PCC), but also result in instabilities due to the series and/or parallel harmonic resonances [4,5].

In [6], through limiting the harmonic and reactive conductance and susceptance, the distorted current caused by nonlinear loads is shared among MFGTIs in coordinated and automatic fashion in accordance with their capacities without communication. This method has the same compensation effect compared with APFs for almost no time delay. However, the topology of DGs and loads is specified as cascade connection and the nonlinear loads must be arranged in the downstream. In [7], a distributed control method for converter-interfaced renewable generation units with active filtering capability is proposed. Agent-based communication makes coordination between the generation units possible.

The prominent issue in power quality problems sharing is communication delay. Even the compensation component is accurately calculated, the tiny delay may deteriorate the compensation performance, sometimes worse. Thus, most

© Springer Nature Singapore Pte Ltd. 2017
K. Li et al. (Eds.): LSMS/ICSEE 2017, Part III, CCIS 763, pp. 247–256, 2017.
DOI: 10.1007/978-981-10-6364-0_25

of the work is devoted to avoid the application of communication channel [8]. However, there are three reasons indicating that it is necessary to take advantage of communication facilities. Firstly, in islanding mode, the information of neighbors is needed to restore the frequency and voltage deviations owing to the secondary control of MFGTIs [9]. Secondly, the cooperation between MFGTIs is inevitable due to the active power and reactive power sharing achieved by economic dispatch in both grid-connected and islanding mode [10]. The third reason is that accurate sharing of compensation task and flexible ancillary services can be reached by overcoming the drawback of communication.

The currents of typical nonlinear loads, such as computers, lamps and air conditioners in which abundant harmonic components are contained are periodic in steady operation which means DFT (Discrete Fourier Transform) can be used to analyze the harmonic components of it. The result can be distributed to MFGTIs as the reference current to ensure the power quality of main grid. If no equipments state changes, then the result will not be sent to MFGTIs and the last reference will be used. Obviously, during the first period of changing state along with communication delay, the harmonic components cannot be compensated by MFGTIs. Since the state of typical nonlinear load will not switch frequently, the compensation task can be achieved after few periods during which the influence of harmonic component can be neglected. In this paper, the compensation of distorted currents through MFGTIs is further studied.

The innovation of this paper can be concluded as:

- A DFT based algorithm is used to generate the compensation reference.
- The compensation task is shared among MFTGIs through communication channel which is avoided in regular papers with the compensation effectiveness demonstrated by simulations.

This paper is arranged as follows. The configuration of MFGTI and proposed method is illustrated in Sect. 2. The generation algorithm of reference current is introduced in Sect. 3. Simulation results are presented in Sect. 4. Section 5 concludes the paper.

2   Configuration

The configuration of MFGTIs topology studied in this paper is depicted in Fig. 1.

Figure 2 illustrates the proposed method of compensation task sharing among MFGTIs.

SDCT is short for sampling, DFT, calculation and transmission. In few periods after the state switches, currents of nonlinear loads are sampled and then the DFT algorithm is used to calculate the amplitude and phase of the relative frequency. The result is transmitted to the MFGTIs as compensation reference. In the steady state, no data packet is transmitted which will notably save the communication resources.

Each MFTGI accumulates the compensation tasks received from n loads and multiply the reference value with the ratio of its own capacity to the sum of all MFTGIs. Thus, the compensation tasks are shared according to their capacities.

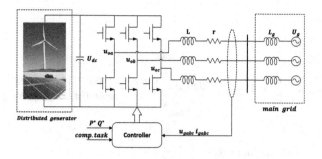

Fig. 1. Configuration of MFGTI

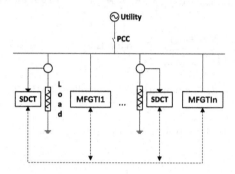

Fig. 2. Compensation task sharing among MFGTIs

3   Generation Algorithm of Reference Current

With the function of Power tracking and harmonic current compensation, the current reference consists of two parts. One is used to track the power generation reference of DG which determined by tertiary control. The other is used to track the harmonic component which is generated by nonlinear loads to ensure the power quality of PCC. The detailed configuration of controller of MFGTI is depicted in Fig. 3.

3.1   Reference Current of Power Generation

To simplify the computation, the Clark transformation is implemented:

$$T_{3s-2s} = \frac{2}{3} \times \begin{bmatrix} 1 & -\frac{1}{2} & -\frac{1}{2} \\ 0 & \frac{\sqrt{3}}{2} & -\frac{\sqrt{3}}{2} \end{bmatrix} \tag{1}$$

whose inverse transformation is written as $T_{2s-3s} = T_{3s-2s}^{T}$. The voltage of MFTGI j, u_{jabc}, can be transformed into $\alpha\beta$ frame by formula (1):

$$\begin{bmatrix} u_{j\alpha} \\ u_{j\beta} \end{bmatrix} = T_{3s-2s} \times \begin{bmatrix} u_{ja} \\ u_{jb} \\ u_{jc} \end{bmatrix} \tag{2}$$

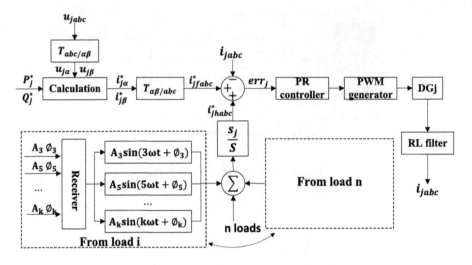

Fig. 3. Configuration of controller of MFGTI

The reference current of power generation $i_{j\alpha}^*$ and $i_{j\beta}^*$ in $\alpha\beta$ frame can be written as:

$$\begin{cases} i_{j\alpha}^* = (u_{j\alpha} * P_j^* + u_{j\beta} * Q_j^*)/(u_{j\alpha}^2 + u_{j\beta}^2) \\ i_{j\beta}^* = (u_{j\beta} * P_j^* - u_{j\alpha} * Q_j^*)/(u_{j\alpha}^2 + u_{j\beta}^2) \end{cases} \tag{3}$$

where P_j^* and Q_j^* are the given active power and reactive power of MFGTI j respectively. Thus the reference current of power generation i_{jfabc}^* in natural abc frame can be calculated by the inverse Clark transformation:

$$\begin{bmatrix} i_{jfa}^* \\ i_{jfb}^* \\ i_{jfc}^* \end{bmatrix} = T_{2s-3s} \times \begin{bmatrix} i_{j\alpha}^* \\ i_{j\beta}^* \end{bmatrix} \tag{4}$$

3.2 Reference Current of Harmonic Compensation

When the compensation reference packet which contains the amplitude and phase of the relative frequency received by MFGTIs, the reference current of harmonic compensation of MFGTIs related to load i is generated by the local function generator with the formula:

$$\sum_{k=3,5,7,\ldots} A_k \sin(k\omega t + \phi_k) \tag{5}$$

where A_k and ϕ_k are amplitude and phase of kth-order harmonics, respectively. The final harmonic reference of MFGTI j, i_{habc}^*, is acquired by multiplying the sum of harmonic references of n loads with the rated capacity $\frac{s_j}{S}$:

$$i_{habc}^* = \frac{s_j}{S} \times \sum_{i=1}^{n} \sum_{k=3,5,7,\ldots} A_k \sin(k\omega t + \phi_k) \tag{6}$$

where s_j is the apparent power of MFTGI j, S is the sum of apparent power of all MFTGIs. Thus, the compensation reference is proportional to the ratio of its capacity to the sum of all MFTGIs, and the compensation tasks are shared according to their capacities.

3.3 Reference Current Tracking

Since the compensation reference in abc frame consists of sinusoidal waves, PI controller will not be able to track the reference within acceptable errors while Proportional Resonant (PR) controller is practical. The model of PR controller is expressed as:

$$G_{PR}(s) = K_p + \sum_{h=1,3,5,7,\ldots} \frac{2K_{rh}\omega_{ch}s}{s^2 + 2\omega_{ch}s + \omega_h^2} \tag{7}$$

where ω_1 and ω_h are the natural angular frequencies of the fundamental and hth-order harmonic resonant terms, ω_{c1} and ω_{ch} are the cut-off frequencies of these terms, K_p and K_{rh} (h = 1, 3, 5, 7) are the proportional and resonant integral gains of the PR controller.

4 Simulation Results

Figure 4 shows the simulation system used to test the validity of proposed compensation task sharing method. The harmonic source is emulated by diode rectifier connected with RL load. A three-phase switch is used to determine whether the MFGTI 2 is put into operation or not. The harmonic component is calculated and distributed to MFGTIs through low-bandwidth channel. The reference of MFTGI consists of active power generation and harmonic compensation.

Key parameters of the simulation model are listed in Table 1.

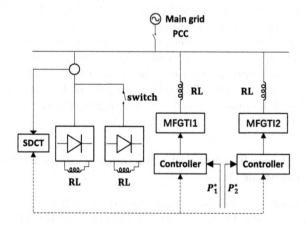

Fig. 4. simulation model of proposed compensation task sharing method

Table 1. Key parameters of the simulation model

Elements	Parameters and values
Main grid	Line-to-line RMS voltage U = 380 V
	Frequency f0 = 50 Hz
RL filter of harmonic source	R = 0.1 Ohms
	L = 0.3 mH
RL filter of MFGTI	R = 0.6 Ohms
	L = 0.5 mH
Harmonic source with breaker	RL loaded by the diode rectifier
	R = 1 Ohms
	L = 0.1 mH
Harmonic source without breaker	RL loaded by the diode rectifier
	R = 2.5 Ohms
	L = 0.1 mH
Three-phase breaker	Initial status = closed
	Switch time = 0.1 s
DC source	Udc = 800 V
PQ reference of MFGTIs	P1 = 40 kw
	Q1 = 0 kvar
	P2 = 60 kw
	Q2 = 0 kvar
Transport delay	Time delay = 0.02 s
Rated power	S1/S2 = 1/2

4.1 Case A: Power Tracking Without Harmonic Compensation

The current waveforms of main grid (I_s), MFGTI 1(I_{g1}), MFGTI 2(I_{g2}) and harmonic source (I_{L1}) are shown in Fig. 5.

Since the reference of MFGTIs incorporate only the active power, the waveforms of I_{g1} and I_{g2} are totally sinusoidal. The harmonic component of I_{L1} is injected into the main grid and degrades the power quality at PCC which is indicated by the distorted waveform. The FFT analysis of I_s is shown in Fig. 6(a). The THD of case A is equal to 9.32% which is bigger than the specified value of IEC61000-2-2 standard, 8%.

4.2 Case B: Power Tracking and Harmonic Compensation Without Communication Delay

As mentioned before, the harmonic compensation reference is absent in the first period of changing state. During the time [0, 20 ms], the MFGTIs generate only the active power which is indicated by the sinusoidal wave. At the same time,

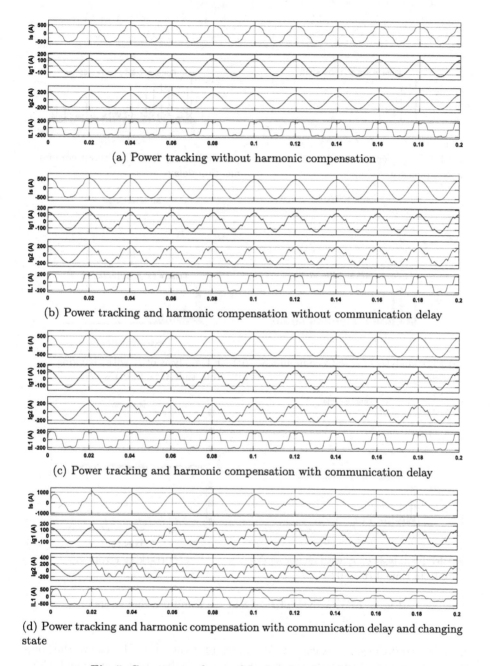

(a) Power tracking without harmonic compensation

(b) Power tracking and harmonic compensation without communication delay

(c) Power tracking and harmonic compensation with communication delay

(d) Power tracking and harmonic compensation with communication delay and changing state

Fig. 5. Current waveforms of I_s, I_{g1}, I_{g2}, I_{L1} of four cases

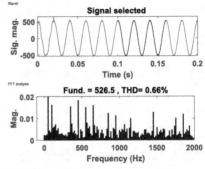

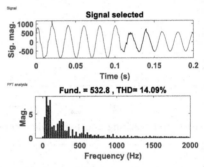

(a) FFT analysis of I_s without harmonic compensation

(b) FFT analysis of I_s without communication delay

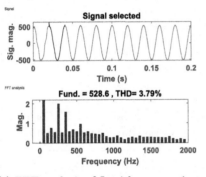

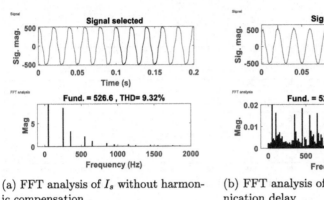

(c) FFT analysis of I_s with communication delay

(d) FFT analysis of I_s with changing state from time 0.11 to 0.15s

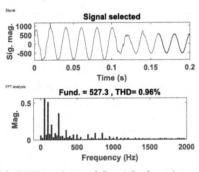

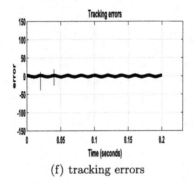

(e) FFT analysis of I_s with changing state from time 0.16 to 0.2s

(f) tracking errors

Fig. 6. FFT analysis of I_s of four cases and tracking errors

I_s is distorted the same as case A. However, after the MFGTIs receive the compensation reference, the output currents I_g inject the harmonic component which is equal to the $-I_{L1h}$, making the I_s being sinusoidal. As shown in Fig. 5(b).

There is a point must be mentioned that the waveform of I_g is distorted with a peak value at t = 20 ms owing to the tracking errors, as shown in Fig. 6(f).

The FFT analysis of I_s is shown in Fig. 6(b). The THD of case B is equal to 0.66% which is obviously an improvement comparing to 9.32%.

4.3   Case C: Power Tracking and Harmonic Compensation with Communication Delay

Communication delay of MFGTI 1 which is equal to 20 ms is introduced while the delay of MFGTI 2 is assumed to be zero for comparison. As shown in Fig. 5(c).

During the time [20, 40 ms], the MFGTI 1 generates only the active power due to communication delay while MFGTI 2 functions as case B. However, as soon as the compensation reference arrives, MFGTI 1 can track the instruction timely as shown in Fig. 5(c) after 40 ms.

The FFT analysis of I_s is shown in Fig. 6(c). The THD of current during time [20,40 ms], 3.79%, is relatively lower comparing with case A due to the compensation of MFGTI 2 while higher comparing with case B. After MFGTI 1 receives the compensation instruction, it is the same case as B.

4.4   Case D: Power Tracking and Harmonic Compensation with Communication Delay and Changing State

At $t = 0.1\,s$, the three-phase breaker switches from closed state to open. During time [0.1, 0.12 s], the compensation reference has not been updated and the current of I_s is distorted due to the mismatched compensation component. However, since the MFGTIs receive the latest instructions, it becomes the case C. As shown in Fig. 5(d).

The FFT analysis of I_s after changing state is shown in Fig. 6(d). The THD of I_s is relatively high after changing state with the value equal to 14.09%.

However, after about 2 periods, the output of MFGTIs track the compensation component after changing state and the THD of I_s decreases to 0.96%, as shown in Fig. 6(e).

In summary, according to the aforementioned simulation results, the proposed DFT based compensation task sharing method is effective for power quality improvement.

5   Conclusions

Considering the typical nonlinear loads, such as computers, lamps and air conditioners whose current waveforms are periodic in steady operation, a DFT based power quality improvement method is proposed. With the communication channel, compensation component is proportionally shared among MFGTIs and the power quality of main grid is ensured. During the steady state, communication sources can be saved without the need to transmit the reference packet. Four cases are provided to illustrate the effectiveness of the proposed method.

Acknowledgments. This work is supported in part by the National Natural Science Foundation of China under Grant Nos. 61533010, 61374055 and 61503193.

References

1. Olivares, D.E., Mehrizi-Sani, A., Etemadi, A.H., Canizares, C.A., Iravani, R., Kazerani, M., Hajimiragha, A.H., Gomis-Bellmunt, O., Saeedifard, M., Palma-Behnke, R., et al.: Trends in microgrid control. IEEE Trans. Smart Grid 5(4), 1905–1919 (2014)
2. Feng, W., Sun, K., Guan, Y., Guerrero, J., Xiao, X.: Active power quality improvement strategy for grid-connected microgrid based on hierarchical control. IEEE Trans. Smart Grid (2016)
3. Peng, C., Zhang, J.: Delay-distribution-dependent load frequency control of power systems with probabilistic interval delays. IEEE Trans. Power Syst. 31(4), 3309–3317 (2016)
4. Enslin, J.H., Heskes, P.J.: Harmonic interaction between a large number of distributed power inverters and the distribution network. IEEE Trans. Power Electron. 19(6), 1586–1593 (2004)
5. Zeng, Z., Yang, H., Zhao, R., Cheng, C.: Topologies and control strategies of multifunctional grid-connected inverters for power quality enhancement: a comprehensive review. Renew. Sustain. Energy Rev. 24, 223–270 (2013)
6. Zeng, Z., Zhao, R., Yang, H.: Coordinated control of multi-functional grid-tied inverters using conductance and susceptance limitation. IET Power Electron. 7(7), 1821–1831 (2014)
7. Macken, K.J., Vanthournout, K., Van den Keybus, J., Deconinck, G., Belmans, R.J.: Distributed control of renewable generation units with integrated active filter. IEEE Trans. Power Electron. 19(5), 1353–1360 (2004)
8. Laaksonen, H.J.: Protection principles for future microgrids. IEEE Trans. Power Electron. 25(12), 2910–2918 (2010)
9. Savaghebi, M., Jalilian, A., Vasquez, J.C., Guerrero, J.M.: Secondary control for voltage quality enhancement in microgrids. IEEE Trans. Smart Grid 3(4), 1893–1902 (2012)
10. Han, Y., Li, H., Shen, P., Coelho, E.A.A., Guerrero, J.M.: Review of active and reactive power sharing strategies in hierarchical controlled microgrids. IEEE Trans. Power Electron. 32(3), 2427–2451 (2017)

A Comprehensive Optimization of PD^μ Controller Design for Trade-off of Energy and System Performance

Ke Zhang[1], Min Zheng[1,2(✉)], Kang Li[3], and Yijie Zhang[1]

[1] School of Mechatronic Engineering and Automation, Shanghai University,
Shanghai 200072, China
zhengmin203@shu.edu.cn
[2] Shanghai Key Laboratory of Power Station Automation Technology,
Shanghai 200072, China
[3] School of Electronics, Electrical Engineering and Computer Science,
Queen's University Belfast, Belfast, UK

Abstract. This paper investigates the optimal trade-off between the system performance and control energy consumption for different settings of the control parameters k_P, k_D, μ and PD^μ controller design, and a comprehensive optimization method is proposed to obtain the optimal PD^μ controller. Detailed correlation analysis between the control performance and energy consumption is presented. The method is applied to the control of a ball-beam system, and the simulation results confirm that the proposed method is practically useful in the analysis and design of the PD^μ controller.

Keywords: Fractional-order PD^μ controller · Control energy consumption · System performance · Ball-beam system · Genetic algorithm

1 Introduction

PID controller has been widely used in industrial process control with the feature of simple structure, robustness and easy to operate. However, conventional PID controllers have an inherent drawback: on meeting a contradictory requirement for both short settling time and small overshoot. The fractional-order PID (FOPID)controller proposed by Podlubny in 1999, which is defined as $PI^\lambda D^\mu$, is more flexible in adjusting the dynamic performance [1]. Compared with the traditional integer-order PID controller, it has two more control parameters λ and μ, therefore the tuning and optimization of the given parameters in the controller is however more difficult. In recent years, the fractional-order controllers have attracted a lot of interest due to their wide spread applications.

Control theory and control engineering has experienced significant progresses in the past century and played a key role in modern industry. It concerns about the automated operation of a machinery or a system to achieve a desired target

© Springer Nature Singapore Pte Ltd. 2017
K. Li et al. (Eds.): LSMS/ICSEE 2017, Part III, CCIS 763, pp. 257–266, 2017.
DOI: 10.1007/978-981-10-6364-0_26

and to avoid unstable or unintended disruptive behavior. In order to cope with more complex industrial control needs, a number of modern control algorithms, such as optimal control, adaptive control, model predictive control, robust control, and networked control have been proposed [2,3]. However, only a few control strategies have considered the relationship between the control energy and system performance [3]. For the traditional optimal control, its main purpose is to seek the optimal control laws to make the performance indexes take the maximum or minimum value under certain constraints. In [4,5], Li et al. for the first time investigated the correlation between the control energy consumption and system performance, However, little has been done so far to consider how the two factors are related to each other and how to choose a proper control target in designing a proper Fractional-order PID (FOPID) controller.

In this note, the Ball-beam system is considered when the Fractional-order PID controller is investigated. Here, we have investigated the correlations between control energy consumption and system performance of the Ball-beam system under the PD^μ control. The nonlinear correlation between the control energy consumption and the system performance are illustrated in the detailed case study. A numerical example demonstrates the efficacy of a comprehensive optimization method to obtain the optimal PD^μ controller in achieving a balanced decision between the system performance and control energy consumption.

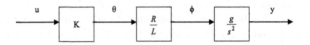

Fig. 1. The architecture of the Ball-beam system.

2   The Architecture of a Ball-Beam System

The Ball-beam system is a typical nonlinear system, where the nonlinear relationships exist between the driver guide motion of the motor spindle and the elevation of the rail, and the transmission of the gear and the guide rail. The system is used to test some advanced control methods.

The mathematical model of a ball-beam system can be approximated as a cascaded system with three parts: the machinery, the angle conversion and the DC motor as shown in Fig. 1. The machinery model connects the location of the ball on the rail $y(t)$ and the angle between the rail and the horizontal line $\varphi(t)$. The angle model associates the elevation angle $\varphi(t)$ of the guide rail with the rotational angle $\theta(t)$ of the motor. The motor model associates the input voltage $u(t)$ of motor control system with $\theta(t)$ which is obtained from position sensor.

A. Machinery Model [6]

The process of rolling the ball on the rails is approximated as a mass rolling on a smooth surface without friction. The force that accelerates the rolling of the ball on the rail is the force of the ball's gravity in the direction parallel to the guide rail and the friction of the ball. According to Newton's theorem, the kinetic equation of the ball rolling on the guide rail is

$$M\ddot{y}(t) = Mg \sin \varphi(t) \tag{1}$$

where $y(t)$ is the position of the ball on the track, M is the mass of the ball, g is the gravitational acceleration, $\varphi(t)$ is the angle between the rail and the horizontal line. Assume $\varphi(t)$ is very small, (1) can be linearized as

$$\ddot{y}(t) = g\varphi(t) \tag{2}$$

The Laplace transfer function of (2) leads to

$$\frac{Y(s)}{\varphi(s)} = \frac{g}{s^2} \tag{3}$$

B. Angle Model

The angle $\varphi(t)$ between the rail and horizontal line is guided by the rotational angle of the DC servo motor. The relationship between $\varphi(t)$ and $\theta(t)$ is non-linear, which is also influenced by the reduction ratio between the big gear and the small gear. (3) can be approximated as follows

$$\frac{\varphi(s)}{\theta(s)} = \frac{\varphi(t)}{\theta(t)} = \frac{R}{L} \tag{4}$$

where R is the radius of the motor plate, L is the length of the rail.

C. Motor Model

The response speed of the motor is usually very fast, and in this case study the motor of the ball-beam system is controlled by a IPM 100 control card, and the time constant of motor rotation $\theta(t)$ to voltage $u(t)$ is very small. So the mathematical model of the motor can be approximated as a gain K.

In order to simplify the calculation process, suppose $KRg/L = 1$. As a result, the transfer function of the whole system therefore can be formulated as

$$\frac{Y(s)}{U(s)} = \frac{KRg/L}{s^2} = \frac{1}{s^2} \tag{5}$$

The z-transform of the ball-beam system model can be described as

$$G_p(s) = \frac{C}{s^2} \Rightarrow g_p(t) = c \cdot t \tag{6}$$

Let $t = nT$, (6) can be reformulated as

$$g(nT) = CnT \tag{7}$$

The z-transform to (7) gives

$$G(z) = \sum_{n=0}^{\infty} CnT \cdot z^{-n} = \frac{CTz}{(z-1)^2} = \frac{Y(z)}{U(z)} \tag{8}$$

Hence, (8) can be written as

$$\frac{CTz}{z^2 - 2z + 1} = \frac{Y(z)}{U(z)} \tag{9}$$

Divided both sides of numerator and denominator of (9) with $z^2 (\neq 0)$, then we have

$$\frac{CTz^{-1}}{1 - 2z^{-1} + z^{-2}} = \frac{Y(z)}{U(z)} \tag{10}$$

Thus

$$CTz^{-1}U(z) = Y(z) - 2z^{-1}Y(z) + z^{-2}Y(z) \tag{11}$$

The difference equation of (11) is

$$CTu(k-1) = y(k) - 2y(k-1) + y(k-2) \tag{12}$$

So, the output sequence of the Ball-beam system becomes

$$y(k) = 2y(k-1) - y(k-2) + CTu(k-1) \tag{13}$$

3   Fractional-Order PID Control

A fractional-order $PI^\lambda D^\mu$ controller (FOPID) can be considered as a generalization of the conventional PID controllers. The transfer function of the FOPID controller has the form

$$C(s) = k_P + \frac{k_I}{s^\lambda} + k_D s^\mu \ (0 < \lambda, \mu < 2) \tag{14}$$

where s is the fractional calculus operator, λ and μ are the integral and derivative order, k_P is proportional gain, k_I is integral gain, and k_D is derivative gain.

The differential-integral operator $_aD_t^\alpha$ is given as

$$_aD_t^\alpha = \begin{cases} \frac{d^\alpha}{dt^\alpha}, & \alpha > 0 \\ 1, & \alpha = 0 \\ \int_a^t (d\tau)^{(-\alpha)}, & \alpha < 0 \end{cases} \tag{15}$$

where α is the fractional order, a and t are the start and end values of the integral, respectively. $_aD_t^\alpha$ is a combined differential and integral operator used in fractional calculus. Some definitions exist for the fractional differential-integral, such as the Caputo, Grunwald-Letnicov, and Riemann-Liouville definitions [8].

The $PI^\lambda D^\mu$ controller in the discrete form is given as

$$u(k) = k_P e(k) + k_I D^{-\lambda} e(t) + k_D D^\mu e(t) \tag{16}$$

Fig. 2. The reedback control block diagram.

4 Optimization of Ball-Beam System with Fractional-Order PD^μ Controller

Consider the block diagram of feedback control loop shown in Fig. 2, where $G_p(s)$ is the transfer function of the ball-beam system, $G_c(s)$ is the fractional-order PD^μ controller, r is the reference input, e is error, u is the output of controller, and y is the output of the loop. The expressions of $G_p(s)$ and $G_c(s)$ are

$$G_p(s) = \frac{C}{s^2} \tag{17}$$

$$G_c(s) = \frac{U(s)}{E(s)} = k_P + k_D s^\mu \ (\mu < 2) \tag{18}$$

Given the above, the difference equations of feedback control loop can be deduced as

$$\begin{cases} e(k) = r(k) - y(k) \\ u(k) = k_p e(k) + k_D D^\mu e(t) \\ y(k) = 2y(k-1) - y(k-2) + CTu(k-1) \end{cases} \tag{19}$$

As the transfer function of the Ball-beam system is a second order system, the theoretical part of the controller needs to be removed. Then, the following incremental PD^μ controller is used.

In this note, the system performance (tracking error) E_y and control energy consumption E_u of the system are defined as

$$E_y = \sum_{k=1}^{t} [r(k) - y(k)]^2 \tag{20}$$

$$E_u = \sum_{k=1}^{t} u(k)^2 \tag{21}$$

$$E_s = E_y + E_u \tag{22}$$

where $y(k)$ is the actual system output, $r(k)$ is the reference input, T is the sampling period. Further, the comprehensive performance is defined as E_s, which is the sum of E_y and E_u. If the value of E_y is becomes smaller, then it indicates that the control error becomes small and the control performance becomes better. If the value of E_u becomes smaller, then it indicates that the control energy consumption becomes small. If the value of E_s becomes smaller, then it indicates that both the control energy consumption and the control performance become better. A proper choice of the PD^μ controller can be made for E_y and E_u which gives the possible best comprehensive control performance for the Ball-beam system.

This paper aims to investigate the system performance E_y and control energy consumption E_u for different settings of the control parameters k_P, k_D and μ in order to achieve the best trade-off between the system performance and control energy consumption.

The optimization method to obtain the optimal control parameters k_P, k_D and μ in achieving a balanced decision between the system performance and control energy consumption is summarized as follows:

Step 1: Using genetic algorithm to search optimal value of μ between μ_1 and μ_2 which make the system performance E_y and the comprehensive performance E_s to the minimum.

Step 2: For a fixed value of $\mu = \mu_1$, plotting the relationship curve between comprehensive performance E_y and the control parameters k_P, k_D on the (k_P, k_D)-plane. When the comprehensive performance E_y is minimum, we choose k_P^1, k_D^1 to be the optimal control parameters.

Step 3: Similar as step 2, for a fixed value of $\mu = \mu_2$, plot the relationship between the performance E_s and control parameters k_P, k_D along the (k_P, k_D) -plane. When the performance E_s is minimum, we choose k_P^2, k_D^2 to be the optimal control parameters.

Step 4: Select a set of optimal from parameters k_P^1, k_D^1 and k_P^2, k_D^2, to meet the system requirements.

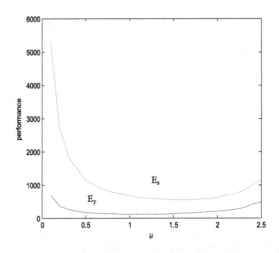

Fig. 3. The minimum value of E_y and E_s with different μ.

5 Simulation Experiments

The genetic algorithm optimization toolbox (GAOT) is used to search for the values of k_P, k_D and μ in this paper. First, the genetic algorithm is used to search the minimum value which makes E_y and E_s to the minimum as shown in Fig. 3.

For each given μ, Fig. 3 shows the optimal value of k_P, k_D to obtain the minimum value of the system performance E_y and comprehensive performance E_s in the feasible region of the control parameters k_P and k_D.

Observation 1: The fractional-order PD^μ controller can make the system achieve better performance (tracking error) and comprehensive performance compared with the traditional integer-order PID controller. It is easily found that E_y can reaches the minimum value when μ is around 1.12, but E_s can reaches the minimum value when μ is around 1.61. So, E_s cannot reaches the minimum value for $\mu = 1.1$ while E_y can reaches the minimum value for $\mu = 1.1$. It is clear that the correlation of the system performance with control energy consumption is nonlinear.

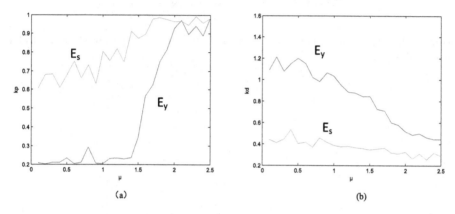

(a)   (b)

Fig. 4. (a) The optimal solution k_P of E_y or E_s with different μ. (b) The optimal solution k_D of E_y or E_s with different μ.

Observation 2: According to Fig. 4, it can be found that the optimal solution k_P of E_y or E_s will increase while the optimal solution k_D for E_y or E_s will decreases as μ increases. The range of optimal solution k_P, k_D of E_y is more variable than E_s. Due to the model characteristics of the Ball-beam system and under different parameters μ, the optimal solutions k_P, k_D of E_y and E_s are within the range: $k_P \leq 1$ and $k_D \leq 1.3$.

Observation 3: In Figs. 5, 6 and 7, the system performance (tracking error) E_y is set to about 160, and the intersection of the three blue curves with the straight line $E_y = 160$ leads to four points E, F, G, H, where the corresponding abscissa values for E and F are approximately $k_D = 0.5$ and $k_D = 0.6$. It can be observed that the corresponding control energy consumption of G, H is larger than E, F. So the control energy consumption required for the corresponding controller parameters at F point is minimum compared with points E, G and H for the $E_y = 160$. This observation shows that the optimal control parameters k_D, k_D and μ in achieving a balanced decision between the system performance

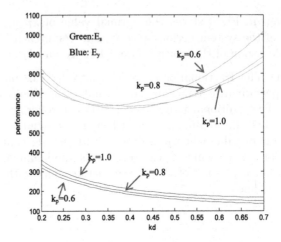

Fig. 5. The relationships between E_y, E_s and the control parameters for μ. (Color figure online)

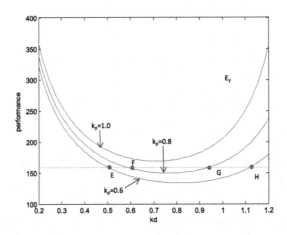

Fig. 6. The relationships between E_y and the control parameters for $\mu = 1.12$ and $k_p \in [0.2, 1.2]$. (Color figure online)

and control energy consumption can be achieved. It shows that the solution that allows the system to obtain optimal control can be found within the feasible domain of the parameters.

Observation 4: Figure 8 show the relationships between the value k_P, k_D and E_s with the different initial value y_1, y_2 when $\mu = 1.12$. The optimal values $k_P = 0.241, k_D = 1.046, E_y = 114.49$ are found when the initial value is $y_1 = 0$, $y_2 = 0.$, and the optimal values $k_P = 0.209, k_D = 0.882, E_y = 110.45$ are found when the initial value is $y_1 = 5, y_2 = 8$. This reveales that the optimal parameters of the PD^μ controller are related to the initial state of the system.

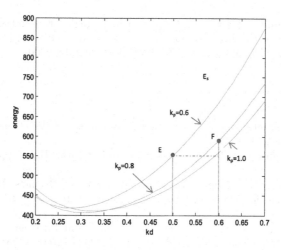

Fig. 7. The relationships between E_s and the control parameters for $\mu = 1.12$ and $k_p \in [0.2, 0.7]$. (Color figure online)

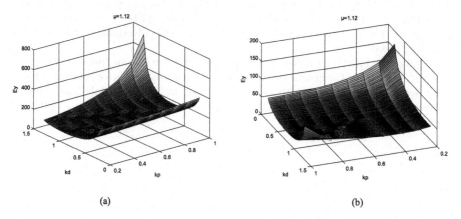

(a)   (b)

Fig. 8. (a) The relationships between the value k_D, k_D and E_s with the initial value $y_1 = y_2 = 0$. (b) The relationships between the value k_D, k_D and E_s with the initial value $y_1 = 5, y_2 = 8$.

6   Conclusion

This paper has investigated the correlations between control energy consumption and system performance of the Ball-beam system using the PD^μ controller design. The nonlinear correlation between the control energy consumption and the system performance for this controller design have been illustrated in detail. Numerical and graphical computation results have shown that the proposed method can be used to analyze and design the PD^μ controller in achieving a balanced decision between the system performance and control energy consumption.

Acknowledgments. This work is supported by Shanghai Key Laboratory of Power Station Technology, the Project-sponsored by SRF for ROCS, SEM, and Shanghai Science Technology Commission No. 14ZR1414800, 14JC1402200.

References

1. Podlubny, I.: Fractional-order systems and -controllers. IEEE Trans. Autom. Control **44**(1), 208–214 (1999)
2. Morari, M., Zafiriou, E.: Robust Process Control. Prentice-Hall, Upper Saddle River (1989)
3. Cloostermand, M.B.G., Hetel, L., Van DeWouwa, N., Heemels, W.P.M.H.: Controller synthesis for networked control systems. Automatica **46**(10), 1584–1594 (2010)
4. Li, K., Wu, Y.L., Li, S.Y., Xi, Y.G.: Energy saving and system performance - an art of trade-off for controller design. In: IEEE 2013 International Conference on Systems, Man, and Cybernetics (SMC), vol. 806, pp. 4737–4742 (2013). doi:10.1109/SMC
5. Wu, Y., Zhao, X., Li, K., et al.: Energy saving-another perspective for parameter optimization of P and PI controllers. Neurocomputing **174**(PA), 500–513 (2016)
6. Ogata, K.T.: Modern Control Engineering, 3rd edn. Electronic Industry Press, Beijing (2000)
7. Podlubny, I.: Fractional Differential Equations. Academic, San Diego (1999)
8. Wang, R.P., Pi, Y.G.: Fractional order proportional and derivative controller design for second-order systems with pure time delay. In: International Conference on Mechatronic Science, Electric Engineering and Computer, China, pp. 1321–1325 (2011)

Hierarchical Time Series Feature Extraction for Power Consumption Anomaly Detection

Zhiyou Ouyang[1,2,3(✉)], Xiaokui Sun[1,2,3], and Dong Yue[1,2,3,4]

[1] School of Automation, Nanjing University of Posts and Telecommunications,
Nanjing 210023, People's Republic of China
netivs@qq.com, medongy@vip.163.com
[2] Hubei Province Collaborative Innovation Center for New Energy Microgrid,
China Three Gorges University, Yichang 443002, People's Republic of China
[3] Institute of Advanced Technology,
Nanjing University of Posts and Telecommunications,
Nanjing 210023, People's Republic of China
[4] Jiangsu Engineering Laboratory of Big Data Analysis and Control for Active
Distribution Network, Nanjing University of Posts and Telecommunications,
Nanjing 210003, China

Abstract. Anomaly of power consumption, particularly due to electricity stealing, has been one of the major concern in power system management for a long time, which may destroy the demand-supply balance and lead to power grid regulating issues and huge profit reduction of electricity companies. One of the essential key to develop machine learning model to solve the above problems is time series feature extraction, which may affect the superior limit of machine learning model. In this paper, a novel systematic time series feature extraction method named hierarchical time series feature extraction is proposed, used for supervised binary classification model that only using user registration information and daily power consumption data, to detect anomaly consumption user with an output of stealing probability. Performance on data of over 100,000 customers shows that the proposed methods are outperforming one of the existing state-of-the-art time series feature extraction library tsfresh [1].

Keywords: Anomaly detection · Power consumption · Electricity stealing · Time series · Feature extraction · Smart gird

1 Introduction

Electric power stealing has been harassing electric power supply enterprises in China. Therefore, accurate detection of abnormal behaviors in power uses is long

---

Z. Ouyang is with the Institute of Advanced Technology and the School of Automation, Nanjing University of Posts and Telecommunications, Nanjing 210003, China. D. Yue is with the Institute of Advanced Technology and the Jiangsu Engineering Laboratory of Big Data Analysis and Control for Active Distribution Network, Nanjing University of Posts and Telecommunications, Nanjing 210003, China.

K. Li et al. (Eds.): LSMS/ICSEE 2017, Part III, CCIS 763, pp. 267–275, 2017.
DOI: 10.1007/978-981-10-6364-0_27

of concern among power supply enterprise. The important evaluation indicators about electricity stealing and leakage include: power anomalies, abnormal load, line loss abnormalities [2]. In order to ensure the normal operation of enterprises and standardize the user's behavior, it is important to obtain accurate detection of abnormal behaviors.

Most existing model-based approach for anomaly detection are firstly constructing profile of normal instances, then identify instances that do not conform to the normal profile as anomalies [3]. There are several methods to solve this problem, such as statistical methods, classification-based methods, and clustering-based methods. Among them, statistical method is a traditional method, and [4] propose the algorithm to deal with two important applications of larger size. It gives accurate results from larger data sets. [5] presents a novel approach to outlier detection based on classification, in an attempt to address two issues: one is the lack of explanation for outlier flagging decisions, and the other is the relatively high computational requirement. [6] presents the cluster-based local outlier algorithm, which is meaningful and provides importance of the local data behavior.

In [7], it demonstrates a way to perform time series anomaly detection via generated states and rules and introduces an algorithm named Gecko for clustering time series data. Nowadays, it is limited to the collection of information about many scenarios. For instance, only user registration information and power consumption data is available, which affects the performance of existing anomaly detection methods. Therefore, we put forward a new algorithm to solve this problem.

In this paper, an anomaly detection model for power consumption based on hierarchical time series feature extraction with supervised binary classification algorithms is proposed, which improve the performance of detection accuracy with only limited data used. The features of this model are as follows:

– Only user information and power consumption data is necessary.
– Hierarchical time series feature extraction used for feature engineering.
– Employing open source tsfresh [1] to automatically generate part of the time series features.
– High performance to win the second prize of 2016 CCF DBCI (rank 3 of 888).

To this end, the major contribution of our work are summarized that a data-driven low-cost solution using only limited data is proposed and avoid any extended Advanced Metering Infrastructure (AMI) hardware. This paper uses hierarchical time series feature extraction with supervised binary classification to build a model that performance well.

The rest of this paper is organized as follows. In Sect. 2, the existing approaches for time series feature extraction are reviewed. In Sect. 3, anomaly detection model based on hierarchical time series feature extraction with supervised binary classification algorithms is described in detail. In Sect. 4, the data used and feature extraction process with performance evaluation is investigated and results based on over 50,000 users' real-world power consumption offered by State Grid of China. In Sect. 5, we conclude this paper.

2   Review of Time Series Feature Extraction

Time series is a set of chronological data, that refers to a set of sequence values obtained at the same time interval and associated with time. It is used in many fields. In [8,9], they propose and analyze methods combining spectral decomposition and feature selection for time series classification problems in discrete cosine transform (DCT) and discrete wavelet transform (DWT) fields. In [10], time series algorithm is used for integration of information from several Electrocardiogram(ECG) features in order to build predictive models for clinical decision-support. [11] presents a feature extraction procedure (FEP) for a brainCcomputer interface (BCI) application where features are extracted from the electroencephalogram (EEG) recorded from subjects performing right and left motor imagery by the time series method. In this paper, the time series method is used to detect of abnormal behaviors of power users.

In abnormal detection field, the mathematical characteristics of time series mainly include random items, period items and trend items [8]. Random items refer to a variety of irregular changes, including strict random changes and irregular effects of sudden changes. Period items refer to a periodic change in a period of time. Trend items refer to a general trend over a longer period of time by some fundamental factor. Then, we build a time series model, like equation (1). $x(t)$ is the input, $O(t)$ is the output. The pulse function of the input to the output is denoted by $f_i(t)$ [1].

$$
\begin{cases}
O(t_1) = & x(t_1) * f_1(t) + d_1(t) \\
O(t_2) = & x(t_2) * f_2(t) + d_2(t) \\
\quad \cdots \\
O(t_n) = & x(t_n) * f_n(t) + d_n(t)
\end{cases}
\tag{1}
$$

where $*$ represents convolution, $d_i(t)$ is the additional noise and $x(t_1), x(t_2), \cdots, x(t_n)$ can be repressed as:

$$
x(t_1) \rightarrow x(t_2) \rightarrow \cdots \rightarrow x(t_v) \rightarrow \cdots \rightarrow x(t_n)
\tag{2}
$$

with $t_{v+1} = t_v + \delta$. As long as the $f_i(t)$ of each transmission channel is obtained, we can detect outlier through the system characteristics of the parameters. $d_i(t)$ can be obtained by the time series model. [1] proposed a feature extraction algorithm for time series, which filters the available features in an early stage of the machine learning pipeline with respect to their significance for the classification or regression task. At the same time, it combines established feature extraction methods with a feature importance filter. The data processing tiers of [1] is illustrated as the following Fig. 1:

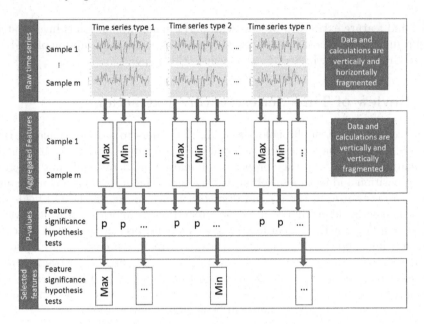

Fig. 1. Data processing tiers of tsfresh

3 Hierarchical Time Series Feature Extraction

For detecting abnormal power energy consumption activities using consumption data, one of the best ways is to take it as a time series classification problem and use supervised classification algorithms, thus, the key of the solution is to abstract time series features. Time series features extracting from existing power energy consumption time series are mainly aimed to describe information about sample distribution and/or domain knowledge that helps to classify anomaly from normal power consumption activities. In this section, a hierarchical time series feature extraction method is proposed to extract systematically time series features of power consumption time series, including summary features, shift features, transform features and decompose features.

Summary features are statistical variables that describe sample distribution of various time periods, so called time-windowed statistical variables, including maximum, minimum, mean, median, standard deviation of daily power consumption and so on. The similarity in those features is that all those features were extracted from global or one time-window piece of global time series, while a time-window piece is part of the historical power consumption that split by a start time and an end time, for example, a time-window can define as [20160101, 20160131] while the given historical consumption times is [20160101, 20161231], then the summary features of this time-window are extracted only to consider power consumption data that from 01/01/2016 to 01/31/2016. The feature extracting process of summary features is illustrated as the following Fig. 2:

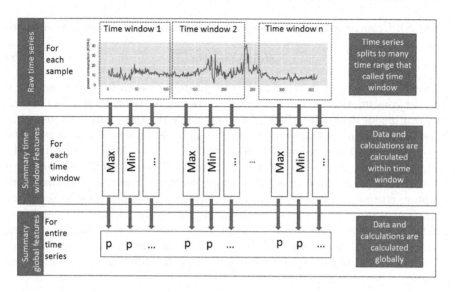

Fig. 2. Summary features extracting process tiers

Some of the summary features are defined as follows, in which S indicates a time-window $[t_1, t_n]$ while s_{t_i} induces a power consumption value at time t_i:

- $maximum(S)$: Maximum value of power consumption within given time-window

$$maximum(S) = max\{s_{t_1}, s_{t_2}, s_{t_3}, \ldots, s_{t_n}\}. \tag{3}$$

- $minimum(S)$: Minimum value of power consumption within given time-window

$$minimum(S) = min\{s_{t_1}, s_{t_2}, s_{t_3}, \ldots, s_{t_n}\}. \tag{4}$$

- $mean(S)$: Arithmetic mean value of power consumption within given time-window

$$mean(S) = \bar{S} = \frac{1}{n}\sum_{i=1}^{n} s_{t_i}. \tag{5}$$

- $median(S)$: Middle of the sorted values of power consumption within given time-window when has an uneven number of samples, while average value of the two middle if even number of samples

$$median(S) = \begin{cases} s_{t_{(n+1)/2}} & : t_n \text{ is uneven,} \\ \frac{1}{2}(s_{t_{(n/2)}} + s_{t_{(n/2+1)}}) & : t_n \text{ is even.} \end{cases} \tag{6}$$

- $var(S)$: Expectation of the squared deviation of power consumption values of given time-window

$$var(S) = \frac{1}{n}\sum_{i=1}^{n_i} (s_{t_i} - \bar{S})^2. \tag{7}$$

– $std(S)$: Uncorrected sample standard deviation of power consumption values

$$std(S) = \sqrt{var(S)}. \tag{8}$$

Shift features are features that extracting from time-shift value comparison of power consumption time series values, including decrement value, decreasing rate, continuous decreasing times and so on, while the time-shift is expressed as "time shift" in daily or time-window, for example, decrement of maximum value of two different time-window [20160101, 20160131] and [20160301, 20160331]. Shift features are designed to describe patterns of comparison distribution of power consumption time series, which is expected to describe abnormal consumption custom changes. Take decreasing rate for example, mostly electricity stealing may result in obvious decrement in power consumption data and may last for some times, thus the decrement value and decreasing rate extracting from two nearly time-window can describe those information and help to identify anomaly and normality. Part but not all of the shift features is defined as follows:

– $decrement(F_t)$: Decrement power consumption attribution F_t of time-window t compare to F_{t-1} at time-window $t-1$, where attribution F is one of the summary features

$$decrement(F_t) = F_t - F_{t-1}. \tag{9}$$

– $decrementRate(F_t)$: Decrement rate of power consumption attribution F_t at time-window t compare to F_{t-1} at time-window $t-1$, where attribution F is one of the summary features

$$decrementRate(F_t) = \frac{F_t - F_{t-1}}{F_{t-1}}. \tag{10}$$

Transform features are features that transform from existing features to another form using different project functions, including $log, sqrt, power, fft$ and so on. The transform features aims to transform features to approach more closely to the real rule of data distribution or to get a more robust form to describe potential data distribution rules. Part but not all transform features are defined as follows:

– $log(F_t)$: Transform exist feature F_t to a new feature by using logarithmic function

$$log(F_t) = log_2(F_t). \tag{11}$$

– $fft(S_t)$: One-dimensional discrete Fourier Transform for given time-windowed power consumption time series S_t

$$fft(S_t) = \sum_{i=t_1}^{t_n} s_i \cdot exp\left(-\frac{2\pi ik(i-1)}{n}\right) \tag{12}$$

Decompose features are features that extracting from decomposition of power consumption time series, and based on the assumption that time series can deconstructs into several components and each representing one of the underlying categories of patterns. Decomposition of time series is a common statistical method and an important technique for all time series analysis, so as to power consumption time series. Decompose features including seasonal features, cyclical features, trend features and so on.

4 Performance Evaluation

In this section, the proposed hierarchical time series feature extraction method is implemented and analyzed in over 54,000 customers' historical power consumption for anomaly detection, with comparison to existing time series feature extraction methods tsfresh [1]. The formation of user historical power consumption data is defined as Table 1.

Table 1. Historical power consumption data

Column name	Format	Sample	Comment
user_id	string	1234567890	Unique id encoded by location and user type
day_index	int	1	Date indexed from 01/01/2014
meter_value	float	12.34	User daily power consumption (KWH)

One of the typical historical power consumption time series is shown as the following Fig. 3, which can be obviously seen that some of the historical power consumption data is missing during period of day index range from 390 to 780, which is also popular in other customers' historical power consumption time series, due to devices error, communication failure, power line modification and so on.

To evaluate the performance of proposed hierarchical time series feature extraction method (named HTSF), features extracting by the open source automatic time series relevant features extraction library tsfresh, introduced in [1], with minimum features (named tsfresh-mini) and full features (named tsfresh-full) is used, as well as one of the most popular and high performance supervised binary algorithms called xgboost is employed, with default parameters used.

The automatic time series feature extracting library tsfresh is a perfect open source python library that can be easily used and satisfy the basic needs of time series feature extracting and filtering, however, the features extracting by tsfresh are designed for common usage and not efficient enough for power consumption anomaly detection comparing to HTSF. And even worse, the feature extracting time of tsfresh with full feature extracting parameter is much larger than acceptable, for example, it even spends more that 36 h in extracting full features from 54,174 customers' power consumption data but HTSF only spends 1 h.

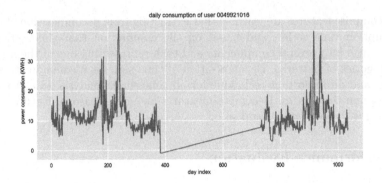

Fig. 3. Summary features extracting process tiers

By using one of the most popular supervised machine learning algorithms xgboost [12], the efficiency of each method is analyzed by comparing the predicted precision. As shown in the following Fig. 4, it shows obviously that the HTSF is more efficient than tsfresh while enough samples is offered. It is also a validation to the view that more samples may increase model precision.

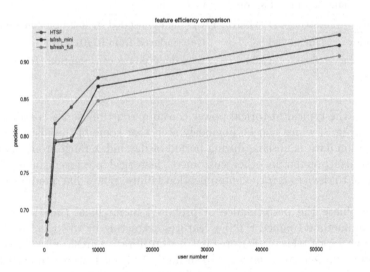

Fig. 4. Feature efficiency comparison

Performance on historical power consumption data with 54,174 labeled users shows that the proposed method is able to obtain a precision of over 93% and outperforms the existing time series feature extraction methods. The proposed method is also the 3rd winning solution to the Customer Abnormal Consumption Activities Detection of 2016 China Computer Federation Big Data and Computational Intelligence competition (2016 CCF DBCI) to predict stealing probability of nearly 100,000 customers.

5   Conclusions

In this paper, we proposed a systematic method of time series feature extraction, including summary features, shift features, transform features and decompose features, for anomaly detection of power consumption. Performance on over 54,000 customers' daily historical power consumption data with user registration shows that the proposed method is outperforming other time series feature extraction algorithms with respect to scalability and achieved accuracy.

Acknowledgments. This work has been partially supported by National Natural Science Foundation (NNSF) of China under Grant 61533010; Open Lab fund of NUPT (2014XSG03).

References

1. Christ, M., Kempa-Liehr, A.W., Feindt, M.: Distributed and parallel time series feature extraction for industrial big data applications. arXiv preprint arXiv: 1610.07717 (2016)
2. Yijia, T., Hang, G.: Anomaly detection of power consumption based on waveform feature recognition. In: 11th International Conference on Computer Science & Education (ICCSE), pp. 587–591. IEEE (2016)
3. Liu, F.T., Ting, K.M., Zhou, Z.H.: Isolation forest. In: Eighth IEEE International Conference on Data Mining, pp. 413–422. IEEE (2008)
4. Rousseeuw, P.J., Driessen, K.V.: A fast algorithm for the minimum covariance determinant estimator. Technometrics **41**(3), 212–223 (1999)
5. Abe, N., Zadrozny, B., Langford, J.: Outlier detection by active learning. In: Twelfth ACM SIGKDD International Conference on Knowledge Discovery and Data Mining, pp. 504–509. ACM, Philadelphia (2006)
6. He, Z., Xu, X., Deng, S.: Discovering cluster-based local outliers. Pattern Recogn. Lett. **24**(9–10), 1641–1650 (2003)
7. Salvador, S., Chan, P., Brodie, J.: Learning states and rules for time series anomaly detection. In: Seventeenth International Florida Artificial Intelligence Research Society Conference, pp. 41–45. IEEE, Miami (2004)
8. Morchen, F.: Time series feature extraction for data mining using DWT and DFT (2003)
9. Batal, I., Hauskrecht, M.: A supervised time series feature extraction technique using DCT and DWT. In: International Conference on Machine Learning and Applications, pp. 735–739. IEEE, Miami (2009)
10. Shandilya, S., Sabouriazad, P., Attin, M., et al.: Time series feature extraction and machine learning for prediction of in-hospital cardiac arrest. Circulation **130**(Suppl 2), A19524–A19524 (2014)
11. Coyle, D., Prasad, G., Mcginnity, T.M.: A time-series prediction approach for feature extraction in a brain-computer interface. IEEE Trans. Neural Syst. Rehabil. Eng. **13**(4), 461–467 (2005)
12. Chen, T., Guestrin, C.: XGBoost: a scalable tree boosting system. In: 22nd SIGKDD Conference on Knowledge Discovery and Data Mining (2016)

Prospect Theory Based Electricity Allocation for GenCos Considering Uncertainty of Emission Price

Yue Zhang[(⊠)] and Shaohua Zhang[(⊠)]

Shanghai Key Laboratory of Power Station Automation Technology,
Department of Automation, Shanghai University, Shanghai, China
yyzhang_mijee@163.com, eeshzhan@126.com

Abstract. Under the electricity market environment, power generation companies (GenCos) can either sell electricity through the spot market or sell them through bilateral contracts. GenCos have to make electricity allocation strategies among different trading choices facing uncertainty of spot market prices. In addition, uncertainty of the emission price is increasing and will become an important risk factor for fossil fuel GenCos. In this paper, we develop a risk decision model for fossil fuel GenCos' electricity allocation based on the prospect theory, which considers GenCos' loss aversion characteristic. Under uncertainties of the electricity spot market price and emission price, the model maximizes the GenCo's overall prospect value through allocating reasonably electricity between the spot market and bilateral contracts. The simulation results show that GenCos' psychological expected profit and loss aversion characteristic have significant effects on their risk decision-making. As uncertainty of the emission price increases, fossil fuel GenCos will increase electricity sale in the spot market.

Keywords: Fossil fuel GenCos · Electricity allocation · Prospect theory · Loss aversion · Carbon emissions

1 Introduction

The market-oriented reform of the power industry provides a variety of trading options for generation companies (GenCos), who can either sell electricity through spot market or sell them through bilateral contracts. Usually, the spot market price is volatile with strong uncertainty, while bilateral contract prices are relatively stable. Due to the different risk characteristics of various trading choices, the fluctuation of various trading prices has different influences on the profits of GenCos. GenCos need to allocate reasonably electricity among various trading markets using risk management in order to achieve a satisfactory risk-return objective [1–5].

Currently the fossil fuel generation is still the main generation form around the world. Fossil fuel GenCos' emissions of carbon dioxide are the main factor leading to global climate change. As global climate change obtains more and more attention, carbon cost becomes a new variable that affects the production decisions of fossil fuel GenCos [6].

© Springer Nature Singapore Pte Ltd. 2017
K. Li et al. (Eds.): LSMS/ICSEE 2017, Part III, CCIS 763, pp. 276–286, 2017.
DOI: 10.1007/978-981-10-6364-0_28

The EU emissions trading system (EU-EST) is the world's first operating carbon trading system with the world's largest carbon trading volume. Beginning in 2013, the EU-ETS stipulates that the carbon emission permits is distributed using the market bidding mechanism instead of the previous free distribution mechanism, which greatly increases the volatility of emissions prices. The uncertainty of emission price becomes an important risk factor in electricity allocation decision-making for fossil fuel GenCos. To date, the risk factors on the income side, such as the uncertainty of the spot market price are considered in most of the relevant studies, while the risk factors on the cost side, such as emission price are ignored [7–9].

The prospect theory was developed by Kahneman and Tversky [10]. Compared with the traditional risk decision-making theory, the prospect theory considers more practical psychological features of real decision-makers such as the bounded rationality, loss aversion and risk preference relying on reference points. The prospect theory has already been applied in the field of power systems. In [11], an optimization of the power system black-start schemes based on prospect theory is introduced. In [12], the prospect theory is applied to the optimal selection of grid planning schemes and an evaluation index system on planning schemes for power grid with high penetration intermittent generations is built. In [13], the problem of optimally charging or discharging of customer-owned storage units in a smart grid is studied taking into account how the customers estimate their utilities with respect to their real-world considerations of gains and losses.

In this paper, a risk decision model is developed for fossil fuel GenCos' electricity allocation based on the prospect theory. GenCos' loss aversion characteristic under uncertainties of the electricity spot market price and emission price is considered. The reasonableness of the proposed method is validated by simulation results.

2 Prospect Theory

The prospect theory reveals the actual individuals' decision-making mechanism under the risk and uncertainty conditions. In the prospect theory, the value function is introduced which has three main characteristics: (1) a decision-maker is risk averse when facing gains; (2) a decision-maker is risk seeking when facing losses; (3) a decision-maker is more sensitive to losses than gains (loss aversion).

The independent variable of the value function is wealth change which compares with personal wealth reference point x_0. The proposed value function is commonly S shaped, concave above the reference point and convex below it. As is illustrated in Fig. 1, when the wealth is greater than the reference point, people in the gain situation tend to choose the deterministic scheme in which risk aversion is dominant. When the wealth is less than the reference point, people in the loss situation tend to choose the risky scheme in which risk seeking is dominant.

Personal reference point plays a key role in the value function. The different reference point will affect the personal judgment on the gain or loss situations, thereby affect personal attitudes towards risk. The individual will be subjective to determine the reference point depending on the current individual wealth level or the wealth level a person seeks to achieve.

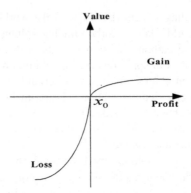

Fig. 1. The value function

In the actual decision-making process, individuals do not use the Bayes' rule to estimate the probability; rather they use the heuristic psychological decision process which includes decision weights of individual emotional factors. Thus in the prospect theory, the weighting function is introduced. The major characteristics of the weighting function are that the low probabilities are overestimated and high probabilities are underestimated.

Risk decision-making in the prospect theory can be divided into the editing process and the evaluation process. In the editing process, the information needs to be collected and processed. In the evaluation process, an individual need to select an appropriate reference point and obtain the prospect value with the combination of the value function and the weigh function. The scheme with the maximum prospect value is chosen as the optimal scheme.

3   GenCos' Electricity Allocation Based on Prospect Theory

3.1   GenCos' Deterministic Electricity Allocation Model

In a certain time period T (1 h), a fossil fuel GenCo can choose to sell electricity through the spot market or sell them through bilateral contracts to avoid the volatility of spot market price. In order to meet the requirements of generating power, the fossil fuel GenCo needs to buy emission permits in carbon market. Therefore, the profit for the fossil fuel GenCo can be divided into two parts: generation cost (fuel cost and carbon emissions cost) and electricity sale income (electricity sale income through spot market and bilateral contract market).

Generation Cost
Generation cost can be calculated as the sum of fuel cost and carbon emission cost. Fuel cost C_F can be expressed as:

$$C_F = \left(aQ_G^2 + bQ_G + c\right)\lambda_F \tag{1}$$

where Q_G is the generating electricity, λ_F is the fuel price, $aQ_G^2 + bQ_G + c$ is fuel consumption for the fossil fuel GenCo.

Assume that carbon emission from the fossil fuel GenCo is calculated as:

$$E = Q_G e_f \tag{2}$$

where e_f is the carbon emission rate for the fossil fuel GenCo.

Emission cost C_E can be expressed as:

$$C_E = Q_G e_f \lambda_E \tag{3}$$

where λ_E is the emission price.

Electricity Sale Income

Electricity sale income can be calculated as the sum of sale income through the spot market and bilateral contracts.

Electricity sale income through the spot market, R_S, can be expressed as:

$$R_S = \lambda_S Q_S \tag{4}$$

where λ_S is the spot market price, Q_S is the electricity sale through the spot market.

Electricity sale income through bilateral contract R_B can be expressed as:

$$R_B = \lambda_B Q_B \tag{5}$$

where λ_B is the bilateral contract price, Q_B is the electricity sale through bilateral contract.

The GenCo's Profit

The fossil fuel GenCo's profit F can be calculated as:

$$F = R_S + R_B - C_F - C_E \tag{6}$$

The objective which aims to maximize the profit of the fossil fuel GenCo can be expressed as:

$$\max F(Q_S, Q_B) = \lambda_S Q_S + \lambda_B Q_B - \lambda_F \left(a Q_G^2 + b Q_G + c\right) - \lambda_E Q_G e_f \tag{7}$$

The following constraint should be considered: The total electricity sale through the spot market and bilateral contract is equal to the total generating electricity:

$$Q_S + Q_B = Q_G \tag{8}$$

3.2 GenCos' Electricity Allocation Model Considering Uncertainties of Spot Market and Emission Price

Assumptions

In a certain time period T (1 h), a fossil fuel GenCo allocates electricity between spot market and bilateral contract under uncertainties of the spot market price and emission price. Decision variable is the electricity sale through bilateral contract Q_B and the electricity sale through the spot market is $Q_G - Q_B$. The probability of the spot market price λ_S based on the historical data obeys a normal distribution, namely, $\lambda_S \sim N(\mu_S, \sigma_S^2)$. The probability of the emission price λ_E satisfies a normal distribution, i.e. $\lambda_E \sim N(\mu_E, \sigma_E^2)$. Here μ and σ are respectively the mean and variance. The correlation between spot market price and emission price is not considered.

From the above assumptions, the GenCo's profit can be calculated as:

$$F(Q_B) = \lambda_S(Q_G - Q_B) + \lambda_B Q_B - \lambda_F(aQ_G^2 + bQ_G + c) - \lambda_E Q_G e_f \qquad (9)$$

According to (9), the probability density distribution of the GenCo's profit $F(Q_B)$ satisfies a normal distribution, i.e. $F(Q_B) \sim N(\mu_F, \sigma_F^2)$, its mean and variance are as follows:

$$\begin{aligned}
\mu_F &= \mu_S(Q_G - Q_B) + \lambda_B Q_B - (aQ_G^2 + bQ_G + c)\lambda_F - \mu_E Q_G e_f \\
\sigma_F^2 &= \sigma_S^2(Q_G - Q_B)^2 + \sigma_E^2 Q_G^2 e_f^2
\end{aligned} \qquad (10)$$

The GenCo's Profit Reference Point and Value Function

The fossil fuel GenCo can set a certain psychological expected profit as profit reference point F_0. Hence, ΔF, as expressed below, can measure the value of deviations from that reference point, i.e. gains and losses.

$$\Delta F = F(Q_B) - F_0 \qquad (11)$$

When the total profit $F(Q_B)$ is greater than the expected profit F_0, i.e. $\Delta F > 0$, the GenCo is satisfied with positive psychological perception. When the profit $F(Q_B)$ is lower than the expected profit F_0, i.e. $\Delta F < 0$, the GenCo is disappointed with negative psychological perception.

According to the prospect theory, the value function has demonstrated to reflect the GenCo's subjective value of income deviation. The value function is concave when the GenCo facing gain prospect, which indicates that the decision maker is more inclined to realize the deterministic profit, namely, risk aversion. The value function is convex when the GenCo facing loss prospect, which indicates that the decision maker is more inclined to be risky, namely, risk seeking. The value function is as follows:

$$v(Q_B) = \begin{cases} (\Delta F)^{\alpha}, & \Delta F \geq 0 \\ -\lambda(-\Delta F)^{\beta}, & \Delta F < 0 \end{cases} \qquad (12)$$

In (12), α and β are respectively the risk aversion parameter and the risk seeking parameter, their value is between 0 and 1. As α decreases, the degree of risk aversion for the GenCo increases. As β decreases, the degree of risk seeking for the GenCo increases. λ is the parameter of loss aversion, $\lambda > 1$. As λ increases, the degree of loss aversion for the GenCo increases.

GenCos' Electricity Allocation Model

When electricity sale through bilateral contract is Q_B, we can calculate the probability distribution of the GenCo's profit which satisfies a normal distribution. By discretizing this distribution, m points are sampled between F_{max} and F_{min} in ascending order, here $F_{max} = \mu_F + 3\sigma_F$, $F_{max} = \mu_F - 3\sigma_F$. The sample points whose profit is smaller than the reference profit F_0 are the loss sample points (suppose there are n loss points). On the contrary, the sample points whose profit is larger than the reference profit F_0 are the gain sample points (suppose there are $m-n$ gain points). The value of each sample point and the conditional expectation value for gains and losses are calculated according to the value function. The overall prospect value for the GenCo is calculated finally when electricity sale through bilateral contract is Q_B.

The objective function for the GenCo's electricity allocation decision-making is as follows, which takes into account the GenCo's psychological gains and losses:

$$\max_{Q_B} V(Q_B) = \sum_{k=1}^{n} p_k^- \cdot v_k(Q_B)^- + \sum_{k=n+1}^{m} p_k^+ \cdot v_k(Q_B)^+ \tag{13}$$

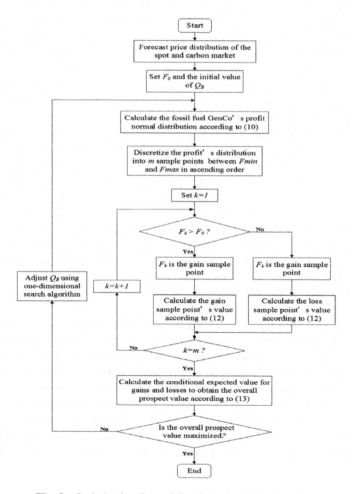

Fig. 2. Optimization flow of GenCos' electricity allocation

In (13), $V(Q_B)$ is the overall prospect value for the fossil fuel GenCo when electricity sale through bilateral contract is Q_B. $v_k(Q_B)^+$ and $v_k(Q_B)^-$ are respectively the samples' value functions for gains and losses. p_k^+ and p_k^- are respectively the samples' probability for gains and losses.

The optimization problem in (13) can be solved using one-dimensional search algorithm. The optimization flow is shown in Fig. 2.

4   Simulation Results and Analysis

The simulation is performed by using data in [14, 15]. In a certain time period T (1 h), we assume that the probability of spot market price obeys a normal distribution, which has the mean value of 36.87 \$/MWh and the standard deviation is 3.54 \$/MWh, namely, $\lambda_S \sim N(36.87, 3.54^2)$. The fuel price λ_F is 2.95 \$/MBtu. The probability of emission price follows a normal distribution, which has the mean value of 12.69 \$/tCO$_2$ and the standard deviation is 0.39 \$/tCO$_2$, namely, $\lambda_E \sim N(12.69, 0.39^2)$. Contract price λ_B is 36 \$/MWh. The total generating electricity Q_G is 3000 MW. In the fuel consumption function, parameters a is 0.00037 MBtu/MWh, b is 4.76 MBtu/MWh and c is 683.91 MBtu [16]. Emission rate e_f is 1.2tCO$_2$/MWh [17]. The loss aversion, risk seeking parameter and risk aversion parameter are determined by Kahneman and Tversky experiment. The loss aversion parameter λ is 2.25, both the risk seeking parameter α and the risk aversion parameter β take a value of 0.88.

Figure 3 shows the optimal prospect value of the GenCo by the proposed model under different psychological expected profits. It can be seen from Fig. 3 that when the expected profit is less than 8000\$, the optimal prospect value of the GenCo is greater than 0, and when the profit is greater than 8000\$, the optimal prospect value of the GenCo is less than 0. According to the prospect theory, the prospect value is equal to the sum of expectation value of gains and losses. A prospect value that is greater than 0 indicates that the expectation value of gains is greater than the expectation value of losses. For prospect values below 0, the expectation value of gains is lower than the expectation value of losses. When the expected profit is less than 8000\$, the optimal prospect value of the GenCo is greater than 0, which indicates that the GenCo is psychologically "gain" at this time and tends to avoid risk to allocate more power to the lower risky contract market. When the expected profit is more than 8000\$, the optimal prospect value of the GenCo is less than 0, which indicates that the GenCo is psychologically "loss" at this time and has a strong risk seeking inclination in decision-making to allocate more electricity to the high risky spot market.

It can be seen from Fig. 4, when the expected profit is low (less than 8000\$), the electricity allocated to contract market increases and the electricity allocated to spot market decreases with increase of the psychological expected profit. It is because when the F_0 is small, the GenCo is psychologically "gain" and tends to avoid risk to allocate more electricity to the lower risky contract market. With increase of the expected profit (more than 8000\$), the electricity allocated to contract market decreases and the electricity allocated to spot market increases. It is because when the F_0 is relatively large, the GenCo is psychologically "loss" and has a strong risk seeking inclination in decision-making to allocate more electricity to the high risky spot market. When the

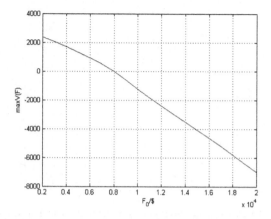

Fig. 3. The optimal prospect values under different expected profits

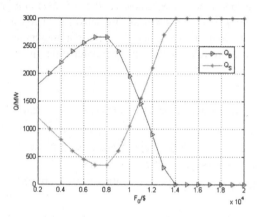

Fig. 4. Electricity allocation under different expected profits

expected profit is large (greater than 14000$), the GenCo will allocate all the electricity into the spot market.

According to the prospect theory, the loss aversion parameter λ reflects the degree of loss aversion of decision maker. As λ ($\lambda > 1$) increases, the degree of loss aversion for the decision maker increases too. Let psychological expected profit equal to 9000$, and the other parameters are consistent with the above. Figure 5 shows the results of the electricity allocation under different loss aversion parameters. It can be seen that with increase of the loss aversion parameter λ, the electricity allocated to contract market increases and the electricity allocated to spot market decreases. It is because, with increase of the loss aversion parameter λ, the GenCo will have greater degree of loss aversion and be more inclined to choose the lower risky contract market and deterministic income. Therefore, the electricity allocated in contract market will increase.

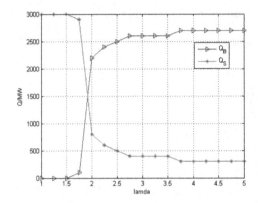

Fig. 5. Electricity allocation under different loss aversion parameters

In addition to the loss aversion parameter λ, the risk aversion parameter α and the risk seeking parameter β can also have an impact on decision-making. Let psychological expected profit equal to 9000$, and the other parameters are consistent with the above. Electricity sales through contract market under different risk factors are shown in Table 1.

Table 1. Electricity sale through contract under different risk factors

α ($\beta = 0.88$)	0.5	0.6	0.7	0.8	0.9	1.0
Q_B/MW	2800	2800	2800	2600	1800	0
β ($\alpha = 0.88$)	0.5	0.6	0.7	0.8	0.9	1.0
Q_B/MW	0	0	2400	2600	2700	2800

It can be seen from Table 1, with the decrease of the risk aversion parameter α, the electricity allocated to the contract market increases. This is because, with decrease of α, the GenCo tends to avoid risk, thus the electricity allocated to the lower risky contract market will increase. With decrease of risk seeking parameter β, the electricity allocated to contract market decrease. The reason is that, with decrease of β, the GenCo tends to seek risk, thus the electricity allocated to high risky spot market will increase. Because of the different characteristics of the decision makers, different strategies will be chosen in the face of the same psychological expected profit. Compared to the traditional risk decision-making theory, which simply regard the decision makers as completely rational decision makers, the value function in the prospect theory can embody different decision makers' personality characteristics. This makes the decision results more consistent with the behaviors of practical decision-makers with bounded rationality.

Table 2 shows the impact of the emission price standard deviation on the electricity allocation decision-making for the GenCo. It can be found that, with increase of the risk in emission price, the GenCo tends to allocate more electricity to high risky spot market in order to reach the psychological expected profit.

Table 2. Electricity sale through contract under different emission price standard deviation

$\sigma_E/\$$	0.1	0.2	0.3	0.4	0.5
Q_B/MWh	2600	2500	2500	2400	2300
$\sigma_E/\$$	0.6	0.7	0.8	0.9	1.0
Q_B/MWh	2200	2100	1900	1900	1800

5 Conclusion

A risk decision model for fossil fuel GenCos' electricity allocation based on the prospect theory is developed in this paper. GenCos' loss aversion characteristic under uncertainties in electricity spot market price and emission price is considered. The electricity allocation strategy between spot market and contract market is determined when the GenCo's overall prospect value is maximized. The simulation results indicate that the GenCo's psychological expected profit and loss aversion characteristic have significant impacts on its allocation strategies. As risk in emission price increases, fossil fuel GenCos will increase electricity sale in the spot market.

References

1. Liu, M., Wu, F.L.: Portfolio optimization in electricity markets. Electr. Power Syst. Res. **77** (8), 1000–1009 (2007)
2. Liu, M., Wu, F.L.: Trading strategy of generation companies in electricity market. Proc. CSEE **28**(25), 111–117 (2008)
3. Wang, R., Shang, J.C., Feng, Y.: Combined bidding strategy and model for power suppliers based on CVaR risk measurement techniques. Autom. Electr. Power Syst. **29**(14), 5–9 (2005)
4. Wang, N., Peng, J.C., Dai, H.C.: A sequence operation theory based capacity allocation strategy for generation companies in uncertain and multi electricity markets. Power Syst. Technol. **30**(23), 77–82 (2006)
5. Liu, Y.A., Xue, Y.S., Guan, X.H.: Optimal allocation in dual markets for price-taker: part two: for generator. Autom. Electr. Power Syst. **28**(17), 12–15 (2004)
6. Dallas, B.: Regulating CO_2 in electricity markets: sources or consumers? Clim. Policy **8**(6), 588–606 (2008)
7. Feng, D., Gan, D.: Supplier asset allocation in a pool-based electricity market. J. Power Syst. IEEE Trans. Power Syst. **22**(3), 1129–1138 (2007)
8. Conejo, A.J., Nogales, F.J., Arroyo, J.M.: Risk-constrained self-scheduling of a thermal power producer. IEEE Trans. Power Syst. **19**(3), 1569–1574 (2004)
9. Amjady, N., Vahidinasab, V.: Security-constrained self-scheduling of generation companies in day ahead electricity markets considering financial risk. Energy Convers. Manag. **65**(6), 164–172 (2013)
10. Kahneman, D., Tversky, A.: Prospect theory: an analysis of decision under risk. Econometrica **47**(2), 263–1131 (1979)
11. Li, R.Q., Tang, L.Q., Ling, W.N.: Optimization of black-start based on prospect theory and grey relational analysis. Power Syst. Prot. Control **41**(5), 103–107 (2013)

12. Gao, S., Wang, S.Z., Li, H.F.: Prospect theory based comprehensive decision-making method of powernetwork planning schemes. Power Syst. Technol. **38**(8), 2029–2036 (2014)
13. Wang, Y., Saad, W.: On the role of utility framing in smart grid energy storage management. In: IEEE International Conference on Communication Workshop, pp. 1946–1951 (2015)
14. European Energy Exchange. http://www.eex.com
15. Nord Pool. http://www.nordpoolspot.com
16. Mathuria, P., Bhakar, R., Li, F.: GenCo's optimal power portfolio selection under emission price risk. Electr. Power Syst. Res. **121**, 279–286 (2015)
17. Hammond, G.P., Akwe, S.O., Williams, S.: Techno-economic appraisal of fossil-fuelled power generation systems with carbon dioxide capture and storage. Fuel Energy Abstr. **36** (2), 975–984 (2011)

Neural-Network-Based Tracking Control of Offshore Steel Jacket Platforms

Zhi-Hui Cai[1(✉)], Bao-Lin Zhang[1(✉)], and Xian-Hu Yu[2]

[1] China Jiliang University, Hangzhou 310018, China
zhcai81@163.com, zhangbl2006@163.com
[2] Ningbo Radio and TV University, Ningbo 315016, China

Abstract. This paper deals with the problem of neural network tracking control for an offshore platform system under external wave forces. A feedforward backpropagation neural-network-based tracking controller (NNTC) is designed to attenuate the displacement response of the offshore platform. In the simulation, the proposed NNTC scheme can effectively improve the stability of the offshore platform. Furthermore, the designed NNTC is more robust than the feedforward and feedback optimal tracking controller (FFOTC) in terms of system parametric perturbations and external wave loads.

1 Introduction

Offshore steel jacket platforms located in 1000+ feet of water usually suffer from continuous vibrations induced by the wind, wave force and earthquakes [1,2]. Over the past few decades, various control methods have been effectively applied to reduce the vibration below the dangerous level. Active and/or passive control methods have been investigated theoretically and experimentally, see [2–10] and the references therein.

Artificial neural networks (ANNs) as a computational model can be used in many problems [11]. Kuźniar et al. [12] show the application of neural networks for determining the natural frequency of an existing structure. Elshafey et al. present a combined method of neural networks and random decrement signature to the damage identification of offshore platforms under random loads [13]. Neural networks have successfully been implemented to control structures under ground excitations by Ghaboussi et al. [14] and Chen et al. [15]. Zhou et al. [16] trained a neural network in order to duplicate a linear quadratic regulator(LQR) controller in a linear offshore structure and showed that the neural controller has the same performance of the LQR controller. Kim et al. [2] have firstly trained a neural network as an optimal controller instead of a linear controller and obtained a neural controller to offshore structures subject to the earthquake. More recently, Uddin et al. [17] show a detailed review on the application of artificial neural network in fixed jacket platforms, Zhang et al. [18] give a summary of references of neural network-based controllers in active control for offshore platforms.

© Springer Nature Singapore Pte Ltd. 2017
K. Li et al. (Eds.): LSMS/ICSEE 2017, Part III, CCIS 763, pp. 287–295, 2017.
DOI: 10.1007/978-981-10-6364-0_29

The artificial neural network has rarely applied in offshore platform under wave loads in most of the previous studies. In the paper [19], the authors obtained an FFOTC by solving some differential equations. In this paper, a neural-network-based tracking controller is developed and applied to an offshore steel jacket platform with active mass damper(AMD) as in [19]. A dynamical system of the offshore structure subjected to wave forces is described. Then, a neural controller is derived. Finally, experiments results show the effectiveness of the proposed neural-network-based tracking controller. The main contributions are as follows:

1. A neuron network tracking controller is proposed for an offshore platform model with AMD under wave loads.
2. Robustness of the designed controller is showed under system parametric perturbations and external wave loadings with small random disturbances.

2   Problem Formulation

An offshore platform model with an AMD device [8], shown in Fig. 1, is used. The dynamical system of the offshore platform can be expressed [20] as follows

$$
\begin{cases}
m_1\ddot{z}_1(t) = -(m_1\omega_1^2 + m_2\omega_2^2)z_1(t) + m_2\omega_2^2 z_2(t) + 2\xi_2\omega_2 m_2\dot{z}_2(t) \\
\qquad\qquad -2(m_1\xi_1\omega_1 + m_2\xi_2\omega_2)\dot{z}_1(t) + f(t) - u(t) \\
m_2\ddot{z}_2(t) = m_2\omega_2^2[z_1(t) - z_2(t)] + 2m_2\xi_2\omega_2[\dot{z}_1(t) - \dot{z}_2(t)] + u(t)
\end{cases}
\tag{1}
$$

where $u(t)$ is the active control force, $f(t)$ is the wave force acting on the model; m_1, $\omega_1(t)$ and $\xi_1(t)$ denote the modal mass, natural frequency and damping ratio of the offshore platform, respectively; m_2, $\omega_2(t)$ and $\xi_2(t)$ denote the mass,

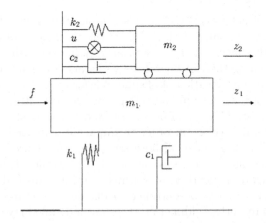

Fig. 1. An offshore platform with an active mass damper [8].

natural frequency and damping ratio of the AMD device, respectively; $z_1(t)$ and $z_2(t)$ denote the displacements of the deck motion of the offshore platform and the AMD device, respectively.

Define the following state variables

$$x_1(t) = z_1(t), \ x_2(t) = z_2(t), \ x_3(t) = \dot{z}_1(t), \ x_4(t) = \dot{z}_2(t) \quad (2)$$

and denote $x(t) = [x_1(t) \ x_2(t) \ x_3(t) \ x_4(t)]^T$. Then the motion Eq. (1) can be expressed as

$$\dot{x}(t) = Ax(t) + Bu(t) + Df(t), \quad x(0) = x_0 \quad (3)$$

where

$$\begin{cases} A = \begin{bmatrix} 0 & 0 & 1 & 0 \\ 0 & 0 & 0 & 1 \\ -(\omega_1^2 + \omega_2^2 \frac{m_2}{m_1}) & \omega_2^2 \frac{m_2}{m_1} & -2(\xi_1\omega_1 + \xi_2\omega_2 \frac{m_2}{m_1}) & 2\xi_2\omega_2 \frac{m_2}{m_1} \\ \omega_2^2 & -\omega_2^2 & 2\xi_2\omega_2 & -2\xi_2\omega_2 \end{bmatrix} \\ \\ B = \left[0, 0, -\frac{1}{m_1}, \frac{1}{m_1} \right]^T, \quad D = \left[0, 0, \frac{1}{m_1}, 0 \right]^T \end{cases} \quad (4)$$

The wave force $f(t)$ in (3) can be expressed as [8]

$$f(t) = \int_0^d p(z,t)\phi(z)dz \quad (5)$$

where z is the vertical coordinate with the origin, $\phi(z)$ is the shape function related to the offshore platform, d is the water depth and $p(z,t)$ is the physical horizontal wave force per unit length and can be approximated as follows

$$p(z,t) = \frac{1}{2}\rho C_d \tilde{D} \sqrt{\frac{8}{\pi}} \sigma_v(z)v(z,t) + \frac{1}{4}\rho\pi C_m \tilde{D}^2 \dot{v}(z,t) \quad (6)$$

where ρ, $\tilde{D}$, C_d and C_m are the water density, diameter of the cylinder, drag, and inertia coefficients, respectively. $v(z,t)$, $\dot{v}(z,t)$ and $\sigma_v(z)$ are the water particle velocity, acceleration, and the standard deviation of the velocity at direction z, respectively.

As shown in [7,19], the wave force $f(t)$ can be derived from the following equations

$$\dot{w}(t) = Gw(t), f(t) = Hw(t) \quad (7)$$

where

$$G = \begin{bmatrix} \tilde{0} & \tilde{I} \\ \tilde{G} & \tilde{0} \end{bmatrix}, \quad H = \tilde{H}[\tilde{I} \ \tilde{0}] \quad (8)$$

$$\tilde{G} = -diag[\omega_1^2, \omega_2^2, \dots, \omega_n^2], \quad \tilde{H} = [1, 1, \dots, 1]\sum_{j=1}^{n} P(\omega_j) \quad (9)$$

and

$$P(\omega_j) = \int_0^d \left[\frac{1}{2}\rho C_d \tilde{D} \sqrt{\frac{8}{\pi}} \sigma_v(z)T_{v\eta}(\omega_j, z) + \frac{1}{4}\rho\pi C_m \tilde{D}^2 T_{\dot{v}\eta}(\omega_j, z) \right] \phi(z)dz$$

The output of system (3) is given as

$$y(t) = Cx(t) \tag{10}$$

where

$$C = \begin{bmatrix} 1 & 0 & 0 & 0 \\ 0 & 0 & 1 & 0 \end{bmatrix} \tag{11}$$

The desired output y_r is obtained as

$$\dot{z}(t) = Mz(t), \quad y_r(t) = Nz(t) \tag{12}$$

where $M \in \mathbb{R}^{r \times r}$ and $N \in \mathbb{R}^{2 \times r}$ are constant matrices.

In the following, a feedforward backpropagation network tracking control law for the system (3) is designed so that $y(t)$ can track the expectation $y_r(t)$ asymptotically.

3   Design of a Neural Network Tracking Control Scheme

3.1   Model Structure

Suppose that the neural network is formed in three layers, namely the input layer, hidden layer and output layer. If a neural network with N_1, N_2, N_3 nodes on each layer has the connection weights $\mathbf{V} \in \mathbb{R}^{N_2 \times N_1}$, $\mathbf{W} \in \mathbb{R}^{N_3 \times N_2}$ between layers and the biases $\mathbf{a} \in \mathbb{R}^{N_2}$, $\mathbf{b} \in \mathbb{R}^{N_3}$ on each layer, then the outputs at the hidden layer and output layer can be described as follows

$$\mathbf{O}_h = s(\mathbf{VI} + \mathbf{a}) \tag{13}$$
$$\mathbf{O}_o = g(\mathbf{Wh} + \mathbf{b}) \tag{14}$$

where $\mathbf{I}$ is the input, $s(\cdot)$ and $g(\cdot)$ are the activation functions [11].

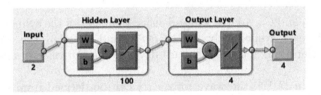

Fig. 2. A neural network structure

In the Simulink of the model of the offshore structure and the neural-network-based tracking controller, it is supposed that the model consists of three blocks: an offshore structure block, a wave force structure block, and a neural network controller block, see Fig. 3.

In the simulation, the offshore structure model with FFOTC [19] is simulated to generate learning data, which is shown in Fig. 4.

In the training process, a neural network, shown in Fig. 2, consists of three layers with 2, 100, 4 nodes on each layer. Here, the activation function $s(\cdot)$ is sigmoid function and $g(\cdot)$ is a linear function.

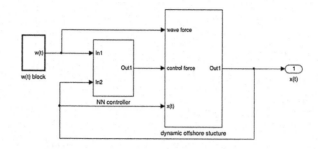

Fig. 3. Simulink model for control simulation.

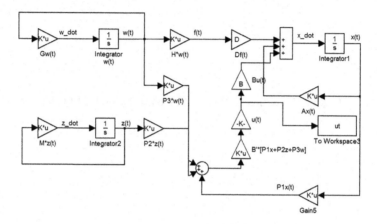

Fig. 4. Simulink model for the offshore platform under FFOTC

3.2 Comparison Analysis

To compare the FFOTC scheme [19] and the NNTC scheme, a Simulink model is designed first. Figure 5(a) shows the external wave force acting on the offshore platform. Then FFOTC $u(t)$ in [19] is obtained, see Fig. 5(b).

The feedforward backpropagation neural network is built and connected to the offshore platform model. The hidden layer of the neural network has 100 nodes. The comparison of the FFOTC and NNTC is demonstrated in Figs. 6(a) and (b), which show the displacement and velocity of the offshore platform under FFOTC and NNTC respectively.

From these two figures, one can see that both of NNTC and FFOTC significantly attenuate the amplitudes of the displacement and the velocity of the offshore platform. It also can be seen that NNTC shows a better control performance than FFOTC.

Assume a small random disturbance $dx(t)$ exists in the state vector $x(t)$, then the corresponding state vector $\tilde{x}(t)$ is changed as follows

$$\tilde{x}(t) = x(t) + dx(t) \tag{15}$$

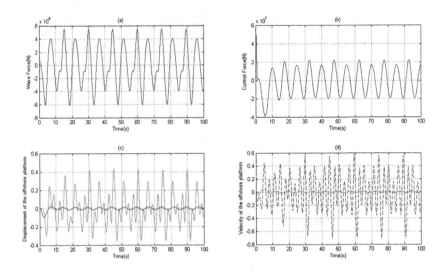

Fig. 5. Simulation of the offshore platform under FFOTC

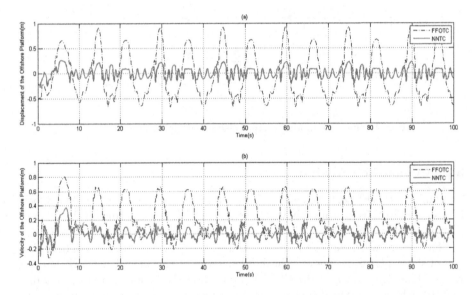

Fig. 6. Simulation of the offshore platform under FFOTC and NNTC

where $dx(t)$ is an uniformly distributed pseudo-random function whose values are in the interval $[-K_1, K_1]$, $K_1 = \lambda_x \min\{x(t)\}$. In the simulation, we set $\lambda_x = 0.1$. Figure 7(a) shows the dynamical displacements of the offshore platform under FFOTC and NNTC in terms of parametric perturbations and wave force, Fig. 7(b) shows the relevant velocities of the offshore platform under FFOTC and NNTC in terms of parametric perturbations and wave force. It shows that NNTC has more robust than FFOTC.

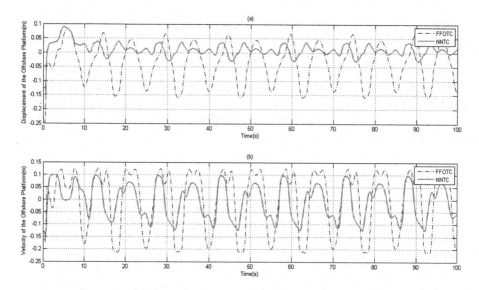

Fig. 7. Simulation of the offshore platform under parametric perturbations.

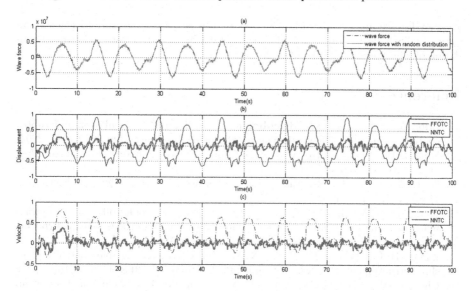

Fig. 8. Simulation of the offshore platform under wave force with a random distribution.

Similarly, suppose a small random disturbance $df(t)$ exists in the wave force model, then the wave force $\tilde{f}(t)$ is changed as follows

$$\tilde{f}(t) = f(t) + df(t) \qquad (16)$$

where $df(t)$ is an uniformly distributed pseudo-random function whose values are in the interval $[-K_2, K_2]$, $K_2 = \lambda_f \min\{f(t)\}$. In the simulation, we set $\lambda_f = 0.1$. Figure 8(a) shows the original wave force and the changed wave force. Figure 8(b) shows the dynamical displacements of the offshore platform with different wave forces, Fig. 8(c) shows the corresponding velocities of the offshore platform subjected to different wave forces. It can be seen that the NNTC scheme can attenuate the vibration amplitudes of the offshore platform subjected to external wave forces with a random distribution.

4   Conclusion

A neural network control scheme for the offshore platform under wave force has been derived. Experiments results show that the proposed NNTC scheme can significantly attenuate the wave-induced oscillation of offshore platform. The proposed NNTC scheme is effective to improve the control performance of the offshore platform. Furthermore, the designed neural-network-based tracking controller is more robust than the FFOTC in terms of system parametric perturbations and external wave loadings with small random disturbances.

Acknowledgments. This work was supported in part by the Natural Science Foundation of Zhejiang Province under Grant LQ12A01025 and the Natural Science Foundation of China under Grant 61379029.

References

1. Zribi, M., Almutairi, N., Abdel-Rohman, M., Terro, M.: Nonlinear and robust control schemes for offshore steel jacket platforms. Nonlinear Dyn. **35**, 61–80 (2004)
2. Kim, D.H.: Neuro-control of fixed offshore structures under earthquake. Eng. Struct. **31**, 517–522 (2009)
3. Abdel-Rohman, M.: Structural control of a steel jacket platform. Struct. Eng. Mech. **4**, 125–138 (1996)
4. Zhang, B.-L., Ma, L., Han, Q.-L.: Sliding mode H_∞ control for offshore steel jacket platforms subject to nonlinear self-excited wave force and external disturbance. Nonlinear Anal.: Real World Appl. **14**, 163–178 (2013)
5. Zhang, B.-L., Han, Q.-L., Zhang, X.-M.: Event-triggered reliable control for offshore structures in network environments. J. Sound Vib. **368**, 1–21 (2016)
6. Chen, C.-W., Shen, C.-W., Chen, C.-Y., Cheng, M.-J.: Stability analysis of an oceanic structure using the Lyapunov method. Eng. Comput. **27**, 186–204 (2010)
7. Ma, H., Tang, G.-Y., Zhao, Y.-D.: Feedforward and feedback optimal control for offshore structures subjected to irregular wave forces. Ocean Eng. **33**, 1105–1117 (2006)
8. Li, H.J., Hu, S.-L.J., Jakubiak, C.: H_2 active vibration control for offshore platform subjected to wave loading. J. Sound Vib. **263**, 709–724 (2003)
9. Ma, H., Hu, W., Wu, J.C., Luo, S.W., Cao, J.M.: Networked vibration control with time delays for offshore platforms under irregular wave forces. In: 42nd Annual Conference of the IEEE Industrial Electronics Society, IECON 2016, pp. 5386–5391 (2016)

10. Nourisola, H., Ahmadi, B.: Robust adaptive sliding mode control based on wavelet kernel principal component for offshore steel jacket platforms subject to nonlinear wave-induced force. J. Vib. Control **22**(15), 3299–3311 (2016)
11. Svozil, D., Kvasnicka, V., Pospichal, J.: Introduction to multi-layer feed-forward neural networks. Chemometr. Intell. Lab. Syst. **39**, 43–62 (1997)
12. Kuźniar, K., Waszczyszyn, Z.: Neural networks and principal component analysis for identification of building natural periods. J. Comput. Civil Eng. **20**, 431–436 (2006)
13. Elshafey, A.A., Haddara, M.R., Marzouk, H.: Damage detection in offshore structures using neural networks. Marine Struct. **23**, 131–145 (2010)
14. Ghaboussi, J., Joghataie, A.: Active control of structures using neural networks. J. Eng. Mech. **121**, 555–567 (1995)
15. Chen, H., Tsai, K., Qi, G., Yang, J., Amini, F.: Neural network for structure control. J. Comput. Civil Eng. **9**, 168–176 (1995)
16. Zhou, Y.J., Zhao, D.Y.: Neural network based active control for offshore platforms. China Ocean Eng. **17**, 461–468 (2003)
17. Uddin, M.A., Jameel, M., Razak, H.A.: Application of artificial neural network in fixed offshore structures. Indian J. Geo-Marine Sci. **44**, 294–403 (2015)
18. Zhang, B.-L., Han, Q.-L., Zhang, X.-M.: Recent advances in vibration control of offshore platforms. Nonlinear Dyn. **89**, 755–771 (2017)
19. Zhang, B.-L., Liu, Y.-J., Han, Q.-L., Tang, G.-Y.: Optimal tracking control with feedforward compensation for offshore steel jacket platforms with active mass damper mechanisms. J. Vib. Control **22**, 695–709 (2016)
20. Zhang, B.L., Feng, A.M., Li, J.: Observer-based optimal fault-tolerant control for offshore platforms. Comput. Electr. Eng. **40**, 2204–2215 (2014)

Intelligent Methods in Developing Electric Vehicles, Engines and Equipment

Short-Term Optimal Scheduling
with the Consideration of Electric Vehicle
Driving Rules

Xiaolin Ge$^{(\boxtimes)}$ and Chenhao Pei

College of Electrical Engineering, Shanghai University of Electric Power,
Shanghai 200090, China
gexiaolin2005@126.com

Abstract. Taking into account the daily driving rules of electric vehicle (EV), a novel short-term optimal scheduling model is proposed. And to describe the driving characteristics of the EV, EVs are divided into four types according to the detailed driving rules. And the stochastic driving time, access time and daily mileage of different types of EVs are simulated by a large number of scenes. Besides, the interaction between the EV and the power systems is added to establish the coupling between the output of units, and the operation of EVs. Due to the complexity of the constraints, the stochastic nonlinear unit commitment model is converted into mixed integer linear programming problem and solved by CPLEX. Case studies show the necessity of considering the stochastic driving rules of EVs, and the classification of EVs can make the dispatching decision more reasonable.

Keywords: Optimal scheduling · Short-term · Vehicle to Grid · Driving rules

1 Introduction

Due to the environmental friendly and energy-efficient advantages, Electric Vehicle (EV) has been developed rapidly in recent years [1]. The production and sales scale of EV are booming. It should be noted that the driving characteristics of EV are uncertain. So when large-scale of EV are integrated into the power system without proper guidance and control, it will increase the difference of peak and valley and even affect the stability of the power grid [2]. Therefore, to reduce negative impact and fully utilize energy storage characteristic of EV, the study of unit commitment with EV has been a hotspot.

Traditionally, the large units with superior economy have the priority to meet the requirement of the base load. While the small units with relative higher cost are used to response to the fluctuation of the load [3]. After the concept of Vehicle to Grid (V2G) is introduced, the traditional unit commitment can be combined with EV. And the reasonable charging and discharging for EV can decrease the cost of the unit commitment by reducing the requirement of small unit [4]. The models of unit commitment with EVs in [5–9], aim at optimizing generation schedule and EV's charging-discharging strategy. The EVs are directly dispatched, and the charging and discharging control of

© Springer Nature Singapore Pte Ltd. 2017
K. Li et al. (Eds.): LSMS/ICSEE 2017, Part III, CCIS 763, pp. 299–308, 2017.
DOI: 10.1007/978-981-10-6364-0_30

the EV is always available in the schedule horizon. However, the driving requirement of EVs is stochastic, causing the charging and discharging of EVs are uncertainty. Thus the directly dispatching model may lead to some errors in practical application, and the stochastic nature of EV should be considered in the scheduling.

In [10], the driving characteristic of EVs is formulated and the charging strategy is optimized, however the inverse discharge capacities of EV is not considered. In [11–14], both of the charging and discharging behaviors are taken into account, and the driving time and mileage of private cars are described. However, the type of EV in these literatures is single. In fact, various types of EV have different driving characteristics [15, 16]. The driving rules of private cars, buses and taxis are analyzed in [17], but the coupling of gird and varies EVs is not established. Therefore, unit commitment model with the consideration of different types of EVs still need to be studied.

In this paper, driving characteristics of various types of EVs are investigated, and a unit commitment model considering the stochastic nature of electric vehicles is proposed. The main contributions are as follows:

(1) To describe the stochastic characteristics of EVs, the access time, departure time and daily driving mileage are simulated as random variables.
(2) The driving rules of various types of EV users including private cars, buses, official vehicles and taxis, are analyzed and formulated.
(3) A stochastic optimization model of unit commitment is developed satisfying various system constraints as well as the driving requirements of EVs.

2 Stochastic Nature of Electric Vehicles

The driving characteristics of the vehicle include the departure time, access time and daily driving distance. These factors will affect the actual scheduling of vehicles. In this paper, four types EVs are considered, and the driving rules are described as follows:

(1) Private car users
    As for the private car users, the statistics of national household driving survey (NHTS) in 2009 are used to analyze the probability density function of initial departure time, initial vehicle access time and daily vehicle mileage. The results show that the access time and the departure time of the vehicle can satisfy the normal distribution, while the daily mileage of the vehicle satisfies the lognormal distribution.
(2) Bus users
    Generally, the bus begins to work at 6:00 am and ends the work at 23:00 pm. Therefore, it is deemed that the bus charges and discharges between 23:00–6:00. The daily mileage of the electric bus satisfies the lognormal distribution.
(3) Taxi
    The taxi, as a kind of operational vehicle, should maximize the operation efficiency and load the passengers immediately after charging. In order to ensure the efficiency of its operation, the taxi is only involved in the charging and not involved in the discharge. Moreover, when the taxi access to charge, it will

immediately be charged. Charging time is generally 0.5–1 h, and then the tax will be immediately put into operation. The start charging time of the vehicle can be uniformly distributed, and the daily mileage can also satisfy the log normal distribution.

(4) Official vehicles

Because the business car is on business at daytime, it is in the driving stage during work and the charging and discharging are not required. They will be charged or discharged at 18:00–8:00. The daily driving mileage of vehicle meets the logarithmic normal distribution.

3 Stochastic Unit Commitment Model with EV

3.1 Objective Function

The objective function is shown as follows:

$$MinF_1^{\cos t} = \sum_{c=1}^{C} \pi(c) \left[\sum_{t=1}^{T} \sum_{i=1}^{N} [C_G(i,t) + C_U(i,t) + C_D(i,t)] \right] \tag{1}$$

where $F_1^{\cos t}$ is total coal consumption; t is the number at the planned period; T is the total number of planned period; i is the unit of Thermal power unit; N is the total number of thermal power unit; c is the number of forecast scene of EV driving; C is the total number of scenes; $\pi(c)$ is the possibility corresponding to scene c; $C_G(i,t)$ is the thermal power unit operating costs; $C_U(i,t)$ is the starting cost; $C_D(i,t)$ is shutdown cost. The operating cost, starting cost and shutdown cost of thermal power unit are solved according to the methods in [18].

3.2 Constrained Conditions

3.2.1 Thermal Power Unit Constraint

For thermal power units, the following constraints are provided:

(1) Load balance constraint:

$$\sum_{i}^{N} P(c,i,t) + \sum_{v=1}^{V} P_{dis}(c,v,t) = D(t) + \sum_{v=1}^{V} P_{ch}(c,v,t) \tag{2}$$

where $P(c,i,t)$ is the power of thermal power unit of scene c under time t; v is the type number of EV, V is all types of EV(Private cars, buses, car rental and official vehicles); $P_{ch}(c,v,t)$ is the charge power of large scale EV; $P_{dis}(c,v,t)$ is the discharge power of large scale EV; $D(t)$ is the maximum load of the system at time t.

(2) System reserve requirement:

$$\sum_i^N \overline{P}(c,i,t) + \sum_{v=1}^V P_{dis}(c,v,t) \geq D(t) + \sum_{v=1}^V P_{ch}(c,v,t) + \overline{K}(t) \tag{3}$$

where $\overline{P}(c,i,t)$ is the upper limit of power of thermal power unit i of scene c at time t; $\overline{K}(t)$ is the upper reserve requirement of load at time t.

$$\sum_i^N \underline{P}(c,i,t) + \sum_{v=1}^V P_{dis}(c,v,t) \leq D(t) + \sum_{v=1}^V P_{ch}(c,v,t) - \underline{K}(t) \tag{4}$$

where $\underline{P}(c,i,t)$ is the lower limit of power of thermal power unit i at time t in scene c; $\underline{K}(t)$ is the lower standby requirement of load at time t.

(3) Thermal units constraint:

This constraint contains power limits, ramping up and down limits, and minimum start-up and shutdown time, the detailed formulation can be found in [21].

3.2.2   EV Correlation Constraints

(1) Driving time constraint:

If $t_{in} \leq t_{out}$, EV must charge and discharge between $[t_{in}, t_{out}]$:

$$\begin{cases} t_{in}(c,v) \leq t \leq t_{out}(c,v), X(c,v,t) + Y(c,v,t) \leq 1 \\ t \leq t_{in}(c,v) \cup t \geq t_{out}(c,v), X(c,v,t) + Y(c,v,t) = 0 \end{cases} \tag{5}$$

where $t_{in}(c,v)$ is the in-grid time of type v vehicle in scene c; $t_{out}(c,v)$ is the off-grid time of type v vehicle in scene c; $X(c,v,t)$ is the charging status of type v vehicle in scene c at time t; $X(c,v,t) = 0$ indicates the vehicle is not under charging status. $X(c,v,t) = 1$ indicates the vehicle is under charging status; $Y(c,v,t)$ is the discharging status of type v vehicle in scene c at time t; $Y(c,v,t) = 0$ indicates the vehicle is in non-discharging status; $Y(c,v,t) = 1$ indicates the vehicle is in charging status.

When $t_{out} \leq t_{in}$, EV only is charged and discharged between $[1, t_{in}] \cup [t_{out}, 24]$:

$$\begin{cases} T_{out}(c,v) \leq t \leq T_{in}(c,v), X(c,v,t) + Y(c,v,t) = 0 \\ t \leq T_{out}(c,v) \cup t \geq T_{in}(c,v), X(c,v,t) + Y(c,v,t) \geq 1 \end{cases} \tag{6}$$

(2) EV charge and discharge power:

$$P_{ch}(c,v,t) = Nev(c,v) \times PV_{ch}(c,v,t) \tag{7}$$

$$P_{dis}(c,v,t) = Nev(c,v) \times PV_{dis}(c,v,t) \tag{8}$$

where Nev(c,v) is the total number of type v vehicles into the grid under scene c. $PV_{ch}(c, v, t)$ is the charging power of type v vehicle in scene c at time t. $PV_{dis}(c, v, t)$ is the discharging power of type v in scene c at time t. It should be noted that because the charging and discharging behavior of EV lack synchronism, in order to scale the charging and discharging behavior of EV users. $P_{ch}(c, v, t)$ and $P_{dis}(c, v, t)$ are used to indicate the charging and discharging power of scale EV. The starting time of driving, the end of driving time and the daily running distance correspond to the characteristics of the corresponding single EV. They are free variable.

(3) Power balance constraint

When EV accesses to grid, namely $t = t_{in}(c, v)$:

$$SOC(c, v, t) = SOC_0(c, v, t) \tag{9}$$

$$SOC_0(c, v, t) = SOCE(c, v) - \frac{d(c, v)M(v)}{100Q(v)} \tag{10}$$

where $SOC(c, v, t)$ is the battery percentage status of vehicle v in scene c at time t; $SOC_0(c, v, t)$ is the full charge status of vehicle v in scene c at time $t = t_{in}(c, v)$. $SOCE(c, v)$ is the expected value of battery charge status of vehicle v in scene c when off grid; $d(c, v)$ is the daily driving mileage of vehicle v in scene c; $M(v)$ is the consumption of vehicle v per 100 km.

It meets EV electricity balance when EV is connected to grid:

$$Q(v)SOC(c, v, t) = Q(v)SOC(c, v, t - 1) + \mu_{EV}^c \times PV_{ch}(c, v, t)\Delta t$$
$$- \frac{1}{\mu_{EV}^d} PV_{dis}(c, v, t)\Delta t \tag{11}$$

where $Q(v)$ is the maximum battery capacity of vehicle v; μ_{EV}^c is the charging efficiency of single vehicle; Δt is the time interval of charging and discharging; μ_{EV}^d is the discharging efficiency of single vehicle.

When EV is out of the grid, namely $t = t_{out}(c, v)$, the battery status of vehicle should meet the expected electricity value of the user. Therefore:

$$SOC(c, v, t) \geq SOCE(c, v, t) \tag{12}$$

(4) EV charging and discharging power limit:

$$PV_{ch}(c, v, t) \leq \overline{PV}_{ch}(v) \times X(c, v, t) \tag{13}$$

$$PV_{dis}(c, v, t) \leq \overline{PV}_{dis}(v) \times Y(c, v, t) \tag{14}$$

where $\overline{PV}_{ch}(v)$ is the upper limit of charging power of vehicle v; $\overline{PV}_{dis}(v)$ is the upper limit of discharging power of vehicle v.

(5) EV Battery status limit:

EV should be charged and discharge without losing the normal life of battery:

$$\underline{SOC} \leq SOC(c,v,t) \leq \overline{SOC} \tag{15}$$

where $\underline{SOC}$ is the lower limit of vehicle under battery load status; $\overline{SOC}$ is the upper limit of vehicle charging status.

(6) Battery charge limit:

The charging capacity of the vehicle shall be no more than the current charging range of the battery:

$$\mu_{EV}^c \times PV_{ch}(c,v,t) \times \Delta t \leq (\overline{SOC} - SOC(c,v,t-1)) \times Q(v) \tag{16}$$

(7) Battery discharge limit:

The discharge of the vehicle shall be no more than the current discharge range of the battery:

$$\frac{1}{\mu_{EV}^d} \times PV_{dis}(c,v,t) \times \Delta t \leq [SOC(c,v,t-1) - \underline{SOC}] \times Q(v) \tag{17}$$

In this paper, the driving time and daily mileage of various types of EV are simulated according to the appropriate sampling method. On this basis, to balance the solution precision and difficulty, scene reduction method based on synchronous back generation technology [19] may eliminate some small probability of the scene and integrate to the similar scene and form the reduced forecast scene.

In order to solve the model of stochastic unit commitment, the linearization technique in [20] indicates the non-linear constraint in the stochastic unit combination as mixed integer linear programming (MILP). Finally, the model after transformation will be simulated on GAMS platform and be solved with CPLEX software package.

4 Case Studies

The application of mentioned model to a concrete case study which contained 10 thermal power units and various types of EV clusters in the region is proposed to illustrate the correctness and effectiveness of the model. The data of the system and the thermal power unit are from the 10-unit test system. Various types of EV parameters and driving parameters are referred to [1]. The number of EV in urban areas is shown in Table 1. In order to obtain the reasonable analysis of travelling behavior of various EV, 10000 scenarios are sampled and reduced to 100.

Table 1. The number of EV in urban areas

Types of EV	Number of EV
Private cars	20000
Buses	2000
Official cars	1000
Taxis	2000

4.1 Considering the Randomness of Travelling or not

The comparison of the charging power with the trip characteristics or not is presented in Fig. 1. It can be seen that the distribution of the charging power occurs in every periods when considering the driving properties of electric vehicles. Because that the electric vehicle is restricted by the charging demanding and the interval of charging time. Thus, there are obvious differences of the distribution of the charging power comparing without considering the stochastic properties. When the electric vehicle is involved in optimal scheduling of unit commitment, it is also needed to follow the vehicle driving properties. Therefore, it is necessary to consider the stochastic properties of electric vehicles when charging for electric vehicles. Otherwise, the optimization strategy is likely to be different from the practical application.

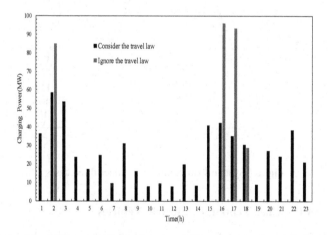

Fig. 1. The comparison of the charging power with or without considering the travel law

4.2 Influence of V2G on Generation Scheduling

To study the influence of V2G on power grid scheduling, three cases are conducted and analyzed as follow:

Case 1: The electric vehicles is charged in disorder, and without the discharging of electric vehicles.

Case 2: The electric vehicles is charged in coordinated way, while without considering the discharging of electric vehicles.

Case 3: Optimal dispatching the electric vehicles by V2G technology.

The load value of the system in each period and the original load of the system without EV under the three cases are present in Fig. 2. It can be concluded that a part of EV in case 1 is charging in peak load which increases the system load during rush hour, since electric vehicles are disorderly charged without charging control. Note that electric vehicles can be charged in the low loan by the coordinated charging of case 2, which could play a certain role in filling the valley. It can be observed that the performance of the load curve of case 3 changes gently compared with case 2 by using V2G. Electric cars can not only charge in low load, and discharge in the peak load, which reduces the load pressure of the system and adjusts the output of each period. Finally, the total cost of system in case 1 is \$591905.706, the total cost of system in case 2 is \$571467.38, the total cost of system in case 3 is \$560129.63, thus, the use of V2G can effectively reduce the cost.

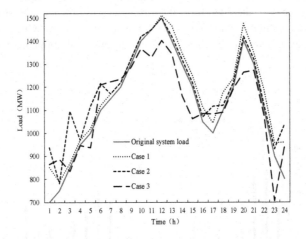

Fig. 2. System load comparison under three cases

4.3 Study on the Classification of Electric Vehicles

In order to study the necessity of classification of electric vehicles, the driving rules of one kind of electric vehicles (private cars) and different types of vehicles are compared. Figure 3(a) shows the charging and discharging power distribution when the driving rules of all the models of vehicles are assumed as private cars, the charging and discharging power distribution of various types of electric vehicles when the driving rules of 4 kinds of different models of vehicles assumed is present in Fig. 3(b). It can be revealed in Fig. 3(b) that all types of vehicles have their own driving rules, such as the bus, it can only be in charged and discharged dispatching within 23:00–6:00. In Fig. 3(a), we assume that the buses are driven with the travelling rules of private cars, charged and discharged power dispatching should be undertaken during their driving time, including 6:00–23:00, which is inconsistent with the actual situation. Due to its unique operative situation of the taxi, fast charging by fixed power after the access to the power grid will be accepted, while discharging will not be accepted, taxi drivers can drive out for timely service should be taken into account. It can be observed in Fig. 3(b) that the taxis are mostly in a state of discharge, which is contrary to the actual situation. Therefore, it is necessary to classify the electric vehicles in the problem of unit commitment.

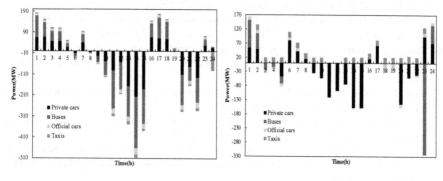

(a) All are considered to be private cars   (b) 4 different types of EVs

Fig. 3. Power distribution scheme.

5 Conclusions

In this paper, a stochastic unit commitment model with the consideration of the characteristics of EV is proposed and the conclusions are summarized as follows:

(1) The stochastic characteristics of electric vehicles should be considered under the reasonable dispatching of electric vehicles. The dispatcher should arrange charging and discharging reasonably in the condition of satisfying the random travel rules of the electric vehicle.
(2) With the advantages of adjusting more flexible of unit commitment and playing a role in peaking load shifting, the V2G technology is adopt and the total cost of the system is optimized.
(3) Different types of electric vehicles have different driving rules, which should be considered in the scheduling model of unit commitment with EV.

Acknowledgments. This work was supported in part by a National Natural Science Foundation of China (No. 51507100), in part by Shanghai Sailing Program (No. 15YF1404600) and in part by the "Chen Guang" project supported by the Shanghai Municipal Education Commission and Shanghai Education Development Foundation (No. 14CG55).

References

1. Luo, Z., Hu, Z., Song, Y., et al.: Study on plug-in electric vehicles charging load calculating. Autom. Electr. Power Syst. **35**(14), 36–42 (2011)
2. Jie, D., Yi, T., Jia, N.: A strategy for distribution of electric vehicles charging load based on user intention and trip rule. Power Syst. Prot. Control **43**(16), 8–15 (2015)
3. Xiaolin, G., Lizi, Z.: Wind-hydro-thermal stochastic unit commitment problem considering the peak regulation constraints. Trans. China Electrotech. Soc. **29**(10), 222–230 (2014)
4. Xue, S., Yuan, Y., Fu, Z.X., et al.: Unit commitment in power system considering vehicle-to-grid. Power Syst. Prot. Control **41**(10), 86–92 (2013)
5. Lu, L., Wen, F., Ledwich, G., et al.: Unit commitment in power systems with plug-in hybrid electric vehicles. Autom. Electr. Power Syst. **23**(7), 1205–1220 (2011)
6. Zhang, S., Zechun, H.U., Song, Y., et al.: Research on unit commitment considering interaction between battery swapping station and power grid. Proc. CSEE **32**(10), 49–55 (2012)
7. Shao, C., Wang, X., Wang, X., et al.: Hierarchical charge control of large populations of EVs. IEEE Trans. Smart Grid **7**(2), 1 (2015)
8. Zakariazadeh, A., Jadid, S., Siano, P.: Multi-objective scheduling of electric vehicles in smart distribution system. Energy Convers. Manag. **79**(3), 43–53 (2014)
9. Saber A.Y., Venayagamoorthy, G.K.: Optimization of vehicle-to-grid scheduling in constrained parking lots. In: Power & Energy Society General Meeting, 2009. PES 2009. IEEE, pp. 1–8. IEEE (2009)
10. Ke, W.U., Wang, C., Sun, H., et al.: Unit commitment in power systems considering large-scale electric vehicles. In: Electric Power Construction (2014)
11. Zhang, X., Juping, G.U., Yue, Y., et al.: Optimum electric vehicle charging model and its approximate solving technique. Proc. CSEE **34**(19), 3148–3155 (2014)
12. Shao, C., Wang, X., Wang, X., et al.: Cooperative dispatch of wind generation and electric vehicles with battery storage capacity constraints in SCUC. IEEE Tran. Smart Grid **5**(5), 2219–2226 (2014)
13. Gaowang, L.I., Qian, B., Shi, D., et al.: Unit commitment problem considering plug-in hybrid electric vehicle. Power Syst. Technol. **37**(1), 32–38 (2013)
14. Tushar, M.H.K., Assi, C., Maier, M., et al.: Smart microgrids: optimal joint scheduling for electric vehicles and home appliances. IEEE Trans. Smart Grid **5**(1), 239–250 (2014)
15. Taylor, J., Maitra, A., Alexander, M. et al.: Evaluation of the impact of plug-in electric vehicle loading on distribution system operations. In: Power & Energy Society General Meeting, 2009. PES 2009, pp. 1–6. IEEE. IEEE Xplore (2009)
16. Luo, Z., Hu, Z., Song, Y., et al.: Study on plug-in electric vehicles charging load calculating. Autom. Electr. Power Syst. **35**(14), 36–42 (2011)
17. Wang, Q.: Research on electric vehicles orderly charging and discharging considering unit commitment and comprehensive benefits. In: North China Electric Power University (2016)
18. Carrión, M., Arroyo, J.M.: A computationally efficient mixed-integer linear formulation for the thermal unit commitment problem. IEEE Trans. Power Syst. **21**(3), 1371–1378 (2006)
19. Li, T., Shahidehpour, M., Li, Z.: Risk-constrained bidding strategy with stochastic unit commitment. IEEE Trans. Power Syst. **22**(1), 449–458 (2007)
20. Ge, X., Zhang, L.: Wind-hydro-thermal stochastic unit commitment problem considering the peak regulation constraints. Trans. China Electrotech. Soc. **29**(10), 222–230 (2014)
21. We, L., Shahidehpour, M., Li, T.: Stochastic security-constrained unit commitment. IEEE Trans. Power Syst. **22**(2), 800–811 (2007)

Dispatching Analysis of Ordered Charging Considering the Randomness Factor of Electric Vehicles Charging

Ling Mao$^{(\boxtimes)}$ and Enyu Jiang

College of Electrical Engineering, Shanghai University of Electric Power,
No. 2588 Changyang Road, Yangpu District, Shanghai 200090, China
maoling2290@126.com

Abstract. With the popularity of Electric Vehicles (EV), the access of electric vehicles makes the load curve of distribution network becomes more and more steep. The increase of EV load brought big impact on the power system and it is necessary to analyze random factors affecting the electric vehicle charging load. Distribution grid dispatching can reduce the gap between peak load and valley load so as to ensure power grid normal operation and power quality. Firstly EV charging load model is established based on the analysis the random factors affecting EV load, and the accumulation method of EV load is given. Secondly the dispatching model of coordination charging is built considering the randomness of EV charging, and the genetic algorithm was used to solve the model. Finally, with the IEEE 33 node test system, the load curve of power grid is obtained in the mode of disorderly charging and order charging dispatching, which proves the validity of the charging dispatching model, and the feasible strategies are provided by the analysis of simulation results.

Keywords: Ordered charging · Monte Carlo method · Electric vehicles · Random factor · Genetic algorithm

1 Introduction

With the development of the times, energy shortage and environmental damage have gradually become a hot issue in modern society, so clean energy has become research focus. The EV which uses electric energy and can realize zero emission has become a developing trend of modern transportation. Therefore, the electric vehicles are also one of the effective ways to reduce air pollution.

With the popularity of electric vehicles, access to a large number of electric vehicles has brought great impact and challenges to distribution network. The negative effects of EV disorder charging bringing to the grid include increasing peak load, increasing power losses, decreasing voltage, decreasing load factor and so on [1–4]. How to optimize and mitigate electric vehicle charging load on the grid has now become one of the key researches. The research of electric vehicle charging load connected to the distribution network mainly focus on the following two aspects: one is on the modeling and prediction of EV charging load, the other is the research on user behavior and control methods of EV charging [5]. The control method of electric

© Springer Nature Singapore Pte Ltd. 2017
K. Li et al. (Eds.): LSMS/ICSEE 2017, Part III, CCIS 763, pp. 309–318, 2017.
DOI: 10.1007/978-981-10-6364-0_31

vehicle charging load is more concentrated. In [6], different charging mode is proposed to regulate electric vehicles effect on the grid. In [7], an ordered charging method of EV based on load forecasting was presented to reduce the negative impact on the distribution network. In [8], a multi-objective optimization model for coordination control electric vehicle including the minimum voltage and power loss are taken as the objective. In [9], the coordination control of electric vehicles was proposed according to user charging requirements. Distribution grid dispatching can reduce the gap between peak load and valley load so as to ensure power grid normal operation and power quality [10–12]. There are also several studies on electric vehicle coordination charging from the economic dispatch such as the regulation of electricity prices [13], and the stochastic economic dispatch model [14]. The above-mentioned studies are mainly to analyze the electric vehicle users, it is imperative to design the dispatching model of coordination charging considering electric vehicle charging uncertain factors. In this paper the dispatching model of electric vehicle coordinated charging is proposed considering these random factors of electric vehicle charging, and the model is solved by genetic algorithm. The developed coordinated control is tested in simulations. The feasible strategies are provided by the analysis of simulation results.

Electric vehicle charging load model is presented in Sect. 2. Section 3 describes the scheduling model and solving method of electric vehicle. A case study is employed in Sect. 4 to illustrate the developed model and method. Results and discussion are presented in Sect. 5.

2 Electric Vehicle Charging Load Model

2.1 The Random Factor

In order to study the load characteristics of electric vehicles and the impact on the distribution network, the random factor in the charging process of electric vehicles is considered such as the type, charging time, charging method and so on, these random factor will affect on the calculation of electric vehicles charge load model. The random factor affecting the EV charging load includes:

(1) Types of electric vehicle

Different type of electric vehicles with different charging properties, the impact on the grid is different. Because the penetration rate of civil private electric vehicles which is in a suspended state most of the day is high, and it is easy to adjust the charging time of electric vehicles in Off-peak electricity, this paper studies objects is the civil private electric vehicles. The type of civil private electric vehicle mainly includes plug-in hybrid electric vehicle (PHEV), battery electric vehicle (BEV), and the mix of PHEV and BEV. The model in this paper takes into account PHEV type. The PHEV is divided into three types according to traveling distance: PHEV30, PHEV40 and PHEV60, which is shown as Table 1.

(2) Charging time of electric vehicle

The charging time of electric vehicle is decided by the user's habit, traveling time, mileage and other decisions. At present, there is no enough and reliable statistical data for the charging behavior of electric vehicles in China. Therefore, in this paper, we use the data from the statistics of the USA car users traveling habits [15]. It is assumed that the electric vehicle users have the same law and the charging time accord with the normal distribution.

(3) Charging mode for electric vehicle

The electric vehicle charging mode are divided into the fast charger mode and slow charger mode. The charging mode is also an important factor of electric vehicle charging load impacting to the power grid. The selection of charging mode is mainly influenced by the electric vehicle user's habit and driving distance.

(4) Charging level of electric vehicle

Different charging degree of electric vehicle battery has a great influence on the distribution network of electric vehicle charging load, and it mainly depends on the user's choice. An investigation found that the 76% of users will choose a full battery state of charge, while 24% of users choose to charge more than 90% of battery capacity.

Table 1. Types of electric vehicles.

Types	Maximum rated power	Battery capacity	Charging power at 0.2 C	Charging power at 1 C	Charging power at 2 C
PHEV30 (MIT)	44	8	1.6	8	16
PHEV40 (USABC)	46	17	3.4	17	34
PHEV60 (EPRI)	90	18	3.6	18	36

2.2   EV Charging Load Calculation Model

The charging power and remaining battery capacity must be known to calculate electric vehicle charging load. The charging power of electric vehicle is different according to the different type, and the remaining battery state of charge (SOC) is decided by the actual use of the user which conform to normal distribution. We can calculate the electric vehicle charging load, and calculate the charging time of electric vehicle.

$$C_i = C_{ie} - C_{is} \qquad (1)$$

$$t_i = \frac{C_i}{P_i} \qquad (2)$$

where C_i is the charging electric quantity of ith electric vehicle i, the final consumption and the initial electric quantity of electric vehicle i is respectively C_{ie}, C_{is}, and P_i is the charging power of electric i, t_i is charging time of the electric vehicle i.

2.3 EV Load Accumulation Method

In order to get daily load curve of an hour time interval, it is need to accumulate the per hour electric vehicles charging power to load point. Specific electric vehicle load accumulation process is as follows: First of all, 24 h a day is divided into 24 intervals. Secondly, the process of charging is divided into multiple interval according to the electric car charging starting time t_{start} and the charging end time t_{end}. Finally, the per hour electric vehicles charging power is accumulated to load point, which is the result of charging power multiplied by corresponding charging power. As shown in Fig. 1, the whole charging process of the electric vehicle i is divided into 3 sections according to the start charging t_{start} and the end charging time t_{end}, the result of the charging interval length multiply by corresponding charging power is accumulated into the hour load. As the calculation process the daily load curve of electric vehicle is obtained. It can be expressed as a formula:

$$P_t = P_{1,t} + P_{2,t} + \cdots + P_{i,t} \tag{3}$$

$$i = [1, N] \tag{4}$$

where $P_{i,t}$ is charging power of electric vehicle i on the t hour, P_t is accumulated charging power of electric vehicles on the t hour.

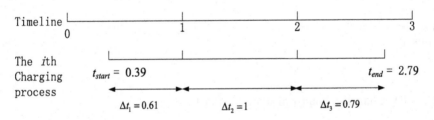

Fig. 1. Electric vehicle charging load accumulating process.

3 Coordination Charging Dispatching Model

3.1 Charging Dispatching Model of Electric Vehicle

Uncoordinated charging of electric vehicle will make power load curve steeper, and threat distribution network security. The increase of EV load brought big impact on distribution network and it is necessary to analyze random factors affecting the electric vehicle charging load. The objective of this paper is to reduce the gap between peak load and valley by coordinated charging dispatching of electric vehicle considering charging random factors. Therefore, the optimal objective function, F, is given as:

$$F = \min(P_{ed,t} + P_{load,t} - \mu) \tag{5}$$

$$\mu = \frac{\sum_{t=1}^{24} P_{ed,t} + P_{load,t}}{24} \tag{6}$$

Where $P_{ed,t}$ is electric vehicle charging load at t time step, $P_{load,t}$ is conventional load at t time step.

At each time step, the electric vehicle charging power is maximized subject to certain constraints in a charging cycle (24 h):

$$\sum_{t=1}^{24} P_{ed,t}\Delta t + C_0 = C_{MAX} \tag{7}$$

Here, C_0 represents the sum of remaining power of all the cars before charging in charging station, C_{MAX} represents the sum of the battery capacity of all electric vehicles which charge in the charging station in the day.

In addition, the charging power of the electric vehicle which can be dispatched is constrained by the number of electric vehicles accessing to grid at each time step, as shown:

$$P_{ed,tMIN} \leq P_{ed,t} \leq P_{ed,tMAX} \tag{8}$$

The next constraint relate to the acceptable voltage. The constraints are as follows:

$$P_{i,t} = \sum_{j \in i} U_{i,t} U_{j,t}(G_{ij} \cos \delta_{ij,t} + B_{ij} \sin \delta_{ij,t}) \tag{9}$$

$$U_{min} \leq U_{i,t} \leq U_{max}(i \in S_B) \tag{10}$$

Here, $P_{ed,tMIN}$ is the minimum permissible charging load of electric at t time step, $P_{ed,tMAX}$ is maximum permissible charging load of electric vehicle at t time step. $P_{i,t}$ is active power of the node i at t time step, $U_{i,t}$ is the voltage the node i, while U_{min}, and U_{max} are the minimum and maximum allowable voltage levels, respectively. $\delta_{ij,t}$ is the angle between node i and node j, G_{ij} and B_{ij}, is the real and imaginary mutual admittance of node i and node j, respectively, S_B is set all nodes in system.

But in the above model, $P_{ed,tMIN}$, $P_{ed,tMAX}$, C_0 and C_{MAX} parameters depends on random factors such as the types of electric vehicles, charging time, charging mode, so the model makes the coordinated charging of electric vehicle become difficult. According to the stochastic optimization theory, the EV ordered charging model is simplified as follows:

$$\min(P_{ed,t} + \overline{P}_{load,t} - \mu) \tag{11}$$

$$\mu = \frac{\sum_{t=1}^{24} P_{ed,t} + \overline{P}_{load,t}}{24} \tag{12}$$

$$\sum_{t=1}^{24} P_{ed,t}\Delta t + C_0 = C_{MAX} \tag{13}$$

$$\overline{P_{ed,tMIN}} \leq P_{ed,t} \leq \overline{P_{ed,tMAX}} \tag{14}$$

$$\overline{P_{i,t}} = \sum_{j \in i} \overline{U_{i,t}U_{j,t}}(G_{ij}\cos\overline{\delta_{ij}} + B_{ij}\sin\overline{\delta_{ij}}) \tag{15}$$

$$U_{\min} \leq \overline{U_{i,t}} \leq U_{\max}(i \in S_B) \tag{16}$$

Where the superscript "-" indicates the expected value of random variables.

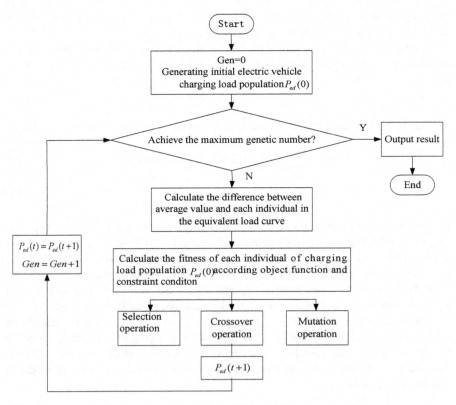

Fig. 2. The flowchart of the solving model process by GA.

3.2   Genetic Algorithm for Solving Model

By using the above model, we can use the genetic algorithm (GA) to solve the EV power flow model (11–13). The detailed solving process is shown in Fig. 2.

Firstly, select the charging power of EV charging station as chromosome individuals. Secondly calculate the fitness of each individual and execute selection operation, crossover operation and mutation operation. Finally get the dispatching result of EV coordinated charging.

4   Example Analysis

In this paper, we use the standard IEEE 33 node 10 kV distribution network system as shown in Fig. 3 to test the model established in this paper. Assume that there are one electric vehicle charging stations connected to node 22. According to the load model and method described in this paper, the charging load of electric vehicle is analyzed.

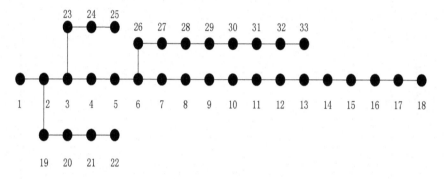

Fig. 3. A standard IEEE 33 node 10 kV distribution network system.

4.1   Uncoordinated Charging Scheduling

Based on the analysis and discussion of the electric vehicle charging load model, the following assumptions:

(1) The number of registered vehicles of Shanghai is assumed to be 8400 until 2017 according to National Household Travel Survey.
(2) Assuming that electric vehicles user charging once a day taking into account the battery life user habits. Once the charging starts, the process will continue until the battery is fully filled.
(3) In an uncoordinated charging mode, electric vehicle charging time is decided by the individual owner. The SOC accord with normal distribution of $SOC \sim N(0.6, 0.12)$, start charging time accord with the normal distribution of $t_{start} \sim N(15.71, 3.725^2)$.

(4) The type ratio of electric vehicles is assumed is 70% PHEV30, 12% PHEV40 and 18% PHEV60. Charging mode is fast and slow charge model 95% and 5%, respectively.

Based on the above assumptions, the disorderly charging of electric vehicles is simulated based on Monte Carlo method. The electric vehicle charging load curve results as shown in Fig. 4. Figure 4 shows the load peak around 16:00 P.M. and the low load around 5:00 A.M. With the popularity of electric vehicles, access to a large number of electric vehicles has brought great impact and challenges to distribution network. If we do not take measures to guide or control user's charging behavior, it will cause the widen difference of peak and valley, increase the peak load, and reduce operation efficiency of grid.

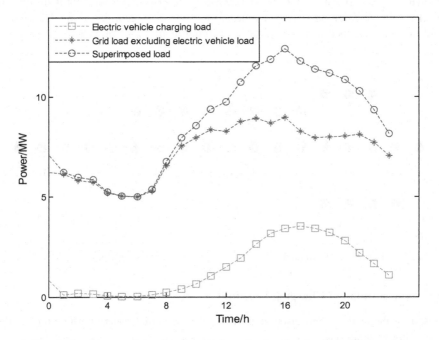

Fig. 4. Load curve of the electric vehicle under the disordered charging mode.

4.2 Coordinated Charging Dispatching

The electric vehicle coordinated dispatched model is simulated based on genetic algorithm. The simulation calculation takes 24 h as the time period to develop the 24 h grid dispatch. The load curve of EV under orderly mode is shown in Fig. 5. Figure 5 illustrated that a large number of electric vehicle charging load is transferred to the night 4:00 A.M. to 7:00 A.M. The EV ordered charging effectively smoothes the load curve, reduces the peak valley difference, and improves the load rate comparing with free charging mode. The feasible strategies are provided by the analysis of simulation

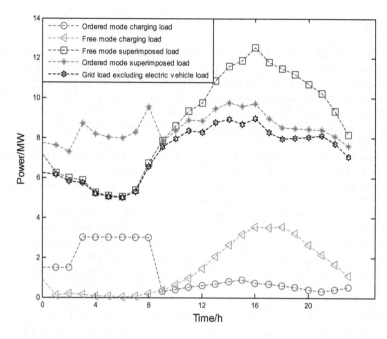

Fig. 5. Load curve of electric vehicle under the mode of orderly charging.

results as follows: (a) the implementation of peak valley price strategy; (b) the limited charging strategy in peak time; (c) reduce the number of charging stations during peak hours.

5   Conclusion

The dispatching model of coordination charging is built considering the randomness of EV charging such as types, charging time, charging mode, and charging level of electric vehicle so on. The disorderly charging of electric vehicles is simulated based on Monte Carlo method. It is found that the peak valley difference increases because of the electric vehicle charging random factors. The dispatching model of coordination charging is built considering the randomness of EV charging, and the genetic algorithm was used to solve the model. Take IEEE 33 as an example to simulate and analyze the ordered charging schedule. Its effectiveness has been demonstrated by comparison EVdisorderly charging. The EV orderly charging effectively smooths the load curve, reduces the peak valley difference, and improves the load rate comparing with free charging mode. The results also show the feasibility of ordered charging scheduling model.

References

1. Wu, D., Aliprantis, D.C., Ying, L.: On the choice between uncontrolled and controlled charging by owners of PHEVs. IEEE Trans. Power Deliv. **26**, 2882–2884 (2011)
2. Clement-Nyns, K., Haesen, E., Driesen, J.: The impact of charging plug-in hybrid electric vehicles on a residential distribution grid. IEEE Trans. Power Syst. **25**, 371–380 (2010)
3. Deilami, S., Masoum, A.S., Moses, P.S.: Real-time coordination of plug-in electric vehicle charging in smart grids to minimize power losses and improve voltage profile. IEEE Trans. Smart Grid **2**, 456–467 (2011)
4. Papadopoulos, P., Skarvelis-Kazakos, S., Grau, I.: Electric vehicles' impact on British distribution networks. IET Electr. Syst. Transp. **2**, 91–102 (2012)
5. Ul-Haq, A., Cecati, C., Strunz, K.: Impact of electric vehicle charging on voltage unbalance in an urban distribution network. Intell. Ind. Syst. **1**, 51–60 (2015)
6. Ahn, C., Li, C.T., Peng, H.: Optimal decentralized charging control algorithm for electrified vehicles connected to smart grid. J. Power Sources **196**, 10369–10379 (2011)
7. Di, G.A., Liberati, F., Canale, S.: Electric vehicles charging control in a smart grid: a model predictive control approach. Control Eng. Pract. **222**, 147–162 (2014)
8. Zakariazadeh, A., Jadid, S., Siano, P.: Multi-objective scheduling of electric vehicles in smart distribution system. Energy Convers. Manag. **79**, 43–53 (2014)
9. Hu, J., You, S., Lind, M.: Coordinated charging of electric vehicles for congestion prevention in the distribution grid. IEEE Trans. Smart Grid **5**, 703–711 (2014)
10. Ma, Z., Callaway, D.S., Hiskens, I.A.: Decentralized charging control of large populations of plug-in electric vehicles. IEEE Trans. Control Syst. Technol. **21**, 67–78 (2013)
11. Binetti, G., Davoudi, A., Naso, D.: A distributed auction-based algorithm for the nonconvex economic dispatch problem. IEEE Trans. Ind. Inf. **10**, 1124–1132 (2014)
12. Gan, L., Topcu, U., Low, S.H.: Optimal decentralized protocol for electric vehicle charging. IEEE Trans. Power Syst. **28**, 940–951 (2013)
13. Rotering, N., Ilic, M.: Optimal charge control of plug-in hybrid electric vehicles in deregulated electricity markets. IEEE Trans. Power Syst. **26**, 1021–1029 (2011)
14. Binetti, G., Davoudi, A., Lewis, F.L.: Distributed consensus-based economic dispatch with transmission losses. IEEE Trans. Power Syst. **29**, 1711–1720 (2014)
15. Saber, A.Y., Venayagamoorthy, G.K.: Intelligent unit commitment with vehicle-to-grid—a cost-emission optimization. J. Power Sources **195**, 898–911 (2010)

A Contract Based Approach for Electric Vehicles Charging in Heterogeneous Networks

Huwei Chen[1], Zhou Su[1]([✉]), Yilong Hui[1], Hui Hui[1], and Dongfeng Fang[2]

[1] School of Mechatronic Engineering and Automation, Shanghai University,
Shanghai, People's Republic of China
zhousuieee@126.com
[2] Department of Electrical and Computer Engineering,
University of Nebraska-Lincoln (UNL), Lincoln, USA

Abstract. With the help of mobile charging stations (MCSs), the charging service of electric vehicles (EVs) can be provided more easily with higher payoff and lower consumption, compared with the fixed charging stations (FCSs). Although many traditional approaches have been used to decide the pricing plans for FCSs, it can not be efficient to design the optimal pricing strategy for MCSs. In this paper, we propose a contract-based scheme to solve the problem of supplying power service to EV users. Firstly, considering quality of service (QoS) and mobility of MCS in the heterogeneous networks, we study and develop the utility function based on the relationship for MCS and EV users. Then, the charging problem for EV users is formulated as an optimization problem through the contract theory. Thirdly, we present the iterative algorithm to achieve the optimal solution. Our simulation results show the effectiveness of the proposed strategy.

1 Introduction

Electric vehicles (EVs) as a significant transportation have attracted a lot of attention and been widely applied with the advantage of higher power efficiency, lower gas emission than conventional vehicles [1,2]. The related report [3] shows that the total number of the world's EV sales has reached 774 thousand with 15.7 billion dollars in 2016, which rises up to 40% in sales compared with that in 2015. It can be predicted that the EV sales will achieve about 310 thousand with 24.3 billion dollars in 2025, and more and more fuel vehicles will be replaced in future.

Due to the limited capacity of batteries in EVs, EV users have to consider the charging problems which affect EV users' driving routing [4,5]. Thanks to the emergence of mobile charging stations (MCSs), it has been recognized as a more beneficially and conveniently available solution to power supply for EV users. Compared with fixed charging stations (FCSs), MCSs could reach dynamic positions based on the specified situation, through which much more time spent on buying charging service can be reduced. However, it brings new challenges

© Springer Nature Singapore Pte Ltd. 2017
K. Li et al. (Eds.): LSMS/ICSEE 2017, Part III, CCIS 763, pp. 319–328, 2017.
DOI: 10.1007/978-981-10-6364-0_32

for MCS in the heterogeneous networks. (1) It needs to study how MCS makes optimal decisions on power supply for EV users, considering quality of service (QoS) in the wireless communication. (2) Due to mobility of MCS, more profits should be achieved based on the optimal strategy offered by MCS.

In order to solve the above challenges, many approaches have been presented and discussed. Liu *et al.* [6] proposed a novel heuristic algorithm to obtain the power allocation with lower impact on power grid for EV users. Karbasioun *et al.* [7] proposed the control policy with the minimal cost for the operator based on the real-time pricing scheme with the dynamic programming. Bayram *et al.* [8] presented the optimal pricing strategy to improve profits of charging station operators during peak hours. Bahrami and Parniani [9] proposed the game-theoretical scheme to schedule the power consumption with the minimum charging cost. However, they do not discuss the requirements on the waiting time and the effect of EV user's type on the charging power as well as batteries' life, when searching the optimal solution. As a result, it could not ensure better performance and charging service for both MCS and EV users.

In this paper, we analyze the interaction between EV users and MCS in the heterogeneous networks. We present the contracted-based approach to make the optimal strategy for MCS, through which more profits can be achieved. MCS sells power to both contracted EV users and non-contracted users, respectively. Firstly, we develop the utility function with bit error ratio (BER) in the heterogeneous networks. Then, in order to make optimal contract items, the charging problem can be formulated as an optimization problem when EV users' types are characterized. Thirdly, through the proposed iterative algorithm, we can obtain the optimal solution. Our simulation results validate the effectiveness of the proposal.

2   System Model and Problem Statement

2.1   Network Model

As shown in Fig. 1, MCS supplied by smart grid can sell power to EV users in the heterogeneous networks, which is composed of macro cell and small cell. Through the heterogeneous network, EV users will exchange information with each other and MCS, including the power price, amount of charging power and waiting time in MCS. Thus, we assume that EV user k with charging request wants to buy power from MCS, $\forall k \in \mathcal{K} = \{1, \ldots, k, \ldots, K\}$, and $K = |\mathcal{K}|$.

In the heterogeneous network, the small cell base station (SCBS) is deployed in Fig. 1, which is used to improve the coverage of macro cell base station (MCBS). BER is considered based on the interferences coming from the wireless infrastructure nearby, which is variable related to its surroundings [10]. We suppose that EV users to be charged by MCS are randomly distributed in the heterogeneous network, and the average BER between MCS and EV user k is denoted by ζ_k.

Considering EV users' driving distance, MCS will decide destination **D** with them [2]. EV users arriving at MCS will wait for being charged, when many

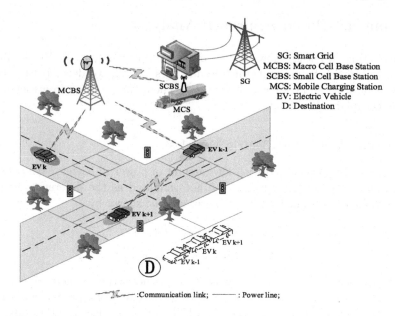

Fig. 1. Example of a charging service model in the heterogeneous networks composed of macro cell and small cell.

EV users need to be charged. In this paper, the $\mathbb{M}/\mathbb{M}/C/S$ queuing model is used to analyze the charging strategy [8]. The service time in MCS follows the exponential distribution with the parameter μ. And EV users arriving at MCS follow a Poisson distribution with the arrival rate λ. We suppose that the average waiting time of EV users in MCS is W.

2.2 Incentive Mechanism

With low state of charge (SOC) in batteries, EV users have to buy the charging service and pay for the power supplied by MCS. As an economic incentive, the contract theory is applied to stimulate EV users to buy larger amount of power [11]. Aiming to maximize its profits, MCS not only supplies power to contracted EV users, but also sells power to non-contracted users with the fixed price decided in advance.

Thus, we suppose that the total amount of power supply Θ in MCS is divided into two kinds: (i) the first part is named as the contracted power supply $\Theta_c = \sum_{k=1}^{K} x_k$ with corresponding payment $Y_c = \sum_{k=1}^{K} y_k$, and the set of contract items is denoted by $\boldsymbol{\Gamma} = \{\Gamma_1, \Gamma_2, \ldots, \Gamma_k, \ldots, \Gamma_K\}$, $\Gamma_k = \{(x_k, y_k), k \in \mathcal{K}\}$, where y_k is EV user k's payment for its own charging power x_k. (ii) the second part is named as the non-contracted power supply $\tilde{\Theta}_{nc} = \Theta - \Theta_c$, which will be sold at the fixed price p_0 decided by MCS in advance. Then, given the total power supply, MCS's total profits mainly depend on the contract item so that how to design the proper contract item is a key issue.

3 Contract-Based Approach Analysis

In this section, we study how to determine the optimal contract items offered by MCS, in order to maximize its profits. First of all, we develop the utility functions of MCS and contracted EV users. Then, the proposed contract-based scheme is explained in detail.

3.1 Utility Model for Electric Vehicles

EV users have to be charged with low amount of power in battery. We consider that the utility function of EV user k is formulated as the difference between the benefit and payment for MCS's power supply, shown by

$$G_k(x_k) = U(x_k) - C(x_k) \tag{1}$$

$$s.t. \ \ x_{k,\min} \leq x_k \leq x_{k,\max} \tag{2}$$

Here, $U(x_k)$ denotes EV user k's benefit, $C(x_k)$ denotes the payment for charging service, x_k denotes the charging power with a constraint of a proper upper limit $x_{k,\max}$ and a proper lower limit $x_{k,\min}$, which is used to decrease the impact of charging service on batteries' lifetime [12].

We suppose that $U(x_k)$ is monotonically increasing on x_k, which can be formulated as a linear function by

$$U(x_k) = a_k \eta_k x_k \tag{3}$$

where a_k is the subsidization coefficient, and η_k is the parameter related to QoS in the heterogeneous network, denoted by $\eta_k = 1 - \zeta_k$.

Since the feature of charging or recharging mainly depends on the battery's life, the cost related to batteries should be taken into account besides the payment for the charging power [12]. Therefore, we have

$$C(x_k) = \frac{\vartheta_k}{\varpi_k} \frac{\tau_k \eta_k x_k}{\delta_k} + \varrho_k \eta_k y_k \tag{4}$$

Here, ϑ_k denotes the expense for EV user k's recharging batteries. ϖ_k is the maximum number of a battery's recharging times used to denote batteries' life. τ_k is the charging efficiency of batteries. δ_k denotes the batteries' capacity, and ϱ_k is the discount coefficient, which is assumed to be monotonically decreasing on the waiting time, and increasing on the proportion of EV user k's driving distance, respectively. Then, it can be described by

$$\varrho_k = \exp\left(\frac{\omega_k \nu_k}{W}\right) \tag{5}$$

where ω_k denotes the adjustment parameter, ν_k is the proportion of driving distance.

Therefore, based on Eqs. (3), (4) and (5), the utility function of EV users in Eq. (1) can be simplified by multiplying both sides by $\frac{1}{\varrho_k \eta_k}$. We have

$$\hat{G}_k(x_k) = A_k x_k - y_k \tag{6}$$

Here,

$$A_k = \left(a_k - \frac{\vartheta_k}{\varpi_k} \frac{\tau_k}{\delta_k} \right) \exp\left(-\frac{\omega_k \nu_k}{W} \right) \tag{7}$$

In addition, $K = |\mathcal{K}|$ EV users to be charged by MCS are characterized into different types shown in Eq. (7). We take the limited number of EV users' types into account, denoted by $\mathcal{I} = \{1, \ldots, i, \ldots, I\}, I = |\mathcal{I}|$. Without loss of generality, we assume that the value of them satisfies $A_1 < A_2 < \cdots < A_I$, and the responding number of EV users with the same type is $M_i, i \in \mathcal{I}$.

According to the above statement, we assume that MCS will sell $\varsigma \tilde{\Theta}_{nc}$ to the contracted EV users based on their power demand, where ς denotes the percent of power supply from the non-contracted power $\tilde{\Theta}_{nc}$. Then, based on EV user k's type, we consider that it will play an important role in the distribution of the power supplied by the non-contracted power, which can be defined as follows,

$$\varepsilon_k = \frac{A_k}{M_k \sum_{i=1}^{I} A_i} \tag{8}$$

where M_k denotes the number of EV users with the same type A_k. Thus, we can know that the total power demand bought by EV user is expressed by

$$X_k = x_k + \varsigma \varepsilon_k \tilde{\Theta}_{nc}$$

$$= x_k + \frac{\varsigma A_k}{M_k \sum_{i=1}^{I} A_i} \left(\Theta - \sum_{j=1}^{K} x_j \right) \tag{9}$$

Based on Eq. (7) and given the power price p_0, we can know that the total payoff of EV users accepting the contract item is expressed by

$$G_{total}(\mathbf{X}) = \sum_{k \in \mathcal{K}} \left\{ A_k x_k - \Pi_k + \frac{\varsigma A_k (A_k - p_0)(\Theta - \sum_{j=1}^{K} x_j)}{M_k \sum_{i=1}^{I} A_i} \right\} \tag{10}$$

Here, $\mathbf{X}$ is the set of total charging power for each EV user, denoted by $\mathbf{X} = \{X_1, X_2, \ldots, X_K\}$.

3.2 Utility Model for Mobile Charging Station

As discussed above, we can know that both power supply and payoff in the contract item offered by MCS play an important role in the total profits, given the total power supply. Then, MCS intends to reduce the power supply with higher payment in the contract item as much as possible. Namely, MCS's profits will monotonically decrease on the power supply, and increase on EV users' payment on the contrary.

In this case, according to [13], we formulate the utility function as the following equation.

$$R = \left(\gamma \Theta - \sum_{k \in \mathcal{K}} x_k \right) \ln \left(1 + \sum_{k \in \mathcal{K}} y_k \right) \tag{11}$$

where γ is a preset parameter given by MCS to implement the elastic contract, satisfying $\frac{\sum_{k \in \mathcal{K}} x_k}{\Theta} < \gamma < 1$.

3.3 Formulation of Contract Items Design

We analyze the interaction between EV users and MCS. Due to the selfishness, MCS aims to maximize its profits without considering EV users' benefit. Moreover, with rationality, they will make decisions to design optimal contracts accepted by EV users with profits. The process of agreement on the contract items will be implemented as follows.

- When EV user k has to be charged with low amount of power in the coverage of heterogeneous networks, it will communicate with MCS for charging service, including the amount of power demand.
- In order to motivate EV user k to buy power, MCS will determine the proper contract item and broadcast it to EV user k.
- After receiving the contract item, EV user k will compute its own payoff. For EV user k, this contract will be accepted and signed when both individual rationality (IR) and incentive compatibility (IC) are simultaneously satisfied. Otherwise, the contract item is equal to zero.
- Receiving EV user k's confirmation, the charging power in the signed contract item will be supplied when EV user k arrives at MCS.

4   Optimal Contract Design

In this section, we use the contract theory to analyze the proposed contract-based scheme. Based on the analysis above, the optimal contract to be accepted simultaneously satisfies the following conditions: (i) IR: it means that EV user i with the type A_i will obtain non-negative profits once the contract item is accepted, i.e. $A_i x_i - y_i \geq 0$. (ii) IC: it means that EV user i will obtain more profits, when it accepts the contract item designed for its own type A_i, i.e. $A_i x_i - y_i \geq A_i x_n - y_n$, $i \neq n$, $i, n \in \mathcal{I}$.

In this case, in order to obtain the optimal solution, this problem can be formulated as the optimization problem, shown by

$$\max R = \left(\gamma \Theta - \sum_{k=1}^{K} x_k \right) \ln \left(1 + \sum_{k=1}^{K} y_k \right) \tag{12}$$

$$s.t. \quad A_i x_i - y_i \geq 0 \quad (IR) \tag{13}$$

Therefore, with the optimal contract item $\Gamma^* = (x_I^*, y_I^*)$ in Eq. (12), MCS could obtain the maximum profits shown as

$$R_{\max} = (\gamma\Theta - M_I x_I^*)\ln(1 + M_I A_I x_I^*) \tag{14}$$

Algorithm 1. An Iterative Search Algorithm

1: Input: Given the total number of EV users K and the responding number M_i, MCS with the total power supply Θ will characterize EV users' types, $i \in \mathcal{I}$. A random x_I is supplied to EV user I at the beginning.
2: Initialization: $x_I^0 = x_I$.
3: **for** $t = 0 : 1 : t_{\max}$ **do**
4:    **if** x_I^t is satisfied based on Ineq. (2) **then**
5:       MCS will update the power supply for EV user I based on the following equations:

$$g^t = x_I^t - \hbar\frac{R'(x_I^t)}{R''(x_I^t)} \tag{15}$$

$$x_I^{t+1} = x_I^t - \frac{R'(x_I^t)}{\frac{1}{\alpha}R''\left[x_I^t + \frac{\alpha}{2}(g^t - x_I^t)\right] + (1 - \frac{1}{\alpha})R''(x_I^t)} \tag{16}$$

$$y_I^{t+1} = A_I x_I^{t+1} \tag{17}$$

6:       Here, $R'(x_I^t) = \frac{\partial R(x_I^t)}{\partial x_I^t}$, $R''(x_I^t) = \frac{\partial^2 R(x_I^t)}{\partial (x_I^t)^2}$. $\hbar$ is the positive step size, and α is bounded by $0 < \alpha < 1$, which are used to adjust the iterative step.
7:    **else**
8:

$$x_I^{t+1} = \min\{\max\{x_I^{t+1}, x_{I,\min}\}, x_{I,\max}\} \tag{18}$$

9:    **end if**
10: **end for**
11: **Until** $\|x_I^t - x_I^{t-1}\|/\|x_I^{t-1}\| \leq \psi_1$.
12: Output the optimal contract item $\Gamma^* = \{x_I^*, y_I^*\}$.

Lemma 1. Knowing exactly information of EV users, the contract item offered by MCS is $\Gamma^* = (x_I^*, y_I^*)$ while the others will be equal to 0, i.e. $x_I > 0$, $y_I > 0$, $x_k = 0$ and $y_i = 0$ ($\forall i \in \{1, 2, \ldots, I-1\}$).

Proof. In order to prove Lemma 1, we use the method of contradiction and assume that the optimal contract exists when the total power demand in the contract item is given and $0 < i < I$.

We suppose that MCS applies the non-negative optimal contract item $\Gamma_i' = (x_i', y_i')$. Thus, the maximum profits can be obtained by

$$\hat{R}_{\max} = \left(\Gamma\Theta - \sum_{i=1}^{I} M_i x_i\right)\ln\left(1 + \sum_{i=1}^{I} M_i A_i x_i\right) \tag{19}$$

Given the total power demand, we can know that

$$M_I A_I x_I^* = \sum_{i=1}^{I} M_i A_I x_i \geq \sum_{i=1}^{I} M_i A_i x_i \iff \ln(1 + M_I A_I x_I^*) \geq \ln\left(1 + \sum_{i=1}^{I} M_i A_i x_i\right)$$
(20)

It implies that $R_{\max} \geq \hat{R}_{\max}$, which contradicts with the assumptions above. Therefore, we prove Lemma 1 and obtain the optimal contract based on Lemma 1. □

From Lemma 1, we can know the optimal contract item is $\Gamma_I^* = (x_I^*, y_I^*)$. In order to obtain the maximum solution, we take the second derivation of Eq. (14) with respect to x_I. It can be described as

$$\frac{\partial^2 R}{\partial x_I^2} = -\frac{A_I M_I^2}{1 + M_I A_I x_I} - \frac{A_I M_I^2(1 + \gamma A_I \Theta)}{(1 + M_I A_I x_I)^2} < 0$$
(21)

Here, Eq. (21) implies that the utility function shown in Eq. (14) is a concave function on x_I and the optimal contract exists. In order to solve the optimization problem, we propose an iterative algorithm to search for the optimal contract item, where ψ_1 is a small threshold value used to obtain precise results. This presented algorithm is shown in **Algorithm 1** in detail.

5   Simulation

5.1   Simulation Scenario

In this section, we verify the performance of the strategy proposed in this paper. Based on [14], we suppose that the average BER is no more than 0.01 in the heterogeneous networks. And, we assume that the total number of EV users' types is 3 in this paper. The other parameters are set as: $a_i(\$/MW) \in \{4, 5.5, 6.5\}$; $\vartheta_i(\$) \in \{600, 800, 1000\}$, $\delta_i(MW) \in \{0.5, 0.8, 1\}$, $\nu_i \in \{0.65, 0.65, 0.7\}$, $\varpi_i \in \{1000, 1200, 1300\}$, $i \in \mathcal{I} = \{1, 2, 3\}$.

5.2   Simulation Results

The total number of EV users is 6, and the responding number of each EV user with the same type, i.e. $M_i \in \{1, 3, 2\}$. We study the relationship between the waiting time and the utility of MCS. From Fig. 2, we can observe that the value of utility of MCS increases with the increase of waiting time. Moreover, we compare the proposed approach with the existing approaches. In uniform distribution scheme (UDS) without different types of users, contracted power supply Θ_c from MCS is supplied to each user with the same amount of charging power. In random distribution scheme (RDS) with different types of users, each user will be supplied with a certain amount of charging power, based on random contract items offered by MCS. And, in UDS with different types of users, each user will be supplied with charging power in the contract item offered by MCS, in which each user's type plays an important role in scheduling the charging power. From the simulation results in Fig. 2, we can obviously see that the proposed contract-based scheme outperforms the other three approaches, where MCS can obtain much better utility than others.

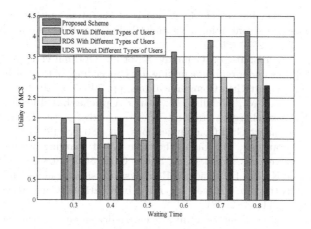

Fig. 2. MCS's utility vs. waiting time, through various schemes.

6   Conclusion and Future Work

This work presents a contract-based approach to provide EV users with charging service from MCS in heterogeneous networks, through which we can obtain the optimal solution based on the proposed iteration algorithm. Considering BER and mobility of MCS in the heterogeneous networks, we define a novel function with consideration of the interaction between EV users and MCS. In order to maximize MCS's payoff, the optimal strategy has been formulated as an optimization problem. Through the proposed iterative algorithm, we achieve the optimal solution. Simulation results have been presented to demonstrate the performance of the proposal. In future research, we will further consider how to design the incentive mechanisms and schedule the charging service task in the wireless communication, where both renewable power and multiple energy trading are considered in the future framework.

References

1. Bayram, I., George, M., Michael, D., et al.: Electrical power allocation in a network of fast charging stations. IEEE J. Sel. Areas Commun. **31**(7), 1235–1246 (2015)
2. Yuan, W., Huang, J., Zhang, Y.: Competitive charging station pricing for plug-in electric vehicles. In: Proceedings of the IEEE International Conference on Smart Grid Communications, pp. 668–673 (2014)
3. http://www.askci.com/news/chanye/2015/11/29/22408603z.shtml

4. Abdelsamad, S., Morsi, W., Sidhu, T.: Impact of wind-based distributed generation on electric energy in distribution systems embedded with electric vehicles. IEEE Trans. Sustain. Energ. **6**(1), 79–87 (2015)

5. Huang, S., He, L., Gu, Y., et al.: Design of a mobile charging service for electric vehicles in an urban environment. IEEE Trans. Intell. Transp. Syst. **16**(2), 787–798 (2015)

6. Liu, N., Chen, Q., liu, J., et al.: A heuristic operation strategy for commercial building microgrids containing EVs and PV System. IEEE Trans. Industr. Electron. **62**(4), 2560–2570 (2014)

7. Karbasioun, M., Ioannis, L., Gennady, S., et al.: Optimal charging strategies for electrical vehicles under real time pricing. In: Proceedings of the IEEE International Conference on Smart Grid Communications, pp. 746–751 (2012)

8. Bayram, I., Ismail, M., Abdallah, M., et al.: A pricing-based load shifting framework for EV fast charging stations. In: Proceedings of the IEEE International Conference on Smart Grid Communications, pp. 680–685 (2014)

9. Bahrami, S., Parniani, M.: Game theoretic based charging strategy for plug-in hybrid electric vehicles. IEEE Trans. Smart Grid. **5**(5), 2368–2375 (2014)

10. Fan, S., Wang, X., Yu, D.: Exact BER analysis for signal code modulation in wireless communication. In: Proceedings of the International Conference on Vehicular Technology, pp. 11–14 (2008)

11. Zhang, K., Mao, Y., Leng, S., et al.: Incentive-driven energy trading in the smart grid. IEEE Access **4**, 1243–1257 (2016)

12. Jeong, S., Jang, Y., Kum, D.: Economic analysis of the dynamic charging electric vehicle. IEEE Trans. Power Electron. **30**(11), 6368–6377 (2015)

13. Lee, J., Guo, J., Choi, J., et al.: Distributed energy trading in microgrids: a game theoretic model and its equilibrium analysis. IEEE Trans. Industr. Electron. **62**(6), 1–10 (2015)

14. Xie, X., Lei, W., Ma, S., et al.: Cognitive and cooperative wireless communication networks. Press (2012). ISBN 978-7-115-27365-9

Review of the Four Ports Electromechanical Converter Used for Hybrid Electric Vehicle

Qiwei Xu[1(✉)], Jing Sun[1], Meng Zhao[1], Xiaobiao Jiang[1], Yunqi Mao[1], and Shumei Cui[2]

[1] Chongqing University, Chongqing 400044, China
xuqw@cqu.edu.cn
[2] Harbin Institute of Technology, Harbin 150001, China

Abstract. Four Ports Electromechanical Converter (FPEC) is a device based on electromagnetic principle to realize speed distribution, torque distribution and electromechanical conversion. The performance of FPEC directly affects the dynamic properties and fuel economy of hybrid electric vehicle (HEV), which can achieve the functions of continuously variable speed, power compensation, brake energy feedback, starter and generator mode by constituting a complete series-parallel hybrid system combining with the internal combustion engine (ICE) and energy storage. This paper has discussed the FPEC used for the hybrid power system and focused on the analysis of magnetic coupling and electrical coupling, which lays the foundation for the further research and practical application of FPEC in the HEV.

Keywords: Hybrid Electric Vehicle (HEV) · Four Ports Electromechanical Converter (FPEC) · Magnetic coupling · Electrical coupling

1 Introduction

Four Ports Electromechanical Converter (FPEC) is a device based on the principle of electromagnetic, which can realize the speed distribution, torque distribution and electromechanical conversion [1]. Traditional motor has an electrical port and a mechanical port, which can be regarded as a two-port device to achieve the conversion of electrical and mechanical energy via the electromagnetic field. The FPEC discussed in this paper has two electrical ports and two mechanical ports. Similarly, the electromagnetic field acts as the coupling medium to achieve the free transmission and conversion of the electrical and mechanical energy between the four ports. The FPEC forms a complete series-parallel hybrid system, which can achieve the functions of continuously variable speed, power compensation, starter and generator mode, and brake energy feedback. The FPEC has the advantages of flexible control, high space utilization, especially in the volume, power density and energy transmission efficiency, which has a broad application prospects in the HEV [2, 3]. This paper has summarized the status quo of magnetic field coupling, speed coupling, and electrical coupling in the FPEC, which has the meanings of theory and practice for the development of FPEC control system used for HEVs.

K. Li et al. (Eds.): LSMS/ICSEE 2017, Part III, CCIS 763, pp. 329–338, 2017.
DOI: 10.1007/978-981-10-6364-0_33

2 Study on the Magnetic Field Coupling

2.1 EVT Based on the Principle of Induction Motor

In 2004, the Prof. Martin Hoeijmakers of Delft Polytechnic University proposed the electrical variable transmission (EVT) based on the principle of induction motor cascade driving [4], as shown in Fig. 1. The winding inner rotor and the inner cage of outer rotor constitute the induction motor, which is named internal machine (IM). The winding stator and the outer cage of outer rotor constitute the outer induction motor, which is called outer machine (OM). The two machines can run as motors or generators.

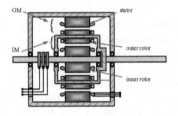

Fig. 1. Structure of EVT

In [5], the magnetic field coupling of IM and OM are seriously. The electromagnetic coupling of EVT with flux-insulation ring or not is studied respectively [6]. The 2D-finite element method (FEM) model and the equivalent magnetic circuit model of them are given in follows (Figs. 2 and 3).

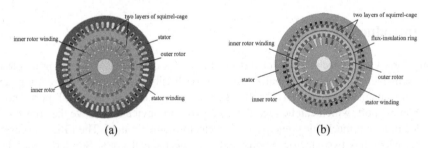

(a)   (b)

Fig. 2. 2D-FEM model of two kinds of EVT

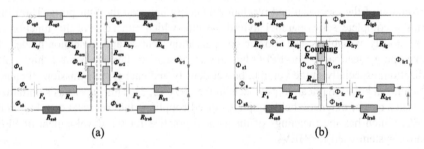

(a)   (b)

Fig. 3. Equivalent magnetic circuit model of two kinds of EVT

2.2 Four Quadrant Transducer

In 2001, the four-quadrant transducer (4QT) was proposed by Prof. Chandur Sadar-angani of the Royal Institute of Technology in Sweden, which also used the radial structure of the stator-outer rotor-inner rotor from outside to inside [7, 8].

In [9], the magnetic coupling in the compound-structure permanent-magnet (CSPM) motor is analyzed, which has the same structure as 4QT. The parameters matching of CSPM motor with THS of the Prius is accomplished to receive the dynamic properties design parameters between them. They both work as the series-parallel power split HEVs to satisfy the same load demand, but the HEV based on CSPM motor is more compact, intelligent, and noiseless. The equivalent magnetic circuit model of the tangential and radial magnetic flux distribution in the CSPM motor's outer rotor are established to explore the internal magnetic coupling rule of CSPM motor, which is shown in Fig. 4.

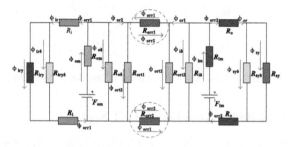

Fig. 4. No-load equivalent magnetic circuit of tangential and radial magnetic fluxes in the outer rotor.

The radial magnetic flux is the coupling flux between DRM and SM, which should be reduced as far as possible under the required electromagnetic energy. By using the network equation method in circuit analysis, the expression of radial coupling magnetic fluxes is given in Eq. (1):

$$\Phi_{\text{orr1}} = \frac{R_{\text{ort1}}}{2R_{\text{i}}} \Phi_{\text{ort1}}$$
$$\Phi_{\text{orr2}} = \frac{R_{\text{ort2}}}{2R_{\text{o}}} \Phi_{\text{ort2}}$$

$$(1)$$

By using the FEM electromagnetic simulation to analysis the magnetic field of CSPM motor in different working conditions, an internal magnetic decoupling scheme is proposed, which meet the requirements of lightweight and miniaturization of CSPM motor. Figure 5 shows the magnetic flux lines distribution of the decoupling CSPM motor.

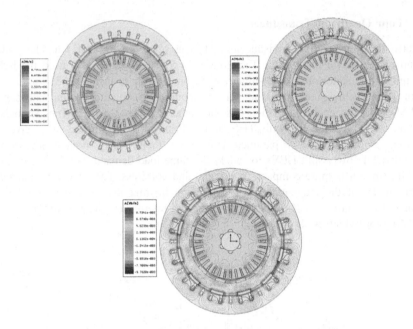

Fig. 5. The magnetic flux lines distribution of the decoupling CSPM motor.

2.3 Hybrid-Exited Radial-Axial Flux E-CVT System

The research group of Hong Kong Polytechnic University proposes a novel hybrid-exited radial-axial flux E-CVT system, which has two stators and two rotors [10]. The Structure of hybrid-exited radial-axial flux E-CVT system is shown in Fig. 6.

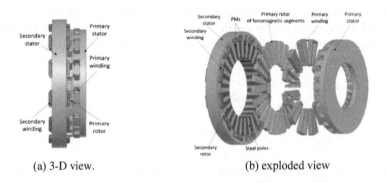

(a) 3-D view.   (b) exploded view

Fig. 6. Structure of hybrid-exited radial-axial flux E-CVT system.

The proposed machine can be seen as a combination of two motor/generators (M/G$_s$). M/G1 is an axial flux modulation machine (AFMM), which is composed of the primary stator and two rotors; M/G2 is a radial flux PM machine (RFPM), which is composed of the secondary stator and secondary rotor. There are some ferromagnetic segments install on the primary rotor, which are arranged between the primary stator and secondary rotor. As can be seen, the secondary rotor is the most commonly utilized, which is mounted on some PMs. All the PMs are magnetized along the azimuthal direction, and every two nearby PMs are magnetized oppositely. In order to achieve the stable torque transmission, the relationship between the pole-pair number of the primary winding (P_{pw}), the magnetic pole-pair number of the secondary rotor (P_{sr}), and the ferromagnetic segment number (N_s) is shown in Eq. (2).

$$P_{pw} = N_s - P_{sr}$$
$$P_{sr}\Omega_{sr} + P_{pw}\Omega_{pw} = N_s\Omega_{pr}$$
(2)

where Ω_{pw}, Ω_{pr}, Ω_{sr} are the rotary speed of the magnetic field generated by the primary winding primary winding, the primary rotor and the secondary rotor.

The hybrid-exited radial-axial flux E-CVT has the advantage of high torque density and torque density, small the volume, weight and copper losses. What's more, due to the hybrid axial and radial flux design, the mutual effect between the primary winding and the secondary winding.

3   Study on the Electrical Coupling

The FPEC used for HEV, which is connected to the energy storage device by superimposing the electrical power of IM and OM. Therefore, the energy storage device can output or feedback the electric power, which causes the forming electrical coupling in the FPEC.

To clearly express the energy flows in this complex system, the ideology of energetic macroscopic representation (EMR) is built to analyze the EVT-based HEV model, as shown in Fig. 7 [11]. The EMR was proposed by Bouscayrol based on the principle of action and reaction in 2000, which in perspective the energy flows.

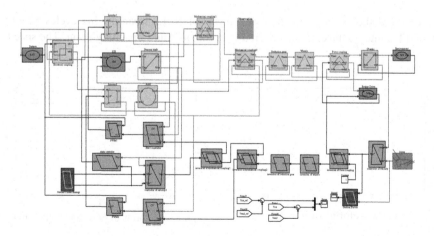

Fig. 7. The simulation model of HEV based on EVT with the ideology of EMR.

The HEV based on FPEC has a new type of power-driven system, which is significantly different from traditional vehicles in power output. Therefore, it is very important to study the control strategy of HEV based on FPEC. By optimizing the control strategies of HEV, the high efficient operation, the improvement of the mileage and the decrease of the emission of HEV are realized. At present, the control strategies of HEV are mainly divided into the rule-based control strategy and the optimized-based control strategy.

3.1 Rule-Based Control Strategy

The rule-based control strategy is mainly divided into target parameter control, switch control, power following control and fuzzy control strategy. The rule-based control strategy regards the driving conditions, powertrain state, engineering experience and experimental test data analysis as the control target to develop the reasonable control rules and constraint conditions.

When using the ON/OFF control strategy to control the HEV based on FPEC, the control states of ICE are only running and stopping. The control block diagram is shown in Fig. 8.

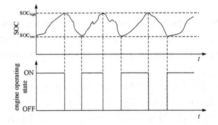

Fig. 8. The simulation model of HEV based on EVT with the ideology of EMR.

Figure 9 shows the simulation waveforms of 10.15 urban driving cycles based on ON/OFF strategy, the engine is operating at the optimum operating point. The speed is

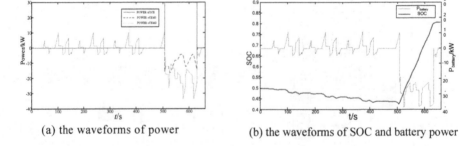

(a) the waveforms of power   (b) the waveforms of SOC and battery power

Fig. 9. The simulation waveforms of 10.15 urban driving cycles based on ON/OFF strategy

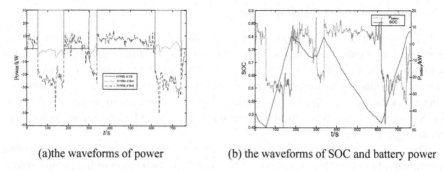

(a)the waveforms of power   (b) the waveforms of SOC and battery power

Fig. 10. The simulation waveforms of HWFET highway driving cycles based on ON/OFF strategy

3000 rpm while the torque is 95 Nm. From this figure it can be concluded that when the vehicle demands less power, the driving force is mainly provided by EVT, and the engine merely runs except the battery SOC reaches the ON/OFF threshold.

Figure 10 shows the simulation waveform of HWFET highway driving cycles. It can be seen that while the demand power of vehicle is significant, the SOC of battery fluctuates violently near the ON/OFF threshold.

When adopting the power follower strategy is used to control the HEV based on FPEC, the output of ICE is adjusted according to the different load demands, which makes the ICE working near the optimal fuel consumption curve. By charging or discharging the battery to provide the output power deviation between ICE and HEV. When the operating point of ICE and the wheel are coincide, the output power of battery is zero. At this time. The FPEC can be regarded as a gear with a gear ratio of 1 during the work process. In the power follower control strategy, in order to improve the fuel economy of vehicles, the characteristic curve of ICE in the efficient region should be determined. Combined the constant power curve of ICE with the constant tangency efficiency curve in the characteristic curve of ICE to compose the optimum operation curve of it, as shown in Fig. 11.

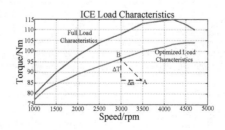

Fig. 11. The optimal operation curve of ICE

3.2 Optimized-Based Control Strategy

Optimized-based control strategy analyzes the overall efficiency of HEV under different working modes according to the working condition of the current and the predict powertrain. The optimization algorithm is used to adjust the vehicle's control rules, then the working state of powertrain in the next time is decided to optimize the efficiencies or fuel economies. At present, the optimized-based control strategy is divided into instantaneous optimization control strategy and global optimization control strategy.

The idea of instantaneous optimization control strategy is to achieve the global optimization based on the current conditions and the optimization of the predicted load demands. This control strategy is less sensitive to the driving conditions. However, the calculation is very large, which needs the comprehensive and accurate data of the car's powertrain in the pre-experimental stage.

3.3 Parameter Sizing in the HEV Based on FPEC

The method of parameter sizing for the powertrain in the HEV is divided into two parts: the parameter primaries and the parameter optimization, which mainly include the matching between the three power devices, i.e., the ICE, FPEC and energy storage device. The parameter primaries is based on the analysis of the constraints between the powertrain in the HEV to meet the basic requirements of driving process, fuel economies and other basic indicators. Through the theoretical analysis and the software simulation to match the parameters of each powertrain. The parameter optimization is based on the parameters' initial value. One of the vehicle's performance parameters is selected as the optimization index, then the optimization objective function is established. The optimization parameters are obtained by simulation with the dynamic parameters.

Prof. Shumei Cui of Harbin Institute of Technology has studied the problem of parameter sizing in the HEV based on FPEC. Through the simulation under a variety of typical conditions, the operation point distribution of EVT is shown in Fig. 12.

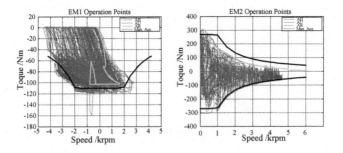

Fig. 12. The operation points distribution of EVT

As shown in Fig. 12, the IM is in the electric state, which has not been fully utilized; however, the area of generation is greater than that of electric in the OM. According to the theory and simulation analysis, the fuel economies of the vehicle can be improved if increase a gear with the fixed speed ratio between the ICE and FPEC, and the designed targets of FPEC are fully utilized, as shown in Fig. 13.

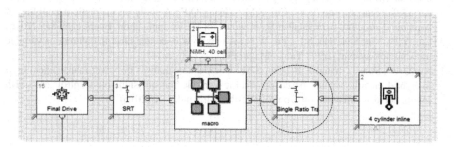

Fig. 13. The power transmission mechanism after increasing fixed ratio gear

4 Conclusion

As a new type of electromechanical energy conversion device, the FPEC is very suitable for use as a dynamic separation device in the hybrid system with many forms of electromechanical energy conversion, which has a wide application prospect. In this paper, the research progresses of magnetic field coupling and electrical coupling in the FPEC used for HEV are summarized detail, which has a dual significance of theory and practice to the development of a control system in the HEV based on FPEC.

References

1. Cheng, Y., Cui, S.M., Song, L.W.: The study of the operation modes and control strategies of an advanced electromechanical converter for automobiles. IEEE Trans. Magn. **43**(1), 430–433 (2007)
2. Kessels, J.T.B.A., Foster, D.L., van den Bosch, P.P.J.: Integrated powertrain control for hybrid electric vehicles with electric variable transmission. In: IEEE Vehicle Power and Propulsion Conference, Dearborn, MI, U.S. (2009)
3. Xu, Q.W., Song, L.W., Tian, D.W., Cui, S.M.: Research on intelligence torque control for the electrical variable transmission used in hybrid electrical vehicle. In: 14th International Conference on Electric Machines and Systems, pp. 1–6, Beijing, China, (2011)
4. Hoeijmakers, M.J., Rondel, M.: The electrical variable transmission in a city bus. In: 35th Power Electronics Specialists Conference, Aachen, Germany (2004)
5. Kim, J., Kim, T., Min, B., Hwang, S., Kim, H.: Mode control strategy for a two-mode hybrid electric vehicle using electrically variable transmission (EVT) and fixed-gear mode. IEEE Trans. Veh. Technol. **60**(3), 793–803 (2011)

6. Xu, Q.W., Sun, J., Zhao, M., Jiang, X.B., Cui, S.M.: Electromagnetic decoupling optimum design of electric variable transmission used for hybrid electric vehicle. Transaction of China Electrotechnical Society (unpublished)
7. Zheng, P., Liu, R.R., Wu, Q.: Magnetic coupling analysis of four-quadrant transducer used for hybrid electric vehicles. IEEE Trans. Magn. **43**(6), 2597–2599 (2007)
8. Zheng, P., Liu, R.R., Thelin, P.: Research on the cooling system of a 4QT prototype machine used for HEV. IEEE Trans. Energy Convers. **23**(1), 61–66 (2008)
9. Xu, Q.W., Sun, J., Luo, L.Y., Cui, S.M., Zhang, Q.F.: A study on magnetic decoupling of compound-structure permanent-magnet motor for HEVs application. Energies **9**, 819 (2016)
10. Liu, Y.L., Niu, S.X., Ho, S.L., Fu, W.N.: A new hybrid-excited electric continuous variable transmission system. IEEE Trans. Magn. **50**(11), 1–4 (2014)
11. Xu, Q.W., Sun, J., Jiang, X.B., Cui, S.M.: Comparison analysis of power management used in hybrid electric vehicle based on electric variable transmission. In: Control 2016-11th International Conference on Control, Northern Ireland, UK (2016)

Research on Parameters Matching of Hybrid Electric Vehicle with Compound-Structure Induction Machine

Qiwei Xu[1(✉)], Xiaobiao Jiang[1], Meng Zhao[1], Xiaoxiao Luo[1],
Weidong Chen[1], Yunqi Mao[1], and Shumei Cui[2]

[1] The State Key Laboratory of Power Transmission Equipment and System
Security and New Technology, Chongqing University,
Chongqing 400044, China
xuqw@cqu.edu.cn
[2] School of Electrical Engineering and Automation,
Harbin Institute of Technology, Harbin 150001, China

Abstract. This paper analyzes the parameters matching problem of hybrid electric vehicle with compound-structure induction machine. According to vehicle dynamics model, the dynamic coupling of compound-structure induction machine can be studied. After the analysis of optimization for degree of hybridization, the research of parameters matching can be carried out. Furthermore, the design requirements of hybrid electric vehicle based on compound-structure induction machine are analyzed. As a result of above analysis, the requirements of three aspects: peak power, continuous power and energy are obtained. Finally, aid from the simulation of Cruise, it is verified that the parameters matching of optimization for degree of hybridization can be feasible.

Keywords: Hybrid electric vehicle · Compound-structure induction machine · Parameters matching · Optimization for degree of hybridization

1 Introduction

The hybrid electric vehicle (HEV) based on the compound-structure induction machine (CSIM) has the electromagnetic power transmission system, and the CSIM as a electric variable transmission (EVT), can replace the function of the continuously variable transmission (CVT). As the theoretical analysis shown, if combined with certain control strategy, the CSIM can replace the planetary gear mechanism of TOYOTA PRIUS series hybrid vehicle, realizing the functions such as the engine idle shutdown, power compensation and regenerative braking [1].

At present, domestic and foreign scholars mainly focus on the electromagnetic design, mathematical modeling, efficiency analysis, cooling system design, brushless double rotor structure and other aspects. The CSIM's body is studied [2–6], which laid a foundation for its further research and application. The current research mainly includes the following aspects: (1) The electromagnetic coupling and control strategy of CSIM: according to the establishment of equivalent magnetic circuit and finite

© Springer Nature Singapore Pte Ltd. 2017
K. Li et al. (Eds.): LSMS/ICSEE 2017, Part III, CCIS 763, pp. 339–350, 2017.
DOI: 10.1007/978-981-10-6364-0_34

element simulation model, the internal magnetic field coupling law of inner motor and outer motor when they work alone and together. (2) Study on electromechanical coupling law of the CSIM: according to energy transfer law of different working modes, the CSIM is regard as the CVT, starter, generator, clutch of traditional hybrid cars to achieve CVT model, starter mode, generator mode, braking mode etc. (3) Study on vehicle control strategy for HEV based on CSIM: according to the fact that the CSIM changes power transmission topology of HEV and the analysis of the application of traditional serial and parallel HEV control strategy in this new structure in hybrid electric vehicle. Through control of the CSIM [7], energy output of engine and storage device can be adjusted.

2   Dynamic Analysis of HEV with CSIM

The EVT architecture is shown in left of Fig. 1, with two mechanical ports and two electrical ports, comprised of the inner rotor, the middle rotor squirrel cage structure and independent double winding stator. The inner rotor is EM1, while the outer rotor is EM2. It is shown in the right of Fig. 1.

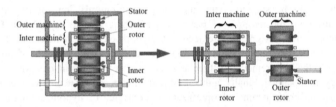

Fig. 1. The structure diagram of compound-structure induction machine

2.1   Vehicle Dynamics Model

The mechanics analysis of the vehicle during driving is shown in Fig. 2.

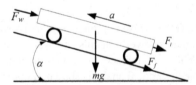

Fig. 2. The dynamics model of a vehicle

According to the Fig. 2, considering the vehicle dynamics, when the vehicle accelerate on the ramp, the demand of traction can be expressed as

$$F_{\text{req}} = F_{\text{f}} + F_{\text{w}} + F_{\text{i}} + F_{\text{j}} \tag{1}$$

where F_{req} is The total driving force;
  F_f is rolling resistance while driving;
  F_w is air resistance while driving;
  F_i is ramp resistance while driving
  F_j is acceleration resistance while driving.
According to the vehicle dynamics analysis, the following equations are given:

$$F_f = mg f_t \cos \alpha \tag{2}$$

$$F_w = 0.047 A c_D (v + v_0)^2 \tag{3}$$

$$F_i = mg \sin \alpha \tag{4}$$

$$F_j = m \delta_b a \tag{5}$$

where g is acceleration of gravity;
  f_t is rolling resistance coefficient;
  α is vehicle climbing angles;
  v is vehicle operating speed;
  δ_b is spinning quality conversion coefficient;
  c_D is coefficient of air resistance;
  A is windward area on front;
  A is the acceleration of vehicle.
  Adding the Eqs. (2)–(5) into the Eq. (1), demand traction is calculated, as shown in Eq. (6).

$$F_{req} = mg f_t \cos \alpha + 0.047 A c_D (v + v_0)^2 + mg \sin \alpha + m \delta_b a \tag{6}$$

Then, the power demand torque T_{req} can be calculated, as shown in Eq. (7).

$$T_{req} = [mg f_t \cos \alpha + 0.047 A c_D (v + v_0)^2 + mg \sin \alpha + m \delta_b a] \cdot r_d \tag{7}$$

Assuming that vehicle under the driving force F_{max} of ultimate adhesion could achieve the maximum speed v_{max}. According to the vehicle dynamic balance relationship, the equation of the vehicle equation is expressed

$$F_{max} = mg \mu_{max} \cos \alpha = F_f + F_w + F_i + F_j \tag{8}$$

where μ_{max} is maximum adhesion coefficient.
  In Eq. (8), v_{max} is written as the quadratic product of v_{max} and v_0.

$$v_{max} = \frac{mg \mu_{max} \cos \alpha - mg f_t \cos \alpha - mg \sin \alpha}{0.141 A c_D v_0} - \frac{m \delta_b a + 0.047 A c_D v_0^2}{0.141 A c_D v_0} \tag{9}$$

2.2   Analysis of Power Output Power Coupling

The HEV with CSIM as shown in Fig. 3. And the split structure of the CSIM is shown in Fig. 4.

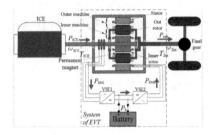

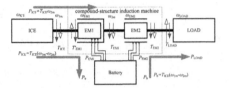

Fig. 3. The structure diagram of hybrid electric vehicle compound-structure induction machine

Fig. 4. The force transmission diagram of HEV based on compound-structure induction machine

As shown in Fig. 4, the mechanical energy output of the engine P_{ICE} is divided into two parts. It can be shown in the Eq. (10).

$$P_{ICE} = P_{EM1} + P_d \tag{10}$$

The electromagnetic torque of EM1 T_{EM1} is balanced with the output torque of motor T_{ICE}. Thus, the power equations can be expressed as follow:

$$\begin{cases} P_{ICE} = T_{ICE}\omega_{ICE} \\ P_{EM1} = T_{EM1}(\omega_{ICE} - \omega_{EM2}) = T_{ICE}(\omega_{ICE} - \omega_{EM2}) \\ P_d = P_{ICE} - P_{EM1} = T_{ICE}\omega_{EM2} = T_{EM1}\omega_{EM2} \end{cases} \tag{11}$$

The shaft output torque T_{2m} of the outer rotor is combined torque of the inner motor and the outer motor is shown in Eq. (12).

$$T_{2m} = T_{EM1} + \frac{P_{EM1}}{\omega_{EM2}} = T_{EM1} + \frac{T_{EM1}(\omega_{ICE} - \omega_{EM2})}{\omega_{EM2}} = \frac{T_{EM1}\omega_{ICE}}{\omega_{EM2}} \tag{12}$$

According to Eq. (6), require power P_{req} can be calculated by the following equation.

$$P_{req} = F_{req\_vhe}v = \frac{1}{3600}[mgf_t \cos\alpha + 0.047Ac_Dv^2 + mg\sin\alpha + m\delta_b a] \cdot v \tag{13}$$

In Fig. 4, since the mechanical power PICE of the engine is shunted by EM1, a part of the power is transmitted to the vehicle energy storage system in the form of output power P_{EM1} via EM1, and the other part of the power P_d is directly transmitted to the wheels to drive the vehicle. The power balance equation when the vehicle is driven, which is shown in Eq. (14).

$$P_{req} = P_d + P_{EM2} \tag{14}$$

In order to meet demand power P_{req}, the CSIM is considered as two motors connected. EM1 and the engine is combined as a power distribution, namely the engine group, as shown in Fig. 5.

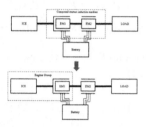

Fig. 5. The schematic of defined engine block at the moment of power distribution

The engine unit outputs the mechanical power P_d and the power generation power P_{EM1}. According to the Eq. (14), the mechanical power P_d meets the power demand to drive the vehicle. The CSIM can realize the CVT function in the traditional vehicle, and the transmission ratio i_{CVT} of the CSIM can be obtained by derivation, as shown in Eq. (15).

$$i_{CVT} = \frac{\omega_{ICE}}{\omega_{EM2}} = \frac{T_{2m}}{T_{EM1}} \tag{15}$$

According to the transmission ratio i_{CVT}, set can be obtained. The relationship among the EM1 output power P_{EM1} in the engine-generator, the direct transmission output power P_d and the engine output power P_{ICE} are obtained.

$$P_{EM1} = \left(\frac{1}{i_{CVT}} - 1\right) P_{ICE} \tag{16}$$

$$P_d = \left(2 - \frac{1}{i_{CVT}}\right) P_{ICE} \tag{17}$$

3 Optimization for Degree of Hybridization of HEV Based on CCSIM

3.1 Calculation of Degree of Hybridization in HEV Based on CSIM

The mixing degree of a hybrid vehicle is the percentage of the electric power P_{elec} in the total drive P_{total}, as shown in Eq. (18).

$$H_S = \frac{P_{elec}}{P_{total}} \times 100\% \tag{18}$$

In the hybrid vehicle, the electric power and the driving total power are as follows:

$$P_{elec} = P_{BAT} = P_{EM2} - P_{EM1} \tag{19}$$

$$\begin{aligned} P_{total} &= P_{ICE} + P_{BAT} = P_{ICE} + P_{EM2} - P_{EM1} \\ &= P_{EM1} + P_d + P_{EM2} - P_{EM1} = P_d + P_{EM2} \end{aligned} \tag{20}$$

(19) and (20) are substituted into the Eq. (18), mixing degree of the hybrid vehicle is obtained as shown in Eq. (21).

$$H_S = \frac{P_{elec}}{P_{total}} \times 100\% = \frac{P_{EM2} - P_{EM1}}{P_{ICE} + P_{EM2} - P_{EM1}} \times 100\% \tag{21}$$

3.2 Optimization for Degree of Hybridization

In the HEV based on the CSIM, by controlling the CSIM, it is possible to adjust the engine to work in the vicinity of the optimal efficiency curve with little change in torque. Therefore, the vehicle demand power, the engine power, the internal motor power and the engine power curve in the full speed range are shown in Fig. 6.

In Fig. 6, the power difference between the vehicle demand power and the engine group is equal to the driving power of the external motor EM2. When the external motor is running at the speed of ω_b, the demand power for the external motor is large, and with the increased vehicle speed, the demand power of the vehicle for the external motor is gradually reduced, as shown in Fig. 7.

When the internal rotor speed is higher than the vehicle speed, the inner motor works in the electric state and the electric power increases with the speed increases, as shown in Fig. 8.

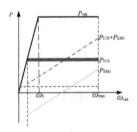

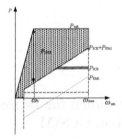

Fig. 6. Power curves of power units in HEV

Fig. 7. The relationship between power of outer machine and vehicle velocity

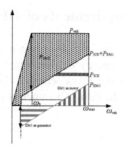

Fig. 8. The relationship between power of inner machine and vehicle velocity

Adjusting the slope of the engine power curve, the motor power generation and electric power can be adjusted. When the generator power and electric power are equal, the internal motor is fully utilized, the design power of the internal motor EM1 is the smallest as shown in Eq. (22).

$$
\begin{aligned}
P_{\text{EM1\_max}} &= T^*_{\text{ICE}} \cdot \omega_{\text{max}} - T^*_{\text{ICE}} \cdot \omega^*_{\text{ICE}} = \frac{P^*_{\text{ICE}}}{\omega^*_{\text{ICE}}} \cdot \omega_{\text{max}} - T^*_{\text{ICE}} \cdot \omega^*_{\text{ICE}} \\
&= \frac{P_{\text{VHE\_max}}}{\omega^*_{\text{ICE}}} \cdot (1 - H_{\text{S}}) \cdot \omega_{\text{max}} - T^*_{\text{ICE}} \cdot \omega^*_{\text{ICE}} \\
&= \frac{T_{\text{VHE\_max}} \cdot \omega_{\text{max}}}{\omega^*_{\text{ICE}}} \cdot (1 - H_{\text{S}}) \cdot \omega_{\text{max}} - T^*_{\text{ICE}} \cdot \omega^*_{\text{ICE}}
\end{aligned}
\tag{22}
$$

According to the engine group, the relationship between motor EM1 and the engine power, the engine power is obtained.

$$
\begin{aligned}
P_{\text{ICE}} &= \frac{i_{\text{CVT}}}{1 - i_{\text{CVT}}} \cdot P_{\text{EM1}} \\
&= \frac{i_{\text{CVT}}}{1 - i_{\text{CVT}}} \cdot \frac{T_{\text{VHE\_max}} \cdot \omega_{\text{max}}}{\omega^*_{\text{ICE}}} \cdot (1 - H_{\text{S}}) \cdot \omega_{\text{max}} - T^*_{\text{ICE}} \cdot \omega^*_{\text{ICE}}
\end{aligned}
\tag{23}
$$

According to the expression of the mixing degree, the power of the external motor EM2 is shown in Eq. (24).

$$
\begin{aligned}
P_{\text{EM2}} &= \frac{H_{\text{S}}}{1 - H_{\text{S}}} \cdot P_{\text{ICE}} + P_{\text{EM1}} = \frac{H_{\text{S}}}{1 - H_{\text{S}}} \cdot \frac{i_{\text{CVT}}}{1 - i_{\text{CVT}}} \cdot P_{\text{EM1}} + P_{\text{EM1}} \\
&= \left(\frac{H_{\text{S}}}{1 - H_{\text{S}}} \cdot \frac{i_{\text{CVT}}}{1 - i_{\text{CVT}}} + 1 \right) \cdot \frac{T_{\text{VHE\_max}} \cdot \omega_{\text{max}}}{\omega^*_{\text{ICE}}} \cdot (1 - H_{\text{S}}) \cdot \omega_{\text{max}} - T^*_{\text{ICE}} \cdot \omega^*_{\text{ICE}}
\end{aligned}
\tag{24}
$$

4 Analysis on Design Requirements of HEV

The design requirements of the HEV mainly include peak power requirements, continuous power requirements and energy requirements. In this paper, the vehicles can achieve the control functions of the Prius series of TOYOTA, therefore, in the calculation of design requirements, the design parameters of Prius series of hybrid vehicle is used, as shown in Table 1.

Table 1. Parameters of hybrid electric vehicle

Type	Parament	Value	Unit
Vehicle parameters	Curb weight	1345	kg
	Load weight	1745	kg
	Windward area	1.746	m^2
	Air resistance coefficient	0.3	
	Transmission system efficiency	0.9	
Tires	Tire radius	0.2929	m
	Rolling resistance coefficient	0.01	

4.1 Analysis of the Peak Power Requirements

The output of dynamic characteristics of the drive system should satisfy the acceleration and climbing driving requirements. For the climb performance requirements, assuming that the vehicle travels on the slope of the road with the speed v_a, the demand power of the drive motor P_i is shown:

$$P_i = \frac{1}{3600\eta_T}\left(mgf_t\cos\alpha + 0.047Ac_Dv_a^2 + mg\sin\alpha\right)v_a \tag{25}$$

where η_T is efficiency of the drive train in a hybrid vehicle.

By analyzing the (25), the relationship between the different vehicle speeds and the power requirements at a given slope can be obtained, as shown in Fig. 9.

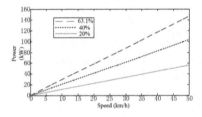

Fig. 9. The relation curves between vehicle velocities and power requirement under different gradients

As known from Fig. 9, the driving power of the HEV is approximately proportional to the vehicle speed, because the proportion of the air resistance component on the low speed climbing of the HEV is small. The acceleration performance depends on the torque control capability of the HEV in the full speed range and will be affected by the external characteristics of the CSIM and the transmission device. In the acceleration process, the required maximum acceleration power is the demand power when the vehicle accelerate to the target vehicle speed.

$$P_i = \frac{1}{3600\eta_T}\left(mgf_t + 0.047Ac_Dv_f^2 + m\delta_b\frac{dv_f}{dt}\right)v_f \tag{26}$$

where v_f is the target speed in the acceleration process.

4.2   Analysis of the Continuous Power Requirements

To meet the requirements of the maximum cruising speed of the HEV in the road, the maximum road driving power demand of hybrid cars is seen as a continuous power demand in this paper. The driving power P_v of the continuous operation region should be satisfied the Eq. (27).

$$P_v \geq \frac{1}{3600\eta_T}(mgf_t + 0.047Ac_Dv_{max}^2)v_{max} \tag{27}$$

where v_{max} is the maximum speed of the road cruise.

By analyzing the Eq. (27), the relationship between the different speeds and the power demands can be gotten when the rolling resistance coefficient is given at 0.01, 0.015 and 0.02 respectively, as shown in Fig. 10.

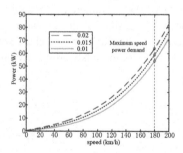

Fig. 10. The relation curves between vehicle velocities and power requirement under different coefficients of rolling resistance

4.3   Analysis of the Energy Requirements

Assuming that the vehicle travels on the flat road when the speed is v and the travel distance is S, the energy demand E_j is shown as Eq. (28):

$$E_{\mathrm{j}} = \frac{k_{\mathrm{b}}S}{3600\eta_{\mathrm{T}}} \left(mgf_{\mathrm{t}} + 0.047Ac_{\mathrm{D}}v^2\right) \tag{28}$$

where k_{b} is energy compensation coefficient during travel.

The relationship between the vehicle speed v and the energy demand E_{j} is shown in Fig. 11.

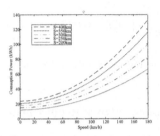

Fig. 11. The relation curves between vehicle velocities and energy requirement under different driving mileage

As the speed increases, the aerodynamic resistance increases nonlinearly, so the energy consumption increases nonlinearly with the increase of the vehicle speed at the same travel distance requirement.

5   Building Models and Simulation Analysis of the System

The establishment of the simulation model is based on the simulation software Cruise. The engine parameters of hybrid vehicle in the Prius series are used in the simulation, focusing on the simulation of matching parameters of the CSIM. In the Cruise simulation model, the condition of US06 is simulated. The simulation results are shown in Fig. 12. In the Fig. 12, the working state of the engine is compared with the power output state of the front end of the main accelerator. The CSIM realizes the function of CVT, and the transmission ratio is continuously adjustable and the fluctuation is small.

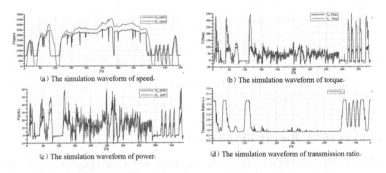

(a) The simulation waveform of speed.   (b) The simulation waveform of torque.

(c) The simulation waveform of power.   (d) The simulation waveform of transmission ratio.

Fig. 12. The simulation results in the driving cycle of US06

The 6 typical driving conditions are the circulation circle UDC of European city, Japanese 1015 circle, American FTP72 circle and high-speed circle US06, the circle of Artemis city and high-speed circle. The simulation results are shown in Figs. 13 and 14.

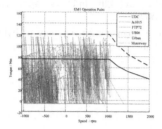

Fig. 13. The operation points distribution of EM1 under different driving cycles

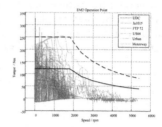

Fig. 14. The operation points distribution of EM2 under different driving cycles

According to the simulation data of Figs. 13 and 14, the matching parameters of EM1 and EM2 can be determined. Because EM1 has a double rotors and it should be used air-cooled, its overload factor can be assumed to be 1.6. EM2 can be designed in accordance with conventional motor, set its overload factor of 2.1. Therefore, the matching parameters of the CSIM are shown in Table 2.

Table 2. The matching parameters of compound-structure induction machine

Category	Parameter name	EM1	EM2
Rated value	Speed (rpm)	1050	1800
	Torque (Nm)	75	120
	Power (kW)	8.25	22.62
Peak value	Speed (rpm)	2100	6000
	Torque (Nm)	120	250
	Power (kW)	13.2	47.12

References

1. Cui, S., Cheng, Y., Chen, Q.: An advanced electrical variable transmission for automobiles. Trans. China Electrotech. Soc. **21**(10), 111–116 (2006)
2. Zheng, P., Liu, R., Thelin, P., Nordlund, E., Sadarangani, C.: Research on the cooling system of a 4QT prototype machine used for HEV. IEEE Trans. Energy Convers. **23**(1), 61–66 (2008)
3. Afsharirad, H., Sharifian, M.B.B.: Improved vector control of dual mechanical port machine for hybrid electric vehicle. In: 2nd Power Electronics, Drive Systems and Technologies Conference, Tehran, Iran, pp. 181–186 (2011)
4. Qiwei, X., Liwei, S., Shumei, C.: Research on force distribution strategy of hybrid electric vehicle based on electric variable transmission. Trans. China Electrotech. Soc. **28**(2), 44–54 (2013)

5. Kessels, J.T.B.A., Foster, D.L., van den Bosch, P.P.J.: Integrated powertrain control for hybrid electric vehicles with electric variable transmission. In: Vehicle Power and Propulsion Conference, Dearborn, MI, USA, pp. 376–381 (2009)
6. Xu, L., Zhang, Y., Wen, X.: Multi-operational modes and control strategies of dual mechanical port machine for hybrid electrical vehicles. In: 42th Industry Application Conference, New Orleans, LA, USA, pp. 1710–1717 (2007)
7. Xu, Q., Cui, S., Song, L., Zhang, Q.: Research on power management strategy of hybrid electric vehicle based on electric variable transmission. Energies 7(2), 934–960 (2014)

Location Model Research of Charging Station for Electric Vehicle Based on Users' Benefit

Fei Xia[1,2(✉)], Zhicheng Wang[1,2], Daogang Peng[1,2], Zihao Li[1],
Zhijiang Luo[1], and Bo Yuan[1,2]

[1] Faculty of Automation Engineering, Shanghai University of Electric Power,
Shanghai 200090, People's Republic of China
xiafeiblue@163.com
[2] Shanghai Engineering Research Center of Intelligent Management and Control
for Power Process, Shanghai 200090, China

Abstract. To improve the electric vehicle charging infrastructure and accelerate the development of electric vehicles, the optimization of electric vehicle charging stations' location and size are studied in this paper. The annual comprehensive cost of society, including user cost and charging station cost, is regarded as the objective function. Weight coefficients are added to the function for increasing the proportion of user cost. An optimized mathematical model for electric vehicle charging stations based on users' benefit is established. The improved Quantum-behaved Particle Swarm Optimization (QPSO) algorithm is adopted to solve this mathematical mode, and result which contain the optimal location and size of electric vehicle charging stations is obtained. Finally, an actual area is taken as the case study to optimize the location and size of electric vehicle charging stations by solving the mathematical model proposed in this paper with the improved QPSO algorithm. Rationality and validity of the model are well improved by the scientific reasonable result.

Keywords: Users' benefit · Electric vehicle · Charging station · Location model · The improved QPSO algorithm

1 Introduction

Under the huge pressure of energy resource and environment preservation, the emergence of electric vehicles can make up for the lack of traditional fuel vehicles in the aspect of energy conservation and environmental protection [1]. Therefore it is very urgent to promote electric vehicles in large-scale. However, the development of electric vehicle industry has been seriously restricted by the insufficient development of electric vehicle charging stations and related infrastructures [2]. Therefore, carrying out the research work of optimization of electric vehicle charging stations' location will have a far-reaching impact on the development of electric vehicles, energy conservation and environmental protection.

At present, there are many scholars have conducted the researches on the optimization design of electric vehicle charging station, such as Ying et al. adopt genetic algorithm, based on minimization comprehensive valve of charging station construction

© Springer Nature Singapore Pte Ltd. 2017
K. Li et al. (Eds.): LSMS/ICSEE 2017, Part III, CCIS 763, pp. 351–361, 2017.
DOI: 10.1007/978-981-10-6364-0_35

and operation cost and user's charging cost as well as the cost of the user to the charging station, to carry out research of optimization of electric vehicle charging stations' location [3]; Li et al. established the minimum annual cost model of electric vehicle charging station, and based on Particle Swarm Genetic Algorithm make allocation plan of electric vehicle charging station [4]; Zhao and Li used differential evolution hybrid particle swarm algorithm for the optimal planning of electric vehicle charging station. During the planning process, the fixed cost of the charging station, the driving cost to the charging station and the waiting cost in the charging station are considered comprehensively [5]. Chen et al. proposed a multi-objective location and capacity planning model for electric vehicle charging station which take carbon emissions into consideration. The optimization goal of the proposed model is to minimize the economic costs of charging station construction and operation, to shorten the time needed for charging electric vehicles and to reduce the carbon emission caused by driving to a charging station. Meanwhile the capacity limit of the charging station is taken as the constraint [6]. However, there are few studies from the user's point of view to optimize the location of charging stations, and the development of electric vehicles can not be separated from the support of users. Although Chu et al. established location model of electric vehicle charging station which use the time satisfaction function to evaluate the quality of service that the charging station provided to the customer [7], but standing on the user's position, only to minimize the cost of users can promote the development of electric vehicle industry more vigorous.

This paper, from the view point of users, an optimized charging station mathematical model based on users' benefit is established which take the annual comprehensive cost of society, including user cost and charging station cost, as the objective function. In order to make sure the model taking more consideration of the interests of the users in the process of optimizing the electric vehicle charging station, weight coefficients are added to the model. Therefore adjusting the weight coefficients can increase the proportion of user cost in the model. The improved QPSO algorithm is adopted to solve the proposed model to getting the most conducive location design of electric vehicle charging station for user. Lastly, taking an area in a city as an example for location optimization of the electric vehicle charging station, the improved OPSO algorithm is used to solve the mathematical model, and the reasonable result verified the scientificity of the mathematical model presented in this paper.

2 Mathematical Model of Charging Station Location

2.1 Objective Function

The annual comprehensive cost of society is an important index to measure the quality of charging station design, which is composed of the user cost and the charging station cost. There are many literatures [3–5] considering the comprehensive cost but few considering the problem from the view point of users and take the users' benefit as the main measure to optimize the charging stations' location. Therefore, a new mathematical model proposed in this paper, which based on comprehensively considering the user cost and the charging station cost, weight coefficients are added to the model for increasing the proportion of user cost in the annual comprehensive cost of society. It makes the optimization of the charging stations' location taking the users benefit as the main measure, so as to achieve the purpose of saving user cost.

$$C = aF_u + bF_s \qquad (1)$$

where C is the annual comprehensive cost of society of electric vehicles, F_u the users cost, F_s the charging station cost, and a, b respectively represent for the weight of users cost and the weight of charging station cost. Then the user cost, charging station cost and constraints will be respectively elaborated.

User Cost. The user cost includes three parts: the user time cost, the charging cost and the charging service cost.

Currently, a series of policies about electric vehicle charging charge were announced, that is, the charging fees containing a certain of charging service fees. Charging fees and charging service fees are seldom considered at the same time in the previous studies. Some improvements have been made in this paper, and the corresponding mathematical model has been established:

$$F_u = f_e + f_t + f_s \qquad (2)$$

where f_e is electric vehicle charging costs, that is, the electricity charges user needs to pay during the charging process, f_t for the time cost, f_s the electric vehicles charging service fees. The users' time cost, charging costs and charging service costs which in the formula (2) will be expatiated in the following parts.

$$f_e = \sum_{i=1}^{m} p \cdot u_i \cdot n_i \cdot 365 \cdot \left(s_i + \frac{\overline{L_i}}{g}\right) \qquad (3)$$

where m is the number of electric vehicle charging station, p the users' charging price, u_i the average times of each user to the i charging station charges every day, n_i the number of electric vehicles in the area covered by the charging station i, s_i average electricity per electric vehicles charged in the i charging station; $\overline{L_i}$ the average distance of electric vehicles to the i charging station; g the mileage per unit electricity.

$$f_s = \sum_{i=1}^{m} q \cdot u_i \cdot n_i \cdot 365 \cdot \left(s_i + \frac{\overline{L_i}}{g}\right) \qquad (4)$$

where q is electric vehicle charging service fees unit price.

$$f_t = \sum_{i=1}^{m} (l_i + w_i) \cdot u_i \cdot n_i \cdot t_p \cdot e \cdot 365 \qquad (5)$$

where l_i is average queue time of electric vehicles every day in the i charging station, w_i the electric vehicles average charging time every day in the i charging station, t_p the electric vehicle owners unit time revenue.

Charging Station Cost. Charging station cost consisted of charging station construction cost and charging station management cost [8].

$$F_s = f_b + f_m \tag{6}$$

where f_b is the charging station construction cost, f_m the charging station management cost.

On the construction cost of electric vehicle charging station, Ying et al. [3] did some researches, but they only studied the initial fixed investment cost and variable cost. In other words, they only studied the charging station construction cost of f_b and f_m, therefore they did not elaborate on the calculation of infrastructure costs E, while Zhao and Li [5] are studied it and presented. Therefore, the above two parts were integrated in this paper and a mathematical calculation model of charging station cost were obtained.

$$E_i = A_i \cdot LH_i + Z_i \cdot N_i + B_i \cdot N_i^2 + G_i \tag{7}$$

where E_i is the infrastructure costs of charging station i, A_i the land area of charging station i, LH_i the land price of the charging station i, Z_i the charging piles purchase price of charging station i, N_i the number of charging piles of charging station i, B_i the transformers, cables and other related costs of charging station i, G_i the fixed investment in business buildings of charging station i.

The method of questionnaire was used to get the expected sojourn time of the user, and according to the formula (8)–(10) the number of charging piles N_i of charging station i were obtained.

$$W_s = \frac{R_i}{H_i(H_i - R_i)} + \frac{1}{H_i} \tag{8}$$

where W_s is the user expected sojourn time(the sum of queueing time and charging time).

$$H_i = \frac{u_i \cdot n_i}{w_i} \tag{9}$$

where H_i is the average service rate of charging station i, and the average service rate refers to the number of people to be served within the unit time (number of people/time).

Due to the number of electric vehicles that go to the charging station for charging are random, and in literature [9] it is turned into the M/M/S queuing model. It is assumed that the number of electric vehicles that arrived the charging station in unit time obey the Poisson Distribution.

$$R_i = \frac{u_i \cdot N_i}{24} \tag{10}$$

where R_i is the average arrival rate of charging station i, and the average arrival rate is equal to the total number of customers divided by the total time.

Thus, the charging station construction cost f_b can be calculated by formula (11):

$$f_b = \sum_{i=1}^{m} E_i \left[\frac{k(1+k)^n}{(1+k)^n - 1} \right] \tag{11}$$

where n is the charging station design operating life, k the investment recovery rate;

Charging station management cost f_m can be calculated by formula (12):

$$f_m = \sum_{i=1}^{m} E_i(1+\delta) \tag{12}$$

In the formula (12) δ convert the charging station staff wages, charging station maintenance cost and other costs into charging station years income coefficient.

2.2 Constraint Condition

According to the size, the charging station is divided into small charging station, medium charging station and large charging station. Different types of charging stations have different requirements for the number of charging piles. Therefore, in this paper the number of charging pile in charging station is constrained.

$$J_{\min} \leq N_i \leq J_{\max} \tag{13}$$

where J_{max} and J_{min} represent the maximum and minimum number of charging piles in the charging station, respectively.

$$M_{\min} = \left[\frac{N}{J_{\max}} \right] \tag{14}$$

$$M_{\max} = \left[\frac{N}{J_{\min}} \right] \tag{15}$$

$$M_{\min} \leq m \leq M_{\max} \tag{16}$$

where $[\cdot]$ is the rounding; M_{max} and M_{min} represent the maximum and minimum number of charging station, respectively.

3 Improved Quantum Particle Swarm Optimization Algorithm

3.1 Algorithm Introduction

In this study, the improved QPSO algorithm is adopted to solve the proposed mathematical model of optimal design of electric vehicle charging station based on users' interest.

The QPSO algorithm is an improved algorithm of classical particle swarm optimization (PSO) algorithm. And it is a quantum behaved PSO algorithm. The algorithm can effectively improve the global search ability. And it has good convergence and stability. This algorithm can make the particles have quantum behavior, so that the particles do not following a fixed route while moving, which can make particles searching through the entire solution space to find the global optimal solution and effectively avoid falling into local optimum [10].

In the QPSO algorithm, during the process of each particles get moved, there exists a DELTA potential well that taking the individual extremum (Pbest) as the center. The position of the particles were updated according to a certain probability to plus or minus.

$$p = a * Pbest(i) + (1 - a) * Gbest \tag{17}$$

$$mbest = 1/M * \sum_{i=1}^{M} Pbest(i) \tag{18}$$

$$b = 1 - generation/\max generation * 0.5 \tag{19}$$

where a is random numbers between 0 and 1, Gbest the global optimum, mbest the intermediate value of individual extreme value (Pbest), M the Population size; generation and max generation respectively represent the currently number of iterations and the maximum number of iterations [11].

In combination with formula (17)–(19) the position of the updated particle swarm is obtained, and the position is shown as the following equation:

$$position = p \pm b * \left| mbest^3/position \right| * \ln(1/u) \tag{20}$$

In the formula (21) u is random numbers between 0 and 1, if $u > 0.5$, then taking "−", if $u < 0.5$, then taking "+" [12].

The improved QPSO algorithm on the basis of QPSO algorithm, and the random number that subject to the Cauchy random distribution was introduced into the process of variation. So it can enhance the global search ability of the algorithm and avoid the optimization process into a local optimum.

According to the formula (17) and formula (20), compared with the QPSO algorithm, the improved QPSO algorithm has the following changes [13]:

$$p = \frac{\alpha * Pbest(i) + \beta * Gbest}{\alpha + \beta} \tag{21}$$

$$position = p \pm b * \left| mbest^3 / position \right| * \ln(1/u) \tag{22}$$

where α, β and u are random numbers obeying Cauchy distribution.

3.2 Model Solution

The improved QPSO algorithm has good global search ability and good convergence and stability, which make the algorithm getting good results in the process of solving. The improved QPSO algorithm is used to solve optimization location model of the charging station that proposed in this paper, the concrete steps as follows:

(1) Initialization. Though setting the size of the population, the number of iterations and other related parameters, inputting the parameters required in the optimization process such as number of electric vehicles, waiting time, time cost, charging station construction costs and so on to initialize the program.

(2) Calculate the number of charging stations. The number of electric vehicle charging stations in the design area is calculated according to the number of electric vehicles in the selected area and combining with the formula (14)–(16).

(3) Fitness calculation. Taking the annual comprehensive cost of society as objective function and according to the mathematical model of optimization location of charging station in the first section to calculate the user cost, charging station cost.

(4) Update the individual extremum and global extreme value. The individual extremum and global extreme value of each individual in the population are recorded and updated constantly in the iteration process.

(5) Renewal the position of particle swarm. The position of all individuals in the population is updated according to the formula (18), (19), (21), (22).

(6) Termination judgment. If the maximum number of iterations is not reached, the step (3)–(5) will be repeated until the maximum iterations get reached. At this time the optimization terminated and the optimal design scheme output.

4 Example Analysis

In this paper, an administrative area of the southern city is taken as an example to carry out study: The area is located in the northeast of the city, and covers an area of 60.61 km^2, with a population of about 1 million 313 thousand, including commercial, industrial and residential areas. According to statistics, In 2013–2014 year the city promotes a total of 11465 new energy vehicles, ranking first in that year. As of the first half of 2015, the city's local electric vehicles are about 30945. Combined with the city's GDP, population, area and other related data of each administrative area, estimating the number of electric vehicles in the area are about 1702.

According to the specific situation of the administrative area, the mathematical model proposed in this paper can be solved according to the steps of the improved QPSO algorithm, which is described in the preceding section. In the process of solving also need a lot of other relevant data. In this paper, a variety of methods are used to obtain the parameters needed in the optimization process:

(1) The relevant data based on users' habits and desires, such as the average charge times per day and the expected price per charge etc., were obtained through the questionnaire survey. Through the statistical analysis of 257 valid questionnaires among the 270 questionnaires, the corresponding data in the model were obtained, as shown in Table 1.

Table 1. Parameters obtained by questionnaire

Name	Symbol	Value	Unit
Charging price	p	0.65	Yuan
Charging service price	q	1.1	Yuan
Average quantity of electricity per charge	s	29.50	Kilowatt hour
Expected sojourn time	W_S	0.35	Hour
Average charging times per day	u	0.8	Times
Mileage of unit electricity	g	6.78	km

(2) Through data collection, field investigation and other forms of research to investigate other parameters in the optimal location model of electric vehicle charging station, such as collecting a number of charging piles price, after filtering and mean calculation, the charging pile price obtained. The model datas are shown in Table 2.

Table 2. Parameters obtained in other ways

Name	Symbol	Value	Unit
Revenue per unit time of owner	t_p	33.6	Yuan
Transformer, cable and other related cost	B	3	Ten thousand yan/year
Charging station design operation life	n	20	Year
Charging station income coefficient	δ	0.3	/
Charging pile unit price	Z	1	Ten thousand yuan
Fixed investment	G	100	Ten thousand yuan/station
Investment recovery	k	0.08	/
Land cost	LH	21140	Yuan per/meter

Based on the obtained model data, the improved QPSO algorithm was applied to solve the proposed mathematical model. The parameters of the algorithm were set as follows: the maximum iteration number is limited to 150 times, and the population size is set to 50. Moreover, in order to verify the feasibility and efficiency of the improved

QPSO algorithm, setting the same parameters, the PSO algorithm is also used to solve the proposed model. In the case of the same number of iterations, the optimization results of the two methods are shown in Figs. 1 and 2, respectively.

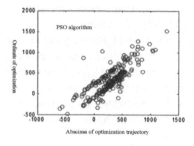

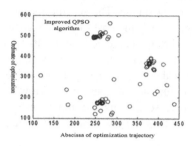

Fig. 1. PSO algorithm iteration 50 times

Fig. 2. Improved QPSO algorithm iteration 50 times

5   Result Analysis

By comparing the results of Figs. 1 and 2, it can be found that, at the 50 iterations, the improved QPSO algorithm starts to appear particle aggregation, while the PSO algorithm is not effective. This shows that the optimal solution can be found more quickly by the improved QPSO algorithm, by contrary, using the PSO algorithm will takes a relatively long time and optimization speed is slower.

The number, size and the specific location of the charging station optimized by the improved QPSO algorithm is shown in Fig. 3a. Figure 3a indicates that the position coordinates of the charging station 1 is (241.5763, 484.4615) which has 5 charging piles. The position coordinates of the charging station 2 is (262.1229, 171.4706) and it includes 6 charging piles. The position coordinates of the charging station 3 is (376.3870, 365.2341) and it contains 7 charging piles. It can be seen from Fig. 3a that the location and scale of the electric vehicle charging station are reasonable. In this case, the comprehensive social cost of optimization is 50783790.79 Yuan.

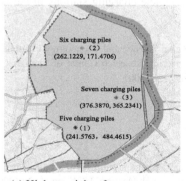

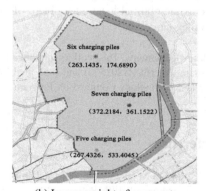

(a) Higher weight of user cost   (b) Lower weight of user cost

Fig. 3. Specific location and scale of the charging stations

At same time, in order to further illustrate the scientificalness of the charging station location scheme based on the user's interests, the user cost in the objective function is compared with the lower weight. Under the circumstance of the user cost with lower weight, solving the objective function get the optimization location of electric vehicle charging station. The specific location and scale of the charging stations are shown in Fig. 3b, and the cost of the objective function is 51538983.71 Yuan.

Through compared the results which obtained by improved QPSO algorithm under the circumstance of the lower weights and higher weights of user cost, in the two case, the scale of the charging station is the same, while there is a small gap of the different charging station coordinates. However, in the situation of the user cost with lower weight, social annual comprehensive cost increased by 755192.92 Yuan. This proofs that the charging station optimization plan shown in Fig. 3a is scientific and reasonable.

6 Conclusion

In this paper, an optimization location model of electric vehicle charging station based on user benefit is proposed. Based on the full consideration of charging station construction cost and a variety of costs in the process of charging station running, the model focuses on the view point of users and increases the user's weight in the proposed optimization location model. So that the optimal location of the electric vehicle charging station can improve the user's convenience for charging electric vehicles and minimize the user's cost, and then increase the number of users of electric vehicles and promote the popularization of electric vehicles. Based on proposing the mathematical model, this paper through the questionnaire survey of the electric vehicle users' interests and consulting relevant information to obtain the practical parameters, then using the improved QPSO algorithm to plan the location of electric vehicle charging stations in an administrative area of a southern city of China, and through compared in the condition of user cost with lower weight, the result shows that the proposed model is scientific and reasonable. In the following work, the proposed model and solution method will be applied to the design of the optimal location software of electric vehicle charging station. The future researches will based on this paper and studies the optimal location problem of electric vehicle charging stations in the more complex situation.

Acknowledgement. This work is supported by Shanghai Science and Technology Commission Program (No.16DZ1202500) and supported by Engineering Research Center of Shanghai Science and Technology Commission Program (No. 14DZ2251100).

References

1. Ferdowsi, M.: Vehicle fleet as a distributed energy storage system for the power grid. In: Proceedings of IEEE Power and Energy Society General Meeting, USA, pp. 1–2 (2009)
2. Chen, L., Zhang, H., Ni, F., Zhu, J.: Present situation and development trend for construction of electric vehicle energy supply infrastructure. Autom. Electr. Power Syst. 35(14), 11–17 (2011)

3. Ying, X., Li, J., Chen, J.: Optimization study of the sitting of the electric vehicle charging stations. Technol. Econ. Areas Commun. **16**(1), 43–46 (2014)
4. Li, L., Tang, C., Li, X.: Layout planning of electrical vehicle charging stations based on particle swarm genetic algorithm. Shanxi Electr. Power **42**(4), 65–69 (2014)
5. Zhao, S., Li, Z.: Optimal planning of charging station for electric vehicle based on PSONDE algorithm. J. North China Electr. Power Univ. **42**(2), 1–7 (2015)
6. Chen, G., Mao, Z., Li, J., et al.: Multi-objective optimal planning of electric vehicle charging stations considering carbon emission. Autom. Electr. Power Syst. **38**(17), 49–53, 136 (2014)
7. Chu, Y., Ma, L., Zhang, H.: Location-allocation and its algorithm for gradual covering electric vehicle charging stations. Math. Pract. Theory **45**(10), 102–106 (2015)
8. Liu, Z., Wen, F., Xue, Y., et al.: Optimal siting and sizing of electric vehicle charging stations. Autom. Electr. Power Syst. **36**(3), 54–59 (2012)
9. Bae, S., Kwasinski, A.: Spatial and temporal model of electric vehicle charging demand. IEEE Trans. Smart Grid **3**(1), 394–403 (2012)
10. Sun, J., Feng, B., Xu, W.: Particle swarm optimization with particles having quantum behavior. In: Congress on Evolutionary Computation, USA (2004)
11. Yu, J., Guo, P.: Quantum-behaved particle swarm optimization algorism with application based on MATLAB. Comput. Digit. Eng. **35**(12), 38–39 (2007)
12. Liu, Z., Zhang, W., Wang, Z.: Optimal planning of charging station for electric vehicle based on quantum PSO algorithm. Chin. Soc. Electr. Eng. **32**(22), 39–45 (2012)
13. Zhao, Y., Zhao, X., Liu, K.: Path planning for UCAV based on Voronoi diagram and quantum-behaved particle swarm algorism. Sci. Technol. Rev. **31**(22), 69–72 (2013)

Research on Double Fuzzy Control Strategy for Parallel Hybrid Electric Bus

Qiwei Xu$^{(\boxtimes)}$, Xiaoxiao Luo, Xiaobiao Jiang, and Meng Zhao

State Key Laboratory of Power Transmission Equipment & System Security and New Technology, Chongqing University, Chongqing, China
xuqw@cqu.edu.cn

Abstract. In this paper, a double fuzzy control strategy (DFLS) for parallel hybrid electric bus (HEB) is proposed. Firstly, the basic parameters of HEB is designed. Then, the single fuzzy logic control strategy (SFLS) is proposed based on the parameters. SOC and torque demand scale factor are taken as the input of fuzzy controller. Combining with the braking energy recovery strategy, this paper proposes a double fuzzy control strategy, which is taken SOC, required torque and bus speed as the input. And this paper makes a comparison of the both strategies in Chinese Bus Driving Cycle (CBDC) based on the simulation results.

Keywords: Hybrid Electric Bus · Single fuzzy control strategy · Double fuzzy control strategy

1 Introduction

Hybrid Electric Bus (HEB) is the transition model between traditional vehicle and electric bus (EB). Comparing with EB, HEB has no limit on driving distance, and comparing with traditional vehicle, HEB has better fuel economy [1, 2].

If the structure and the characteristic of HEB are given, the fuel economy is largely determined by HEB control strategy. Therefore, energy control strategy is highly influence the performance of HEB. And energy control strategy includes: precise logic control strategy, fuzzy logic control strategy, global optimum control strategy and instantaneous optimal control strategy [3]. The control system of HEB is a time-varying and highly nonlinear system. And fuzzy logic control strategy has good real-time performance, adaptability and robustness. Therefore, fuzzy logic is adopted to design the energy control strategy in this paper [4–6].

There are lots of research on fuzzy logic control strategy for HEB. Lee designed a fuzzy controller to restrict the emission of NOX, but the fuzzy controller cannot keep the balance of SOC [7]. So, Lee improved the previous design, proposed a fuzzy control strategy with two fuzzy controller, one for predicting diver's instruction, the other for keeping the balance of SOC [8]. Bauman proposed a fuzzy control strategy based on rule-based control strategy (RBCS) [9]. Schouten designed a fuzzy control strategy which taken driver's instruction, motor speed and SOC as the input of fuzzy controller [10]. However, the researches only focus on the energy management or energy recovery strategy. In this paper, the parameters of HEB is matching firstly.

© Springer Nature Singapore Pte Ltd. 2017
K. Li et al. (Eds.): LSMS/ICSEE 2017, Part III, CCIS 763, pp. 362–370, 2017.
DOI: 10.1007/978-981-10-6364-0_36

According to this, SFLS is designed. Considering the energy recovery strategy, DFLS is established by adding energy recovery fuzzy control strategy for further optimization of vehicle fuel economy.

2 Component Selection

Before the energy control strategy design, it is necessary to match the parameters of HEB. The most important of the parameters matching is to select the appropriate power for engine, motor and battery.

In this paper, keep a traditional bus's external parameters unchanged, transforming its power system into parallel hybrid electric system (Table 1).

Table 1. Parameters of hybrid electric vehicle.

Power source	Parameter	Power (kw)
Engine	Four-stroke diesel engine	110
Motor	AC motor	40
Battery	NI-MH	52

3 SFLS Design

The main advantage of fuzzy control is its self-tuning and adaptive ability. The energy distribution of HEB is a responsible non-linear, multi-energy coupled system, and fuzzy control based on fuzzy reasoning, does not require the vehicle energy consumption of the precise mathematical model. Only according to the operator's experience and the robustness of operational data control system, it is applicable to solve non-linear, time-varying and hysteresis systems that are difficult to solve by conventional control. It can not only optimize the engine, but also take into account the motor, battery, transmission and other components of the work efficiency, controlling the engine as much as possible in the efficient area. The battery at the beginning and end of the basic balance, achieve the overall efficiency of the hybrid drive system.

3.1 Torque Distribution

The energy control strategy is used to get better fuel economy. And engine work in high efficiency area can improve the fuel economy. It means that the strategy controls the engine work in the high efficiency area by using motor to work together. Therefore, how to distribute the output torque is the key to the strategy.

In this paper, motor is used to compensate engine output torque to make engine work in high efficiency area. So, the rules of torque distribution can be defined:

1. When SOC is small, engine and motor priority for battery charging.
2. When required torque is too small, close engine and motor outputs torque to drive bus.

3. When required torque is a little bit small, the motor outputs negative torque as the load to improve the engine efficiency.
4. When required torque is already make engine work in high efficiency area, close motor and engine output all the torque.
5. When required torque is a little bit large, motor outputs positive torque to drive bus with engine.
6. When required torque is larger than the maximum output torque of engine, engine and motor work together to achieve the requirement without considering engine efficiency.

3.2 Fuzzy Controller Design

The first step for fuzzy controller design is input and output variables selection. According to torque distribution rules, it is obviously to know that required torque, SOC, motor and engine output torque are the important variables in the fuzzy control. Therefore, SOC and scaling factor (required torque divided by engine optimal torque) are selected as the input of fuzzy controller. The scaling factor of motor target torque k is selected as the output. The fuzzy controller is shown in Fig. 1.

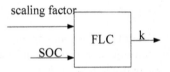

Fig. 1. Fuzzy controller

The next step is to build the membership function. According to torque distribution rules and fuzzy logic theory, the membership function can be built. Then based on theory and simulation results analysis, adjust the function to optimize the fuzzy control energy strategy. The membership function is shown in Fig. 2.

The last step is to determine the rule base of fuzzy controller. The rule base is shown in Table 2. In different condition, according to torque distribution rules, the rules can be defined as below:

If (u is LL) and (SOC is LL) then (k is NL).

3.3 Fuzzy Controller Design

The motor has high work efficiency so that it is used to improve the engine work efficiency based on the torque distribution rules. Under the condition with stable SOC, motor outputs appropriate torque to make engine output torque close to its optimal torque curve. According to this, the engine work efficiency can be improved.

Therefore, the energy control system can be designed. The topological figure of it is shown in Fig. 3.

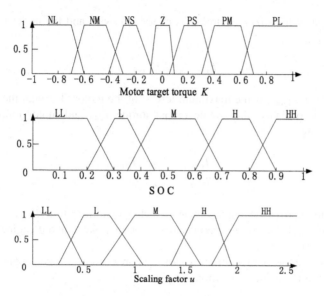

Fig. 2. Membership function

Table 2. Rule base.

Motor target torque k		Torque require scaling factor u				
		LL	L	M	H	HH
SOC	LL	NL	NL	NM	NM	NS
	L	NL	NM	NS	PM	PM
	M	Z	Z	PS	PM	PL
	H	PS	PS	PM	PL	PL
	HH	PM	PS	PM	PL	PL

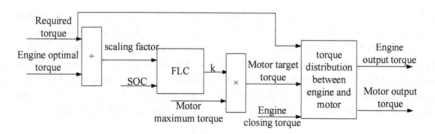

Fig. 3. Energy control system

It should be noted that only positive torque calculated in this system. When the torque is negative, the energy should be calculated in energy recovery module and recovery the energy based on energy recovery strategy. The strategy is described in detail in Sect. 4.

The output of the fuzzy controller is expressed by k, and the target torque of the motor is T_{mot_obj}:

$$T_{mot\_obj} = kT_{mot\_max} \tag{1}$$

In Eq. (1), T_{mot_max} is the maximum torque of the motor. Through the distribution module, in the ordinary, the engine output torque T_{eng} and motor torque T_{mot} are expressed below:

$$T_{eng} = T_{req} - T_{mot}$$
$$T_{mot} = T_{mot\_obj} \tag{2}$$

In order to improve the engine efficiency, when required torque T_{req} is very low, the engine should be closed. So the engine closing torque is set to avoid engine work in the low efficiency area.

When the engine output torque is lower than closing torque, engine should be closed and motor provides the whole torque.

$$T_{eng} = 0$$
$$T_{mot} = T_{req} \tag{3}$$

When the required torque is larger than the engine maximum out torque T_{eng_max}, engine outputs maximum torque and the rest is provided by motor.

$$T_{eng} = T_{eng\_max}$$
$$T_{mot} = T_{mot\_obj} \tag{4}$$

4 SFLS Design

Brake energy recovery is one of the major sources of energy savings for HEB compared to traditional fuel vehicles. Through the use of regenerative braking, the energy can be converted into electrical energy stored in the battery to prepare for use, to improve the vehicle's energy efficiency. But at the same time, the control of friction braking and regenerative braking between the distribution ratio must be coordinated, in order to ensure the stability of braking under the premise of as much as possible to recover the braking energy in order to form a safe, stable and efficient braking system.

4.1 Brake Force Distribution

Considering the braking force requirement, the strategy needs to recovery energy as much as possible. Therefore, it is important to distribute braking force. The more force distributed to regenerative braking system, the more energy recovered. According to this, the rules can be defined below.

1. When required braking force is small, the mechanical braking system does not work and motor transformed into generator to recover energy.
2. When required braking force is large, mechanical braking system and regenerative braking system work together to provide braking force.

4.2   Brake Force Distribution

The factors that affect the braking energy recovery include motor power generation, battery capacity and so on. Therefore, the required input information must include the SOC, the bus speed and the braking force required by the driver. The braking torque distribution scaling factor is taken as the output. The fuzzy controller is shown in Fig. 4.

According to this, required braking force, SOC and bus speed are selected as the input.

Then, the membership function and rule base should be determined. The method to build these is same as the SFLS. The membership function is shown in Fig. 5, the energy recovery system is shown is Fig. 6, and the rule base is shown in Table 3.

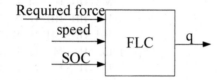

Fig. 4. Fuzzy controller

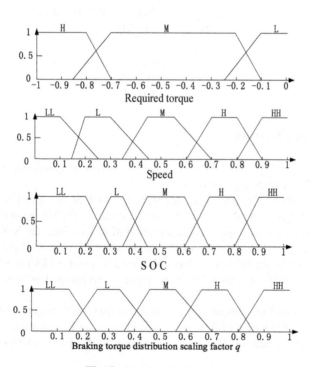

Fig. 5. Membership function

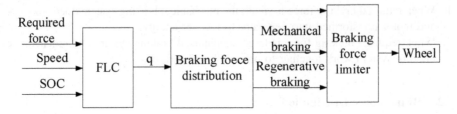

Fig. 6. Energy recovery system

Table 3. Rule base.

Braking torque distribution scaling factor q			SOC				
			LL	L	M	H	HH
Required torque L	Speed	LL	H	M	L	L	LL
		L	H	H	M	L	LL
		M	HH	H	M	L	LL
		H	HH	HH	H	M	L
		HH	HH	HH	H	H	M
	M Speed	LL	H	M	M	L	LL
		L	H	H	M	L	LL
		M	HH	H	M	M	L
		H	HH	HH	H	M	L
		HH	HH	HH	H	M	M
	H Speed	LL	HH				
		L					
		M					
		H					
		HH					

5 Simulation Results

In order to verify SFLS and DFLS designed in this paper, simulating in ADVISOR with CBDC. The curves are shown below.

Figure 7(a) shows that the both strategies are well fit the speed requirement of driving cycle. Figure 7(b) shows the engine speed curves are almost same because of the same transmission and speed requirement. And in Fig. 7(c), the SOC of the both strategies can keep stable in the reasonable range which is 0.2–0.8. Therefore, the both strategies are suit for the HEB. It is obviously to see that SOC of DFLS is higher than it of SFLS in the whole simulation. It means there is more energy has been recovered by using DFLS.

Comparing with Fig. 7(d) and (e), the motor outputs more negative torque at 210–220 s and 800–1000 s of DFLS so that more energy can be recovered. And at 610–650 s, the required torque of HEB is large, it is same as the rule.5 of torque distribution.

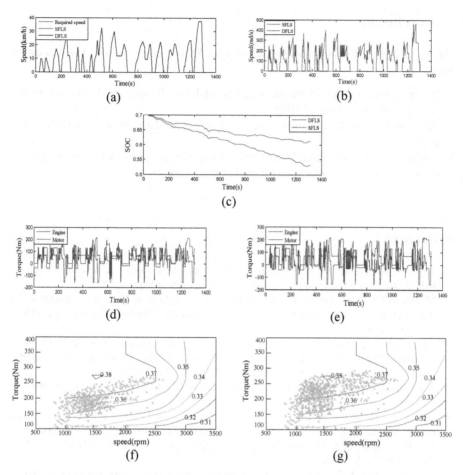

Fig. 7. Simulation results: (a) speed follow curve; (b) engine speed; (c) battery SOC change curve; (d) engine and motor torque of SFLS; (e) engine and motor torque of DFLS; (f) engine working area of SFLS; (g) engine working area of SFLS

Therefore, the motor should provide positive torque to drive HEB with engine. The motor of DFLS provides more positive torque to make engine works in high efficiency area.

In Fig. 7(f) and (g), the numerical values present the engine efficiency. The greater value is; the higher work efficiency of engine is. Comparison between with Fig. 7(f) and (g), the engine working points of DFLS is more concentrate on the high work efficiency area so that DFLS has better fuel economy as shown in Table 4.

Table 4. Fuel consumption.

	SFLS	DFLS
Fuel consumption (L/100 km)	33.6	31.4

6   Conclusion

This paper aims at the fuzzy control strategy for HEB. Following conclusion could be made:

1. Fuzzy control has good robustness and adaptability. It has a good performance for energy control strategy of HEB.
2. According to simulation results, the two fuzzy controllers are suit for HEB in this paper.
3. Comparing with SFLS and DFLS, after adding energy recovery system, DFLS can recover more energy and improve engine work efficiency.

References

1. Bayer, J., Koplin, M., Butcher, J.A., et al.: Optimizing the university of wisconsin's parallel hybrid-electric aluminum intensive vehicle. SAE Technical Papers (2000)
2. Montazeri-Gh, M., Poursamad, A., Ghalichi, B.: Application of genetic algorithm for optimization of control strategy in parallel hybrid electric vehicles. J. Frankl. Inst. **343**(4), 420–435 (2006)
3. Sun, K., Shu, Q.: Overview of the types of battery models. **1416**(1), 3644–3648 (2011)
4. Bostanian, M., Barakati, S.M., Najjari, B., et al.: A genetic-fuzzy control strategy for parallel hybrid electric vehicle. Int. J. Automot. Eng. **3**(3), 483–495 (2013)
5. Mohammad, E.F., Said, F.: Optimization and control of a HEV. In: ICEEOT, pp. 388–391 (2016)
6. Sorrentino, M., Rizzo, G., Arsie, I.: Analysis of a rule-based control strategy for on-board energy management of series hybrid vehicles. Control Eng. Pract. **19**(12), 1433–1441 (2011)
7. Lee, C.C.: Fuzzy logic in control systems: fuzzy logic controller. I. IEEE Trans. Syst. Man Cybern. **20**(2), 404–418 (1990)
8. Lee, H.D., Sul, S.K.: Fuzzy-logic-based torque control strategy for parallel-type hybrid electric vehicle. IEEE Trans. Ind. Electron. **45**, 625–632 (1998)
9. Baumann, B.M., Washington, G., Glenn, B.C., et al.: Mechatronic design and control of hybrid electric vehicles. IEEE/ASME Trans. Mech. **5**(1), 58–72 (2000)
10. Schouten, N.T., Salman, M.A., Kheir, N.A.: Fuzzy logic control for parallel hybrid vehicle. IEEE Trans. Control Syst. Technol. **10**, 460–468 (2002)

Optimal Battery Charging Strategy Based on Complex System Optimization

Haiping Ma[1,2(✉)], Pengcheng You[1], Kailong Liu[3], Zhile Yang[3],
and Minrui Fei[4]

[1] State Key Laboratory of Industrial Control Technology,
Department of Control, Zhejiang University, Hangzhou, China
Mhping1981@126.com, pcyou@zju.edu.cn
[2] Department of Electrical Engineering, Shaoxing University,
Shaoxing, Zhejiang, China
[3] School of Electronics, Electrical Engineering and Computer Science,
Queen's University Belfast, Belfast, UK
{kliu02,zyang07}@qub.ac.uk
[4] Shanghai Key Laboratory of Power Station Automation Technology,
School of Mechatronic Engineering and Automation,
Shanghai University, Shanghai, China

Abstract. This paper proposes a complex system optimization method to obtain an optimal battery charging strategy. First, a real-world lithium-ion battery charging model is built as a complex system problem, which includes electric subsystem and thermal subsystem. The optimization objectives of electric subsystem includes battery charging time and energy loss, and the optimization objectives of thermal subsystem includes battery internal temperature rise and surface temperature rise. Then a called biogeography-based complex system optimization (BBO/Complex) algorithm is introduced, which is a heuristic method for complex system optimization. Finally, BBO/Complex is applied to the complex system of battery charging strategy, and the results show that the proposed method is a competitive algorithm for solving batter charging problem studied in this paper.

Keywords: Battery charging · Complex system · Heuristic algorithm · BBO/Complex

1 Introduction

In recent years, portable electronic devices have been widely used in many domains. Lithium-ion batteries are tending to replace the traditional rechargeable batteries such as lead-acid batteries used in these devices, because they show some outstanding performance such as high power and energy densities, broad operating temperature range, long-life cycles, and low self-discharge rate [1]. These merits of lithium-ion batteries make them become a very promising primary power source for electronic devices in the future. Therefore, for the applications of lithium-ion batteries, a well-designed battery charger plays a vital role for sustaining battery performance and lifespan, and the key is to obtain a proper battery charging strategy including the

© Springer Nature Singapore Pte Ltd. 2017
K. Li et al. (Eds.): LSMS/ICSEE 2017, Part III, CCIS 763, pp. 371–378, 2017.
DOI: 10.1007/978-981-10-6364-0_37

selection of charging current pattern, the control and termination of charging process, and the safety and behavior of the battery.

In the past years, a lot of approaches have been developed to improve the battery charging performance. Some of the approaches involve computation intelligence including neural networks [2, 3], grey prediction [4], and fuzzy control [5]. Some of strategies take the battery charging behaviors as an optimization problem which is further solved using heuristic methods. In [6], genetic algorithm (GA) is used to manage online battery charging state for electric and hybrid vehicle applications. In [7], particle swarm optimization (PSO) is employed to obtain optimum battery energy storage system considering dynamic demand response for micro grids. But these studies only consider battery charging performance as a single-objective or multi-objective optimization problem. In fact, the battery temperatures including the surface and internal temperatures also consist of a system optimization problem, and they are important factors during the battery charging process, because too high or low temperature would be harmful to the battery charging safety and behavior. Undoubtedly, it becomes more complex than ever before, and the optimization becomes more difficult under considering battery temperatures. In this situation, battery charging strategy cannot be treated as a typical single-objective or multi-objective optimization problem any long. Strictly, it is taken as a complex system, which contains multiple subsystems, each of which contains multiple objectives, multiple constraints and multiple variables. So it is necessary to build new heuristic methods to tackle the battery charging problem under new circumstances.

Complex system optimization is a class of optimization methods dedicated to solving complex problems with multiple subsystems, multiple objectives, and multiple constraints. Traditional complex system optimization methods includes multidisciplinary feasible (MDF), individual discipline feasible (IDF) and collaborative optimization (CO), which are popular in engineering domain [8, 9]. But these methods only provide conceptual frameworks without involving the detailed algorithms, which are usually specified based on the user's preference. Recently, a heuristic method, called biogeography-based complex system optimization (BBO/Complex) is proposed by Simon and Du [10] to solve complex problems. Some literatures showed that BBO/Complex had obtained good performance for the virtual machine placement [11], the economic emission load dispatch [12], the speed reducer problem, the power converter problem, the heart dipole problem and the propane combustion problem [10].

For battery charging management, an important but challenging problem is to achieve optimal charging performance considering various factors including efficiency, reliability and safety. Motivated by these considerations, this paper adopts BBO/Complex to obtain the optimal battery charging strategy. The remainder of this paper is organized as follows. Section 2 builds a battery charging model for complex system optimization. Section 3 reviews BBO/Complex as a complex system optimization method. Section 4 applies BBO/Complex to solve battery charging model and presents optimization results. Section 5 provides conclusions and suggests directions for future work.

2 Problem Formulation of Battery Charging

The complex system model of battery charging is formulated as two subsystems, each of which includes two objective functions. One is the electric subsystem with two objectives of battery charging time and energy loss. Another is the thermal subsystem with two objectives of battery internal temperature rise and surface temperature rise.

2.1 Battery Electric Subsystem

In the electric subsystem, the charging time during battery charging is an important optimization indicator. Generally, the shorter the charging time is, the better the performance is. Another important optimization indicator is the battery energy loss. The smaller energy loss is, the higher the battery charging efficiency is.

The objective functions of the battery charging time and energy loss are defined as

$$J_{CT} = t_s \cdot k_{tf} \tag{1}$$

$$J_{EL} = t_s \cdot \sum_{k=0}^{k_{tf}} \left(i^2(k) \cdot R(k) + \frac{V_1^2(k)}{R_1(k)} + \frac{V_2^2(k)}{R_2(k)} \right) \tag{2}$$

where t_s is the sampling time interval during the battery charging process, k_{tf} is the number of sample when the capacity of battery reaches its target, i is the charging current, which remains constant during a given sample time internal, R, R_1 and R_2 are the battery diffusion resistances, and V_1 and V_2 are the battery RC network voltages.

2.2 Battery Thermal Subsystem

In the thermal subsystem, the battery internal temperature rise and surface temperatures rise are key performance indicators during the charging process. The higher the temperature is, the more serious the damage is for the service life of the battery.

The objective functions for the battery internal temperature rise and surface temperature rise can be defined as

$$J_{ITR} = t_s \cdot \sum_{k=0}^{k_{tf}} T_{IT}(k) \tag{3}$$

$$J_{STR} = t_s \cdot \sum_{k=0}^{k_{tf}} T_{ST}(k) \tag{4}$$

where T_{IT} and T_{ST} represent the battery internal and surface temperature respectively.

2.3   Optimization and Constraints

Optimal battery charging strategy is to find the appropriate charging current i to simultaneously minimize objective functions J_{CT} and J_{EL} in battery electric subsystem and objective functions J_{ITR} and J_{STR} in battery thermal subsystem. It is defined as

$$\text{minimize } \{J_{CT},\ J_{EL},\ J_{ITR},\ J_{STR}\}. \tag{5}$$

Furthermore, during the battery charging process, some constraints and updates need to be satisfied for the battery parameters such as voltage and current, which are described as follows:

$$
\begin{aligned}
V_1(k) &= a_1 \cdot V_1(k-1) - b_1 \cdot i(k-1) \\
V_2(k) &= a_2 \cdot V_2(k-1) - b_2 \cdot i(k-1) \\
V(k) &= V_1(k) + V_2(k) + i(k) \cdot R(k) \\
T_{IT}(k) &= (1 - t_s \cdot k_1/D_1) \cdot T_{IT}(k-1) + (t_s \cdot k_1/D_1) \cdot T_{ST}(k-1) \\
&\quad + t_s \cdot R(k-1) \cdot i^2(k-1)/D_1 \\
T_{ST}(k) &= (t_s \cdot k_1/D_2) \cdot T_{IT}(k-1) + (1 - t_s \cdot (k_1+k_2)/D_2) \cdot T_{ST}(k-1)
\end{aligned}
\tag{6}
$$

and

$$
\begin{aligned}
T_{IT}(0) &= 0, \quad T_{ST}(0) = 0 \\
a_j &= \exp(-t_s/R_j), \ j = 1,\ 2 \\
b_j &= R_j \cdot (1 - a_j) \\
i_{\min} &\le i(k) \le i_{\max} \\
V_{\min} &\le V(k) \le V_{\max}
\end{aligned}
\tag{7}
$$

where k_1, k_2, D_1 and D_2 are pre-defined parameters, $i_{\min}$ and $i_{\max}$ are the minimum and maximum values of charging current i, $V_{\min}$ and $V_{\max}$ are the minimum and maximum values of the voltage V.

3   Biogeography-Based Complex System Optimization

This section provides an overview of BBO/Complex for complex system [10]. Before we introduce the details of BBO/Complex, there are some notations we need to clarify. BBO/Complex is an extension to the standard BBO, but it is different to BBO. Standard BBO is a single-objective or multiple-objective optimization algorithm, which is suitable to a single system. BBO/Complex is a complex system optimization algorithm, and it is suitable to a complex system with multiple subsystems, each of which contains multiple objectives and multiple constraints. On the other hand, some definitions and operating strategies of standard BBO, including migration and mutation, are reserved, which are not described repeatedly in this paper.

Now we introduce some new BBO/Complex notations which are different with standard BBO. Let $P = \{A_1, A_2, A_3, \ldots\}$ denote an ecosystem that is comprised of archipelagos, each of which corresponds to one subsystem. $A_h = \{O_{h1}, O_{h2}, O_{h3}, \ldots;$ $C_{h1}, C_{h2}, C_{h3}, \ldots; I_{h1}, I_{h2}, I_{h3}, \ldots\}$ represents an arbitrary archipelago, which is comprised of objective O_{hi}, constraints C_{hi} and candidate solutions I_{hi}. $I_{hi} = \{S_{hi1}, S_{hi2}, S_{hi3}, \ldots\}$ represents an arbitrary candidate solutions, which is comprised of independent variables S_{hij}.

Based on the original paper [10], the framework of BBO/Complex is shown in Fig. 1, which includes within-subsystem migration, cross-subsystem migration and mutation.

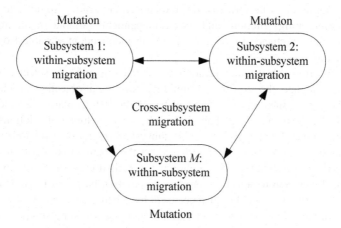

Fig. 1. Framework of BBO/Complex, including within-subsystem migration, crossover-subsystem migration, and mutation.

The main steps of BBO/Complex are depicted as follows.

Step 1: Initialize the population and parameters;
Step 2: Perform within-subsystem migration for each subsystem;
Step 3: Perform cross-subsystem migration for selected subsystem pairs;
Step 4: Perform probabilistic mutation for each candidate solution;
Step 5: Terminate if the termination condition is satisfied, otherwise, generate the next population and go to Step 2.

In step 1, BBO/Complex parameters include the number of subsystems, the number of candidate solutions in each subsystem, the maximum immigration rate and emigration rate, the mutation probability and stopping criterion.

In step 2, within-subsystem migration is very similar to standard BBO migration. For standard BBO, the calculation of migration rates is based on the solution fitness, and for BBO/Complex, the calculation of migration rates is based on the solution rank. Note that solution rank in BBO/Complex combines all information of objectives and constraints to calculate the migration rates, and the calculation method is the same to non-dominated sorting [13]. The process of within-subsystem migration shows as follows: first, probabilistically choose the immigrating solution based on immigration rate, and use roulette-wheel selection based on emigration rates to select the emigrating solution. Immigration rate and emigration rate linearly related to the solution rank, which are calculated as

$$\lambda = \frac{k}{K}, \quad \mu = 1 - \lambda \tag{8}$$

where λ and μ are the immigration rate and emigration rate respectively, k and K are the solution rank and total number of solutions in a subsystem respectively.

Finally, migration is performed from the chosen emigrating solution to the corresponding immigrating solution, and each independent variable in an immigrating solution will have a chance to be replaced by an independent variable from an emigrating solution.

In step 3, cross-subsystem migration is carried out only on selected subsystem pairs. First, calculate the constraint similarity level and objective similarity level between every two subsystems, which is based on fast similarity level calculation (FSLC) [14]. Next calculate Euclidian distance between each pair of solutions from two selected subsystems. Finally perform cross-subsystem migration: probabilistically find suitable pair of subsystems to migrate based on the obtained similarity levels. After that, we need to choose emigrating solution for each immigrating solution. We use roulette-wheel selection to select the emigrating solution based on Euclidian distances of solutions. Solutions with better distances will have better chance to be selected as the emigrating solution. Each independent variable in an immigrating solution will have a chance to be replaced by an independent variable from an emigrating solution.

In step 4, probabilistically perform mutation on each solution based on the mutation probability, which is the same as that in the standard BBO algorithm.

Based on the above description, we find that the two most important components are within-subsystem migration and cross-subsystem migration for BBO/Complex. In standard BBO, migration is a simple operator because only one subsystem evolves in the entire system. But in complex system, it has multiple subsystems. We need to combine all information within and cross subsystems, including objectives, constraints, and solution variables, to determine to migration.

4   Simulation Results

In this section, we use BBO/Complex to solve the proposed battery charging problem. The purpose of this simulation is to show the feasibility and effectiveness of BBO/Complex to solve real-world complex systems. So we compare BBO/Complex with CO, MDF, and IDF [10], which are well-known traditional complex system optimization methods. But we do not compare it with other evolutionary algorithms such as GAs, PSO and so on.

The battery charging parameters are set as follows: $t_s = 1s$, $R = 0.0152\Omega$, $R_1 = 0.0037\Omega$, $R_2 = 0.0034\Omega$, $k_1 = 1.6423$, $k_2 = 0.3102$, $D_1 = 286.35$, and $D_2 = 30.9$. In addition, the minimum and maximum values of charging current and voltage are $i_{min} = -30A$, $i_{max} = 0A$, $V_{min} = 2.6V$, $V_{max} = 3.65V$ respectively. The more detail of the parameters of battery charging model refers to [15]. The performance criteria is based on the cost values of battery charging time, energy loss, internal temperature rise and surface temperature rise, and the optimization goal is to find the minimum values of these costs.

The parameters of BBO/Complex have been manually tuned for optimal performance. For BBO/Complex and the complementary methods in CO, MDF, and IDF, the size of population is 10, mutation rate is 0.01 per independent variable in solution, and the number of Monte Carlo simulations is 20, with a maximum number of function evaluations equal to 1000 for each Monte Carlo simulation. The optimization results are shown in Table 1.

Table 1. The optimization results of the battery charging model for CO, MDF, IDF and BBO/Complex.

Objective functions	Complex system optimization			
	CO	MDF	IDF	BBO/Complex
Charging time J_{CT}	1342	1388	1524	1252
Energy loss J_{EL}	17875	17912	18807	16908
Internal temperature rise J_{ITR}	10245	10107	11823	9805
Surface temperature rise J_{STR}	3428	3473	3612	3349

From Table 1, we see that BBO/Complex has the smallest cost value of charging time, the smallest cost value of energy loss, the smallest cost value of internal temperature rise and smaller surface temperature rise. That is, BBO/Complex has better performance than traditional complex system optimization methods including CO, MDF, and IDF. This is because BBO/Complex improves the diversity of solutions by employing cross-subsystem migration and within-subsystem migration to enhance optimization performance. According to these results, we conclude that BBO/Complex has good complex system optimization performance for battery charging problem studied in this paper.

5 Conclusions

In this paper, we propose a model of real-world lithium-ion battery charging, which is formulated as a complex system with two subsystems, each of which includes two objectives. Then, we introduce BBO/Complex, which includes within-subsystem migration, cross-subsystem migration and mutation, to satisfy the structure of complex systems. Finally, we apply BBO/Complex to the proposed battery charging model, and the simulation results demonstrate that BBO/Complex can effectively obtain the optimal battery charging strategy, which shows it is a competitive complex system optimization algorithm.

This paper shows that BBO/Complex has good optimization performance for solving battery charging problem, but it still opens other research directions for additional development and empirical investigation. First, we consider some real-world charging circumstance constraints into battery charging model, which are important factors for charging performance. Second, we consider adjusting cross-subsystem migration strategy to improve BBO/Complex optimization performance for complex system problems.

Acknowledgments. This research was supported by the National Natural Science Foundation of China under Grant Nos. 61640316 and 61633016, and the Fund for China Scholarship Council under Grant No. 201608330109.

References

1. Vo, T., Chen, X., Shen, W., Kapoor, A.: New charging strategy for lithium-ion batteries based on the integration of Taguchi method and state of charge estimation. J. Power Source **273**, 413–422 (2015)
2. Xu, L., Wang, J., Chen, Q.: Kalman filtering state of charge estimation for battery management system based on a stochastic fuzzy neural network battery model. Energy Convers. Manag. **53**(1), 33–39 (2012)
3. Sun, M., Ni, S., Ma, H.: Delay-dependent robust H∞ control for time-delay systems with polytopic uncertainty. In: Proceedings of the 48th IEEE Conference on Decision and Control with the 28th Chinese Control Conference, pp. 280–284. Shanghai, China (2009)
4. Chen, L., Hsu, R., Liu, C.: A design of a grey-predicted Li-ion battery charge system. IEEE Trans. Ind. Electron. **55**(10), 3692–3701 (2008)
5. Li, C., Liu, G.: Optimal fuzzy power control and management of fuel cell/battery hybrid vehicles. J. Power Sources **192**(2), 525–533 (2009)
6. Chen, Z., Mi, C., Fu, Y., Xu, J., Gong, X.: Online battery state of health estimation based on genetic algorithm for electric and hybrid vehicle applications. J. Power Sources **240**, 184–192 (2013)
7. Kerdphol, T., Qudaih, Y., Mitani, Y.: Optimum battery energy storage system using PSO considering dynamic demand response for micro-grids. Int. J. Electr. Power Energy Syst. **83**, 58–66 (2016)
8. Allison, J.: Complex system optimization: a review of analytical target cascading, collaborative optimization, and other formulations. M.S. Thesis, University of Michigan, Ann Arbor, MI (2004)
9. Ma, H., Su, S., Simon, D., Fei, M.: Ensemble multi-objective biogeography-based optimization with application to automated warehouse scheduling. Eng. Appl. Artif. Intell. **44**, 79–90 (2015)
10. Du, D., Simon, D.: Complex system optimization using biogeography-based optimization. Math. Probl. Eng. **2013**, 17 p. (2013). Article ID 456232
11. Zheng, Q., Li, R., Li, X., Shah, N., Zhang, J., Tian, F., Chao, K.-M., Li, J.: Virtual machine consolidated placement based on multi-objective biogeography-based optimization. Future Gener. Comput. Syst. **54**, 95–122 (2016)
12. Ma, H., Yang, Z., You, P., Fei, M.: Complex system optimization for economic emission load dispatch. In: Proceedings of 11th UKACC International Conference Control, pp. 1–6. Belfast, UK (2016)
13. Deb, K., Pratap, A., Agarwal, S., Meyarivan, T.: A fast and elitist multi-objective genetic algorithm: NSGA-II. IEEE Trans. Evol. Comput. **6**(2), 182–197 (2002)
14. Hathaway, R., Bezdek, J.: Fuzzy c-means clustering of incomplete data. IEEE Trans. Syst. Man Cybern. **31**(5), 735–744 (2001)
15. Liu, K., Li, K., Yang, Z., Zhang, C., Deng, J.: An advanced Lithium-ion battery optimal charging strategy based on a coupled thermoelectric model. Electrochim. Acta **225**, 330–344 (2017)

Experimental Research on Power Battery Fast Charging Performance

Jinlei Sun[1(✉)], Lei Li[1], Fei Yang[1], Qiang Li[1], and Chao Wu[2]

[1] School of Automation, Nanjing University of Science and Technology,
Nanjing 210094, China
jinlei.sun@hotmail.com,
{lileinjust,yangfei,chnliqiang}@njust.edu.cn
[2] Department of Electrical Engineering, Luoyang Institute of Science
and Technology, Luoyang 471023, China
shiningi@163.com

Abstract. Lithium-ion battery fast charging issues have become a crucial factor for the promotion of consumer interest in commercialization, such as mobile devices and electric vehicles (EVs). This paper focuses on the experimental research on fast charging. A battery thermal model is introduced to investigate the temperature variation at high charging current rates 1C, 3C, 4C, 5C. And charging experiments are taken at these current rates respectively. The results show that high charging current rates could effectively reduce charging time. Besides, batteries can be charged to 77.5%, 76.2%, 72.5% of the capacity at 3C, 4C, 5C current rates respectively. The maximum temperature rises during charging are 4.5 °C, 5.5 °C, 6.6 °C respectively.

Keywords: Battery · Fast charging · Temperature variation · Thermal model

1 Introduction

Nowadays, with the rapid development of portable electronic devices and EVs/HEVs, Lithium-ion batteries have attracted more and more attention. Lithium-ion batteries have been extensively used for energy storage in many fields, owing to the advantages, such as high energy density, high power density and long life [1]. However, the slow charging speed limits the further development in applications. Lithium-ion battery fast charging challenges constitute a principle bottleneck of EVs/HEVs applications. Therefore, battery fast charging control is an area of research deserve attention in the field of EVs/HEVs.

The demand of long operation time for commercial devices and long running distance for EVs/HEVs drives large battery cell capacity to achieve high power and energy requirement [2]. Therefore, the charging speed, energy efficiency, and thermal safety are critical for battery system. Traditional charging method of Constant-Current and Constant-Voltage (CC-CV) is widely used to avoid overcharging [3]. Battery is charged at a constant current until the upper voltage limit is reached, and a constant voltage is used until the current falls below a predetermined value. The CC-CV method effectively avoids overcharging and make battery fully charged; however, it is not

© Springer Nature Singapore Pte Ltd. 2017
K. Li et al. (Eds.): LSMS/ICSEE 2017, Part III, CCIS 763, pp. 379–385, 2017.
DOI: 10.1007/978-981-10-6364-0_38

suitable to fast charging because the constant voltage stage increases the charging time. Fast charging typically involves high current rates, high energy throughputs and high temperatures, all of these effects force the deterioration of a battery's electric characteristics and affect its functionality [4].

To accelerate the charging speed, several studies have been proposed in the last decade [5–8]. Reference [5] used ant colony system based optimization method to determine the optimal charging current among five charging states. Reference [6] proposed a pulse-charge strategy using a phase-locked battery charger to reduce charging time. In reference [7], Zhang proposed a polarization based charging time and temperature rise optimization strategy to search for optimal charging current trajectories. In reference [8], Liu proposed a coupled thermoelectric model to optimal charging. The previous references mainly focus on charging time, energy loss, however, the thermal behavior during charging, therefore, fast charging with high current rates considering the effects above worth further investigation.

This paper is a fundamental study for fast charging strategy, the battery thermal model is proposed and the battery temperature variation is estimated at 1C–5C current rates accordingly. Experiments are taken at high charging current rates to achieve fast charging, the charging capacity, temperature rise and charging time are evaluated. The results are of significant importance to further study on fast charging.

2 Battery Thermal Model

A thermal model is applied to investigate battery temperature changes during charging. This model is also used in reference [9]. To facilitate the analysis, the average battery surface temperature is taken as battery temperature in this model. The thermal balance equalization is as follows.

$$mC\frac{dT_s}{dt} = Q_g - Q_d \tag{1}$$

where, m(g) is the weight, $C(Jg^{-1} K^{-1})$ is the heat capacity, Ts(K) is the battery temperature, Qg(W) is the heat generation power, Qd(W) is the transferred heat power to the surroundings.

Heat generation is usually described as irreversible Joule heat and reversible entropy change heat. In this study, the entropy change part is ignored. Thus, Qg can be expressed as follows.

$$Q_g = I(U - V_{OCV}) \tag{2}$$

where I is the charge/discharge current (positive for charge, negative for discharge), V_{OCV} is the open circuit voltage, U is the cell terminal voltage, The OCV-SOC curve is obtained every 10% SOC interval after 3 h resting. The OCV-SOC curve is shown in Fig. 1.

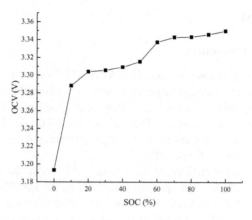

Fig. 1. OCV-SOC curve for battery charging.

The heat dissipation is taken place by conduction through the cell body and by convection and radiation from the cell surface to the environment. In this case, the radiation part is ignored. The heat dissipation power Q_d can be described as (3).

$$Q_d = hA(T_s - T_a) \tag{3}$$

where $h(Wm^{-2} K^{-1})$ is the heat transfer coefficient, $A(m^2)$ is the cell surface area, $Ts(K)$ is the cell average temperature, $Ta(K)$ is the ambient temperature.

According to Eqs. (1)–(3), the linear differential equation can be obtained, and the next sample time temperature can be calculated using laplasse transform and inverse transform after discretization. The discrete time formula is shown in (4).

$$T_s((k+1)T_{sample}) = e^{-\frac{IT_s\frac{\partial V_{OCV}}{\partial T_s}+hA}{mC}} \left(T_s(kT_{sample}) + \frac{I(kT_{sample})(V_{OCV(kT_{sample})} - U(kT_{sample}))}{mC} + \frac{hAT_a}{mC} \right) \tag{4}$$

where T_{sample} is the sampling time, k is the number of sampling.

The charging process is common in EVs/HEVs applications, and the charging safety should be taken seriously, especially for fast charging. It can be seen from the heat generation formula that overpotential is related to SOC. Besides, OCV and SOC has nonlinear relationship. The charging current plays an important role in heat generation especially for large charging current. Therefore, experiments are taken to investigate the temperature distribution for high current rate charging.

The physical parameters required are shown in Table 1, and the heat capacity is provided by the manufacturer.

Table 1. Battery physical parameters

Item	Value
Heat capacity C/Jg^{-1}K^{-1}	1.06
Heat transfer coefficient h/Wm^{-2}K^{-1}	6
Ambient temperature Ta/K	292.6

3 Experimental

3.1 Battery Test Procedures

A multichannel battery testing system NEWARE-BTS4000 (5 V, 200 A, four chan-nels) was used to charge and discharge the battery and record the voltage and current. The battery testing system could measure voltage with accuracy of ±0.1% and has four measurement range, 0.05 A–10 A, 10 A–50 A, 50 A–100 A, 100 A–200 A with accuracy of ±0.1% for each measurement range. Temperature sensor records battery temperature with accuracy of ±0.1 °C.

The experiments were performed on a single cell under the condition of room temperature (23 °C ± 5 °C). The specification of the power LiFePO4 battery (Made in Shanghai, China) is shown in Table 1. The cell was charged at 4 current rates, 1C, 3C, 4C, 5C. The initial state is 0% state of charge (SOC), which is achieved by discharging at 1C current rate until the discharge cutoff voltage 2.5 V mentioned in Table 2 was met. The charging process was performed at constant current rate and the process was terminated when charging cutoff voltage 3.65 V was reached. During the charging test at each current rate, the temperature variation was monitored and recorded with tem-perature sensor.

Table 2. Basic features of target battery.

Item	Value
Capacity	8 Ah
Charging cutoff voltage	3.65 V
Discharging cutoff voltage	2.5 V
Maximum charging current	10C-rate
Maximum discharging current	30C-rate

3.2 Results and Discussion

Several charging tests have been taken to investigate the charging performance with different current rates, 1C, 3C, 4C, 5C. Figure 2 shows the charging voltage curve at different current rates.

In Fig. 2, 1C current rate is the recommended charging current rate provided by the manufacture, and 3C, 4C, 5C are current rates to achieve fast charging. The detailed charging performance at different charging current rates is shown in Table 3.

It can be seen in Table 3 that, charging at 1C current rate could be charged to 96.25%, because of the polarization characteristic. When the charging current rate increases, the charging time decreases dramatically, which is desired for fast charging.

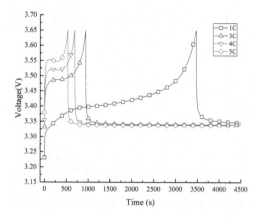

Fig. 2. Charging voltage curve for battery at 1C, 3C, 4C, 5C current rates.

Table 3. Battery charging performance at different charging current rates.

Current rate	Charging capacity (Ah)	Temperature rise (°C)
1C	7.7	1.5
3C	6.2	4.5
4C	6.1	5.5
5C	5.8	6.6

However, the drawbacks of fast charging are the decrease of charging capacity and temperature rise. That means only 77.5%, 76.2%, 72.5% of the battery capacity could be charged at 3C, 4C, 5C current rates, respectively. Besides, battery performance is sensitive to temperature, and high temperature operation environment accelerates aging. Charging and discharging at high current rate causes high temperature rise, owing to the chemical reaction inside the battery. In the fast charging tests, the temperature rises greatly at the target charging current rates, although the charging time is shorter than the recommended charging current rate. Therefore, the temperature should be monitored during charging, Fig. 3 shows the calculated and measured temperature variation at different charging current rates.

In Fig. 3, the temperature rise at the target current rates are obvious. And the battery temperature can be calculated accurately according to the proposed model and battery temperature. And the maximum temperature error at 3C, 4C, 5C during charging are 0.3 °C, 0.6 °C, 0.7 °C, respectively.

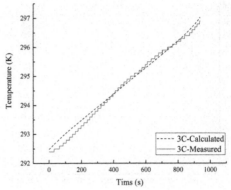

a) 3C charging current rate temperature comparison

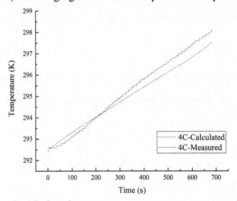

b) 4C charging current rate temperature comparison

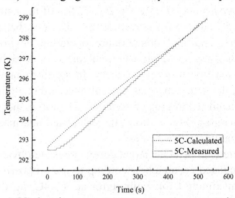

c) 5C charging current rate temperature comparison

Fig. 3. Comparison between calculated and measured temperature at 3C, 4C, 5C current rates.

4   Conclusions

The fast charging performance at 1C, 3C, 4C, 5C current rates are investigated experimentally. The battery thermal model is built, and the battery temperature is estimated during charging process. Fast charging results show that battery could be charged to more than 70% of the capacity with charging time less than 1000 s. Also, the temperature variation is estimated during charging at target current rates. The maximum temperature rise during charging are 4.5 °C, 5.5 °C, 6.6 °C respectively, and the estimation error is within 1 °C. The future work will focus on charging strategy taking energy loss and temperature variation into consideration.

Acknowledgments. This research was supported by "Natural Science Foundation of Jiangsu Province", No. BK20160837 and "Henan scientific and technological research program" Research and development of battery management system for new energy vehicles under complex operating conditions (152102210120).

References

1. Ansean, D., Dubarry, M., Devie, A., Liaw, B.Y., Garcia, V.M., Viera, J.C., Gonzalez, M.: Fast charging technique for high power LiFePO4 batteries: a mechanistic analysis of aging. J. Power Sources **321**, 201–209 (2016)
2. Thanh, T.V., Xiaopeng, C., Weixiang, S., Kapoor, A.: New charging strategy for lithium-ion batteries based on the integration of Taguchi method and state of charge estimation. J. Power Sources **273**, 413–422 (2015)
3. Zheng, C., Bing, X., Mi, C.C., Rui, X.: Loss-minimization-based charging strategy for lithium-ion battery. IEEE Trans. Ind. Appl. **51**(5), 4121–4129 (2015)
4. Ansean, D., Gonzalez, M., Viera, J.C., Garcia, V.M., Blanco, C., Valledor, M.: Fast charging technique for high power lithium iron phosphate batteries: a cycle life analysis. J. Power Sources **239**, 9–15 (2013)
5. Liu, Y.H., Luo, Y.F.: Search for an optimal rapid-charging pattern for Li-ion batteries using the Taguchi approach. IEEE Trans. Ind. Electron. **57**(12), 3963–3971 (2010)
6. Chen, L.R., Chen, J.J., Chu, N.Y., Han, G.Y.: Current-pumped battery charger. IEEE Trans. Ind. Electron. **55**(6), 2482–2488 (2008)
7. Zhang, C., Jiang, J., Gao, Y., Zhang, W., Liu, Q., Hu, X.: Charging optimization in lithium-ion batteries based on temperature rise and charge time. Appl. Energy **194**, 569–577 (2017)
8. Liu, K., Li, K., Yang, Z., Zhang, C., Deng, J.: An advanced lithium-ion battery optimal charging strategy based on a coupled thermoelectric model. Electrochim. Acta **225**, 330–344 (2017)
9. Rad, M.S., Danilov, D.L., Baghalha, M., Kazemeini, M., Notten, P.H.L.: Adaptive thermal modeling of Li-ion batteries. Electrochim. Acta **102**, 183–195 (2013). doi:10.1016/j.electacta.2013.03.167

A Novel RBF Neural Model for Single Flow Zinc Nickel Batteries

Xiang Li[1], Kang Li[1(✉)], Zhile Yang[1], and Chikong Wong[2]

[1] School of Electronics, Electrical Engineering and Computer Science,
Queen's University Belfast, Belfast, UK
{xli25,k.li,zyang07}@qub.ac.uk
[2] Department of Electrical and Computer Engineering, The University of Macau,
Zhuhai, Macau, China
ckwong@umac.mo

Abstract. As a popular type of Redox Flow Batteries (RFBs), single flow Zinc Nickel Battery (ZNB) was proposed in the last decade without requiring an expensive and complex ionic membrane in the battery. In this paper, a Radial Basis Function (RBF) neural model is proposed for modelling the behaviours of ZNBs. Both the linear and non-linear parameters in the model are tuned through a new feedback-learning phase assisted Teaching-Learning-Based Optimization (TLBO) method. Besides, the fast recursive algorithm (FRA) is applied to select the proper inputs and network structure to reduce the modelling error and computational efforts. The experimental results confirm that the proposed methods are capable of producing ZNB models with desirable performance over both training and test data.

Keywords: Zinc Nickel Batteries (ZNBs) · Radial Basis Function (RBF) · Teaching-Learning-Feedback-Based Optimization (TLFBO)

1 Introduction

Large-scale energy storages are indispensable for addressing the integration issue of a large amount of intermittent and variable renewable energy. Differing from conventional energy storage technologies, the active materials bearing the energy in Redox Flow Batteries (RFBs) are stored as the flowing electrolyte, driven by pumps, in two separated reservoirs, and do not rely on the battery structural container [1]. However, the two non-stopping auxiliary pumps reduce the holistic system efficiency and BRFs have to face the challenges of encompassing huge manufacturing expense on ion-separators and generating toxic chemicals [1,2].

A novel Zinc Nickel Battery (ZNB) was proposed in [3]. The mechanical configurations are simplified to having only one flowing passage and the manufacturing costs are significantly reduced without using expensive membrane and raw materials [3,4]. A bench-scale ZNB was first fabricated to show an aggressive performance [4]. It was shown that electrodes are too vulnerable to sustain

© Springer Nature Singapore Pte Ltd. 2017
K. Li et al. (Eds.): LSMS/ICSEE 2017, Part III, CCIS 763, pp. 386–395, 2017.
DOI: 10.1007/978-981-10-6364-0_39

extreme cases such as over-charging and over-discharging. Thereby, to develop an accurate battery model becomes a vital step for safe and secure operation of ZNBs. One mathematical model [5] was proposed to interpret the reactions within ZNBs using the conservation laws of mass, energy and momentum transport of electrolytes. Whereas this model uses complex partial differential equations and a number of non-measured quantities which need to be determined. On the other hand, the equivalent circuit model (ECM) [6] represents the system using electronic components and the parameters are identified from available measurements. However, the model accuracy varies with different structure of ZNB models.

The Radial Basis Function (RBF) neural network has a simple structure for modelling non-linear systems, which has been intensively researched and used to model other battery types [7]. A key issue to build an RBF neural model is to optimize the non-linear parameters in the basis functions. In this note, a variant of heuristic approach, namedly, Teaching-Learning-Feedback-Based Optimization (TLFBO) [8] is utilized to optimise the linear and non-linear parameters simultaneously in the RBF neural model to improve the model accuracy. Moreover, in order to filter out less relevant model inputs, a fast recursive algorithms (FRA) [9] is applied to select the model structure as well as the model inputs.

This paper is organized as follows. The fundamentals of ZNBs are reviewed in Sect. 2. The RBF neural model construction, the FRA method and the TLFBO optimizer are presented in Sects. 3, 4 and 5 respectively. The experimental study of TLFBO based RBF neural modelling is given in Sects. 6 and 7 respectively. Section 8 concludes the paper.

2 Novel Single Flow Zinc Nickel Battery

A ZNB flow battery system consists of a stack, an electrolyte reservoir which can be either separated or integrated with battery stack as illustrated in Fig. 1 [6], a pump, simplified hydraulic pipes, and the system is free of membranes. The stack is assembled by a series of paralleled single cells that are constructed by nickel oxide positive electrodes, and zinc negative current collectors. The principal reactions are briefly given as follows. Cathode chemical reaction:

$$2NiOOH + 2H_2O + 2e^- \xrightarrow[charging]{discharging} 2Ni\left(OH\right)_2 + 2OH^-$$

Anode chemical reaction:

$$Zn + 2OH^- \xrightarrow[charging]{discharging} Zn\left(OH\right)_2 + 2e^-$$

Overall chemical reaction:

$$Zn + 2H_2O + 2NiOOH \xrightarrow[charging]{discharging} 2Ni\left(OH\right)_2 + Zn\left(OH\right)_2$$

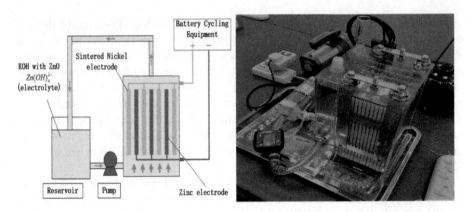

Fig. 1. schematic of ZNBs system (a) and the experimental 200 Ah apparatus (b)

3   Radial Basis Function Neural Networks

The RBF neural networks have been widely used to model non-linear systems, herein, a general multiple-input-single-output RBF neural network is considered. The outputs are formulated as follows.

$$y\left(t\right) = \sum_{i=1}^{n} \omega_i \cdot \phi_i(X) \tag{1}$$

where $y\left(t\right)$ is the RBF neural model output representing the system outputs in discrete sampling time. ω_i is the linear parameter and denotes the output weight for the corresponding $i-th$ RBF node. X is the inputs of the ZNB system in terms of the readily measurable terminal voltage and applied current signals. The output of the hidden node is given as follows.

$$\phi_i(X) = exp(-\tfrac{1}{2\sigma_i^2}\left\|X - c_i\right\|),\ i = 1, 2, \cdots, n \tag{2}$$

where σ_i and c_i are the non-linear parameters in the Gaussian function ϕ_i and denote values of width and centre vectors for the corresponding $i-th$ node.

When the RBF network is used to model the ZNBs, three essential phases in terms of inputs selection, network structure determination involving the number of hidden neurons, and optimization of the linear/non-linear parameters, need to be considered. In this paper, the FRA method is first applied to select both the model structure and identify the significant input terms. The novel TLFBO is proposed to optimize the parameters in the RBF network simultaneously.

4   Fast Recursive Algorithm

The Fast Recursive algorithm (FRA) is a powerful fast forward method to both select model structure and estimate model parameters. In this work the battery model is considered as a discrete non-linear dynamic system $f(.)$ including 20 input items and single output in terms of the t instant terminal voltage outlined as follows.

$$y(t) = f\left(V(t-1), \cdots, V(t-l_{10}), I(t-1), \cdots, I(t-l_{10})\right) \tag{3}$$

According to Eq. (1), the pre-set 20 input terms as the inputs X for the RBF NN will cause significant computational expenses. Besides, the number of RBF hidden nodes needs to be decided. It is clear that the modelling accuracy and the computational efforts increase as the number of hidden nodes increases. According to [10], the number of RBF hidden nodes can be selected based on a simpler ARX model built from the system input and output data. Thereby, the FRA method is applied to pre-select the RBF neural inputs X and the number of hidden nodes. According to [9], a recursive matrix $\mathbf{M}_k$ and a residual matrix $\mathbf{R}_k$ are defined as the basis for the FRA method.

Assuming that

$$\mathbf{\Phi} = \{V(t-1), \cdots, V(t-l_{10}), I(t), \cdots, I(t-l_9)\} \tag{4}$$

$$\mathbf{M}_k \triangleq \mathbf{\Phi}_k^T \mathbf{\Phi}_k \quad \mathbf{M}_k \in \mathfrak{R}^{N \times k}, k = 1, \cdots, n$$
$$\mathbf{R}_k \triangleq I - \mathbf{\Phi}_k \mathbf{M}_k^{-1} \mathbf{\Phi}_k^T, \quad \mathbf{R}_0 \triangleq I \tag{5}$$

where $V(t-i), i = 1, \cdots, 10$ and $I(t-i), i = 0, \cdots, 9$ are the battery voltage and current at time instant i, and these are candidate neural inputs need to be selected. $\mathbf{\Phi}_k = [\varphi_1, \cdots, \varphi_k]$, $k = 1, \cdots, p$ represents the selected items from the regression matrix $\mathbf{\Phi}$ and $\varphi_i = [\varphi_i(1), \cdots, \varphi_i(N)]$. Further, the form of recursive matrix is defined as follows.

$$\mathbf{R}_{k+1} = \mathbf{R}_k - \frac{\mathbf{R}_k \varphi_{k+1} \varphi_{k+1}^T \mathbf{R}_k^T}{\varphi_{k+1}^T \mathbf{R}_k \varphi_{k+1}}, \quad k = 0, 1, \cdots, n-1 \tag{6}$$

Based on [11], $\mathbf{R}$ has the following properties.

$$\mathbf{R}_k^T = \mathbf{R}_k; \quad (\mathbf{R}_k)^2 = \mathbf{R}_k, \quad k = 0, 1, \cdots, n \tag{7}$$

$$\mathbf{R}_i \mathbf{R}_j = \mathbf{R}_j \mathbf{R}_i = \mathbf{R}_i, \quad i \geq j; \ i, j = 0, 1, \cdots, n \tag{8}$$

$$\mathbf{R}_k \varphi_i = 0, \quad \forall i = 0, 1, \cdots, n \tag{9}$$

E_k denotes the modelling error

$$E_k = \mathbf{y}^T \mathbf{R}_k \mathbf{y} \tag{10}$$

Applying Eqs. (6), (10) and the fundamental properties of recursive matrix from Eqs. (7), (8) and (9), the net contribution to the cost function by the $(k+1)th$ element can be expressed as.

$$\Delta E_{k+1} = \mathbf{y}^T (\mathbf{R}_k - \mathbf{R}_{k+1}) \mathbf{y} = \frac{\mathbf{y}^T \mathbf{R}_k \varphi_{k+1} \varphi_{k+1}^T \mathbf{R}_k \mathbf{y}}{\varphi_{k+1}^T \mathbf{R}_k \varphi_{k+1}} \tag{11}$$

Using the FRA method, the most significant terms with the maximum net contributions are selected to further refine the candidates pools and the number of applied hidden nodes of RBF network. In this paper, five important terms with appropriate time lags are chosen, including $V(t-1), V(t-5), V(t-10), I(t),$ $I(t-7)$ and 6 effective hidden nodes are chosen from 20 candidate hidden neurons to predefine the network structure.

5   Teaching-Learning-Feedback-Based Optimization

The Teaching-Learning-based-Optimization (TLBO) method mimics the process of knowledge sharing in the class to optimize non-linear dynamic systems. Inspired by the supervised learning, a Feedback learning phase is added to the original TLBO in [8] to increase the converging speed. Followed by the learner phase, the last global optima denoted as the previous teacher collaborates with the newly selected teacher to provide collective feedbacks of the learning results. This new TLBO variant is called TLFBO. In this work, the linear and non-linear parameters of the RBF network are optimized by this novel proposed TLFBO method.

5.1   Teacher Phase

A teacher is expected to circulate knowledge to improve the mean solution of the whole class. Students learn from the differences between teacher and the mean solution. The updating procedure is formulated below.

$$DM_i = rand_1 * (Teacher_i - T_F * M_i). \tag{12}$$

where M_i is the mean solution, and $Teacher_i$ denotes the teacher who has the best solution. The teaching factor T_F is ranged as 1 or 2.

$$T_F = round(1 + rand_2(0,1)) \tag{13}$$

$$St_i^{new} = St_i^{current} + DM_i \tag{14}$$

where the St_i^{new} is the new generated population after obtaining the information from the teacher. As aforementioned, the most knowledgeable particle will be the updated teacher for the following populations.

5.2   Learner Phase

The learner phase is a mutually learning stage where students exchange their knowledge randomly. The optimum solution is achieved through two steps expressed as follows.

$$St_k^{new} = \begin{cases} St_k^{current} + rand_3(St_k - St_j) iff(St_k < St_j) \\ St_k^{current} + rand_3(St_j - St_k) iff(St_j < St_k) \end{cases} \tag{15}$$

where according to the fitness function or objective function, the marks of students St_k and St_j are compared randomly.

5.3 Feedback Learning Phase

Contradictory to the well-known Particle Swarm Optimization (PSO) method [12], though TLBO is a precise exploration, and the convergence speed is slow. Therefore, an extra learning phase is employed to gain feedbacks from the previous teacher (the previous optima) and current teacher to accelerate the convergence speed as shown in [8]. This procedure can be is formulated as follows.

$$St_i^{new} = St_i^{current} + l_1 * w * D_{last} + l_2 * w * D_{current} \tag{16}$$

$$D_{last} = Tr_i^{last} - St_i^{current} \tag{17}$$

$$D_{current} = Tr_i^{current} - St_i^{current} \tag{18}$$

$$w = (G_{total} - G_{current}/G_{total}) \tag{19}$$

where w is the mutation weight to restraint the exploring scope. l_1, l_2 range between 0 to 1 and $l_1 + l_2 = 1$, they are used to tune the feedback weights from the selected teachers. G_{total} is the predefined total generations; $G_{current}$ is the current generation index.

6 TLFBO Based RBF Network Modelling

The Root Mean Squared Error (RMSE) is used as the objective function.

$$RMSE = \sqrt{\frac{1}{N} \cdot \sum_{i=1}^{N} (\hat{y} - y)} \tag{20}$$

Based on the Eqs.(1), (2), the modelling estimates $\hat{y}$ is calculated. The procedures of building the RBF neural model are given below.

6.1 Input Selection

As shown in Sect. 4, the past battery voltage and current measurements with up to 10 time lags are used as the candidate neural model inputs. Then, an ARX model with 20 terms is constructed, and the contribution of each term is computed. Applying the FRA method, the most 5 significant terms are selected, from which the following 5 significant inputs $V(t-1), V(t-5), V(t-10), I(t), I(t-7)$ are selected to build the RBF neural model.

6.2 Network Construction

At beginning, the number of hidden neurons was chosen to be 20 by trial and error, where the initial parameters of the RBF neurons are generated randomly. Then, using the FRA method, 6 largest contributors from the original 20 neurons are chosen to build the RBF neural model, and the 6 hidden neurons are further optimized using the TLFBO method as given below.

6.3 Network Parameters Optimization

The non-linear and linear variables including σ_i, c_i and ω_i in the radial function are optimized simultaneously using the TLFBO given in Sect. 6. The optimization procedures are outlined as follows,

1. Initialization:
   (a) Configure the inputs X and RBF structure in terms of the numbers of neurons h_n;
   (b) Pre-set the numbers of generations $G_m = 50$, population size $N_p = 30$ and the upper/lower bounds of each solution as $St_{up} = 5$, $St_{low} = 0$;
   (c) Randomize the first population St_1 where the dimension of the parameters is $D = 3 * 6 = 18$, as there are 3 parameters in one neuron to be optimized.
   (d) Check the constraints to adjust the position of all the particles X_i to avoid violating system constraints;
2. Teacher Phase:
   (a) Compute the objective function f to select the teacher T_i;
   (b) Calculate the mean M_i in column-wise;
   (c) Apply the different performance between teacher T_i and mean value M_i denoted as DM_i according to Eq.(14) to exert the influence of teacher;
   (d) Update the generation St_i^{new}.
3. Learner Phase:
   (a) Compute the objective function f and students mutually exchange the knowledge to improve the solutions described in Eq. (15);
   (b) Choose most knowledgeable one as the next Teacher;
4. Feedback Learning Phase:
   (a) Calculate the differences between students and the previous teacher/ current teacher denoted as D_{last}, $D_{current}$;
   (b) Pre-set the learning weight values as $l_1 = 0.3$, $l_2 = 0.7$ apply the different performance between teacher T_i and mean value M_i denoted as DM_i according to Eq.(14) to exert the influence of two teachers;
   (c) Update the generation St_i^{new} and choose the most knowledgeable one as the next Teacher.

7 Experimental Results

The training data was collected from a bench-scale 3.7 Ah ZNB [6,13]. While the validation data was collected from a commercialized large-scale 200 Ah ZNBs. The battery tests were performed using a battery testing system CT-3008W (ShenZhen Neware Corp., China). For the handmade 3.7 Ah battery, the charging/discharging current was 1 C. For the 200Ah battery shown in Fig. 1, the applied current is 0.25 C. The upper and lower threshold voltages were predefined as 2.05 V and 1.2 V respectively. During the testing procedure, the temperature was fixed as the room temperature. In order to capture and demonstrate the non-linear relationship, 10 intervals representing different state of charge (SOC) points were used in the experiments. Eight points between SOC 10% and 90% were evenly chosen during charging processing. After each pulse current, the duration of relaxation was up to 30 min.

7.1 Training Error

The training errors of normal RBF network and within FRA selection are shown in Figs. 2 and 3 The RMSE modelling errors associated with the two cases are extremely small as $6.77E^{-4}$ and $2.77E^{-5}$ respectively. It is clear that the proposed FRA selection method not only improved the modelling performance, but also reduced the computational efforts significantly.

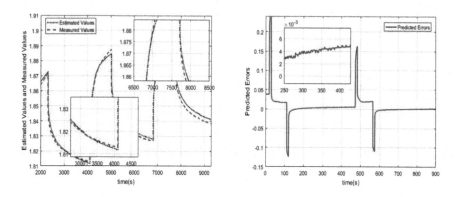

Fig. 2. TLFBO based RBF modelling without FRA selection

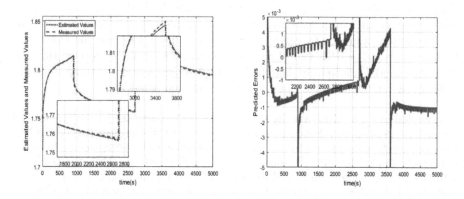

Fig. 3. TLFBO based RBF modelling with FRA selection

7.2 Validation Error

The validation errors of the two methods are shown in Fig. 4 The RMSE values are quite similar on these two optimizers as $3.74E^{-2}$ and $2.67E^{-2}$ respectively. However, it can be seen that the proposed TLFBO method outperforms the PSO method in addressing the over-fitting and local optimum issues. The TLFBO is more efficient to optimize the linear and non-linear parameters in the RBF neural model simultaneously. More details about the description of the TLFBO method and the corresponding processing can be referred to the previous work [8].

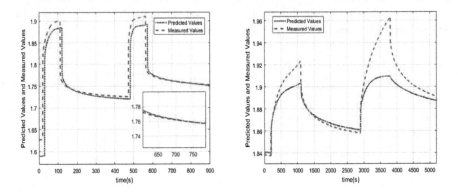

Fig. 4. validation results of TLFBO based RBF modelling (a) and PSO based RBF modelling (b)

8 Conclusion

In this paper, a new TLBO variant, namely, TLFBO was used to optimize the non-linear parameters of a novel ZNB model using the RBF neural network. To further reduce the computational efforts and simplify the model structure, the FRA selection approach was used to both select the inputs and the number of hidden nodes in the RBF neural model. The experimental results show that the developed model can well predict the battery terminal voltage outputs, in terms of both the training accuracy and the generalization capability.

Acknowledgments. X Li and CK Wong would like to thank the Macao Science and Technology Development Fund (FDCT) 's support with the project (111/2013/A3)-Flow Battery Storage System Study and Its Application in Power System. The paper is partially funded by EPSRC under grant EP/L001063/1, the NSFC under grant 61673256, NSFC under grant 61633016, and NSFC under grant 61533010 and by China State Key Laboratory of Alternate Electrical Power System with Renewable Energy Source under LAPS17018.

References

1. Weber, A.Z., Mench, M.M., Meyers, J.P., Ross, P.N., Gostick, J.T., Liu, Q.: Redox flow batteries: a review. J. Appl. Electrochem. **41**(10), 1137 (2011)
2. Wang, W., Luo, Q., Li, B., Wei, X., Li, L., Yang, Z.: Recent progress in redox flow battery research and development. Adv. Funct. Mater. **23**(8), 970–986 (2013)

3. Cheng, J., Zhang, L., Yang, Y.S., Wen, Y.H., Cao, G.P., Wang, X.D.: Preliminary study of single flow zinc-nickel battery. Electrochem. Commun. **9**(11), 2639–2642 (2007)
4. Ito, Y., Nyce, M., Plivelich, R., Klein, M., Steingart, D., Banerjee, S.: Zinc morphology in zinc-nickel flow assisted batteries and impact on performance. J. Power Sources **196**(4), 2340–2345 (2011)
5. Gomadam, P.M., Weidner, J.W., Dougal, R.A., White, R.E.: Mathematical modeling of lithium-ion and nickel battery systems. J. Power Sources **110**(2), 267–284 (2002)
6. Li, Y.X., Wong, M.C., Ip, W.F., Zhao, P.C., Wong, C.K., Cheng, J., You, Z.Y.: Modeling of novel single flow zinc-nickel battery for energy storage system. In: 2014 IEEE 9th Conference on Industrial Electronics and Applications (ICIEA), pp. 1621–1626. IEEE (2014)
7. Charkhgard, M., Farrokhi, M.: State-of-charge estimation for lithium-ion batteries using neural networks and EKF. IEEE Trans. Industr. Electron. **57**(12), 4178–4187 (2010)
8. Li, X., Li, K., Yang, Z.: Teaching-learning-feedback-based optimization. In: Tan, Y., Takagi, H., Shi, Y. (eds.) ICSI 2017. LNCS, vol. 10385. Springer, Cham (2017). doi:10.1007/978-3-319-61824-1_8
9. Li, K., Peng, J.X., Irwin, G.W.: A fast nonlinear model identification method. IEEE Trans. Autom. Control **50**(8), 1211–1216 (2005)
10. Li, K., Peng, J.X.: Neural input selection-a fast model-based approach. Neurocomputing **70**(4), 762–769 (2007)
11. Li, K., Peng, J.X., Bai, E.W.: Two-stage mixed discrete-continuous identification of radial basis function (RBF) neural models for nonlinear systems. IEEE Trans. Circ. Syst. I: Regul. Pap. **56**(3), 630–643 (2009)
12. Kennedy, J.: Particle swarm optimization. In: Sammut, C., Webb, G.I. (eds.) Encyclopedia of Machine Learning, pp. 760–766. Springer, Heidelberg (2011). doi:10.1007/978-0-387-30164-8_630
13. Li, X., Wong, C., Yang, Z.: A novel flowrate control method for single flow zinc/nickel battery. In: International Conference for Students on Applied Engineering (ISCAE), pp. 30–35. IEEE (2016)

State-of-Charge Estimation of Lithium Batteries Using Compact RBF Networks and AUKF

Li Zhang[1]([✉]), Kang Li[2]([✉]), Dajun Du[1], Minrui Fei[1], and Xiang Li[2]

[1] School of Mechatronics and Automation, Shanghai University,
Shanghai 200072, China
zl_qee@163.com, ddj@shu.edu.cn, mrfei@staff.shu.edu.cn
[2] School of Electronics, Electrical Engineering and Computer Science,
Queen's University Belfast, Belfast, UK
{k.li,xli25}@qub.ac.uk

Abstract. A novel framework for the state-of-charge (SOC) estimation of lithium batteries is proposed in this paper based on an adaptive unscented Kalman filters (AUKF) and radial basis function (RBF) neural networks. Firstly, a compact off-line RBF network model is built using a two-stage input selection strategy and the differential evolution optimization (TSS_DE_RBF) to represent the dynamic characteristics of batteries. Here, in the modeling process, the redundant hidden neurons are removed using a fast two-stage selection algorithm to further reduce the model complexity, leading a more compact model in line with the principle of parsimony. Meanwhile, the nonlinear parameters in the radial basis function are optimized through the differential evolution (DE) method simultaneously. The method is implemented on a lithium battery to capture the nonlinear behaviours through the readily measurable input signals. Furthermore, the SOC is estimated online using the AUKF along with an adaptable process noise covariance matrix based the developed RBF neural model. Experimental results manifest the accurate estimation abilities and confirm the effectiveness of the proposed approach.

Keywords: State-of-charge (SOC) · Two-stage selection (TSS) · Radial basis function (RBF) · Differential evolution (DE) · Adaptive unscented Kalman filter (AUKF)

1 Introduction

The lithium batteries have been playing an important role in electric vehicles (EVs) and hybrid electric vehicles (HEVs) in recent years [1]. The battery management system (BMS) is indispensable to maintain the safety and efficient operation. Though the BMS has been studied for a number of years, the battery states estiamtion, especially the real-time SOC estimation is still a big challenge.

Various SOC estimation methods have been proposed so far. The well-adopted model-free methods [1] such as the Ampere-hour method [2], the open

© Springer Nature Singapore Pte Ltd. 2017
K. Li et al. (Eds.): LSMS/ICSEE 2017, Part III, CCIS 763, pp. 396–405, 2017.
DOI: 10.1007/978-981-10-6364-0_40

circuit voltage (OCV)-based method [3], etc. are sensitive to the underlying accumulating measurement errors. Besides, the OCV curve is subject to a flat segement, which means that any slightly introduced error over the measurements would lead to a significant deviation in the terminal values. On the other hand, model-based methods, including the white-box, the grey-box, and the black-box, are capable of capturing the representative battery behaviours and producing accurate enough estimations of the SOC [4–6].

Generally speaking, the white-box models are often unsuitable for real-time applications due to their computational complexity. Grey-box models, such as the equivalent electric circuit models (EECMs) are widely used in the engineering area. Zhang et al. [1,4] proposed an evolutionary EECM using the teaching learning based optimization (TLBO) to identify the model parameters. The black-box models use readily measured signals to infer the battery states. Meng et al. [6] implemented the least-square support vector machine (LSSVM) to build the measurement equations.

On the other hand, the RBF network is capable of approximating nonlinear systems and capturing the complicated nonlinear relationships from the raw data. The critical factors to construct a good fitting RBF network include optimizing the centres and the widths of the hidden nodes and the weights in relation to the output nodes [7]. In this paper, the differential evolution (DE) [8] is applied to optimize the nonlinear and linear parameters simultaneously. Further, according to the principle of parsimony, a compact network is preferable to a more complex one under the similar approximation and generalization performance. Li et al. proposed a fast recursive algorithm (FRA) [9] and a fast two-stage selection algorithm (TSS) [10] combining forward selection and backward refinement. This paper uses the TSS selection method to improve the compactness of the RBF network model for SOC estimation. Generally speaking, the Kalman filter and its extended techniques, i.e. extended Kalman filter (EKF) [11] and AUKF [6] are widely adopted for battery state estimation. In the scope of this paper, the AUKF method is adopted to estimate the SOC values based on a compact RBF neural model. The process noise covariance matrix and the measurement noise covariance matrix are adjusted to improve the estimation accuracy significantly.

The reminder of this paper is organized as follows. A brief introduction of the TSS_DE_RBF method is presented in Sect. 2. Section 3 introduces the established battery TSS_DE_RBF model. Further, the AUKF based SOC estimation approach is described in Sect. 4. The experimental and simulation results are compared in the Sect. 5. Finally, Sect. 6 concludes this paper.

2 TSS_DE_RBF Algorithm

According [7,9,10], the TSS_DE_RBF method can not only build a compact RBF model but also optimize the parameters of the activation functions in the hidden nodes. All candidate neurons are reviewed by the TSS algorithm to select the most significant hidden nodes in building the new compact model. Meanwhile, the DE optimizer is used to determine the optimal values of centers and widths in the radial basis function hidden nodes.

2.1   RBF Neural Networks

A canonical nonlinear dynamic system can be described using a multi-input-single-output (MISO) RBF network as follows:

$$y(t) = \sum_{k=1}^{n} \theta_k \varphi_k(X(t); c_k; \sigma_k) + \varepsilon(t) \tag{1}$$

where $y(t)$, $X(t) \in R^m$ and $\varepsilon(t)$ are output, input and model error at time instant t respectively. Herein, m and n denote the number of inputs and hidden nodes respectively. And $\varphi_k(X(t); c_k; \sigma_k)$ is the Gaussian function as the activation function for the hidden nodes. $c_k \in R^m$ is the centre vector and σ is the RBF width, θ_k represents the output linear weights.

A set of N samples including the $X(t)$, $y(t)$, $t = 1, \cdots, N$ are used as the training data for the RBF network construction, therefore, Eq. (1) can be reformulated in a matrix form as:

$$\mathbf{y} = \Phi\theta + \mathbf{e} \tag{2}$$

where $\Phi = [\phi_1, \cdots, \phi_n]^T \in R^{N \times n}$ is known as the output matrix of the hidden nodes. $\phi_i = [\varphi_i(X(1)), \cdots, \varphi_i(X(N))]^T$, $i = 1, \cdots, n$, $\mathbf{y} = [y(1), \cdots, y(N)]^T \in R^N$ is the output vector. Besides, $\theta = [\theta_1, \cdots, \theta_n]^T \in R^n$ denotes the output weights.

2.2   Two Stage Selection Method

In order to construct a parsimonious RBF neural model with a minimal number of hidden nodes, a two-stage selection method is employed in this paper. During the first stage, the FRA is used to build a compact model [9]. Then, the already selected important model terms (nodes) are reviewed in the second stage. The optimised model with significantly improved performance possesses a fast convergence rate and strong generalization capability.

A. Stage 1 Forword Selection
The forward selection method is used to construct a compact RBF neural network, where the biggest net contribution neurons are stored in a regression matrix $\mathbf{P}$ and rearranged in Φ. For a RBF neural network with hidden neurons randomly generated, it is a linear-in-the-parameter model. For such a model with i model terms (neurons), the corresponding optimal linear parameters (output weights in the RBF neural network) are given below using the least square algorithm (LS)

$$\hat{\theta}_i = (\Phi_i^T \Phi_i)^{-1} \Phi_k^T \mathbf{y} \tag{3}$$

The cost function with the optimal output weights is formulated as

$$\mathbf{J}_i = \mathbf{y}^T \mathbf{R}_i \mathbf{y} \tag{4}$$

where $\mathbf{R}_i = I - \Phi_i[\Phi_i^T \Phi_i]^{-1}\Phi_i^T$ is the residue matrix. Then, the net contribution of a new model term φ_{i+1} at the $i + 1^{th}$ iteration is expressed as

$$\Delta \mathbf{J}_{i+1}(\varphi_{i+1}) = ((\mathbf{y}^{(i)})^T \cdot \varphi_{i+1}^{(i)})^2 / (\varphi_{i+1}^{(i)})^T \cdot \varphi_{i+1}^{(i)} \tag{5}$$

where φ_{i+1} is reordered one, $\mathbf{y}^{(i)} = \mathbf{R}_i\mathbf{y}$, $\varphi_{i+1}^{(i)} = \mathbf{R}_i\varphi_{i+1}$, $k = 0, \cdots, n-1$.

If k model terms are selected to build the compact model, the intermediate matrices $A = [a_{i,j}]_{k\times n}$, $A_y = [ay_i]_{n\times 1}^T$ and $B = [b_i]_{n\times 1}$ are defined as

$$
a_{i,j} = \begin{cases}
0, & j < i \\
(\varphi_i^{(i-1)})^T\varphi_i^{(i-1)} = (\varphi_i)^T\varphi_i - \sum_{h=1}^{i-1} a_{h,i}^2/a_{h,h}, & j = i \\
(\varphi_i^{(i-1)})^T\varphi_j^{(i-1)} = (\varphi_i)^T\varphi_j - \sum_{h=1}^{i-1}(a_{h,i}a_{h,j})/a_{h,h}, & j > i
\end{cases} \tag{6}
$$

$$
ay_i = \begin{cases}
\varphi_i^T\mathbf{R}_{i-1}\mathbf{y} = \varphi_i^T\mathbf{y} - \sum_{h=1}^{i-1}(a_{h,i}ay_h)/a_{h,h}, & 1 \le i < k \\
\varphi_i^T\mathbf{R}_k\mathbf{y} = \varphi_i^T\mathbf{y} - \sum_{h=1}^{k-1}(a_{h,i}ay_h)/a_{h,h}, & k \le i \le n
\end{cases} \tag{7}
$$

$$
b_i = \begin{cases}
(\varphi_i^{(i-1)})^T\varphi_i^{(i-1)} = (\varphi_i)^T\varphi_i - \sum_{h=1}^{i-1} a_{h,i}^2/a_{h,h}, & 1 \le i < k \\
\varphi_i\mathbf{R}_k\varphi_i = (\varphi_i)^T\varphi_i - \sum_{h=1}^{k-1} a_{h,i}^2/a_{h,h}, & k \le i \le n
\end{cases} \tag{8}
$$

Therefore, the net contribution of the $k+1^{th}$ node can be expressed as

$$
\Delta\mathbf{J}_{k+1}(\varphi_{i+1}) = \frac{(\mathbf{y}^T\varphi_{i+1} - \sum_{h=1}^{k}(ay_h a_{h,i+1}/a_{h,h}))^2}{(\varphi_{i+1})^T\varphi_{i+1} - \sum_{h=1}^{k} a_{h,i+1}^2/a_{h,h}}, \quad k \le i \le n \tag{9}
$$

B. Stage 2 Backward Refinement

The backward refinement stage is used to optimize the model constructed by the forward selection approach as the model built at the first stage is not optimal. In brief, the unselected items in the initial set of nodes that have more significant net contributions will replace insignificant ones in the selected nodes.

The forward regression matrix comprising k selected nodes is denoted as $\mathbf{P}_k = [p_1, \cdots, p_k]$. Then the adjacent terms are interchanged ($\hat{p}_q = p_{q-1}$ & $\hat{p}_{q-1} = p_q$). Therefore, some changes occur in the corresponding residue matrix and intermediate matrices which constitute the regression context defined in [10]. According to [10], the residue matrix $\hat{\mathbf{R}}_q$ is derived as

$$
\hat{\mathbf{R}}_q = \mathbf{R}(p_1, \cdots, p_{q-1}, \hat{p}_q) = \mathbf{R}_{q-1} - \mathbf{R}_{q-1}\hat{p}_q\hat{p}_q^T\mathbf{R}_{q-1}^T/\hat{p}_q^T\mathbf{R}_{q-1}\hat{p}_q \tag{10}
$$

besides, the interchanged $A_{i,q:q+1}$ are given by

$$
\begin{cases}
\hat{A}(1:q-1,q) = A(1:q-1,q+1) \\
\hat{A}(1:q-1,q+1) = A(1:q-1,q)
\end{cases} \tag{11}
$$

The position interchanging in the row is expressed as

$$
\begin{cases}
\hat{A}(q,q) = A(q+1,q+1) + A^2(q,q+1)/A(q,q) \\
\hat{A}(q,q+2:n) = A(q+1,q+2:n) + \frac{A(q,q+1)A(q,q+2:n)}{A(q,q)} \\
\hat{A}(q+1,q+1) = A(q,q) - A^2(q,q+1)/\hat{A}(q,q) \\
\hat{A}(q+1,q+2:n) = A(q,q+2:n) - \frac{A(q,q+1)\hat{A}(q,q+2:n)}{\hat{A}(q,q)}
\end{cases} \tag{12}
$$

while $B(q:q+1)$ can be rewritten as

$$
\hat{B}(q) = \hat{A}(q,q), \hat{B}(q+1) = \hat{A}(q+1,q+1) \tag{13}
$$

similarly, $A_y(q : q + 1)$ is updated as

$$\begin{cases} \hat{A}_y(q) = A_y(q+1) + A(q, q+1)A_y(q)/A(q,q) \\ \hat{A}_y(q+1) = A_y(q) - A(q, q+1)\hat{A}_y(q)/\hat{A}(q,q) \end{cases} \quad (14)$$

And then, the remaining candidates will be moved to the kth position, i.e. $\hat{p}_k = \varphi_i$. Its net contribution is computed as follows

$$\Delta J_k(\varphi_i) = (\mathbf{y}^T\varphi_i - \sum_{h=1}^{k-1} \hat{a}y_h \hat{a}_{h,i}/\hat{a}_{h,h})^2 / (\varphi_i)^T \varphi_i - \sum_{h=1}^{k-1} \hat{a}_{h,i}^2/\hat{a}_{h,h} \quad (15)$$

where $\hat{\mathbf{R}}_{k-1}$, $\hat{a}_{.,\cdot}$ and $\hat{a}y_{.}$ are the $(k-t)^{th}$ updated values in Step 1. And t is the initial position index of the reviewed elements.

Subsequently, the corresponding weights will be estimated as

$$\hat{\theta}_j = (a_{j,y} - \sum_{h=j+1}^{k} \hat{\theta}_h a_{j,h})/a_{j,j}, \quad j = k, k-1, \cdots, 1 \quad (16)$$

2.3   DE Optimization

Due to the highly nonlinear nature of the lithium battery model, the DE method is used to optimize he centers and the widths of the hidden nodes for the battery RBF neural model.

A nonlinear vector is defined as follows

$$\mathbf{X} = [x_1, x_2, \cdots, x_n]^T \quad (17)$$

where $x_i \in [x_{i,min}, x_{i,max}]$, specifically, in this paper, $\mathbf{X} = [\mu, \sigma]^T$.

The procedures of the DE algorithm are summarised as follows:

Step 1 Initialization. A population with N real-valued vectors(solutions) is generated. Furthermore, the upper/lower bounds $[X_{min}, X_{max}]$, the maximum number of iterations T, the mutation parameter $F \in [0, 2]$ and the crossover factors $CR \in [0, 1]$ are predetermined. Hence, the ith vector at the tth ($t = 0$ initialization) generation is expressed as

$$\mathbf{X}_{i,t} = [x_{i,t}^1, x_{i,t}^2, \cdots, x_{i,t}^n]^T \quad (18)$$

Step 2 Evaluation. An individual vector is evaluated using Eqs. (9) and (15).

Step 3 Mutation. Firstly, 3 integers $(r_1, r_2, r_3 \in 1, \cdots, N$ and $r_1 \neq r_2 \neq r_3 \neq i)$ are obtained randomly. Then, the mutated vector is derived as

$$\mathbf{V}_{i,t} = \begin{cases} \mathbf{X}_{r_1,t} + F(\mathbf{X}_{r_2,t} - \mathbf{X}_{r_3,t}), & \mathbf{V}_{i,t} \in [X_{min}, X_{max}] \\ \mathbf{X}_{min} + F(\mathbf{X}_{max} - \mathbf{X}_{min}), & otherwise \end{cases} \quad (19)$$

Step 4 Crossover. Some parameters in the existing target vector are updated, leading to a trial vector. The updated corresponding element is formulated as

$$\mathbf{u}_{i,t}^j = \begin{cases} v_{i,t}^j & rand_j \leq CR \quad or \quad j = randn_i \\ x_{i,t}^j & rand_j > CR \quad and \quad j \neq randn_i \end{cases} \quad j = 1, 2, \cdots, n \quad (20)$$

where $rand_j \in [0, 1]$ and $randn_i \in 1, 2, \cdots, n$, $v_{i,t}^j$ donotes the element of $V_{i,t}$.

Step 5 Selection. Evaluated the vectors $X_{i,t}$ and $U_{i,t}$, the best one will be selected in the next generation The process is described as follows

$$X_{i,t+1} = \begin{cases} U_{i,t} & net(U_{i,t}) > net(X_{i,t}) \\ X_{i,t} & net(U_{i,t}) \leq net(X_{i,t}) \end{cases} \quad i = 1, 2, \cdots, N \quad (21)$$

where $net(\cdot)$ is the net contribution calculated by Eqs. (9) and (15).

Step 6 Judgement. A loop from **step2** to **step6** will be performed until the pre-set criteria is met.

3   Battery SOC Estimation Framework

In the paper, the proposed battery SOC estimation framework has two parts. The RBF neural model is developed to reflect the nonliear behaviours of batteies. Then, the AUKF is used for real-time estimation of SOC based on the built RBF model, while considering the measurement noises corrupted into the measured signals. The schematic is illustrated in Fig. 1(a).

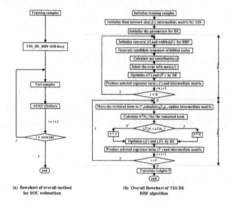

(a) flowchart of overall method for SOC estimation

(b) Overall flowchart of TSS DE RBF algorithm

Fig. 1. Overall method and modeling flowchart diagram for battery SOC estimation

3.1   State Equation

Battery SOC is defined as the ratio of the reserved charge to the nominal capacity. Assuming Q_n, Q_t, Q_0 are the nominal capacity, the remaining capacity and the initial capacity respectively. Thus the expression of SOC is formulated as:

$$SOC_t = Q_t/Q_n = (Q_0 - \int_0^t \eta \cdot i(\tau)\, d\tau)/Q_n = SOC_0 - \int_0^t \eta \cdot i(\tau)\, d\tau/Q_n \quad (22)$$

where $i(\tau)$ is the momentary current at time instant τ, and $\eta = 1/3600$.

Afterwards, the new formation of Eq. (22) in the time-discrete form is shown as follows

$$SOC_k = SOC_{k-1} - \eta \cdot \Delta t \cdot i_k / Q_n \tag{23}$$

where SOC_k and SOC_{k-1} represent SOC at the time instants k and $k-1$, respectively; i_k denotes the discharging current, and Δt is the time interval during the sample rate.

3.2   Measurement Equation

According to Eq. (23), the discharging current at the time instants k and $k-1$, i.e. i_k and i_{k-1}, are selected as the inputs of for the measurement equation. The output of the model is the terminal voltage (V_k) at time instant k. Thus the measurement equation can be derived from Eq. (1) as follows

$$V_k = h(x_k, i_k, i_{k-1}) = \sum_{i=1}^{m} \theta_i exp(-\|u_k - uc_i\|^2 / (2\sigma_i)^2) + \varepsilon_k \tag{24}$$

where $u_k = [i_k, i_{k-1}]$ is the control variable in the AUKF method, the Gaussian function $exp(-\|u_k - uc_i\|^2 / (2\sigma_i)^2)$ is the output of the ith hidden nodes, and m is the total number of hidden nodes in the RBF neural model.

The specific processes are illustrated in Fig. 1(b).

4   SOC Estimation Based on AUKF

After the battery model is built using the method discussed above, battery behavior can be formulated as a state-space equation. Then, SOC will be estimated online using AUKF method [6]. The SOC estimation procedure of AUKF is detailed in Table 1.

5   Experimental Results

A 5-Ah LiFePo4 battery was tested at the room temperature $(25°C)$, using an Arbin BT2000 battery test system. The samples are taken from Hybrid Pulse Power Characterization (HPPC) discharging tests.

Numerous simulation tests in relation to the different sizes of the RBF networks were conducted and compared with the proposed compact battery model as shown in Fig. 2(a), (b). It is clear that even only 10 selected neurons using the proposed method can achieve a highly accurate approximation, while retaining the computational efficiency.

In order to further improve the accuracy as showed in Fig. 2(a), (b), an extra input item $i_{(t-1)}$ was added into the model. Then, a comparison of results obtained from different methods are also indicated in Fig. 2(a), (b). One method is TSS_DE_RBF_LOO which computes the net contribution using leave-one-out(LOO) cross validation. The other is RBF neural network construction method provided by the Matlab. Thereby, the proposed TSS_DE_RBF method

Table 1. SOC estimation procedure using the AUKF

---

1. Problem formulation

state equation: $x_{k+1} = f(x_k, i_k) + w_k = x_{k-1} - \eta \cdot \Delta t \cdot i_k / Q_n + w_k$

output equation: $Z_{k+1} = h(x_k, i_k, i_{k-1}) + v_k = \sum_{i=1}^{k} \theta_i exp(\frac{-\|u - uc_i\|^2}{(2\sigma_i)^2}) + \varepsilon_k + v_k$

where the state equation is formulated based on Ah method and the output equation
is built using TSS_DE_RBF. $u = [x_k, i_k, i_{k-1}]^T$, $x_k = SOC_k$, Z_{k+1}
denotes terminal voltage V_{k+1}, w_k and v_k are the process
and measurement noise respectively. $E(w_k w_k^T) = Q_k$, $E(v_k v_k^T) = R_k$

2. Initialize

$\hat{x}_0 = E[x_0]$, $\hat{w}_0 = E[w_0]$, $\hat{v}_0 = E[v_0]$, $P_0 = E[(x_0 - \hat{x}_0)(x_0 - \hat{x}_0)^T]$, $Q_0 = E(w_0 w_0^T)$,

$R_0 = E(v_0 v_0^T)$, $\alpha = 1$ determines the distribution of sigma points, $\beta = 2$ is used to
reduce the high-order error, $t = 0$ ensures the covariance matrix is positive.

---

3. For $k = 1, 2, \cdots$

1) state extension:

$\hat{X}_{k-1} = [\hat{x}_{k-1}, \hat{w}_{k-1}, \hat{v}_{k-1}]$, $P_{X,k-1} = diag[P_{x,k-1}, Q_{k-1}, R_{k-1}]$,

2) unscented transformation:

sigma points: $X_{k-1,0} = \hat{X}_{k-1}$, $X_{k-1,i} = \hat{X}_{k-1} + (\sqrt{(L+\lambda)P_{X,k-1}})_i$ $i = 1, \cdots, L$,

$$X_{k-1,i} = \hat{X}_{k-1} - (\sqrt{(L+\lambda)P_{X,k-1}})_i \quad i = L+1, \cdots, 2L$$

weighting coefficients: $c_0^{(m)} = \lambda/(L+\lambda)$, $c_0^{(c)} = \lambda/(L+\lambda) + (1 - \alpha^2 + \beta)$

$$c_i^{(m)} = c_i^{(c)} = \lambda/2(L+\lambda) \qquad i = 1, \cdots, 2L$$

where $L = length(\hat{X}_{k-1})$, $\lambda = \alpha^2(L+t) - L$, $(\sqrt{(L+\lambda)P_{X,k-1}})_i$ is the ith row
of $(\sqrt{(L+\lambda)P_{X,k-1}})$.

3) prediction:

$\hat{x}_{k|k-1} = \sum_{i=0}^{2L} c_i^{(m)} [f(X_{k-1,i}^x, u_{k-1}) + X_{k-1,i}^w] = \sum_{i=0}^{2L} c_i^{(m)} X_{k|k-1,i}^x$

$P_{x,k|k-1} = \sum_{i=0}^{2L} c_i^{(c)} (X_{k|k-1,i}^x - \hat{x}_{k|k-1})(X_{k|k-1,i}^x - \hat{x}_{k|k-1})^T$

$\hat{Z}_{k|k-1} = \sum_{i=0}^{2L} c_i^{(c)} [h(X_{k|k-1,i}^x, u_{k-1}) + X_{k-1,i}^v] = \sum_{i=0}^{2L} c_i^{(c)} Z_{k|k-1,i}$

where $X_{k|k-1,i}^x$, $X_{k-1,i}^w$, $X_{k-1,i}^v$ are x, w, v component of the vector $X_{k-1,i}$
respectively.

4) correction:

$\hat{x}_k = \hat{x}_{k|k-1} + K_k(z_k - \hat{Z}_{k|k-1})$, $K_k = P_{xz,k} P_{z,k}^{-1}$, $P_{x,k} = P_{x,k|k-1} - K_k P_{z,k} K_k^T$

where z_k is the actual terminal voltage.

5) adjustment process:

$\mu_k = z_k - h(\hat{x}_k, u_k)$, $F_k = \mu_k \mu_k^T$, $Q_k = K_k F_k K_k^T$,

$R_k = (F_k + \sum_{i=0}^{2L} c_i^{(c)} (Z_{k|k-1,i} - z_k)(Z_{k|k-1,i} - z_k)^T / 2)$

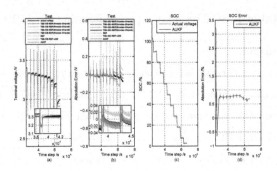

Fig. 2. Test results based on using different methods

is demonstrated to be an effective approach in addressing the over-fitting issue. Through a closer look into the holistic simulation results in Table 2, the developed simple model has been successfully able to produce a more parsimonious network with faster convergence speed and satisfactory accuracy ($\pm 0.1V$).

Table 2. Performance for different algorithms

	TSS_DE_RBF	TSS_DE_RBF	TSS_DE_RBF	TSS_DE_RBF	TSS_DE_RBF_LOO	RBF
Number of hidden nodes	10	30	50	30	30	30
Number of hidden inputs	2	2	2	3	3	3
Mean of time(s)	99.4067	353.5123	648.5841	345.7022	320.3859	120.3198
MSE	7.1634e-4	3.7689e-4	2.7573e-4	4.1294e-4	4.4237-3	3.2219e-3

The simulation results of SOC estimation based on TSS_DE_RBF and AUKF approach are shown in Fig. 2(c), (d). It is worth mentioning that the estimated errors of SOC estimation are less then 3.5%, while the majority is within the range of 1%. The results are proved to be superior to most results published in the literature. Meanwhile compared to sampling time $1s$, the computation time less than 6ms indicates that AUKF can be used for online SOC estimation.

6   Conclusion

The RBF network has a simple structure and has been intensively used to learn the dynamic behaviours of complex systems and perform well in comprehensive tasks. In this paper, an accurate lithium battery model has been built using the RBF network. Both the modelling performance and computational efficiency are significantly improved by the proposed TSS_DE_RBF method. Compared with results published in the literature, the proposed combination i.e. TSS_DE_RBF based AUKF has achieved a desirable performance in SOC estimation. Future work will investigate real-time construction of compact RBF neural model online because of the computational advantage of the TSS_DE_RBF method.

Acknowledgments. This paper was partially funded by the NSFC under grant 61673256, 61633016 and 61533010. State Key Laboratory of Alternate Electrical Power System with Renewable Energy Source under grant LAPS17018, and Shanghai Science Technology Commission No. 14ZR1414800.

References

1. Zhang, C., Li, K., Pei, L., Zhu, C.: An integrated approach for real-time model-based state-of-charge estimation of lithium-ion batteries. J. Power Sources **283**, 24–36 (2015)
2. Jeong, Y.M., Cho, Y.K., Ahn, J.H., Ryu, S.H., Lee, B.K.: Enhanced coulomb counting method with adaptive SOC reset time for estimating OCV. In: Proceedings of the IEEE Energy Conversion Congress and Exposition, Pittsburgh, PA, USA, 14–18 September, pp. 1313–1318 (2014)
3. Petzl, M., Danzer, M.A.: Advancements in OCV measurement and analysis for lithium-Ion batteries. IEEE Trans. Energy Convers. **28**(3), 675–681 (2013)
4. Zhang, C., Li, K., Deng, J., Song, S.: Improved realtime state-of-charge estimation of LiFePO4 battery based on a novel thermoelectric model. IEEE Trans. Ind. Electron. **240**(1), 184–192 (2013)
5. Yang, Z., Li, K., Foley, A., Zhang, C.: A new self-learning TLBO algorithm for RBF neural modelling of batteries in electric vehicles. In: IEEE Congress on Evolutionary Computation (CEC), Beijing, China, 6–11 July (2014)
6. Meng, J., Luo, G., Gao, F.: Lithium polymer battery state-of-charge estimation based on adaptive unscented kalman filter and support vector machine. IEEE Trans. Power Electron. **31**(3), 2226–2238 (2016)
7. Deng, J., Li, K., Harkin-Jones, E., Fei, M., et al.: A novel two stage algorithm for construction of RBF neural models based on a-optimality criterion. In: 2013 Ninth International Conference on Natural Computation (ICNC) (2013)
8. Cheng, S.-L., Hwang, C.: Optimal approximation of linear systems by a differential evolution algorithm. IEEE Trans. Syst. Hum. Cybern. **31**(6), 698–707 (2001)
9. Li, K., Peng, J., Irwin, G.: A fast nonlinear model identification method. IEEE Trans. Autom. Control **50**(8), 1211–1216 (2005)
10. Li, K., Peng, J.X., Bai, E.W.: A two-stage algorithm for identification of nonlinear dynamic systems. Automatica **42**(7), 1189–1197 (2006)
11. Chen, Z., Fu, Y.H., Mi, C.C.: State of charge estimation of Lithium-Ion batteries in electric drive vehicles using extended Kalman filtering. IEEE Trans. Veh. Technol. **62**(3), 1020–1030 (2013)

Intelligent Computing and Control in
Power Systems

Design of Adaptive Predictive Controller for Superheated Steam Temperature Control in Thermal Power Plant

Hong Qian[1,2], Yu-qing Feng[1(✉)], and Zi-bin Zheng[1]

[1] Shanghai University of Electric Power, Shanghai, China
hades450313863@163.com
[2] Shanghai Key Laboratory of Power Station Automation Technology,
Shanghai, China

Abstract. In this paper, an adaptive model predictive controller for overheating steam temperature control of thermal power plants is designed, which is based on the control object with large delay, large inertia, nonlinearity and strong time-varying properties. Through the on-line identification and control of different models, compared with predictive controllers in a general model, in terms of adjusting the superheat steam temperature, it can shorten adjusting time drastically, reduce even eliminate the overshoot and improve the dynamic performance greatly when applying in adaptive model predictive controller. The results show that the adaptive model predictive controller, because of its simple implementation, can be used in power plants, and also can be applied to solve similar problems, which has a broad application prospects.

Keywords: Thermal power plants · Superheated steam temperature control · Online identification algorithm · Adaptive control · Model predictive control · Controller design

1 Introduction

The quality of the superheated steam temperature in the thermal power plant will affect the operating efficiency of the power plant unit. At present, the thermal power plant can not achieve the satisfactory effect on the regulation of the superheated steam temperature. The reason is that the overheated steam temperature control object has the characteristics of large delay and large inertia, and because of the frequent changes in power plant load causes the accused object has nonlinear and time-varying nature, the conventional control system has been unable to meet the quality requirements of the superheated steam temperature regulation of the power plant, so it needs to adopt advanced control algorithm to solve.

At present, a large number of studies have been carried out on researchers of superheated steam temperature control [1–8], and [9] proposed a method to adjust PID cascade main steam temperature control strategy using radial basis (RBF) neural network. And the defect of this method is that the establishment of the neural network requires a lot of training data and it is not easy to realize. Fuzzy control is a universal

© Springer Nature Singapore Pte Ltd. 2017
K. Li et al. (Eds.): LSMS/ICSEE 2017, Part III, CCIS 763, pp. 409–419, 2017.
DOI: 10.1007/978-981-10-6364-0_41

nonlinear feature domain controller, which is suitable for the control of nonlinear and time-varying systems. In [10] cascade PID control and fuzzy control are combined to design a cascade fuzzy controller. However, the fuzzy rules and membership functions of fuzzy control are derived from the experience, the controller design is more complex, difficult to implement and popularize. In [11], a fuzzy B-spline neural network can be used to design an adaptive fuzzy controller, and a suitable training algorithm is proposed to adjust the superheated steam temperature. In the literature [12] and [13], the proposed algorithm can improve the dynamic performance and anti-jamming ability of the system by using the constrained predictive control algorithm, which is suitable for the problem of large delay, Large inertia of the controlled object, but its dependence on the controlled object model is strong, in a nonlinear, strong denaturation of the controlled object on the weak adaptability.

Therefore, through the research of on-line identification algorithm and the design of adaptive predictive controller, an adaptive predictive controller for overheating steam temperature control of thermal power plant is developed. Finally, the control capability of the adaptive predictive controller is verified by the system simulation test, which achieves the better control performance requirement.

2   Design of Adaptive Model Predictive Controller

2.1   System Structure

The system structure of the adaptive model predictive controller designed for the control of superheated steam temperature in thermal power plant is shown in Fig. 1.

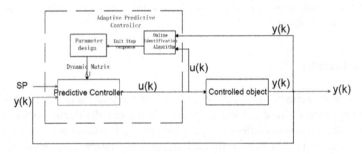

Fig. 1. Structure chart of adaptive model predictive control system

The adaptive predictive control algorithm of the adaptive predictive controller is a dynamic matrix control (DMC) algorithm based on the step response model. The controller samples and stores the control law u (k) and the adjusted amount y (k) at the current time at each sampling time k. The online identification algorithm module calculates the dynamic matrix A1 of the currently controlled object online using the stored control law sequence and the adjusted sequence. The original dynamic matrix is replaced by the calculated dynamic matrix A1 to achieve the purpose of dynamic matrix adaptation.

2.2   Design of Controller

The adaptive model predictive controller consists of three modules: model predictive controller, on-line identification algorithm module and predictive controller parameter design module. The model predictive controller is the control layer, the online identification algorithm module and the predictive controller parameter design module form the online identification layer.

The control layer and the online identification layer work in parallel. In the control layer, the model predictive controller calculate and output the control law with the superheated steam temperature set point and the actual value as input value The online identification layer samples and stores the control law u(k) and the adjusted real time value y(k) at each time to update the control law sequence and the modulated sequence. At each time, the online identification module uses the updated control law sequence and the adjusted sequence to identify and obtains the unit step response sequence of the current controlled object. The unit step response sequence will be used by the predictive controller parameter design module as the new dynamic matrix A1 to replace the original dynamic matrix A of the model predictor controller to participate in the iterative operation of the next time prediction control algorithm.

The performance optimization indicators of the model predictive control algorithm at time k:

$$\min J(k) = \omega_p(k) - \hat{y}_{p0}(k) - A\Delta U_M(k)_Q^2 \tag{1}$$

The necessary conditions for finding extreme values of performance optimization indicators:

$$\frac{dJ(k)}{d\Delta U_M(k)} = 0 \tag{2}$$

The optimal values of the amount of control law change can be obtained from Eqs. (1) and (2):

$$\Delta U_M(k) = \begin{bmatrix} \Delta u(k) \\ \Delta u(k+1) \\ \Delta u(k+2) \\ \vdots \\ \vdots \\ \Delta u(k+M-1) \end{bmatrix} = (A^T Q A + R)^{-1} A^T Q [\omega_p(k) - \hat{y}_{p0}(k)] \tag{3}$$

The instantaneous control effect of the moment k + 1 is given by $\Delta u\ (k + 1)$:

$$\Delta u(k+1) = d^T \times [\omega_p(k+1) - \hat{y}_{p0}(k+1)] \tag{4}$$

The initial prediction vector at (k + 1) is shown in Eq. (5):

$$\hat{y}_{p0}(k+1) = [I_{P\times P} \quad 0]_{P\times N} \times S_{N\times N} \times \hat{y}_{cor}(k+1) \tag{5}$$

$$\hat{y}_{cor}(k+1) = [\hat{y}_{N0_{N\times 1}}(k) + A \times \Delta u(k) + h_{N\times 1} \times [y(k+1) - [\hat{y}_0(k+1|k) + a_1 \times \Delta u(k)]]] \tag{6}$$

Formula (4)–(6) shows the iterative relationship between $\Delta u(k+1)$ and $\Delta u(k)$. The core of the iterative relation is the dynamic matrices A, which determined the relation between $\Delta u(k + 1)$ and $\Delta u(k)$. The dynamic matrix A is the prediction model of the predictive control algorithm. And the model predictive control algorithm is highly dependent on the prediction model, and the accuracy of the prediction model is directly related to the adjustment effect of the controller. Therefore, the main function of the adaptive model predictive controller is to dynamically identify the current controlled object by online identification algorithm and update the dynamic matrix A of the predictive control algorithm in real time, which can avoid the occurrence of deviation between the predicted model and the actual controlled object.

3   Online Identification Algorithm

3.1   The Structure of Online Identification Algorithm

The structure of the online identification algorithm is shown in Fig. 2. The online identification algorithm module has two inputs and one output. At each sampling time, the online identification algorithm module uses the control law increment and the adjusted amount y as input, and output the required prediction model (the dynamic matrix A).

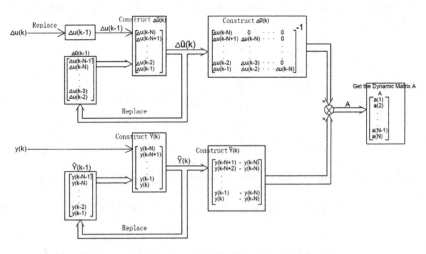

Fig. 2. Structure chart of online identification algorithm

At the sampling time k, the on-line identification module samples the control law increment Δu (k) and the adjusted amount y (k). Δu (k) replaces the control law increment Δu (k − 1) at the previous sampling time to be stored for online identification at the next time. Δu (k − 1) and the control law increment vector $\Delta \hat{u}(k - 1)$ together constitute the vector $\Delta \hat{u}$(k).

$$\Delta\hat{u}(k-1) = [\Delta u(k-N-1)\Delta u(k-N)\cdots\Delta u(k-3)\Delta u(k-2)]^T \tag{7}$$

All elements of the vector $\Delta\hat{u}$(k − 1), except the first element, move one bit up, and Δu (k−1) is the last element. The resulting new vector is $\Delta\hat{u}(k)$.

$$\Delta\hat{u}(k) = [\Delta u(k-N)\Delta u(k-N+1)\cdots\Delta u(k-2)\Delta u(k-1)]^T \tag{8}$$

$\Delta\hat{u}(k)$ replace the original $\Delta\hat{u}(k - 1)$, for the next moment of online identification. N control law increment construct N-dimensional matrix $\Delta\bar{u}$(k).

$$\Delta\bar{u}(k) = \begin{bmatrix} \Delta u(k-N) & 0 & \cdots & 0 \\ \Delta u(k-N+1) & \Delta u(k-N) & \cdots & 0 \\ \vdots & \vdots & \cdots & 0 \\ & & \cdots & 0 \\ \Delta u(k-2) & \Delta u(k-3) & \cdots & \cdots & 0 \\ \Delta u(k-1) & \Delta u(k-2) & \cdots & \Delta u(k-N) \end{bmatrix}^{-1} \tag{9}$$

At the sampling time k, the other sample value is the adjusted amount y(k). The N + 1 former adjusted amount and y(k) make up vector $\hat{Y}(k)$. All elements of the vector $\hat{Y}(k - 1)$, except the first element, move one bit up, and y(k) is the last element. The resulting new vector is $\hat{Y}(k)$. Similarly, $\hat{Y}(k)$ replace the original $\hat{Y}(k - 1)$, for the next online identification.

$$\hat{Y}(k-1) = [y(k-N-1)y(k-N)\cdots y(k-1)]^T \tag{10}$$

$$\hat{Y}(k) = [y(k-N)y(k-N+1)\cdots y(k-1)y(k)]^T \tag{11}$$

The historical value of the N + 1 adjusted amount of Vector $\hat{Y}(k)$ will be used to construct the N-dimensional vector $\bar{Y}$(k).

$$\bar{Y}(k) = \begin{bmatrix} y(k-N+1) - y(k-N) \\ y(k-N+2) - y(k-N) \\ \vdots \\ \vdots \\ y(k-1) - y(k-N)\vdots \\ y(k) - y(k-N) \end{bmatrix} \tag{12}$$

Multiply the N-dimensional matrix $\Delta \bar{u}(k)$ by the N-dimensional vector $\bar{Y}(k)$. And the resulting N-dimensional vector is the required forecasting model (the dynamic matrix A).

3.2   The Expression of Online Identification Algorithm

At k moments, vector $\Delta \hat{u}(k)$ and vector $\hat{Y}(k)$ are saved in online identification algorithm module.

$$\hat{Y}(k) = [y(k - N)y(k - N + 1) \cdots y(k - 1)y(k)]^T \tag{13}$$

$$\Delta \hat{u}(k) = [\Delta u(k - N) \cdots \Delta u(k - 2)\Delta u(k - 1)]^T \tag{14}$$

The dynamic matrix A that needs to be calculated by the online identification algorithm is:

$$A = [a_1 a_2 a_3 \cdots a_{N-1} a_N]^T \tag{15}$$

The relationship among $\hat{Y}(k)$ $\Delta \hat{u}(k)$ and A can be expressed by the following expression:

$$y(k - N + 1) = y(k - N) + a_1 \times \Delta u(k - N)$$
$$y(k - N + 2) = y(k - N) + a_1 \times \Delta u(k - N + 1) + a_2 \times \Delta u(k - N)$$
$$y(k - N + 3) = y(k - N) + a_1 \times \Delta u(k - N + 2) + a_2 \times \Delta u(k - N + 1) + a_3 \times \Delta u(k - N)$$

$$\vdots \qquad \vdots \qquad \vdots \qquad \vdots$$

$$y(k) = y(k - N) + a_1 \times \Delta u(k - 1) + a_2 \times \Delta u(k - 2) + \cdots + a_N \times \Delta u(k - N)$$
$$\tag{16}$$

Scilicet:

$$y(k - N + i) = y(k - N) + \sum_{j=1}^{N} a_j \times \Delta u(k - N + (i - j)), i = 1, 2, 3, \cdots, N j \leq i \tag{17}$$

Change the above expression into a matrix equation:

$$
\begin{bmatrix} y(k-N+1) \\ y(k-N+2) \\ y(k-N+3) \\ \vdots \\ \vdots \\ y(k) \end{bmatrix} =
\begin{bmatrix} y(k-N) \\ y(k-N) \\ y(k-N) \\ \vdots \\ \vdots \\ y(k-N) \end{bmatrix} +
\begin{bmatrix} \Delta u(k-N) & 0 & \cdots & 0 \\ \Delta u(k-N+1) & \Delta u(k-N) & \cdots & 0 \\ \Delta u(k-N+2) & \Delta u(k-N+1) & \cdots & 0 \\ \vdots & \vdots & \cdots & 0 \\ \vdots & \vdots & \cdots & 0 \\ \Delta u(k-1) & \Delta u(k-2) & \cdots & \Delta u(k-N) \end{bmatrix} \times
\begin{bmatrix} a_1 \\ a_2 \\ a_3 \\ \vdots \\ \vdots \\ a_N \end{bmatrix}
$$
$$\tag{18}$$

Solve the matrix equation to obtain the required dynamic matrix A:

$$
A = \begin{bmatrix} a_1 \\ a_2 \\ a_3 \\ \vdots \\ \vdots \\ a_N \end{bmatrix} = \begin{bmatrix} \Delta u(k-N) & 0 & \cdots & 0 \\ \Delta u(k-N+1) & \Delta u(k-N) & \cdots & 0 \\ \Delta u(k-N+2) & \Delta u(k-N+1) & \cdots & 0 \\ \vdots & \vdots & \cdots & 0 \\ \vdots & \vdots & \cdots & \cdots & 0 \\ \Delta u(k-1) & \Delta u(k-2) & \cdots & \Delta u(k-N) \end{bmatrix}^{-1} \times \begin{bmatrix} y(k-N+1)-y(k-N) \\ y(k-N+2)-y(k-N) \\ y(k-N+3)-y(k-N) \\ \vdots \\ \vdots \\ y(k)-y(k-N) \end{bmatrix}
$$

(19)

The resulting dynamic matrix A is the dynamic matrix A in Fig. 2.

$$
A = \begin{bmatrix} a_1 \\ a_2 \\ a_3 \\ \vdots \\ \vdots \\ a_N \end{bmatrix} = \Delta \bar{u}(k) \times \bar{Y}(k)
$$

(20)

The N model coefficients of the dynamic matrix A:

$$
a_i = \begin{cases} \dfrac{y(k-N+1)-y(k-N)}{\Delta u(k-N)}, i = 1 \\[2ex] \dfrac{[y(k-N+j)-y(k-N)]-\left[\sum_{j=1}^{i-1} a_j \times \Delta u(k-N+i-j)\right]}{\Delta u(k-N)}, i = 2,3,4,\cdots N \end{cases}
$$

(21)

The dynamic matrix A obtained by online identification is a new predictive model of predictive control algorithm. The dynamic matrix A is used instead of the original dynamic matrix to complete a model online identification.

4 Simulation Example

4.1 The Establishment of the Controlled Object Model of Superheated Steam Temperature Control System

The change of the flow rate of the spraying water is the main factor that causes the main steam temperature change. The control system with superheated steam temperature regulation by spraying water is simple, and it is advantageous for safe operation of superheater. It is a widely used superheated steam temperature adjustment method.

The dynamic characteristics of the controlled object of the superheated steam temperature are identified by the test of the steam temperature change caused by the disturbance of spraying water flow. Figure 3 is the superheated steam temperature controlled object structure. The system controls the main steam temperature y by changing the reducing-temperature valve u.

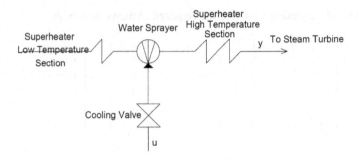

Fig. 3. Structure chart of superheated steam temperature controlled object

A step disturbance test of the valve opening of the 300 MW unit of the power plant is done. Identify the mathematical model of the controlled object, through the input and output data.

According to the five different load conditions, the controlled object model is identified and processed, and the five models are shown in Table 1 below. In this paper, the simulation test of the adaptive predictive controller is carried out for the above five models. And the test shows the advantage of the adaptive predictive controller when the controlled object changes, compared to the conventional predictive controller.

Table 1. Controlled object mathematical model

Model	Model 1 $G_1(s)$ (original)	Model 2 $G_2(s)$	Model 3 $G_3(s)$	Model 4 $G_4(s)$	Model 5 $G_5(s)$
$G(s) = \frac{\Delta y}{\Delta u}\,(^\circ C/\%)$	$\frac{1.2486}{(123s+1)^2}$	$\frac{1.2619}{350s+1}$	$\frac{1.2619}{150s+1}$	$\frac{1.6486}{(123s+1)^2}$	$\frac{1.9486}{(123s+1)^2}$

4.2 Simulation of Adaptive Predictive Controller

The simulation test platform is built by Visual Studio2010 and MATLAB/Simulink simulation software. The function of the adaptive predictive controller is implemented on Visual Studio 2010. The adaptive predictive controller is connected with MATLAB/Simulink simulation platform through OPC interface.

The MATLAB/Simulink simulation model is shown in Fig. 4 below. The adaptive control controller calculates the control law through the OPC interface to the OPC Read module named "U_MPC" in the following figure and acts on the controlled object The adjusted amount will be return to the adaptive predictive controller by the OPC Write module which is called "Actual_Temper" to participate in calculating the control law.

The simulation test will be carried out on the models shown in Table 1. The initial temperature of the superheater was 523°C and the set temperature was 540°C. Comparing the characteristic curves of the adaptive predictive controller and the general predictive controller under the same model. It is ensured that the initial prediction model of the adaptive predictive controller is consistent with the initial prediction model (dynamic matrix A) of the general predictive controller, using the unit step

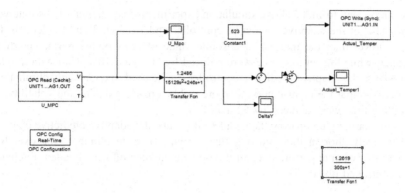

Fig. 4. Simulation chart of adaptive model predictive controller

response sequence of the initial model (model 1). The four models of the model 2 to the model 5 were used as the controlled objects to the experiments. Finally, the control curves were displayed by the oscilloscope. The results are shown in Figs. 5, 6, 7 and 8.

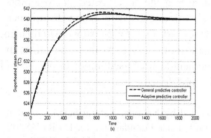

Fig. 5. Curve comparison between general predictive controller and adaptive predictive controller control in model 2.

Fig. 6. Curve comparison between general predictive controller and adaptive predictive controller control in model 3

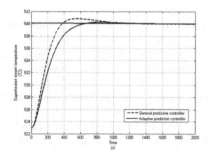

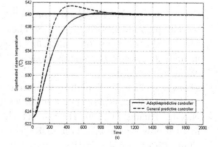

Fig. 7. Curve comparison between general predictive controller and adaptive predictive controller control in model 4

Fig. 8. Curve comparison between general predictive controller and adaptive predictive controller control in model 5

It can be seen from the above simulation experiments that integral of absolute value of error [14] of the adaptive predictive controller under model 2 and model 3 is 17% and 3% lower than the integral of absolute value of error of the normal predictive controller, while the settling time were reduced by 50% and 10%. Under the model 4 and model 5, the adaptive predictive controller can be used to completely eliminate the overshoot, so that the control curve is more stable, greatly reducing the volatility, and the settling time were reduced by 52 and 43%.

In summary, by comparing the simulation results, the adaptive predictive controller can effectively shorten the adjusting time, reduce or even eliminate the overshoot, improve the dynamic performance of the system, and can effectively adapt to changes of the controlled object.

5 Conclusion

In this paper, the design and research of the adaptive predictive controller for the control of the superheated steam temperature of the thermal power plant is carried out. Through the research of the on-line identification algorithm and the reasonable design of the adaptive predictive controller, the problem that the superheated steam temperature regulation can not achieve the satisfactory effect caused by some disturbances caused by the change of the power plant load is solved, making the superheated steam temperature tracking set value faster, the process more stable. At the same time, the online identification algorithm is simple to implement, can be used in power plants, and also to promote the application to solve similar problems.

Acknowledgements. This work was partially supported by National Natural Science Foundation of China, Grant No. 61503237, Shanghai Natural Science Foundation (No. 15ZR1418300), Shanghai Key Laboratory of Power Station Automation Technology (No. 13DZ2273800).

References

1. Lin, J., Shen, J., Li, Y.: Adaptive predictive control for super heated steam temperature based on multiple models switching. J. Southeast Univ. (Nat. Sci. Ed.) **38**(1), 69–74 (2008)
2. Ma, L., Shi, Z.: Study on boiler superheated steam temperature compensation control with neural network inverse process models. J. North China Electr. Power Univ. (Nat. Sci. Ed.) **38**(5), 70–75 (2011)
3. Zhang, L., Chen, F.: Superheated steam temperature control system based on adaptive smith predictive compensation. Inf. Control **44**(5), 513–518 (2015)
4. Nie, L., Huang, H., Liu, H., Han, H., Lv, P.: Superheated steam temperature control of thermal power plant based on ADRC combined with neural network strategy. Control Eng. China **17**(S2), 45–48 (2010)
5. Wu, D., Zhang, L., Wang, J.: Study on boiler superheated steam temperature control system based on TDFMD PID. J. Chin. Soc. Power Eng. **35**(12), 970–974 (2015)

6. Sun, L., Dong, J., Li, D.: Active disturbance rejection control for superheated steam boiler temperatures using the fruit fly algorithm. J. Tsinghua Univ. (Sci. Technol.) **54**(10), 1288–1292 (2014)
7. Wang, F.: Networked predictive control of main steam temperature for once-through boiler based on reverse transfer strategy. Proc. CSEE **35**(19), 4981–4990 (2015)
8. Qian, J., Zhao, J., Xu, Z.: Predictive Control, 1st edn, pp. 49–68. Chemical Industry Press, Beijing (2007)
9. Wang, J., Jiang, G., Wang, S.: Main steam temperature control using RBF neural network based on hybrid learning algorithm. Therm. Power Gener. **38**(2), 28–32 (2009)
10. Zhang, B.: The cascade fuzzy controller in the application of the superheat vapor temperature. Boil. Technol. **41**(2), 5–9 (2010)
11. Pu, W., Chen, L.: Designing adaptive fuzzy controllers based on B-spline neural networks. Control Theory Appl. **13**(4), 448–454 (1996)
12. Wang, X., Feng, X.: Application of constrained predictive model to power plant superheated steam temperature control. J. Chin. Soc. Power Eng. **30**(3), 192–195 (2010)
13. Yang, Z.: Control System Optimization of Typical Low Nitrogen Combustion AGC Unit Based on Model Technology. Shanghai University of Electric Power, Shanghai (2016)
14. Wang, Q.: Performance Assessment to Predictive PI Control System Based on Relay Feedback. Shenyang University of Technology, Shenyang (2008)

Extended State Space Predictive Control of Gas Turbine System in Combined Cycle Power Plant

Guolian Hou[1(⊠)], Tian Wang[1], Huan Du[1], Jianhua Zhang[1],
and Xiaobin Zheng[2]

[1] School of Control and Computer Engineering,
North China Electric Power University, Beijing 102206, China
{hgl,zjh}@ncepu.edu.cn, wangtianncepu@163.com,
duhuanncepu@163.com
[2] Beijing Hualian Electric Power Engineering Supervision Company,
Building 40 No. 188, South 4th Ring Road, Fengtai District,
Beijing 100070, China
zhengxiaobin@heesc.com

Abstract. In this paper, an extended state space predictive control (ESSPC) strategy has been applied to gas turbine in combined cycle power plant. This proposed predictive control needn't solve Diophantine equation online. In addition, in order to overcome the shortcoming of the conventional state space predictive control (SSPC) which only takes output errors into consideration, the objective function of ESSPC includes both output errors and the variation of the system states. In the rolling optimization part, the quadratic program (QP) method is applied to deal with the limitations on the inputs of system. Simulation results show that the proposed algorithm has better tracking ability and stability compared with conventional state space predictive controller in the same condition.

Keywords: Gas turbine · Combined cycle power plant · Extended state space model · Predictive control

1 Introduction

In recent years, the gas turbines have played a critical role in the field of thermal power generation due to the rapid transmission of electricity, small size and cleanness of natural gas. The gas turbine is a key factor in determining the efficiency of combined cycle power supply. For gas turbine control system, the main task is to command the power output to satisfy the power generation need of power plants when kept the fuel flow and inlet guide vane position within given thresholds. Gas turbine system is extremely complex because of its high order, strong coupling, which may lead to enormous challenge to control strategies. The traditional PID control is difficult to achieve the expected control effect. Up to now, various control methods have been studied by researchers for the control of gas turbines.

© Springer Nature Singapore Pte Ltd. 2017
K. Li et al. (Eds.): LSMS/ICSEE 2017, Part III, CCIS 763, pp. 420–428, 2017.
DOI: 10.1007/978-981-10-6364-0_42

Robust control and fuzzy control have been favored by some scholars. Considering the fuel flow and inlet guide vane position limits, reference [1] proposes a robust controller for the speed and temperature control of gas turbine. Although the simulation result shows the settling time is extremely fast, the turbine speed is out of the desired interval. In [2], a fuzzy logic controller is designed for gas turbine. The overshoot of the simulation result is acceptable, irrespective of the settling time is a little long. In [3], the piecewise-fuzzy theory is introduced into the gas turbine speed control system. The appropriate fuzzy group and inferential rules are selected according to different dynamic characteristics of the gas turbine, in addition that the controller parameters are adjusted to meet the dynamic performance indexes of the gas turbine control adaptively.

In power plant control, the model predictive control (MPC) has been extensively used as an advanced control strategy. Reference [4] has proposed a predictive controller based on a nonlinear model in a gas turbine system. The purpose of the controller is to track the compressor discharge demand to adjust the compressor outlet pressure. This controller has better performance in terms of constraint handling and disturbance rejection and can be triumphantly applied to a nonlinearized gas turbine system. Reference [5] applies model predictive control to a gas turbine system and shows great superiorities compared with the PI controller.

In order to improve the abilities of optimization, the model predictive control algorithm has been applied by a variety of researchers to deal with the control problem of multi-variable systems which have strong coupling. MPC is chosen to apply to gas turbine system because of its advantages of simple principle, short operation time and excellent optimization ability. Nevertheless, the traditional MPC also has some limitations, so it is necessary to propose an improved state space predictive controller for a better control performance.

In the present paper, an extended state space predictive controller which based on conventional predictive control [6] is proposed to regulate the power and speed of gas turbine system in gas-steam combined cycle power plant. The algorithm of ESSPC extends the output errors to state variables. It not only takes the input and output variables, but also their past values and output errors as new state variables, furthermore the cost function includes the change of the system state variation. Compared with the predictive control based on process input-output transfer function model, the predictive control based on the state space model can perform state feedback which take the state variable constraints into consideration. More importantly, it does not need to solve the Diophantine equation, which greatly reduces the amount of computation and run time. A quadratic program (QP) problem is used to deal with the input constraints for the sake of ensuring the safe operation of the system. The gas turbine model used in this paper is based on Rowen model investigated in [7, 8].

The rest of this paper is organized as follows. In the next section, the model of gas turbine has been introduced briefly. After that, the extended state space predictive controller is proposed in Sect. 3. In Sect. 4, the simulation results using the proposed algorithm are given. Eventually, conclusions are drawn in Sect. 5.

2 Description of Gas Turbine System

The model of combined cycle power plant can be divided into three components: gas turbine, heat recovery steam generator (HRSG) and steam turbine. Furthermore, each subsystem will be divided into several parts according to the object structure, heat transfer and flow of the working medium. The whole dynamic process follows the rules of conservation of energy, mass and momentum. Figure 1 shows the structure of the gas turbine.

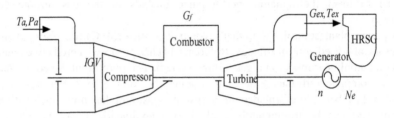

Fig. 1. Structure diagram of the gas turbine.

From the above picture, N_e is power. T_a and P_a denote environmental temperature and pressure. G_{ex} and T_{ex} represent turbine exhaust gas flow and temperature. G_f, IGV and n are respectively fuel flow, compressor inlet guide vanes and rotor speed. Gas turbine control system mainly includes speed control system, acceleration control system, power control system, temperature control system and the compressor inlet guide vane control system. The gas turbine system has the peculiarities of multi-variable, strong coupling and frequent disturbance. The working principle of the gas turbine is described as follows. The compressor draws air from the surroundings and compresses the air to higher pressure gas. In this process, the temperature of the air rises as it passes through the axial compressor. The compressed high-pressure gas is injected into the combustor and mixed with the fuel. Then the compound would become high temperature, high pressure gas by combusting. The gas with high temperature and high pressure expands in the gas turbine and devotes itself to operation, turns the turbine around, drives the compressor and generator into high speed rotation. As a result, the gas turbine converts the chemical energy portion of the gas or liquid fuel into mechanical energy. In Fig. 1, Gf and IGV are control inputs which are determined by the controller; N_e and n are measured outputs.

The dynamics of the gas turbine system working in rated condition can be described as following MIMO model:

$$x_m(k+1) = A_m x_m(k) + B_m u(k)$$
$$y(k) = C_m x_m(k)$$

$$(1)$$

where, $x_m \in R^p$ is state variable, $y \in R^n$ is measured output, $u \in R^m$ is input variable, $A_m \in R^{p \times p}$, $B_m \in R^{p \times m}$ and $C_m \in R^{n \times p}$ are real matrices, where,

$$A_m = \begin{bmatrix} 0.7775 & -2.0028 & 0 & 0 & 0 & 0 & 0 & 0 \\ 0.0279 & 0.7813 & 0 & 0 & 0 & 0 & 0 & 0 \\ 0 & 0 & 0.7775 & -2.0028 & 0 & 0 & 0 & 0 \\ 0 & 0 & 0.0279 & 0.7813 & 0 & 0 & 0 & 0 \\ 0 & 0 & 0 & 0 & 0.7775 & -2.0028 & 0 & 0 \\ 0 & 0 & 0 & 0 & 0.0279 & 0.7813 & 0 & 0 \\ 0 & 0 & 0 & 0 & 0 & 0 & 0.7775 & -2.0028 \\ 0 & 0 & 0 & 0 & 0 & 0 & 0.0279 & 0.7813 \end{bmatrix}$$

$$B_m = \begin{bmatrix} 0.0279 & 0.0030 & 0 & 0 & 0.0279 & 0.0030 & 0 & 0 \\ 0 & 0 & 0.0279 & 0.0030 & 0 & 0 & 0.0279 & 0.0030 \end{bmatrix}^T$$

$$C_m = \begin{bmatrix} 0.0021 & 17.0175 & 0.0139 & 108.5789 & 0 & 0 & 0 & 0 \\ 0 & 0 & 0 & 0 & 0.0544 & 0 & 0.3456 & 0 \end{bmatrix}$$

The input variable u_1 is the inlet guide vane position, u_2 is the fuel flow. The output y_1 is the produced power of gas turbine, y_2 is the speed of turbine. The model is obtained by transforming the model from [9].

Generally, complex industrial processes involve multifarious constraints, such as the limitations imposed by inherent properties of system and economic benefits. Gas turbine system needs to give priority to the following constraints: the change rate of the fuel flow (Δu_1) should be controlled within the scope of the limit of the real accelerate rate of each device to avoid the breakdown caused by the excessive output; the increasing rate of the inlet guide vane position (Δu_2) should be lower than the acceleration and deceleration limit of the servo valve controller in the gas turbine control system. Therefore, the constraint conditions of control input variables and their derivatives are given as follows:

$$
\begin{aligned}
u_{1min} \leq u_1 \leq u_{1max} \\
u_{2min} \leq u_2 \leq u_{2max} \\
\Delta u_{1min} \leq \Delta u_1 \leq \Delta u_{1max} \\
\Delta u_{2min} \leq \Delta u_2 \leq \Delta u_{2max}
\end{aligned}
\tag{2}
$$

3 Extended State Space Predictive Control

Considering the discrete state space model described in Eq. (1), the difference equation of the state variable can be expressed as:

$$\Delta x_m(k+1) = x_m(k+1) - x_m(k) \tag{3}$$

The deviation of control variable is:

$$\Delta u(k) = u(k) - u(k-1) \tag{4}$$

These are the increments of the variables $x_m(k)$ and $u(k)$. Thus, the deviation of the state space equation is:

$$\Delta x_m(k+1) = A_m \Delta x_m(k) + B_m \Delta u(k) \tag{5}$$

Supposing the anticipant output is $r(k)$, and the output tracking error can be defined as: $e(k) = y(k) - r(k)$. Then $e(k+1)$ can be denoted by:

$$e(k+1) = e(k) + C_m A_m \Delta x_m(k) + (C_m B_m + D_m)\Delta u(k) - \Delta r(k+1) \tag{6}$$

where $\Delta r(k+1) = r(k+1) - r(k)$.

Extend the state variable $x_m(k)$ so that it can contain the output tracking error $e(k)$. Therefore, the extended state variable is defined as:

$$z(k) = [\Delta x_m(k) \quad e(k)]^T \tag{7}$$

The extended state space model can be represented as:

$$z(k+1) = Az(k) + B\Delta u(k) + C\Delta r(k+1) \tag{8}$$

where $A = \begin{bmatrix} A_m & 0 \\ C_m A_m & I_q \end{bmatrix}$ $B = \begin{bmatrix} B_m \\ C_m B_m \end{bmatrix}$ $C = \begin{bmatrix} 0 \\ -I_q \end{bmatrix}$, and 0 is a zero matrix, I_q is a unit matrix.

The cost function of the algorithm is described as follow:

$$J = \sum_{j=1}^{P} z^T(k+j)Q_j z(k+j) + \sum_{j=1}^{M} \Delta u^T(k+j)L_j \Delta u(k+j) \tag{9}$$

where P is the prediction horizon, M denotes the control horizon, Q_j represents the weighted matrix, L_j is the weighted factor of control input increments, generally Q_j is taken as: $Q_j = diag\{Q_{jx}, Q_{je}\}$. Otherwise, the weighted factor of state variable Q_{jx} is defined as $Q_{jx} = diag\{q_{j,x1}, q_{j,x2}, \ldots, q_{j,xn}\}$ and the weighted factor of output error $Q_{je} = diag\{q_{j,e1}, q_{j,e2}, \ldots, q_{j,eq}\}$.

The future state variables corresponding with sampling instant k can be denoted by:

$$Z = Fz(k) + \Phi\Delta U + S\Delta R \tag{10}$$

where

$$Z = \begin{bmatrix} z(k+1) \\ z(k+2) \\ \vdots \\ z(k+P) \end{bmatrix} \quad \Delta U = \begin{bmatrix} \Delta u(k) \\ \Delta u(k+1) \\ \vdots \\ \Delta u(k+M-1) \end{bmatrix} \quad \Delta R = \begin{bmatrix} \Delta r(k+1) \\ \Delta r(k+2) \\ \vdots \\ \Delta r(k+P) \end{bmatrix} \quad F = \begin{bmatrix} A \\ A^2 \\ \vdots \\ A^P \end{bmatrix}$$

$$\Phi = \begin{bmatrix} B & 0 & 0 & \cdots & 0 \\ AB & B & 0 & \cdots & 0 \\ A^2B & AB & B & \cdots & 0 \\ \vdots & \vdots & \vdots & \ddots & \vdots \\ A^{P-1}B & A^{P-2}B & A^{P-3}B & \cdots & A^{P-M}B \end{bmatrix} \quad S = \begin{bmatrix} C & 0 & 0 & \cdots & 0 \\ AC & C & 0 & \cdots & 0 \\ A^2C & AC & C & \cdots & 0 \\ \vdots & \vdots & \vdots & \ddots & \vdots \\ A^{P-1}C & A^{P-2}C & A^{P-3}C & \cdots & C \end{bmatrix}$$

The cost function is rewritten as:

$$J = Z^T Q Z + \Delta U^T L \Delta U \tag{11}$$

where $Q = block\ diag\{Q_1, Q_2, \ldots, Q_P\}, L = block\ diag\{L_1, L_2, \ldots, L_M\}$. Let

$$\frac{\partial J}{\partial \Delta u} = 0 \tag{12}$$

The optimal control law can be expressed as follow:

$$\Delta U = -\left(\Phi^T Q \Phi + L\right)^{-1} \Phi^T Q(Fz(k) + S\Delta R) \tag{13}$$

$$K_S = \left(\Phi^T Q \Phi + L\right)^{-1} \Phi^T Q F, \quad K_R = \left(\Phi^T Q \Phi + L\right)^{-1} \Phi^T Q S \tag{14}$$

Afterwords, the increment of control input at sampling instant k can be expressed as:

$$\Delta u(k) = -k_S z(k) - k_R \Delta R \tag{15}$$

where k_S and k_R are the first P lines of K_S and K_R respectively.

The resulting optimization problem would be a quadratic program problem when considering the constraint condition. The cost function can be further expressed as:

$$J = \Delta U^T (\Phi^T Q \Phi + L) \Delta U + 2(Fz(k) + S\Delta R)^T \Phi \Delta U \tag{16}$$

Then the input constraints are needed to be expressed as follows:

$$\begin{bmatrix} I & 0 & \cdots & 0 \\ I & I & 0 & \vdots \\ \vdots & \vdots & \ddots & 0 \\ I & I & I & I \end{bmatrix} \Delta U_k \geq \begin{bmatrix} u_{min} - u_{k-1} \\ u_{min} - u_{k-1} \\ \vdots \\ u_{min} - u_{k-1} \end{bmatrix} \tag{17}$$

$$\begin{bmatrix} I & 0 & \cdots & 0 \\ I & I & 0 & \vdots \\ \vdots & \vdots & \ddots & 0 \\ I & I & I & I \end{bmatrix} \Delta U_k \leq \begin{bmatrix} u_{max} - u_{k-1} \\ u_{max} - u_{k-1} \\ \vdots \\ u_{max} - u_{k-1} \end{bmatrix} \tag{18}$$

After this, the optimal control vector can be calculated in every sampling period.

4 Simulation Results

The sampling time of the gas turbine system is determined as 2 s. The prediction horizon P is chosen as 15 while the control horizon M is selected as 5. The weighted factor Q_j consists of Q_{jx} and Q_{je}. The value of weighted factor is taken as follows:

$$Q_{jx} = diag\{0.1, 0.1, \ldots, 0.1\}, \ Q_{je} = diag\{2, 1\}, \ L_j = diag\{1, 1\} \tag{19}$$

The constraint conditions are given as follows:

$$52 \leq u_1 \leq 65, \ 14.5 \leq u_2 \leq 27.5$$
$$-2 \leq \Delta u_1 \leq 2, \ -2 \leq \Delta u_2 \leq 2 \tag{20}$$

The working conditions of the system have been changed as follows. The load is decreasing from 200 MW to 180 MW at time 10 s, then the set-point value increases from 180 MW to 200 MW at time 50 s. The Figs. 2 and 3 show the comparison of conventional state space predictive control with the extended state space predictive control. The blue solid lines show the inputs and outputs of the system equipped with the extended state-space predictive controller, while red dotted lines show the inputs and outputs of the system with conventional state space predictive control in Fig. 2.

Figure 2(a) shows the variation of the power and speed. As the figure indicates, response overshoot of power is acceptable and the controller could bring the response to the desired value within settling time of 15 s approximately. It indicated that the ESSPC algorithm has higher tracking speed. As manifested in Fig. 2(a), the overshoot of the speed response curve is less than 0.2%, which fully meets the requirements of the speed variation. It can be obviously find that the change of turbine speed which using improved algorithm is lower than conventional algorithm.

The upper and lower parts of Fig. 2(b) show the change of input variables (IGV and fuel flow) respectively. The figure also demonstrates obviously that the speed limit of the input variables that ensure the safe operation of the system has a preferable effect.

Figure 3 show the state changes of two different algorithms. As it can be seen, the variation of the state variables is quite small and can be obtained quickly. The states of ESSPC have less setting time than the other.

The simulation results indicate the controller proposed in this paper has good tracking ability compared with conventional state space predictive control. In addition that it also can handle the constraint problems perfectly.

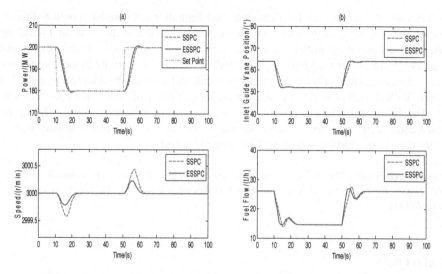

Fig. 2. Output and input curves when set point of power decreases by 10%

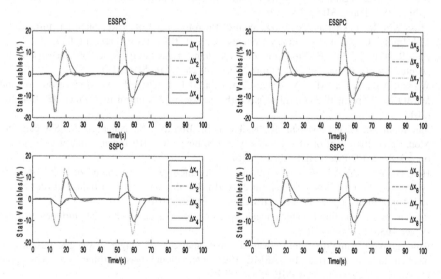

Fig. 3. State curves of ESSPC and SSPC

5   Conclusion

In this paper, an extended state space predictive controller is designed for a gas turbine system. The advantage of the proposed ESSPC approach is that the state and output errors are contained in the extended state variable. In addition, the cost function not only includes output errors, but also includes the change of the system state variation. A quadratic program algorithm is used to handle the restriction of variables. Simulation results indicate that the controller can guarantee limitations of variables could be limited within the allowable range which highlights the advantage of the constraint handling ability of proposed controller. As can be seen from the simulation results, the output variables have higher tracking speed and little overshoot. According to the comparison between simulation results using different algorithms, the extended state space predictive controller has been proved to possess anticipant capability of power and speed control.

The future work will focus on the improvement of predictive control theory by researching new predictive control algorithm. In addition, the accomplishment of a more precise gas turbine model will be further studied.

Acknowledgments. This work is supported by the National Science Foundation of China (61374052, 61511130082) and the Fundamental Research Funds for the Central Universities 2016ZZD03.

References

1. Najimi, E., Ramezani, M.H.: Robust control of speed and temperature in a power plant gas turbine. ISA Trans. **51**(2), 304–308 (2012)
2. Mosaferin, R., Fakharian, A.: Fuzzy control for rotor speed of power plant gas turbine. In: 13th Iranian Conference on Fuzzy Systems, pp. 1–4. IEEE Press, Qazvin (2013)
3. Zhu, R., Wu, Y., Yin, F.: Study on control system for gas turbine based on piecewise-fuzzy adaptive theory (Chinese). Gas Turbine Technol. **26**(2), 48–51 (2013)
4. Wiesea, A.P., Bloma, M.J., Manziea, C., Breara, M.J., Kitchenerb, A.: Model reduction and MIMO model predictive control of gas turbine systems. J. Control Eng. Pract. **45**, 194–206 (2015)
5. Surendran, S., Chandrawanshi, R., Kulkarni, S., Bhartiya, S., Nataraj, P.S., Sampath, S.: Model predictive control of a laboratory gas turbine. In: 2016 Indian Control Conference, pp. 79–84. IEEE Press, Hyderabad (2016)
6. Hou, G., Yang, Y., Yin, F., Li, Q., Zhang, J.: Nonlinear predictive control of a boiler-turbine system based on T-S fuzzy model. In: 2016 IEEE 11th Conference on Industrial Electronics and Applications, pp. 2159–2163. IEEE Press, Hefei (2016)
7. Rowen, W.I.: Simplified mathematical representations of heavy-duty gas turbines. J. Eng. Power **105**, 865–869 (1983)
8. Rowen, W.I.: Simplified mathematical representations of single shaft gas turbines in mechanical drive service. In: ASME 1992 International Gas Turbine and Aeroengine Congress and Exposition. ASME Press (1992)
9. Niu, L., Liu, X.: Multivariable generalized predictive scheme for gas turbine control in combined cycle power plant. In: 2008 IEEE Conference on Cybernetics and Intelligent Systems, pp. 791–796. IEEE Press, Chengdu (2008)

Decentralized H$_\infty$ Load Frequency Control for Multi-area Power Systems with Communication Uncertainties

Yanliang Cui[1($\boxtimes$)], Guangtian Shi[1], Lanlan Xu[2], Xiaoan Zhang[1], and Xue Li[1]

[1] School of Mechanical Engineering, Lanzhou Jiaotong University,
Lanzhou 730070, Gansu, China
cyl1600@126.com
[2] School of Civil Engineering, Lanzhou Jiaotong University,
Lanzhou 730070, Gansu, China

Abstract. This paper investigates the distributed load frequency control (LFC) for multi-area power systems with communication switching topologies and data transmission time-delays. For stabilizing the power flow frequency while encompassing situations of seldom subsystem disconnections, a decentralized Markov switching control scheme is proposed. To further reduce conservative of the controller, a time-delay equipartition technique is developed. In addition, the distributed cooperative control (DCC) scheme is also discussed and proved to be unsuitable as a LFC strategy. Finally, illustrative examples are provided to validate effectiveness of the proposed methods.

Keywords: Load frequency control · Multi-area power systems · Switching topology · Time-delay

1 Introduction

Smart grid (SG) is conceived as modernized power system, wherein new and sustainable energy resource, transportation, distributed usage will be integrated by incorporating pervasive sensing, communications and control functionalities [1,2]. One of the most prominent characteristics of SG is to provide bidirectional power and information flows for facilitating effective and reliable power delivery, more significantly, alleviating the rapid growth of pressures upon load consumption and greenhouse gas emission [3].

As a large scale system, the SG can be conceptually separated as multiple sub-power systems that are dispersed in different geographical locations and interconnected via specific networks. However, the integration between subsystems with overall utility grid is not flexible yet because their inter complementary features are still not be fully revealed. One fundamental problem is known as the power flow frequency deviation, such a phenomena is induced when load sudden changes or disturbed by unexpected events and uncertainties, consequently, it will result in deterioration of the power exchange efficiency [5]. To overcome

© Springer Nature Singapore Pte Ltd. 2017
K. Li et al. (Eds.): LSMS/ICSEE 2017, Part III, CCIS 763, pp. 429–438, 2017.
DOI: 10.1007/978-981-10-6364-0_43

the above issue, a popular remedy method is called the load frequency control (LFC), its objective is to accommodate variant load demands by maintaining the power frequency to the given tolerance and keeping power flows within scheduled limits. For example, an improved internal model control scheme is advocated for single power system area in [6]. A robust LFC scheme for multiple wind power system is proposed in [7]. A decentralized sliding mode robust LFC for multi-interconnected power systems is presented in [8]. An H_∞ tracking LFC strategy for multi-area power system is researched in [9].

In the above-mentioned literatures, the employed communication networks are supposed to be trustworthy. However, some communication imperfections may degrade the LFC performance or even deteriorate system stability [10]. For enhancing system adaptiveness and robustness against such defections, an event-triggered LFC is proposed for reducing data transmission amount in [11]. Consider communication switching topologies, a distributed gain scheduling strategy of LFC is proposed in [12]. For reducing data transmission burden, an event-triggering LFC method is advocated in [13].

With the aforementioned motivations, this paper investigates distributed LFC for multi-area power systems with communication uncertainties. The contributions of this work are listed as follows. (i) A decentralized LFC design method is mainly proposed. (ii) To further reduce the conservative, a time-delay equipartition technique is proposed. (iii) The typical DCC is also studied and proved to be unsuitable as a LFC scheme.

The remainder of the paper is organized as follows. Problem is formulated in Sect. 2. Main result is presented in Sect. 3. Section 4 gives simulation examples. Section 5 outlines conclusion remark.

The following notations are given which will be used throughout in the literature. Let $\mathscr{R}$, $\mathscr{N}$ and $\mathscr{C}$ denote the real, integer and complex numbers, respectively. The operator notation '$\odot$' denotes Hadamard product. The notation $I \in \mathscr{R}^{4N \times 4N}$ denotes a identity matrix, I_i denotes a column block matrix with i^{th} element is I and the rests are zero matrices.

2   Problem Formulation

The overall power system is consisted by N heterogeneous subsystems. The structure of ith subsystem is illustrated as Fig. 1, where $i \in \{1, 2, \ldots, N\}$.

As shown in Fig. 1, the dynamic of ith subsystem can be represented as:

$$\dot{x}_i(t) = A_i x_i(t) - \sum_{j=1, j \neq i}^{N} A_{ij} x_j(t) + B_i u_i(t) + \tilde{B}_i \omega_i(t). \tag{1}$$

where $x_i(t) = (\Delta f_i(t), \Delta p_{mi}(t), \Delta p_{vi}(t), \Delta p_{tie}^{ij}(t))^T$, $u_i(t) = \Delta P_{ci}(t)$,

$$A_i = \begin{pmatrix} -\frac{D_i}{M_i} & \frac{1}{M_i} & 0 & -\frac{1}{M_i} \\ 0 & -\frac{1}{T_{chi}} & \frac{1}{T_{chi}} & 0 \\ -\frac{1}{R_i T_{gi}} & 0 & -\frac{1}{T_{gi}} & 0 \\ \sum_{j=1, j \neq i}^{N} T_{ij} & 0 & 0 & 0 \end{pmatrix}, \quad A_{ij} = \begin{pmatrix} 0 & 0 & 0 & 0 \\ 0 & 0 & 0 & 0 \\ 0 & 0 & 0 & 0 \\ T_{ij} & 0 & 0 & 0 \end{pmatrix}, \quad B_i = \begin{pmatrix} 0 \\ 0 \\ \frac{1}{T_{gi}} \\ 0 \end{pmatrix},$$

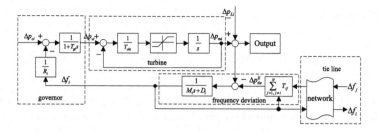

Fig. 1. Structure of ith sub power system

$\tilde{B}_i = diag(-\frac{1}{M_i}, 0, 0, 0)$, $\omega_i(t) = (\Delta P_{Li}^T(t), \Delta P_{Li}^T(t), \Delta P_{Li}^T(t), \Delta P_{Li}^T(t))^T$, $\Delta p_{vi}(t)$, $\Delta f_i(t)$, $\Delta p_{ci}(t)$, $\Delta p_{Li}(t)$ and $\Delta p_{tie}^{ij}(t)$ denote the governor value position deviation, frequency deviation, reference load set-point, grid load deviation and the tie line power flow between i and j, respectively; R_i, T_{gi}, Δp_{mi}, T_{chi}, D_i, M_i and T_{ij} are the speed droop coefficient, time constant of governor, generator mechanical power deviation, constant of turbine, damping coefficient, moment of inertia of generator and the synchronizing power coefficient.

The power load $\Delta P_{Li}(t)$ is modelled as an energy-bounded disturbance, namely, $\Delta P_{Li}(t) \in L_2[0, +\infty)$, where $L_2[0, +\infty)$ is the energy-bounded functions space, each function $f(t)$ in $L_2[0, +\infty)$ satisfies $\int_0^{+\infty} f^*(t)f(t)dt < \infty$.

2.1   Distributed Feedback Control

The trend of frequency in subsystem indicates not only the mismatch between the actual values with the pre-specified frequency standard but also the deviations with other correlated subsystems. Therefore, a decentralized control law is preferred as:

$$u_i(t) = -K_i x_i(t). \tag{2}$$

Remark 1. A typical distributed cooperative control (DCC) can be expressed as $u_i(t) = -K_{i1} x_i(t) - K_{i2} \sum_{j=1, j\neq i}^{N} x_j(t)$. It can be discover that $B_i K_{i2}$ always has at least triple zero eigenvalues. It implies that the close-looped system has multiple zero eigenvalues as well. In this case, it is more appropriate for DCC law is excluded in LFC practice.

2.2   Data Transmission Delay

The communication network of power system is mostly builded by so-called medium speed networks, the unavoidable data transmission time-varying delay fluctuates in a time interval of $3 \sim 100\,\mathrm{ms}$ [4]. By (1) and (2), the close-looped system is obtained as:

$$\dot{x}_i(t) = (A_i - B_i K_i)x_i(t) - \sum_{j=1, j\neq i}^{N} A_{ij}x_j(t - \tau(t)) + \tilde{B}_i\omega_i(t). \tag{3}$$

It is assumed that $0 \leq \tau(t) = \max(\tau_{ij}(t), j \in N_i) < \tau, \dot{\tau}(t) \leq \mu, \mu > 0$, where τ and μ denote the upper boundary and increment of $\tau(t)$, respectively.

2.3   Switching Topology

For the interconnected subsystems, their state information are conveniently captured via communication networks by a non-weighted direction graph $\mathscr{G}$. Due to communication imperfections, its Laplacian matrix $\mathscr{L}$ will be unavoidably changed to a time-varying value as $\mathscr{L}(t)$ [12,14,15]. Therefore, $\mathscr{L}(t)$ is modeled as $\mathscr{L}_{\sigma(t)}$ with a witching sequence of $\sigma(t)$, $\sigma(t) \in S = \{1, 2, \ldots, s\}$. To stabilize the electric flow frequency, two kind of control strategies are proposed as follows.

(i) Global graph method (GGM). Regardless of whether subsystems are switched out or not, the global graph of $\mathscr{L}_{\sigma(t)}$ is employed in the control scheme and $\sigma(t)$ is described as one unified probability sequence.

(ii) Maximal connected subgraph method (MCSM). When subsystem i is switched off from the power system, one can intuitively deduce from A_i that such a system is naturally stable. Therefore, only the maximal connected subgraph of $\mathscr{L}_{\sigma(t)}$ should be concerned.

As an effective method, the topology switching dynamics can be modeled as a Markov chain [16,17]. Let $r(t) \in S$, $S = (1, 2, \ldots)$ denotes the Markov chain, where S denotes the finite set including s possible switching topologies. Let $\Pi = (\pi_{ij}), i, j \in S$ denotes the state transform probability matrix, and it satisfies $P\{r(t+\xi = j)|r(t) = i\} = \begin{cases} \pi_{ij}\xi + o(\xi), & i \neq j, \\ 1 + \pi_{ii}\xi + o(\xi), & i = j, \end{cases}$, where $P\{\bullet\}$ is conditional probability, ξ is the ith subsystem activation time and $o(\xi)$ is the corresponding higher order infinitesimal to ξ, it satisfies $\lim_{\xi \to 0} \dfrac{o(\xi)}{\xi} = 0$. The transition probabilities in Π are partly unknown.

Accordingly, (3) is changed as:

$$\dot{x}(t) = \bar{A}_{(1,r(t))}x(t) + \bar{A}_{(2,r(t))}x(t - \tau(t)) + \tilde{B}\omega(t), \tag{4}$$

where $\bar{A}_{(1,r(t))} = diag(\tilde{A}_{(i,r(t))} - B_i K_{(i,r(t))})$, $\bar{A}_{(2,r(t))} = \mathscr{A}_{r(t)} \odot A$,

$$\tilde{A}_{(i,r(t))} = \begin{pmatrix} -\frac{D_i}{M_i} & \frac{1}{M_i} & 0 & -\frac{1}{M_i} \\ 0 & -\frac{1}{T_{chi}} & \frac{1}{T_{chi}} & 0 \\ -\frac{1}{R_i T_{gi}} & 0 & -\frac{1}{T_{gi}} & 0 \\ \sum_{j=1, j\neq i}^{N} l_{(ij,r(t))}T_{ij} & 0 & 0 & 0 \end{pmatrix},$$

$$\mathscr{A}_{r(t)} = \begin{pmatrix} 0 & l_{(12,r(t))} & \cdots & l_{(1N,r(t))} \\ l_{(21,r(t))} & 0 & \cdots & l_{(2N,r(t))} \\ \cdots & \cdots & \cdots & \\ l_{(N1,r(t))} & l_{(N2,r(t))} & \cdots & 0 \end{pmatrix}, \, i = 1, \ldots, N.$$

As a traditional method, H_∞ control is ordinarily employed to reduce the effects of exogenous disturbances [18,19]. Commonly, the H_∞ performance can be written as an equivalent linear quadratic index (LQI) as:

$$\int_0^\infty (x^T(t)x(t) - \gamma^2 \omega^T(t)\omega(t))dt \le 0. \tag{5}$$

3 Main Result

For clarity, the controller design method is presented at first.

Theorem 1. Given positive scalars $\tau > 0$, $\mu > 0$ and $0 < \gamma < 1$, integer number n, if there exist positive definite matrices $\bar{P}_\sigma = \bar{P}_\sigma^T > 0$, $\bar{P}_\sigma = diag(\tilde{P}_\sigma, \ldots, \tilde{P}_\sigma)$, $\sigma \in S$, semi-positive definite matrices $\bar{Q}_{ij} \ge 0$, $\bar{R}_j \ge 0$ and any appropriate dimension matrices $\bar{M}_1$, $\bar{M}_{kj}$; $i = 1, 2$, $j = 1, \ldots, N$, $k = 2, 3$; such that:

$$\bar{\Gamma} = \begin{pmatrix} \bar{\Phi} + \bar{\Omega} & \bar{B} & \alpha\tau \bar{f}_\sigma^T & \alpha\tau\bar{M} & \cdots & \alpha\tau\bar{M} \\ * & -\gamma^2 I & \bar{P}_\sigma \bar{B}^T & 0 & \cdots & 0 \\ * & * & \alpha\tau(\sum_{j=1}^{n} \bar{R}_i - 2\bar{P}_\sigma) & 0 & \cdots & 0 \\ * & * & * & -\alpha\tau\bar{R}_1 & \cdots & 0 \\ * & * & * & * & \cdots & 0 \\ * & * & * & * & * & -\alpha\tau\bar{R}_n \end{pmatrix} < 0 \tag{6}$$

hold, then (4) achieves asymptotical stability with γ performance under controller (2); the controller gains are given as $K_{(k,\sigma)} = \bar{K}_{(k,\sigma)}\tilde{P}_\sigma^{-1}$ and $k = 1, 2, \ldots, N$.

where:

$$\alpha = \frac{1}{n}, \bar{f}_\sigma^T = \Gamma_{(g,\sigma)}^T I_1 - \Gamma_{(p,\sigma)}^T I_{2n+1}, \bar{B}^T = ((\tilde{B}\bar{P}_\sigma)^T, \underbrace{0, \ldots, 0}_{2n})_{4(2n+1)N \times 4N}$$

$$\bar{\Phi} = \begin{pmatrix} \Phi_1 & 0 & 0 & 0 & 0 & 0 & 0 & 0 & \Phi_2 \\ * & \Phi_3 & 0 & 0 & 0 & 0 & 0 & 0 & 0 \\ * & * & \cdots & 0 & 0 & 0 & 0 & 0 & 0 \\ * & * & * & \Phi_4 & 0 & 0 & 0 & 0 & 0 \\ * & * & * & * & \Phi_5 & 0 & 0 & 0 & 0 \\ * & * & * & * & * & \Phi_6 & 0 & 0 & 0 \\ * & * & * & * & * & * & \cdots & 0 & 0 \\ * & * & * & * & * & * & * & \Phi_7 & 0 \\ * & * & * & * & * & * & * & * & \Phi_8 \end{pmatrix}_{4(2n+1)N \times 4(2n+1)N},$$

$$\Phi_1 = \Gamma_{(g,\sigma)} + \Gamma_{(g,\sigma)}^T + \sum_{j=1}^{s} \tilde{\pi}_{\sigma j} \bar{P}_j + \bar{Q}_{11} + \bar{Q}_{21} + I, \Phi_2 = \Gamma_{(p,\sigma)}, \Phi_3 = \bar{Q}_{12} - \bar{Q}_{11},$$

$$\Phi_4 = \bar{Q}_{1,n} - \bar{Q}_{1,n-1}, \Phi_5 = -\bar{Q}_{1n}, \Phi_6 = -(1 - \alpha\mu)\bar{Q}_{22},$$

$$\Phi_7 = -(1 - (n-1)\alpha\mu)(\bar{Q}_{2,n-1} - \bar{Q}_{2n}), \Phi_8 = -(1 - \mu)\bar{Q}_{2n}, \bar{\Omega} = \bar{M}I_x + I_x^T \bar{M},$$

$$\Gamma_{(g,\sigma)} = diag(\tilde{A}_{(1,\sigma)}, \ldots, \tilde{A}_{(N,\sigma)})\bar{P}_\sigma - diag(B_1\bar{K}_{(1,\sigma)}, \ldots, B_N\bar{K}_{(N,\sigma)}), \Gamma_{(p,\sigma)} = \mathscr{A}_\sigma\bar{P}_\sigma,$$

$$\tilde{A}_{(i,\sigma)} = \begin{pmatrix} -\frac{D_i}{M_i} & \frac{1}{M_i} & 0 & -\frac{1}{M_i} \\ 0 & -\frac{1}{T_{chi}} & \frac{1}{T_{chi}} & 0 \\ -\frac{1}{R_i T_{gi}} & 0 & -\frac{1}{T_{gi}} & 0 \\ \sum_{j=1,j\neq i}^{N} l_{(ij,\sigma)} T_{ij} & 0 & 0 & 0 \end{pmatrix}, \mathscr{A}_\sigma = \begin{pmatrix} 0 & l_{(12,\sigma)} & \cdots & l_{(1N,\sigma)} \\ l_{(21,\sigma)} & 0 & \cdots & l_{(2N,\sigma)} \\ \cdots & \cdots & \cdots & \cdots \\ l_{(N1,\sigma)} & l_{(N2,\sigma)} & \cdots & 0 \end{pmatrix},$$

$$\bar{M} = (\bar{M}_1^T, \underbrace{\bar{M}_{21}^T, \ldots, \bar{M}_{2n}^T}_{n}, \underbrace{\bar{M}_{31}^T, \ldots, \bar{M}_{3n}^T}_{n})^T, I_x = (I, \underbrace{I, -I, \ldots, I, -I}_{n}, \underbrace{I, -I, \ldots, I, -I}_{n}).$$

Remark 2. Note that $\mathscr{L}_\sigma$ is supposed to be a directed graph, it is convenient for Theorems 1 to be extended to undirected circumstances. Moreover, $\mathscr{L}_\sigma$ probably has no spinning tree.

Remark 3. Given large number n, the conservative of proposed controller should be reduced. However, the negative side is that the computation complexity of (6) will be dramatically increased. Therefore, Theorem 1 provides a trade-off between controller conservative and computational burden.

Proof. Dividing τ and $\tau(t)$ into n equal fragmented segments as $\tau = \bigcup_{i=1}^{i=n} [\frac{i-1}{n}\tau, \frac{i}{n}\tau)$ and $\tau(t) = \bigcup_{i=1}^{i=n} [\frac{i-1}{n}\tau(t), \frac{i}{n}\tau(t))$, respectively. Denote $r(t) = \sigma$ and $\alpha = \frac{1}{n}$, a fragmented Lyapunov functional is designed for (4) as:

$$V(t) = x^T(t) P_\sigma x(t) + \sum_{i=1}^{n} \int_{-i\alpha\tau}^{-(i-1)\alpha\tau} \int_{t+\theta}^{t} \dot{x}^T(s) R_i \dot{x}(s) ds$$

$$+ \sum_{i=1}^{n} \int_{t-i\alpha\tau}^{t-(i-1)\alpha\tau} x^T(s) Q_{1i} x(s) ds + \sum_{i=1}^{n} \int_{t-i\alpha\tau(t)}^{t-(i-1)\alpha\tau(t)} x^T(s) Q_{2i} x(s) ds,$$

where $P_\sigma = P_\sigma^T > 0$, $P_\sigma \in \mathscr{R}^{N\times N}$, $Q_i = Q_i^T \geq 0$, $R_i = R_i^T \geq 0$, $Q_i, R_i \in \mathscr{R}^{N\times N}$ and $i = 1, 2, \ldots, n$.

Calculating differential of $V(t)$ along the trajectory of (4), define $\eta(t) = (x^T(t), \underbrace{x^T(t-\alpha\tau), \ldots, x^T(t-\tau)}_{n}, \underbrace{x^T(t-\alpha\tau(t)), \ldots, x^T(t-\tau(t))}_{n})^T$, add the follows into $\dot{V}(t)$): $2\eta^T(t) M \left(\sum_{i=1}^{n} (x(t-i\alpha\tau) - x(t-(i-1)\alpha\tau) - \int_{t-i\alpha\tau}^{t-(i-1)\alpha\tau} \dot{x}(s) ds \right)$, where $M = (M_1^T, \underbrace{M_{21}^T, \ldots, M_{2n}^T}_{n}, \underbrace{M_{31}^T, \ldots, M_{3n}^T}_{n})^T$, $M_1, M_{2i}, M_{3i} \in \mathscr{R}^{N\times N}$ denote any appropriate dimension matrices, $i = 1, \ldots, n$. Then one obtains:

$$\dot{V}(t) \leq \eta^T(t)\{I_1^T (P_\sigma \bar{A}_{(1,\sigma)} + \bar{A}_{(1,\sigma)}^T P_\sigma + \sum_{j=1}^{s} (\pi_{\sigma j} P_j) + Q_{11} + Q_{21}) I_1 - I_2^T Q_{11} I_2$$

$$+ I_2^T Q_{12} I_2 - I_3^T Q_{12} I_3 + \ldots - I_{n+1}^T Q_{1n} I_{n+1} - (1-\alpha\mu) I_{n+2}^T Q_{21} I_{n+2} + \ldots$$

$$+ (1-(n-1)\alpha\mu) I_{2n}^T (Q_{2n} - Q_{2,n-1}) I_{2n} - (1-\mu) I_{2n+1}^T Q_{2n} I_{2n+1}$$

$$+ \sum_{j=1}^{n} \alpha\tau M R_i^{-1} M^T + M I_x + I_x^T M^T\}\eta(t) + \sum_{j=1}^{n} \alpha\tau f^T R_i f$$

$$- \sum_{i=1}^{n} \int_{t-i\alpha\tau}^{t-(i-1)\alpha\tau} (\eta^T(t) M + \dot{x}^T(s) R_i) R_i^{-1} (M^T \eta(t) + R_i \dot{x}(s)),$$

where:

$$f = \bar{A}_{(1,\sigma)}x(t) + \bar{A}_{(2,\sigma)}x(t - \tau(t)) + \tilde{B}\omega(t),$$
$$I_x = (\underbrace{I, I, -I, \ldots, I, -I}_{n}, \underbrace{I, -I, \ldots, I, -I}_{n}).$$

Define sets $U^\sigma = U_k^\sigma \bigcup U_{uk}^\sigma$, $\sigma \in S$, $U_k^\sigma = \{\pi_{\sigma j}|j \in S\}$, $U_{uk}^\sigma = \{\tilde{\pi}_{\sigma j}|j \in S\}$ and $\pi_k^\sigma = \sum_{b \in U_k^\sigma} \pi_{\sigma b}$, then one has $\sum_{j=1}^{s} \pi_{\sigma j}P_j \leq \sum_{j=1}^{s} \tilde{\pi}_{\sigma j}P_j$, where $\tilde{\pi}_{\sigma i} = \pi_{\sigma i}, \forall \pi_{\sigma i} \subset U_k^\sigma$, otherwise, $\tilde{\pi}_{\sigma i} = 1 - \pi_k^\sigma$. Hence, one obtains $I_1^T \sum_{j=1}^{s} \pi_{\sigma j}P_j I_1 \leq I_1^T \sum_{j=1}^{s} \tilde{\pi}_{\sigma j}P_j I_1$.

Regulate $P_\sigma = diag(\bar{P}_\sigma, \bar{P}_\sigma, \bar{P}_\sigma, \bar{P}_\sigma)$, where $\bar{P}_\sigma = \bar{P}_\sigma^T > 0$, $\bar{P}_\sigma \in \mathcal{R}^{n \times n}$ and $i = 1, 2, 3, 4$; perform congruence transformations by $J = \begin{pmatrix} \underbrace{P_\sigma^{-1}, \ldots, P_\sigma^{-1}}_{3n+2} \end{pmatrix}$, then denote $\bar{Q}_{ij} = P_\sigma^{-1}Q_{ij}P_\sigma^{-1}$, $\bar{M}_{ij} = P_\sigma^{-1}M_{ij}P_\sigma^{-1}$, $\bar{R}_i = P_\sigma^{-1}R_iP_\sigma^{-1}$, $\bar{P}_i = P_\sigma^{-1}P_iP_\sigma^{-1}$, $i = 1, 2$, $j = 1, 2, 3, 4$; $\bar{M}_1 = P_\sigma^{-1}M_1P_\sigma^{-1}$; $\bar{K}_{(k,\sigma)} = K_{(k,\sigma)}\bar{P}_\sigma^{-1}$, $k = 1, 2, \ldots, N$, $\Gamma_{(g,\sigma)} \doteq \bar{A}_{(1,\sigma)}\bar{P}_\sigma^{-1} =$, $\Gamma_{(p,\sigma)} \doteq \bar{A}_{(2,\sigma)}\bar{P}_\sigma^{-1}$ and $\tilde{P}_\sigma \doteq \bar{P}_\sigma^{-1}$, the original NLMI can be feasibly decoupled to an inequivalent LMI.

By LQI (5) gives $\int_0^\infty \left(\eta^T(t)\Gamma\eta(t) + x^T(t)x(t) - \gamma\omega^T(t)\omega(t)\right) dt < 0$. Define an augment state as $\bar{\eta}(t) = (\eta(t), \omega(t))^T$, the above inequality can be written as $\int_0^\infty \bar{\eta}^T(t)\bar{\Gamma}\bar{\eta}(t)dt < 0$, where $\bar{\Gamma}$ is given in (6) thus completes the proo.

4   Numerical Example

The effectiveness of proposed GGM and the MCSM are evaluated at first. Afterwards, some experiment proofs are also given to show that the DCC is an unappropriate LFC scheme.

The overall power system is consisted by four interconnected subsystems as shown in Fig. 2, wherein $\sigma \in S$ denote the current switching case. It is assumed that the transmission probability of switching signal satisfies $\Pi = \begin{pmatrix} 0.3 & ? & ? & ? \\ ? & 0.3 & ? & ? \\ ? & ? & ? & 0.4 \\ ? & 0.1 & ? & ? \end{pmatrix}$. The mark ? denotes the corresponding information is unknown.

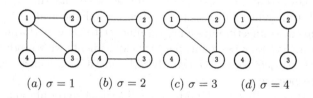

(a) $\sigma = 1$ (b) $\sigma = 2$ (c) $\sigma = 3$ (d) $\sigma = 4$

Fig. 2. Switching cases of $\mathscr{L}$ with switching signal of σ

The state trajectories of the close-looped system under GGM and MCSM are shown as Figs. 3 and 4, respectively.

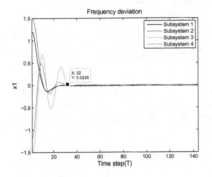
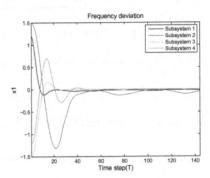

Fig. 3. Frequency deviation of GGM   **Fig. 4.** Frequency deviation of MCSM

Figures 3 and 4 show that the state trajectories are exponentially attenuated and converge to the equilibrium point gradually. Compared with the former GGM performance, MCSM takes more longer period to achieve the objective. Although the desired stability is achieved, the dynamic behavior of MCSM is relatively worse than GGM.

4.1   DCC Performance

One can easily discover that Theorem 1 can be feasibly extended to DCC scheme of

$$u(t) = -K_{i1}x_i(t) - K_{i2} \sum_{j=1,j\neq i}^{N} x_j(t - \tau(t))$$ by substitute K_i with K_{i1} and be app-

end an extra item $\begin{pmatrix} 0 & l_{(12,\sigma)}B_1\bar{K}_{(12,\sigma)} & \dots & l_{(1N,\sigma)}B_1\bar{K}_{(1N,\sigma)} \\ l_{(21,\sigma)}B_2\bar{K}_{(21,\sigma)} & 0 & \dots & l_{(2N,\sigma)}B_2\bar{K}_{(2N,\sigma)} \\ \dots & \dots & \dots & \dots \\ l_{(N1,\sigma)}B_N\bar{K}_{(N1,\sigma)} & l_{(N2,\sigma)}B_N\bar{K}_{(N2,\sigma)} & \dots & 0 \end{pmatrix}$

into $\Gamma_{(p,\sigma)}$. Then, the controllers can be connivently solved by $K_{(i1,\sigma)} = \bar{K}_{(i1,\sigma)}$ $\tilde{P}_\sigma^{-1}$ and $K_{(i2,\sigma)} = \bar{K}_{(i2,\sigma)}\tilde{P}_\sigma^{-1}$, $i = 1, 2, \ldots, N$, respectively.

By the DCC law, the state trajectories of the close-looped system are shown as Fig. 5.

Figure 5 shows that the wave crests emerge twice and the curves fluctuate violently each time when subsystem 4 is switched off the grid. It indicates that the subsystems disconnection has a great influence to DCC performance. To further evaluate its dynamic behavior, perform experiments with random switching signal, then the trajectories of GGM, MCSM and DCC are shown as Fig. 6 (a), (b) and (c), respectively.

Figure 6 (a) show that the GGM curves fluctuate smoothly and converge to the equilibrium point quickly. Figure 6 (b) illustrate that the desired stability can be achieved by MCSM, however, the dynamic behavior is worse than GGM. As

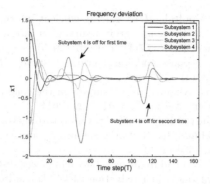

Fig. 5. State trajectories of DCC

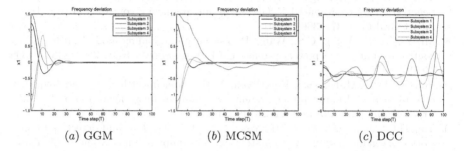

(a) GGM   (b) MCSM   (c) DCC

Fig. 6. Frequency deviations under random switching signal

shown in Fig. 6 (c), the DCC fails apparently. It indicates that the DCC method can not guarantee stability under a fast switching signal. It indicates that the DCC should not be engaged in LFC due to risk of system stability.

5 Conclusion

This article investigates the notion of LFC for multi-area power systems with communication drawbacks. A time-delay equipartition technique is designed and a feasible decentralized Markov switching control scheme is proposed. The advocated control law not only guarantees the frequency stability for each local subsystem but also synchronize with overall power system. Furthermore, the DCC method is also discussed and it is concluded that DCC endanger system stability counterproductively.

Acknowledgments. This work was supported by the Open Project of State Key Laboratory of Traction Power, Southwest Jiaotong University (TPL1604).

References

1. Fang, X., Misra, S., Xue, G.L., Yang, D.J.: Smart grid? The new and improved power grid: a survey. IEEE Commun. Surv. Tutor. **14**(4), 944–980 (2012)

2. Erol-Kantarci, M., Mouftah, H.T.: Energy-efficient information and communication infrastructures in the smart grid: a survey on interactions and open issues. IEEE Commun. Surv. Tutor. **17**(1), 179–197 (2015)
3. Siano, P.: Demand response and smart grids - a survey. Renew. Sustain. Energy Rev. **30**, 461–478 (2014)
4. Wang, W.Y., Xu, Y., Khanna, M.: A survey on the communication architectures in smart grid. Comput. Netw. **55**, 3604–3629 (2011)
5. Pandey, S.K., Mohanty, S.R., Kishor, N.: A literature survey on load frequency control for conventional and distribution generation power systems. Renew. Sustain. Energy Rev. **25**, 318–334 (2013)
6. Saxena, S., Hote, Y.V.: Load frequency control in power systems via internal model control scheme and model-order reduction. IEEE Trans. Power Syst. **28**(3), 2749–2757 (2013)
7. Vachirasricirikul, S., Ngamroo, I.: Robust LFC in a smart grid with wind power penetration by coordinated V2G control and frequency controller. IEEE Trans. Smart Grid **5**(1), 371–380 (2014)
8. Mi, Y., Fu, Y., Wang, C.S., Wang, P.: Decentralized sliding mode load frequency control for multi-area power systems. IEEE Trans. Power Syst. **28**(4), 4301–4309 (2013)
9. Yousef, H.A., AL-Kharusi, K., Albadi, M.H., Hosseinzadeh, N.: Load frequency control of a multi-area power system: an adaptive fuzzy logic approach. IEEE Trans. Power Syst. **29**(4), 1822–1830 (2014)
10. Yan, H.C., Qian, F.F., Zhang, H., Yang, F.W., Guo, G.: H_∞ fault detection for networked mechanical spring-mass systems with incomplete information. IEEE Trans. Ind. Electron. **63**(9), 5622–5631 (2014)
11. Peng, C., Li, L.C., Fei, M.R.: Resilient event-triggered HN load frequency control for networked power systems with energy limited DoS attacks, IEEE Trans. Power Syst. (2017, in press)
12. Liu, S.C., Liu, X.P.P., Saddik, A.E.: Modeling and distributed gain scheduling strategy for load frequency control in smart grids with communication topology changes. ISA Trans. **53**, 454–461 (2014)
13. Wen, S., Yu, X.H., Zeng, Z.G., Wang, J.J.: Event-triggering load frequency control for multi-area power systems with communication delays. IEEE Trans. Ind. Electron. **63**(2), 1308–1317 (2016)
14. Shamsi, P., Fahimi, B.: Stability assessment of a DC distribution network in a hybrid micro-grid application. IEEE Trans. Smart Grid **5**(5), 2527–2534 (2014)
15. Alobeidli, K.A., Syed, M.H., Moursi, M.S.E., Zeineldi, H.H.: Novel coordinated voltage control for hybrid micro-grid with sslanding capability. IEEE Trans. Smart Grid **6**(3), 1116–1127 (2015)
16. Patrinos, P., Sopasakis, P., Sarimveis, H., Bemporad, A.: Stochastic model predictive control for constrained discrete-time Markovian switching systems. Automatica **50**(10), 2504–2514 (2014)
17. Li, W.Q., Wu, Z.J.: Output tracking of stochastic high-order nonlinear systems with Markovian switching. IEEE Trans. Autom. Control **58**(6), 1585–1590 (2013)
18. Hu, S.L., Yue, D., Xie, X.P., Du, Z.P.: Event-triggered H_∞ stabilization for networked stochastic systems with multiplicative noise and network-induced delays. Inf. Sci. **299**, 178–197 (2015)
19. Tian, E.G., Wong, W.K., Yue, D.: Robust H_∞ control for switched systems with input delays: a sojourn-probability-dependent method. Inf. Sci. **283**, 22–35 (2014)

Cyber Security Against Denial of Service of Attacks on Load Frequency Control of Multi-area Power Systems

Yubin Shen[1], Minrui Fei[1(✉)], Dajun Du[1], Wenjun Zhang[2], Srdjan Stanković[3], and Aleksandar Rakić[3]

[1] Shanghai Key Laboratory of Power Station Automation Technology, Department of Automation, School of Mechatronical Engineering and Automation, Shanghai University, Shanghai 200072, China
shenyb616@163.com, mrfei@staff.shu.edu.cn
[2] School of Communication and Information Engineering, Shanghai University, Shanghai, China
wjzhang@mail.shu.edu.cn
[3] School of Electrical Engineering, University of Belgrade, Belgrade, Serbia
stankovic@etf.rs

Abstract. While open communication infrastructures are embedded into multi-area power systems to support data transmittion, it make communication channels vulnerable to cyber attacks, reliability of power systems is affected. This paper studies the load frequency control (LFC) of multi-area power systems under DoS attacks. The state space model of power systems under DoS attacks is formulated, where event-triggered control scheme is integrated for the multi-area power systems under DoS attacks. By utilizing average dwell time design approach, exponential stability and L_2-gain of the multi-area power systems can be obtained for event-triggered LFC scheme under DoS attacks, if choosing an unavailability rate of communication channels for DoS attacks properly. Finally, the example shows that the convergences of frequency deviation of three-area power systems are compared under different DoS attack scenarios, when the proportion of the total time of DoS attacks can obtain the result properly.

Keywords: Denial-of-Service (DoS) attacks · Multi-area power systems · Load frequency control (LFC) · Event-triggered mechanism

1 Introduction

An open communication infrastructure is used to support the increasingly decentralized property of Load Frequency Control (LFC) services in the modern deregulated power systems. Security of the construction multi-area power systems has attracted considerable attention. LFC is an important control problem for the dynamical operation of power systems in the last few decades [1]. The purpose

© Springer Nature Singapore Pte Ltd. 2017
K. Li et al. (Eds.): LSMS/ICSEE 2017, Part III, CCIS 763, pp. 439–449, 2017.
DOI: 10.1007/978-981-10-6364-0_44

of LFC scheme is to return the frequency to its nominal value and minimize unscheduled tie-line power flows between neighboring control areas, when there is any change in load [2]. But, if power systems with LFC scheme suffer to attack from the malicious adversary, the economy and people's life will suffer major negative impacts. Security of cyber and communication networks is fundamental to the reliable operation of the power systems [3]. The purpose of malicious attacks is to destabilize the power systems, the dynamic performance and stability of multi-area power systems with LFC scheme are degraded for cyber-attacks [4,5]. Denial-of-Service (DoS) attack, data injection and deception attack are three major kinds of attack behavior for the security issues [6]. DoS attacks may be one of the most detrimental that affects the packet delivery [7]. Malicious DoS attacker can send a great quantities of inauthentic packets to cause network congestion, the communication channel is briefly interrupted. So, stability of multi-area power systems with LFC under DoS attacks motivate our research.

2   Dynamic Model of the Multi-area Power System with LFC Scheme

2.1   System Model

Model of Area-i Power System with LFC. A multi-area power system is consisted of a number of interconnected control areas which are connected by tie-lines. The area-i of power systems may be consisted of the different types of generating units like nonreheat turbine and renewable generation. Due to the use of open communication channels, time delays will arise, when telemetered signals are sent from remote terminal units (RTUs) to control center, and control signals are transmitted from the control center to individual units.

The model of area-i power system is a fourth order generating units model and is given by [8].

$$
\begin{bmatrix} \Delta\dot{\omega}_i \\ \Delta\dot{P}_{mech_i} \\ \Delta\dot{P}_{v_i} \\ \Delta\dot{P}_{tie,i} \end{bmatrix} = \begin{bmatrix} -\frac{D_i}{M_i^a} & \frac{1}{M_i^a} & 0 & -\frac{1}{M_i^a} \\ 0 & -\frac{1}{T_{CH_i}} & \frac{1}{T_{CH_i}} & 0 \\ -\frac{1}{R_i^f T_{G_i}} & 0 & 0 & -\frac{1}{T_{G_i}} \\ \sum_{i\neq j,j=1}^{N} 2\pi T_{ij} & 0 & 0 & 0 \end{bmatrix} \begin{bmatrix} \Delta\omega_i \\ \Delta P_{mech_i} \\ \Delta P_{v_i} \\ \Delta P_{tie,i} \end{bmatrix}
$$
$$
+ \begin{bmatrix} 0 \\ 0 \\ 0 \\ -\sum_{i\neq j,j=1}^{N} 2\pi T_{ij} \end{bmatrix} \Delta\omega_j + \begin{bmatrix} 0 \\ 0 \\ \frac{1}{T_{G_i}} \\ 0 \end{bmatrix} \Delta P_{ref_i} + \begin{bmatrix} -\frac{1}{M_i^a} \\ 0 \\ 0 \\ 0 \end{bmatrix} \Delta P_{L_i}.
$$

(1)

With the state vector $x_i(t) = \begin{bmatrix} \Delta\omega_i \ \Delta P_{mech_i} \ \Delta P_{v_i} \ \Delta P_{tie,i} \end{bmatrix}^T$, control input $u_i(t) = \Delta P_{ref_i}$, disturbance input $w_i(t) = \Delta P_{L_i}$, and output $y_i(t) = ACE_i$. $\Delta\omega_i, \Delta P_{mech_i}, \Delta P_{v_i}$ and $\Delta P_{tie,i}$ represent the deviation in the frequency, the generator mechanical output, the valve position and total tie-line power flow in area-i, respectively.

The state-space model of the areas-i power system with LFC scheme can be represented as follows:

$$\dot{x}_i(t) = A_{ii}x_i(t) + \sum_{i\neq j,j=1}^{N} A_{ij}x_j(t) + B_iu_i(t) + F_iw_i(t),$$
$$y_i(t) = C_ix_i(t)$$
(2)

where system matrices are defined as

$$A_{ii} = \begin{bmatrix} -\frac{D_i}{M_i^a} & \frac{1}{M_i^a} & 0 & -\frac{1}{M_i^a} \\ 0 & -\frac{1}{T_{CH_i}} & \frac{1}{T_{CH_i}} & 0 \\ -\frac{1}{R_i^f T_{G_i}} & 0 & 0 & -\frac{1}{T_{G_i}} \\ \sum_{i\neq j,j=1}^{N} 2\pi T_{ij} & 0 & 0 & 0 \end{bmatrix}, A_{ij} = \begin{bmatrix} 0 & 0 & 0 & 0 \\ 0 & 0 & 0 & 0 \\ 0 & 0 & 0 & 0 \\ -2\pi T_{ij} & 0 & 0 & 0 \end{bmatrix}, B_i = \begin{bmatrix} 0 \\ 0 \\ \frac{1}{T_{G_i}} \\ 0 \end{bmatrix}^T,$$

$$F_i = \begin{bmatrix} -\frac{1}{M_i^a} & 0 & 0 & 0 \end{bmatrix}^T, C_i = \begin{bmatrix} \beta_i & 0 & 0 & 1 \end{bmatrix}.$$

Dynamic of the Multi-area Power System with LFC. Based on the state space model of areas-i power system with LFC, the dynamic model of the multi-area power systems is described as the following form:

$$\dot{x}(t) = Ax(t) + Bu(t) + Fw(t),$$
$$y(t) = Cx(t)$$
(3)

where $x(t) = \begin{bmatrix} x_1(t) & x_2(t) & \cdots & x_n(t) \end{bmatrix}^T, y(t) = \begin{bmatrix} y_1(t) & y_2(t) & \cdots & y_n(t) \end{bmatrix}^T, u(t) = \begin{bmatrix} u_1(t) & u_2(t) & \cdots & u_n(t) \end{bmatrix}^T, A = \begin{bmatrix} A_{11} & A_{12} & \cdots & A_{1n} \\ A_{21} & A_{22} & \cdots & A_{2n} \\ \vdots & \vdots & \ddots & \vdots \\ A_{n1} & A_{n2} & \cdots & A_{nn} \end{bmatrix}, B = diag\{B_1, B_2, \cdots, B_n\},$

$F = diag\{F_1, F_2, \cdots, F_n\}, C = diag\{C_1, C_2, \cdots, C_n\}$. $x(t) \in \mathbb{R}^n$ is plant state vector, $u(t) \in \mathbb{R}^m$ is control input vector, $w(t)$ is disturbance vector and $y(t) \in \mathbb{R}^p$ is measurement output vector, respectively. For the equation as $\sum_{i=1}^{N} \Delta P_{tie,i} = 0$, the net tie-line power exchange must be satisfies among the multi-area power systems. For (3), the input control component of the multi-area power systems with LFC scheme as $u(t) = Ky(t) = KCx(t)$, where $K = diag\{K_1, K_2, \cdots, K_n\}$.

2.2   DoS Attacks Analysis

DoS as the attack phenomenon can affect measurement and control transmission channels respectively, the available data may lose in the present DoS attacks. The nth DoS time-interval is represented

$$D_n = \{\xi_n\} \cup (\xi_n, \xi_n + \varsigma_n),$$
(4)

ξ_n denote the time instant at nth DoS attack preventing transition from absence to presence, ς_n is the length of time that nth DoS attack interval. It is very obvious that there are the time instants of DoS attacks $t_0 < t_{\xi_1} < \cdots < t_{\xi_n} < \cdots < t$ during $[t_0, t]$.

For the whole time interval $[t_0, t)$, the index set $\aleph(t_0, t)$ is divided into two subsets $\aleph_S(t_0, t)$ and $\aleph_D(t_0, t)$, which represent the set of time intervals where communication channels are allowed and denied, respectively.

$$\aleph(t_0, t) = \aleph_S(t_0, t) \cup \aleph_D(t_0, t). \tag{5}$$

$$|\aleph_S(t_0, t)| + |\aleph_D(t_0, t)| = t - t_0. \tag{6}$$

Given $t_0, t \in \mathbb{N}_0$ with $t \geq t_0$, let

$$\aleph_D(t_0, t) = \bigcup_{n \in \mathbb{N}} D_n \cap [t_0, t] \quad or \quad \aleph_D(t_0, t) = \sum_{l=1}^{N} [t_{\xi_l}, t_{\xi_l} + \varsigma_l) \tag{7}$$

and

$$\aleph_S(t_0, t) = \aleph(t_0, t) \backslash \aleph_D(t_0, t) \quad or \quad \aleph_S(t_0, t) = \sum_{l=0}^{N} [t_{\xi_l} + \varsigma_l, t_{\xi_l}), \tag{8}$$

where $t_{\xi_0} = t_0$ and $\varsigma_0 = 0$.

There are many of DoS attacks before the current time t, let $n(t)$ denote the current time instant of present DoS attacks,

$$n(t) = \begin{cases} -1, & t < t_0 \\ \sup\{n \in N | \xi_n < t)\} & otherwise \end{cases}, \tag{9}$$

since

$$\aleph_D(t_0, t) = \left\{ \bigcup_{n=0}^{n(t)-1} D_n \right\} \cup [\xi_{n(t)}, \min\{\xi_{n(t)} + \zeta_{n(t)}, t\}), \tag{10}$$

represent the sum of DoS intervals for interval $[t_0, t)$.

It is convenient that the following definitions are necessary.

Definition 1. *There is a certain $\delta \geq 0$, it is defined as:*

$$\frac{|\aleph_D(t_0, t)|}{|\aleph_S(t_0, t)|} \leq \delta, \forall t \in \mathbb{N}_{\geq 0}. \tag{11}$$

Definition 2. *Let $n(t_0, t)$ denote the number of DoS from the absence to the presence transmissions occuring on the interval $[t_0, t)$. There exists constants $\vartheta \in \mathbb{N}_{\geq 0}$ and $\tau_D \in \mathbb{N}_{\geq T}$ such that*

$$n(t_0, t) \leq \vartheta + \frac{t - t_0}{\tau_D} \tag{12}$$

for all $t_0, t \in \mathbb{N}_{\geq 0}$ with $t \geq t_0$. $n(t_0, t)$ can be denote the sequences of DoS intervals. τ_D provides the average dwell time between any consecutive DoS intervals on the time interval $[t_0, t)$.

2.3   Event-Triggered Communication Mechanism

The successful transmitted of sampling instants are described by t_k, the transmission instants of sampling sequence are described as $M_{t_k} = \{t_k h, t_k = 0, 1, \ldots, \}$, satisfying $M_{t_k} \subset M_k$. Thus, The measurement output of the multi-area power system with LFC is presented as $y(t) = y(t_k h) = Cx(t_k h), t \in [t_k, t_{k+1})$.

Considering the effects of the total network-induced delay τ_{t_k}, the output of the ZOH can be modeled as $\bar{y}(t) = y(t_k h)$, the control center receives the data $\bar{y}(t)$ at the time $t_k h + \tau_{t_k}$.

$$u(t) = Ky(t) = K\bar{y}(t) = Ky(t_k h), t_k h + \tau_{t_k} \leq t < t_{k+1}h + \tau_{t_{k+1}}, k \in \mathbb{N}. \quad (13)$$

The next time instant for releasing data is determined from the following communication scheme:

$$t_{k+1}h = \min\{t_k h + ih | i \in N,$$
$$(y(t_k h + ih) - y(t_k h))^T \Phi(y(t_k h + ih) - y(t_k h)) > \sigma y(t_k h)^T \Phi y(t_k h)\}. \quad (14)$$

$$e_k(t) = y(t_k h + ih) - y(t_k h) = C[x(t_k h + ih) - x(t_k h)], \quad (15)$$

$e_k(t)$ is the error measurement data between the latest successful transmission time instant and the current real-time measurement time instant.

2.4   Modeling of the Power System Under DoS Attack

The time axis is divided into many of intervals in the data transmission process, but other measurement output data can not be transmitted in other intervals, due to the occurrence of DoS attacks. The model for multi-area power systems with LFC scheme under DoS attacks has been modeled as time-varying delay switched system for all time intervals. The regions of unstable subsystem are determined for value of τ_D.

Based on the above analysis, the actual control input applied to power system is devoted by

$$u(t) = Ky(t_{k(t)}h). \quad (16)$$

$$k(t) = \sup\{k \in \mathbb{N} | t_k \notin \aleph_D(t_0, t)\}. \quad (17)$$

Therefore, the model of the multi-area power system with LFC scheme (3) can be rewritten as

$$\dot{x}(t) = Ax(t) + BKCx(t_{k(t)}h) + Fw(t). \quad (18)$$

The L_2-gain property of the system (3) can be applied in this paper, $\omega(t)$ is the disturbance that belongs to $L_2[0, \infty)$, under zero initial condition, it holds

$$\int_0^\infty y^T(u)y(u)du \leq \gamma^2 \int_0^\infty \omega^T(u)\omega(u)du \quad (19)$$

where $\gamma > 0$.

3   Design of Event-Trigger LFC Under DoS Attacks

The following definition and lemmas are very useful in the proof of the main results.

Definition 3. *The initial solution of system* (3) *is said to be exponentially stability, if there exists* $\theta > 0$ *and* $\mathfrak{M} \geq 0$ *such that for any initial data* $x_{t_0} = \phi$

$$\|x(t, t_0, \phi)\| \leq \mathfrak{M} \|\phi\|_\iota e^{-\theta(t-t_0)}, \tag{20}$$

where $(t_0, \phi) \in \mathbb{N}^+ \times PC([-\iota, 0], \mathbb{R}^n)$, θ *is called the exponential convergence rate.*

Lemma 1 [9]. *Consider a given matrix* $R > 0$. *Then, for all continuous function* ω *in* $[a, b] \to \mathbb{R}^n$ *the following inequality holds:*

$$\int_a^b \omega^T(u) R\omega(u) du \geq \frac{1}{b-a} \left(\int_a^b \omega(u) du \right)^T R \left(\int_a^b \omega(u) du \right) + \frac{3}{b-a} \Omega^T R\Omega. \tag{21}$$

where $\Omega = \int_a^b \omega(s) ds - \frac{2}{b-a} \int_a^b \int_a^s \omega(r) dr ds$.

Lemma 2. *For an* $n \times n$ *matrix* $R > 0$, *a scalar* $\alpha \geq 0$, *and a vector function* $x \in C([-\eta_2, -\eta_1], \mathbb{R}^n)$, *the following inequality holds:*

$$(\eta_2 - \eta_1) \int_{t-\eta_2}^{t-\eta_1} e^{2\alpha(u-t)} \dot{x}^T(u) R\dot{x}(u) du \geq$$
$$\left(\int_{t-\eta_2}^{t-\eta_1} e^{\alpha(u-t)} \dot{x}(u) du \right)^T R \left(\int_{t-\eta_2}^{t-\eta_1} e^{\alpha(u-t)} \dot{x}(u) du \right) + 3\tilde{\Omega}^T R\tilde{\Omega}. \tag{22}$$

where

$$\tilde{\Omega} = e^{-\alpha\eta_1} x(t-\eta_1) - e^{-\alpha\eta_2} x(t-\eta_2) - \alpha \int_{t-\eta_2}^{t-\eta_1} e^{\alpha(u-t)} x(u) du, \tag{23}$$

$$-\frac{2}{\eta_2 - \eta_1} [(\eta_2 - \eta_1) x(t) - \int_{t-\eta_2}^{t-\eta_1} e^{\alpha(u-t)} x(u) du - \alpha \int_{-\eta_2}^{-\eta_1} \int_{t+u}^t e^{\alpha(r-t)} x(r) dr du]. \tag{24}$$

3.1   State of System Without Denial-of-Service Attacks

There are not DoS attacks about in the set of the time intervals $\aleph_s(t_0, t), t \in \mathbb{N}_0$,

$$e_k(t) = y(t_{k(t)} h) - y(t_{k(t)} h + ih),$$

represent the error of measurement data between the last successful control update time and the current time.

$$\tau(t) = t - t_{k(t)} h - ih, \quad i \in \mathbb{N}.$$

We have $\eta_1 \leq \tau(t) < \eta_2$ due to $\tau_{t_k} \leq \tau(t) < h + \tau_M$, where $\eta_1 = \tau_m$ and $\eta_2 = h + \tau_M$, then

$$y(t_{k(t)}h) = y(t - \tau(t)) + e_k(t). \tag{25}$$

It is clear that (18) can be rewritten as

$$\dot{x}(t) = Ax(t) + BKCx(t - \tau(t)) + BKe_k(t) + Fw(t). \tag{26}$$

Lemma 3. *Consider the system* (3), *give scalars* $a_1 > 0, 0 \leq \sigma < 1, \eta_1 > 0$ *and* $\eta_2 > 0$. *If there exist matrices* $P_1 > 0, Q_i > 0, R_i > 0$ *and* $Z_j > 0$ $(i = 1, 2; j = 1, 2, 3)$, Φ *is positive definite symmetric matrices, such that the following LMI holds,*

$$\begin{bmatrix} \widehat{\Xi} & \begin{bmatrix} \widetilde{\Gamma} \\ 0 \end{bmatrix} \\ * & \widetilde{\Theta} \end{bmatrix} < 0, \tag{27}$$

then along the trajectory of the system (26), *it follows that:*

$$V_1(t) \leq e^{-2a_1(t-t_0)} V_1(t_0) - \int_{t_0}^{t} e^{-2a_1(u-t)} L(u) du. \tag{28}$$

3.2   State of System with Denial-of-Service Attacks

The event-triggered transmission data may be attacked by DoS which are frequently encountered in the set of the time intervals $\aleph(t_0, t), t \in \mathbb{N}_0$. We can obtain the model of power system.

$$\dot{x}(t) = Ax(t) + BKCx(t_{k(t)}h) + Fw(t). \tag{29}$$

Due to that,

$$\begin{aligned} e_k^T(\xi_n) \Phi e_k(\xi_n) &= [y(t_{k(\xi_n)}h) - y(\xi_n)]^T \Phi[y(t_{k(\xi_n)}h) - y(\xi_n)] \\ &\geq y^T(t_{k(\xi_n)}h) \Phi y(t_{k(\xi_n)}h) - y^T(\xi_n) \Phi y(\xi_n). \end{aligned} \tag{30}$$

It is gained that $y(\xi_n) \leq y(t_{k(t)}h)$ because of exponentially stability (28) when $t \in [t_{k(t)}h, \xi_n + \varsigma_n)$. As a result,

$$e_k^T(t) \Phi e_k(t) \leq \sigma(1 + \sigma) y^T(\xi_n) \Phi y(\xi_n) \leq \sigma(1 + \sigma) y^T(t_{k(t)}h) \Phi y(t_{k(t)}h). \tag{31}$$

Lemma 4. *Consider the system* (3), *give scalars* $a_2 > 0, 0 \leq \sigma < 1, \eta_1 > 0$ *and* $\eta_2 > 0$. *if there exist matrices* $P_2 > 0, Q_i > 0, R_i > 0$ *and* $Z_j > 0$ $(i = 3, 4; j = 4, 5, 6)$, Φ *is positive definite symmetric matrices, such that the following LMI holds,*

$$\begin{bmatrix} \widehat{\Xi} & \begin{bmatrix} \widehat{\Gamma} \\ 0 \end{bmatrix} \\ * & \widehat{\Theta} \end{bmatrix} < 0, \tag{32}$$

then along the trajectory of the system (29), *it follows that:*

$$V_2(t) \leq e^{2a_2(t-t_0)} V_2(t_0) - \int_{t_0}^{t} e^{2a_2(t-u)} L(u) du. \tag{33}$$

3.3 Stability Analysis

We investigate the condition of exponentially stability of multi-area power systems with LFC under DoS attacks by using the average dwell time approach based on the above analysis.

Theorem 1. *Consider the system* (3), *give scalars* $a_1 > 0, a_2 > 0, 0 \leq \sigma < 1$, $\tau_D > 0, \gamma > 0, \eta_1 > 0$ *and* $\eta_2 > 0$. ϑ *be any arbitrary constant, if there exist matrices* P_i, Q_j, R_j *and* Z_k $(i = 1, 2; j = 1, 2, 3, 4; k = 1, 2, 3, 4, 5, 6)$, Φ *is positive definite symmetric matrices, such that LMI* (27) *and* (32) *hold, the system* (3) *is exponentially stability with* L_2-*gain, and the following inequality holds:*

$$\|x(t)\| \leq \mathfrak{H} \|x(t_0)\| e^{-\mathfrak{G}(t-t_0)}, \tag{34}$$

where

$$\mathfrak{H} = \left(\tfrac{\beta}{\alpha}\rho^\vartheta\right)^{\frac{1}{2}}, \mathfrak{G} = \tfrac{1}{2}\left[\tfrac{(2a_1 - 2a_2\delta)}{1+\delta} - \tfrac{\ln\rho}{\tau_D}\right],$$

the ratios δ *between* $|\aleph_D(t_0, t)|$ *and* $|\aleph_S(t_0, t)|$ *satisfies following the conditions:*

$$\delta < \frac{2a_1\tau_D - \ln\rho}{2a_2\tau_D + \ln\rho}, \tag{35}$$

$\rho \geq 1$, *satisfies* $P_i \leq \rho P_j, Q_m \leq \rho Q_n, R_m \leq \rho R_n, Z_p \leq \rho Z_q$, $(i, j = 1, 2; m, n = 1, 2, 3, 4; p, q = 1, 2, 3, 4, 5, 6)$.

4   Simulatin Example

In this section, a three-area power systems with LFC is used to evaluate the impacts of DoS attacks, where all generators are assumed to be equipped with a nonreheat turbine, multiple generators are modeled by a single equivalent generator in each Area. There are the open communication channels in the power systems, which measurement data will be transmitted from remote terminal units (RTUs) to control center and from control center to generators. We assumed DoS attacks exist in the control loop of the three-area power system. Similar to [10], all the corresponding parameters of the three-area power systems models (3) are given in Table 1.

Table 1. Model parameters for a three-area power systems model with LFC.

Parameters	M_i^a	R_i^f	D_i	$T_{CH_i}^f$	T_{G_i}	β_i	T_{ij}
Area 1	3.5	0.03	2	50	40	1	7.54
Area 2	4.0	0.07	2.75	10	25	1	7.54
Area 3	3.75	0.05	2.4	30	32	1	7.54

Our purpose is to find out the adequate proportion of the activation time of the present and absent DoS attacks, such that the frequency deviation of

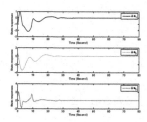
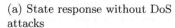
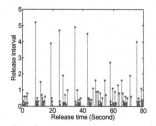

(a) State response without DoS attacks

(b) Release instants and release interval of measured data without DoS attacks

Fig. 1. State response (a) release instants and release time interval (b) of measured data without DoS.

the three-area is exponentially stable with L_2-gain γ. There are three cases for different attack proportions:

Case 1: There is the absence of DoS attacks in the power systems, where $\delta = 0$. We can obtain the delay upper bound of the three-area power systems with LFC scheme under event-triggered mechanism to achieve zero steady states for frequency deviations. Hence, given $a_1 = 0.25, \sigma = 0.2, \gamma = 3.5$ and $\eta_1 = 0.01$, by utilizing the LMI Toolbox, it follows from Lemma 3 that we can obtain $\eta_2 = 0.314$.

The frequency deviation of the multi-area power systems with LFC is shown in Fig. 1(a), and it is said to be exponentially stabilization, it can be found that the frequency deviation of three-area could reach to zero in about 30 s. The measurement data of system are transmitted using event-triggered scheme via communication channels, which the number of transmission data can largely reduce, release instants and release intervals of measurement data are shown in Fig. 1(b).

Case 2: The frequency deviation of the three-area power systems with LFC had been change according to different DoS attacks launching time, where $\delta \geq 3$, given $a_2 = 0.2, \rho = 2$. Figure 2(a) shows that the dynamic of the three-area power systems with LFC is seriously effected by DoS attacks. In the Fig. 2(a), it is enlarged to see the difference color of background, the time intervals are painted white owing to the absent DoS attacks, the red violet intervals indicates that the three-area power systems are being subjected to DOS attacks, it is reasonable for the adversary to launch DoS attacks. The frequency deviation of the three-area power systems still can not guarantee convergence. Before the simulation ends, the available event-triggered data are so small in Fig. 2(b) that the frequency deviation of the three-area power systems can only be caused fluctuations in the unstability state.

Case 3: The number and duration of DoS attacks are greatly reduced in Fig. 3(a), when $\delta = 0.1$. The number of unavailable event-triggered data are

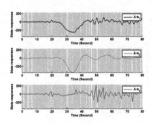 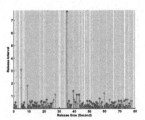

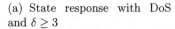

(a) State response with DoS and $\delta \geq 3$

(b) Release instants and release interval of measured data with DoS and $\delta \geq 3$

Fig. 2. State response (a) release instants and release time interval (b) of measured data with DoS and $\delta \geq 3$. (Color figure online)

Table 2. Values of τ_D by different values of δ.

δ	0.1	0.25	0.5	0.75	0.9
τ_D	1.7073	2.1875	3.5	6.125	9.5

much smaller than the case 1. Since, there are more available event-triggered data that are transmitted to the control center in Fig. 3(b). The frequency deviation of the three-area power systems after short-period fluctuations can still guarantee exponentially stabilization in Fig. 3(a).

According to values of various ratio δ between $|\aleph_D(t_0, t)|$ and $|\aleph_S(t_0, t)|$, the dynamic performance of the three-area power systems can be obtain for different values of $0 \leq \delta \leq 1$, there are different values of average dwell time τ_D for different values of δ in Table 2. The three-area power systems can maintain stability to be attack by malicious adversary, when the total time of $\aleph_D(t_0, t)$ are less than or equal to the total time of $\aleph_S(t_0, t)$ in this paper.

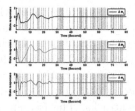 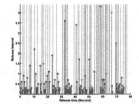

(a) State response with DoS and $\delta = 0.1$

(b) Release instants and release interval of measured data with DoS and $\delta = 0.1$

Fig. 3. State response (a) release instants and release time interval (b) of measured data with DoS and $\delta = 0.1$.

5   Conclusion

This paper considered the problem that DoS attacks in the open communication infrastructures can affect the dynamic performance of the multi-area power systems. Event-triggered mechanism has been utilized to transmit the data at its event time, in order to reduce the communication burdens and lowers the control updating frequency. By formulating DoS attacks as a switch action on communication channels, L_2-gain problem of the time-varying delay system with the disturbance input had been extend in the main result, it obtained constraint condition for finding a ratio of the total time intervals of the present DoS attack and the absent DoS attack. In case studies, the dynamics of a three-area power systems remain exponentially stability with different attack-launching instants.

Acknowledgments. This work was supported by the National Key Scientific Instrument and Equipment Development Project under Grant No. 2012YQ15008703, National Science Foundation of China under Grant No. 61633016, and the Key Project of Science and Technology Commission of Shanghai Municipality under Grant No. 15220710400.

References

1. Pourmousavi, S.A., Nehrir, M.H.: Introducing dynamic demand response in the LFC model. IEEE Trans. Power Syst. **29**(4), 1562–1572 (2014)
2. Kundur, P.: Power System Stability and Control. McGraw-Hill Inc., New York (1994)
3. Amin, S.M.: Electricity infrastructure security: toward reliable, resilient and secure cyber-physical power and energy systems. In: 2010 IEEE Power and Energy Society General Meeting, pp. 1–5. IEEE Press (2010)
4. Sargolzaei, A., Yen, K.K., Abdelghani, M.N.: Delayed inputs attack on load frequency control in smart grid. In: 2014 Innovative Smart Grid Technologies Conference, pp. 1–5. IEEE Press, Washington, USA (2014)
5. Sargolzaei, A., Yen, K.K., Abdelghani, M.N.: Control of nonlinear heartbeat models under time-delay-switched feedback using emotional learning control. Int. J. Recent Trends Eng. Technol. **10**(2), 85–91 (2014)
6. Amin, S., Schwartz, G.A., Sastry, S.S.: Security of interdependent and identical networked control systems. Automatica **49**(1), 186–192 (2013)
7. Amin, S., Cárdenas, A.A., Sastry, S.S.: Safe and secure networked control systems under Denial-of-service attacks. In: Majumdar, R., Tabuada, P. (eds.) HSCC 2009. LNCS, vol. 5469, pp. 31–45. Springer, Heidelberg (2009). doi:10.1007/978-3-642-00602-9_3
8. Kanchanaharuthai, A., Ngamsom, P.: Robust H∞ load frequency control for interconnected power systems with D-stability constraints via LMI approach. In: Proceedings of the 2005, American Control Conference, pp. 4387–4392. IEEE Press, Portland, USA (2005)
9. Seuret, A., Gouaisbaut, F.: Wirtinger-based integral inequality: application to time-delay systems. Automatica **49**(9), 2860–2866 (2013)
10. Ma, M., Chen, H., Liu, X., Allgwer, F.: Distributed model predictive load frequency control of multi-area interconnected power system. Int. J. Electr. Power Energy Syst. **62**, 289–298 (2014)

Detecting Replay Attacks in Power Systems: A Data-Driven Approach

Mingliang Ma[1], Peng Zhou[1]([envelope]), Dajun Du[1], Chen Peng[1], Minrui Fei[1], and Hanan Mubarak AlBuflasa[2]

[1] Shanghai University, Shanghai, China
pzhou@shu.edu.cn
[2] University of Bahrain, Zallaq, Bahrain

Abstract. Detecting replay attacks in power systems is quite challenging, since the attackers can mimic normal power states and do not make direct damages to the system. Existing works are mostly model-based, which may either suffer from a low detection performance or induce negative side effects to power control. In this paper, we explore purely data-driven approach for good detection performance without side effects. Our basic idea is to learn a classifier using a set of labelled data (i.e., power state) samples to detect the replayed states from normal ones. We choose the Support Vector Machine (SVM) as our classifier, and a self-correlation coefficient as the data feature for detection. We evaluate and confirm the effectiveness of our approach on IEEE bus systems.

Keywords: Power system security · Replay attacks · Support vector machine

1 Introduction

Modern power systems are now opening to the Internet for resource optimization and operation efficiency, but meanwhile induce potential security threats from cyber space. To tackle such threats, many defenses (such as intrusion detection systems [1], network firewalls [2] and trust management solutions [3] etc.) have been deployed [4] and try to secure power systems in the Internet paradigm. These defenses are usually detecting intrusions by checking whether power system runs on visible faults, hence being able to catch the attacks when or after real damages (lead to system faults) are made on the system. However, this kind of detection is too late if the damage is deadly and irreversible. In fact, many cyber attacks intrude into the system much earlier than they make damages. They conceal themselves by replaying normal power states (named replay attack) and thus can wait for a right moment to launch sudden destruction (in order to maximize the impacts). A real world example of such intrusion is Stuxnet virus [5]. This virus performs replay attacks to stealthily hide in Iran nuclear power system after intrusion [6], and can execute malicious control commands to destroy the nuclear centrifuges at the moment pre-defined by attackers.

© Springer Nature Singapore Pte Ltd. 2017
K. Li et al. (Eds.): LSMS/ICSEE 2017, Part III, CCIS 763, pp. 450–457, 2017.
DOI: 10.1007/978-981-10-6364-0_45

Unlike other cyber attacks, replay attacks are very insidious since they do not make any direct damages to the system, and thus are quite challenging to detect. Despite some preliminary works [6–8] have been proposed in the literature, they either have a low detection performance or may affect normal control process. For example, Mo et al. [7] have conducted a pioneer work to detect replay attacks using a system residual threshold chosen by a χ^2 failure detector. This method has an additional input drawn from i.i.d. Guassian distribution as authentication signal, and detects replay attacks by assuming the replayed states can be linearly classified (the classification line is determined by the threshold) from the normal ones in case the authentication single cannot be replayed. But actually, this assumption does not hold well, hence leading to a very poor detection performance (the detection rate is less than 40% reported by [7]). The authors of [7] acknowledged this weakness and proposed an optimal watermarked signal to improve the detection rate in their later research [6]. The idea is to obtain the authentication signal by maximizing the expected Kullback-Leibler divergence between the replayed and normal residuals. Despite the watermarked signal can achieve a better detection performance than the i.i.d. signal, they both are additional model inputs and necessarily add noises and uncertainties to the system to reduce the control performance. Moreover, Hoehn and Zhang [8] added an additional control element to the control loop for replay attack detection. Such solution is costly, and may incur re-configuration to power control.

In this paper, we break through the linear classification assumption on model residual (go beyond [6,7]) and design purely data-driven approach to avoid side effects (e.g., additional inputs or control elements) on power control (go beyond [6–8]). Our main contributions are two folds. First, we design self-correlation coefficient, a new set of data feature that is more discriminative than the model residual to differentiate the replayed power states from the normal ones. Second, we build an non-linear classification data-driven approach, specifically the Support Vector Machine (SVM) with a Gaussian kernel, on a set of labelled data samples (through the data features) for replay attack detection. We have evaluated our solution on IEEE bus systems using MATPOWER platform [9], and successfully confirmed the effectiveness of our design.

2 System Model and Threat Model

2.1 System Model

In general, a power system is a distributed network which connects with power plants, transmission lines, electrical transformers and consumers [10]. To control the whole progress of electric distribution, the power system usually deploys sensors to measure the voltages and flows (i.e., power states) from geographically distributed power plants over time, a state estimator to filter measurement noises, and a controller to take actions upon state changing. In this paper, we follow prior research [7] to consider the power system linear time invariant (LTI) and model it on a sequence of discrete time intervals. In particular, we model the power state transition over time as:

$$s_{k+1} = F \cdot s_k + B \cdot i_k + e_{1k}, \tag{1}$$

where, $s_k \in \mathbb{R}^n$ is the vector of power state variables at the time interval k (the power system is assumed to have n power states). i_k and e_{1k} are the vectors of control inputs and system internal noises at time k, respectively. F is the power system matrix and B is the power control matrix. Besides, we model the power state measurement as:

$$z_k = H \cdot s_k + e_{2k}, \tag{2}$$

where, $z_k \in \mathbb{R}^m$ is the vector of $m \leq n$ measurements of power states at time k, H is measurement matrix and e_{2k} represents the measurement noises. We also consider a Kalman filter as the state estimator to avoid the uncertainty caused by noise (which includes e_{1k} and e_{2k}). We show a diagram to illustrate the power system model in Fig. 1.

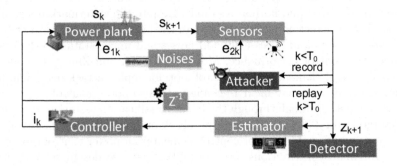

Fig. 1. An example of power system model.

2.2 Threat Model

We consider the replay attackers have compromised some measurement sensors in the power system. We use an indicator vector $a \in \mathbb{R}^m$ to represent compromised sensors. That is, if the j-th sensor is compromised, $a(j) = 1$, otherwise, $a(j) = 0$. We follow the typical replay attack model proposed by [7], in which the attackers can record the power states by compromised sensors from time interval T_s to $T_s + T - 1$, and then replay these recorded states in the following time windows $[T_s + p \cdot T, T_s + (p+1) \cdot T - 1]$. Under attack, the measured states monitored by the power system detector is $z_k = z_k^a + (I - a)z_k$, and z_k^a can be expressed as:

$$z_k^a = \begin{cases} a \cdot z_k & T_s \leq k \leq T_s + T - 1 \\ a \cdot z_{k-pT} & T_s + pT \leq k \leq T_s + (p+1)T - 1, \end{cases} \tag{3}$$

where, $p = 1, 2, 3, \cdots, N$. The replay attack is started at time T_s and lasted for the consequent $N \cdot T$ time intervals. N and T are both attacking parameters set by the replay attackers. T defines the replay period and NT defines the total duration of replay attack.

3   New Detection Feature and Method

3.1   Data Feature for Detection - Self-correlation Coefficient

In this paper, we solve the replay attack detection problem without the need to estimate the system's model (i.e., F and B in Eq. (1)) for residual calculation (i.e., $r_k = z_k - H(F \cdot s_{k-1} + B \cdot i_{k-1})$), and do not induce any additional control inputs and elements, in order to not disturb the power system's normal operation. Specifically, we propose a purely data-driven solution by analyzing only the measurement data results (i.e., z_ks) for detecting the replayed states (i.e., z_k^a). To facilitate the data-driven detection, we design a new set of data features. The insight is from the fact that: the replayed states always show strong periodicity on measured data (as described in Eq. (3), the z_k^a is repeated in the period of T) while the normal ones may not. We thus propose self-correlation coefficient (or SCC in short) as the data feature for detection. That is, given a time sequence $\{z_k(j)\}_{k\in[1,M]}$, where $z_k(j)$ represents the j-th element in vector z_k, we calculate the corresponding SCC as a vector of $\beta_j \in \mathbb{R}^{\frac{M}{2}}$ (M is the length of time sequence that we can determine for detection), and compute β_j's t-th element $\beta_j(t)$ as:

$$\beta_j(t) = \sum_{x=1}^{t} \sum_{y=1}^{\frac{M}{t}-1} |z_{x+(y-1)t}(j) - z_{x+yt}(j)|, \tag{4}$$

where, $|*|$ is a function which returns an absolute value. For the replayed states, if $t \bmod T$ equals 0, we can have $\beta_j(t) = 0$ by the definition of Eq. (4). While for the normal states, β_j have a very low probability to include $\beta_j(t) = 0$ conditions (due to the measurement noises and system uncertainties). Therefore, we can use these $\beta_j(t) = 0$s in SCC feature as detection tricks to discriminate the replayed states from normal ones. Figure 2 presents an example for such detection tricks captured by SCC.

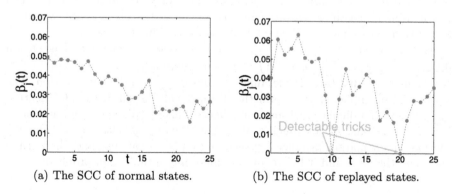

(a) The SCC of normal states.   (b) The SCC of replayed states.

Fig. 2. An example of SCC for normal and replayed ($T_s = 0, T = 10$) states.

3.2   Data-Driven Detection Method - SVM Approach

As another major contribution, we propose a non-linear data-driven classification method to ensure the detection performance. In prior work [6,7], the authors implemented a model-based detection method for replay attack detection. This method relies on a pre-defined residual threshold that is selected according to system model, and hence implicitly assumed the replayed states can be linearly classified from the normal ones in a one dimensional feature space (i.e., the system residual feature space). However, this assumption cannot well reflect the truth. In fact, replay attackers can purposely choose different attacking parameters T_s and T, and record different fragments of normal states to replay, hence making a very complex distribution of the replayed states. Therefore, we cannot simply consider the discrimination of the replayed states from the normal ones as a simply linear classification problem (even from the view of system model), and should design new detection method from non-linear classification perspective.

In this paper, we extend the detection problem from the single feature of model-based residual threshold to a more powerful data-driven feature space (i.e., self-correlation coefficient or SCC in short), and thus can apply a non-linear classifier to solve it. In particular, we use the typical support vector machine (SVM in short) with a Gaussian kernel [11] to address this non-linear classification problem. The basic idea is to learn a sequence of support vectors from a training set of labelled replayed and normal data (i.e., state) samples over the SCC feature space (which is a $\frac{M}{2}$ dimensional space). We formalize the training set with n samples as $(f^{(i)}, \ell^{(i)})$ for $i = 1, 2, \cdots, n$, where $f^{(i)} = \beta_j$ is the SCC feature of the sample (i) and $\ell^{(i)}$ is a label indicating whether (i) is an attack sample $(\ell^{(i)} = -1)$ or a normal one $(\ell^{(i)} = 1)$. The SVM employs a kernel trick technique to map the non-linear classification problem to a higher dimensional space for solving. The key insight is that a non-linear distribution in a lower dimensional space can possibly be linearly distributed in a higher dimensional space, and the kernel trick can approximate such distribution without requiring to re-build the space explicitly. Considering the replay attack detection problem is a non-linear classification problem even in the SCC feature space, we adopt kernel trick technique and choose a Gaussian kernel in our design.

Given training set $(f^{(i)}, \ell^{(i)})$ for $i = 1 : n$, we write the classifier's Lagrangian dual as $\ell = g(f) = \sum_{i=1}^{n} w_i \ell^{(i)} \kappa(f^{(i)}, f) + b$, where $\kappa(f^{(i)}, f) = e^{-||f^{(i)} - f||^2 / 2\sigma^2}$ is a Gaussian radial basis kernel function and $|| * ||^2$ is a 2-norm function. The $\kappa(f^{(i)}, f)$ can implicitly transform the distance calculation in the SCC feature space to a higher dimensional space, and hence make the distance computation in higher dimensional space explicitly. We expect the optimal classifier $g(f)$ having the maximum margin distance between the replayed samples and the normal ones, and thus can find it by solving the following optimization problem:

$$\max_{w} \sum_{i=1}^{n} w_i - \frac{1}{2} \sum_{i,j=1}^{n} w_i w_j \ell^{(i)} \ell^{(j)} \kappa(f^{(i)}, f^{(j)}), s.t., \ell^{(i)} g(f^{(i)}) \geq 1, \qquad (5)$$

where, the samples $(f^{(i)}, \ell^{(i)})$ that make $\ell^{(i)} g(f^{(i)}) = 1$ are support vectors, and the maximum margin classifier $g(f)$ can be found by solving the w in Eq. (5). We use this optimal $g(f)$ as our replay attack detector.

4   Evaluation

We show the effectiveness of our approach by running experiments on MAT-POW platform [9]. In particular, we choose a typical IEEE bus test systems, specifically the IEEE 14 bus (we choose this bus since it is perhaps the most popular IEEE bus used in the literature), in our experiments, and implement state estimation using a state forecasting method [12]. We run each power state estimation with 50 time intervals and calculate the SCC feature on this time window. We consider the noises $e_{1k}, e_{2k} \sim N(0, 0.01)$. We execute replay attacks with random T_s and T, since the smart attackers can purposely choose any values for these two parameters to evade the detection. In our evaluation, we have 1000 times of power state estimation processes in total (500 times for normal and 500 times for replay attacks). We train SVM classifier (implemented by libSVM in Matlab) by randomly selecting 250 normal data and 250 attacked ones, and use the reminder data for test. We regard each SVM training and testing process as one experiment, and run 100 times of such experiments in our evaluation. We report our results in Fig. 3. As can be seen, our approach can achieve good detection performance, in around 80% detection rates with no more than 15% false positives. Through the experimental results, we have confirmed an important point of view: a purely data-driven approach without side effects on power control is possible to detect replay attacks with good detection performance. This point sheds light the potential for future research in this direction.

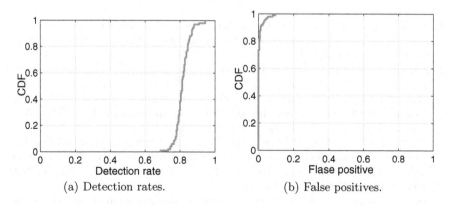

(a) Detection rates.   (b) False positives.

Fig. 3. The CDF of detection results on IEEE 14 bus.

5   Conclusion and Future Work

In this paper, we propose a data-driven approach (specially the SVM with a Gaussian kernel) to detect replay attacks on a new set of self-correlation coefficient data feature space. Our approach can achieve good detection performance, and more importantly, does not introduce any additional model inputs and control devices to the power control, which is required by state-of-the-art counterparts (i.e., [6–8]).

We acknowledge there still exists some potential space to improve the effectiveness of data-driven approaches for replay attack detection, and the future works can do this improvement by two research directions. The first is to apply much more robust classifiers, such as deep learning architecture and reinforcement learning strategies, for the detection. The second is to consider the arms races between the detector and attacker. That is, if the attackers have some knowledge on how the detection is performed, they will do some necessary countermeasures to avoid the detection. As a result, the detector should take into account the attackers' response during the training phase. A promising solution is to apply a deep learning architecture [13] or adversarial learning methods for classifier training [14].

Acknowledgments. This work was partially supported by the National Natural Science Foundation of China (Nos. 61502293, 61633016 and 61673255), the Shanghai Young Eastern Scholar Program (No. QD2016030), the Young Teachers' Training Program for Shanghai College & University, the Science and Technology Commission of Shanghai Municipality (Nos. 17511107002 and 15411953502) and the Shanghai Key Laboratory of Power Station Automation Technology.

References

1. Zhang, Y., Wang, L., Sun, W., Green, R.C., Alam, M.: Distributed intrusion detection system in a multi-layer network architecture of smart grids. IEEE Trans. Smart Grid **2**(4), 796–808 (2011)
2. Ericsson, G.N.: Cyber security and power system communicationessential parts of a smart grid infrastructure. IEEE Trans. Power Deliv. **25**(3), 1501–1507 (2010)
3. Zhou, P., Jiang, S., Irissappane, A., Zhang, J., Zhou, J., Teo, J.C.M.: Toward energy-efficient trust system through watchdog optimization for WSNs. IEEE Trans. Inf. Forensics Secur. **10**(3), 613–625 (2015)
4. Wei, D., Lu, Y., Jafari, M., Skare, P.M., Rohde, K.: Protecting smart grid automation systems against cyberattacks. IEEE Trans. Smart Grid **2**(4), 782–795 (2011)
5. Falliere, N., Murchu, L.O., Chien, E.: Symantec Corp. Secur. Response. W32. stuxnet dossier **5**, 6 (2011). White paper
6. Mo, Y., Weerakkody, S., Sinopoli, B.: Physical authentication of control systems: designing watermarked control inputs to detect counterfeit sensor outputs. IEEE Control Syst. **35**(1), 93–109 (2015)
7. Mo, Y., Sinopoli, B.: Secure control against replay attacks. In: 47th Annual Allerton Conference on Communication, Control, and Computing, pp. 911–918. IEEE (2009)

8. Hoehn, A., Zhang, P.: Detection of replay attacks in cyber-physical systems. In: American Control Conference (ACC), pp. 290–295. American Automatic Control Council (AACC) (2016)
9. Zimmerman, R.D., Murillo-Sánchez, C.E., Thomas, R.J.: Matpower: steady-state operations, planning, and analysis tools for power systems research and education. IEEE Trans. Power Syst. **26**(1), 12–19 (2011)
10. Gonen, T.: Electric Power Distribution Engineering. CRC Press, Boca Raton (2016)
11. Burges, C.J.: A tutorial on support vector machines for pattern recognition. Data Min. Knowl. Disc. **2**(2), 121–167 (1998)
12. Da Silva, A.L., Do Coutto Filho, M., De Queiroz, J.: State forecasting in electric power systems. In: IEE Proceedings C (Generation, Transmission and Distribution), Vol. 130, pp. 237–244. IET (1983)
13. Zhou, P., Gu, X., Zhang, J., Fei, M.: A priori trust inference with context-aware stereotypical deep learning. Knowl.-Based Syst. **88**, 97–106 (2015)
14. Laskov, P., Lippmann, R.: Machine learning in adversarial environments. Mach. Learn. **81**(2), 115–119 (2010)

A Novel Dynamic State Estimation Algorithm in Power Systems Under Denial of Service Attacks

Mengzhuo Yang$^{(\boxtimes)}$, Xue Li$^{(\boxtimes)}$, and Dajun Du

School of Mechatronical Engineering and Automation, Shanghai University,
Shanghai 200072, China
ymz108@163.com, lixue@i.shu.edu.cn

Abstract. The paper is concerned with a dynamic state estimation algorithm in power systems under denial of service (DoS) attacks. Firstly, the character of data packet losses caused by DoS attacks is described by Bernoulli distribution, and the dynamic model of power system is reconstructed. Using Holt's two-parameter exponential smoothing and extended Kalman filtering techniques, a dynamic state estimation algorithm is proposed, where the recursion formula of the parameter identification, state prediction and state filtering contain the statistical properties of data packet losses. Simulation results confirm the feasibility and effectiveness of the proposed algorithm.

Keywords: Power system · Dynamic state estimation · Denial of service attack · Extended Kalman filter

1 Introduction

Supervisory Control and Data Acquisition (SCADA) systems are widely used in the modern power systems, which collect a large number of signals and the state estimators are extensively used to filter the measurement noise and detect gross errors [1]. The optimal estimate of the power system is then employed to achieve some advanced operation functions of the energy management system (EMS), e.g., the optimal power flow analysis, the automatic generation control, and the contingency analysis.

However, such vulnerabilities in the SCADA system as the communication links, the Remote Terminal Units (RTUs) and the software and databases in the control center are usually challenged by various cyber-attacks such as DoS attack [2]. It compromises the availability of resources by jamming the communication channel, attacking network protocols, and flooding the network traffic [3,4], leading to measurement packets losses. To solve this problem, some efforts have been stimulated. For example, two queueing models are proposed to describe the stochastic process of packet transmission under DoS attacks [5]. Security control for discrete-time linear dynamical systems is investigated, where the sensor and control packets losses are caused by a random or a resource-constrained attacker in [6]. Taking packet losses into account in [7], an improved recursive least squares

© Springer Nature Singapore Pte Ltd. 2017
K. Li et al. (Eds.): LSMS/ICSEE 2017, Part III, CCIS 763, pp. 458–466, 2017.
DOI: 10.1007/978-981-10-6364-0_46

algorithm is proposed. Different from most existing works under given attack patterns, the optimal attack schedules are developed in [8]. The nonlinear characteristic of power systems is analyzed, and a stochastic extended Kalman filter technique is proposed to obtain optimal estimates of power networks in [9]. The random DoS attack model is described by a Bernoulli sequence in [10]. Therefore, thus far the state estimation problem for power systems under denial of service attacks has seldom been reported.

Motivated from the above observations, the paper mainly investigate dynamic state estimation algorithm in power systems under DoS attacks. The rest of this paper is organized as follows: problem formulation as well as the dynamic model of a power system is presented in Sect. 2. Section 3 presents dynamic state estimation algorithm including the parameter identification, state prediction and state filtering. Simulations are given in Sect. 4, following the conclusions in Sect. 5.

2   Problem Formulation

The state of a power network is composed of complex nodal voltages in the entire system, when the network topology and parameters are completely known. The dynamic model of a power system can be expressed by

$$
\begin{aligned}
x_{k+1} &= F_k x_k + u_k + w_k, \\
z_k &= h(x_k) + v_k,
\end{aligned}
\tag{1}
$$

where $k = 1, 2, \ldots, N$ is time horizon, $x_k \in \Re^{n \times 1}$ is the state vector at time instant k, $x_{k+1} \in \Re^{n \times 1}$ is the state vector at time instant $k + 1$, $z_k \in \Re^{m \times 1}$ is the measurement vector at time k, $F_k \in \Re^{n \times n}$ is the state transition matrix, $u_k \in \Re^{n \times 1}$ is the state transition vector, $w_k \in \Re^{n \times 1}$ is the model noise with zero mean and covariance matrix Q_k, $v_k \in \Re^{m \times 1}$ is the measurement noise with zero mean and covariance matrix R_k. It is worth noting that the model noise w_k and the measurement noise v_k are uncorrelated, and the initial state of the power system denoted as x_0 is independent of w_k and v_k under the condition that $E[x_0] = m_0$ and $E\left[(x_0 - m_0)(x_0 - m_0)^T\right] = P_0$.

With the rapid development of communication and information technology, these measurement outputs z_k are transmitted to the estimator via the communication network. When the network environment is considered to be perfect, there exist no data packet dropouts. The linear discrete-time state transition equation in (1) can be directly solved by the proposed dynamic state estimation method [11]. As a large amount of smart sensors are deployed in power network, these devices are exposed in an unencrypted communication environment, causing the increasing severe cyber security issues. This paper focuses on the measurement outputs attacked by DoS, and data packet dropouts are inevitably introduced. Therefore, the measurement outputs z_k are becoming $\hat{z}_k$, it follows that

$$
\hat{z}_k = \gamma_k z_k,
\tag{2}
$$

where $\gamma_k = 0$ if the measurement outputs z_k are lost under DoS attack, otherwise $\gamma_k = 1$. Therefore, (1) can be rewritten as

$$
\begin{aligned}
x_{k+1} &= F_k x_k + u_k + w_k, \\
z_k &= h\left(x_k\right) + v_k, \\
\hat{z}_k &= \gamma_k z_k.
\end{aligned}
\tag{3}
$$

Using the same method [6], the packet loss is modeled as the Bernoulli process γ_k with the following probability distribution law:

$$
\begin{aligned}
&\Pr\left(\gamma_k = 0\right) = \rho, \Pr\left(\gamma_k = 1\right) = 1 - \rho, \\
&Var\left(\gamma_k\right) = \rho\left(1 - \rho\right).
\end{aligned}
\tag{4}
$$

To obtain an optimal estimate in (3), the following cost function which indicates the filter error covariance matrix should be minimized, i.e.,

$$
P_{k|k} = E\left[e_{k|k} e_{k|k}^T\right],
\tag{5}
$$

where $e_{k|k} = x_{k|k} - x_k$ is the error at time instant k, E is the expectation based on the state vector x_k, the noise w_k and v_k and the parameter γ_k.

3　Dynamic State Estimation

Dynamic state estimation of a power system usually consists of three stages, i.e., parameter identification, state prediction and state filtering. These stages are thereinafter explained in detail respectively.

3.1　Parameter Identification

According to (1), the dynamic model at time instant k can be given by

$$
x_k = F_{k-1} x_{k-1} + u_{k-1} + w_{k-1},
\tag{6}
$$

where F_{k-1} and u_{k-1} are unknown state transition parameters identified by online or offline techniques. They can be calculated by one of the most commonly used identification methods, i.e., Holt's two-parameter exponential smoothing method [12]. Firstly, the predicted state vector (i.e., $x_{k|k-1}$) at time instant k need be calculated by

$$
x_{k|k-1} = a_{k-1} + b_{k-1},
\tag{7}
$$

$$
a_{k-1} = \alpha x_{k-1|k-1} + (1 - \alpha) x_{k-1|k-2},
\tag{8}
$$

$$
b_{k-1} = \beta\left(a_{k-1} - a_{k-2}\right) + (1 - \beta) b_{k-2},
\tag{9}
$$

where $a_{k-1} \in \Re^{n \times 1}$ is the estimate of the level of the series at time instant $k-1$, $b_{k-1} \in \Re^{n \times 1}$ is the estimate of the slope of the series at time instant $k-1$, α and β is the smoothing parameters with the fixed values between 0 and 1.

The desired transition parameters are then given by

$$F_{k-1} = \alpha(1 + \beta)I, \tag{10}$$

$$u_{k-1} = (1 + \beta)(1 - \alpha)x_{k-1|k-2} - \beta a_{k-2} + (1 - \beta)b_{k-2}, \tag{11}$$

where $I \in \Re^{n \times n}$ is an identity matrix.

Once parameters F_{k-1} and u_{k-1} are obtained, the state prediction can then be implemented in the following section.

3.2   State Prediction

The prediction equation can be obtained by performing conditional expectation on (6) with the previous state estimate $x_{k-1|k-1}$, i.e.,

$$x_{k|k-1} = F_{k-1}x_{k-1|k-1} + u_{k-1}. \tag{12}$$

Define the prediction error $e_{k|k-1} = x_{k|k-1} - x_k = F_{k-1}e_{k-1|k-1} - w_{k-1}$, then the prediction error covariance matrix is expressed as

$$
\begin{aligned}
P_{k|k-1} \\
&= E\left[e_{k|k-1}e_{k|k-1}^T\right] \\
&= E\left\{\left[F_{k-1}e_{k-1|k-1} - w_{k-1}\right]\left[F_{k-1}e_{k-1|k-1} - w_{k-1}\right]^T\right\} \\
&= F_{k-1}E\left[e_{k|k-1}e_{k|k-1}^T\right]F_{k-1}^T - F_{k-1}E\left[e_{k-1|k-1}w_{k-1}^T\right] \\
&\quad - E\left[w_{k-1}e_{k-1|k-1}^T\right]F_{k-1}^T + E\left[w_{k-1}w_{k-1}^T\right] \\
&= F_{k-1}P_{k-1|k-1}F_{k-1}^T + Q_{k-1},
\end{aligned}
\tag{13}
$$

where $P_{k-1|k-1}$ denotes the filtering error covariance matrix at time instant $k-1$.

3.3   State Filtering

According to the received measurements set at time instant k, the estimate of the power system $x_{k|k}$ is obtained by the state filtering process. The filter equation is expressed as

$$x_{k|k} = x_{k|k-1} + K_k\left[\hat{z}_k - h\left(x_{k|k-1}\right)\right], \tag{14}$$

where $K_k \in \Re^{n \times m}$ is the gain matrix.

Taylor series expansion is employed in EKF algorithms to process nonlinear measurement vectors. Therefore, $\hat{z}_k$ in (2) can be further given by

$$\hat{z}_k = \gamma_k z_k = \gamma_k\left[h\left(x_{k|k-1}\right) + H_k\left(x_k - x_{k|k-1}\right) + v_k\right], \tag{15}$$

where H_k represents Jacobian matrix of $h(\cdot)$ at time instant k.

According to (14) and (15), the filter error $e_{k|k} = x_{k|k} - x_k$ is calculated by

$$
\begin{aligned}
e_{k|k} &= e_{k|k-1} + K_k\left\{\gamma_k\left[h(x_{k|k-1}) + H_k(x_k - x_{k|k-1}) + v_k\right] - h\left(x_{k|k-1}\right)\right\} \\
&= (I - K_k\gamma_k H_k)e_{k|k-1} - K_k(1 - \gamma_k)h(x_{k|k-1}) + K_k\gamma_k v_k.
\end{aligned}
\tag{16}
$$

Furthermore, $P_{k|k}$ in (5) can be calculated as

$$
\begin{aligned}
P_{k|k} = P_{k|k-1} &- K_k \sigma H_k P_{k|k-1} - P_{k|k-1} H_k^T \sigma K_k^T \\
&+ K_k[\sigma^2 H_k P_{k|k-1} H_k^T + (1-\sigma)^2 h\left(x_{k|k-1}\right) h(x_{k|k-1})^T + \sigma^2 R_k]K_k^T,
\end{aligned}
\tag{17}
$$

where $\sigma = E\left[\gamma_k\right] = 1 - \rho$.

Minimizing the error covariance matrix, partial derivative of $P_{k|k}$ with regard to K_k is zero, i.e.,

$$
\frac{\partial P_{k|k}}{\partial K_k} = \frac{\partial Trace\left\{E\left[e_{k|k}e_{k|k}^T\right]\right\}}{\partial K_k} = 0.
\tag{18}
$$

The optimal filter gain is derived as

$$
\begin{aligned}
K_k = P_{k|k-1}H_k^T \sigma \\
\times \left[\sigma^2 H_k P_{k|k-1} H_k^T + (1-\sigma)^2 h\left(x_{k|k-1}\right) h(x_{k|k-1})^T + \sigma^2 R_k\right]^{-1}.
\end{aligned}
\tag{19}
$$

Once the optimal gain matrix K_k is obtained, the filtered state and the corresponding covariance matrix can be then computed by using (14) and (17) respectively.

4   Simulation Example

To verify the effectiveness of the proposed algorithm, the proposed algorithm is operated on an IEEE14 power system to illustrate our proposed algorithm. The following indices are employed to evaluate simulation results, i.e.,

$$
\varepsilon_{v_k} = \sum_{i=1}^{n} \left|v_{k|k}\left(i\right) - v_k\left(i\right)\right| /n \times 100\%,
\tag{20}
$$

$$
\varepsilon_{\theta_k} = \sum_{i=1}^{n} \left|\theta_{k|k}\left(i\right) - \theta_k\left(i\right)\right| /n \times 100\%,
\tag{21}
$$

$$
J_k = \sum_{i=1}^{m} \left|z_{k|k}\left(i\right) - z_k\left(i\right)\right| / \sum_{i=1}^{m} \left|\hat{z}_k\left(i\right) - z_k\left(i\right)\right|.
\tag{22}
$$

To analyze the performance of the proposed algorithm under DoS attacks, two experiments are carried out.

1. Under DoS attacks, the traditional EKF algorithm and the proposed algorithm are employed respectively. Figure 1 presents the data packet losses sequence when $\rho = 0.05$. Figures 2 and 3 illustrate the estimation error (ε_{v_k}, ε_{θ_k}) and the performance index (J_k). It is clearly seen from Fig. 1 that the system is attacked at the 12th, 28th, 47th, 61th and 93th sampling time, resulting in a sudden increase in the estimation error. The performance index is

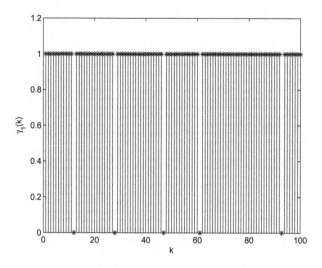

Fig. 1. The data packet losses sequence under DoS attacks ($\rho = 0.05$)

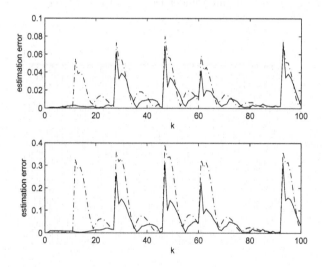

Fig. 2. The state estimation error using EKF algorithm (dashed line) and proposed algorithm (solid line) under DoS attacks ($\rho = 0.05$)

also deteriorated, as described in Fig. 3. These results confirm that the normal estimation process is significantly affected if the power system is under DoS attack. It can also be seen from the estimation error comparison of these two algorithms in Fig. 2 and the performance index comparison in Fig. 3 that the proposed algorithm provides significantly better performance.

2. Under DoS attacks, the above two algorithms are also performed when $\rho = 0.1$. Figure 4 shows the corresponding data packet losses sequence.

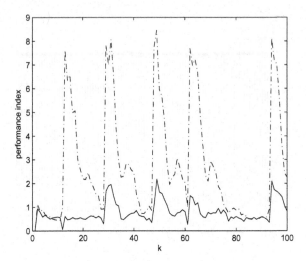

Fig. 3. The state estimation performance index using EKF algorithm (dashed line) and proposed algorithm (solid line) under DoS attacks ($\rho = 0.05$)

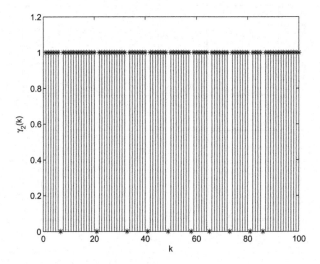

Fig. 4. The data packet losses sequence under DoS attacks ($\rho = 0.1$)

Figures 5 and 6 illustrate the estimation error (ε_{v_k}, ε_{θ_k}) and the performance index (J_k). The power system is attacked at the 7th, 21th, 33th, 41th, 49th, 58th, 65th, 73th, 81th and 86th sampling time, as described in Fig. 4. It can be seen from the estimation error comparison in Fig. 5 and the performance index comparison in Fig. 6 that the estimation performance is obviously improved using the proposed algorithm.

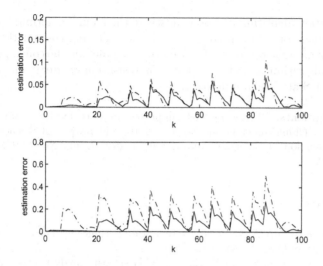

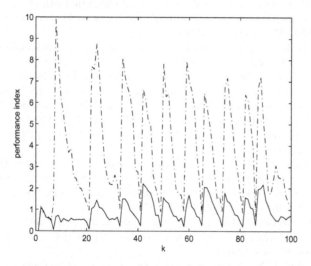

Fig. 5. The state estimation error using EKF algorithm (dashed line) and proposed algorithm (solid line) under DoS attacks ($\rho = 0.1$)

Fig. 6. The state estimation performance index using EKF algorithm (dashed line) and proposed algorithm (solid line) under DoS attacks ($\rho = 0.1$)

5   Conclusion

In this paper, a novel optimal state estimation algorithm has been proposed for power networks under denial of service attacks. The character of data packet losses caused by DoS attacks is described by Bernoulli distribution, and the dynamic model of power system is reconstructed. Using Holt's two-parameter exponential smoothing and extended Kalman filtering techniques, a dynamic state estimation algorithm is proposed, where the recursion formula of the parameter identification, state prediction and state filtering contain the statistical properties of data packet losses. Future work may include the convergence analysis for the algorithm as well as taking compensation strategy and power constraints of DoS attacks into account.

Acknowledgments. This work was supported in part by the National Science Foundation of China under Grant No. 61533010, and project of Science and Technology Commission of Shanghai Municipality under Grants No. 14JC1402200 and 15JC1401900.

References

1. Abur, A., Exposito, A.G.: Power System State Estimation: Theory and Implementation. Marcel Dekker, New York (2004)
2. Teixeira, A., Amin, S., Sandberg, H.: Cyber security analysis of state estimators in electric power systems. In: 49th IEEE Conference on Decision and Control, pp. 5991–5998. IEEE Press, Atlanta (2010)
3. Pasqualetti, F., Dorfler, F., Bullo, F.: Control-theoretic methods for cyberphysical security: geometric principles for optimal cross-layer resilient control systems. IEEE Control Syst. **35**(1), 110–127 (2015)
4. Xu, W., Ma, K., Trappe, W.: Jamming sensor networks: attack and defense strategies. IEEE Netw. **20**(3), 41–47 (2006)
5. Long, M., Wu, C.H., Hung, J.Y.: Denial of service attacks on network-based control systems: impact and mitigation. IEEE Trans. Ind. Inf. **1**(2), 85–96 (2005)
6. Amin, S., Cárdenas, A.A., Sastry, S.S.: Safe and secure networked control systems under denial-of-service attacks. In: Majumdar, R., Tabuada, P. (eds.) HSCC 2009. LNCS, vol. 5469, pp. 31–45. Springer, Heidelberg (2009). doi:10.1007/978-3-642-00602-9_3
7. Du, D., Shang, L., Zhao, W.: An online recursive identification method over networks with random packet losses. In: Fei, M., Peng, C., Su, Z., Song, Y., Han, Q. (eds.) LSMS/ICSEE 2014. CCIS, vol. 462, pp. 449–458. Springer, Heidelberg (2014). doi:10.1007/978-3-662-45261-5_47
8. Zhang, H., Cheng, P., Shi, L.: Optimal DoS attack policy against remote state estimation. In: 52nd IEEE Conference on Decision and Control, pp. 5444–5449. IEEE Press, Firenze (2013)
9. Su, C.L., Lu, C.N.: Interconnected network state estimation using randomly delayed measurements. IEEE Trans. Power Syst. **16**(4), 870–878 (2001)
10. Sinopoli, B., Schenato, L., Franceschetti, M.: Kalman filtering with intermittent observations. IEEE Trans. Autom. Control **49**(9), 1453–1464 (2004)
11. Debs, A.S., Larson, R.E.: A dynamic estimator for tracking the state of a power system. IEEE Trans. Power Appar. Syst. PAS **89**(7), 1670–1678 (1970)
12. Hyndman, R.J., Athanasopoulos, G.: Forecasting: Principles and Practice. OTexts, London (2014)

Small-Signal Refinement of Power System Static Load Modelling Techniques

Gareth McLorn and Seán McLoone$^{(\boxtimes)}$

School of Electronics, Electrical Engineering and Computer Science,
Queen's University Belfast, Belfast, Northern Ireland, UK
{gmclorn02, s.mcloone}@qub.ac.uk

Abstract. Loads are often represented as a weighted combination of constant impedance (Z), current (I) and power (P) components, so called ZIP models, by various power systems network simulation tools. However, with the growing need to model nonlinear load types, such as LED lighting, ZIP models are increasingly rendered inadequate in fully representing the voltage dependency of power consumption traits. In this paper we propose the use of small-signal ZIP models, derived from a neural network model of appliance level consumption profiles, to enable better characterizations of voltage dependent load behavior. Direct and indirect approaches to small-signal ZIP model parameter estimation are presented, with the latter method shown to be the most robust to neural network approximation errors. The proposed methodology is demonstrated using both simulation and experimentally collected load data.

Keywords: ZIP models · Exponential models · Load modelling · Neural networks

1 Introduction

This paper proposes an enhancement of existing static load modelling techniques to facilitate greater accuracy in the characterization of active and reactive power consumption, as functions of applied voltage, in power systems studies. This research is pertinent, given the emergence of many modern load types for which conventionally understood behaviors, in response to imposed variations in voltage, do not readily apply. Conventional modelling practices ought to be reviewed so that the characteristics of modern loads are reproduced in simulation with greater fidelity and their aggregated influence upon electricity networks may become better quantified.

Conservation Voltage Reduction (CVR) is a noteworthy energy conservation application, for which the accuracy of load models is intrinsic to its effectiveness. Utilities tend to employ CVR to lower service level voltages as a proxy for reducing electricity demand, especially during peak hours. Its objectives are typically achieved via the coordinated control of assorted distribution network assets, such as on load tap-changing (OLTC) transformers and switched capacitor banks. The effectiveness of enacting CVR within a targeted electricity network may be encapsulated by its CVR factor [1], values which aim to quantify the responsiveness (in percentage terms) of the energy consumed by an electrical appliance, household, feeder or entire network, ΔE,

© Springer Nature Singapore Pte Ltd. 2017
K. Li et al. (Eds.): LSMS/ICSEE 2017, Part III, CCIS 763, pp. 467–476, 2017.
DOI: 10.1007/978-981-10-6364-0_47

to an applied percentage change in the voltage supply, ΔV. For changes in active energy, this paper qualifies the corresponding CVR factor scalar with a 'p' subscript, as in CVR_p. Similarly, for reactive power the notation CVR_q is used. In the medium to long term the CVR_p scalar may be used to predict the total energy savings attributable to CVR. In the short term steady-state, it may also be used by utilities to predict the demand reduction impact of CVR during the daily peak [2, 3].

$$CVR_p = \frac{\Delta E(\%)}{\Delta V(\%)} \tag{1}$$

CVR_p ratios are in effect concise, aggregated representations of the steady-state parameters that individual loads present to the network upon connection. Ultimately, it is the aggregated combination of the many different loads within each household (modelled as a lumped load) and the composite CVR_p quantity therein that will determine the extent by which customers save energy (if at all) within a CVR scheme. The aggregation concept may be extended further so that CVR factors determined at the appliance, customer and feeder levels are used to form a CVR factor for the entire network. Other research [4, 5] has sought to overcome the complexity of extrapolating a deluge of individual load traits to the network level by introducing a linearized approximation to the established definition for CVR_p outlined in (1).

This paper explores new ways of interpreting load behaviors in the presence of voltage fluctuations, beyond that of established modelling practice, with the intention of assisting the analysis and predictability of network voltage optimization tools, such as CVR. Specifically, a small-signal load characterization methodology is proposed in which an accurate nonlinear model of the load behavior is first generated using a Multilayer Perceptron (MLP) neural network, and then voltage dependent small-signal load model parameters derived analytically as function of the MLP parameters.

The remainder of the paper is organized as follows. Section 2 introduces the two most widely used static load models (ZIP and exponential), and discusses their weaknesses. The proposed small-signal neural network based methodology is then described in Sect. 3. Results demonstrating the efficacy of the methodology are presented in Sect. 4 and finally Sect. 5 concludes the paper.

2   Static Load Modelling

2.1   ZIP Models

The ZIP load modelling approach [7] seeks to approximate active or reactive power draw as a function of applied voltage, expressed in the form of a quadratic polynomial. The coefficients of the model correspond to constant impedance (Z), constant current (I) and constant power (P) consumption terms. An equality constraint applies to the model, whereby the values of the three coefficients must exactly sum to 1. The ZIP formulation for active power, $P(V)$, is given by the following equation, in which P_0 corresponds to the active power observed at the nominal voltage level, V_0.

$$P(V) = P_0 \cdot \left[Z_p \cdot \left(\frac{V}{V_0} \right)^2 + I_p \cdot \left(\frac{V}{V_0} \right) + P_p \right] \qquad (2)$$

Cursory application of (2) shows that for a purely resistive load, i.e. one for which $Z_p = 1$, $I_p = 0$ and $P_p = 0$, the active power consumption scales with the square of the applied voltage level. Thus, the application of CVR to such loads is highly conducive in terms of reducing the power consumption of end users. The classic interpretation of a constant impedance load is the incandescent lamp [6], which until very recent times was a ubiquitous feature of distribution networks and a major component of aggregate demand. Load types of this nature are not expected to benefit from the application of CVR. Conversely, for purely constant power loads, i.e. where $Z_p = 0$, $I_p = 0$ and $P_p = 1$, active power consumption is wholly independent of applied voltage fluctuations.

Moreover, the act of reducing the voltages applied to such devices is, by definition, consistent with increasing the currents drawn and so the impact of CVR may be to increase the level of network losses incurred in delivering power to end users.

Given some measured data sets, $\mathbf{P}$ and $\mathbf{V}$, obtained for a load type that is subject to characterization, two approaches exist to estimate the resulting ZIP model:

$$\mathbf{P} = [P_1 \quad P_2 \quad \cdots \quad P_m]^T, \quad \mathbf{V} = [V_1 \quad V_2 \quad \cdots \quad V_m]^T, \qquad (3)$$

where m is the number of measurements (data points). The first step is to convert the data to per-unit form (i.e. $\mathbf{P}_{pu} = \mathbf{P}/P_0$, $\mathbf{V}_{pu} = \mathbf{V}/V_0$) and to then employ constrained least squares regression to estimate the ZIP parameters [9]. This method assumes that the nominal power (P_0) is known a priori. The second approach fits a quadratic model to the raw data. From this the nominal power (P_0) level corresponding to the nominal voltage (V_0) is derived and the corresponding per-unit ZIP parameters can be deduced analytically. The advantage of the latter approach is that the sum to unity constraint on ZIP parameters, outlined in [8], is inherently satisfied. Hence, parameter estimation can be performed via conventional unconstrained least squares regression.

2.2 Exponential Models

The exponential load model approximates the load power-voltage curve as

$$P(V) = P_0 \cdot \left(\frac{V}{V_0} \right)^{n_p} \qquad (4)$$

In contrast to the ZIP model, in (4) a load's relationship with voltage is encapsulated by a single exponent term, n_p, rather than three individual scalars. For small variations in applied voltage, it is possible to show, algebraically, that n_p offers a reasonable approximation to CVR_p. Synergies with the ZIP model become apparent by once again considering the attributes of constant impedance and constant power example load types. For a constant impedance load, (2) and (4) are observed to be equivalent when $n_p = 2$; whereas for constant power loads, $n_p = 0$. More generally, by computing a quadratic approximation to the exponential model at the nominal voltage, the relationship between the ZIP and exponential parameters can be expressed as

$$Z_p = \frac{1}{2}n_p(n_p - 1), \quad I_p = n_p(2 - n_p), \quad \text{and} \quad P_p = \frac{1}{2}\left(n_p^2 - 3n_p + 2\right) \quad (5)$$

Therefore, each n_p maps to an equivalent set of ZIP parameters.

The exponential model can also be estimated from measured data, taken in either raw or per unit form. A linear least squares estimate can be obtained by taking the log of per-unit data, giving

$$n_p = \frac{[\log(\mathbf{V}_{pu})]^T [\log(\mathbf{P}_{pu})]}{[\log(\mathbf{V}_{pu})]^T [\log(\mathbf{V}_{pu})]} \quad (6)$$

When working with raw data, expressed in base units rather than per unit, both P_0 and n_p must be estimated. Again logarithms are taken, which allows the parameter estimates to obtained from $n_p = \theta_1$ and $P_0 = \exp(\theta_2)$, where $\mathbf{X}^+$ denotes the Moore-Penrose pseudoinverse of $\mathbf{X}$, and

$$\boldsymbol{\theta} = \begin{bmatrix} \theta_1 \\ \theta_2 \end{bmatrix} = [\log(\mathbf{V}) \quad 1]^+ [\log(\mathbf{P})] \quad (7)$$

2.3   Load Modelling Challenges

The application of both the ZIP and exponential load models is best suited to the characterization of those load types that exhibit simple relationships with applied voltage. As such, their ongoing applicability has arguably become more dubious as the adoption of complex, power electronic circuitry within everyday consumer appliances has proliferated. Many such devices exhibit strongly non-linear or piecewise voltage response characteristics, against which the quadratic composition of the classic ZIP load model (2) can often only muster a rudimentary fit.

An example of piecewise active power consumption behavior is presented in Fig. 1(a) for a modern, Light Emitting Diode (LED) based lamp. Lamp performance has been tested under strict laboratory conditions across a range of applied voltage levels within the statutory range. The ensuing active power characteristic is not readily resolved to a quadratic function through ZIP modelling techniques. As such, the equivalent ZIP-fitted curve (depicted by the continuous line trace) resolves to a linearly sloping power against voltage relationship, across all voltage levels. Conversely, the raw measured data (depicted by the scattered points) portrays a piecewise, constant power trend, within which different constant power levels are defined across different bands of applied voltage. A similar characteristic, across the same range of voltages, is presented for an incandescent bulb in Fig. 1(b). In this instance the ZIP-fitted curve offers a much better approximation to the measured data.

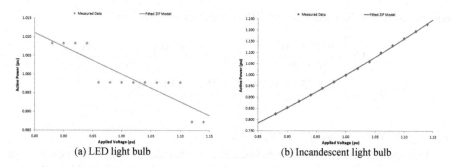

Fig. 1. Per-unit active power consumption profile versus per-unit applied voltage, for: (a) a modern 9 W LED rated lamp, and (b) a traditional 60 W rated incandescent lamp

The divergent characteristics exhibited for the LED lamps are indicative of why new innovations in load modelling are increasingly required, as the limitations of fitting modeled behaviors to simplistic quadratic or exponential functions becomes more glaring within network level, power systems studies. Many load flow simulation tools invoke ZIP load modelling techniques. Amongst the most compelling arguments for continuing this approach is the convention that each of the ZIP coefficients, Z_p, I_p and P_p, pertain to a physical property (impedance, current and power) that is easy for electrical engineers to comprehend and thus convenient for analysis. However, aside from potential increases in computational effort, there are few clear justifications for precluding the use of more complex models to represent load characteristics in simulation, especially if they are able to impart greater levels of accuracy.

3 Small-Signal Neural Network Based Load Modelling

As discussed previously, the use of a stationary set of ZIP or exponential parameters to model load characteristics, when fitted across a relatively wide range of voltage variations, fails to render a precise characterization for many modern loads. We propose an alternative framework, in which model coefficients are resolved dynamically with respect to the prevailing voltage level, across the full range of applied voltage. This approach ensures that a higher level of accuracy is obtained, as the characteristic can be more readily fitted against localized perturbations and piecewise functionality. A general nonlinear modelling paradigm $\hat{P} = f(V, \mathbf{W})$ is employed, where the model parameters, $\mathbf{W}$, are chosen so as to minimize $\left\| \mathbf{P} - \hat{\mathbf{P}}_2 \right\|$. From a Taylor series expansion of $f(V, \mathbf{W})$, small-signal ZIP and exponential models are derived to describe load behaviors with respect to small changes in voltage, about a localized reference point. The small-signal exponential model can be obtained as

$$n_p(V) = V \frac{f'(V, \mathbf{W})}{f(V, \mathbf{W})}, \tag{8}$$

and the corresponding small signal ZIP parameters are given by

$$Z_p(V) = \frac{V^2 f''(V, \mathbf{W})}{2f(V, \mathbf{W})}, \quad I_p(V) = \frac{V f'(V, \mathbf{W}) - V^2 f''(V, \mathbf{W})}{f(V, \mathbf{W})},$$
$$P_p(V) = \frac{f(V, \mathbf{W}) + 0.5 V^2 f''(V, \mathbf{W}) - V f'(V, \mathbf{W})}{f(V, \mathbf{W})}, \tag{9}$$

where $f'(V, \mathbf{W}) = \frac{\partial f}{\partial V}$ and $f''(V, \mathbf{W}) = \frac{\partial^2 f}{\partial V^2}$.

Various approaches exist for creating general nonlinear models of the power versus voltage load profile, such as high order polynomials, B-splines, SVMs and neural networks. In this instance, single hidden layer Multilayer Perceptron (MLP) neural networks are chosen as the model structure. These have universal function approximation capabilities and provide smooth fits to nonlinear functions. By employing appropriate training and cross-validation procedures the number of hidden layer neurons (N_h) and network weights ($\mathbf{W}$) can be optimized. This choice of model is also known to handle discontinuities well. The MISO MLP with one hidden layer, sigmoidal activation functions in the hidden layer neurons and a linear activation function in the output layer is defined as

$$y = \sum_{i=1}^{N_h} c_i \mathrm{sig} \left(\sum_{j=1}^{N_I} w_{ij} u_j + b_i \right) + d \tag{10}$$

where $\{c_i, w_{ij}, b_i, d\}$ are the network weights, constituting the model parameters $\mathbf{W}$, and $\mathrm{sig}(x) = \frac{1}{1+e^{-x}}$ is the sigmoid function. The Jacobian $\frac{\partial y}{\partial \mathbf{u}} = \left[\frac{\partial y}{\partial u_1} \quad \cdots \quad \frac{\partial y}{\partial u_{N_I}} \right]^{\mathrm{T}}$ of the MLP is computed as

$$\frac{\partial y}{\partial u_p} = \sum_{i=1}^{N_h} c_i w_{ip} \mathrm{sig}' \left(\sum_{j=1}^{N_I} w_{ij} u_j + b_i \right) \tag{11}$$

and the Hessian matrix $\mathbf{H} = \left[h_{pq} \right]$ of second derivatives $\frac{\partial^2 y}{\partial u_p \partial u_q}$ is given by

$$\frac{\partial^2 y}{\partial u_p \partial u_q} = \sum_{i=1}^{N_h} c_i w_{ip} w_{iq} \mathrm{sig}'' \left(\sum_{j=1}^{N_I} w_{ij} u_j + b_i \right) \tag{12}$$

where the first and second derivatives of $\mathrm{sig}(x)$ are given by

$$\mathrm{sig}'(x) = \mathrm{sig}(x)(1 - \mathrm{sig}(x))$$
$$\mathrm{sig}''(x) = \mathrm{sig}'(x)(1 - 2\mathrm{sig}(x)) \tag{13}$$

For the load characterization approach considered here, the MLP model reduces to

$$f(V, \mathbf{W}) = \sum_{i=1}^{N_h} c_i \mathrm{sig}(w_i V + b_i) + d \tag{14}$$

and the corresponding Jacobian and Hessian matrix reduce to

$$f'(V, \mathbf{W}) = \sum_{i=1}^{N_h} c_i w_i \mathrm{sig}'(w_i V + b_i) \tag{15}$$

$$f''(V, \mathbf{W}) = \sum_{i=1}^{N_h} c_i w_i^2 \mathrm{sig}''(w_i V + b_i) \tag{16}$$

4 Results

In this section, the methodology introduced in Sect. 3, Eqs. (8) to (16) in particular; is applied to estimate load models for: (a) a simulated load characteristic corresponding to a 5 kΩ resistance; (b) measurement data for a 60 W incandescent light bulb; and (c) measurement data for a modern 9 W LED light bulb. In all cases active power measurements are recorded for applied voltages in the range -12% to $+14\%$ of the nominal voltage of 230 V in increments of 2%. The MLP networks are trained using the BFGS training algorithm [10] with leave-one-out cross validation used to optimize the number of neurons, N_h. The results obtained for the three loads are presented in rows (a), (b) and (c) of Figs. 2 and 3, respectively.

In Fig. 2, the plots in the column denoted (i) show the MLP approximation of the P-V characteristic achieved, while plots of the small-signal exponential model coefficient, n_p, (computed according to (8)) are displayed in column (ii). In Fig. 3, the plots in column (i) show the small-signal ZIP parameter estimates determined directly from the MLP model, using (9); with column (ii) depicting plots of the small-signal ZIP parameter estimates determined indirectly from the n_p estimate recovered in (5).

As an ideal constant impedance example, the expected value of n_p for the 5 kΩ resistor over all voltage levels is 2, and the corresponding ZIP model values are $Z_p = 1$, $I_p = 0$, $P_p = 0$. Figure 2 confirms that the proposed methodology yields small-signal ZIP and exponential model parameter estimates consistent with these theoretical values. It is noteworthy that the indirect estimation of the ZIP parameters is more stable than direct estimation which is more susceptible to model approximation errors. This is a consequence of the extra degrees of freedom in the ZIP model and its reliance on local curvature information for parameter estimation.

The limitations of the direct small-signal ZIP model parameter estimation are even more apparent in Fig. 3, which shows the results for the incandescent light bulb. Here the stationary (global) exponential load model estimate is $n_p = 1.54$ and the stationary (global) ZIP load model of the device is $Z_p = 0.72$, $I_p = 0.11$, $P_p = 0.17$. One might expect for this device to have a n_p value of 2. However, in practice the resistance of the bulb is strongly temperature dependent, and thus increases with applied voltage level, yielding a n_p value closer to 1.5. Comparing the stationary model coefficients with the small-signal variants, it is clear that the small-signal exponential model provides consistent estimates – varying between 1.38 and 1.58 with an average of 1.53. In contrast, the directly estimated small-signal ZIP model parameters vary widely and do not correlate with the stationary ZIP model parameters. The indirectly estimated, small-signal ZIP

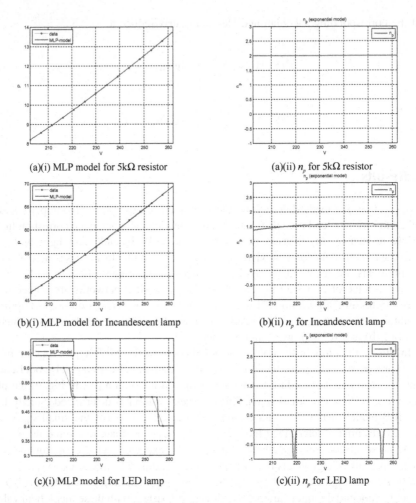

(a)(i) MLP model for 5kΩ resistor

(a)(ii) n_p for 5kΩ resistor

(b)(i) MLP model for Incandescent lamp

(b)(ii) n_p for Incandescent lamp

(c)(i) MLP model for LED lamp

(c)(ii) n_p for LED lamp

Fig. 2. Plots of the P-V data and MLP model approximation and small-signal exponential model estimates for each of the sample loads.

parameters are more stable, but it is noted that they differ substantially from the conventional, stationary model with mean values of 0.41, 0.71 and −0.12 for Z_p, I_p, and P_p, respectively. This underscores the differences that exist between a single, stationary model covering the full device operating envelope and locally valid, small-signal models.

Figure 2(c)(i) shows how the LED bulb switches between three piece-wise, constant power intervals over the voltage range investigated. However, the conventional stationary ZIP model fit of this load profile, plotted in Fig. 1(a), does not adequately capture this behavior. It is clear that the MLP model provides a much superior fit to the load profile than the downwards sloping, close to linear approximation observed within Fig. 1(a). In addition, the small-signal exponential and ZIP models correctly identify the load as being constant power over most of the voltage range, except in the vicinity of the discontinuities, where they break down.

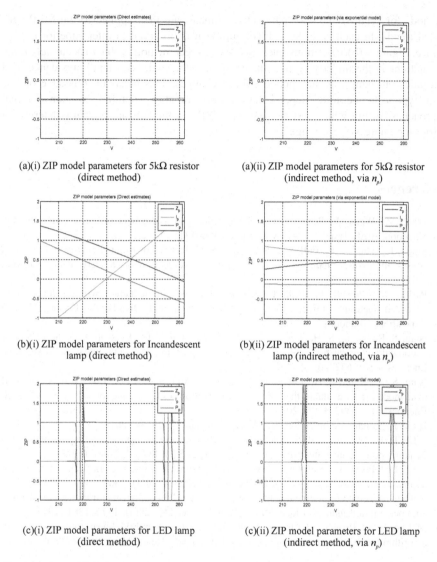

(a)(i) ZIP model parameters for 5kΩ resistor
(direct method)

(a)(ii) ZIP model parameters for 5kΩ resistor
(indirect method, via n_p)

(b)(i) ZIP model parameters for Incandescent
lamp (direct method)

(b)(ii) ZIP model parameters for Incandescent
lamp (indirect method, via n_p)

(c)(i) ZIP model parameters for LED lamp
(direct method)

(c)(ii) ZIP model parameters for LED lamp
(indirect method, via n_p)

Fig. 3. Plots of small-signal ZIP parameters estimates (directly from MLP model); and small-signal ZIP parameter estimates, indirectly derived from the exponential model estimate for each of the sample loads

5   Conclusions

This paper introduces a preliminary method for producing high accuracy, load model parameter estimates for modern load types. An MLP-based approach has been shown to offer a high fidelity fit to the P-V characteristics determined for the featured load types, from experimental measurements. This is particularly evident in the case of the piecewise functionality exhibited by an LED lamp in Fig. 2(c). Direct and indirect methods for small-signal, ZIP parameter estimates are then applied to the MLP generated trends. In each case the indirect approach is observed to offer superior performance. Figure 3 (b)(ii) demonstrates how the indirect method is able to accurately track fluctuations in the underlying impedance of an archetypal, "constant" impedance load (incandescent lamp), as the applied voltage changes. This behavior, reflective of this load type's temperature dependency, is not reliably captured by conventional approaches, within which a stationary set of ZIP coefficients are typically determined across the entire voltage range.

References

1. Wang, Z., Wang, J.: Review on implementation and assessment of conservation voltage reduction. IEEE Trans. Power Syst. **29**(3), 1306–1315 (2014)
2. Singh, R., Tuffner, F., Fuller, J., Schneider, K.: Effects of distributed energy resources on conservation voltage reduction (CVR). In: IEEE Power and Energy Society General Meeting (2011)
3. Peskin, M., Powell, P., Hall, E.: Conservation voltage reduction with feedback from advanced metering infrastructure. In: Transmission and Distribution Conference and Exposition, Orlando, FL (2012)
4. Nam, S.-R., Kang, S.-H., Lee, J.-H., Ahn, S.-J., Choi, J.-H.: Evaluation of the effects of nationwide conservation voltage reduction on peak-load shaving using SOMAS data. Energies **6**(12), 6322–6334 (2013)
5. Nam, S.-R., Kang, S.-H., Lee, J.-H., Choi, E.-J., Ahn, S.-J., Choi, J.-H.: EMS-data-based load modeling to evaluate the effect of conservation voltage reduction at a national level. Energies **6**(8), 3692–3705 (2013)
6. Hunt, J.: Voltage optimization. IET Wiring Matters **47**, 29–34 (2013)
7. IEEE Task Force on Load Representation for Dynamic: Performance: load representation for dynamic performance analysis. IEEE Trans. Power Syst. **8**(2), 472–482 (1993)
8. IEEE Task Force on Load Representation for Dynamic: Performance: load representation for dynamic performance analysis. IEEE Trans. Power Syst. **10**(3), 1302–1313 (1995)
9. Bokhari, A., et al.: Experimental determination of the ZIP coefficients for modern residential, commercial, and industrial loads. IEEE Trans. Power Del. **29**(3), 1372–1381 (2014)
10. McLoone, S., Brown, M., Irwin, G., Lightbody, G.: A hybrid linear/nonlinear training algorithm for feedforward neural networks. IEEE Trans. Neural Netw. **9**(4), 669–684 (1998)

H_∞ Prediction Triggering Control of Multi-area Power Systems Load Frequency Control Under DoS Attacks

Zihao Cheng[1]([✉]), Dong Yue[2], Xinli Lan[3], Chongxin Huang[2], and Songlin Hu[2]

[1] School of Computer Science,
Nanjing University of Posts and Telecommunications, Nanjing 210023, China
2016040229@njupt.edu.cn
[2] Institute of Advanced Technology,
Nanjing University of Posts and Telecommunications, Nanjing 210023, China
[3] School of Automation, Nanjing University of Science and Technology,
Nanjing 210094, China

Abstract. This paper is concerned with load frequency control of multi-area power system under DoS attacks. We introduce inner virtual event triggering mechanism under framework of model based predictive control to design load frequency control. Then Lyapunov stability theory is used to analysis H_∞ predictive control problem. Sufficient condition is derived in form of linear matrix inequalities which guarantees the closed loop system is stable with H_∞ performance. Simulation of a two-area power system is given to illustrate the effectiveness of the proposed method in dealing with long duration DoS attack.

Keywords: Load frequency control (LFC) · Denial-of-service (DoS) attacks · Model predictive control · Event-triggered control · Power system

1 Introduction

With more open IT infrastructure embedded, power system is vulnerable to cyber threat from communication network. Recently, cyberattack of networked system investigated has been mainly classified into two classes. One is denial-of-service (DoS) attack which prevents signals from exchanging between specially distributed notes of system i.e. plant and remote controller. The other one is deception attack which affecting the value of transmitted signals. Some interesting results has been reported in literatures. De Persis in [1] established a novel DoS model by frequency and duration and give constraint condition between trigger time and DoS under which system is input-to-state stability. Yuan in [2] studied the resilient control of network control system under DoS attack via a unified game approach to get optimal solution both defender and attacks. Considering DoS attack as one kind of network constraint conditions, Ding et al. in [3] and [4] investigated security problem of stochastic system concentrating on the random behavior of both DoS attack and deception attack. And Zhang in [5] introduced time-delay system method for one kind of DoS attack with maximum upper bounded.

K. Li et al. (Eds.): LSMS/ICSEE 2017, Part III, CCIS 763, pp. 477–487, 2017.
DOI: 10.1007/978-981-10-6364-0_48

Amin et al. in [6] presented an optimal feedback control method minimizing a given objective function with energy constraint DoS attack. Other optimal methods has been studied in [7, 8] from the viewpoint of attacker. For multi-agent system under DoS attacks, Amullen et al. in [9] investigated the secured formation control problem based on continuous on-line system identification.

For multi-area power system, load frequency control (LFC) is an effective mechanism to balance frequency derivation and power exchange among other area system with the existence of load change. And this problem has been attracted most popular researchers and some valuable control methods has been provided to design LFC including Proportional-Integral control [10], state feedback control [11, 12], model predictive control [13].

Furthermore, in networked control system there are two thoughts dealing with system control problem with physical network constraint. One positive strategy is predictive control method which could compensate date dropt and date delay to accomplish system performance like local control [14, 15]. From other aspect, event trigger mechanism is one of most effective method of saving network transmission resources [17–20]. It also indicates a fact that system performance could be obtained with less control inputs which has an obvious advantage when lacks chances of control input like happening DoS attack.

In this paper, we consider attack scenario of DoS launched on communication channels in the control loop of power systems. If attacked succeed, control signals would be dropped before obtained by actuator. This situation could be viewed as a case of open loop control of system. And the more duration of DoS attack, the more malicious influence of system [8]. Hence, we aim to solve the long duration of DoS attack in this note. And we not only utilize model based predictive control method to compensate control date packet loss induced by DoS attack but also introduce virtual event-trigger into controller to improve compensation length.

The remainder of this paper is organized as follows. In Sect. 2, the power system is established by form of linear discrete time system. Then, we introduce observer and model based predictor with event trigger in Sect. 3. System analysis and control parameters design is in Sect. 4. Simulation results are presented in Sect. 5. Finally, Sect. 6 concludes this paper.

2   LFC Model of Multi-area Power Systems

In this section, the classical model of load frequency control of multi-area interconnected power system [10, 11] is established with DoS attacks occurring in control channel shown in Fig. 1. From Fig. 1, the MPC controller sends control signals to actuator through communication channels. And the adversaries launch DoS attacks in customer-edge router to cause transmission channel break. Then without feedback control inputs, system operation is in open loop which seriously damage dynamic performance of power system. Hence, we plan to utilize predictive control scheme which send out prediction control signals to network DoS compensator (NDC) which provides control inputs during DoS attack. The following model describes the dynamic of LFC in area i.

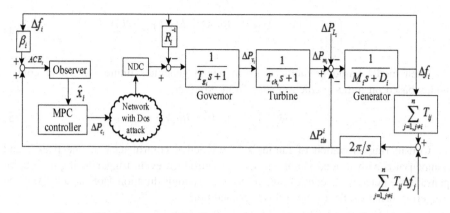

Fig. 1. Block diagram of the ith control area under network DoS attack

$$\dot{\Delta f_i} = \frac{1}{M_i}\left(-D_i\Delta f_i + \Delta P_{m_i} - \Delta P_{tie}^{ij} - \Delta P_{L_i}\right), \quad \dot{\Delta P}_{m_i} = \frac{1}{T_{ch_i}}\left(-\Delta P_{m_i} + \Delta P_{v_i}\right)$$

$$\dot{\Delta P}_{v_i} = \frac{1}{T_{g_i}}\left(-\frac{1}{R_i}\Delta f_i - \Delta P_{v_i} + \Delta P_{c_i}\right), \quad \dot{\Delta P}_{tie}^i = \sum_{j=1,j\neq i}^{n} 2\pi T_{ij}\left(\Delta f_i - \Delta f_j\right) \quad (1)$$

$$ACE_i = \beta_i\Delta f_i + \Delta P_{tie}^i$$

For the whole multi-area power system, system model is given by

$$\begin{cases} x(k+1) = Ax(k) + Bu(k) + Fw(k) \\ y(k) = Cx(k) \end{cases} \quad (2)$$

where h is sample time, $A = e^{A_c h}, B = \int_0^h e^{A_c \tau}d\tau B_c, F = \int_0^h e^{A_c \tau}d\tau F_c$.

And we employ state feedback control as strategy of LFC

$$u(k) = Kx(k) \quad (3)$$

where K is state feedback gain to be designed.

3 Model-Based Predictive Triggering Control Under DoS Attack

Because states of system could not be measured directly in actual system. For utilizing states feedback control scheme, it is essential to design an observer at the plant sides. And observer function is given as follows

$$\begin{cases} \hat{x}(k+1) = A\hat{x}(k) + Bu(k) + L(\hat{y}(k) - y(k)) \\ \hat{y}(k) = C\hat{x}(k) \end{cases} \tag{4}$$

Then model based predictor is given which generates future states and control inputs based on initial prediction state from observer.

$$\hat{x}(k+1) = A\hat{x}(k) + B\hat{u}(k) \tag{5}$$

Considering limitation of numbers of predictive control data in one packet and compensator with limited storage space, we introduce event-trigger to improve compensation efficiency of control signals to against long duration DoS attack. And the triggering instant satisfies the following condition.

$$k_{s_i+m+1} = k_{s_i+m} + \min_{\tau}\{\tau | [\hat{x}(k_{s_i+m}+\tau) - \hat{x}(k_{s_i+m})]^T \Phi [\hat{x}(k_{s_i+m}+\tau) - \hat{x}(k_{s_i+m})]$$
$$> \mu \hat{x}(k_{s_i+m}+\tau)^T \Phi \hat{x}(k_{s_i+m}+\tau)\}$$

where k_{s_i} represents the successful transmitted time of control data packet and k_{s_i+m} is virtual triggering time in controller with $m = 0, 1, \ldots, N_c - 1$, and $0 < \mu < 1$ is a given scalar parameter, Φ is a positive definite symmetric matrix to be designed.

Remark 1. It is clear that $\{k_{s_i}\} \subseteq \{k\}$ and $k_{s_{i+1}} = k_{s_i} + 1$ meaning of without DoS attack. And this event-trigger is called virtual trigger because it is just calculated in controller program instead of occurring in actual network transmission.

Remark 2. If DoS attack happens, actuator would pick out predictive control signal according to timestamp $k_{s_i+m}, m \in 0, 1, 2, \ldots N_c - 1$. Otherwise, only first control signal $\hat{u}(k_{s_i})$ would be implemented in actuator.

Furthermore, DoS attack is one kind of cyber accident which damage the usability of control input from time. It is different from random data missing caused by network itself physically. Because attacker with limited energy is ambitious to lead to more damage to control system. Hence successive attack is one of most serious strategy which leads to long time of successive packet dropout causing system out of tolerant stability zone. Therefore it is urgent to compensate system during DoS attack.

In our work, we consider one serious attack scenario that DOS attack with limited energy is implemented in forward loop channel between controller and actuator. And only attack duration as the most fatal characteristic of DoS attack is considered with

$$\lambda_D \leq T_D$$

And to simplify analysis process, we set $T_D = Mh$ and $T_D < k_{s_i+N_c} - k_{s_i}$.

Assumption 1. The constraints between timestamp of prediction control signals and DoS attack duration satisfies

$$k_{s_{i+1}} = k_{s_i} + T_D + 1, k_{s_{i+1}} \leq k_{s_i+N_c}, k_{s_{i+1}-1} = k_{s_i+N_c-1}$$

which means that the predictive length is longer than DoS duration.

For the convenience of analysis, we divided interval into subintervals as $k \in [k_{s_i}, k_{s_{i+1}}) \cap [k_{s_i+m}, k_{s_i+m+1})$, $m = 0, 1, \ldots, s_{i+1} - s_i - 1$. Further, system dynamic with predictive error is represented according to the above function

$$x(k+1) = (A+BK)x(k) - BK\left(e_p(k) + \hat{\delta}(k)\right) + Fw(k) \tag{6}$$

where $e_p(k) = x(k) - \hat{x}(k|k_{s_i})$, $\hat{\delta}(k) = \hat{x}(k|k_{s_i}) - \hat{x}(k_{s_i+m}|k_{s_i})$, $e_p(k), \hat{\delta}(k)$ are states error of prediction and event trigger error respectively.

Observer error function is given by

$$e_o(k+1) = (A - LC)e_o(k) + Fw(k) \tag{7}$$

Furthermore, defining $X(k) = \left[x^T(k), e_o^T(k)\right]^T$ and $W(k) = \left[e_p^T(k), \hat{\delta}^T(k), w^T(k)\right]^T$, according to (6) and (7) the closed-loop system can be described as

$$X(k+1) = \overline{A}X(k) + \overline{F}W(k)$$
$$Y(k) = \overline{C}X(k) \tag{8}$$

where

$$\overline{A} = \begin{bmatrix} A+BK & 0 \\ 0 & A-LC \end{bmatrix}, \quad \overline{F} = \begin{bmatrix} -BK & -BK & F \\ 0 & 0 & F \end{bmatrix}, \quad \overline{C} = [C \quad 0]$$

Remark 3. According to the event triggered condition, the trigger errors satisfy $\hat{\delta}^T(k)\Phi\hat{\delta}(k) \le \mu\left(x(k) - e_p(k)\right)^T \Phi\left(x(k) - e_p(k)\right)$. It could be further described by

$$W^T(k)\overline{\Phi}W(k) \le \mu[I_1X(k) - I_2W(k)]^T \Phi[I_1X(k) - I_2W(k)] \tag{9}$$

Definition 1. For a given scalar $\gamma > 0$, system (8) is said to be stabilization with an H_∞ performance if there exists a state feedback law $U(k_{s_i})$ such that

(1) The closed-loop system is asymptotically stable when $W(k) = 0$.
(2) Under zero initial condition, the controlled output satisfies

$$\sum_{k=0}^{\infty} \|Y(k)\| < \gamma^2 \sum_{k=0}^{\infty} \|W(k)\|$$

for any $W(k) \in l_2(0, \infty)$.

From above analysis, power system LFC in open loop control caused by DoS attack has been transformed into a disturbance attenuate problem. The disturbances include not only load deviation but also state prediction error and prediction states triggered error. And robust H_∞ control method can be used to eliminate the two deviations. Besides the proposed method has a new meaning that timeliness problem of control caused by DoS attack is translated to value problem of state deviation because of introducing model-based predictive control strategy.

4 H_∞ Predictive Controller Design

Theorem 1. For given parameter $0 < \mu < 1, \gamma > 0$, if there exist matrices $P > 0, \Phi > 0$ and matrix K, L with appropriate dimensions satisfying the following inequality

$$\begin{bmatrix} \bar{A}^T P \bar{A} - P + \bar{C}^T \bar{C} + \mu \Phi_1 & \bar{A}^T P \bar{F} - \mu \Phi_{12} \\ * & \bar{F}^T P \bar{F} + \mu \Phi_2 - \bar{\Phi} - \gamma^2 I \end{bmatrix} < 0 \qquad (10)$$

system (2) is asymptotically stable with an H_∞-norm bound γ.

Proof. Choose a Lyapunov functional candidate as

$$V(k) = X^T(k) P X(k)$$

where P is a symmetric positive definite matrix. For $k \in \left[k_{s_i}, k_{s_{i+1}}\right) \cap \left[k_{s_i + m}, k_{s_i + m + 1}\right)$ with $w(k) = 0$, calculating the difference of $V(k)$ along the trajectory of (8) yields

$$\Delta V(k) = V(k+1) - V(k) = X^T(k)\left(\bar{A}^T P \bar{A} - P\right) X(k)$$

Form (10), it indicates that $\Delta V(k) < 0$, which guarantees that the closed-loop system (8) is asymptotically stable.

Then, we prove the stable system having H_∞ performance with considering the event triggered condition in (9)

$$\begin{aligned} & \Delta V(k) + Y^T(k) Y(k) - \gamma^2 W^T(k) W(k) - \hat{\delta}^T(k) \Phi \hat{\delta}(k) \\ & \quad + \mu \left(x(k) - e_p(k)\right)^T \Phi \left(x(k) - e_p(k)\right) \\ & = \begin{bmatrix} X^T(k) & W^T(k) \end{bmatrix} \begin{bmatrix} \bar{A}^T P \bar{A} - P + \bar{C}^T \bar{C} + \mu \Phi_1 & \bar{A}^T P \bar{F} - \mu \Phi_{12} \\ * & \bar{F}^T P \bar{F} + \mu \Phi_2 - \bar{\Phi} - \gamma^2 I \end{bmatrix} \begin{bmatrix} X(k) \\ W(k) \end{bmatrix} \end{aligned}$$

where $\Phi_1 = I_1^T \Phi I_1, \Phi_{12} = I_1^T \Phi I_2, \Phi_2 = I_2^T \Phi I_2$.

According to (9) and (10), it follows that

$$\Delta V(k) + Y^T(k) Y(k) - \gamma^2 W^T(k) W(k) < 0 \qquad (11)$$

Under zero initial condition, adding both sides of (11) from $k = 0$ to $k = \infty$, we can obtained that

$$\sum_{k=0}^{\infty} \|Y(k)\| < \gamma^2 \sum_{k=0}^{\infty} \|W(k)\|$$

Thus, the H_∞-norm in Definition 1 of the closed-loop system (8) is less than γ. The proof is completed.

Theorem 1 gives a sufficient condition which guarantees system stable with H_∞ disturbance attenuate lever γ. Further we will convert matrix inequality condition (10) into LMIs by solving which controller and observer parameters could be obtained.

Theorem 2. For given parameter $0 < \mu < 1$, $\alpha, \beta, \lambda, \gamma > 0$, the system (8) is asymptotically stable with H_∞ norm bound γ if there exist matrices $\tilde{P}, \hat{P} > 0, \tilde{\Phi} > 0$ and matrix $\tilde{K}, M$ with appropriate dimensions satisfying the following LMI

$$
\begin{bmatrix}
-\tilde{P} + \mu\tilde{\Phi} \\
0 & -\alpha\tilde{P} \\
-\mu\beta\tilde{\Phi} & 0 & \Pi_1 \\
0 & 0 & 0 & \Pi_2 \\
0 & 0 & 0 & 0 & -\gamma^2 I \\
\Pi_3 & 0 & -B\tilde{K} & -B\tilde{K} & F & -\tilde{P} \\
0 & \Pi_4 & 0 & 0 & F & 0 & -\alpha\tilde{P} \\
C\tilde{P} & 0 & 0 & 0 & 0 & 0 & 0 & -I
\end{bmatrix} < 0 \qquad (12)
$$

where

$$\Pi_1 = \mu\beta^2\tilde{\Phi} - \beta^2\gamma^2\hat{P}, \Pi_2 = -\lambda^2\tilde{\Phi} - \lambda^2\gamma^2\hat{P}, \Pi_3 = A\tilde{P} + B\tilde{K}, \Pi_4 = \alpha(A\tilde{P} - MC)$$

Proof. The inequality (10) in Theorem 1 could be rewritten by

$$
\begin{bmatrix} \bar{A} & \bar{F} \\ \bar{C} & 0 \end{bmatrix}^T \begin{bmatrix} P & 0 \\ 0 & I \end{bmatrix} \begin{bmatrix} \bar{A} & \bar{F} \\ \bar{C} & 0 \end{bmatrix} + \begin{bmatrix} -P + \mu\Phi_1 & -\mu\Phi_{12} \\ * & \mu\Phi_2 - \bar{\Phi} - \gamma^2 I \end{bmatrix} < 0 \qquad (13)
$$

Applying schur complement lemma and pre- and post-multiplying both sides of (13) with $diag\{P^{-1}, I, I, I\}$, we have

$$
\begin{bmatrix}
-P^{-1} + \mu P^{-1}\Phi_1 P^{-1} \\
-\mu\Phi_{12}^T P^{-1} & \mu\Phi_2 - \bar{\Phi} - \gamma^2 I & * \\
\bar{A}P^{-1} & \bar{F} & -P^{-1} \\
\bar{C}P^{-1} & 0 & 0 & -I
\end{bmatrix} < 0 \qquad (14)
$$

Then letting $P = diag\{\tilde{P}^{-1}, \alpha^{-1}\tilde{P}^{-1}\}$ and pre- and post-multiplying (14) with $diag\{I, I, \beta\tilde{P}, \lambda\tilde{P}, I, I, I, I\}$, where $\tilde{P}$ is also a positive definite matrix and α, β, λ are positive parameters that can be tuned. By denoting $\tilde{\Phi} = \tilde{P}\Phi\tilde{P}, \tilde{K} = K\tilde{P}, \hat{P} = \tilde{P}P$ and $MC \triangleq LC\tilde{P}, NC \triangleq C\tilde{P}$, we can obtain (12). According to Theorem 1, it can be obtained that if the LMI (12) holds, then the system (8) is asymptotically stable with H_∞ performance. This proof is completed.

In the above proof, we introduce two terms of variable substitution to make (14) LMIs including $MC \triangleq LC\tilde{P}, NC \triangleq C\tilde{P}$ and $\hat{P} = \tilde{P}\tilde{P}$. Here, to figure out observer parameter L and deal with the linear term of definition, we introduce two additional LMIs respectively as followings.

$$\left\{ \begin{bmatrix} -\varepsilon I & * \\ NC - C\tilde{P} & -I \end{bmatrix} < 0 \quad \text{and} \quad \left\{ \begin{bmatrix} -\hat{P} - \eta I & * \\ \tilde{P} & -I \end{bmatrix} < 0 \atop \eta \to 0 \right. \tag{15}$$

From above analysis, we finish system synthesis section. The parameters of controller $K = \tilde{K}\tilde{P}^{-1}$ with event-trigger $\Phi = \tilde{P}^{-1}\tilde{\Phi}\tilde{P}^{-1}$ and observer $L = MN^{-1}$ can be obtained by solving LMIs (12) and (15).

5 Numerical Example

The effectiveness of the proposed predictive triggering control (PTC) will be demonstrated through a numerical simulation of two-area power system [10] under DoS attack. One of the two-area system diagram is shown in Fig. 1, where area-1 has two generators and area-2 is with four generators. The generators in both area-1 and area-2 are model by a single equivalent generators. The plant and controller parameters are listed in as following.

Area-1: parameters of plant

$$M_1 = 10, \quad D_1 = 1, \quad T_{ch1} = 0.3\,\text{s}, \quad T_{g1} = 0.1\,\text{s}, \quad R_1 = 0.1\,\text{s},$$
$$\beta_1 = 2/R_1 + D_1, \quad T_{12} = 0.2,$$

Area-2: parameters of plant

$$M_2 = 12, \quad D_2 = 1.5, \quad T_{ch2} = 0.17\,\text{s}, \quad T_{g2} = 0.4\,\text{s}, \quad R_2 = 0.05\,\text{s},$$
$$\beta_2 = 4/R_2 + D_2.$$

And load changes in form of impulse happen with times and values as followings.

$$t = \begin{bmatrix} 10\,\text{s} & 20\,\text{s} & 30\,\text{s} & 40\,\text{s} & 50\,\text{s} & 60\,\text{s} & 70\,\text{s} & 80\,\text{s} \end{bmatrix}$$
$$\Delta P_{L_1} = \begin{bmatrix} -0.8\,\text{pu} & -0.5\,\text{pu} & -0.6\,\text{pu} & -1\,\text{pu} & 0.3\,\text{pu} & 0.5\,\text{pu} & 0.6\,\text{pu} & 1\,\text{pu} \end{bmatrix}$$
$$\Delta P_{L_2} = \begin{bmatrix} -0.8\,\text{pu} & 0.5\,\text{pu} & -0.6\,\text{pu} & -0.2\,\text{pu} & 0.3\,\text{pu} & 0.7\,\text{pu} & 0.6\,\text{pu} & -0.3\,\text{pu} \end{bmatrix}$$

The simulation interval is $[0, 100\,\text{s}]$ and DoS attack is launched during $[10\,\text{s}, 60\,\text{s}]$. Then we set the discrete time system sampling interval $h = 10^{-3}\,\text{s}$, H_∞ performance index $\gamma = 4$, and virtual event-trigger parameter $\mu = 0.8$. Furthermore, applying Theorem 2 with given parameters $\alpha = 1.8, \beta = 3, \lambda = 10, \varepsilon = 10^{-6}, \eta = 10^{-6}$, the feedback control gain K is obtained as follows.

$$K = 1 \times 10^4 \begin{bmatrix} -1.0113 & -0.0263 & -0.0100 & 0.0163 & 0.2469 & 0.0027 & -0.0000 & 0.0142 \\ -0.0948 & -0.0011 & -0.0000 & -0.0111 & -2.2009 & -0.0155 & -0.0400 & 0.0233 \end{bmatrix}$$

The frequency deviation dynamic of two-area power system and load changes are shown in Figs. 2 and 3. From Fig. 2, it is shown that system states are approached to zero after load change which means the closed loop power system is stable. By a simple calculation, one can obtain $\|y\|_2 = 0.0751$ and $\|w\|_2 = 6.27$, which yields the H_∞-norm $\gamma^* = 0.012$. It is obvious that $\gamma^* < \gamma = 4$ indicating the system with H_∞ disturbance attenuate performance under DoS attack within the framework of proposed control method.

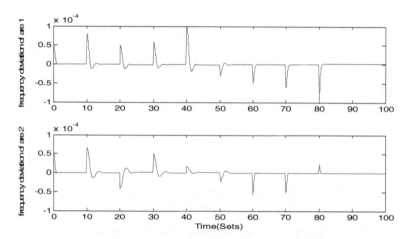

Fig. 2. The frequency derivation of two-area power system under DoS attack

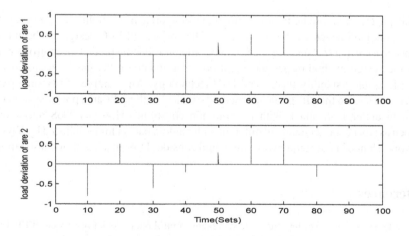

Fig. 3. The load change of two-area power system

From the other aspect, it is shown in Fig. 4 that the plant is operated by using the predictive control inputs in compensator. And the control update is based on the timestamp of control inputs stuck by the virtual event trigger in predictive controller. The statistical number of times of control update is 59 accounting for 0.0012 of DoS attack duration [10 s, 60 s]. And comparing with the results in [21] which the maximum number of packet losses induced by DoS attack is 3 h (h = 0.002 s), the method in this paper has an advantage in dealing with long DoS attack.

This result verify our original thought that predictive triggering control could compensate long DoS attack under transmission constraint in practice that the number of control signals in one packet transmitted is very limited rather than infinite.

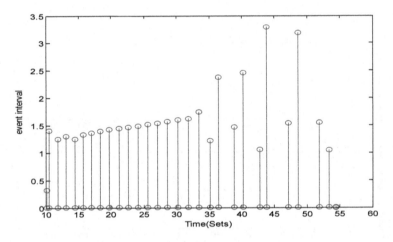

Fig. 4. Event interval during DoS attack interval

6 Conclusion

In this paper, a H_∞ predictive triggering control approach for multi-area power system LFC under DoS attack has been presented. The problem of LFC design has been transformed to a normal H_∞ state feedback control after introducing predictive control scheme. And the design method of parameters of observer and controller has been provided by using Lyapunov stability theory and LMI technology. An example of two-area power system has shown that the proposed method can deal with multi-area power system LFC problem under DoS attack with transmission constraint. However DoS attack often happen as stochastic accident but different with random packet losses induced by physical network. Hence in our future work, we would consider DoS attack as a stochastic model.

References

1. De Persis, C., Tesi, P.: Input-to-state stabilizing control under denial-of-service. IEEE Trans. Autom. Control **60**(11), 2930–2944 (2015)

2. Yuan, Y., Yuan, H., Guo, L., Yang, H., Sun, S.: Resilient control of networked control system under DoS attacks: a unified game approach. IEEE Trans. Ind. Inf. **12**(5), 1786–1794 (2016)
3. Ding, D., Wang, Z., Wei, G.: Event-based security control for discrete-time stochastic systems. IET Control Theory Appl. **10**(15), 1808–1815 (2016)
4. Zhang, J., Peng, C., Masroor, S., Sun, H., Chai, L.: Stability analysis of networked control systems with denial-of-service attacks. In: 2016 UKACC 11th International Conference on Control, Belfast, pp. 1–6 (2016)
5. Hu, L., Wang, Z., Naeem, W.: Security analysis of stochastic networked control systems under false data injection attacks. In: 2016 UKACC 11th International Conference on Control, Belfast, pp. 1–6 (2016)
6. Amin, S., Cárdenas, A.A., Sastry, S.S.: Safe and secure networked control systems under denial-of-service attacks. In: Majumdar, R., Tabuada, P. (eds.) HSCC 2009. LNCS, vol. 5469, pp. 31–45. Springer, Heidelberg (2009). doi:10.1007/978-3-642-00602-9_3
7. Zhang, H., Cheng, P.: Optimal denial-of-service attack scheduling with energy constraint. IEEE Trans. Autom. Control **60**(11), 3023–3028 (2015)
8. Zhang, H., Cheng, P., Shi, L., Chen, J.: Optimal DoS attack scheduling in wireless networked control system. IEEE Trans. Control Syst. Technol. **24**(3), 843–852 (2016)
9. Amullen, E.M., Shetty, S., Keel, L.H.: Secured formation control for multi-agent systems under DoS attacks. In: 2016 IEEE Symposium on Technologies for Homeland Security (HST), Waltham, MA, pp. 1–6 (2016)
10. Bevrani, H., Hiyama, T.: Robust decentralized PI based LFC design for time-delay power system. Energy Convers. Manag. **49**, 193–204 (2008)
11. Dotta, D., Silva, A.S.: Wide-area measurements-based two–level control design considering signal transmission delay. IEEE Trans. Power Syst. **24**(1), 208–216 (2009)
12. Dey, R., Ghosh, S., Ray, G.: H_∞ load frequency control of interconnected power systems with communication delays. Int. J. Electr. Power Energy Syst. **42**(1), 672–684 (2012)
13. Shiroei, M.: Supervisory predictive control of power system load frequency control. Int. J. Electr. Power Energy Syst. **61**, 70–80 (2014)
14. Liu, G.P.: Predictive controller design of networked systems with communication delays and data loss. IEEE Trans. Circuits Syst. II Express Briefs **57**(6), 481–485 (2010)
15. Yin, X., Yue, D., Songlin, H., Peng, C., Xue, Y.: Model-based event-triggered predictive control for networked systems with data dropout. SIAM J. Control Optim. **54**(2), 567–586 (2016)
16. Yin, X., Yue, D., Hu, S.: Model-based event-triggered predictive control for networked systems with communication delays compensation. Int. J. Robust Nonlinear Control **25**, 3572–3595 (2015)
17. Hu, S., Yue, D.: Event-triggered control design of linear networked systems with quantizations. ISA Trans. **51**, 153–162 (2012)
18. Yue, D., Tian, E., Han, Q.: A delay system method for designing event-triggered controllers of networked control systems. IEEE Trans. Autom. Control **58**(2013), 475–481 (2013)
19. Garcia, E., Antsaklis, P.: Model-based event-triggered control for systems with quantization and time-varying network delays. IEEE Trans. Autom. Control **58**(2013), 422–434 (2013)
20. Heemels, W., Donkers, M.: Model-based periodic event-triggered control for linear systems. Autom. J. IFAC **49**(2013), 698–711 (2013)
21. Peng, C., Li, J., Fei, M.R.: Resilient event-triggered H_∞ load frequency control for networked power systems with energy-limited DoS attacks. IEEE Trans. Power Syst. **PP**(99), 1 (2016)

New Framework Mining Algorithm Based Main Operation Parameters Optimization in Power Plant

Wencheng Huang[1], Li Jia[1(✉)], and Daogang Peng[2]

[1] School of Mechatronics Engineering and Automation, Shanghai Key
Laboratory of Power Station Automation Technology,
Shanghai University, Shanghai, China
jiali@staff.shu.edu.cn
[2] College of Automation Engineering, Shanghai Key Laboratory of Power
Station Automation Technology, Shanghai University of Electric Power,
Shanghai, China

Abstract. Association rule mining algorithm based on support-confidence framework is widely applied to the optimization of main operating parameters value in thermal power plant. But some important potential knowledge is easy to be overlooked by the framework in the actual mining process. Moreover, the simulation experiments show that there is a great relationship between mining results and a given minimum support threshold. Thus a dynamic interestingness-support framework mining algorithm based on metarules guided is proposed by which parameters for multidimensional association rules can be determined. The new framework reduces the redundancy of results by metarule-guided mining. And it mainly screens association rules with the index of interestingness except support, so as to weaken the dependence between mining results and the minimum support threshold. What is more, a new similarity criterion is introduced in dividing production process condition, to avoid the single spherical cluster determined by the Euclidean distance. Therefrom overcome the shortness of traditional dividing. The simulation results show that the algorithm proposed in this paper can effectively tap out the rules. And the rules can correctly reflect the knowledge of the unit and improve the accuracy of main operation parameters value in thermal power plant.

Keywords: New framework · Metarules · Main operation parameters optimization · Production process condition

1 Introduction

In order to ensure the safety, stability and efficiency of the thermal power plant, the operation parameters optimization in thermal power plant becomes one of the research fields. Conditions emerging in high frequency are not always the optimal ones. The frequency of production process condition is closely related to the unit operation level and other external factors [1]. Under the same external constraint conditions, the unit is operating under different stable operation conditions due to differences in the operation

© Springer Nature Singapore Pte Ltd. 2017
K. Li et al. (Eds.): LSMS/ICSEE 2017, Part III, CCIS 763, pp. 488–496, 2017.
DOI: 10.1007/978-981-10-6364-0_49

level. The power plant's production process condition is determined by a group of parameters that consist of environmental factors, fuel characteristics and the load of unit and other corresponding operating parameters which are not easy to be controlled in power plant. For thermal power plant's operators, the optimal parameters under certain conditions have guidance significance. The optimal steady-state production process condition is the core issues in the study of operation parameters optimization in the thermal power plant.

Recently, the operation parameters optimization research is focused on the following aspects. Using rough set theory [2] to optimal operation parameters which are combined with data mining to complete attributes reduction and optimal target value by association rules mining. The data discretization is another one in these important studies, such as equal width binning which transforms continuous attributes to discrete ones. Fuzzy association rules mining with fuzzy discretization algorithm extract the knowledge rules from the historical database [3]. In addition, the research of dynamic data mining in thermal power plant is also an improvement of the on-line operation optimization of the unit based on the current mining algorithm [4]. Furthermore, several methods for improving the performance of k-means algorithm are used to divide the production process conditions including parameters of coal quality, circulating water inlet temperature and unit power load. In short, the above research results are not aware of the weakness of the support-confidence framework and the artificial divisions for production process condition [6].

2 Framework of New Algorithm

In operation parameters optimization research, traditional association rules mining framework will product large amounts of mining results due to its own weakness. Users can not directly get the interested knowledge from the association rules, and it needs spend a lot of time to filter out the interesting part of the whole results (Fig. 1).

2.1 The Condition Division Based on a New Similarity

In the condition division section, we propose a new definition of the similarity of two points to enhance the ability of traditional similarity in processing high dimensional data with more information on both distance and angle to avoid the single spherical cluster in traditional algorithm [8]. To this end, the following similarity $s_{i,q}$, is defined:

$$s_{i,q} = \gamma \frac{1}{\sqrt{2\pi}\sigma} e^{-\frac{d(x_i,x_q)}{2\sigma^2}} + (1-\gamma)\cos(\theta_{i,q}), \quad if \ \cos(\theta_{i,q}) \geq 0 \tag{1}$$

where $\gamma \in (0,1)$ is a weight parameter and is constrained between 0 and 1, σ is mean of quadratic norm of x_i and x_q. And the Euclidian between x_i and x_q can be defined as:

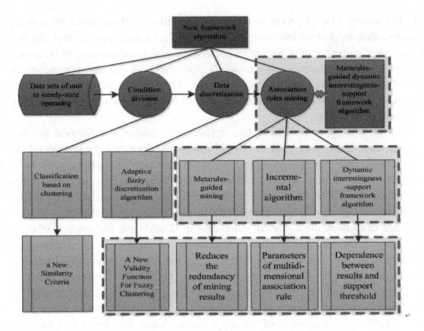

Fig. 1. Architecture diagram of new algorithm

$$d(x_i, x_q) = \sqrt{(x_i - x_q)'(x_i - x_q)} \tag{2}$$

And the angle $\theta_{i,q}$, for point x_i and x_q defined as:

$$\cos(\theta_{i,q}) = (x_i' x_q)/(\|x_i\| \|x_q\|) \tag{3}$$

where $\|x_i\| = \sqrt{\|x_i\|' \|x_i\|}$.

It is important to note that Eq. (1) will not be used to compute the similarity particle s_i between x_i and x_q, if $\cos(\theta_{i,q})$ is negative. For simplicity, this point is illustrated in the 2-dimensional space, where x_q denotes the vector perpendicular to x_q. It is clear that x_p which lies in the right of $\bar{x}_q$ is more similar to x_q then x_q which lies in the left of $\bar{x}_q$. The use of cosine function to discriminate the directionality between x_i and x_q indicates that these two vectors are dissimilar if cosine is negative, and hence the x_q in the reference group is discarded.

2.2   Adaptive Fuzzy Discretization

At present, the methods mainly used in the discretization of the historical operation data set are those such as the equal width method, the equal frequency method, the clustering algorithm and so on. Method of equal width will divide the attribute range partition into sections with the same width which generally contains the same number of samples. Number of discretization interval is determined by the user, but if there is a very serious deviation of the data in the discretization interval, the result is not credible. The method of equal-frequency discretization is similar to the equal-width discretization method. Discretization methods mentioned above is a hard division which is a strict division for the power plant operation parameters into each cluster. But their values are not strictly divided to the clusters actually. The researchers introduced fuzzy set theory and effectively solved the problem of data divided excellently [3].

An adaptive fuzzy discretization algorithm is proposed and applied in thermal power plant operation parameters value discretization procedure to exclude previous fuzzy discretization parameter which is artificially given with a validity function for fuzzy clustering introduced in this paper [7].

In geometry, good partition should satisfy two requirements [7]: (a) divergence: The inter-cluster distances should be as bigger as possible; (b) compactness: The intra-cluster distances should be as smaller as possible. The value of the ratio of the compactness and the divergence can be the criterion of the clustering validity. According to this guideline and the statistic variable, we constructed the following new validity function as formula (2).

The numerator of $L(c)$ denotes the sum of the distances between classes and the denominator of L(c) denotes the sum of the intra-distances of all the clusters. So the bigger L(c) is, the more reliable the result of clustering is. The cluster number c is the best one when L(c) reaches its maximum value [7].

Firstly calculate the central vector of the overall data, where $x_j (j = 1, 2, \cdots, n)$ is a pattern in data set X, $v_j (j = 1, 2, \cdots, c)$ is the prototype of the *jth* cluster:

$$\bar{x} = \frac{\sum_{i=1}^{c} \sum_{j=1}^{n} u_{ij}^m x_j}{n} \tag{4}$$

New validity function of cluster numbers c:

$$L(c) = \frac{\sum_{i=1}^{c} \sum_{j=1}^{n} u_{ij}^m \|v_i - \bar{x}\|^2 \Big/ (c - 1)}{\sum_{i=1}^{c} \sum_{j=1}^{n} u_{ij}^m \|x_j - v_i\|^2 \Big/ (n - c)} \tag{5}$$

where *m* is the fuzzy weighting exponent,c is the number of clusters to be explored, *n* is the card of dataset *X*.

The value of m can affect the convexity of the validity function and the convergence of the algorithm. It commonly determined by m = 2, which has a good performance in most cases. The adaptive function for the number of clusters, and the condition $\theta > 0$ to stop the clustering iteration, the initial clustering number $c = 2$ and the initial center matrix $V^{(0)}$ are given. The following is the adaptive process of the algorithm [7]:

STEP 1. Calculate the partition matrix:

$$u_{ij}^{(k)} = \frac{1}{\sum\limits_{r=1}^{c} (\frac{d_{ij}^{(k)}}{d_{rj}^{(k)}})^{\frac{2}{m-1}}}$$

(6)

If there exist j, r, so that $d_{rj}^{(k)} = 0$, then $u_{ij}^{(k)} = 1$ and for $i \neq r$, $u_{ij}^{(k)} = 0$.

STEP 2. Calculate the prototypes:

$$v_i^{(k+1)} = \frac{\sum\limits_{j=1}^{n} (u_{ij}^{(k)})^m x_j}{\sum\limits_{j=1}^{n} (u_{ij}^{(k)})^m}$$

(7)

STEP 3. Calculate the variation of partition matrix:

$$\left\| V^{(k+1)} - V^{(k)} \right\| \leq \theta$$

(8)

If the inequality is satisfied, then stop the iteration, else go to STEP 1, with $k = k + 1$.

STEP 4. Calculate L(c), with $L(1) = 0$, under $c > 1$ and $c < n$. Get the local maximum point, and keep the corresponding center matrix and the membership degree matrix. $c = c + 1$, if $c \leq n - 1$, go to STEP 1, else stop the iteration. Select the best c and the corresponding cluster center point matrix V_{ij} and fuzzy partition matrix U_{ij} as the result of the adaptive fuzzy discretization algorithm.

2.3   Metarule-Guided Mining of Association Rules

An example of such a metarule is

$$P_1 \wedge P_2 \wedge \ldots \wedge P_l \Rightarrow Q_1 \wedge Q_2 \wedge \ldots \wedge Q_r$$

(9)

where $P_i(i = 1, 2, \cdots, l)$ and $Q_j(j = 1, 2, \cdots, r)$ are either instantiated predicates or predicate variables. To find interdimensional association rules satisfying the template, Typically, a user will specify a list of attributes to be considered for instantiation with P1 and P2. Otherwise, a default set may be used [5].

2.4   Incremental Algorithm of Interestingness Threshold

This paper presents a dynamic adjustment incremental algorithm to determine the threshold of interestingness shown in formula (10).

$$\min\_conf_{fp\_r} = 2 \times (\frac{1 - \min\_conf_{fp\_2}}{1 + e^{-r}} + \frac{\min\_conf_{fp\_2} - 1}{2}) + \min\_conf_{fp\_2}$$

(10)

Thus, the dependence between the mining results and a given minimum support threshold is reduced. The potential knowledge in the history of the power plant is more fully excavated than before.

3 New Framework's Association Rule Mining Algorithm

This paper combines the metarule-guided mining technology to the new mining framework. The following is the specific operation process of the association rules algorithm based on the dynamic interestingness-support framework:

STEP 0. Algorithm initialization: given the minimum interestingness threshold of 2-dimensional association rules. Determine the instantiated predicate sets L_{P_1} and L_{Q_1} from Q_r;

STEP 1. Based on the metarules template, the single element from L_{Q_1} and all the single element's subsets from L_{P_1} is constructed as the 2-dimensional association rules named t;

$$P_i \Rightarrow Q_j \tag{11}$$

STEP 2. All 2-dimensional association rules whose fuzzy confidence is larger than $min\_conf_{fp\_2}$ are selected. Reconstruct L_{P_2} based on the selected results and calculate of higher dimensional association rules according to the incremental formula for minimum confidence threshold. It is a monotone increasing function with its Two order Derivative less than zero, which means it will increasing more and more slowly.

STEP 3. All r-dimensional subsets were composed of value whose corresponding numeric attribute only appears at most once in each rule. Subset of L_{P_r} and elements in L_{Q_1} constitute a r + 1-dimensional association rules. Then calculate the fuzzy confidence and use $min\_conf_{fp\_r+1}$ to generate $L_{P_{r+1}}$.

STEP 4. If $L_{P_{r+1}}$ is not an empty set, go to STEP 3. Otherwise replace the subset element of L_{Q_1} and go to STEP 1, if the single element is completed at the end of the mining process, the process stops.

4 The Application of Mining Algorithm in Operation Optimization

Based on the history data of a 300 MW power plant unit, the typical parameters which are related to operation optimization are analyzed. The dataset contains 2880 actual steady-state data for a certain month with load, main steam pressure, flue gas temperature and flue gas oxygen content parameters included. In this paper, the confidence is chosen as the interestingness index. And the result is compared with the original support-confidence framework algorithm [9–11].

4.1   Condition Division Based on a New Similarity

Condition division based on SIS system is the foundation of consumption difference analysis, operation optimization, fault diagnosis and other important applications. Previous typical condition division mainly sources from the statistical analysis methods and some clustering algorithms. The purpose of this paper is to explore a new similarity of the condition points from the historical database with the large amounts of operating data included to efficiently improve the rationality of the condition division compared to the artificial division of the attributes.

4.2   Mining Results of the Original Mining Framework Under Different Support Thresholds

For the former support-confidence framework, the minimum fuzzy support threshold is set to 0.01, 0.05, and 0.1 respectively. If the minimum fuzzy support threshold is given at 0.1, it is 51 frequent 1-itemsets and 0 frequent 2-itemsets, which shows there is no rule that is mined from the fuzzy dataset. If the minimum fuzzy support threshold is given at 0.05, it is 116 frequent 1-itemsets, 21 frequent 2-itemsets and 3 frequent 3-itemsets, illustrating that there is a few rules mined from the fuzzy dataset. If it is given at 0.01, it is 116 frequent 1-itemsets, 457 frequent 2-itemsets and 493 frequent 3-itemsets, indicating that there are large quantities of rules that are mined from the fuzzy dataset and it requires a lot of time to screen the interesting rules [13–17]. It can be concluded that the minimum support threshold has a great impact on the mining results (Table 1).

Table 1. Optimization value analysis of main operating parameters by each algorithm

Algorithm	Load	Main steam pressure	Flue gas temperature	B-oxygen content	Power economical parameter	Fuzzy confidence
Support-confidence framework	219.49 MW	[15.12, 15.31] MPa	105.42 °C	4.20%	0.41	100%
With average of interval	219.00 MW	[15.12, 15.31]	105.45 °C	4.15%	0.42	100%
With midpoint of the equal-width discretization interval	218.00 MW	[14.80, 15.70] MPa	107.41 °C	4.02%	0.44	100%
New framework with new similarity criterion in condition division in this paper	219.34 MW	15.25 MPa	105.78 °C	4.22%	0.41	99.34%

4.3   Comparison of the Optimal Values by the Different Mining Algorithm

When operation data is in a skewed distribution, the results of algorithm proposed were similar with the confidence-support framework with the value determined by median and average of the discretization interval by the equal-width discretization whose results can reflect the distribution of the original data. Mining results of the support-confidence framework with median of the equal-width discretization interval is most similar with the mining results by the algorithm proposed in this paper. Also, results under the original and new framework with cluster centers of the adaptive fuzzy discretization are basically the same, but the main steam pressure is valued by the new framework only. The value of the main steam pressure is in the specific interval determined by the mining algorithm under original framework. It is proved that the algorithm proposed in this paper is easy to mine the potential rule knowledge which is easy to be ignored by the previous framework so as to improve the accuracy of optimal operation parameters in power plant.

In addition, the new framework determines an association rule at 99.34% confidence which not only confirm the parameters in the table above, but also the other attributes including main steam flow, make-up water, air volume and the steam temperature in air intake of intermediate-pressure cylinder. In addition, flue gas oxygen content parameter which has a great influence on the economical performance of the unit is also the most easily adjusted parameter in the operating process of boiler under variable conditions. It has the widest range and strong coupling relationship with other economical parameters in the power plant [12].

The design value of this parameter in this unit is 4.16%. The mining result in this paper is 4.22%. Although there is a certain gap between the two values, it represents the best stable operation level where the unit actually was ever operated.

5   Conclusion

According to the characteristics of historical operation data sets in thermal power plant, the metarule-guided dynamic interestingness-support framework algorithm based on a new similarity criterion in dividing production process condition is proposed in this paper. Firstly, a new similarity is introduced in dividing production process condition. It is designed to avoid the single spherical cluster determined by the Euclidean distance that can overcome the shortness of former dividision in high dimensional space. Another advantage is that the adaptive fuzzy discretization algorithm is used to complete the discretization of historical data. The mining algorithm under new framework makes several improvements to determine optimal parameters of multidimensional association rules. And it reduces the redundancy of mining results by metarule-guided technology and the index of interestingness except support. It weakens the dependence between mining results and the minimum support threshold. Simulation results show that the new framework can mine more useful and reliable rules compared to the former support-confidence framework. What's more, it improves the accuracy of optimal operation parameters in power plant.

References

1. Liu, B., He, J., Zeng, X.: The application of nested data mining in power plant operating condition analysis. Power Syst. Eng. **30**(5), 13–17 (2014). (in Chinese)
2. Chen, D.: Data Analysis and Operation Optimization in Power Plant Based on Rough Set Theory, pp. 1–60. North China Electric Power University, Beijing (2013). (in Chinese)
3. Qiuping, W., Zhiqiang, C., Hao, W.: The summary of optimal operation parameters in power station based on the data mining. Electr. Power Sci. Eng. **31**(7), 19–24 (2015). (in Chinese)
4. Ran, P.: Research on Operation Optimization of Power Plant Thermal System by Dynamic Data Mining Approaches, pp. 1–124. North China Electric Power University, Beijing (2015). (in Chinese)
5. Han, J., Kamber, M.: Data Mining: Concepts and Techniques. Morgan Kaufmann Publishers, San Francisco (2001)
6. Li, J., Niu, C., Gu, J., Liu, J.: Application of data mining in optimal parameters value of power plant. Baoding J. North China Electr. Power Univ. **35**(4), 53–56 (2008). (in Chinese)
7. Li, Y., Yu, F.: A new validity function for fuzzy clustering. In: International Conference on Computational Intelligence and Natural Computing, pp. 462–465 (2009)
8. Zhai, S., Huang, X., Liu, J.: Data mining to economic norms of power plant based on condition division. Electr. Power **42**(7), 68–71 (2009). (in Chinese)
9. Luo, C.: Introduction to Fuzzy Sets. Beijing Normal University Press, Beijing (2005)
10. Li, J.: The Research and Application of Data Mining in Power Plant Operation Optimization, pp. 1–119. North China Electric Power University, Baoding (2006). (in Chinese)
11. Li, J.Q., Niu, C.L., Gu, J.J., et al.: Energy loss analysis based on fuzzy association rule mining in power plant. In: International Symposium on Computational Intelligence and Design, Wuhan, China, 17–18 October, pp. 186–189 (2008)
12. Lai, C.: The Application of Data Mining Technology in Power Plant Supervisory Information System, pp. 1–59. Shanghai Jiao Tong University, Shanghai (2014). (in Chinese)
13. Hong, J., Cui, Y., Bi, X., et al.: A unit on-line operation optimization system and determination of real-time optimum operation mode. Autom. Electr. Power Syst. **31**(6), 86–90 (2007). (in Chinese)
14. Liu, B., He, J., Liu, G.: A Study of Working Conditions Based on Cluster Analysis. In: APPEEC (2010)
15. Ren, H., Li, W., et al.: The analyze of operation index for the power unit under different loads. Proc. CSEE **19**(9), 50–56 (1999). (in Chinese)
16. Agrawal, R., Imielinski, T., Swami, A.: Mining association rules between sets of items in large databases. In: Proceedings of the ACM SIGMOD Conference on Management of Data, Washington D.C., pp. 207–216 (1993)
17. Li, W., Liu, C., Sheng, D., et al.: Status and development trend of economic operation and optimization system in domestic coal-fired power plant. Power Syst. Eng. **20**(1), 59–61 (2004). (in Chinese)

A Consensus-Based Distributed Primal-Dual Perturbed Subgradient Algorithm for DC OPF

Zhongyuan Yang$^{(\boxtimes)}$, Bin Zou, and Junmeng Zhang

School of Mechatronical Engineering and Automation, Shanghai University,
Shanghai 200072, China
yzll23772@sina.cn, zoubin@shu.edu.cn,
stzhangjm@163.com

Abstract. In this paper, an consensus-based distributed primal-dual perturbed subgradient algorithm is proposed for the DC Optimal Power Flow (OPF) problem. The algorithm is based on a double layer multi-agent structure, in which each generator bus and load bus in electric power grid is viewed as bus agent and connects with the grid by a network agent. In particular, network agents employ the average consensus method to estimate the global variables which are necessary for bus agents to update their generation using a local primal-dual perturbed subgradient method. The proposed approach is fully distributed and realizes the privacy protection. The employment of primal-dual perturbation method ensuring the convergence of the algorithm. Simulation results demonstrate the effectiveness of the proposed distributed algorithm.

Keywords: Distributed optimization · Multi-agent systems · Consensus protocols · DC optimal power flow · Smart grid

1 Introduction

The distributed operation and control of the power system has been widely concerned since the beginning smart grid research. In fact, the amount of distributed energy resources and adjustable demands such as renewable energy resources, energy storage systems is expected to significant increase in smart grid [1]. Hence, the coordination of distributed energy resources is becoming a challenging problem. Furthermore, in the future power system, neither energy management system nor electric plants are willing to reveal all of their economic and control data to each other for their own economic reasons [2]. These factors provide motivation for power system to move from the current highly centralized control structure towards a more distributed one.

In this paper, a novel distributed algorithm with double layer multi-agent system is proposed to solve DC-OPF problem. In this algorithm, each generator bus and load bus is designed as bus agent, which connects to the transmission grid through a network agent. Bus agents and network agents constitute the bus agent layer and the network agent layer of the double layer multi-agent system, respectively.

The proposed algorithm handle the DC-OPF problem iteratively, at each iteration, network agents exchange information with its neighbors to estimate the global information using consensus algorithm, meanwhile, bus agents exchange information with

© Springer Nature Singapore Pte Ltd. 2017
K. Li et al. (Eds.): LSMS/ICSEE 2017, Part III, CCIS 763, pp. 497–508, 2017.
DOI: 10.1007/978-981-10-6364-0_50

the connecting network agent and update their local variables using primal-dual per-turbed subgradient method [3]. Instead of reveal all global information of the grid to bus agents directly, the network agent encrypts the information and sends it to the connecting bus agent to calculate the optimization problem locally. The algorithm realizes the information isolation between bus agents and the grid, which is benefit to protect the privacy.

2 DC Optimal Power Flow Problem

The objective of the DC OPF problem is to seek the minimum cost generation dispatch to supply a given load with system operational constraints, which may be formulated as

$$\min \left(\sum_{\substack{i \in S_G \\ i \neq 1}} f_i(p_{G_i}) + f_1 \left(\sum_{i \in S_D} p_{D_i} - \sum_{\substack{i \in S_G \\ i \neq 1}} p_{G_i} \right) \right) \tag{1}$$

s.t.

$$p_i^{SP} = p_{G_i} - p_{D_i} \quad \forall i \in \{1, \ldots, N\} \tag{2}$$

$$P_{G_i}^{\min} \leq p_{G_i} \leq P_{G_i}^{\max} \quad \forall i \in \Omega_G \tag{3}$$

$$|p_{ij}| < P_{ij}^{U} \quad ij \in S_L, i \neq j \tag{4}$$

$$\theta_1 = 0 \tag{5}$$

where p_{G_i} is the real power production of generator bus i, N is the the number of buses in the system, I is the number of generator bus, p_i^{SP} denotes the real power injection at bus i, S_i is the set of buses physically connected to bus i, the voltage angle at bus i and bus j are denoted by θ_i and θ_j, respectively, p_{D_i} is the load at bus i. It is assumed that each generator bus has only one generator, S_G is the set of generators in the system, $P_{G_i}^{\max}$ and $P_{G_i}^{\min}$ are lower and upper limits on generation for the generator at bus i, which corresponds to the bus in which it is located. P_{ij}^{U} is the transmission capacity of the line connecting buses i and j, S_L denotes the set of transmission lines. Note that, by convention, bus $i = 1$ is taken to be the reference for which the voltage angle is set to zero.

The cost function of generator i is commonly assumed to be quadratic to real power generation p_{G_i} with cost parameters a_i, b_i, c_i such that

$$f_i(p_{Gi}) = a_i p_{Gi}^2 + b_i p_{Gi} + c_i \tag{6}$$

To keep the total power balance of the system, the reference bus $i = 1$ is set to be the balancing bus that satisfies the following condition

$$p_{G_1} = \sum_{d \in S_D} p_{D_d} - \sum_{\substack{i \in S_G \\ i \neq 1}} p_{G_i} \qquad (7)$$

thus the objective function (2-1) can be written as (8)

$$\min\left(\sum_{\substack{i \in S_G \\ i \neq 1}} f_i(p_{G_i}) + f_1\left(\sum_{i \in S_D} p_{D_i} - \sum_{\substack{i \in S_G \\ i \neq 1}} p_{G_i} \right) \right) \qquad (8)$$

meanwhile, the real power injection at bus i can be determined by

$$p_i^{SP} = \sum_{j \in S_i} \frac{\theta_i - \theta_j}{x_{ij}} \quad \forall i \in \{1, \ldots, N\} \qquad (9)$$

In this paper, bus $i = 1$ is defined as a reference bus with $\theta_1 = 0$, the voltage angle of the bus can be determined as (10) based on (2)

$$\theta = B^{-1} \cdot P^{SP} \qquad (10)$$

where $\theta = [\theta_1, \ldots, \theta_N]_{N \times 1}^T$ is the angle phase vector, $P^{SP} = [p_1^{SP}, \ldots, p_1^{SP}]_{N \times 1}^T$ is the real power injection vector, the admittance matrix of the grid can be written as

$$B^{-1} = \begin{bmatrix} 0 & 0 & \cdots & 0 \\ 0 & \begin{bmatrix} B_{22}' & \cdots & B_{2N}' \\ \vdots & \ddots & \vdots \\ B_{N2}' & \cdots & B_{NN}' \end{bmatrix}^{-1} \end{bmatrix}_{N \times N} \qquad (11)$$

B_{ii}' and B_{ij}' are self-admittance of bus i and mutual admittance between bus i and j, respectively.

Thus, the Lagrangian function of the DC OPF problem may be written as (12).

$$L = \sum_{\substack{i \in S_G \\ i \neq 1}} f_i(p_{Gi}) + f_1\left(\sum_{i \in S_D} p_{Di} - \sum_{\substack{i \in S_G \\ i \neq 1}} p_{Gi} \right) + \sum_{ij \in S_L} \mu_{ij}^+ \left(p_{ij} - P_{ij}^U \right) + \sum_{ij \in S_L} \mu_{ij}^- \left(-p_{ij} - P_{ij}^U \right)$$

$$(12)$$

where $\mu_{ij}^+ \geq 0$ and $\mu_{ij}^+ \geq 0$ $(ij \in S_L)$ are the Lagrangian multipliers correspond to the transmission line constraint, the vector of these Lagrangian multipliers can be shown as $\mu_i^+ = [\mu_{ij}^+]^T$, $i = 1, \ldots, N$, $j = 1, \ldots, N$, $i \neq j$ and, respectively. The $\eta^+ \geq 0$ and $\eta^- \geq 0$ are Lagrangian multipliers which correspond to the upper and lower generation limits of the reference generator bus, respectively.

3 Centralized Primal-Dual Perturbed Algorithm

The DC OPF problem can be solved by a primal-dual perturbed (PDP) subgradient method in which the perturbation points is applied to relax the concave and convex constraints of (12) [4]. More precisely, at iteration k, the PDP subgradient method preforms

$$P_{G_r}^{(k)} = P_\chi \left(P_{G_r}^{(k-1)} - a_{k1} L_{P_{G_r}} \left(P_{G_r}^{(k-1)}, \hat{\beta}^{(k)} \right) \right)$$

$$P_{G_1}^{(k)} = P_\chi \left(\sum_{i \in S_D} p_{D_i} - \sum P_{G_r}^{(k)} \right) \tag{13}$$

$$\lambda^{(k)} = \left(\lambda^{(k-1)} + a_{k2} L_\lambda \left(\hat{\alpha}^{(k)}, \lambda^{(k-1)} \right) \right)^+$$

where $P_{G_r}^{(k)} \in R^{I-1}$ is the generation vector of non-reference buses, P_{G_i} is the vector of reference bus, P_χ is the projection function that ensure $p_n^{(k)} \in \left[P_{G_n}^{min}, P_{G_n}^{max} \right], \forall n \in \Omega_G$, $\lambda^T = [\mu_{1j}^{+\,(k)}, \ldots, \mu_{Nj}^{+\,(k)}, \ldots, \mu_{ij}^{-\,(k)}, \ldots, \mu_{Nj}^{-\,(k)}, \lambda^{+\,(k)}, \lambda^{-(k)}]^T$ is the vector of Lagrangian multiplier, $\hat{\alpha}^{(k)}$ and $\hat{\beta}^{(k)}$ are perturbation points, $a_{k1} > 0$, $a_{k2} > 0$ are constant step sizes, and the perturbation point updates are

$$\hat{\alpha}^{(k)} = P_\chi \left(P_{G_r}^{(k-1)} - \rho_1 L_{P_{G_r}} \left(P_{G_r}^{(k-1)}, \lambda^{(k-1)} \right) \right)$$

$$\hat{\beta}^{(k)} = \left(\lambda^{(k-1)} + \rho_2 L_\lambda \left(P_{G_r}^{(k-1)}, \lambda^{(k-1)} \right) \right)^+ \tag{14}$$

where $\rho_1 > 0$ and $\rho_2 > 0$ are constants, and

$$L_{P_{G_r}} \left(P_{G_r}^{(k-1)}, \lambda^{(k-1)} \right) = \left[\frac{\partial L}{\partial p_{G_2}}, \ldots, \frac{\partial L}{\partial p_{G_I}} \right]^T_{\left(P_{G_r}^{(k-1)}, \lambda^{(k-1)} \right)} \tag{15}$$

where

$$\left. \frac{\partial L}{\partial p_{G_i}} \right|_{\left(P_{G_r}^{(k-1)}, \lambda^{(k-1)} \right)} = \begin{bmatrix} 2a_i p_{G_i} + b_i \\ -2a_1 \left(\sum_{i \in S_D} p_{Di} - \sum_{\substack{i \in S_G \\ i \neq 1}} p_{Gi} \right) - b_1 \\ -\eta^+ + \eta^- \\ + \sum_{ij \in S_L} \mu_{ij}^+ \left(\frac{\partial p_{ij}}{\partial p_{G_i}} \right) - \sum_{ij \in S_L} \mu_{ij}^- \left(\frac{\partial p_{ij}}{\partial p_{G_i}} \right) \end{bmatrix}_{\left(P_{G_r}^{(k-1)}, \lambda^{(k-1)} \right)} \tag{16}$$

and

$$L_\lambda\left(P_{G_r}^{(k-1)}, \lambda^{(k-1)}\right) = \left[\frac{\partial L}{\partial \mu_{ij}^+}, \frac{\partial L}{\partial \mu_{ij}^-}, \frac{\partial L}{\partial \eta^+}, \frac{\partial L}{\partial \eta^-}\right]^T_{\left(P_{G_r}^{(k-1)}, \lambda^{(k-1)}\right)} \tag{17}$$

where

$$\begin{aligned}
\frac{\partial L}{\partial \mu_{ij}^+} &= P_l - P_l^U \\
\frac{\partial L}{\partial \mu_{ij}^-} &= -P_l - P_l^U \\
\frac{\partial L}{\partial \eta^+} &= \sum_{i \in S_D} p_{D_i} - \sum_{\substack{i \in S_G \\ i \neq 1}} p_{G_i} - P_1^{\max} \\
\frac{\partial L}{\partial \eta^-} &= P_1^{\min} - \sum_{i \in S_D} p_{D_i} + \sum_{\substack{i \in S_G \\ i \neq 1}} p_{G_i}
\end{aligned} \tag{18}$$

(15) and (17) represent the subgradients of Lagrangian function at $\left(P_{G_r}^{(k-1)}, \lambda^{(k-1)}\right)$ with respect to P_{G_r} and λ, respectively.

The convergence analysis of the PDP subgradient method is shown in [3].

4 Distributed Algorithm

In this section, a double layer multi-agent system is developed to distribute the optimization algorithm of DC OPF problem and protect the privacy. Meanwhile, a modified consensus algorithm with auxiliary variable is introduced in which network agents exchange information only with their neighboring network agents to obtain the estimation of global voltage angle vector θ. Then, the procedure of the proposed Consensus-based Distributed Primal-dual Perturbed Algorithm is illustrated.

4.1 Double Layer Multi-agent System

A power system is commonly modeled by a multi-agent system in which the geographically distributed buses are designed as agents with local sensing, communication and computation abilities. Different from the common multi-agent system, a double layer multi-agent structure is developed in this paper to achieve information isolation and privacy protection.

Here we set a 6-bus power system as an example. The communication topology of the system is shown in Fig. 1(a) in which the communication path that two buses can exchange information is represented by the edge among agents. The double layer multi-agent system structure that corresponds to the 6-bus power system is shown in Fig. 1(b). Functions of network agent and the bus agent is illustrated in Table 1. It is

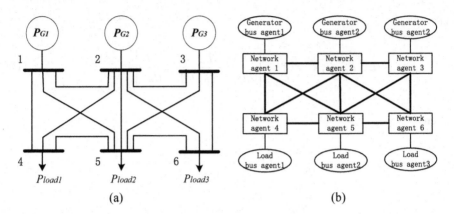

Fig. 1. (a) The communication topology of the 6-bus power system, (b) The structure of the double layer multi-agent system

Table 1. Functions of network agent and bus agent

Function	Network agent	Generator/Load bus Agent
Information storage	• Store the impedance of all transmission lines • Store the admittance matrix of the grid • Store the local estimation of the global voltage angle	• Store its local cost function and constraints
Information receiving	• Receive local voltage angle estimation from the neighboring network bus agents • Receive real power generation or demand information from the connecting bus agent	• Receive information concerning local variable updates from the neighboring network agent
Information sending	• Send information concerning local variables update to the connecting bus agent • Send local voltage angle estimation to the neighboring network agent	• Send real power generation or demand information to the connecting network agent

obviously that the double layer multi-agent system can efficiently protects the privacy and distributes the PDP subgradient algorithm.

4.2   Graph Theory

The graph theory is commonly adopted to describe the characteristic of the multi-agent system in which the network agent system can be represented by a weighted graph G (V, E, A) [5], V is the set of elements called nodes, E is the set of distinct nodes called edges, and A is the associated adjacency matrix.

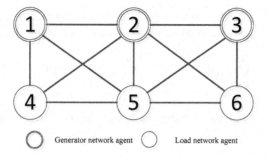

Fig. 2. The network agents graph of the 6-bus power system

In this paper, graph nodes represent the network agents of the power system, edges represent the communication path among agents, and the adjacency matrix reflects the weight of edge. The element a_{ij} of the adjacency matrix is positive if the corresponding edge exists, otherwise, it is zero. The graph of the network agents in Fig. 1(a) can be easily depicted in Fig. 2.

4.3 Consensus Protocols

To have a distributed optimization algorithm for solving the DC OPF problem, a modified consensus algorithm is applied. Let x_i denotes the state of node i, we call the graph achieves consensus if and only if (19) holds.

$$x_1 = x_2 = \cdots = x_N \tag{19}$$

The element a_{ij} of the adjacency matrix A in i-th row and j-th column is defined as

$$a_{ij} = \begin{cases} 1/d_i & j \in N_i \\ 0 & j \notin N_i \end{cases} \tag{20}$$

where N_i is the set of the neighbors of node i (including node i) and d_i is its cardinality. A is called a doubly stochastic matrix when each row-sum and column-sum of the adjacency matrix A are both equal to 1, while, the structure of real power system is always too complex to meet the double stochastic condition, for example, the communication structures in Fig. 2, whose adjacency matrix is just a stochastic matrix for only the row-sums of A are equal to 1. In this condition, a modified consensus algorithm should be employed [6].

In the procedure of the modified consensus algorithm, two auxiliary variables $y_i^{(0)}$ and $z_i^{(0)}$ are introduced as follows

$$y_i^{(0)} = 1/d_i, \quad z_i^{(0)} = x_i^{(0)}/d_i \tag{21}$$

where $y_i^{(0)}$ and $z_i^{(0)}$ are the initial values of the two auxiliary variables, d_i is the cardinality of N_i, in each iteration the state of each node is updated as follows

$$y_i^{(k)} = \sum_{j=1}^{N} a_{ij} y_j^{(k-1)}, \quad z_i^{(k)} = \sum_{j=1}^{N} a_{ij} z_j^{(k-1)}, \quad x_i^{(k)} = z_i^{(k)} / y_i^{(k)} \tag{22}$$

where N is the number of nodes in the graph, with the increasing of iteration, the result will converges to

$$\lim_{k \to \infty} x_i^{(k)} = \frac{1}{N} \sum_{i=1}^{N} x_i^{(0)} \tag{23}$$

thus, each node can also get the estimation of the average or sum of all nodal initial states, thereby the global information of the system is accessible to all nodes locally. This is the key for the proposed distributed algorithm to realize distribution. The convergence property of the modified consensus algorithm had been proved in [6].

4.4   Consensus-Based Distributed Primal-Dual Perturbed Algorithm

A Distributed Primal-dual Perturbed Algorithm for solving (12) is proposed basing graph theory and the modified consensus algorithm in 4.3. In this iterative procedure, each network agent firstly obtain its real power injection using (24) since the real power generation and demand of the connecting bus agent is accessible to it:

$$p_i^{SP(k)} = p_{G_i}^{(k)} - p_{D_i}^{(k)} \tag{24}$$

where k denotes the iteration counter, then, we define the initial state vector as fellows

$$\hat{\theta}_i^{(k)} = [\hat{\theta}_{i1}^{(k)}, \cdots, \hat{\theta}_{iN}^{(k)}]_{N \times 1}^T \tag{25}$$

where

$$\hat{\theta}_{ij}^{(k)} = [B^{-1}]_{ij} p_i^{SP(k)} \tag{26}$$

in which $[B^{-1}]_{ij}$ is the element of matrix B^{-1} in i-th row and j-th column. It is obvious that for all network agents, there has

$$\hat{\theta}^{(k)} = \left[\hat{\theta}_1^{(k)}, \cdots, \hat{\theta}_N^{(k)} \right]_{N \times N}$$

$$= \begin{bmatrix} \hat{\theta}_{11}^{(k)} & \cdots & \hat{\theta}_{N1}^{(k)} \\ \vdots & \ddots & \vdots \\ \hat{\theta}_{1N}^{(k)} & \cdots & \hat{\theta}_{NN}^{(k)} \end{bmatrix}_{N \times N} \tag{27}$$

To obtain the local estimation of all bus agents in the system, the modified consensus algorithm proposed in 4.3 is applied. Let the elements of i-th row of $\hat{\theta}^{(k)}$, i.e., $[\hat{\theta}^{(k)}_{1i}, \ldots, \hat{\theta}^{(k)}_{Ni}]_{1 \times N}$, correspond to $x_i^{(0)}$, $i = 1, \ldots, N$, and initialize $y_i^{(0)}$ and $z_i^{(0)}$ as (21) and update them as (22), with the increasing of iteration, the state of each network agent will reaches consensus on $\tilde{\theta}_i$, where

$$\tilde{\theta}_i = \frac{1}{N} \sum_{j=1}^{N} \hat{\theta}^{(k)}_{ij} \tag{28}$$

thus, the voltage angle estimation of i-th network agent θ_i can be determined locally for all network agents in the system:

$$\theta_i = N \cdot \tilde{\theta}_i \quad i = 1, \ldots N \tag{29}$$

therefore, the estimation of voltage angle vector $\theta = [\theta_1, \ldots, \theta_N]^T_{N \times 1}$ is obtained at each network agent, then $L_\lambda \left(P^{(k-1)}_{G_r}, \lambda^{(k-1)} \right)$ and the update of Lagrangian multipliers can be calculated in each network agent locally.

Moreover, the proposed distributed algorithm achieves privacy protection. In iteration, each network agent calculates (30) and sent IG_i to the connecting bus agent. According to (16), it is obvious that the generator bus agent can computes $L_{P_{G_i}} \left(P^{(k-1)}_{G_r}, \lambda^{(k-1)} \right)$ locally by (31), thus the generators p_{G_i} can be locally updated using the primal-dual perturbed algorithm. With iteration increasing, the optimal generator output solution of the system is achieved

$$IG_i = \left[\begin{array}{c} -2a_1 \left(\sum\limits_{i \in S_D} p_{Di} - \sum\limits_{\substack{i \in S_G \\ i \neq 1}} p_{Gi} \right) - b_1 \\ -\eta^+ + \eta^- \\ + \sum\limits_{2j \in S_L} \mu_{2j}^+ \left(\frac{\partial p_{2j}}{\partial p_{G_2}} \right) - \sum\limits_{2j \in S_L} \mu_{2j}^- \left(\frac{\partial p_{2j}}{\partial p_{G_2}} \right) \end{array} \right]_{\left(P^{(k-1)}_{G_r}, \lambda^{(k-1)} \right)} \tag{30}$$

$$L_{P_{G_i}} \left(P^{(k-1)}_{G_r}, \lambda^{(k-1)} \right) = 2a_i p_{G_i} + b_i + IG_i \tag{31}$$

5   Simulation Results

In this section, the comparison with the centralized PDP subgradient method shows the effectiveness of the proposed distributed algorithm. Several case studies are discussed to show the performance of the algorithm.

The 6-bus power system in [5] with 3 loads, 3 generators and 11 transmission lines is considered and set 1-th bus as the reference. The single-line diagram of the power system is shown in Fig. 1, while the corresponding communication graph is shown in Fig. 1(b). The characteristic of the generators are given in Table 2, while the reactance for the transmission lines are given in the archive of Matpower [7]. The initial load demand is 300 MW, with three loads of 100 MW for each load bus. The real power capacity constraint on each transmission line is $P_{ij}^U = 100$ MW. The constants ρ_1, ρ_2 are 0.6, 0.3, respectively, and the step sizes a_{k1}, a_{k2} are 1.3, 0.3, respectively.

Table 2. Generator characteristics of the 6-bus system

Bus	c_i [$]	b_i [$/MW]	a_i [$/MW^2]	p_i^{min} [MW]	p_i^{max} [MW]
1	213.1	11.669	0.00533	50	200
2	200	10.333	0.00889	37.5	150
3	240	10.833	0.00741	45	180

The comparison between the centralized PDP subgradient algorithm and the proposed distributed algorithm is carried out utilize the 6-bus power system above, optimal power flow solutions are shown in Table 3, which illustrates the effectiveness of the proposed distributed algorithm.

Table 3. Best solution by centralized algorithm and distributed algorithm

Transmission line	Power flow solution of the centralized PDP sugradient algorithm [MW]	Power flow solution of the consensus-based distributed algorithm [MW]
1–2	4.71	4.71
1–4	37.70	37.70
1–5	30.23	30.23
2–3	−2.59	−2.58
2–4	65.97	65.97
2–5	27.08	27.09
2–6	32.83	32.84
3–5	33.75	33.74
3–6	72.16	72.14
4–5	3.82	3.82
5–6	−5.19	−5.19

Case Study 1: Time-Varying Demand. This case study considers the 6-bus power system above to evaluate the capability of the proposed algorithm to handle automatically changes in the load demand. The generator output is assumed to stay minima when iteration $k = 0$, at iteration $k = 400$, the load demand in each load bus is increased to 130 MW, while at iteration $k = 800$, the load demand in each load bus is reduced to 100 MW, the real power output can be adjusted automatically as shown in Fig. 3(a).

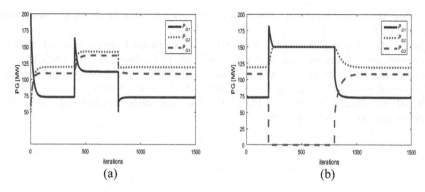

Fig. 3. (a) Power outputs with time-varying demand, (b) power outputs with plug and play capability

Case Study 2: Plug and Play Capability. In this case study, the plug and play adaptability of the proposed distributed algorithm is discussed. The system is assumed to hold steady as the real power outputs in the end of case study 1 until iteration $k = 200$, the generator bus 3 is removed from the system, while at iteration $k = 800$, the generator bus 3 is connected again, simulation result indicates that the system can always achieve power balance, as shown in Fig. 3(b).

6 Conclusion

A consensus-based distributed PDP subgradient algorithm for solving DC OPF problem is proposed in this paper. This algorithm employs a double layer multi-agent structure to model the power system, which achieves the isolation between local information and global information. A modified consensus method is employed to estimates the global voltage angle locally and distributes the PDP subgradient algorithm. The effectiveness of the proposed method is demonstrated by numerical simulations on several test systems with transmission constraints, time-varying demand and plug and play capacity. Future works will make further comparison between the proposed algorithm and other distributed algorithms and extend it to AC OPF problem.

References

1. Mohammadi, J., Hug, G., Kar, S.: Agent-based distributed security constrained optimal power flow. IEEE Trans. Smart Grid **PP**(99), 1 (2016)
2. Disfani, V.R., Fan, L., Miao, Z.: Distributed DC optimal power flow for radial networks through partial primal dual algorithm. In: 2015 IEEE Power & Energy Society General Meeting, pp. 1–5. IEEE (2015)
3. Chang, T.H., Nedic, A., Scaglione, A.: Distributed constrained optimization by consensus-based primal-dual perturbation method. IEEE Trans. Autom. Control **59**(6), 1524–1538 (2014)

4. Kallio, M.J., Ruszczynski, A.: Perturbation methods for saddle point computation. Work. Pap. **119**(35), 37–59 (1994)
5. Binetti, G., Davoudi, A., Lewis, F.L., Naso, D.: Distributed consensus-based economic dispatch with transmission losses. Power Syst. IEEE Trans. **29**(4), 1711–1720 (2014)
6. Olshevsky, A., Tsitsiklis, J.N.: Convergence speed in distributed consensus and averaging [reprint of mr2480125]. Siam J. Control Optim. **48**(1), 33–55 (2006)
7. Wood, A.J., Wollenberg, B.F., Sheblé, G.B.: Power generation, operation and control. IEEE Power Ene. Mag. **37**(3), 195 (2003)

Model Predictive Control Based
on the Dynamic PLS Approach to Waste Heat
Recovery System

Jianhua Zhang$^{(\boxtimes)}$, Haopeng Hu, Jinzhu Pu, and Guolian Hou

School of Control and Computer Engineering,
North China Electric Power University, Beijing 102206, China
{zjh,hgl}@ncepu.edu.cn, hu94hao@qq.com,
pjz111111@126.com

Abstract. This paper investigates model predictive control scheme based on PLS latent space for CO_2 transcritical power cycle based waste heat recovery system. First, a control-oriented model is developed for the transcritical CO_2 power cycle system. For the sake of solving multi-variable and strong coupling problems of the transcritical CO_2 cycle system, model predictive control scheme based on the dynamic PLS approach is adopted and applied to this waste heat recovery system. The experimental results show that the adopted control method shows better performance in disturbance rejection and set-point tracking than PLS-PID control scheme for the CO_2 transcritical power cycle system.

Keywords: CO_2 · Transcritical power cycle · Waste heat recovery · Model predictive control · Partial least squares

1 Introduction

As the energy consumption increases, the traditional fossil energy has dried up. Therefore, it has important significance to enhance the energy efficiency and develop new kinds of energy sources.

In order to enhance the energy efficiency, several kinds of new power cycle are developed to replace traditional steam cycle in waste heat recovery systems (WHRSs). As a new type of thermodynamic cycle, transcritical CO_2 cycle has excellent performance in reusing low-grade waste energy. That is because the CO_2 working fluid has so many advantages. For example, the supercritical state of CO_2 can be reached easily in the power cycle process [1], which the critical pressure is 7.38 MPa and the critical temperature is 31.1 °C. CO_2 is in the supercritical state all the time in the evaporator, so there is no phase change in the evaporator. In addition, CO_2 is non- poisonous, non-explosive, inexpensive and luxuriant in the world [2]. It is well known that CO_2 has potentially favorable thermodynamics and transports properties in the supercritical area.

Some investigations have been done on transcritical CO_2 power cycle during the past decades. Most of the investigations are concerned with thermodynamic analysis,

© Springer Nature Singapore Pte Ltd. 2017
K. Li et al. (Eds.): LSMS/ICSEE 2017, Part III, CCIS 763, pp. 509–518, 2017.
DOI: 10.1007/978-981-10-6364-0_51

performance analysis and performance optimization [3, 4]. However, no publication has focused on modeling and control of CO_2 power cycle systems so far.

Reference [5] built a model for transcritical CO_2 compression systems in which the CO_2 working fluid in gas cooler operated in a supercritical state. Considering transcritical CO_2 compression cycle is opposite to transcritical CO_2 power cycle, following reference [5], a control-oriented physical model is established for transcritical CO_2 power cycle based waste heat recovery systems in this work.

CO_2 transcritical power cycle system is a multi-variable, complex system with strong coupling. In order to deal with this problem, there are many kinds of approaches developed by researchers, such as partial least squares (PLS). Partial Least Squares is one kind of data driven technology which is used to decoupling dimension reduction, noise remove and solve the collinearity problem [6]. In 1992, Kaspar and Ray [7] put forward the PLS control scheme for the first time, regarding the load matrix of PLS algorithm as pre-compensator and post-compensator to connect the PLS hidden space with the pristine physical space. In this paper, model predictive controller rather than PID controller is applied to transcrtical CO_2 power cycle based waste heat recovery systems in PLS latent space.

The residual of the article is designed as follows. The physical model of a CO_2 transcritical power cycle system is established in Sect. 2. The dynamic PLS framework is presented in Sect. 3. The GPC controller design is presented in Sect. 4. The experimental results of proposed control system is illustrated in Sect. 5. At last, the summary of the whole paper is presented in Sect. 6.

2 Modeling of a CO_2 Transcritical Power Cycle Based WHRS

In this section, a physical model is obtained for the carbon dioxide transcritical power cycle system after building the model of each component. The diagram of the 10 kW waste heat recovery power cycle system is presented in Fig. 1. The waste heat is

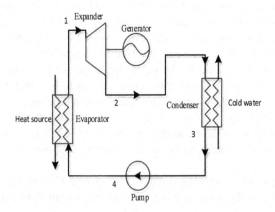

Fig. 1. Diagram of CO_2 transcritical power cycle system.

transferred to the evaporator in which the working fluid CO_2 is heated, the evaporated CO_2 drives the expander for power generation. Then, the working fluid CO_2 is cooled into liquid state in a condenser. The CO_2 from the condenser is then pressurized by the pump and sent back to the evaporator.

2.1 Evaporator

The CO_2 working fluid in evaporator is operating in a supercritical state, the lumped parameter model is used for describing the model of the evaporator.

The double-pipe heat exchanger is used as the evaporator in this system, the heat transfer coefficients on the air side and on the carbon dioxide side are calculated using Petukhov correlation [8] as follows:

$$\alpha = \frac{k}{D} \left[\frac{\frac{f}{8} \cdot Re \cdot Pr}{12.7 \cdot \left(\frac{f}{8}\right)^{0.5} \cdot \left(Pr^{\frac{2}{3}} - 1\right) + 1.07} \right]. \tag{1}$$

2.2 Condenser

Moving boundary lumped parameter model is used for describing the model of the condenser in which both the inside and outside of the wall are considered. The inside part is divided into three zones: an overheated zone, a two phase zone and a sub-cooling zone.

The shell-tube heat exchanger is used as the condenser in this system, the heat transfer coefficients can be calculated as follows:

On the cold water side and for the single-phase carbon dioxide Eq. (7) is used. For the two-phase region of the carbon dioxide, Cavallini's correlation [9] is used as follows

$$\alpha = 0.05 \cdot Re_{eq}^{0.8} \cdot Pr_{satliq}^{0.33} \cdot \frac{k_{satliq}}{D_i}. \tag{2}$$

where the equivalent Reynolds number is calculated by the following formulas

$$Re_{eq} = Re_{vap} \cdot \frac{\mu_{satvap}}{\mu_{satliq}} \cdot \left(\frac{\rho_{satliq}}{\rho_{satvap}}\right)^{0.5} + Re_{liq}. \tag{3}$$

$$Re_{liq} = \frac{\dot{m}}{A} \cdot (1 - \bar{\gamma}) \cdot \frac{D_i}{\mu_{satliq}}. \tag{4}$$

$$Re_{vap} = \frac{\dot{m}}{A} \cdot \bar{\gamma} \cdot \frac{D_i}{\mu_{satvap}}. \tag{5}$$

2.3   Expander

Compared with the dynamics of the evaporator and condenser, a steady-state model can be established for the scroll expander.

Losses caused by internal leakage, friction, heat transfers and supply pressure drop can be lumped into a mechanical efficiency η_{exp}, thus the mechanical work of expander can be formulated as follows [10]

$$W_{exp} = \dot{m}_{exp} \cdot ((h_{exp,sup} - h_{exp,i}) + v_{exp,i}(P_{exp,i} - P_c))\eta_{exp}. \tag{6}$$

The mass flow rate entering the expander can be formulated

$$\dot{m}_{exp} = \frac{ff \cdot V_s \cdot N_{exp}}{60 \cdot v_{exp,i}}. \tag{7}$$

where ff is the filling factor and V_s the swept volume of expander. N_{exp} is the rotating speed of the scroll expander and $v_{exp,i}$ is the inlet specific volume of carbon dioxide.

2.4   Pump

The mass flow rates of working fluid CO_2 according to different rotation speeds can be obtained based on similarity principle [11]

$$\frac{\dot{m}_{p1}}{\dot{m}_{p2}} = \frac{N_{p1}}{N_{p2}}. \tag{8}$$

For given pump efficient coefficient η_p and average specific volume $\bar{v}_p$, the outlet enthalpy of the pump is calculated by

$$h_{p,o} = h_{p,i} + \frac{\bar{v}_p(P_{po} - P_{pi})}{\eta_p}. \tag{9}$$

2.5   Overall Model

The overall model of the CO_2 transcritical power cycle waste heat recovery system is established by linking the models of each subcomponent. The input and output of this overall model are $u = [N_p, N_{rot}, \dot{m}_c]^T$ and $y = [P_e, T_{su}, T_c]^T$ respectively. Consequently, the overall model of the whole process can be formulated by the following state space equation

$$\begin{aligned} \dot{x} &= f(u,t) \\ y &= g(u,t). \end{aligned} \tag{10}$$

3 Dynamic PLS Framework for Controller Design

In modern industrial processes, there are a lot of multivariable, strong coupling and complex industrial systems, such as the CO_2 transcritical power cycle waste heat recovery system. It brings big challenge for controller design due to the interaction between input variables and output variables. For such industrial systems, a model predictive control scheme based on the PLS approach is shown.

3.1 Partial Least Squares

In standard PLS, the given data is divided into two blocks, input data matrix $X = (x_{kn})_{K \times N}$ and output data matrix $Y = (y_{km})_{K \times M}$, where K, N and M represent the numbers of observations, the dimension of inputs and outputs. The inputs and outputs are scaled and mean-centered first and then broken down into a sum of several component matrices, which is called outer relations

$$X - E = \sum_{r=1}^{R} t_r p_r^T = TP^T. \tag{11}$$

$$Y - F = \sum_{r=1}^{R} v_r q_r^T = VQ^T. \tag{12}$$

where R represents the value of retained latent variables, which can be defined by some heuristic and other statistical methods. T and V is the score matrices of X and Y, E and F are residual matrices, which are useless to describe the input variables and the output variables. P and Q show the influences of X and Y respectively. Now the connection between inputs and outputs is constructed by the inner relation

$$v_r = b_r t_r. \tag{13}$$

$$b_r = \frac{v_r^T t_r}{t_r^T t_r}. \tag{14}$$

where b_r is the linear regression coefficient. Thus, the relationship between the blocks of data is contained in the final PLS model

$$Y - F = b_1 t_1 v_1^T + b_2 t_2 v_2^T + \cdots + b_R t_R v_R^T = TBV^T. \tag{15}$$

3.2 Dynamic PLS Model and Control Framework

The PLS control framework is shown in Fig. 2. In this picture, W_x and W_y are the scaling matrices added before using the PLS method. P is the loading matrix for the input variables. Q represents the loading matrix for the output variables, and Q^+ represents the appropriate inverse of Q. As in general process control nomenclature, G_c, G_p and G_d represent the transfer function matrices for the controller, controlled object and disturbances respectively.

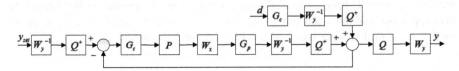

Fig. 2. Block diagram for PLS compensation.

Because the conventional PLS modeling methods only consider the steady state relationship, and always ignore the dynamic relationships in the actual processes, in this case, Kaspar and Ray put forward the dynamic filters method, which becomes some connections of the inner PLS model [12]. The merit of this method is that there are no lagged variables exist in the outer PLS model. This can lead to a simple adaptation of the PLS model for controller design. In this article, this approach is chosen.

Thus, the inner PLS model is a matrix in diagonal form using this method, which can be shown as

$$
\mathrm{BH} = \begin{bmatrix} G_{p1}(s) & \cdots & 0 \\ \vdots & \ddots & \vdots \\ 0 & \cdots & G_{pR}(s) \end{bmatrix}. \tag{16}
$$

Accordingly, the controller design in SISO system can be directly used to each loop in the PLS latent space to realize the appropriate performance.

4   Model Predictive Control in PLS Framework

In order to improve the performance of system, model predictive control method in dynamic PLS framework is designed [13].

Generalized predictive control (GPC) is one of the most effective and comprehensive MPC methods nowadays. It can deal with many complex control issues. The predictive model using this method is named the controlled auto-regressive integrated moving average (CARIMA) model

$$
A(z^{-1})Y(t) = B(z^{-1})X(t-1) + \frac{1}{\Delta}C(z^{-1})\varepsilon(t). \tag{17}
$$

where $A(z^{-1}) = I + A_1 z^{-1} + \cdots + A_{n_a} z^{-n_a}$, $B(z^{-1}) = B_0 + B_1 z^{-1} + \cdots + B_{n_b} z^{-n_b}$, $C(z^{-1}) = C_0 + C_1 z^{-1} + \cdots + C_{n_c} z^{-n_c}$, $\Delta = 1 - z^{-1}$.

The multi-loop model predictive control method within the PLS frame is shown in Fig. 3.

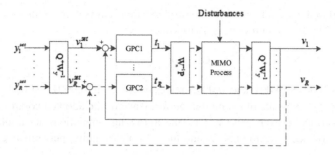

Fig. 3. Model predictive control method within the dynamic PLS frame.

In Fig. 3, the initial MIMO controller design is transformed to a number of individual SISO controllers design. Each predictive controller is designed under each latent subspace of PLS inner model. For each GPC, the cost function is

$$J^i(k) = \sum_{j=1}^{Np} ||v_{sp}(k+j) - v_i(k+j)||_2^2 + \lambda_i \sum_{j=1}^{M} ||\Delta t_i(k+j-1)||_2^2. \tag{18}$$

In other words, the control effect is calculated in the PLS latent space. The system output will be converted to score variables again to contrast with the score set-points. When the control system performance of each subsystem is reached, the whole control system performance can be achieved.

5   Simulation Results

In this chapter, we use the 10 kW CO_2 transcritical power cycle system established in Sect. 2 as the controlled plant to examine the ability of this control approach. The control objective is to keep the three controlled variables in a safe and steady region by tuning the three manipulated variables.

In order to construct the PLS model, 500 sets of related data are gathered by a series of stimulating pseudo random signal to input variables to excite all the correlative dynamic models. Measurable disturbances with a signal-to-noise ratio of 10 are applied to controlled variables in order to be closer to the scene of the actual industrial processes. After identification, the inner PLS model using dynamic filter method is gotten as follows

$$\mathbf{BH} = \begin{bmatrix} \frac{35.62}{s+0.0434} & 0 & 0 \\ 0 & \frac{34.02}{s+0.1547} & 0 \\ 0 & 0 & \frac{23.48}{s+0.2188} \end{bmatrix}. \tag{19}$$

The next phase is to design the GPC in each SISO loop under PLS framework. In this simulation, the sampling period T_s is 2 s. The GPC control parameters of each loop are prediction horizon $Np = 10$ and control horizon $M = 4$. To investigate the ability of the proposed method, the PID controllers are designed in contrast to the GPC controllers. The PID controllers is as follows

$$G_{PI} = k_p + k_i/s. \tag{20}$$

where k_p is 0.0018, 0.001 and 0.0095 respectively, k_i is 0.0001, 0.0002 and 0.0043 respectively. For the sake of testing the performance of the adopted control method for a CO_2 transcritical power cycle system, the following simulations are implemented to test the disturbance rejection ability and the set-point tracking performance of system.

5.1   Set-Point Tracking Test

In this test, the set-point of evaporating pressure is added from 15 MPa to 15.05 MPa at 1000 s; decreased by 0.2 MPa at 2000 s. The set-point of evaporating outlet temperature is added from 130 °C to 132 °C at 1000 s; decreased by 1 °C at 2000 s. The set-point of condenser outlet temperature is decreased by 2 °C at 1000 s, then is increased by 1 °C at 2000 s. The simulation results are presented in Fig. 4.

The experimental results demonstrate that both two control method have good performance in tracking the set-point, but the GPC with dynamic PLS model has better control performance and has less overshoot and settling time. Figure 4 shows that the deviation of manipulated variables is less than PLS-PID control method.

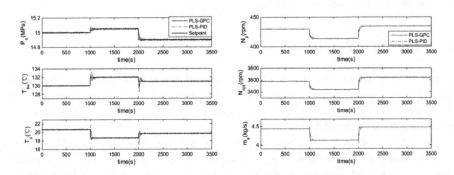

Fig. 4. Evolutions of controlled variables and manipulated variables.

5.2   Disturbance Rejection Test

The disturbance is imposed to the manipulated variables respectively. First, a step range of 5 in the pump rotation is applied at 500 s; a step range of −5 is made at 1000 s. Then, a step range of 50 in the expander rotation is applied at 1500 s; a step range of –50 is made at 2000 s. Finally, a step range of 0.1 in the mass flow rate of cooled water is applied at 2500 s; a step range of –0.1 is made at 3000 s. The evolutions of controlled variables and manipulated variables are presented in Fig. 5.

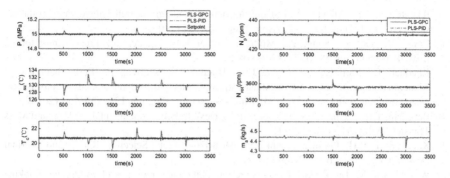

Fig. 5. Evolutions of controlled variables and manipulated variables.

In Fig. 5, all the manipulated variables have a slight deviation because of the change of each manipulated variables, but the controlled variables can quickly restore to their set-points. Moreover, the change amplitude of manipulated variables using GPC method is smaller than that using PLS-PID control method. Thus, the performance of PLS-GPC control method is better than PLS-PID control method.

6 Conclusion

In this paper, CO_2 transcritical power cycle system model is established. In order to solve the multivariable and strong coupling problems of the CO_2 transcritical power cycle system, GPC control scheme based on the PLS framework is presented. Some simulation experiments are carried out to test the performance of the control strategy. The results demonstrate that proposed control scheme shows better performance in disturbance rejection and set-point tracking than PLS-PID control scheme.

Acknowledgments. The article was supported by China National Science Foundation under Grant (61374052 and 61511130082). These are gratefully acknowledged.

References

1. Chen, Y.: Novel cycles using carbon dioxide as working fluid. Licenciate thesis, School of Engineering and Management, Stockholm (2006)
2. Clementoni, E.M., Cox, T.L., King, M.A.: Off-nominal component performance in a supercritical carbon dioxide brayton cycle. J. Eng. Gas Turbines Power **138**(1), 011703 (2016)
3. Chen, Y., Lundqvist, P., Platell, P.: Theoretical research of carbon dioxide power cycle application in automobile industry to reduce vehicle's fuel consumption. Appl. Therm. Eng. **25**(14), 2041–2053 (2005)
4. Walnum, H.T., Neksa, P., Nord, L.O., Andresen, T.: Modelling and simulation of CO_2 (carbon dioxide) bottoming cycles for offshore oil and gas installations at design and off-design conditions. Energy **59**, 513–520 (2013)

5. Rasmussen, B.P., Alleyne, A.G.: Control-oriented modeling of transcritical vapor compression systems. J. Dyn. SYST-T. Asme **126**(1), 54–64 (2004)
6. Cong, Y., Ge, Z., Song, Z.: Multirate partial least squares for process monitoring. IFAC-PapersOnLine **48**(8), 771–776 (2015)
7. Khedher, L., Ramirez, J., Gorriz, J.M., Bramin, A., Segovia, F.: Early diagnosis of Alzheimer's disease based on partial least squares, principal component analysis and support vector machine using segmented MRI images. Neurocomputing **151**, 139–150 (2015)
8. Bergman, T.L., Lavine, A.S., Incropera, F.P., DeWitt, D.P.: Fundamentals of Heat and Mass Transfer. Wiley, Hoboken (2011)
9. Kakac, S., Liu, H., Pramuanjaroenkij, A.: Heat Exchangers: Selection, Rating, and Thermal Design. CRC Press, Boca Raton (2012)
10. Wei, D., Lu, X., Lu, Z., Gu, J.: Dynamic modeling and simulation of an Organic Rankine Cycle (ORC) system for waste heat recovery. Appl. Therm. Eng. **28**(10), 1216–1224 (2008)
11. Zhang, J., Zhang, W., Hou, G., Fang, F.: Dynamic modeling and multivariable control of organic Rankine cycles in waste heat utilizing processes. Comput. Math Appl. **64**(5), 908–921 (2012)
12. Li, G., Qin, S.J., Zhou, D.: Geometric properties of partial least squares for process monitoring. Automatica **46**(1), 204–210 (2010)
13. Kaspar, M.H., Ray, W.H.: Chemometric methods for process monitoring and high performance controller design. AIChE J. **38**(10), 1593–1608 (1992)

Optimized Control of Ship DC Electric Propulsion System with Energy Storage Unit

Feng Ding[1], Shuofeng Wang[2(✉)], and Shaohua Zhang[1]

[1] Key Laboratory of Power Station Automation Technology,
Department of Automation, Shanghai University, Shanghai 200072, China
shudf222@163.com
[2] Shanghai Marine Equipment Research Institute, Shanghai 200040, China
wsf704@163.com

Abstract. The frequent load fluctuations caused by the marine environmental variability and the operational requirements of the ship itself will have adverse impacts on the economics and reliability of the ship power grid. To alleviate these adverse impacts, the energy management technology is adopted and the super capacitor is employed as the energy storage unit in the ship DC electric propulsion system. In addition, the smooth fluctuation power control method is used, and the particle swarm optimization algorithm is applied to optimize the cut-off frequency of the low-pass filter and the capacity of super capacitor. As results, the fuel consumption cost of the diesel generator and energy storage cost can be minimized, and the negative impact caused by the ship load fluctuations can be mitigated. Finally, the simulation results show that the proposed methods can effectively improve the performance of ship propulsion system.

Keywords: Ship power grid · Electric propulsion · Super capacitor · Smooth fluctuation · Particle swarm optimization

1 Introduction

With development of power electronics technology and intensifying of energy crisis, ship propulsion technology is also involved in a rapid development stage. The shift from traditional diesel propulsion to electric propulsion technology has become an unstoppable trend. To guarantee the consistency of voltage, frequency and phase of the AC grid, ship electric propulsion system with AC power grid requires the generator to operate at the rated speed. Based on the analysis of the fuel consumption characteristic curve of diesel generators, it is not ideal for the fuel consumption rate when the diesel generator operates with the rated speed but with a low load [1]. If the diesel generator can operate with a low speed and a high torque according to the actual situation of the load, the diesel fuel consumption rate can be reduced. In DC electric propulsion system, no special requirement is needed to the generator speed because the diesel generator is linked in parallel to DC busbar. Besides, diesel generator speed can be adjusted according to the actual situation of the load [2].

Due to the complexity and variability of the marine environment and the operational requirements of the ship itself, electric propulsion ship has been faced with a

© Springer Nature Singapore Pte Ltd. 2017
K. Li et al. (Eds.): LSMS/ICSEE 2017, Part III, CCIS 763, pp. 519–528, 2017.
DOI: 10.1007/978-981-10-6364-0_52

technical problem: the load of engineering ship, such as icebreaker, dredger, changes rapidly in a short time, resulting in frequency disturbance that has a huge impact on the ship power grid [3]. On the one hand, load changes rapidly in a short time. But the start and shutdown of diesel generators will need a time period, which leads to a poor capability of the system dynamic response. Frequent start-shutdown will also shorten the lifespan of the diesel generators and increase the maintenance cost. In the process of frequent start-shutdown, the fuel efficiency is low [4]. On the other hand, there is need to have large reserve capacity for the diesel generator to deal with fluctuations in the load, which leads to lower utilization of the generator.

There are some solutions to address the negative impact of load fluctuations on the ship power grid. A battery-based energy storage control system is employed to smooth the grid-injected power in [5]. Based on the smooth fluctuation power control method, the energy storage control method of variable filter time constant is proposed in [6]. Based on the secondary low-pass filter, reference [7] proposes a filter time constant control method of the hybrid energy storage according to the spectrum diagram and modifies the filtering output power considering the battery charge and discharge power limitation. Reference [8] presents the multiple energy storage elements usability for ships using a passive hybrid topology.

Given the above background, in order to solve the negative impact of ship load fluctuations on the ship itself, this paper uses the super capacitor as the energy storage unit to stabilize the ship load fluctuations. Then, in order to minimize the fuel consumption cost of diesel generators and the cost of energy storage unit, this paper uses the smooth fluctuation power control method and the particle swarm optimization (PSO) algorithm to optimize the cut-off frequency of low-pass filter and the capacity of super capacitor. Finally, the validity and feasibility of the proposed methods are verified by the experimental data.

2   System Structure and Optimized Control Strategy

2.1   The Structure of Ship DC Electric Propulsion System

The main component in the power plant of ship power grid is diesel generator, which is the main energy source of the system. The energy storage unit is composed of super capacitor which is used to provide or absorb the energy when the load fluctuates. The load side is mainly composed of propulsion load and hotel load. The DC bus is connected to the rectifier and inverter of frequency converter. The power transmission mode of the whole system is roughly the same as that of the AC power propulsion system. The difference is that the diesel generator and propulsion motor are connected with the DC bus through the AC/DC converter or DC/AC converter respectively. As the energy storage unit, the super capacitor connected to the DC bus through the bidirectional DC/DC converter. Figure 1 shows the structure of the ship DC electric propulsion system [9].

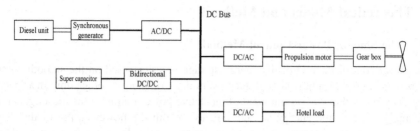

Fig. 1. Structure of ship DC electric propulsion system

2.2 Optimized Control Strategy

In this paper, the super capacitor as a storage unit is applied in the ship DC power propulsion system. The energy storage unit can absorb or provide energy to alleviate load fluctuations. Since the charge and discharge control of the super capacitor is mainly controlled by the bidirectional DC/DC converter, the bidirectional DC/DC converter is optimized to control the charge and discharge of the super capacitor and the power output of diesel generators. Firstly, the power output of diesel generators is predicted by the load fluctuation under the operating condition of the ship. Then, the output spectrum of the diesel generators is calculated by using the fast Fourier transform (FFT). This paper uses the smooth fluctuation power control method, and the high-frequency component of the diesel generator output is separated by the first-order low-pass filter. Because of some advantages of the super capacitor, such as power density, long cycle lifespan and high efficiency, the high-frequency components are real-timely compensated by the super capacitor. Finally, the reference depth of charge and discharge of super capacitor and the reference output power of diesel generators is calculated by using the smooth fluctuation power control method. The PSO is used to further optimize the cut-off frequency and capacity of super capacitor. On the basis of minimizing the fuel consumption cost and energy storage cost, the diesel generator can be maintained in the best working point and the energy storage unit can solve the negative impact caused by the load fluctuations of ship. The block diagram of the smooth power control method based on the particle swarm algorithm is presented in Fig. 2.

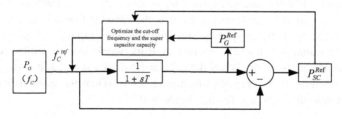

Fig. 2. Block diagram of optimized control strategy

3   Theoretical Model and Methods

3.1   The Smooth Power Control Method

The power control method of the energy storage inverter based on the smooth control refers to the real-time adjustment of the power output of energy storage unit. In a specific frequency band, the wave component of the active power output of the diesel generator is compensated by the energy storage unit to smooth the power output of the diesel engine power output. At first, P_G, the power output of the diesel generator, is predicted by the load fluctuation under the operating condition of the ship. Then, P_G^{Ref}, the reference signal of the active power of the diesel generator, is obtained by the first-order low-pass filter. At last, the reference signal of the active power of super capacitor will be obtained by $P_G - P_G^{Ref}$. The positive value indicates that the super capacitor is discharged and the negative value indicates that a super capacitor is charged [10, 11].

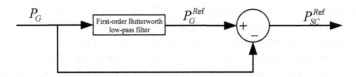

Fig. 3. Block diagram of the smooth control strategy

The smooth power control method used in this paper is described as follows:

$$P_G^{Ref}(s) = \frac{1}{1+sT}P_G(s) \tag{1}$$

$$P_{SC}^{Ref}(s) = P_G^{Ref}(s) - P_G(s) = \frac{sT}{1+sT}P_G(s) \tag{2}$$

where P_G^{Ref} is the reference output power of diesel generators using the smooth power control method; P_G is the output power of diesel generators predicted by the load fluctuation under the operating condition of the ship; P_{SC}^{Ref} is the reference depth of charge and discharge of super capacitor.

A first-order Butterworth low-pass filter is adopted in this method. The transfer function under s domain is transformed into the transfer function in the frequency domain. According to the frequency characteristic and amplitude frequency characteristic, the amplitude-frequency function of Butterworth low-pass filter is a monotonically decreasing function. When the angular-frequency is equal to 0, the maximum value of amplitude of the transfer function is set 1. When the angular-frequency is equal to the cut-off frequency, the amplitude is 0.707.

$$f_C = \frac{1}{2\pi T} \tag{3}$$

where T is the filter time constant; f_C is the cut-off frequency.

3.2   Optimization of Cut-off Frequency and Capacity of Super Capacitor

Objective Function. The establishment of the model is the key to the schedule of the ship energy storage unit and diesel engine. The main task is to make decisions according to different optimization objectives and constraints. In a given period, combining the situations of load demand, both of the diesel generator output and smooth conditions of energy storage unit are optimized and predicted. Therefore, the economics and stability of the DC electric propulsion system are subsequently achieved.

$$\min C_{sum} = \sum_{t=1}^{T}(COST_{SC} + COST_G) \tag{4}$$

The Fuel Consumption Cost of Diesel Generators. Assume that the diesel generator is a variable speed unit. The fuel consumption characteristics of the diesel generator, F_G, depends on the power output of diesel generators according to the following formula [12]:

$$COST_G = \sum_{t=1}^{T} F_G(t) \cdot q_G \tag{5}$$

$$F_G(t) = a \cdot P_G^{Ref}(t) + b \cdot (P_G^{Ref}(t))^2 + c \tag{6}$$

where C_{sum} is the total cost including the fuel consumption cost of diesel generator and the super capacitor cost of energy storage unit; $COST_G$ is the fuel consumption cost of diesel generators; q_G is the price of the diesel; $P_G^{Ref}(t)$ is the power output of diesel generators; a and b are the coefficients of fuel consumption characteristic curve.

The Cost Model of the Super Capacitor. The cost of energy storage unit mainly includes the investment cost and the operation and maintenance cost. The calculation model of costs is as follows [13, 14].

$$COST_{SC} = \frac{P_{SC}^{Ref}}{P_r} \cdot C_{sb} \cdot \left(\frac{r(1+r)^l}{(1+r)^l - 1}K_I + K_{OM}\right) \tag{7}$$

where C_{sb} is the capacity of super capacitor; K_I is the investment cost per unit; K_{OM} is the per unit operation and maintenance cost of super capacitor; r is the construction interest rate of the energy storage; l is the lifespan of super capacitor; P_r is the rated discharge and charge depth of super capacitor.

Equality and Inequality Constraints

Active Power Balance Equation. Due to the load fluctuations of the ship power grid, the power balance will reduce the influence of the uncertain factors on power quality and voltage frequency.

$$P_G(t) + P_{SC}(t) = P_{load}(t) \tag{8}$$

where $P_{SC}(t)$ is the charge and discharge power of the super capacitor. When $P_{SC}(t) < 0$, the super capacitor is charged; when $P_{SC}(t) < 0$, the super capacitor is discharged. $P_{load}(t)$ is the total load including propulsion load and hotel load.

Minimum and Maximum Power Limits. It is necessary to keep the diesel generator to operate in a certain range.

$$P_G^{\min} \leq P_G(t) \leq P_G^{\max} \tag{9}$$

where $P_G^{\max}$ and $P_G^{\min}$ are the maximum and minimum active outputs of the diesel generators, respectively (Fig. 3).

Energy Storage Unit Constraints. The energy storage system can effectively be used to mitigate the volatility of ship power grid and stabilize the load fluctuations through the energy storage unit [15].

$$E_{SC}(t) = \begin{cases} E_{SC}(t-1) + P_{SC}(t)\Delta t \eta_{SC}^{ch} & P_{SC}(t) > 0 \\ E_{SC}(t-1) + P_{SC}(t)\Delta t / \eta_{SC}^{dis} & P_{SC}(t) < 0 \end{cases} \tag{10}$$

$$P_{SC,ch}^{\max}(t) \leq P_{SC}(t) \leq P_{SC,dis}^{\max}(t) \tag{11}$$

$$SOC_{SC}^{\min} \leq SOC_{SC}(t) \leq SOC_{SC}^{\max} \tag{12}$$

where $E_{SC}(t)$ is the stored energy in the super capacitor in the period t; $SOC_{SC}^{\max}$ and $SOC_{SC}^{\min}$ are respectively the maximum and minimum state of charge; $P_{SC,ch}^{\max}$ ($P_{SC,dis}^{\max}$) is the largest depth of charge and discharge; η_{SC}^{ch} (η_{SC}^{dis}) is the efficiency of charge and discharge.

The above optimization problem is solved by the PSO algorithm.

4 Simulation Analysis

Consider a set of load data that fluctuates in a time period. The sampling frequency is 1 Hz. Figure 4 shows the load fluctuation of a vessel in a time period of 1 min. Table 1 shows the basic parameters.

As seen in the Fig. 5, according to the conditions of load fluctuations, the diesel generation is forecasted and the power spectrum of diesel generation is obtained by using the FFT in the discrete Fourier transform method.

By using the smooth fluctuation power control method based on particle swarm optimization, this paper optimize the cut-off frequency of low-pass filter and super capacitor under the premise of minimizing cost of fuel consumption and energy storage unit. Figure 6 shows the convergence of PSO algorithm.

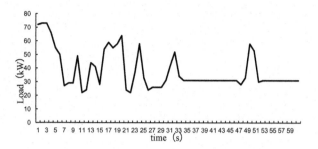

Fig. 4. Load profile

Table 1. Basic parameters

Diesel generator	c (g/h)	8488.1	PSO algorithm	C_1	1
	a (g/kWh)	115.65		C_2	1
	b (g/kWh)	0.202		maxgen	100
	q_G ($/g)	6.04		popsize	50
Super capacitor	K_I ($/kWh)	600		W_{max}	0.9
	K_{OM} ($/kWh)	20		W_{min}	0.4
	L	0.06			
	η	60%			
	r (year)	10			

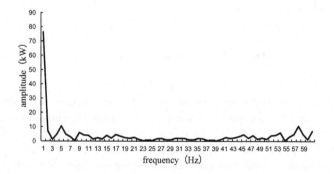

Fig. 5. Power spectrum of diesel generation

Figure 7 shows the output of diesel generators and the charge and discharge depth of super capacitor before and after using the smooth power control strategy. Due to the limitation of charge state of the super capacitor, the rated charge and discharge depth of the super capacitor is not more than 20% of its own capacity, so the maximum discharge capacity of the super capacitor is calculated as 175 kW. As seen in the Fig. 7, the fluctuations of diesel generation are significantly mitigated by using the super capacitor. Due to the existence of energy storage unit, the diesel generator can be

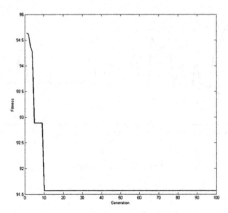

Fig. 6. Convergence curve of PSO

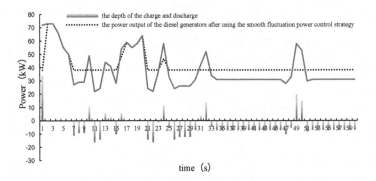

Fig. 7. Power output of diesel generators

controlled in a stable operating point in most of time. Besides, the stable operating point is calculated by the particle swarm optimization algorithm to guarantee the minimum fuel and storage cost. Therefore, the smooth fluctuation power control strategy based on particle swarm optimization satisfies the minimization of the cost, as well as the optimal cut-off frequency of the low-pass filter. Meanwhile, the rated capacity of the super capacitor is also determined by particle swarm optimization.

Figure 8 shows the SOC of super capacitor. Under normal circumstances, the minimum capacity of super capacitor is set to 10% of its own capacity in this paper. If below this state, super capacitor is deeply discharged. In addition, the initial energy state of super capacitor is set to 10% of its own capacity. Based on the particle swarm optimization algorithm, the cut-off frequency of the low-pass filter is 0.014 Hz. Under the operating condition within 1 min, the cost of the fuel consumption and the super capacitor using smooth fluctuation power control strategy based on the particle swarm optimization algorithm are 91.57 $/h. The conventional cost of the fuel consumption is 116.35 $/h according to the fuel consumption characteristics of the ship. Therefore, the

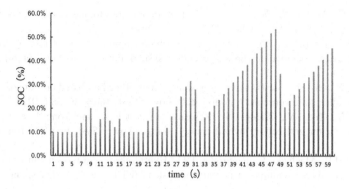

Fig. 8. State of charge

smooth fluctuation power control strategy based on particle swarm optimization can not only reduce the negative impact of the load fluctuation on the ship power grid, but also save fuel and storage costs.

5 Conclusions

This paper employed the DC electric propulsion system instead of the AC electric propulsion system. In this system, the diesel generators can directly control the speed according to different load. In order to reach the purpose of costs reduction of fuel and energy storage, this paper uses the smooth fluctuation power control method and the particle swarm optimization algorithm to optimize the cut-off frequency of low-pass filter and capacity of the super capacitor. Simulation results show that the proposed methods can effectively improve the performance of ship propulsion system.

At present, the energy storage unit has become an indispensable part of the development of the ship. With the improvement of the large-capacity lithium battery and super capacitor technology, multi-energy hybrid ship will become an unstoppable trend. As the multi-energy hybrid ships often use wind power, photovoltaic and other renewable energy, the issues caused by intermittence and volatility are needed to be addressed. Therefore, the energy storage unit will play a more important role in the future ship electric propulsion system.

References

1. Hansen, F., Lindtjørn, J.O., Vanska, K.: Onboard DC grid for enhanced DP operation in ships. In: Dynamic Positioning Conference, Norway, pp. 1–8 (2011)
2. Zhang, P., Shi, Y.: Research on simulation of ship's power system based on DC power grid. Mar. Electr. Electron. Eng. **36**(9), 53–56 (2016)
3. Chen, C., Wang, X., Feng, H.: Application of li-battery and super capacitor in electric propulsion ship. Ship Eng. **38**(2), 186–190 (2016)

4. Chen, C., Wang, X., Xiao, J.: Application of energy storage devices in ship electric propulsion system. Navig. China **37**(4), 25–29 (2014)
5. Qiu, P., Ge, B., Bi, D.: Battery energy storage-based power stabilizing control for grid-connected photovoltaic power generation system. Power Syst. Prot. Control **39**(3), 29–33 (2011)
6. Zhang, Y., Guo, L., Jia, H.: An energy storage control method based on state of charge and variable filter time constant. Autom. Electr. Power Syst. **36**(6), 36–38 (2012)
7. Sang, B., Tao, Y., Zheng, G.: Research on topology and control strategy of the super-capacitor and battery hybrid energy storage. Power Syst. Prot. Control **42**(2), 1–6 (2014)
8. Trovão, J.P., Machado, F., Pereirinha, P.G.: Hybrid electric excursion ships power supply system based on a multiple energy storage system. IET Electr. Syst. Transp. **6**(3), 190–201 (2016)
9. Xu, L., Chen, D.: Control and operation of a DC microgrid with variable generation and energy storage. IEEE Trans. Power Deliv. **26**(4), 2513–2522 (2011)
10. Jiang, Q., Wang, H.: Two-time-scale coordination control for a battery energy storage system to mitigate wind power fluctuations. IEEE Trans. Energy Convers. **28**(1), 52–61 (2013)
11. Sadegh, S., Reza, K.P.: Reduction of energy storage system for smoothing hybrid wind-PV power fluctuation. In: Proceedings of 2012 11th International Conference on Environment and Electrical, Venice, Italy, pp. 115–117 (2012)
12. Li, J., Jin, X., Xiong, R.: Multi-objective Optimal energy management stragegy and economic analysis for an range-extended electric bus. Energy Procedia **88**, 814–820 (2016)
13. Li, P., Duo, X., Zhou, Z., Lee, W.-J., Zhao, B.: Stochastic optimal operation of microgrid based on chaotic binary particle swarm optimization. IEEE Trans. Smart Grid **7**(1), 66–73 (2016)
14. Hu, X., Liu, T., He, C.: Multi-objective optimal operation of microgrid considering the battery loss characteristics. Proc. CSEE **36**(10), 2674–2681 (2016)
15. Carpinelli, G., Mottola, F., Proto, D., Russo, A.: A multi-objective approach for microgrid scheduling. IEEE Trans. Smart Grid. 1–10 (2016)

The Application of the Particle Swarm Algorithm to Optimize PID Controller in the Automatic Voltage Regulation System

Jing Wang[1], Naichao Song[2], Enyu Jiang[3(✉)], Da Xu[3], Weihua Deng[3], and Ling Mao[3]

[1] College of Information Technology, Shanghai Ocean University,
Shanghai, People's Republic of China
[2] Xu Chang Electric Power Supply Company, Xu Chang City,
Henan Province, People's Republic of China
[3] College of Electrical Engineering, Shanghai University of Electric Power,
Shanghai, People's Republic of China
enyu_1981@163.com

Abstract. Automatic voltage regulation (AVR) is a system that used to adjust the voltage stability and balance reactive power and also for regulating power plant generator. Focusing on the traditional PID automatic voltage regulation system, this paper investigated the effect of particle swarm optimization (PSO) algorithm in optimizing the parameters of PID controller in AVR system, and compared with genetic algorithm (GA) for PID parameters optimization. The simulation results showed that the AVR system optimized by PSO had more stability and robustness, which indicated the good application prospect of the proposed method.

Keywords: Automatic voltage regulation · Particle swarm optimization · PID · Genetic algorithm

1 Introduction

In the automatic voltage control system, the automatic voltage regulation (AVR) system is composed of the equipments involved in regulating terminal voltage of power systems, especially the excitation system [1, 2]. The AVR system consists of amplifiers, exciter, generator and sensor. According to the stability requirement of the power system, a controller for improving the response speed need to be added, and the common PID controller used in industry is often adopted to solve this problem [3–6].

Control performance of PID controller is determined by the PID parameter. Therefore, the improvement of the PID controller's performance is to find the effective settings of PID parameters [7, 8]. At present, commonly used method of adjusting PID parameters are Z-N method, critical proportion method, inverse curve method and so on [9–11]. The application of these methods of setting PID controller often cause larger amount of overshoot, longer oscillation and so on [12].

© Springer Nature Singapore Pte Ltd. 2017
K. Li et al. (Eds.): LSMS/ICSEE 2017, Part III, CCIS 763, pp. 529–536, 2017.
DOI: 10.1007/978-981-10-6364-0_53

Therefore, this paper discussed the affect of PID controller parameters on the performance of the AVR system. Meanwhile the particle swarm optimization (PSO) algorithm was proposed for optimizing the parameters of the PID controller, and the effect of the proposed method was compared with that of the genetic algorithm optimized PID controller.

2   The Automatic Voltage Regulation System

AVR system is used to guarantee the stability of the final voltage at the power plant machine in the same specific level under system fault, the voltage fluctuation. A simple AVR system includes four main parts—amplifier, exciter, generator and sensor. In order to meet the requirements of power system stability, we added a PID controller. This four parts continuous run under the condition of normal operation, you can ignore the amount of saturated or nonlinear characteristics, the four parts can be as linear elements. These parts reasonable transfer function can be respectively performance is as follows:

- Amplifier Model

   Amplifier model uses an amplification coefficient and a time constant.

$$\frac{V_R(s)}{V_e(s)} = \frac{K_A}{1 + \tau_A s} \tag{1}$$

   The typical values of K_A is from 10 to 400. The range of time constant τ_A is very small, range 0.02 to 0.1 s.

- Exciter Model

   Exciter of transfer function is shown as an amplification coefficient and a time constant.

$$\frac{V_F(s)}{V_R(s)} = \frac{K_E}{1 + \tau_E s} \tag{2}$$

   The typical values of K_E is from 0.5 to 1. The range of time constant τ_E is very small, from 0.4 to 1.0 s.

- The Generator Model

   In linearized model, the transfer function of the generator reflects the relationship between voltage of the generator and voltage of magnetic field. It can be characterized by a amplification coefficient and a time constant.

$$\frac{V_t(s)}{V_F(s)} = \frac{K_G}{1 + \tau_G s} \tag{3}$$

These constant size is determined with load, the range of K_G is from 0.7 to 1, and the range of τ_G is from 1.0 to 2.0.

- The Model of Sensor

The model of sensor is expressed as a simple linear transfer function:

$$\frac{V_S(s)}{V_t(s)} = \frac{K_R}{1 + \tau_R s} \tag{4}$$

The range of τ_R is from 0.001 to 0.06 s.

- PID Controller Model:

PID controller transfer function is as follows:

$$\frac{V_{in}(s)}{V_{out}(s)} = k_p + k_i/s + k_d s \tag{5}$$

3   Particle Swarm Optimization

The particle swarm optimization algorithm is originated from artificial life and predatory birds' behavior research, which is a global search strategy, and a competitive neural network learning algorithm. The search space dimension is D, particle population scale is S, the i-th particle's position vector X_i and velocity V_i is expressed $X_i = (x_{i1}, x_{i2}, \ldots, x_{iD})^T$ and $V_i = (v_{i1}, v_{i2}, \ldots, v_{iD})^T$, $i = 1, 2, \ldots, S$. The i-th particle's optimal position vector is P_i, and optimal position vector of all particles is P_g, and both are D vector. The evolution rules are as follows [5]:

$$V_{id}^{t+1} = \omega V_{id}^t + c_1 r_1 \left(P_{id} - X_{id}^t \right) + c_2 r_2 \left(P_{gd} - X_{id}^t \right) \tag{6}$$

$$X_{id}^{t+1} = X_{id}^t + V_{id}^{t+1} \tag{7}$$

Here $d = 1, 2, \ldots, D$. c_1 is cognitive learning factor, it is acceleration term weight of the i-th particle's optimal position vector P_i. c_2 is social learning factor, and it is acceleration term weight of the optimal position vector of all particles P_g. Their values are between 0 and 2. r_1 and r_2 are uniformly distributed random numbers in $[0, 1]$. ω is the momentum factor and it is non-negative. While large value is easier to search the global optimal solution, the partial convergence is poorer. Generally $\omega_{max} = 0.9$, $\omega_{min} = 0.4$. In the evolutionary process, to ensure the algorithm's convergence, particles' speed limit V_{max} should be set. Before using formula (6) to update particle velocity value, whether $V_{id} \in [-V_{max}, V_{max}]$ holds should be judged, if it holds, value could be updated using formula (6), otherwise using

$$V_{id} = \text{sgn}(V_{id})V_{\max} \tag{8}$$

In formula (8), $\text{sgn}()$ is sign function. The iterative termination conditions are chosen as maximum iterating times or satisfying specified standard error.

Penalty factor in formula (3) is important for sample classification penalty and calculation accuracy. Kernel function width in formula (6) is important for recognizing ability of the Kernel function and generalization ability. The traditional way such as cross validation test method are very complex, since they need certain experience through repeated try to determine the right parameter value. The proposed way using PSO algorithm is shown as follows:

1. Extract feature vector by Wavelet packet analysis of colonic contractions measured in colon end. Define training sample as $T = \{(x_i, y_i)|i = 1, 2, \ldots, n\}$, and test sample as $T' = \{(x'_i, y'_i)|i = 1, 2, \ldots, m\}$. Initialize penalty factor and kernel function width and particle's velocity vector V. According to physiological characteristics of rectal pressure signal, rectal pressure signal category and number of wavelet packet layer, SVM structure can be determined;
2. After initialization, input the training sample set into the network, assess each particle's fitness according to formula (9)

$$J = \frac{1}{N} \sum_{i=1}^{N} \left(h(x)^d - h(x) \right)^2 \tag{9}$$

Here $h(x)$ is chosen from the SVM classification function $f(x) = \text{sgn}\{h(x)\}$. In PSO algorithm, if the current particle' fitness is better than its parents, the current particle value is P_{best}, if in the whole group, the fitness of another particle in rest particles is better, that particle value is G_{best};

3. Calculate velocity vector and position vector of the new generation by substituting the updated P_{best} and G_{best} into formula (6) and (7);
4. If the current iteration times reached the predefined maximum number or the minimum target error, output the final value according to the SVM decision function, otherwise turn to step 1.

After taking in a particle, it is necessary to evaluate the advantages and disadvantages of the whole control system. Under the condition of the step, a control system can reflect the response characteristics of time domain evaluation index including overshoot M_p, rise time t_τ, stability time t_s, and steady-state error E_{ss}. In order to reflect the four characteristics in a data, four data will be compounded for a $W(K)$, as shown in formula (8).

$$W(K) = (1 - e^{-\beta}) \bullet (M_P + E_{ss}) + e^{-\beta} \bullet (t_s - t_\tau) \tag{10}$$

$W(K)$ represents the overall performance of the control system under step input, while a smaller value of $W(K)$ indicates a better control.

β in Eq. (8) represents the weighting coefficient of which the change of value can be used to satisfy different requirements of performance. For instance, by setting the value of β greater than 0.7, the overshoot and static steady-state error can be reduced, while the effect is more distinguishable with high value of β. On the other hand, setting the value of β smaller than 0.7 reduces the rising time and convergence time. According to the system properties of AVR and the need for steady voltage of power grid, β is set between 0.8 and 1.5, minimizing overshoot and E_{ss}.

4 Simulation Experiment

In the simulation experiment, K_A is set to 100, K_E is set to 1, τ_E is set to 0.4, K_G is set to 1, τ_G is set to 1, K_R is set to 1, τ_R is set to 0.1, k_p, k_i, k_d are the 3 dimensions used in the PSO. The block diagram of AVR system is shown in Fig. 1.

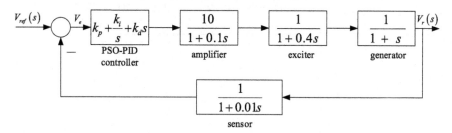

Fig. 1. Block diagram of an AVR system with a PSO-PID controller

The effect of the PSO based simulation was compared with the other simulations optimized by GA and manual setting approach in order to evaluate the performance. The step response of the AVR with no controller installed was shown in Fig. 2. The figure indicated that the AVR with no controller installed had existed more excessive overshoot, longer oscillation time and larger oscillation amplitude, thus the AVR system with no controller was unable to maintain in a stable voltage.

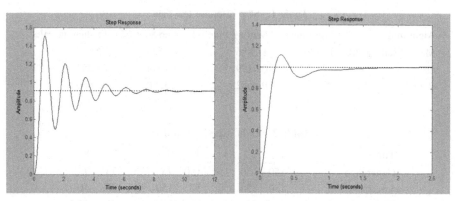

Fig. 2. Step response of the AVR with no controller installed

Fig. 3. Step response of the AVR with PID parameters manually configured

The step response of the AVR with PID parameters manually configured is shown in Fig. 3. The performance of it had been greatly improved compared to the one with no controller installed, though the problem of overshoot and stabilization period being too long were not eliminated.

The step response of the AVR with PID parameters of PSO approach configured is shown as the real line in Fig. 4. The step response of the AVR with PID parameters of GA approach configured is shown as the dotted line in Fig. 4. In order to facilitate the application of PSO algorithm and GA algorithm, we defined an evaluation function $f = \frac{1}{W(k)}$, while a smaller value of wk indicates a better control. It can be observed from the figure that both algorithms are able to provide significant improvements to the response characterics of AVR systems, and to satisfy the need of stable voltage in the power systems.

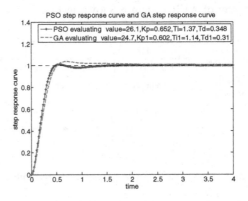

Fig. 4. Step response curve of PSO algorithm and GA algorithm

As shown in the Tables 1 and 2, during the 8 simulations, PSO algorithm kept evaluating value higher than that of GA algorithm with changing iteration time, group size and search region, therefore showed better optimization than GA algorithm.

Table 1. Step response test data

Algorithm	Rise time (s)	Settling time (s)	Overshoot (%)	Evaluating value
Manual setting PID	0.1476	1.2886	11.3628	2.917
GA	0.2775	0.4199	1.0054	25.255
PSO	0.3031	0.4652	0.3586	25.669

Table 2. Eight experimental data

Times	1	2	3	4	5	6	7	8
Generation	150	150	150	150	75	75	75	75
Swarm	150	150	50	50	150	150	50	50
PSO evaluating value	25.7	26.0	25.6	26.0	25.7	25.7	25.7	25.6
GA evaluating value	25.2	25.1	25.3	25.4	25.3	25.6	25.2	25.0

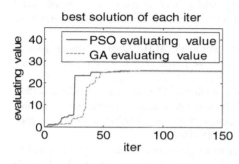
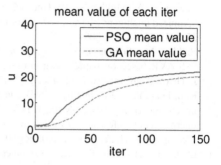

Fig. 5. Best particle evaluating value vary with generation

Fig. 6. Average evaluating value vary with generation

The convergence speed of PSO algorithm and GA algorithm were also compared, comparing the evaluating value of best particle in every generation, as is shown in Fig. 5. Average evaluating value of all particles for each generation is defined as μ, as is shown in Fig. 6. In the Figs. 5 and 6, PSO algorithm is described as real line and GA algorithm is described as dotted line. The results showed in the Figs. 5 and 6 indicated that the convergence speed of PSO algorithm was greater than that of GA algorithm.

5   Conclusion

This paper discussed the importance of maintaining the voltage stability and the power balance by the AVR system in the AVC system, and used the PSO algorithm to set controller parameters of AVR system, in order to improve the corresponding features of AVR system, and further to improve voltage stability of the area. It was proved by experiments that the capability of maintaining the voltage stability for AVR system by using PSO algorithm had been improved greatly. Furthermore, compared with the GA commonly used in the parameter setting of PID controller, the proposed PSO optimized method had the greater effect on improving the maintenance of voltage stability and power balance the AVR system.

Acknowledgements. This work was supported by the Shanghai Sailing Program (16YF1415700); the 2015 Doctoral Scientific Research Foundation of Shanghai Ocean University (A2-0203-00-100348).

References

1. Anbarasi, S., Muralidharan, S.: Enhancing the transient performances and stability of AVR system with BFOA tuned PID controller. Control Eng. Appl. Inform. **18**(1), 20–29 (2016)
2. Aidoo, I.K., Sharma, P., Hoff, B.: Optimal controllers designs for automatic reactive power control in an isolated wind-diesel hybrid power system. Int. J. Electr. Power Energy Syst. **81**, 387–404 (2016)

3. Chatterjee, A., Mukherjee, V., Ghoshal, S.P.: Velocity relaxed and craziness-based swarm optimized intelligent PID and PSS controlled AVR system. Int. J. Electr. Power Energy Syst. **31**(7–8), 323–333 (2009)
4. Haddin, M., Soebagio, S., Soeprijanto, A., Purnomo, M.H.: Optimal setting gain of PSS-AVR based on particle swarm optimization for power system stability improvement. J. Theor. Appl. Inf. Technol. **42**(1), 42–47 (2012)
5. Kim, D.H., Park, J.I.: Intelligent PID controller tuning of AVR system using GA and PSO. In: Huang, D.-S., Zhang, X.-P., Huang, G.-B. (eds.) Advances in Intelligent Computing. International Conference on Intelligent Computing, ICIC 2005. LNCS, vol. 3645, pp. 366–375. Springer, Heidelberg (2005). doi:10.1007/11538356_38
6. Li, C., Li, H., Kou, P.: Piecewise function based gravitational search algorithm and its application on parameter identification of AVR system. Neurocomputing **124**, 139–148 (2014)
7. Wong, C.-C., Li, S.-A., Wang, H.-Y.: Optimal PID controller design for AVR system. Tamkang J. Sci. Eng. **12**(3), 259–270 (2009)
8. Rahimian, M., Raahemifar, K.: Optimal PID controller design for AVR system using particle swarm optimization algorithm. In: 2011 Canadian Conference on Electrical and Computer Engineering, CCECE 2011, Niagara Falls, Canada, pp. 337–340 (2011)
9. Miavagh, F.M., Miavaghi, E.A.A., Ghiasi, A.R., Asadollahi, M.: Applying of PID, FPID, TID and ITID controllers on AVR system using particle swarm optimization (PSO). In: 2nd International Conference on Knowledge-Based Engineering and Innovation, KBEI 2015, Tehran, pp. 866–871 (2015)
10. Mukherjee, V., Ghoshal, S.P.: Intelligent particle swarm optimized fuzzy PID controller for AVR system. Electr. Power Syst. Res. **77**(12), 1689–1698 (2007)
11. Ramezanian, H., Balochian, S., Zare, A.: Design of optimal fractional-order PID controllers using particle swarm optimization algorithm for automatic voltage regulator (AVR) system. J. Control Autom. Electr. Syst. **24**(5), 601–611 (2013)
12. Bourouba, B., Ladaci, S.: Comparative performance analysis of GA, PSO, CA and ABC algorithms for fractional (PID mu)-D-lambda controller tuning. In: Proceedings of 2016 8th International Conference on Modelling, Identification and Control ICMIC 2016, pp. 960–965, Algiers (2016)

Research on the Bio-electromagnetic Compatibility of Artificial Anal Sphincter Based on Transcutaneous Energy Transfer

Peng Zan[1], Chundong Zhang[1], Suqin Zhang[2], Yankai Liu[1], and Yong Shao[1(✉)]

[1] Shanghai Key Laboratory of Power Station Automation Technology,
School of Mechatronics Engineering and Automation,
Shanghai University, Shanghai, China
shaoyong@shu.edu.cn
[2] Naval Aeronautical University Qingdao Campus, Qingdao, China

Abstract. For the treatment of anal incontinence, a new type of artificial anal sphincter is designed. The artificial anal sphincter system based on transcutaneous energy transfer mainly consists of sensor execution subsystem, wireless communication control subsystem and transcutaneous energy supply subsystem. Aim at the energy supply problem and the electromagnetic compatibility of the device, the energy transmission circuit is designed and optimized. At the same time, the three-dimensional model of the transmitting coil is constructed, and the high precision electromagnetic model of human body is carried out by using the finite difference time domain method. The distribution of specific absorption rate of different tissues is obtained. The safety analysis is carried out according to the International standard for electromagnetic safety of human body. The simulation results show that the artificial anal sphincter can stably and reliably supply energy to the internal device and it features favorable bio-electromagnetic compatibility. This study makes a firm theoretical foundation for the application of artificial anal sphincter.

Keywords: Artificial anal sphincter · Transcutaneous energy supply · Finite difference time domain · Biocompatibility

1 Introduction

Anal incontinence, also known as fecal incontinence, is a kind of common symptoms of the human body to lose the ability of defecation and bring the psychological burden [1]. According to relevant reports, fecal incontinence prevalence rate was 8.3% and the number of anal incontinence in elderly is much higher [2]. The main methods used in the treatment of anal incontinence at home and abroad are drug therapy, biofeedback therapy and surgical treatment. However, there are many defects, such as recurrent, severe infection and long treatment cycle, and the treatment effect is not ideal. In order to obtain an effective method for the treatment of anal incontinence, domestic and foreign academicians have conducted extensive research. Artificial anal sphincter is

© Springer Nature Singapore Pte Ltd. 2017
K. Li et al. (Eds.): LSMS/ICSEE 2017, Part III, CCIS 763, pp. 537–546, 2017.
DOI: 10.1007/978-981-10-6364-0_54

widely used as an effective way and its function and structure have been constantly improved [3–5]. But there are two problems have not yet been resolved. First, the use of conventional wire dragging methods to provide energy to the body device is easy to cause serious wound infection in patients with poor results. The transcutaneous energy transmission mode has some disadvantages such as low efficiency and poor stability. Second when the transcutaneous energy transmission system provides energy to the device in the body, the human body will be exposed to alternating electromagnetic fields and that may cause harm to human tissues and organs. However, there is no in-depth analysis of the electromagnetic compatibility of the artificial anal sphincter system, lack of relevant reports.

In view of the above problems, a new artificial anal sphincter is proposed based on transcutaneous energy supply. The artificial anal sphincter is designed, and the efficiency of energy transfer is optimized. The system realizes stable and efficient energy supply to the in vivo device. The simulation model of three-dimensional coil and the electromagnetic model of human body are established. By using the finite difference time domain method (FDTD), the distribution of the specific absorption rate of the emitter to the human body is calculated. Compared with international standards for electromagnetic safety of human body, the electromagnetic safety of the artificial anal sphincter is verified.

2 System Overview

The artificial anal sphincter system is composed of three modules, which are sensor execution module, wireless communication control module and transcutaneous energy supply module. The overall framework of the system is shown in Fig. 1. The sensor execution module is mainly responsible for the collection of intestinal pressure signals and the realization of artificial anal sphincter prosthesis. Wireless communication control module is used for signal acquisition, signal processing and signal transmission, at the same time to achieve the alarm signal and select the sphincter of the operating

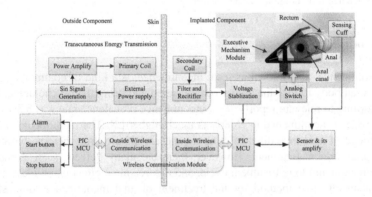

Fig. 1. Outline of artificial anal sphincter system

mode. The transcutaneous energy transmission module is used to satisfy the power supply demand of the sensing execution module and the wireless communication control module. A new actuator is designed, which is composed of a DC push-pull electromagnet, a bionic sphincter, a slide block structure, a flexible baffle and two small silica gels. The function of closed anus and defecation control is realized.

3   Transcutaneous Energy Transfer System

The transcutaneous energy transfer system provides a stable and reliable energy supply to the device. It replaces the traditional way of drag wire or battery, so that the practical significance of artificial anal sphincter increased significantly. The equivalent circuit diagram of the loosely coupled transformer is shown in Fig. 2. In order to overcome the skin barrier, the primary coil of the loosely coupled transformer is closely attached to the external skin, and the secondary coil is implanted into the skin tissue. In this way, the power supply of the device can be realized by the principle of electromagnetic induction. According to Fig. 2 the following equations can be obtained:

$$V_2 = \frac{j\omega M Z_2}{\omega^2(M^2 - L_1 l_2) + j\omega(L_1 R_2 + L_2 R_1 + L_1 Z_2) + R_1 R_2 + R_1 Z_2} \tag{1}$$

$$\eta = \frac{\omega^2 M^2 R}{R_1[(R_2 + R)^2 + (\omega L_2 + X)^2] + \omega^2 M^2 (R_2 + R)} \tag{2}$$

R and X are the real and imaginary parts of Z_2, respectively. The Eq. (2) shows that the coupling efficiency reaches the maximum, $\omega L_2 = -X$. This can be achieved by adding a shunt capacitor at the input impedance Z_2 of the secondary coil. At the same time, this paper use MATLAB to optimize the structure of the coil, such as proximity effect and parasitic capacitance, and the frequency of the work is set to 358 kHz. The parameters of the coil settings are shown in Table 1.

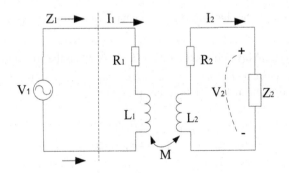

Fig. 2. Inductive coil link principle

Table 1. Coil specifications

Parameters	Primary coil	Secondary coil
Outer diameter	36 mm	30 mm
Inner diameter	26 mm	22 mm
Thickness	10 mm	6 mm
Inductance	28 μH	18 μH
Equivalent resistance	2.5 Ω	5.8 Ω

4 The Model of Transmitter Coil and the Electromagnetic Model of Human Body

4.1 The Model of Transmitter Coil

Because the current of the transmitting coil is much larger than the receiving coil, the electromagnetic field produced by the receiving coil has little effect on the human tissue. In order to simplify the analysis, the three-dimensional model of the transmitting coil is established. The three-dimensional model of the transmitting coil is established by means of simplification in this paper [7]. In order to correctly calculate the interaction between the transmitter coil and the human body, the finite-length solenoid model is presented according to the characteristics of the transmitting coil, as shown in Fig. 3.

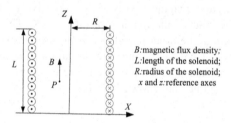

B:magnetic flux density;
L:length of the solenoid;
R:radius of the solenoid;
x and z:reference axes

Fig. 3. Section drawing of the limited close-wound solenoid

According to the Maxwell equation $\nabla \times E = \frac{\partial B}{\partial t}$, when the solenoid is connected to a low frequency current, $I = I_0 e^{-j\omega\tau}$ the expressions of the electric field strength and magnetic field strength of the solenoid are presented:

$$B = \frac{2\pi R n^2 \mu_0 I}{\sqrt{(2\pi R n)^2 + 1}} \tag{3}$$

$$I = I_0 e^{-j\omega\tau} \tag{4}$$

$$E = -j\omega I_0 \frac{x\pi R n^2 \sqrt{\mu_0 \varepsilon_0}}{\sqrt{(2\pi R n)^2 + 1}} e^{-j\omega\tau} \tag{5}$$

where R is the radius of the solenoid, n is the number of turns of solenoid, μ_0 is the vacuum permeability, ω is the angular frequency, τ is the time, ε_0 is the relative dielectric constant, I_0 is the effective value of the current I. The simulation coil is located in the abdomen of the human body and the structure and dimensions of the transmitting coil are shown in Fig. 4.

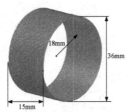

Fig. 4. The model of transmitter coil

4.2 The Electromagnetic Model of Human Body

The numerical calculation method is used to solve the electromagnetic effect of the transmitting coil on the human body. First, it needs to set up the electromagnetic calculation model of human body. The external shape of the human body is irregular and its internal structure and electromagnetic properties is very complex. Most of the studies on the distribution of human biological dose are two or three layers simplified model in early stage. The results are not accurate because of the coarse partition and low resolution. In order to obtain more accurate and detailed conclusions, a high precision electromagnetic calculation model of human body is established. By using digitized data in the form of transverse color images from the VHP (Visible Human Project, USA), the three-dimensional electromagnetic model of human body is constructed [8]. The whole modeling process is shown in Fig. 5(a). In order to reduce the amount of computation and workload, this paper chooses 375 pictures in VHP images. Then, the images are segmented and the distance between the two images is 5 mm. The human body model has 52 kinds of human tissues and the resolution of is 5 mm 5 mm × 5 mm. After the color image is grayed, the mapping relationship between gray value and the organization is established. The obtained images are arranged in a vertical direction and the human body structure voxel model is set up. Through the use of 3D-Doctor software for three-dimensional image modeling, the structural model of human body with body weight of 90 kg and height of 180 cm is got.

In the numerical calculation, it is necessary to define the electromagnetic properties of each tissue and organ of the human body structure, including the electrical conductivity and relative permittivity [9]. The dielectric parameters can be calculated by the following formula:

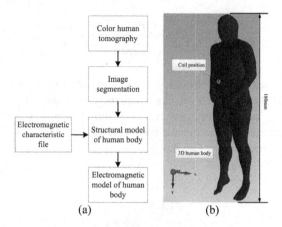

(a)   (b)

Fig. 5. (a) Model building process. (b) The electromagnetic model of human body (Color figure online)

$$\hat{\varepsilon}(\omega) = \varepsilon_\infty + \sum\nolimits_{n=1}^{4} \frac{\Delta\varepsilon_n}{1 + (j\omega\tau_n)^{1-\alpha_n}} + \frac{\sigma_i}{j\omega\varepsilon_0} \tag{6}$$

where ε_n is the dielectric constant, ε_0 is the vacuum dielectric constant, σ_i is the static ionic conductivity and τ_n is the relaxation time. By means of formula (6), the electromagnetic properties of different tissues at different frequencies can be acquired and then an electromagnetic characteristic file can be formed. The electromagnetic model of human body is shown in Fig. 5b. According to formula (6), the FCC and IFAC-CNR sites provide the dielectric constant and conductivity of different tissues or organs in the frequency range of 100 GHz. The electromagnetic parameters of some human tissues at 358 kHz frequency are shown in Table 2.

Table 2. Electromagnetic parameters of some tissues of human body ($f = 385$ kHz)

Tissue name	Conductivity [S/m]	Relative permittivity
Blood	0.73185	4483.4
Muscle	0.42485	4458.7
Heart	0.26656	3884.7
Fat	0.024758	38.829
Bone Marrow	0.0037492	54.061
Skin wet	0.16097	4690.8
Stomach	0.54812	2178.8

5 Research Method

5.1 SAR Calculation

In the discussion of human exposed to electromagnetic fields on the human body, the specific absorption rate is the most important parameter. At present, SAR is usually used to describe the degree of interaction between electromagnetic wave and human body in the world, and the maximum value of electromagnetic radiation can be determined by the physical quantity [10]. Specific absorption rate (SAR) can be defined as the electromagnetic power absorbed or consumed by a unit of mass in a unit of time, which can be expressed by:

$$SAR = \frac{d}{dt}\left(\frac{dW}{dm}\right) = \frac{d}{dt}\left(\frac{dW}{\rho dV}\right) \tag{7}$$

where m is the quality of biological tissue, t is the time, W is the radiation power, V is the volume of biological tissue, and ρ is the biological tissue conductivity. The average specific absorption rate can be used to describe the size of the absorbed dose in the electromagnetic environment [11]. The average specific absorption rate is calculated as follows:

$$SAR = \frac{1}{V}\int vSARdV = \frac{1}{V}\int v\frac{\sigma}{\rho}|E|^2 dV \tag{8}$$

The most commonly used average SAR is the calculation of 1 g (SAR_{1g}) or 10 g (SAR_{10g}) tissue.

5.2 Finite Difference Time-Domain Method

The commercial software HFSS based on finite difference time domain method is used to calculate the specific absorption rate distribution. The FDTD method provides a direct method for solving the Maxwell's equation in the time domain, which is convenient for the calculation of electromagnetic radiation in irregular and inhomogeneous media [12]. Firstly, the human body model is imported into HFSS, and the excitation parameters of the transmitter coil are set up. Then the mesh is divided, so that the output can be calculated and processed. When the excitation parameters of the transmitting coil are set, its current is a sinusoidal signal of 1A, the number of turns of the coil is 25 turns, and the working frequency is 358 kHz.

6 Simulation Results and Discussion

6.1 Simulation Results

This section will show the SAR distribution of the human body on the vertical section. Prior to the analysis of SAR, an axis coordinate system is established and the center of the computational model as the origin (0, 0, 160). Figure 6 shows the SAR value distribution along the Y-Z axis by analyzing the midpoint of the model. At the same time, this paper carries on the sampling curve to the human body model, and the curve coincides with the axis of the transmitting coil. The SAR curve is shown in Fig. 7.

It can be seen from Fig. 6 that the distributions of SAR_{1g} and SAR_{10g} are similar, only the values are different. The distribution of the specific absorption rate is larger in the vicinity of the transmitting coil, with the increase of the distance between the human tissue and the coil, the value of the absorption rate decreases rapidly. The maximum value appears near the transmitting coil. The average value of SAR in the 1 g tissue is 8.012 mW/kg, when the transcutaneous energy transmission system is working. The average value of SAR in the 10 g tissue is 1.472 mW/kg, and the whole body averaged SAR value is 9.046 mW/kg.

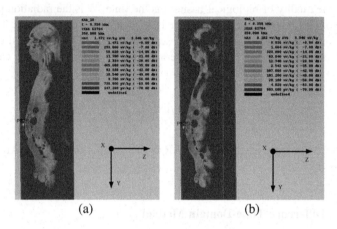

(a)   (b)

Fig. 6. Distribution of the SAR in human body: (a) SAR_{1g} and (b) SAR_{10g}

As shown in Fig. 7, the value of average SAR (1 g and 10 g) in the human body model with the sampling curve can be got. The trend of SAR_{1g} curve and SAR_{10g} curve are basically the same. With the increase of the distance from the transmitting coil, SAR value increases first and then decreases rapidly. This is due to the different dielectric properties of different tissues of the human body that the conductivity of the

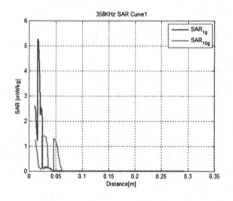

Fig. 7. 1D representation of SAR_{1g} and SAR_{10g} along the curve

muscle is greater than the skin and the fat. At the same time, with the increase of the distance from the source of electromagnetic wave, the radiation power decreases rapidly. The maximum value of SAR_{1g} is 5.2 mW/kg and the maximum value of SAR_{10g} is 1.5 mW/kg. The value is close to zero when the distance is close to 6 cm.

6.2 Discussion

The limits of international and European adoption are that the average SAR value of 10 g biological tissue is less than 2 W/kg in the uncontrolled electromagnetic environment. The United States provides a more stringent standard that the average SAR value of the 1 g biological tissue is less than 1.6 W/kg. The simulation results show that the maximum value of SAR_{1g} is 5.2 mW/kg and the maximum value of SAR_{10g} is 1.5 mW/kg, as well as the average SAR value of the whole body is 9.046 mW/kg. The values are far less than the safety limits set by the International Commission on Non-ionizing radiation Protection (ICNIRP) and the Institute of electrical and Electronics Engineers (IEEE). From the SAR value of the human body and the plot of the curve change, the main region of the electromagnetic radiation of the human body is located near the transmitting coil, and the trend is significantly reduced with the increase of the distance from the source. The maximum values of the curves at the same time are all within the safety limits. In conclusion, the electromagnetic field generated by transcutaneous energy transmission system in the human tissue SAR within the safety limits. Therefore, the transcutaneous energy transmission system is safe for human body. The new artificial anal sphincter system has favorable bio-electromagnetic compatibility.

7 Conclusion

In this paper, the electromagnetic safety of artificial anal sphincter is analyzed and the radiation dose of human body is studied in the process of transcutaneous energy transmission system. In order to simulate the electromagnetic radiation damage received by the real human body, the high precision electromagnetic model of human body is established. By using the finite difference time domain method, the electromagnetic field produced by the transmitting coil in different tissues of human body is solved. The distribution of the specific absorption rate is obtained, which provides a theoretical basis for the design and optimization of transcutaneous energy transmission system. The results show that the artificial anal sphincter based on transcutaneous energy supply is in accordance with the international ICNIRP standard in terms of radiation dose safety, which provides a safe verification for the electromagnetic compatibility of artificial anal sphincter.

However, there are still some deficiencies in this study. Firstly, this paper only considers the influence of the transcutaneous energy transmission system on the electromagnetic radiation of the human body at fixed frequency, and the study on the electromagnetic dose absorbed by human body under different intensity and frequency is not carried out. Secondly, transcutaneous energy transmission system may increase the temperature of the human tissue. In the following study, the distribution of specific absorption rate of transcutaneous energy transmission system under different intensity and frequency will be studied. And the temperature rise safety will be studied.

Acknowledgments. This work was supported by National Natural Science Foundation of China (No. 31570998)

References

1. Magnus, H., Nicholas, J.T.: Fecal incontinence: mechanisms and management. Curr. Opin. Gastroenterol. Deliv. **28**(1), 57–62 (2012)
2. Duelundjakobsen, J., Worsoe, J., Lundby, L., et al.: Management of patients with faecal incontinence. Ther. Adv. Gastroenterol. **9**(1), 86 (2016)
3. Fattorini, E., Brusa, T., Gingert, C., et al.: Artificial muscle devices: innovations and prospects for fecal incontinence treatment. Ann. Biomed. Eng. **44**(5), 1355–1369 (2016)
4. Mantoo, S., Meurette, G., Podevin, J., et al.: The magnetic anal sphincter: a new device in the management of severe fecal incontinence. Expert Rev. Med. Devices **9**, 483–490 (2012)
5. Lei, K., Yan, G., Wang, Z., et al.: Power flow control of TET system for a novel artificial anal sphincter system. J. Med. Eng. Technol. **39**(1), 9 (2014)
6. Aldhaher, S., Luk, C.K., Bati, A., et al.: Wireless power transfer using class E inverter with saturable DC-feed inductor. IEEE Trans. Ind. Appl. **50**(4), 2710–2718 (2014)
7. Arai, K., Ninomiya, A., Ishigohka, T., et al.: Acoustic emission during DC operations of the ITER Central Solenoid model coil. IEEE Trans. Appl. Supercond. **12**(1), 504–507 (2002)
8. Spitzer, V., Ackerman, M.J., Scherzinger, A.L., et al.: The visible human male: a technical report. J. Am. Med. Inf. Assoc. Jamia **3**(2), 118–130 (1996)
9. Gabriel, S., Lau, R.W., Gabriel, C.: The dielectric properties of biological tissues: III. Parametric models for the dielectric spectrum of tissues. Phys. Med. Biol. **41**(11), 2271–2293 (1996)
10. Vinding, M.S., Guérin, B., Vosegaard, T., et al.: Local SAR, global SAR, and power-constrained large-flip-angle pulses with optimal control and virtual observation points. Magn. Reson. Med. **84**(5), 679–686 (2015)
11. Bamba, A., Joseph, W., Vermeeren, G., et al.: A formula for human average whole-body SARwb under diffuse fields exposure in the GHz region. Phys. Med. Biol. **59**(23), 7435–7456 (2014)
12. Ozgun, O., Kuzuoglu, M.: Transformation-based metamaterials to eliminate the staircasing error in the finite difference time domain method. Int. J. RF Microw. Comput. Aided Eng. **22**(4), 530–540 (2012)

The Role of Intelligent Computing in Load Forecasting for Distributed Energy System

Pengwei Su, Yan Wang, Jun Zhao[✉], Shuai Deng, Ligai Kang,
Zelin Li, and Yu Jin

Key Laboratory of Efficient Utilization of Low and Medium Grade Energy,
Ministry of Education of China, Tianjin University, Tianjin 300350, China
zhaojun@tju.edu.cn

Abstract. The integration of renewable energy into the distributed energy system has challenged the operation optimization of the distributed energy system. In addition, application of new technologies and diversified character-istics of the demand side also impose a great influence on the distributed energy system. Through a literature review, the load forecasting technology, which is a key technology inside the optimization framework of distributed energy system, is reviewed and analyzed from two aspects, fundamental research and appli-cation research. The study presented in this paper analyses the research methods and research status of load forecasting, analyses the key role of intelligent computing in load forecasting in distributed energy system, and realizes and explores the application of load forecasting in practical energy system.

Keywords: Distributed energy system · Load forecasting · Intelligent computing · Fundamental research · Applied research

1 Introduction

With the rapid development of the economy, global warming induced by the con-sumption of large amounts of fossil fuel has begun to threaten the ecological balance of the planet, which has caused every country all over the world to seek low-carbon energy systems. As an efficient energy supply system, distributed energy system is becoming more and more attractive because it can utilize multi-energy as resource effectively. However, the increasing of the proportion of renewable energy in energy system, the energy storage technology integrated in energy system and the various characteristics of the demand side affect the efficient operation of the distributed energy system a lot.

Load forecasting is used to predict load data at a specific moment in the future, then the schedule of energy supply system could be modulated in real-time. So the load forecasting is considered as an effective application in operation strategy optimization of distributed energy system. In recent years, intelligent computing (IC) is applied in various fields more often. The calculating efficiency and accuracy could be improved in load forecasting by using intelligent computing as it can help to solve the complex relationship between the forecast results and the influencing factors.

© Springer Nature Singapore Pte Ltd. 2017
K. Li et al. (Eds.): LSMS/ICSEE 2017, Part III, CCIS 763, pp. 547–555, 2017.
DOI: 10.1007/978-981-10-6364-0_55

The study presented in this paper analyses the research methods and research status of load forecasting, from two aspects, fundamental research and applied research, analyses the key role of intelligent computing in load forecasting in distributed energy system. In the face of the existing research mostly stay in the theoretical stage, the application of load forecasting in practical energy system is realized and explored based on the actual project.

2   Load Forecasting

2.1   Fundamental Research

2.1.1   Concept and Classification

Load forecasting is used to predict load data at a specific moment in the future based on various factors such as the operation characteristics, capacity expansion decisions, natural conditions, and social influences of the system when a certain amount of precision is needed [1]. It is considered an effective supporting method that complements decision-making, which can be used to adjust the supply of the energy system and precisely adjusts the load by changing the operation strategy.

Load forecasting can be divided into long-term load forecasting, medium-term load forecasting, and short-term load forecasting according to the length of the forecasting duration. Long-term load forecasting typically refers to forecasts made 1 to 10 years in advance, and is often used as basis for determining future energy demand and planning policies. The medium-term load forecasting often refers to forecasts made for a few weeks to a few months in advance, which is used to guide the companies to make plans. Short-time load forecasting includes forecasts made for a few hours to a few days in advance, which can be used to supervise the operation and control of the distributed energy system.

2.1.2   Research Methods

The steps such as energy supply, transmission, and utilization in the distributed energy system can be expressed using matrix (1).

$$L = (C \ \ S) \cdot \begin{pmatrix} P \\ A \end{pmatrix} \tag{1}$$

Here, L represents the building load, which includes building demands such as cooling, heating and power; P represents the energy supply, which includes traditional energy resources and renewable energy resources; A represents the energy storage supply, which includes energy storage such as cooling, heating and power; C is a performance parameter of the energy supply system; S represents the performance parameter of the energy storage device.

The matrix (2) describes the major content of the research in load forecasting, the left side of the equation shows the load at the moment $t + 1$, the F matrix in the right side of the equation is a matrix of coefficients, and L_t is the load at the moment t.

$$
\underbrace{\begin{pmatrix} L_{t_1}+1 \\ L_{t_2}+1 \\ \vdots \\ L_{t_m}+1 \end{pmatrix}}_{L_{t+1}} = \underbrace{\begin{pmatrix} f_{11} & f_{12} & \cdots & f_{1n} \\ f_{21} & f_{22} & \cdots & f_{2n} \\ \vdots & \vdots & \ddots & \vdots \\ f_{m1} & f_{m2} & \cdots & f_{mn} \end{pmatrix}}_{F} \underbrace{\begin{pmatrix} L_{t_1} \\ L_{t_2} \\ \vdots \\ L_{t_m} \end{pmatrix}}_{L_t} \tag{2}
$$

Here, F represents the parameters of load influencing factors, which is a matrix consisting of influencing factors such as outdoor weather conditions, number of personnel, and holidays. This matrix is combined with Eq. (1) to revise the load matrix in the left side of the equation. Thus, as shown in Fig. 3, the difference between supply and demand is minimized. With the reduction in the duration of the forecast, the matching may be fulfilled in a smaller time scale, which approaches the ideal condition, of instant matching.

$$
\min \Delta E = \int_{\tau_1}^{\tau_2} \Delta E dt \tag{3}
$$

$$
\Delta E = E_{\text{sup}} - E_{dem} = (C \ S)\begin{pmatrix} P \\ A \end{pmatrix} - L \tag{4}
$$

Normally, when applying load forecasting to research into distributed energy systems, the process is often performed using the following steps: first data preparation, and the major content of the data preparation includes data cleaning, data transformation, and data reduction; then data mining and modeling are performed by effectively supervising and studying the data, and the relationship between variables and the load is established; then it comes to error analysis, by analyzing the error between the modeling forecast value and the actual value, the model is modified, and the precision of the model is improved [2]. Later, the predicted error is determined based on the conditions of forecast and application, and actual application is carried out after the conditions are satisfied. Otherwise, the process is modified during data mining and modification, as shown in Fig. 1.

2.1.3 Research Progress

With the development of distributed energy systems, the spatial scale is getting bigger and bigger, from the family housing to the community and park [3–5], the increase of system spatial scale makes the load influencing factors more complicated. Typical studies on the load forecasting in recent years are summarized in Table 1. With reference to review papers in a related research area published prior to 2013, it can be seen that the research in this field is expanding from pure electricity load forecasting towards the cooling and heating load forecasting [6, 7]. The time scale is changed from month, day to hour, and the prediction of the small time scale will help the system to adjust precisely for the load, and the research is transitioning towards the more challenging short-time forecast.

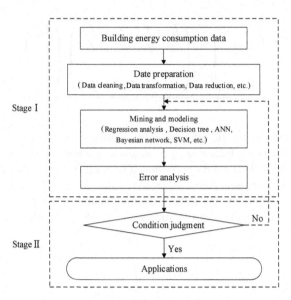

Fig. 1. Flow chart of load forecasting

Table 1. Typical studies on load forecasting

Year	Author	Load type	Time scale	Intelligent computing
2016	Gupta et al. [8]	Electricity	Hour	Wavelet neural network
2016	Idowu et al. [9]	Heating	Hour	SVM
2016	Deb et al. [10]	Cooling	Day	ANN
2015	Chitsaz et al. [11]	Electricity	Hour	Self-recurrent wavelet neural network
2015	Protic et al. [12]	Heating	Hour	SVM-Wavelet
2015	Abdoos et al. [13]	Electricity	Hour	SVM
2014	Chou and Bui [14]	Cooling and heating	Hour	SVM-ANN
2014	Rodrigues et al. [15]	Electricity	Hour	ANN
2013	Xue et al. [16]	Electricity	Month	GA

2.1.4 Application of IC in Load Forecasting

There are many influencing factors of load in distributed energy system, mainly include the ambient temperature, solar radiation, humidity, occupation rate of the room, behavior characteristics and the envelope of building. As the relationship between load demands and the influencing factors is nonlinear, the kernel of the load forecasting is to find the relationships between load demands and the influencing factors. With the development of distributed energy system, the higher requirement of load forecast is needed. the complexity of the load influencing factors, the diversification of the load types and the reduction of the forecasting time scale increase the difficulty of load forecasting a lot.

In order to acquire a more accurate load forecasting data, artificial neural network (ANN), genetic algorithm (GA) and support vector machine (SVM) et al. were applied in load forecasting. As shown in Table 1, the application of intelligent computing in load forecasting helps to solve how to find the complex relationships between the forecast results and the influencing factors which improved the accuracy in load forecasting.

2.2   Application Research

2.2.1   Existing Application Researches

Currently, applied research into the load forecasting in the distributed energy system is subjected to factors such as the high cost of regulation and control, and thus is still in its initial stage. Existing representative findings are shown in Table 2, while most research remains in stage I, as shown in Fig. 1, and few studies reach stage II.

Table 2. Current situation of research and application of load forecasting

Year	Authors	Accuracy (MAPE)	Load type	Purpose	Stage I	II
2016	Li et al. [17]	3.43%	Electricity	Save electricity	√	
2016	Fang and Lahdelma [18]	8.40%	Heating	Energy conservation	√	
2016	Papakonstantinou et al. [19]	–	Heating	Energy conservation		√
2015	Lee et al. [20]	2.61%	Electricity	Save electricity	√	
2015	Lahouar and Slama [21]	2.30%	Electricity	Save electricity	√	
2015	Lou and Dong [22]	4.56%	Electricity	Save electricity	√	
2014	Vaghefi et al. [23]	3.10%	Electricity and cooling	Save electricity	√	
2015	Qiao and Chen [24]	2.59%	Gas	Energy conservation	√	
2014	Amini et al. [25]	1.44%	Electric vehicle	Reduce the pressure of the grid	√	
2011	Chen et al. [26]	–	Electricity	improve power quality of microgrid		√

Nikolaos et al. proposed an algorithm aims to provide more heat energy to the difficult consumers when they need it the most [19]. The required input information are the short term weather forecast, the supply hot water temperature propagation delays of the district heating grid as a function of the grid load level and consumption profiles.

The methodology is applied on a simplified case study of a 120 MW district heating grid. The results showed that within a specific supply water temperature range the performance of the grid in terms of minimum pressure difference at the consumers over a year was significantly better using the proposed proactive algorithm compared to simple reactive and constant temperature control strategies. Chen et al. proposed an active control strategy of the micro-grid energy storage system based on short-term load forecast [26], when performing short-time forecast of the load in the micro-grid, this strategy would actively control the charging and discharging of the storage system, optimize the load curve of the micro-grid, and realize peak load shifting under when the capacity of the batteries, the charging-discharging characteristics, and limitations in the charging-discharging numbers are considered. Thus, the stable operation of the distributed power source can be ensured, and the electric energy quality of the micro-grid can be improved.

2.2.2 Existing Demonstration of IC in Load Forecasting

Based on a flagship project of China & Singapore governments, "Eco-city district energy station system optimization and its dispatch system based on weather forecast", our project team has studied the application of load forecasting in energy system. The optimization of scheduling of the operation for the No. 2 energy station in Sino-Singapore Tianjin Eco-city was carried out based on the load forecasting results in this project. It primarily pre-determined the load using weather forecast data, and the predictive model is shown in Fig. 2. Moreover, the distributed energy system is shown in Fig. 3, which consists of combined cooling-heating-and-power (CCHP), ground source heat pump machine group, electric chillers, and energy storage device. Based on data such as the historical load, environmental temperature, weather conditions, and type of day, the load predictive model was established to obtain the load values in the future 24 h. These were entered into the operation platform in the energy station based on weather forecasts. Further, the platform controlled and adjusted related machine

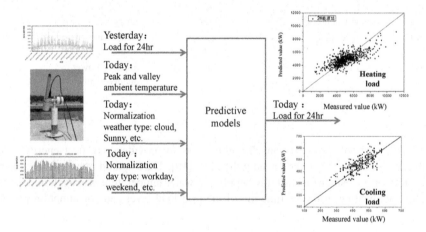

Fig. 2. Model for the prediction

group according to the results of load forecasting to optimize the operation of the system. Work related to this project successfully demonstrated the feasibility of applying the load forecasting technology to the distributed energy system.

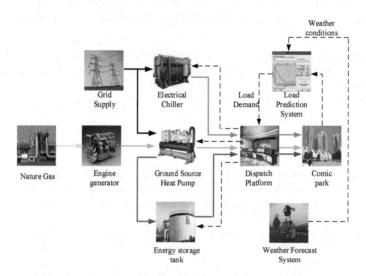

Fig. 3. The system structure diagram of energy station

3 Summary Remarks

With the development of distributed energy system, the increasing of the proportion of renewable energy in energy system, the energy storage technology integrated in energy system and the various characteristics of the demand side affect the efficient operation of the distributed energy system a lot. As a key technology in distributed energy system operation optimization, the calculating efficiency and accuracy could be improved in load forecasting by using intelligent computing. the load forecasting technology is reviewed and analyzed from two aspects, fundamental research and applied research. The study presented in this paper analyses the key role of intelligent computing in load forecasting in distributed energy system, and realizes and explores the application of load forecasting in practical energy system. Some conclusions can be made:

1. Load forecasting can be used to adjust the supply of the energy system and pre-cisely adjusts the load by changing the operation strategy, the complexity of the load influencing factors, the diversification of the load types and the reduction of the forecasting time scale increase the difficulty of load forecasting a lot. The load forecasting is carried out by intelligent computing method, which solves the problem of complex relationship between load and influencing factors, and improves the calculation efficiency and forecast accuracy.

2. Current studies on the load forecasting mostly remain at the theoretical stage, and there is much more work to do in order to apply the predictive model in actual energy system. For this reason, future work should focus on the development of technical solutions and products by integrating big data and the cloud storage technology, the government may guide and carry out demonstration applications of load forecasting, thus the distributed energy system can be better developed.

Acknowledgments. This study was financially supported by National High Technology Research and Development Program ("863" program) of China under Grant Number 2015AA050403.

References

1. Nihuan, L.: Review of the short-term load forecasting methods of electric power system. Power Syst. Protect. Control. **39**(1), 147–152 (2011). (in Chinese)
2. Liangjun, Z., Tan, Y., Gang, X.: MATLAB data analysis and data mining. China Machine Press, Beijing (2015). (in Chinese)
3. Molin, A., Schneider, S., Rohdin, P., et al.: Assessing a regional building applied PV potential—spatial and dynamic analysis of supply and load matching. Renew. Energy **91**, 261–274 (2016)
4. Väisänen, S., Mikkilä, M., Havukainen, J., et al.: Using a multi-method approach for decision-making about a sustainable local distributed energy system: a case study from Finland. J. Clean. Prod. **137**, 1330–1338 (2016)
5. Guarino, F., Cassarà, P., Longo, S., et al.: Load match optimization of a residential building case study: a cross-entropy based electricity storage sizing algorithm. Appl. Energy **154**, 380–391 (2015)
6. Singh, A.K., Ibraheem, I., Khatoon, S., et al.: Load forecasting techniques and methodologies: a review. In: International Conference on Power, Control and Embedded Systems, pp. 1–10 (2012)
7. Powell, K.M., Sriprasad, A., Cole, W.J., et al.: Heating, cooling, and electrical load forecasting for a large-scale district energy system. Energy **74**(5), 877–885 (2014)
8. Gupta, S., Singh, V., Mittal, A.P., et al.: Weekly load prediction using wavelet neural network approach. In: Second International Conference on Computational Intelligence and Communication Technology, pp. 174–179 (2016)
9. Idowu, S., Saguna, S., Åhlund, C., et al.: Applied machine learning: forecasting heat load in district heating system. Energy Build. **133**, 478–488 (2016)
10. Deb, C., Eang, L.S., Yang, J., et al.: Forecasting diurnal cooling energy load for institutional buildings using artificial neural networks. Energy Build. **121**, 284–297 (2016)
11. Chitsaz, H., Shaker, H., Zareipour, H., et al.: Short-term electricity load forecasting of buildings in micro grids. Energy Build. **99**, 50–60 (2015)
12. Protić, M., Shahaboddin, S., et al.: Forecasting of consumers heat load in district heating systems using the support vector machine with a discrete wavelet transform algorithm. Energy **87**(3), 343–351 (2015)
13. Abdoos, A., Hemmati, M., Abdoos, A.A.: Short term load forecasting using a hybrid intelligent method. Knowl. Based Syst. **76**, 139–147 (2015)

14. Chou, J.S., Bui, D.K.: Modeling heating and cooling loads by artificial intelligence for energy-efficient building design. Energy Build. **82**, 437–446 (2014)
15. Rodrigues, F., Cardeira, C., Calado, J.M.F., et al.: The daily and hourly energy consumption and load forecasting using artificial neural network method: a case study using a set of 93 households in Portugal. Energy Procedia **62**, 220–229 (2014)
16. Xue, B., Geng, J., Zheng, Y., et al.: Application of genetic algorithm to middle-long term optimal combination power load forecast. In: IEEE Region 10 Annual International Conference, pp. 1–4 (2013)
17. Li, S., Goel, L., Wang, P.: An ensemble approach for short-term load forecasting by extreme learning machine. Appl. Energy **170**, 22–29 (2016)
18. Fang, T., Lahdelma, R.: Evaluation of a multiple linear regression model and SARIMA model in forecasting heat demand for district heating system. Appl. Energy **179**, 544–552 (2016)
19. Papakonstantinou, N., Savolainen, J., Koistinen, J., et al.: District heating temperature control algorithm based on short term weather forecast and consumption predictions. In: IEEE International Conference on Emerging Technologies and Factory Automation, ETFA (2016)
20. Lee, W.J., Hong, J.: A hybrid dynamic and fuzzy time series model for mid-term power load forecasting. Int. J. Electr. Power Energy Syst. **64**, 1057–1062 (2015)
21. Lahouar, A., Slama, J.B.H.: Day-ahead load forecast using random forest and expert input selection. Energy Conv. Manag. **103**, 1040–1051 (2015)
22. Lou, C.W., Dong, M.C.: A novel random fuzzy neural networks for tackling uncertainties of electric load forecasting. Int. J. Electr. Power Energy Syst. **73**, 34–44 (2015)
23. Vaghefi, A., Jafari, M.A., Bisse, E., et al.: Modeling and forecasting of cooling and electricity load demand. Appl. Energy **136**, 186–196 (2014)
24. Qiao, W., Chen, B.: Hourly load prediction for natural gas based on Haar wavelet tansforming and ARIMA-RBF. Shiyou Huagong Gaodeng Xuexiao Xuebao/J. Petrochem. Univ. **28**(4), 75–80 (2015)
25. Amini, M.H., Kargarian, A., Karabasoglu, O.: ARIMA-based decoupled time series forecasting of electric vehicle charging demand for stochastic power system operation. Electric Power Syst. Res. **140**, 378–390 (2016)
26. Chen, Y., Zhang, B., Wang, J.: Active control strategy for microgrid energy storage system based on short-term load forecasting. Power Syst. Technol. **35**(08), 35–40 (2011). (in Chinese)

Intelligent Control Methods of Demand Side Management in Integrated Energy System: Literature Review and Case Study

Yan Wang, Pengwei Su, Jun Zhao$^{(\boxtimes)}$, Shuai Deng, Hao Li, and Yu Jin

Key Laboratory of Efficient Utilization of Low and Medium Grade Energy, Ministry of Education of China, Tianjin University, Tianjin 300350, China
zhaojun@tju.edu.cn

Abstract. Demand side management (DSM) would become an important method to guarantee the stability and reliability of the innovative energy structure model, have received increasing attention. DSM is regarded as an integrated technology solution for planning, operation, monitoring and management of building utility activities. However, there are several problems and technical challenges on the research level of fundamental methodology, which causes difficulties for the practical application of intelligent DSM control strategy. Therefore, optimization would play a vital role in the implementation process of DSM. A real case study was presented to demonstrate how to relieve and solve the existing technical challenges by the application effective optimization strategy and methods of DSM. At last, several possible research directions, that application of intelligent methods in the development of DSM optimization techniques, were presented.

Keywords: Integrated energy system · Demand side management · Optimization algorithm · Intelligent control methods

1 Introduction

Since the 21st century, the rising demand for energy has posed a contradiction between economic growth and environmental protection. A large-scale application of non-fossil energy and a synergetic development of conventional energy and renewable energy are becoming the critical pathway to revolution of energy generation and consumption in the near- and middle-term. On the other hand, the information and communication technology (ICT) sector represented by computer, communication and network technology has been enormous innovation since the 1950s, which offered powerful technical support to the development of energy sector. These driven factors are shaping the new structures of energy system and network: Smart Grid (SG), Integrated Energy System (IES) and Energy Internet (EI), as innovative concepts of energy structure model, have emerged in the energy field. These concepts express a vision of the future energy systems integrating advanced intelligent control technologies at transmission and distribution sides to maximize energy and resources value in an intelligent way.

© Springer Nature Singapore Pte Ltd. 2017
K. Li et al. (Eds.): LSMS/ICSEE 2017, Part III, CCIS 763, pp. 556–565, 2017.
DOI: 10.1007/978-981-10-6364-0_56

Compared with traditional energy system, IES can achieve a bidirectional transmission of energy and information flow, which means it would be capable of responding to changes, resulted from supply to demand side. The emphasis of IES is to restore the merchandise property of energy, and improve energy market trading mechanism. Thus, the end-user requirements on the demand side would become the main guidance for energy production in IES. Meanwhile, high penetration rate of renewable energy source and the plug-in-and-play of distributed generation are accessed, would influence the stability of both source and demand sides in energy system [1]. Consequently, a typical strategy measures on demand side is demand side management (DSM), which is a significant method to guarantee the stability and reliability of IES, have received increasing attention.

Buildings accounts for about a third of total energy consumption in China, which play an important role in DSM of IES. The building demand management aims at minimize the impact of real-time price, critical peak price and time-of-use rates on the service quality of buildings [2], to relieve the pressure of power imbalance in energy production, transmission and utilization processes. Meanwhile, the advanced intelligent control methods and technologies of buildings DSM, such as building automation systems, intelligent monitoring and smart meters, could be implemented to better realize bidirectional operation mode between supply and demand side (i.e., buildings). Considering most studies on intelligent control method of DSM are conducted without a definite technological boundary, a relatively comprehensive review, which includes literature research and case study of building DSM, is introduced in this paper briefly.

2 Overview of Demand Side Management

2.1 Definition and Classification of DSM

Demand side management (DSM), which is a bi-directional link between supply side and end-users, is defined as an integrated technical solution for planning, implementation and monitoring of utility activities at demand side. Be similar to demand side response (DSR) comes from power industrial field, and both concepts were driven by administrative and economic benefit measure, to achieve the aim of energy-efficient on demand side. DSR is one of the main solutions for DSM, and it should be design at specific time, and implemented within a short period time. In addition, there are several differences between DSM and DSR, according to their participator, leading role, acting time and main purpose (see Table 1).

Table 1. Differences between DSM and DSR

Aspects	DSM	DSR
Participator	Demand side	Supply and demand side
Leading role	Consumer	Grid
Acting time	Short-term	Long-term (the whole life cycle)
Main purpose	Grid protection	Grid protection, energy-saving, etc.

Therefore, it can be clearly seen that DSM is to encourage the consumer to adjust their own consumption pattern on energy consumption not only during the peak period of power demand, but also during a longer period of time. From this, the consumer behavior is emphasized in this concept, and is generally assumed to be driven by the level of economic, education and social development, etc.

DSM generally divided into six load shifting categories according to the change of load shape: peak clipping, valley filling, load shifting, strategic conservation, strategic load growth, and flexible load shape, as shown in Fig. 1 [3].

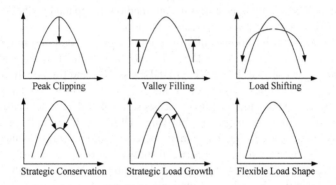

Peak Clipping   Valley Filling   Load Shifting

Strategic Conservation   Strategic Load Growth   Flexible Load Shape

Fig. 1. DSM category [3]

Mathematical modeling to described the definition of DSM, is originally proposed in reference [1], based on the energy balance method between supply and demand side. This mathematical equation used a coupling matrix related to DSM, and through the reshaped and shifted load, to adapt the change from the supply side. In addition, for the on-site renewable energy system, as shown in Fig. 2, DSM can work through adjust the profile of energy consumption to achieve a better match with the profile of renewable energy generation.

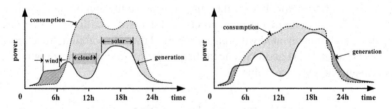

Fig. 2. DSM for on-site renewable power generation system [4]

2.2 Technical Challenge

There are several problems as follows on the research level of fundamental methodology, which causes difficulties for implementation of intelligent DSM control strategy.

2.2.1   Time and Space Scales are Indefinite

According to size difference of covered regional area, the space scales of IES can be divided into local-area [5], wide-area [6, 7], national [8] and global [9]. From a realistic respective, the global IES with large space scale highlights transmission and delivery [8], while the wide-area IES with a relative small space scale focus more on distribution and consumption [6, 7]. In addition, the high penetration rate of renewable energy source (RES) is regarded as an important feature of IES which is different with traditional energy supply network [10]. Meanwhile, RES have characteristics of random volatility and strong intermittency with an excessive dependence on topographic factors and natural conditions, which led to its temporospatial complementarities across a wide area [11, 12]. Thus, the utilization of the time difference (or called time lag) in the areas with deferent scales can make the renewable energy become practical to realize flexible peak-valley allocation for global energy consumption [13].

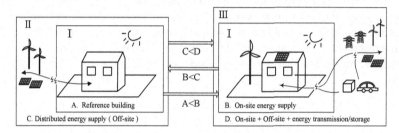

Fig. 3. Influence of space scale on performance assessment of generalized IES

It can be found in Fig. 3 that there are three kinds of so-called "hierarchy" space boundary from I to III for the same reference building. When the space scale was set as the same boundary I in case A and B, relative to the reference building in case A without on-site renewable energy supply, case B applies available on-site renewable energy (such as rooftop photovoltaic and small wind turbine) to provide energy products for the building. Although the supply capacity from on-site renewables is limited by local climate, such design definitely affect the overall performance of the energy system. Using specific assessment indicators, such as solar fraction or permeability of renewable resources, the performance of case B was superior to that of case A. Furthermore, when the space scale was extended to II, a distributed power/energy system can be proposed for the reference building. Thus, the ratio of off-site wind/photovoltaic power to energy demand of building can be increased. Considering multi-aspect performance indicators such as the fraction of renewable energy, carbon footprint and environmental impact, case C shows a better performance than case B.

Third, with a further extend of space scale to III, the physical boundary could even be extended into a virtual economical boundary. The owner of reference building can purchase green power from remote renewable energy farm for a better performance on environmentally-friendly. Also, it can be realized by above-mentioned wide-area IES and global IES, because such large-scale IES can accesses urban power, heat and gas networks as well as other controllable resources such as batteries and electric vehicles to shape a optimized configuration of energy supply across a large space scale. Thus, case D has the potential capacity to obtain a better performance than case C by strengthening supply capacity of IES with an acceptable cost.

Based on the above-mentioned comparison, it can be found that a clear clarification of the time and space scale would have a direct influence on the performance evaluation of early-stage design, optimal control and follow-up assessment of the specific project.

2.2.2 Complexity of Assessment Indicators System

IES performance is difficult to evaluate through the conventional assessment indicators, because such new energy system involves several industrial sectors including energy, information, finance and etc. Meanwhile, IES, as a coupled and shared system of multi-network, covered so many intelligent technologies that are lack of clear criteria for quantitative performance. Likewise, the screening of assessment indicator is a significant step for the optimization of DSM control strategy. Thus, considering unclear technology boundary, time and space scales, some innovative assessment indicators are proposed in recent years, and two of the typical cases, namely load matching (LM) and grid interaction (GI) indicators, are developed to evaluate the coupling performance among grid, renewable energy of supply side and building load of demand side (see Fig. 4).

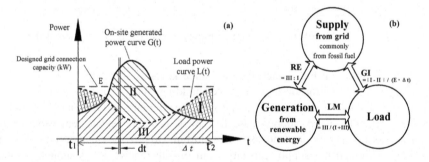

Fig. 4. Relationship diagram of LM and GI: (a) Graphical version (the drawing is adapted from reference [14]); (b) simplified version (GI definition is derived from reference [15])

A ratio to which the building load demand compares with the renewable energy supply is commonly used to evaluate the performance level of LM. It can enhance from two aspects: adjusting the energy supply side to the demand side, and adjusting the demand side to the energy supply side which is also called DSM [15]. However, the emphasis of LM only on the relationship between demand and supply inside the footprint of building, but it could not describe the interaction between building load and grid. Thus, the definition of GI is required to be quantified as a ratio of generation and load demand profiles within an annual cycle, and pay attention to the unmatched parts of them [16].

3   Existing Optimization Models and Methods for DSM

3.1   Classification of DSM Optimization Model

Optimization deals with the problem of minimizing or maximizing a certain function in a finite dimensional Euclidean space over a subset of that space, which is determined by functional inequalities. There are a variety of methods to achieve an optimized, including linear optimization, convex optimization, global optimization, discrete optimization, and so on. Through a survey of DSM definition, classification and mathematical modeling, optimization methods and models for DSM can be classified as three types (see Fig. 4) [17].

3.2   Objective Functions of the DSM Optimization Model

The benefits from applying DSM in IES are mutual for both the supply side and demand side. For supply side, utilities will have better utilization of the available system capacity to improve energy efficiency. For demand side, that all customers main concerns that the improvement in the electrical service quality, as well as the amount of monthly electric bill will be decreased. Therefore, two categories for the objective functions of DSM optimization problem are contributed, which include maximized load factor on energy supply side, and minimized the cost for the consumers on demand side. Although there exist two categories of objective functions for six load shifting categories mentioned above, the related constraints for these DSM methods are slightly different, affected by demand type at control variables (Fig. 5) [18].

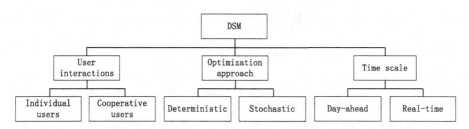

Fig. 5. Classification of DSM optimization model [17]

4   Case Study

Buildings, as the primary energy consumers at demand side, could play an important role in DSM of IES. For residential buildings, the way of DSM is mostly to optimize the schedule of equipment operation to reduce the electricity consumption [19, 20]. However, for commercial buildings during on-peak period, DSM would achieve economic benefits for both demand and supply side of IES. Therefore, load shifting becomes one of major means for DSM in commercial buildings. Load shifting defines the shifting loads from on-peak to off-peak periods, to take advantage of real-time pricing and time-of-use rates. In addition, several typical technologies could be used to support the implementation of load shifting, including energy storage system [21], building thermal mass [23, 24], and so on.

4.1 Design of DSM Load Control Strategies

As a case study, an IES model with energy supply, generation and utilization side on TRNSYS. The system configuration of this IES is shown in Fig. 6. The system can be divided into three parts: energy supply system, thermal storage system, and building service system. Besides, two types of DSM load control strategies are designed used for air-conditioning systems of commercial building in IES:

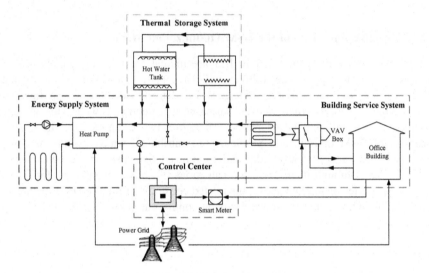

Fig. 6. Configuration of IES on TRNSYS

(1) Demand-based indirect load control strategy (with thermal storage system). Making the most of electric power cost between on-peak and off-peak, GSHP using water storage technology to store heat at night, and to satisfy the higher electric load during on-peak period in daytime.

(2) Demand-based direct load control strategy (without thermal storage system). When there is an urgent request in on-peak period, the control center would adjust the heat supply of each room on demand side, according to a status queue of all rooms. The status queue obtained from the energy consumption characteristics and thermal comfort index in each room. In this way can reduce power consumption to realize the "peak clipping".

4.2 Possible Application of Intelligent Methods in Optimization of DSM

In the last thirty years, a great progress has been made in the modeling of DSM, and its optimization algorithms have been introduced into the optimization solution of the intelligent control strategy for DSM. In return, such application researches also promoted the development of software and simulation modules, based on the optimization algorithms. Apart from Transient Energy System Simulation Program (TRNSYS) [26], there is some other software, the Hybrid Optimization Model for Electric Renewables (HOMER) [27], the Distributed Energy Resources Customer Adoption Model (DER-CAM) [28], and Generic Optimization Program (GenOpt) are frequently applied as widely-recognized tools.

GenOpt is a generic optimization program, which is designed for finding the values of user-selected design parameters that minimize an objective function, leading to optimal operation of a specific energy system. Furthermore, the objective function can be calculated by some external simulation platform, such as TRNSYS [29].

In the field of evolutionary computation, the genetic algorithm (GA) is an optimization technique based on natural genetics, and to simulate growth and decay of living organisms in a natural environment [30]. The Particle Swarm Algorithms (PSO) is an evolutionary computation is an efficient stochastic global optimization technique, developed in 1995. Compared with GA, PSO is easily achieved, and only need to set few parameters [31].

In addition, there are many other algorithms are widely available to the optimization of DSM, such as the tabu search (TS), differential evolution algorithms (DE), simulated annealing (SA), artificial neural network (ANN), ant colony algorithm (ACO) and so on. Hence more and more algorithms can also be integrated to achieve optimal results in a flexible way, and contributed to the evaluation of DSM control method from different perspectives.

5 Summary Remarks

As the innovative concepts of energy structure model, such as IES emerged in the energy field, DSM would become an important measure to guarantee the stability and reliability of IES, have received increasing attention. However, there are several problems on the research level of fundamental methodology, including indefinite of time and space scales, complexity of assessment indicators system and other technical challenge, which causes difficulties for implementation of intelligent DSM control strategy.

Based on the above challenges, optimization would play a vital role in the implementation process of DSM. After a scoping review on optimization methods and models of DSM, a real case study was presented to demonstrate how to relieve and solve the existing technical challenge by the application effective optimization strategy and methods of DSM.

At last, several possible research directions, that application of intelligent methods in the present development of DSM optimization methods, were presented. Most optimization methods of DSM are for single and separately managed customers on demand side. Therefore the energy storage, load match and grid interaction can further integration with DSM technologies should do more innovation efforts and attempts in future studies. Also comprehensive approach combined two or more algorithms for different time and space scales would turn to become one of possible future work directions.

References

1. Kienzle, F., Ahcin, P., Andersson, G.: Valuing investments in multi-energy conversion, storage, and demand-side management systems under uncertainty. IEEE Trans. Sustain. Energy **2**(2), 194–202 (2011)
2. Wang, S., Tang, R.: Supply-based feedback control strategy of air-conditioning systems for direct load control of buildings responding to urgent requests of smart grids. Appl. Energy (2016)
3. Qureshi, W.A., Nair, N.K.C., Farid, M.M.: Impact of energy storage in buildings on electricity demand side management. Energy Convers. Manag. **52**(5), 2110–2120 (2011)
4. Kolokotsa, D., Rovas, D., Kosmatopoulos, E., Kalaitzakis, K.: A roadmap towards intelligent net zero- and positive-energy buildings. Sol. Energy **85**(12), 3067–3084 (2011)
5. Rui, L., Chen, L., Yuan, T., Chunlai, L.: Optimal dispatch of zero-carbon-emission micro energy internet integrated with non-supplementary fired compressed air energy storage system. J. Mod. Power Syst. Clean Energy **4**(4), 566–580 (2016)
6. Perry, S., Kleme, J.í., Bulatov, I.: Integrating waste and renewable energy to reduce the carbon footprint of locally integrated energy sectors. Energy **33**, 1489–1497 (2008)
7. Peng, Y.L., Walmsley, T.G., Alwi, S.R.W., Manan, Z.A., Klemeš, J.J., Varbanov, P.S.: Integrating district cooling systems in locally integrated energy sectors through total site heat integration. Appl. Energy **167**, 155–157 (2016)
8. Yan, B., Wang, B., Zhu, L., et al.: A novel, stable, and economic power sharing scheme for an autonomous microgrid in the energy internet. Energies **8**(11), 12741–12764 (2015)
9. Liu, Z.: Global Energy Internet. China Electric Power Press, Beijing (2015)
10. Yang, Y., Wu, K., Long, H., Gao, J., Xu, Y., Kato, T., et al.: Integrated electricity and heating demand-side management for wind power integration in china. Energy **78**, 235–246 (2014)
11. Kempton, W., Pimenta, F.M., Veron, D.E., Colle, B.A.: Electric power from offshore wind via synoptic-scale interconnection. Proc. Natl. Acad. Sci. **107**(16), 7240–7245 (2010)
12. Oswald, J., Raine, M., Ashraf-Ball, H.: Will British weather provide reliable electricity? Energy Policy **36**(8), 3212–3225 (2008)
13. Hoicka, C.E., Rowlands, I.H.: Solar and wind resource complementarity: advancing options for renewable electricity integration in ontario, canada. Renew. Energy **36**(1), 97–107 (2011)
14. Cao, S., Hasan, A., Kai, S.: On-site energy matching indices for buildings with energy conversion, storage and hybrid grid connections. Energy Build. **64**(5), 423–438 (2013)
15. Salom, J., Marszal, A.J., Widén, J., Candanedo, J., Lindberg, K.B.: Analysis of load match and grid interaction indicators in net zero energy buildings with simulated and monitored data. Appl. Energy **136**, 119–131 (2014)
16. Voss, K., Sartori, I., Napolitano, A., Geier, S., Gonzalves, H., Hall, M., et al.: Load matching and grid interaction of net zero energy buildings. In: Eurosun 2010, Graz (2010)
17. Livengood, D.J.: The Energy Box: Comparing Locally Automated Control Strategies of Residential Electricity Consumption Under Uncertainty. Massachusetts Institute of Technology, Cambridge (2011)
18. Esther, B.P., Kumar, K.S.: A survey on residential demand side management architecture, approaches, optimization models and methods. Renew. Sustain. Energy Rev. **59**, 342–351 (2016)
19. Attia Hussein, A.: Mathematical formulation of the DSM problem and its optimal solution. In: Proceedings of the 14th international middle east power systems conference (MEPCON 2010), Paper ID314. Cairo University, Egypt, 19–21 December 2010
20. Toronto Hydro Electric System Peak Saver (2010). http://www.torontohydro.com/peaksaver

21. Ji, H.Y., Bladick, R., Novoselac, A.: Demand response for residential buildings based on dynamic price of electricity. Energy Build. **80**, 531–541 (2014)
22. Fukai, J., Hamada, Y., Morozumi, Y., Miyatake, O.: Improvement of thermal characteristics of latent heat thermal energy storage units using carbon-fiber brushes: experiments and modeling. Int. J. Heat Mass Transf. **46**(23), 4513–4525 (2003)
23. Kousksou, T., Rhafiki, T.E., Omari, K.E., Zeraouli, Y., Guer, Y.L.: Forced convective heat transfer in supercooled phase-change material suspensions with stochastic crystallization. Int. J. Refrig. **33**(8), 1569–1582 (2010)
24. Xue, X., Wang, S., Sun, Y., Fu, X.: An interactive building power demand management strategy for facilitating smart grid optimization. Appl. Energy **116**(3), 297–310 (2014)
25. Kirby, B., Kueck, J., Laughner, T., Morris, K.: Spinning reserve from hotel load response. Electr. J. **21**(10), 59–66 (2008)
26. University of Wisconsin-Madison. A TRaNsient SYtems Simulation Program. http://sel.me. wisc.edu/trnsys/
27. HOMER Energy LLC. HOMER. http://www.homerenergy.com/
28. Laurence Berkeley National Laboratory. Distributed Energy Resources Customer Adoption Model (DER-CAM). https://www.bnl.gov/SET/DER-CAM.php
29. Wetter, M.: Design Optimization with Genopt. Building Energy Simulation User News (2000)
30. Sanaye, S., Hajabdollahi, H.: Thermo-economic optimization of solar CCHP using both genetic and particle swarm algorithms. J. Sol. Energy Eng. **137**(1), 011001 (2014)
31. Crossland, A.F., Jones, D., Wade, N.S.: Planning the location and rating of distributed energy storage in LV networks using a genetic algorithm with simulated annealing. Int. J. Electr. Power Energy Syst. **59**(7), 103–110 (2014)

Optimal Design and Operation of Integrated Energy System Based on Supply-Demand Coupling Analysis

Qiong Wu and Hongbo Ren[✉]

College of Energy and Mechanical Engineering,
Shanghai University of Electric Power, Shanghai 200090, China
tjrhb@163.com

Abstract. In this study, a bottom-up energy system optimization model is developed to assist the decision-making towards a sustainable energy system in the local area, while accounting for both supply-side and demand-side measures. The demand-side energy efficiency measures have been modeled as virtual energy generators, so as to be considered within a uniform optimization framework. The optimization model can provide feasible system configuration of both supply-side and demand-side appliances, as well as corresponding operating strategies, in terms of either economic performance or environmental benefit. As an illustrative example, a residential area located in Kitakyushu, Japan, is employed for analysis. The simulation results suggest that the combination of distributed energy resources and energy efficiency measures may result in better economic, energy and environmental performances. Moreover, it is technically and economically feasible to achieve more than 40% reduction in CO_2 emissions within the local area.

Keywords: Integrated energy system · Optimization model · Energy efficiency · Distributed energy resources

1 Introduction

As a "no-regrets" option for CO_2 emissions mitigation, energy efficiency measures (EEM) are expected to play the most important role in a sustainable energy system. On the other hand, as a "less-regrets" mitigation option, distributed energy resources (DER) with satisfied energy and environmental performances are paid more and more attention. However, a local energy system integrating EEM and DER may result in a higher degree of complexity, while determining the optimal combination of energy supply and utilization technologies. To support such a complex decision-making problem, a mathematical model based optimization procedure has proved its validity.

In the past few years, a considerable number of studies have been reported on the optimization of DER systems, especially the combined cooling, heating and power (CCHP) systems [1, 2]. On the other hand, the optimal selection of energy efficiency measures is also paid enough attention. Penna et al. [3] developed a multi-objective model for the determination of optimal retrofit solutions in terms of either economic performance or energy benefit. Karmellos et al. [4] developed a multi-objective

© Springer Nature Singapore Pte Ltd. 2017
K. Li et al. (Eds.): LSMS/ICSEE 2017, Part III, CCIS 763, pp. 566–575, 2017.
DOI: 10.1007/978-981-10-6364-0_57

optimization model for the prioritization of EEMs in buildings. According to the above discussions, previous studies generally focused on either the supply side or the demand side of the local energy system. As the two pillars of sustainable energy policy, EEM and DER should be analyzed in an integrated framework.

In this study, a bottom-up energy system optimization model covering both supply and demand sides is developed to assist the decision-making towards a sustainable energy system in the local area.

2 Methodology

2.1 Research Object

Problem Definition. As shown in Fig. 1, by means of the connected energy flows, primary energy sources, energy conservation options, energy demand forms and end-use appliances have been included in the local energy system [5]. The supply-side consists of various DER options, and the final energy consumptions are always realized by operating various appliances. Thus, the energy-saving potential can be exploited for each component satisfying various energy forms. Therefore, the right part of the figure illustrates various types of EEM for energy substitution and conservation. The decision-makers should determine the optimal combination of DER and EEM for the local area based on the supply-demand coupling analysis.

In this study, a model framework proposed in previous studies has been extended and enhanced to include not only DERs but also EEMs [6]. In addition, in order to be comparable with the DERs within a consistent framework, the EEMs have been modeled as virtual energy supply appliances, which may reduce the energy demand instead of providing more efficient or clean supply.

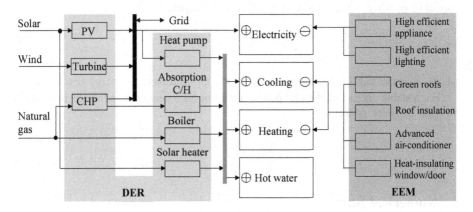

Fig. 1. Sample layout of an integrated local energy system.

2.2   Model Formulation

Objective Function. In this study, the optimization model aims at determining the optimal combination of technologies for energy supply and conversion, and to some extent, for final energy demand. Firstly, the system is optimized by using total annual cost of the entire energy system over the whole study period as the objective function to be minimized. In a simplified form the objective function can be written as:

$$[\text{min}] \quad f_C = C_f + C_v = f(I_i, R_i) + f(Q_{i,j}) \tag{1}$$

where, C_f and C_v indicate the fixed and variable costs, respectively. The fixed cost is usually the initial investment which should be considered from the life-cycle viewpoint. It is a function of the equipment size R_i, as well as an integer variable I_i which indicates the existence or number of each equipment. The variable cost is described as a function of real output ($Q_{i,j}$, the virtual output of energy efficiency measures as well) of the ith technology at the jth time interval.

Alternatively, the environmental objective function is described in Eq. (2). In this study, the environmental performance is assessed based on total CO2 emissions.

$$[\text{min}] \quad f_E = CO_e + CO_f = f(E_j) + f(Q_{i,j}) \tag{2}$$

where, CO_e and CO_f indicate the CO_2 emissions from utility electricity and on-site fuel consumption, respectively. E_j indicates the imported amount of utility electricity.

Main Constraints. Generally, equality constraints in terms of energy balances and inequality constraints in terms of technical restrictions should be considered.

Energy generation and consumption balance is formulated following the proposed system model, which includes utility grid, distributed generation, power export, and energy conservation thanks to the introduction of energy efficiency measures. Corresponding equality constraints have the following expressions:

$$\sum_i Q_{i,j} + E_j - S_j = D_j \quad \forall j \tag{3}$$

where, S_j is the exported amount of electricity. D_j indicates the energy demands.

For a stable operation, the real power output of each generator is limited in terms of its rated capacity as follows:

$$Q_{i,j} \le R_i \quad \forall i,j \tag{4}$$

3   Numerical Study

In this study, the developed optimization model is applied to a residential area which is composed of detached house, terraced house and apartment.

3.1   Energy Demands

In this study, to illustrate the fluctuation of energy demands, three days have been selected as typical days representing winter, mid-season, and summer, respectively [7, 8]. As shown in Fig. 2, the winter's day, 1st January, has relatively high thermal requirement, and the daily average heat-to-power ratio is about 2.3. Both thermal and electric demands occur from morning until night with relatively high coincidence. As to the typical day in the mid-season (1st May), due to the absence of cooling and heating loads, the thermal demand is relatively low and the daily heat-to-power ratio is only 0.6. Furthermore, the summer's day (1st Sep) has intermediate thermal requirement, which occurs from 17:00 in the evening until mid-night.

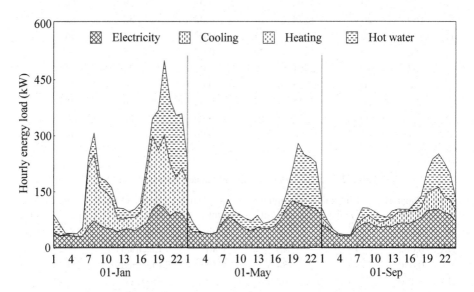

Fig. 2. Hourly load profiles on three typical days.

3.2   Technical Options

While accounting for the characteristics of residential buildings, various DERs and EEMs available in the market have been selected as the candidates. In detail, the DERs considered include: fuel cell (FC), gas engine (GE), micro-turbine (MT), and PV cells, as shown in Table 1. Except the PV system, all other technologies are fired by natural gas, with heat recovery for heating and cooling demands.

On the other hand, multiple EEMs are assumed, as shown in Table 2. While considering the breakdown of residential energy consumption, the examined EEMs can be summarized as high-efficient lighting (HL), high-efficient refrigerator (HR), high-efficient television (HT), high-efficient air-conditioner (HA), wall insulation (WI, with materials of polystyrene and plastic fiber for WI-1, WI-2, respectively), heat-insulated door (ID), heat-insulated window (IW), green roof (GR) and solar water

Table 1. Parameters of distributed generators [10].

Type	Capacity (kW)	Lifetime (year)	Capital cost ($/kW)	Fixed O&M cost ($/kW)	Variable O&M cost ($/kWh)	Power efficiency (%)	Thermal efficiency (%)
FC	1	10	12436	20	0.04	37	50
	5	12	8700	17	0.04	32	36
	10	12	6500	15	0.03	35	35
	125	20	4000	10	0.03	35	27
GE	1	20	10819	28	0.03	20	65
	30	20	2029	23	0.02	28	64
	60	20	1851	19	0.02	29	62
	75	20	1796	18	0.02	29	61
	100	20	1774	17	0.02	30	62
MT	28	10	3046	23	0.02	23	54
	60	10	2420	20	0.02	25	56
	67	10	2201	16	0.02	25	45
	76	10	2225	17	0.02	24	48
	100	10	2015	14	0.02	26	45
PV	1	20	7450	12	0.00	12	0
SW	1	15	250	5	0.00	0	40

Table 2. Parameters of energy efficiency measures [11–13].

Type	Capacity (kW)	Lifetime (year)	Capital cost ($/Unit)	Reference unit	Reduced load type	Applicable period (hour)
HL	0.096	25	13	Unit	Electricity	18–24
HR	0.082	15	210	Unit	Electricity	0–24
HT	0.050	20	174	Unit	Electricity	18–24
HA	0.190	15	362	Unit	Heating, cooling	0–24
WI-1	0.003	30	293	m^3	Heating, cooling	0–24
WI-2	0.004	30	439	m^3	Heating, cooling	0–24
ID	0.030	30	293	m^2	Heating, cooling	0–24
IW	0.067	30	37	m^2	Heating, cooling	0–24
GR	0.030	30	164	m^2	Heating, cooling	0–24

heater (SW) [9]. As aforementioned, the EEMs have been modeled as virtual energy supply appliances. Therefore, the capacity of each EEM is assumed to be hourly energy savings due to the introduction of a single unit of the measure.

3.3 Scenario Description

In this study, in terms of the energy supply strategy for the residential area, four scenarios have been assumed according to the predetermined technical combination and user preference.

Scenarios 1–3 focus on the economic optimization of the local energy system because economic performance always plays the main role in the final decision-making, especially for the end-users. However, the alternative technical options are different for three scenarios. In detail, in Scenario 1, only EEMs are included; in Scenario 2, only DERs are accounted; and in Scenario 3, both of the former two options are considered. Similar with Scenario 3, Scenario 4 considers EEM and DER simultaneously, while taking annual CO_2 emissions as the objective function to be minimized.

On the other hand, besides the aforementioned four scenarios, as a base scenario for reference, a conventional system without DER or EEM adoption is also assumed, in which external grid meets the electric demand and gas boiler serves the thermal load.

4 Results and Discussion

Based on the input data, optional design and operation of the integrated energy system is determined, by executing the developed optimization model. In the following, besides the discussions on optimal design and operation results of various scenarios, a cost-benefit analysis of the low-carbon energy system is also included.

4.1 Optimal Adoption Results

Table 3 shows the deduced system combinations and corresponding economic, energy and environmental performances for various scenarios. Generally, it can be found that both EEM and DER can contribute a lot to the local energy system from either economic, energy or environmental viewpoint. Especially in Scenarios 1–3, for the economic objective is considered, the introduction of proposed local energy system may achieve win-win-win benefits, resulting in reduced total costs, CO_2 emissions and primary energy consumptions. For example, compared with the base scenario, Scenario 3 with both DER and EEM adoption achieves reasonable reduction ratios of total cost, CO_2 emissions and primary energy consumption with values of 17.7%, 18.3% and 21.3%, respectively. In addition, it is interesting to notice that Scenario 2 (with only DER adoption) enjoys better economic performance but worse energy and environmental performances than Scenario 1 (with only EEM introduction).

Moreover, as the environmental issues are paid the main attention (Scenario 4), most of EEMs (except WI-2 and GR) are selected, whereas the total capacity of DER is less than that of Scenario 2 and Scenario 3. Correspondingly, local CO_2 emissions, as expected, is reduced by 59.4% (compared with the base scenario), while total cost increases dramatically (97% more than the base case).

Table 3. Optimal adoption results of various scenarios.

Scenario	DER capacity (kW)	EEM capacity (kW)	System combination		Total cost (Thousand $)	Total CO$_2$ emissions (Ton)	Total primary energy consumption (MJ)
			DER	EEM			
Base Scenario	0	0	–	–	141.6	314.2	7.9
Scenario 1	0	54	–	HL: 306; HR: 69; SW: 58	130.4	278.1	7.0
Scenario 2	400	0	GE-75: 1; PV: 325	–	126.4	291.5	7.1
Scenario 3	400	23	GE-75:1; PV: 325	HL: 306; HR: 69	116.6	256.7	6.2
Scenario 4	366	815	FC-1: 19; GE-100:2; PV: 147	HL: 306; HR: 69; HT: 69; HA: 237; WI-1: 50; ID: 214; IW: 573; SW: 1316	278.9	127.6	3.4

4.2   Annual Energy Balances

Electric Balance. As shown in Fig. 3, the introduction of DER and EEM results in reduced power input from the utility grid. In detail, the shares of power supply from DERs and EEMs are 18%, 38%, 54% and 76% for scenarios 1–4, respectively. Generally, DERs may serve most of the local electric requirement if necessary, while the EEMs with virtual power supply also cannot be neglected. For example, in Scenario 1, EEMs serve nearly 18% of total electricity load. Among which, HL and HR take shares of 8% and 9%, respectively. Again, when DER option is paid the only attention (Scenario 2), gas engine unit takes a share of about 38% of total annual electric demand. Furthermore, it is interesting to notice that although PV array is introduced, it supplies no electricity for the local requirements. This is because all the power generation from the PV system is sold back to the grid due to the assumed high buy-back price. Detailed discussion will be illustrated in the following subsection. Moreover, as mentioned above, when the environmental aspect is focused (Scenario 4), most of local electric demand is served by the on-site generation (including EEM based virtual generation). As to the EEMs, HL, HR and HT supply 8%, 9% and 1% of total electricity load, respectively. On the other hand, the shares of power from FC, GE and PV are 9%, 20% and 28%, respectively.

Thermal Balance. Figure 4 illustrates the share of various thermal sources for local thermal demands. With the introduction of DERs or EEMs, the direct gas and electricity consumption for local thermal demands is reduced more or less. The detail information is introduced as follows.

In Scenario 1, only solar water heater is introduced for the reduction of hot-water load, and it takes up about 5.1% of annual total thermal demand. As to the scenario with DER option (Scenario 2), the recovered heat from gas engine contributes most (about 59%) of local thermal demand, while the shares of electricity and city gas are 24.4% and 16.6%, respectively. In Scenario 3, although both DER and EEM are considered for optimization, the shares of various thermal sources are similar to the

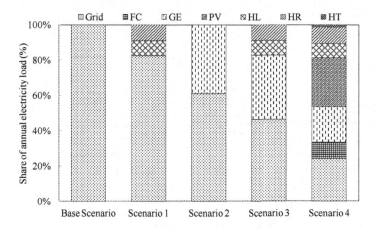

Fig. 3. Annual electrical balance of various scenarios.

situation in Scenario 2 because only EEM with virtual power generation are selected (see Table 3). As to Scenario 4, due to the relatively high attention paid to the environmental performance, the EEMs (HA, WI-1, ID, IW, SW) with virtual thermal energy supply about 23.1% of total thermal demand, and DERs (GE and FC) take a share of 42.0%.

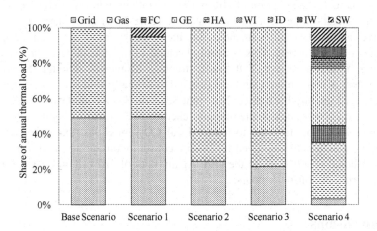

Fig. 4. Annual thermal balance of various scenarios.

4.3 Hourly Operating Strategies

Hourly operating schedules are of vital importance to the understanding of the technical characteristics and corresponding performances of different system alternatives. As an example, Fig. 5 shows the hourly electric balances of the examined four scenarios on a summer day.

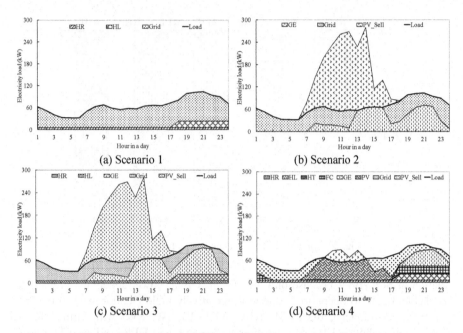

Fig. 5. Hourly electrical balance of various scenarios (Summer).

In Scenario 1, the adoption of HR and HL results in reduced electric consumption during the specific time period as indicated in Table 3. As to Scenario 2, gas engine unit serves all the electricity demand during the on-peak period, and is out of operation during the off-peak period. This is mainly because of the fluctuating hourly tariff rates according to the time-of-use (TOU) tariff structure. Moreover, although the power out of PV system is much more than local demand, it is totally sold back to the grid due to the relatively high buy-back price. Note that Scenario 3 is a combination of Scenario 1 and Scenario 2. Nevertheless, the running schedule of Scenario 4 is quite different with other three scenarios. In order to reduce local CO_2 emissions, the power out of PV system is preferentially employed for on-site consumption during the daytime period, and the residual is sold back to the grid. Again, both of fuel cell and gas engine units operate during the off-peak hours.

5   Conclusions

In this study, an optimization model has been developed to investigate the feasibility of introducing DERs and EEMs in a local energy system. As an illustrative example, a residential area has been employed for analysis. According to the simulation results, the conclusions can be deduced:

(1) As the economic objective is optimized, annual system cost, CO_2 emissions and primary energy consumption of the scenario with both DER and EEM adoption are reduced by 17.7%, 18.3% and 21.3%, respectively. On the other hand, when

the environmental issues are focused, the system may enjoy better environmental performance but is at cost of worse economic performance.

(2) Local energy prices may have great effect on the design and operation of the local energy system. For example, due to the fluctuating hourly tariff rate, the gas fired DER unit usually operates during the on-peak period but is out of operation during the off-peak hours.

Acknowledgements. This work was supported in part by National Natural Science Foundation of China (No. 71403162) and Shanghai Sailing Program (No. 17YF1406800).

References

1. Hajabdollahi, H., Ganjehkaviri, A., Jaafar, M.N.M.: Assessment of new operational strategy in optimization of CCHP plant for different climates using evolutionary algorithms. Appl. Therm. Eng. **75**, 468–480 (2015)
2. Karami, R., Sayyaadi, H.: Optimal sizing of Stirling-CCHP systems for residential buildings at diverse climatic conditions. Appl. Therm. Eng. **89**, 377–393 (2015)
3. Penna, P., Prada, A., Cappelletti, F., Gasparella, A.: Multi-objectives optimization of energy efficiency measures in existing buildings. Energy Build. **95**, 57–69 (2015)
4. Karmellos, M., Kiprakis, A., Mavrotas, G.: A multi-objective approach for optimal prioritization of energy efficiency measures in buildings: model, software and case studies. Appl. Energy **139**, 131–150 (2015)
5. Ren, H., Zhou, W., Gao, W., Wu, Q.: A mixed-integer linear optimization model for local energy system planning based on simplex and branch-and-bound algorithms. In: Li, K., Fei, M., Jia, L., Irwin, George W. (eds.) ICSEE/LSMS -2010. LNCS, vol. 6328, pp. 361–371. Springer, Heidelberg (2010). doi:10.1007/978-3-642-15621-2_40
6. Ren, H., Gao, W.: A MILP model for integrated plan and evaluation of distributed energy systems. Appl. Energy **87**(3), 1001–1014 (2010)
7. Center for Global Environmental Research: National Greenhouse Gas Inventory Report of JAPAN. Japan: Ministry of the Environment (2015)
8. Kashiwagi T.: Natural Gas Cogeneration Plan/Design Manual. Japan Industrial Publishing Co. Ltd (2002)
9. Ministry of Economy: Trade and Industry: Catalog of Energy Conservation Performance. Ministry of Economy, Trade and Industry, Tokyo (2010)
10. Firestone, R.: Distributed Energy Resources Customer Adoption Model Technology Data. Berkeley Lab, Berkeley (2004)
11. Castleton, H.F., Stovin, V., Beck, S.B.M., Davison, J.B.: Green roofs: building energy savings and the potential for retrofit. Energy Build. **42**(10), 1582–1591 (2010)
12. Dicorato, M., Forte, G., Trovato, M.: Environmental-constrained energy planning using energy-efficiency and distributed-generation facilities. Renew. Energy **33**(6), 1297–1313 (2008)
13. Zheng, G., Jing, Y., Huang, H., Shi, G., Zhang, X.: Developing a fuzzy analytic hierarchical process model for building energy conservation assessment. Renew. Energy **35**(1), 78–87 (2010)

Modeling, Simulation and Control in Smart Grid and Microgrid

Control Strategies for the Microgrid Control System with Communication Delays

Weihua Deng[1], Pengfei Chen[1], Kang Li[2], and Chuanfeng Li[3(✉)]

[1] College of Electrical Engineering, Shanghai University of Electric Power,
Shanghai, China
[2] School of Electronics, Electrical Engineering and Computer Science,
Queen's University Belfast, Belfast, UK
[3] School of Computer and Information,
Luoyang Institute of Science and Technology, Luoyang, China
dwh197859@126.com

Abstract. In this paper, two kinds of microgrid system architectures and the control approaches are studied. The proposed architectures are designed above a kind of communication network. The communication network characteristic is mainly described by network-induced delays which have greater influences on the control system performance. The network-induced delays in this paper is depicted by the inverse Gaussian distribution function. The proposed control strategies are implemented depending on the achitectures of themselves. The principle of event-triggered and droop-based approach are employed to restrain the different disturbances such as the break of main grid and insertion of new load. Some numerical examples are used to illustrated the effectiveness of the control approaches in this paper.

Keywords: Microgrids · Network-induced delays · Inverse Gaussian distribution

1 Introduction

Microgrid [7] R2R3 is a combination of many factors such as the distributed generations (DGs) and variable loads and energy storage battery. Distributed generation can be either the renewable energy source or conventional fossil energy. In microgrid control system, there is another important component that is the communication network. Above communication networks, the control strategies can be applied into the control of microgrids. The networks generally are divided into three types. The first one is at the lowest level, which connect the sensors, controllers and actuators of every DG. The second type of network is among the distributed controllers of DGs. And the last layer is used to transmit information between multiple microgrids. The above components construct a complicated

---

C. Li–This work was supported by Natural Science Foundation of Shanghai under Grant 15ZR1417500.

K. Li et al. (Eds.): LSMS/ICSEE 2017, Part III, CCIS 763, pp. 579–586, 2017.
DOI: 10.1007/978-981-10-6364-0_58

microgrid control system. The energy can be transferred among the microgrids according to their demands.

In the operation of microgrid control system, the communication networks has its advantage as stated in the above and has also some disadvantages such as delays and dropouts. At present, majority of research mainly focus on the design control system by use of the advantages of communication networks. And the work of considering the communication delays in microgrid control system performance is rare. The Zigbee network delay is analyzed and concluded that there is a great impact on stability in [6]. And the network-induced delays are modelled networked control method is used to analyze the system stability in [2,3].

This paper will analyze microgrid system stability based on a kind of communication delay proposed in [8]. The kind of delay is modelled by use of an inverse Gaussian-distributed function which can depict a physical IEEE 802.11b wireless channel. The microgrid control system stability is researched mainly from the point of simulation view.

2   The Delay Model

The random delay is modelled as an inverse Gaussian-distributed function. Its main parameters are μ and λ, namely, $\tau(k) \sim IG(\mu, \lambda)$. The cumulative density function (CDF) of the inverse Gaussian distribution is shown in the following

$$
\begin{aligned}
F(\tau(k)) = &\Phi(\sqrt{\tfrac{\lambda}{\tau(k)}}(\tfrac{\tau(k)}{\mu} - 1)) \\
&+ \exp(\tfrac{2\lambda}{\tau(k)})\Phi(-\sqrt{\tfrac{\lambda}{\tau(k)}}(\tfrac{\tau(k)}{\mu} + 1))
\end{aligned}
\tag{1}
$$

The influences of this kind of delays on the microgrid stability will be analyzed in this paper.

3   Control System Structure

Two kinds of control system structure will be researched. CB means circuit breaker in the following paper.

3.1   The Distributed Control Structure of the Islanded Microgrid

The islanded microgrid system based on droop control is shown in Fig. 1. The microgrid is composed of one DG and three loads. Load 0 is located on DG and others are far away from the DG.

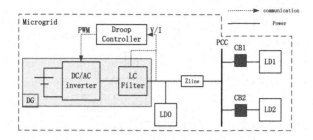

Fig. 1. The distributed control architecture under islanded microgrid.

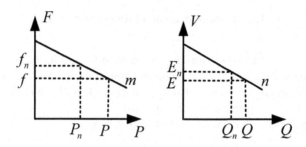

Fig. 2. Droop characteristic curves.

The Droop-Based Control Strategy. Droop control can achieve DGs such as wind or solar emulating the principle of the governor of a conventional synchronous generator through the linear droop characteristic of frequency to active power $(F-P)$ and voltage magnitude to reactive power $(U-Q)$ as show in Fig. 2. The voltage and frequency droop characteristics are given by

$$
\begin{aligned}
f &= f_n + m\,(P_n - P) \\
E &= E_n + n\,(Q_n - Q)
\end{aligned}
\tag{2}
$$

where m and n are the droop coefficients that are selected based on the active and reactive power rating of DG. f_n, E_n, P_n and Q_n are the control references. f and E are the amplitude and frequency of the DG generated. P and Q are the measured active and reactive power at the DGs terminal

$$
\begin{aligned}
P &= \tfrac{\omega_c}{s+\omega_c}\,(u_{od}i_{od} + u_{oq}i_{oq}) \\
Q &= \tfrac{\omega_c}{s+\omega_c}\,(u_{od}i_{oq} - u_{oq}i_{od})
\end{aligned}
\tag{3}
$$

where ω_c represents the cut-off frequency of low-pass filter; s is the Laplace transform factor; $_{od}$, $_{oq}$, i_{od} and i_{oq} are the output voltages and currents in the dq reference frame.

The controller is formed by three loops as show in Fig. 3: (1) power controller, used to set the frequency and magnitude for the fundamental voltage of the inverter according to the droop coefficients. (2) voltage control, yield close control of the output voltage and synthesize the reference current vector. The

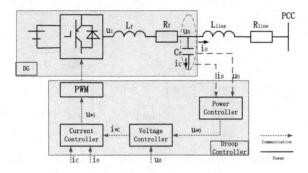

Fig. 3. Block diagram of droop controller.

voltage control loop employs a standard proportional and integral (PI) regulator. (3) current controller, rapidly response to the filter-capacitor current, generate voltage vector signal to pulse-width modulation (PWM) module. The current loop employs only proportional regulator in [5].

The voltage controller model is shown as follow

$$
\begin{aligned}
i_{cd}^* &= \left(K_{PV} + \frac{K_{IV}}{s}\right)\left(u_{od}^* - u_{od}\right) - \omega C_f u_{oq} \\
i_{cq}^* &= \left(K_{PV} + \frac{K_{IV}}{s}\right)\left(u_{oq}^* - u_{oq}\right) + \omega C_f u_{od}
\end{aligned}
\tag{4}
$$

where K_{PV}, K_{IV} are the proportional and integral gains of voltage, respectively; C_f is the per-phase capacitance of LC filter ω represents the frequency of microgrid. The current controller model is shown as follow

$$
\begin{aligned}
u_{id}^* &= K_{PI}\left(i_{cd}^* - i_{cd}\right) - \omega L_f \left(i_{cq} + i_{oq}\right) \\
u_{iq}^* &= K_{PI}\left(i_{cq}^* - i_{cq}\right) + \omega L_f \left(i_{cd} + i_{od}\right)
\end{aligned}
\tag{5}
$$

where K_{PI} is the proportional of current; L_f is the per-phase inductor of LC filter; i_{cd}, i_{cq} are the capacitance currents in dq reference frame.

3.2 Event-Triggered Coordinated Control Structure

The Coordinated Control Structure. The coordinated control architecture is shown in Fig. 4. This kind of architecture is used to connect the multi-microgrids. Based on it the power can be dispatched among the multi-microgrids according to their requirements. The coordination controller (CC) is responsible for the higher level of control such as coordination amongst the microgrids. While the above mentioned center and distributed control are responsible for the regulation inside one microgrid. The mechanism of event-triggered is also used in this control strategy. The proposed coordinated control architecture is satisfied with the following assumptions:

- Assumption 1: the microgrids control system consists of two microgrids and one coordination controller. Every microgrid includes one DG and one load and one MCC;

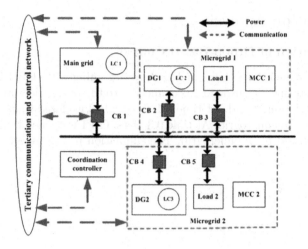

Fig. 4. Coordinated control architecture.

- Assumption 2: at the beginning the two microgrids work normally, namely, CB 2, CB 3, CB 4 and CB 5 are closed, and CB 1 is open;
- Assumption 3: the rated power of DG 2 may meet the two load demand or may not;
- Assumption 4: here CB is of the function of measure and control and actuation.
- Assumption 5: DG 1 is disconnected at 1 s because of some faults, namely, CB 2 is open. The DG 2 is disconnected at 2.5 s because of some faults, namely, CB 5 is open.

The Coordinated Control Approach. The control will make load 1 and load 2 run at rated power when the microgrids meet some disturbances. The coordinated control algorithm are as follows:

Algorithm 1:

- Step 1: at 0.01 s, CB 2 open and controller generates a signal and transmits the information to CC;
- Step 2: CC accepts the signal from CB 2 and generates a trigger signal transmit it to DG 2;
- Step 3: LC 3 accepts the signal from CC and makes the active power of DG 2 increased;
- Step 4: at 0.025 s, CB 4 open and generates a signal and transmits it to CC;
- Step 5: CC accepts the signal from CB 4 and transmits a trigger signal to CB 1;
- Step 6: CB 1 accepts the signal from CC and CB 1 will be closed.

4   Simulation Analysis

4.1   Simulation of Droop-Based Control

The simulink time is 0.5 s. Look at Fig. 2, to beginning with, CB 1 is closed and
CB 2 is opened that means load 1 is connected to the microgrid and load 2 is not
connected to the microgrid. At time 0.1–0.2 s, CB 2 is closed that means load 2
is connected to the microgrid. At time 0.3–0.4 s, CB 1 is opened that means load
1 is not connected to the microgrid. The simulation results are given in Figs. 5,
6 and 7.

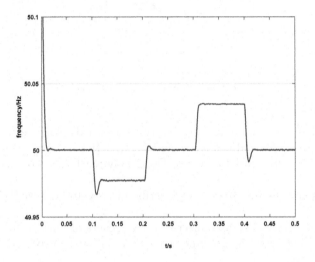

Fig. 5. The frequency of islanded microgrid.

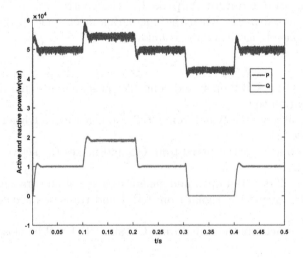

Fig. 6. The output active/reactive power of DG.

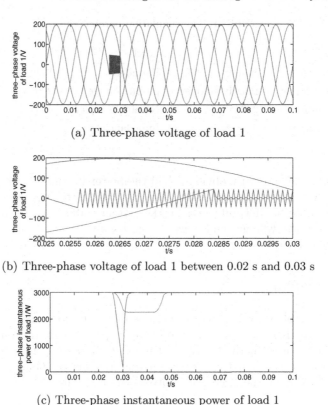

(a) Three-phase voltage of load 1

(b) Three-phase voltage of load 1 between 0.02 s and 0.03 s

(c) Three-phase instantaneous power of load 1

Fig. 7. The case of coordinated control architecture with IGD delays.

From Fig. 6, we can see that the microgrid normal frequency is 50 Hz. At time
0.1–0.2 s, the load 2 is connected to the microgrid and DG needs to increase the
output active power. According to the principle of droop control, the frequency
is decreased. At time 0.3–0.4, the situation is inverse. Figure 6 is the output
active/reactive power of DG. The output of DGs normal active power is 50 kw
and normal reactive power is 10 kvar. At time 0.1–0.2 s, the load 2 is connected
to the microgrid and active power is increased to 55 kw and reactive power is
increased to 1.9 kvar. At time 0.3–0.4 s, the load 1 is disconnected to the microgrid and active power is decreased to 43 kw and reactive power is de increased
to 0 var.

4.2   Simulation of coordinated control

The related parameters are given in Table 1, which are used to simulate the case
of coordinated control. The results are shown in Fig. 7. The microgrid control
systems in Fig. 5 are disturbed at 0.01 s and 0.025 s by the disconnecting of CB 2
and CB 4. The changes of voltages are shown in Fig. 7(a) under that disturbances

Table 1. The simulation parameters.

DG		Load	
Parameter	Value	Parameter	Value
Phase sequence	Positive-sequence	ph-ph voltage	200 (V)
Amplitude	200 (V)	Nominal frequency	60 Hz
Frequency	60 (Hz)	Impedance	$19.2 + j76.5\,(\Omega)$

and the change is not very obvious. And the power in Fig. 7(c) dropped at the corresponding disturbance moments. But from Fig. 7(b), we can see that the IGD delays have heavier influences on the system stability between 0.025 s and 0.03 s. Finally the voltages and power all go back to their stability states after coordination control.

5   Conclusion

The two kinds of control system structures of microgrid based on the inverse Gaussian delay model are investigated in this paper. The droop control and the coordination control methods are implemented on their system constructions from the aspect of simulation. It shown that the inverse Gaussian delay has greater influences on the microgrid control system stability.

References

1. Ipakchi, A., Albuyeh, F.: Grid of future. IEEE Power Energy Mag. **7**, 52–62 (2009)
2. Singh, A., Singh, R., Pal, B.C.: Stability analysis of networked control in smart grids. IEEE Trans. Smart Grid **6**, 381–390 (2015)
3. Peng, C., Zhang, J.: Delay-distribution-dependent load frequency control of power systems with probabilistic interval delays. IEEE Trans. Power Syst. **31**, 3309–3317 (2016)
4. Farhangieh, H.: The path of smart grid. IEEE Power Energy Mag. **8**, 18–28 (2010)
5. Yu, K., Ai, Q., Wang, S.Y., Ni, J.M., Lv, T.G.: Analysis and optimization of droop controller for microgrid system based on small-signal dynamic model. IEEE Trans. Smart Grid **7**, 695–705 (2016)
6. Setiawan, M.A., Shahnia, F., Rajakaruna, S.: ZigBee-based communication system for data transfer within future microgrids. IEEE Trans. Smart Grid **6**, 2343–2355 (2015)
7. Hatziargyriou, N., Asano, H., Iravani, R.: Microgrid. IEEE Power Energy Mag. **5**, 78–94 (2007)
8. Deng, W.H., Li, K., Irwin, G.W., Fei, M.R.: Identification and control of Hammerstein systems via wireless networks. Int. J. Syst. Sci. **44**, 1613–1625 (2013)

Secondary Voltage Control of Microgrids with Distributed Event-Triggered Mechanism

Jing Shi$^{(\boxtimes)}$, Dong Yue, and Shengxuan Weng

School of Automation and the Institute of Advanced Technology,
Nanjing University of Posts and Telecommunications, Nanjing 210046, China
mejingshi@126.com

Abstract. This paper presents a secondary voltage control scheme with distributed event-triggered mechanism for multiple distributed generators in microgrids. First, to mitigate the over-provisioning of communication resources in microgrids, a distributed event-triggered mechanism is proposed. Then, based on the proposed triggering scheme, distributed secondary controllers are designed for distributed generators. Finally, simulation results demonstrate that with the adoption of the control strategy, the voltages of distributed generators are synchronised to their nominal values.

Keywords: Microgrids · Secondary voltage control · Distributed event-triggered

1 Introduction

Nowadays, to deal with the energy crisis and environmental degradation issues, a growing number of distributed generators (DGs) are integrated into the microgrid systems [1], which run in either grid-connected or islanded mode. As in islanded mode, microgrids should maintain frequency and voltage stability automonously. How to regulate output voltages and frequencies of DGs to their prescribed values in microgrids is called secondary control problem [2]. The conventional secondary control relies on the microgrid centralized control center which requires complex communication networks. However, due to the fact that a vast range of DGs are widely dispersed geographically, it calls for reliable communication networks and the centralized control scheme may reduce the reliability of the microgrid systems [3,4]. Different from the centralized control, distributed control with sparse communication network makes it hard that failures happen and improves the system performance by letting local controllers to exchange information with their neighbors'. During the past decades, some fruitful research on the distributed secondary control of microgrids have been proposed, for instance [5,6], where the implementation of the distributed control requires the local information transmission between DGs.

Note that among these results, it is assumed that the information of each DG is continuously transmitted over communication network. In addition, the

© Springer Nature Singapore Pte Ltd. 2017
K. Li et al. (Eds.): LSMS/ICSEE 2017, Part III, CCIS 763, pp. 587–596, 2017.
DOI: 10.1007/978-981-10-6364-0_59

traditional periodic sampling control scheme may yield conservative results as the sampling rate is selected ideally [7,8]. For saving communication resources between DGs, it is desirable to design effective controllers that can work in event-driven environments and update their values only when needed. As a result, the unnecessary redundant information can be reduced [9–11].

This paper introduces a distributed event-triggered mechanism for secondary controllers of microgrids with multiple DGs. In order to compensate the voltage deviation caused by the primary control and restore them to their prescribed values, a distributed control law for DGs is built with a event-triggered mechanism. The main contributions of the paper are given as follows.

1. The event-triggered mechanism is adopted when designing the microgrid secondary controller. This mechanism can reduce the amount of data packet transmission, and ease communication pressure.
2. Both the designed event-triggered mechanism and the implementation of the secondary voltage controller require the neighbors' information. This indicates that the proposed distributed secondary control can avoid the drawbacks of centralized scheme as mentioned above.

This paper is organized as follows: Sect. 2 provides the problem formulation. The secondary voltage control with distributed event-triggered mechanism is presented in Sect. 3. An experimental test of the proposed scheme in an islanded MG is given in Sect. 4. Section 5 concludes this paper.

2   Problem Formulation

This section will study a microgrid operating in islanded mode which involves the inverter-interfaced DGs and the loads. As a primary DC source (e.g., wind system, PV array, etc.), DGs are connected by voltage source inverters, while the loads are connected through an LC filter and coupling inductance. To stabilize the output voltage of DGs when MG switches to islanded mode, the primary control is implemented locally in the internal control loops of DGs by using the notable droop technique, which does not need any communication link. As is well known, the primary control consists of three controllers, e.g. the power, voltage and current controllers, and one can refer to the detailed descriptions of the three controllers in [3,12].

Droop technique is a decentralized strategy, which shares active and reactive powers among DGs and maintains the output levels of voltage and frequency within acceptable ranges. At the ith DG terminal, desired relationships for inductive lines between reactive power and voltage can be given as follows

$$v_{omagi}^* = E_{ni} - n_i Q_i \qquad (1)$$

where E_{ni} are the references in primary control, v_{omagi}^* is the output voltage magnitude reference value calculated by power controller, n_i are the droop coefficients, Q_i are the measured reactive power.

The magnitude of the DG output voltage v_{omagi} is obtained as $v_{omagi} = \sqrt{v_{odi}^2 + v_{oqi}^2}$ and the primary voltage control strategy aligns the voltage magnitude of each DG on the direct axis of its reference frame. Then the primary voltage control strategy is expressed as follows

$$\begin{cases} v_{odi} = E_{ni} - n_i Q_i \\ v_{oqi} = 0 \end{cases} \tag{2}$$

To compensate the voltage deviation caused by the primary control, the secondary controller is applied to tune the voltage amplitude of every DG to their normal values by setting the primary references E_{ni}. Therefore, the secondary control selects E_{ni} such that the voltage amplitude of each DG synchronises to their prescribed value, that is, $v_{odi} \rightarrow v_{ref}$.

3   Preliminaries and Main Results

Secondary control of microgrids is essentially a tracking synchronization problem, where all DGs need to be synchronized with the command generator. Thus, each DG communicates and exchanges the information with neighbouring DGs.

The details of the distributed secondary coordination control scheme with event-triggered mechanism for the islanded MG is shown in Fig. 1.

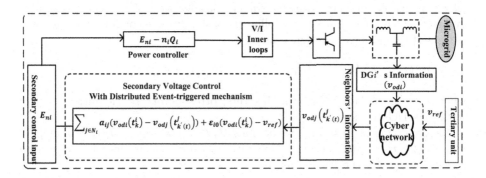

Fig. 1. The block diagram of the distributed secondary coordination control scheme.

3.1   Preliminaries

A fixed undirected graph $\mathcal{G} = (\mathcal{V}, \mathcal{E}, A)$ is used to describe the communication topology for DGs. $\mathcal{V} \triangleq (v_1, v_2, \ldots, v_N)$ is the node set where v_i signifies the ith DG. $\mathcal{E} \triangleq \{(v_j, v_i), if \ j \rightarrow i\}$ denotes the edge set in which $j \rightarrow i$ represents that the information of the jth DG can be transmitted to the ith one. $A \triangleq \{a_{ij}\} \in R^{N \times N}$ is the weighted adjacency matrix of graph $\mathcal{G}$, where $a_{ij} = 1$ if and only if $(v_j, v_i) \in \mathcal{E}$, otherwise $a_{ij} = 0$, and it is assumed that $a_{ii} = 0$. The neighbouring set of node v_i is denoted by $N_i \triangleq \{v_j \in \mathcal{V} \mid (v_j, v_i) \in \mathcal{E}\}$, and the cardinal number of N_i is represented as $|N_i|$. The 'undirected' means that $(v_i, v_j) \in \mathcal{E}$ if and only if $(v_j, v_i) \in \mathcal{E}$, and the 'fixed' means all elements in $\mathcal{E}$ are constant. Defining the in-degree matrix $D \triangleq diag\{d_1, \ldots, d_N\}$ where $d_i = \sum_{j \in N_i} a_{ij}$, the Laplacian matrix of graph $\mathcal{G}$ is given as $L = D - A$. L has all row sums equal to zero, that is $L1_N = 0$, with 1_N being the vector of ones with the length of N.

To ease the communication pressure, the event-triggered idea is adopted in the following. Each DG transmits its own information, namely, the output voltage to its neighbors at its triggering instant. This triggering instant is determined by a triggering condition which will be given in the following. Denote the triggering times of the output voltage of the ith DG by t_k^i for $k = 0, 1, 2, \ldots$.

3.2 Secondary Voltage Control with Distributed Event-Triggered Mechanism

In this section, the secondary voltage control with distributed event-triggered mechanism is designed such that the voltage magnitudes of DGs v_{odi} is synchronized to the reference voltage v_{ref}. We need to select control inputs E_{ni} for the secondary voltage control in (2).

Therefore by using (2) and the input−output feedback linearisation method [14], a dynamic system for obtaining the input E_{ni} can be obtained as

$$\dot{v}_{odi} = \dot{E}_{ni} - n_i \dot{Q}_i = u_i \tag{3}$$

where u_{1i} is an auxiliary control. Under consideration of the effectiveness of event-triggered mechanism, the following distributed secondary voltage controllers of the ith DG is designed as

$$u_i(t) = -k\left(\Sigma_{j \in N_i} a_{ij}(v_{odi}(t_k^i) - v_{odj}(t_{k'(t)}^j)) + \varepsilon_{i0}(v_{odi}(t_k^i) - v_{ref}) \right) \tag{4}$$

where $t_{k'(t)}^i$ is the latest triggering instants corresponding to the output voltage of the jth DG at instant t.

Before giving main results, we define $d_i(t) \triangleq v_{odi}(t) - v_{ref}$, $\hat{d}_i(t) \triangleq v_{odi}(t_k^i) - v_{ref}$, $t \in [t_k^i, t_{k+1}^i)$ and the utilization ratio measurement errors with respect to voltage output for the ith DG as

$$e_i(t) \triangleq v_{odi}(t) - v_{odi}(t_k^i), t \in [t_k^i, t_{k+1}^i) \tag{5}$$

Then, combining the definitions yields $e_i(t) = d_i(t) - \hat{d}_i(t)$ and

$$\dot{d}_i(t) = -k\left(\sum_{j \in N_i} a_{ij}(\hat{d}_i(t) - \hat{d}_j(t)) + \varepsilon_{i0}\hat{d}_i(t) \right) \tag{6}$$

Theorem 1. *It is assumed that at least one ε_{i0} equals to 1 for $i = 1, \ldots, n$, then under the voltage controller (4) with the following event-triggered condition*

$$|e_i(t)| > \frac{\beta_i h_i(t)^2}{\sum_{j=1}^n a_{ij}|h_j(t)| + (\varepsilon_{i0} + |N_i|)|h_i(t)|} \tag{7}$$

where

$$h_i(t) = \sum_{j \in N_i} a_{ij}(v_{odi}(t_k^i) - v_{odj}(t_{k'(t)}^j)) + \varepsilon_{i0}(v_{odi}(t_k^i) - v_{ref}), \tag{8}$$

and $0 < \beta_i < 1$, $\varepsilon_{i0} = 1$ if and only if the ith DG can receive the information of the leader, otherwise $\varepsilon_{i0} = 0$, then the output voltage magnitude of all DGs asymptotically achieve consensus, that is, the voltage magnitude $v_{odi}, i = 1, \ldots, N$ synchronizes to the reference value v_{ref}.

Proof. Let $d(t) = [d_1(t), d_2(t), \ldots, d_N(t)]^T$, and choose a Lyapunov function candidate as

$$V(t) = \frac{1}{2}\left(d(t)^T L d(t) + \sum_{i=1}^{n} \varepsilon_{i0} d_i(t)^2\right) \tag{9}$$

By the property of the Lapacian matrix L described in Preliminaries, the Lyapunov function $V(t)$ is positive. Then, the derivative of V with respect to t along with the solution of system (3) is

$$\dot{V} = \sum_{i=1}^{n}\left(\sum_{j=1}^{n} l_{ij} d_j + \varepsilon_{i0} d_i\right) \dot{d}_i$$

$$= \sum_{i=1}^{n}\left(\sum_{j=1}^{n} l_{ij}(e_j + \hat{d}_j) + \varepsilon_{i0}(e_i + \hat{d}_i)\right) \dot{d}_i \tag{10}$$

$$= \sum_{i=1}^{n}\left(\sum_{j=1}^{n} l_{ij} e_j + \varepsilon_{i0} e_i\right) \dot{d}_i + \sum_{i=1}^{n}\left(\sum_{j=1}^{n} l_{ij} \hat{d}_j + \varepsilon_{i0} \hat{d}_i\right) \dot{d}_i$$

where l_{ij} is the element in the Laplacian matrix L. From (6) and (8), we obtain

$$\sum_{i=1}^{n}\left(\sum_{j=1}^{n} l_{ij} \hat{d}_j + \varepsilon_{i0} \hat{d}_i\right) \dot{d}_i = -k \sum_{i=1}^{n} h_i^2 \tag{11}$$

On one hand, for the undirected graph G, we have

$$\sum_{i=1}^{n}\left(\sum_{j=1}^{n} l_{ij} e_j + \varepsilon_{i0} e_i\right) \dot{d}_i = \sum_{i=1}^{n}\left(\sum_{j=1}^{n} a_{ij}(e_i - e_j) + \varepsilon_{i0} e_i\right) \dot{d}_i$$

$$\leq \sum_{i=1}^{n}\left(\sum_{j=1}^{n} a_{ij}|e_j| + (\varepsilon_{i0} + |N_i|)|e_i|\right)|\dot{d}_i| \tag{12}$$

$$= \sum_{i=1}^{n}\left(\sum_{j=1}^{n} a_{ij} k|h_j| + (\varepsilon_{i0} + |N_i|)k|h_i|\right)|e_i|$$

On the other hand, for the given condition (7), we can obtain that

$$|e_i(t)| \leq \frac{\beta_i h_i(t)^2}{\sum_{j=1}^{n} a_{ij}|h_j(t)| + (\varepsilon_{i0} + |N_i|)|h_i(t)|} \tag{13}$$

Combining (11)–(13) yields

$$\dot{V}_1 \leq \sum_{i=1}^{n} \left(\sum_{j=1}^{n} a_{ij} k |h_j| + (\varepsilon_{i0} + |N_i|) k |h_i| \right) |e_i| - k_1 \sum_{i=1}^{n} h_i^2$$
$$\leq - \sum_{i=1}^{n} (1 - \beta_i) h_i^2 \tag{14}$$

Applying LaSalle's invariance principle, one can obtain the output voltage v_{odi} of (3) converges to the reference value v_{ref}. This completes the proof. □

Theorem 1 implies that the event-triggered secondary controller (4) can achieve the voltage control objective effectively. As mentioned above, the designed controller is able to reduce the redundant information transmission sharply as the information is transmitted only when needed. However, this may result in Zeno behaviour problem [15] since the proposed event-triggered scheme (7) is continuously monitoring. For this problem, we are seeking the lower bounds of the event time intervals $t_{k+1}^i - t_k^i$ for $k = 1, 2, \ldots$.

Theorem 2. *Consider the conditions as in Theorem 1, and the event time interval of the conditions (7) is lower bounded as*

$$t_{k+1}^i - t_k^i > \frac{\beta_i h_i^2(t_{k+1}^i)}{M_i \sum_{j=1}^{n} a_{ij} |h_j(t_{k+1}^i)| + (\varepsilon_{i0} + |N_i|) |h_i(t_{k+1}^i)|} \tag{15}$$

where M_i is a positive constant, therefore the Zeno behavior can be avoided.

Proof. Consider (6) and the triggering condition (7), we have

$$\frac{d}{dt} |e_i(t)| \leq |\frac{d}{dt} e_i(t)| = |\frac{d}{dt} (v_{odi}(t) - v_{odi}(t_k^i))| = |\frac{d}{dt} v_{odi}(t)| \leq k |h_i(t)| \tag{16}$$

for $t \in [t_k^i, t_{k+1}^i)$. Since the continuously differentiable functions $v_{odi}(t)$ converge to some constant values, the function $h_i(t)$ has an upper bound by some positive constants for each $i = 1, \ldots, N$. Thus, there are positive constants $M_i \geq k |h_i(t)|$ such that $\frac{d}{dt} |e_i| \leq M_i$, which implies $e_i(t) \leq M_i(t - t_k^i)$ for $t \in [t_k^i, t_{k+1}^i)$ by using $e_i(t_k^i) = 0$. For the trigging condition (7), the next triggering instant $t = t_{k+1}^i$ for the ith DG happens when $|e_i(t)| > \frac{\beta_i h_i^2(t)}{\sum_{j=1}^{n} a_{ij} |h_j(t)| + (\varepsilon_{i0} + |N_i|) |h_i(t)|}$, which indicates (15) holds. Two cases are analyzed here: (1) $h_i(t_{k+1}^i) \neq 0$. In this case, the inter-event time interval $t_{k+1}^i - t_k^i$ is strictly positive, which means the ith DG implements the $k + 1$th information transmission in a finite time about the output voltage magnitude $v_{odi}(t)$ after the kth one; (2) $h_i(t_{k+1}^i) = 0$. From (6), we obtain that $v_{odi}(t_{k+1}^i)$ achieves the equilibrium point. By a similar analysis in [13], the event is not necessarily be triggered at $t = t_{k+1}^i$. By the above analysis, we can ensure that the Zeno behavior does not happen for the closed-loop system (3). □

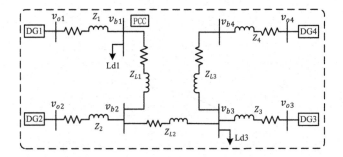

Fig. 2. The microgrid test system.

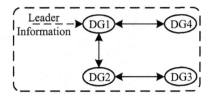

Fig. 3. The communication digraph.

4   Simulation Results

In order to verify the effectiveness of distributed secondary controllers with event-triggered mechanisms designed in Sects. 3, some simulations for the

Table 1. System parameters

System Quantities	Values	System Quantities	Values
Microgrid frequency	50 HZ	*DC voltage*	800 V
Leader information		*Line impedance*	
ω_{ref}	314 rad/s	Z_{L1}	0.23 Ω and 0.35e−3 H
v_{ref}	311 V	Z_{L2}	0.35 Ω and 1.847e−3 H
		Z_{L3}	0.23 Ω and 0.35e−3 H
Output connector		*Loads*	
$Z_1 = Z_2 = Z_3 = Z_4$	0.03 Ω and 0.35e−2 H	L_{d1}	20 kW and 12 kVar
		L_{d3}	15.3 kW and 7.6 kVar
Voltage controller		*Current controller*	
$K_{pv1} = K_{pv2}$	0.1	$K_{pc1} = K_{pc2}$	15
$K_{pv3} = K_{pv4}$	0.05	$K_{pc3} = K_{pc4}$	10.5
$K_{iv1} = K_{iv2}$	420	$K_{ic1} = K_{ic2}$	20000
$K_{iv3} = K_{iv4}$	390	$K_{ic3} = K_{ic4}$	16000
Droop coefficients		*Droop coefficients*	
$m_1 = m_2$	9.4e−5	$n_1 = n_2$	1.3e−3
$m_3 = m_4$	12.5e−5	$n_3 = n_4$	1.5e−3

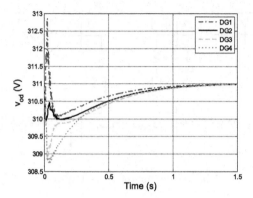

Fig. 4. Distributed generator voltages.

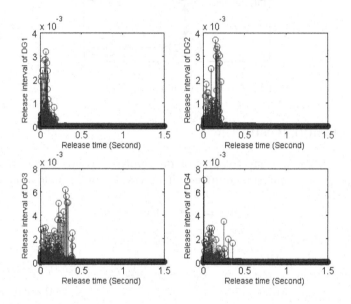

Fig. 5. Broadcast periods of the four generators with event-triggered mechanism.

microgrid test system of 380 V (per phase RMS), 50 Hz (314 rad/s) in have been implemented through MATLAB/Simulink. The test system contains four DGs from Fig. 2. The detailed parameters of the system are summarized in Table 1. And a fixed communication network in Fig. 3 is considered in this paper. Besides, we select DG1 as the leader in the system. In the following, the proposed voltage control scheme tunes the voltage amplitudes of the microgrid to their prescribed values.

The simulation results under the voltage controllers (4) with event-triggered mechanism (7) is given in this section. The parameters are set as $k = 20$, $\beta = [0.2; 0.5; 0.3; 0.4]$. It is assumed that the microgrid disconnects form the main grid at $t = 0$. Figure 4 shows the variations of output voltage amplitudes of all the four DGs with respect to time. The four generators are chosen to show their broadcast periods of d-axis voltage in Fig. 5, where the x-coordinates of crosses in this plot signify the triggering time, and the y-coordinates of them denote the elapsed time. The controller events under (7) for the third generator are triggered 14297, which shows that the redundant information is obviated for the control purpose.

5 Conclusion

The issue of microgrid secondary control has been investigated in this paper. To ease the communication pressure, the distributed event-triggered mechanism is adopted to design the secondary controller. By using the control strategy, the voltages and frequencies of distributed generators are synchronised to their nominal values, which has been demonstrated in the simulation results. Moreover, non-ideal signal transmission, such as time delay and packet loss, particularly in practical communication network will be considered in our future works.

Acknowledgments. This work is supported in part by the National Natural Science Foundation of China under Grant Nos. 61533010, 61374055 and 61503193.

References

1. Guerrero, J.M., Chandorkar, M., Lee, T.L., Loh, P.C.: Advanced control architectures for intelligent microgrids, part I: decentralized and hierarchical control. IEEE Trans. Ind. Electron. **60**(4), 1254–1262 (2013)
2. Yazdanian, M., Mehrizi-Sani, A.: Distributed control techniques in microgrids. IEEE Trans. Smart Grid **5**(6), 2901–2909 (2014)
3. Bidram, A., Lewis, F.L., Davoudi, A.: Distributed control systems for small-scale power networks: using multiagent cooperative control theory. IEEE Control Syst. **34**(6), 56–77 (2014)
4. Weng, S., Yue, D., Shi, J.: Distributed cooperative control for multiple photovoltaic generators in distribution power system under event-triggered mechanism. J. Frankl. Inst. **353**(14), 3407–3427 (2016)
5. Bidram, A., Davoudi, A., Lewis, F.L., Qu, Z.: Secondary control of microgrids based on distributed cooperative control of multi-agent systems. IET Gener. Transm. Distrib. **7**(8), 822–831 (2013)
6. Lu, X., Yu, X., Lai, J., Guerrero, J., Zhou, H.: Distributed secondary voltage and frequency control for islanded microgrids with uncertain communication links. IEEE Trans. Ind. Inform. (2016)
7. Peng, C., Zhang, J.: Delay-distribution-dependent load frequency control of power systems with probabilistic interval delays. IEEE Trans. Power Syst. **31**(4), 3309–3317 (2016)
8. Zhang, J., Peng, C., Fei, M.R., Tian, Y.C.: Output feedback control of networked systems with a stochastic communication protocol. J. Frankl. Inst. **354**(9), 3838–3853 (2017)
9. Wang, X., Lemmon, M.D.: Event-triggering in distributed networked control systems. IEEE Trans. Autom. Control **56**(3), 586–601 (2011)

10. Guo, G., Ding, L., Han, Q.L.: A distributed event-triggered transmission strategy for sampled-data consensus of multi-agent systems. Automatica **50**(5), 1489–1496 (2014)
11. Zhang, J., Peng, C.: Event-triggered H_∞ filtering for networked Takagi-Sugeno fuzzy systems with asynchronous constraints. IET Signal Process. **9**(5), 403–411 (2015)
12. Shafiee, Q., Guerrero, J.M., Vasquez, J.C.: Distributed secondary control for islanded microgrids novel approach. IEEE Trans. Power Electron. **29**(2), 1018–1031 (2014)
13. Li, H., Liao, X., Huang, T., Zhu, W.: Event-triggering sampling based leader-following consensus in second-order multi-agent systems. IEEE Trans. Autom. Control **60**(7), 1998–2003 (2015)
14. Bidram, A., Davoudi, A., Lewis, F.L., Guerrero, J.M.: Distributed cooperative secondary control of microgrids using feedback linearization. IEEE Trans. Power Syst. **28**(3), 3462–3470 (2013)
15. Dimarogonas, D.V., Frazzoli, E., Johansson, K.H.: Distributed event-triggered control for multi-agent systems. IEEE Trans. Autom. Control **57**(5), 1291–1297 (2012)

Frequent Deviation-Free Control
for Micro-Grid Operation Modes Switching
Based on Virtual Synchronous Generator

Yan Xu[1], Tengfei Zhang[1(✉)], and Dong Yue[2]

[1] College of Automation, Nanjing University of Posts and Telecommunications,
Nanjing 210023, China
xy265134@163.com, tfzhang@126.com
[2] Jiangsu Engineering Laboratory of Big Data Analysis and Control for Active
Distribution Network, Nanjing 210023, China

Abstract. The virtual synchronous generator (VSG), which overcomes the impact of the traditional inverter without the moment of inertia to the power grid, improves the stability of the power system and has received extensive attention in Micro-grid. However, since the VSG uses the traditional active power-frequency droop control, there is a frequency deviation in island mode, which will adversely affect the load in Micro-grid. A frequent deviation-free control strategy based on VSG is proposed, i.e., the frequency proportional-integral (PI) module feedback is used to replace the traditional damping module. It will eliminate the frequency deviation of Micro-grid in island mode and realize the Micro-grid inverter to work in multi-mode control. The simulation results show that the effectiveness of the presented VSG based frequent deviation-free control strategy.

Keywords: Micro-grid · Virtual synchronous generator · Smooth switching · Frequent deviation-free

1 Introduction

With the continuous consumption of fossil fuels as well as the severe environmental pollution, traditional power generations from fossil fuels may not be able to meet the requirements of sustainable development [1]. Thus the distributed generation (DG) is getting more and more attentions [2]. Generally, DG is connected to the distribution network through inverter. However, the inverter is static device with small inertia or damping [3], it is difficult to participate in grid regulation, thus it will not provide the corresponding voltage and frequency support to distribution network contained with DG. It is not possible to provide the necessary damping for Micro-grid with relatively poor stability [4, 5], which lacks a mechanism that can be "synchronized" to the distribution network and Micro-grid [3]. If there is a large number of DG based on power electronics inverter interface, it will affect the dynamic response and stability of the power system.

The rotor of synchronous generator has the mechanical inertia which can contain a large amount of kinetic energy. Based on the experience of traditional power system

© Springer Nature Singapore Pte Ltd. 2017
K. Li et al. (Eds.): LSMS/ICSEE 2017, Part III, CCIS 763, pp. 597–606, 2017.
DOI: 10.1007/978-981-10-6364-0_60

operation, the grid connected inverter has similar operating characteristics of synchronous generator, which can realize the friendly access of distributed generation and improve the stability of power system. At the same time, the relevant control strategies and theoretical analysis methods of traditional synchronous generators can also be effectively referenced [6]. Thus the domestic and foreign scholars proposed the VSG technology [7–9]. By using the VSG technology, the inverter obtains a similar inertia to the synchronous generator [10], which facilitates the smooth switching of Micro-grid between grid connected and island mode, thus it will play an important role in the future development of the Micro-grid.

In literature [11], the VSG technology is proposed to provide the inertia and damping for the system, and the mode switching of Micro-grid island mode/grid connected is realized by simulating the synchronous device of the synchronous generator, but the adaptability of VSG in grid connected is not considered. Literature [12, 13] proposed a droop control strategy of the grid connected inverter in island mode, where the deviation of voltage and frequency is introduced in the active and reactive power of grid connected inverter. It can make the grid connected inverter share the load power of the power grid in accordance to and its rated capacity and the deviation of frequency and voltage of Micro-grid. The mechanical and electrical equations of the synchronous generator are used to control the grid connected inverter, thus the grid connected inverter can be compared with the synchronous generator in both the mechanism and the external characteristics [14, 15]. It is suitable for the connection between energy storage device and the distribution network. The VSG technology can be used to realize smooth switching of Micro-grid between grid connected and island mode [8]. But in island mode, the frequency deviation problem exists in the algorithm of VSG. The angle droop is used to replace the frequency droop, which can eliminate the frequency deviation, but a common corner frequency reference is needed for the multi inverter [16].

The VSG in island mode uses active power-frequency droop control, which belongs to a differential control. It will cause a large frequency deviation of the system when the large area power has changed, so VSG still exist a problem of frequency deviation. For the problem, this paper studies and proposes a kind of active power-frequency control algorithm of frequent deviation-free. In the motion equation of rotor, the traditional damping module is replaced by the frequency PI module feedback, which eliminates the frequency deviation in island mode, and can also realize the multi-mode control of the inverter.

2 The Basic Principles of VSG

2.1 The Mathematical Model of VSG

2.1.1 Active Power-Frequency Control

The active power-frequency droop control of VSG is actually the governor of synchronous generator. The output of virtual mechanical torque control and adjust the corresponding frequency by detecting power deviation-free. The change of output power is controlled by the damping coefficient of VSG when the frequency is changed. The motion equation of rotor of VSG is:

$$\begin{cases} J\frac{d\omega}{dt} = T_m - T_e - D\Delta\omega = \frac{P_m}{\omega} - \frac{P_e}{\omega} - D(\omega - \omega_0) \\ \frac{d\varphi}{dt} = \omega - \omega_0 \end{cases} \qquad (1)$$

where J is virtual inertia, T_m and T_e are virtual mechanical torque and electromagnetic torque, D is constant damping coefficient, P_m is mechanical power, P_e is electromagnetic power, ω is rotor angular velocity, ω_0 is no-load rotor angular velocity, φ is power angle. Because of the existence of J, the grid connected inverter has inertia in the dynamic process of power and frequency. Due to the existence of D, the grid connected power generation device also has the ability to damp grid power oscillation.

The mechanical power equation of VSG is:

$$P_m = P_{ref} + K_\omega(\omega_0 - \omega) \qquad (2)$$

where P_{ref} is active power reference, K_ω is adjustment coefficient. According to the above analysis, the droop characteristics exist between the active power and the frequency, and the VSG has the characteristics of self-synchronization of the conventional generator.

By the analysis of the literature and control block diagram are as follows [17]:
Active-power-frequency control equation of inverter is:

$$\frac{1}{2Hs}\left[\frac{1}{D_p}\left(\omega_{ref} - \omega_g\right) + P_{ref} - P_{avg} - K_d\left(\omega_1 - \omega_g\right)\right] = \omega_1 \qquad (3)$$

where H is virtual inertial time constant, ω_{ref}, ω_g, ω_1 are angular velocity reference, angular velocity of common bus and inverter angular velocity respectively, P_{ref} and P_{avg} are active power reference and average active power after first order filtering, K_d is damping coefficient, D_p is frequency droop coefficient. Active power-frequency control block diagram of VSG is as Fig. 1.

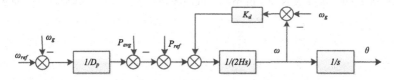

Fig. 1. The active power-frequency control block diagram of VSG.

2.1.2 Reactive Power-Voltage Control

The reactive power-voltage control of VSG is used to simulate the excitation regulation function of synchronous generator, which is used to realize the droop characteristic of reactive power and voltage amplitude. The reactive power-voltage control plays a different role in two different operation modes, which are grid connected and island mode. In grid connected mode, the control purpose is to transmit the specified reactive power to the power grid, while in island mode, the output reactive power is determined by the load, and its main purpose is to control the output voltage of the inverter.

In steady state, the inverter maintains reactive power-voltage droop control. The terminal voltage reference U_{ref} expression is:

$$U_{ref} = U_n + k_m(Q_{ref} - Q_{avg}) \tag{4}$$

where U_{ref} is terminal voltage amplitude reference, U_n is nominal voltage. k_m is reactive power-voltage droop coefficient, Q_{ref} is reactive power reference, Q_{avg} is reactive power output of VSG. Reactive power-voltage control system structure is as shown in Fig. 2, k_e is controller parameters.

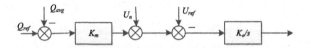

Fig. 2.   Reactive power-voltage control block diagram of VSG

2.2   Voltage and Current Double Loop Control

Double loop controller can not only improve the power quality of power grid, but also track the upper VSG controller timely and accurately, thus to simulate the characteristics of synchronous generator better. This paper selects the LC filter current which can real-time adjust the capacitance current as the inner ring current feedback, the output voltage is also got timely adjustment duo to the differential of capacitive current, it make the load capacity of inverter better. Voltage and current double loop control block diagram is as shown in Fig. 3 [18].

In Fig. 3, where U_{odq}^* is input voltage, i is inner ring input current, k is gain coefficient, k_p is transfer function of inverter, U_0 is inverter output voltage, C and L are filter capacitance and inductance respectively, i_C and i_L are capacitor and inductor current respectively, i_{abc} is bus current Micro-grid, U_{odq} is output control voltage.

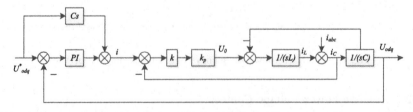

Fig. 3.   Voltage current double loop control block diagram

3   Frequent Deviation-Free Control of Micro-Grid in Island Mode

3.1   Frequent Deviation-Free Analysis

Literature [19] proposed that the rotor motion equation of VSG can be represented as follows when ignore damping component:

$$P_m - P_e = J\omega \frac{d\omega}{dt} \approx J\omega_0 \frac{d\omega}{dt} \tag{5}$$

The algorithm is by controlling the inverter to simulate the rotor motion equation synchronous generator, thus obtaining the similar frequency of inertia of synchronous generator.

Although the literature has pointed out that damping part has been neglected, in the actual situation, compared to Fig. 1, the damping part still exists, and when work in island mode, public bus frequency and the frequency of the inverter power supply are equal, i.e., $\omega_g = \omega$, the dynamic response of frequency is controlled by frequency modulation controller. Therefore, VSG in island mode adopted droop control algorithm can make the problem of frequency deviation.

3.2   Frequent Deviation-Free Control Strategy

Frequent deviation-free control strategy is proposed in this paper based on literature [19] to solve the problem of frequency deviation, where use PI module $(K_p + K_i/s)$ to replace damping module K_d in Fig. 1 to realize frequent deviation-free control.

When $\Delta\omega = \omega - \omega_0 \neq 0$: if $\Delta\omega > 0$, because of the output of negative feedback branch is big duo to integral instruments have a cumulative, the input of controller K_i/s will decrease and the output frequency ω will also decrease. When the steady state is reached, the inverter output frequency is equal to the given frequency $\omega = \omega_0$.

Frequent deviation-free active power-frequency control block diagram is shown in Fig. 4.

From Fig. 4, the transfer function between $\Delta\omega$ and ΔP is:

$$G(s) = \frac{\Delta\omega}{\Delta P} = \frac{K_\omega s}{s^2 + K_p K_\omega s + K_i K_\omega} \tag{6}$$

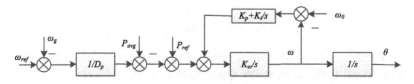

Fig. 4. Frequent deviation-free active power-frequency control block diagram

The initial value theorem of unit power step excitation:

$$\Delta\omega_0 = \lim_{s\to\infty}[sG(s)(1/s)] = 0 \qquad (7)$$

The final value theorem of unit power step excitation:

$$\Delta\omega_\infty = \lim_{s\to0}[sG(s)(1/s)] = 0 \qquad (8)$$

By formulas (6), (7) and (8), it is the second order transfer function relationship between $\Delta\omega$ and ΔP, and the unit step excitation power $\lim_{s\to\infty}\Delta\omega_0 = 0$, $\lim_{s\to0}\Delta\omega_0 = 0$, $\omega = \omega_0$.

Above all, frequent deviation-free control strategy can realize frequent deviation-free control of VSG in the steady state. Both in grid connected and island mode, the controller model is the same, which is conducive to multi-mode control.

4   The Frequent Deviation-Free Switching Control Strategy of Micro-Grid

4.1   Switching Control Strategy of Grid Connected/Island Mode

In the planned island mode or unplanned island mode, VSG can still maintain the potential and the phase of the virtual generator under the condition of grid connection when power grid is removed. There will not be obvious transient process while switching, and can successfully achieve grid connected/island mode smooth switch.

4.2   Switching Control Strategy of Island Mode/Grid Connected

When the VSG is running on island mode, due to the effect of adjustment of the voltage and frequency, there will be a certain deviation between voltage amplitude, frequency of Micro-grid and voltage amplitude, frequency of the power grid. If switching the Micro-grid to grid connected directly at this point, there will cause a large current shock and lead to a failure switching.

Taking C phase as an example, u_c is terminal voltage of VSG in island mode, u_{gc} is power grid voltage, U_v is voltage amplitude of Micro-grid in island mode, U_g is power grid voltage amplitude.

$$u_{gc} = U_g \sin(\omega_g t + \theta_g) \qquad (9)$$

$$u_{vc} = U_v \sin(\omega_v t + \theta_v) \qquad (10)$$

Usually, the difference of voltage amplitude are not big between Micro-grid with island mode and power grid, it can approximately be $U_g \approx U_v = U$. Thus from formulas (9) and (10), the instantaneous deviation between the two voltages is:

$$\Delta u = u_{gc} - u_{vc} = U_g \sin(\omega_g t + \theta_g) - U_v \sin(\omega_v t + \theta_v)$$
$$\approx 2U \sin\left[\frac{\omega_g - \omega_v}{2} + \frac{\theta_g - \theta_v}{2}\right] \cdot \cos\left[\frac{\omega_g + \omega_v}{2} + \frac{\theta_g + \theta_v}{2}\right] \qquad (11)$$

It can be seen that the frequency deviation, phase deviation and amplitude deviation exist between the Micro-grid voltage and the power grid voltage, and the instantaneous voltage deviation between the point of couple connection (PCC) can reach to $2U$. So it need to ensure that the inverter output voltage amplitude, phase and frequency are equal to voltage amplitude, phase and frequency of power grid before the inverter switching. Otherwise, the power grid will have a greater impact current and voltage, which will result in island mode/grid connected switching failure. The control algorithm proposed in this paper will not produce the frequency deviation at the steady state, thus it avoids the frequency adjustment process in the process of pre-synchronization.

5 Simulation Studies

The VSG has been simulated by using the proposed control strategy in MATLAB/ Simulink. The simulation results are presented to verify the effectiveness and feasibility of the proposed control strategy. The parameters are set as follows: Micro-grid side bus voltage is 60 V, power grid side bus voltage is 380 V, active power of load 1 is 100 kW, reactive power of load 1 is 10 kvar, active power of load 2 is 7 kW, reactive power of load is 1 kvar, active power load 3 is 2500 W, reactive power of load 3 is 1000 var; DC inverter source voltage is 120 V, rated active power is 5000 W, rated reactive power is 1500 var.

The simulation result of Fig. 5 shows: in the 0–0.45 s, the Micro-grid operates in island mode. In the 0–3 s, DC source use the frequent deviation-free control based on VSG, and in the 0.3–0.45 s, the pre-synchronization controller has received the grid connected signal and starts to work. Switch is closed at 0.45 s, the Micro-grid is incorporated into the power grid, and the DC source becomes PQ control. The Micro-grid is out of the power grid again at 0.7 s, and the DC source is still controlled by PQ. While detecting island mode signal at 0.75 s, the DC inverter is still switched into the frequent deviation-free control based on VSG.

The Fig. 6 shows that at 0.3 s, pre-synchronous controller starts to work, the active output of DC source almost has no change, at 0.45 s, the Micro-grid is incorporated into the power grid, there is a switch of control strategy between Micro-grid and power grid, but active output fluctuation is small. The Micro-grid is out of the power grid at 0.75 s, the active power output fluctuation is still small, smooth switching can be there. From Fig. 7, the reactive power output of the DC source is relatively small when the corresponding control strategy is switched, which can maintain the stable operation of the Micro-grid, and further verify the effectiveness of the control strategy.

Figure 8 shows that the frequency variation in the whole process is in the range of 1 Hz, which meet the frequency change indicators of small power Micro-grid. After the pre-synchronization controller worked, the Micro-grid side bus voltage can quickly track the power grid side bus voltage, and complete the grid connection condition.

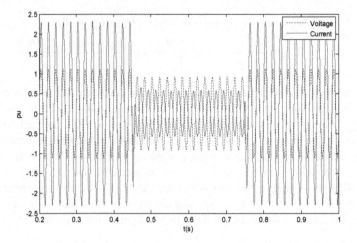

Fig. 5. Integrated control chart of Micro-grid smooth switching

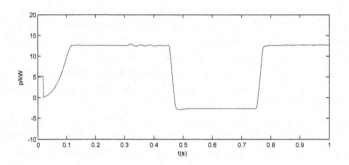

Fig. 6. Active output of Micro-grid

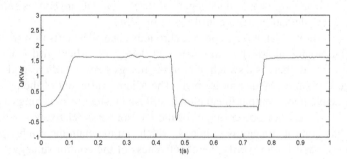

Fig. 7. Reactive power output of Micro-grid

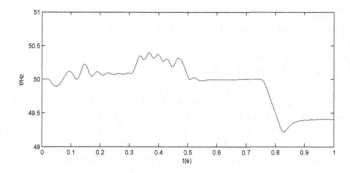

Fig. 8. Frequency variation of Micro-grid

6 Conclusion

Aiming at the problem of frequency deviation of active power-frequency control algorithm based on VSG in steady state, this paper proposed a frequent deviation-free active power-frequency droop control algorithm. Compared with the traditional VSG active power droop control algorithm, the proposed algorithm has two obvious advantages: (1) The PI module replace traditional damping module can solve the problem of frequency deviation. (2) Both grid connected mode and island mode are controlled by the same active power-frequency controller, which does not need to change the operating mode of the controller. It is helpful to realize multi-mode operation and smooth switching. The simulation results show that the Micro-grid has no large transient oscillation in island mode, grid-connected and mode switching, it realize multi-mode operation and smooth switching of the Micro-grid. The simulation results verify the effectiveness and feasibility and of the algorithm.

Acknowledgments. This research was supported by National Natural Science Foundation of China (61533010, 61105082), Qing Lan Project of Jiangsu Province (QL2016), '1311 Talent Plan' of Nanjing University of Posts and Telecommunications (NY2013), and Natural Science Foundation of NUPT (Grant No. 215149).

References

1. Yuan, Y., Li, Z., Feng, Y., et al.: Development purposes, orientations and prospects of micro grid in China. Autom. Electr. Power Syst. **34**(1), 59–63 (2010)
2. Lasseter, R.H.: Micro grids and distributed generation. J. Energy Eng. **133**(3), 144–149 (2007)
3. Zhipeng, L., Wanxing, S., Qingchang, Z., et al.: Virtual synchronous generator and its application in micro grid. Proc. CSEE **34**(16), 2591–2603 (2014)
4. Borges, C.L.T., Martins, V.F.: Multistage expansion planning for active distribution networks under demand and distributed generation uncertainties. Int. J. Electr. Power Energy Syst. **36**(1), 107–116 (2012)

5. Li, Y., Vilathgamuwa, D.M., Loh, P.C.: Design, analysis, and real-time testing of a controller for multibus micro-grid system. IEEE Trans. Power Electron. **19**(5), 1195–1204 (2004)

6. Tianwen, Z., Laijun, C., et al.: Technology and prospect of virtual synchronous generator. Autom. Electr. Power Syst. **21**, 026 (2015)

7. Xiang-zhen, Y., Jian-hui, S., Ming, D., et al.: Control strategy for virtual synchronous generator in micro-grid. In: 4th International Conference on Electric Utility Deregulation and Restructuring and Power Technologies (DRPT), pp. 1633–1637. IEEE (2011)

8. Zhong, Q.C., Weiss, G.: Synchronverters: inverters that mimic synchronous generators. IEEE Trans. Industr. Electron. **58**(4), 1259–1267 (2011)

9. D'Arco, S., Suul, J.A., Fosso, O.B.: Control system tuning and stability analysis of virtual synchronous machines. In: Energy Conversion Congress and Exposition (ECCE), pp. 2664–2671. IEEE (2013)

10. Dan, Z., Su, J., Wu, B.: Research on control method of micro-grid based on virtual synchronous generator. Autom. Electr. Power **32**(4), 59–62 (2010)

11. Shi, R., Zhang, X., Liu, F., et al.: A dynamic voltage transient suppression control strategy for micro grid inverter. In: 2014 International Power Electronics and Application Conference and Exposition (PEAC), pp. 205–209. IEEE (2014)

12. Chung, I.Y., Liu, W., Cartes, D.A., et al.: Control methods of inverter-interfaced distributed generators in a micro grid system. IEEE Trans. Industr. Appl. **46**(3), 1078–1088 (2010)

13. Mohamed, Y.A.R.I., El-Saadany, E.F.: Adaptive decentralized droop controller to preserve power sharing stability of paralleled inverters in distributed generation micro-grids. IEEE Trans. Power Electron. **23**(6), 2806–2816 (2008)

14. Chen, Y., Hesse, R., Turschner, D., et al.: Improving the grid power quality using virtual synchronous machines. In: 2011 International Conference on Power Engineering, Energy and Electrical Drives (POWERENG), pp. 1–6. IEEE (2011)

15. Loix, T.: Participation of inverter-connected distributed energy resources in grid voltage control. KU Leuven **6** (2011)

16. Majumder, R., Chaudhuri, B., Ghosh, A., et al.: Improvement of stability and load sharing in an autonomous micro grid using supplementary droop control loop. IEEE Trans. Power Syst. **25**(2), 796–808 (2010)

17. Meng, J., Wang, Y., Shi, X.C., et al.: Control strategy and parameter analysis of distributed inverter based on virtual synchronous generator. Proc. Electrotech. Soc. **29**(12), 1–10 (2014)

18. Zhang, T., Li, X.: Island/ containing the photovoltaic source network smooth switching control strategy. Power Syst. Technol. **39**(4), 904–910 (2015)

19. Shi, R., Zhang, X., Xu, H., et al.: Seamless switching control strategy of micro-grid based on virtual synchronous generator. Autom. Electr. Power Syst. **40**(10), 16–23 (2016)

A Novel Data Injection Cyber-Attack Against Dynamic State Estimation in Smart Grid

Rui Chen, Dajun Du$^{(\boxtimes)}$, and Minrui Fei

School of Mechatronical Engineering and Automation, Shanghai University,
Shanghai 200072, China
ddj559@163.com

Abstract. Dynamic state estimation is usually employed to provide real-time operation and effective supervision of smart grid (SG), which has been also found vulnerable to a typical data injection cyber-attack submerged into big data. The attacks against dynamic state estimation can purposely manipulate online measurements to mislead state estimates without posing any anomalies to the bad data detection (BDD). Aiming at Kalman filter estimation, a novel data injection cyber-attack is proposed in this paper. Unlike the previous injection attack perfectly escaping the BDD, an imperfect attack targeting state variables is firstly investigated, and these targeted state variables are then determined by a new search approach, i.e., a ε-feasible injection attack strategy. Simulation results confirm the feasibility of the proposed attack strategy.

Keywords: Smart grid · Kalman filter estimation · Bad data detection · Data injection cyber-attack

1 Introduction

The smart grid (SG), as a typical smart cyber-physical system (CPS), uses two-way flows of electricity and information to create a widely distributed automated energy delivery network [1,2]. Meanwhile, with a large number of sensors deployed in the grid, the state estimation based on the measurements and the grid model is usually employed to provide real-time and effective supervision for the operation of the whole system, which greatly improves the level of comprehensive automation and management. Since SG highly relies on wired/wireless network to achieve the data exchange and communication, it inevitably introduces cyber-attacks emerging from the network, such as denial of service (DoS) attacks, reply attacks, and false data injection attacks, etc., leading to potentially significant damages to power networks [3,4]. Therefore, the cyber security of SG is an important and open problem.

As traditional power systems exclusively operate in an isolated physical environment, their security mainly focuses on the random failures of the systems while cyber security issues are not taken into account [5]. The addition of wired/wireless network in SG results in increasing complexity and potentially

© Springer Nature Singapore Pte Ltd. 2017
K. Li et al. (Eds.): LSMS/ICSEE 2017, Part III, CCIS 763, pp. 607–615, 2017.
DOI: 10.1007/978-981-10-6364-0_61

more security threats, it is important to understand and defend such cyber-attacks. Although bad data detection (BDD) schemes based on the measurement data and state estimation have been used to identify and detect anomaly data, these BDD schemes may not solve cyber security issues and the state estimators have been found vulnerable to those cyber-attacks. Typically, the data injection attacks can purposely craft sparse measurements to perturb the results of state estimation while successfully evading the BDD [6].

Recently, the data injection attacks against state estimation in the grid have attracted great research interests. The data injection attack against static state estimation was first discussed in [6]. Based on the alternating direction method of multipliers, a target attack and a strategic attack were presented in [7]. The possibility of implementing the data injection attack was examined when the operator used the more practical dynamic state estimation [8]. Vulnerability assessment of nonlinear state estimation with respect to data injection cyber-attacks was discussed in [9].

The existing research work is mostly the data injection cyber-attacks based on a linear measurement model, while only a little work emphasizes such attacks against dynamic state estimation. In fact, the grid operator usually employs the more practical dynamic kalman filter estimation, and dynamic estimates based on online measurements often provide real-time and effective supervision for operation of the system, leading to an increasing need to investigate such attacks against kalman filter estimation. Therefore, how to develop a novel data injection attack against dynamic kalman estimation is the main motivation of this study. In view of this, the paper investigates one kind of the data injection attack against kalman filter estimation. The main contributions of the paper include: (1) The ε-feasible injection attack aiming at state variables is proposed, where the value ε can be designed by the attackers. (2) A new search approach is designed to determine these targeted state variables.

The rest of the paper is organized as follows. Section 2 presents the mathematical modeling and dynamic state estimation. The ε-feasible injection attack is analyzed in Sect. 3. Simulation results are given in Sect. 4, followed by the conclusion section in Sect. 5.

2   Mathematical Modeling and Dynamic State Estimation

2.1   Dynamic Model

Consider the non-linear dynamic characteristics of real power systems, the corresponding state and measurement equations are expressed as

$$x(k + 1) = f(x(k)) + w(k), \tag{1}$$

$$z(k + 1) = h(x(k + 1)) + v(k + 1), \tag{2}$$

where $x(k) = [V(k), \theta(k)] \in \Re^n$ is a state vector including bus voltage magnitudes and phase angles, and $z(k) = [P(k), Q(k), V(k), \theta(k)] \in \Re^m$ is a measurement vector including active and reactive powers, bus voltage magnitudes and phase angles; $f(x(k))$ and $h(x(k+1))$ are nonlinear equations of state variables; $w(k) \sim N(0,Q)$ is the system noise with zero mean and diagonal covariance matrix Q and $v(k+1) \sim N(0,R)$ is the measurement noise.

Since Kalman filter is carried out based on a linear model, according to Holt's two-parameter linear exponential smoothing method in [10], $f(x(k))$ in (1) can be expressed as

$$\tilde{x}(k+1/k) = S(k) + b(k), \tag{3}$$

$$S(k) = \alpha x(k) + (1-\alpha)\tilde{x}(k/k-1), \tag{4}$$

$$b(k) = \beta(S(k) - S(k-1)) + (1-\beta)b(k-1), \tag{5}$$

where α, β are preset constants, $x(k)$ is the real state at time instant k, and $\tilde{x}(k+1/k), \tilde{x}(k/k-1)$ are forecast states at time instant k and $k+1$, respectively.

According to (3)–(5), (1) can be rewritten as

$$x(k+1) = F(k)x(k) + G(k) + w(k), \tag{6}$$

where $F(k) = \alpha(1+\beta), G(k) = (1+\beta)(1-\alpha)\tilde{x}(k/k-1) - \beta S(k-1) + (1-\beta)b(k-1)$.

Furthermore, using the first-order Taylor expansion, the measurement Eq. (2) can be linearized as

$$\begin{aligned} z(k+1) &= h(x(k+1)) + v(k+1) \\ &= h(x_0) + H(k+1)(x(k+1) - x_0) + v(k+1) \\ &= H(k+1)x(k+1) + l(k+1) + v(k+1) \end{aligned} \tag{7}$$

where x_0 is the point of linearization, $H(k+1) = \partial h(x(k+1))/\partial x(k+1)|_{x(k+1)=x_0}$, $l(k+1) = h(x_0) - H(k+1)x_0$.

2.2 Kalman Filter Estimation

According to (6)–(7), the extended kalman filter can be used to obtain the state estimates $\hat{x}(k+1)$ at time instant $k+1$. Firstly, by performing the conditional expectation operation on (6), the forecasted state vector $\tilde{x}(k+1/k)$ with its forecasted error covariance matrix $P(k+1/k)$ are obtained as

$$\tilde{x}(k+1/k) = F(k)\hat{x}(k) + G(k), \tag{8}$$

$$P(k+1/k) = F(k)P(k)F^T(k) + Q. \tag{9}$$

Then, by incorporating the online measurements $z(k+1)$, the filtering estimates $\hat{x}(k+1)$ with its estimated error covariance matrix $P(k+1)$ are as follows

$$K(k+1) = P(k+1/k)H^T(k+1)\left[H(k+1)P(k+1/k)H^T(k+1)+R\right]^{-1} \tag{10}$$

$$\hat{x}(k+1) = \tilde{x}(k+1/k) + K(k+1)\left[z(k+1) - h(\tilde{x}(k+1/k))\right], \quad (11)$$

$$P(k+1) = \left[I - K(k+1)H(k+1)\right]P(k+1/k), \quad (12)$$

where $K(k+1)$ is the kalman gain, and the initial conditions are defined as $\hat{x}(0) = x(0), P(0) = P(0)$.

According to (10)–(12), the system states are estimated at each time instant with the online measurements to monitor the operation of the whole system in real time. However, with the wide application of information and communication technology in the SG, the online measurements obtained from the intelligent sensors are subject to cyber-attacks. These attacks can purposely manipulate online measurements to perturb the results of dynamic state estimation without being detected by the BDD (i.e., $\|z(k+1) - \hat{z}(k+1)\| \leqslant \tau, \tau$ is preset), which will inevitably bring serious security threats to the SG [11].

3 The ε-feasible Injection Attack Against Kalman Filter Estimation

According to the data injection attack against linear state estimation, when the system configuration information is obtained by the attackers, the system measurement matrix is also available. This, in turn, provides the information of which measurement is dependent on which state variable due to a linear relationship between the measurements and states. Consequently, the attackers can purposely design the strategic attack and the goal of the attack to contaminate the results of linear state estimation [6,7].

However, there exist some differences for the injection attack against dynamic state estimation. When the state variable is acted as the attacked target like linear state estimation, it would be found that the corresponding required measurement changes can not be directly determined since the associated state variable values are also supposed to be known by the attackers. Similarly, when the measurement is chosen to be corrupted, the attackers are also required to determine which state variable will be attacked and to know other related state variable values [8].

Therefore, with regards to the injection attack against dynamic state estimation, the attackers need to be equipped with more abilities. This is because the attacker needs to know the system state estimation process as well as the system configuration information to pass the BDD of the dynamic state estimator. According to the BDD mechanism, the norm of measurement residuals with the attack can be expressed as

$$\begin{aligned}
&\|z_a(k+1) - \hat{z}_a(k+1)\| \\
&= \|h(x(k+1)) + a(k+1) + v(k+1) - h(\hat{x}_a(k+1))\| \\
&\leqslant \|H(k+1)(x(k+1) - \hat{x}(k+1)) + v(k+1)\| \\
&\quad + \|a(k+1) + h(\hat{x}(k+1)) - h(\hat{x}_a(k+1))\| \\
&= \|z(k+1) - \hat{z}(k+1)\| + \|a(k+1) + h(\hat{x}(k+1)) - h(\hat{x}_a(k+1))\|. \quad (13)
\end{aligned}$$

To achieve a hidden attack against the nonlinear state estimation, a necessary condition can be derived as

$$\|a(k+1) + h(\hat{x}(k+1)) - h(\hat{x}_a(k+1))\| \leqslant \varepsilon, \tag{14}$$

where ε is a small positive value, $\varepsilon = 0$ indicates that such a data injection attack can perfectly evade the existing dynamic BDD. This is, the required change values in measurements (i.e., the attack vector), can be formulated as $a(k+1) = h(\hat{x}_a(k+1)) - h(\hat{x}(k+1))$ with $\hat{x}_a(k+1)$ being attacked estimates at time instant $k+1$.

From (11), the attacked filtering estimation can be written as

$$
\begin{aligned}
\hat{x}_a&(k+1) \\
&= \tilde{x}(k+1/k) + K(k+1)\left[z(k+1) + a(k+1) - h(\tilde{x}(k+1/k))\right] \\
&= \tilde{x}(k+1/k) + K(k+1)[z(k+1) - h(\tilde{x}(k+1))] + K(k+1)a(k+1) \\
&= \hat{x}(k+1) + c(k+1).
\end{aligned}
\tag{15}
$$

where $c(k+1) = K(k+1)a(k+1)$ is the required change values in the estimated state variables at time instant $k+1$.

To make the attack perfectly evade the dynamic BDD, the required changes $c(k+1)$ in (15) is further written as

$$c(k+1) = K(k+1)[h(\hat{x}(k+1) + c(k+1)) - h(\hat{x}(k+1))]. \tag{16}$$

Unfortunately, there exists no such $c(k+1)$ to satisfy (16), this is, it is infeasible to develop a perfect injection attack to make the norm of attacked residuals remain the same as the original value (i.e., $\|z_a(k+1) - \hat{z}_a(k+1)\| = \|z(k+1) - \hat{z}(k+1)\|$). Therefore, the ε-feasible injection attack is considered, i.e., $\|a(k+1) + h(\hat{x}(k+1)) - h(\hat{x}_a(k+1))\| \leqslant \varepsilon, 0 < \varepsilon < \tau$.

To achieve the goal of the attacker, the attackers do not intend to contaminate the system measurements directly, instead, one kind of the injection attack aiming at state variables is taken into account. If the targeted state variables relate to as few system measurements as possible, it will take less effort for the attacker to implement the ε-feasible injection attack. Thus, a new search approach is given here to determine these targeted state variables:

(1) Construct a connection matrix of the measurements and state information, where the index entry (i, j) represents the branch $i - j$ measurement information associated with the state variables of the node i and node j, and each element of the connection matrix is initialized as zero;
(2) Assign the weight (i.e., measurement times) for the element in the connection matrix according to the collected measurements, e.g., the element for the index entry (i, j) represents the number of the branch $i - j$ measurement and the element (i, i) represents the number of the node i measurement;
(3) Calculate the total number of the damaged measurements with attacking each state of the node i, i.e., calculate the sum of nonzero elements in the $i - th$ row and $i - th$ column of the connection matrix;
(4) Based on the results of procedure 3, select k nodes to attack, whose states associate with fewer measurements, where k is set by the attacker.

4 Numerical Simulations

A modified IEEE 14-bus system is employed to illustrate the feasibility of the ε-feasible injection attack, and the detailed IEEE 14-bus system can be referred to [12]. The system measurements ($m = 59$) include active and reactive bus injection powers, active and reactive branch powers and bus voltage magnitudes, and bus measurement noises follow $v_i \sim N(0, 0.02^2)$ and branch measurement noises follow $v_i \sim N(0, 0.01^2)$. State variables including bus voltage magnitudes and phase angles ($n = 27$) are estimated with reference bus phase angle $\theta_1 = 0$, and each system noise follows $w_i \sim N(0, 10^{-4})$. The initial condition is set as $\hat{x}(0) = x(0)$ from the power flow calculation, and the initial covariance is defined as $P(0) = 10^{-4} \times eye(n)$, where 'eye(n)' is the n-by-n identify matrix. The constants are set as $\alpha = 0.8$ and $\beta = 0.5$.

According to the proposed search method in Sect. 3, the number of the attacked measurements against each state variable and the corresponding attacked measurements are listed in Table 1. To save space, only the prior 5 nodes associated with fewer measurements are presented. The state variables θ_3, θ_8, V_3 and V_8 are chosen as the attacked targets from Table 1 when $k = 2$, then the required changes (i.e., $\hat{\theta}_3 \times 0.2, \hat{\theta}_8 \times 0.2, \hat{V}_3 \times 0.1$ and $V_8 \times 0.1$) are added to these targeted state variables, respectively. Firstly, ε is set as 0.2, and the online measurements are only attacked at certain time instant. The state estimation errors of θ_3, θ_8, V_3 and V_8 without and with the attack are shown in Fig. 1, and the norm of measurement residuals are shown in Fig. 2. Figure 1 shows that the proposed injection attack aiming at state variables can perturb the results of state estimators. As shown in Fig. 2, the deviation of the norm of measurement residuals without and with attack remains within the set threshold value ε. It confirms the ε-feasible injection attack is feasible because it can evade the BDD of dynamic state estimation.

Table 1. The number of attacked measurements against each state variable and the corresponding attacked measurements

The attacked nodes	8	3	11	12	10
The number of attacked meas	4	6	8	8	10
The attacked measurements	7, 7–8	4, 2–3, 3–4	10, 11, 6–11, 10–11	12, 13, 6–12, 12–13	9, 10, 11, 9–10, 10–11

Furthermore, when the value k is set as 5, i.e., the state variables of buses 3, 8, 10, 11, 12 are chosen as attacking targets in Table 1. The value ε is set as 2, and the online measurements are corrupted by the injection attack at different time instants, and other factors remain unchanged as above. The norm of measurement residuals without and with attack are shown in Fig. 3, different

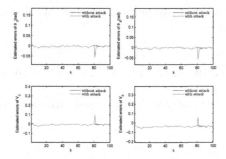

Fig. 1. State estimation errors of the targeted state variables without and with attack

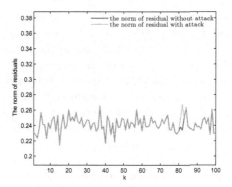

Fig. 2. The norm of measurement residuals without attack and with attack

attack time instants and the corresponding norm of measurement residuals are further presented in Table 2. From Fig. 3 and Table 2, it is obviously seen that the ε-feasible injection attack can evade the existing BDD when the injection attack occurs at different time instants. Simulation example further demonstrates that the ε-feasible injection attack is feasible.

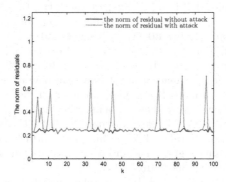

Fig. 3. The norm of measurement residuals without attack and with attack at different attack time instants

Table 2. Different attack time instants and the corresponding measurement residuals without and with attack

Time instants	4	6	11	33	45	70	83	96
Residuals without attack	0.2370	0.2413	0.2478	0.2419	0.2188	0.2405	0.2499	0.2497
Residuals with attack	0.5240	0.4288	0.5926	0.6666	0.6380	0.6648	0.7074	0.7057

5 Conclusion

Aiming at the problem that dynamic state estimation is vulnerable to the cyber-attack, the paper has proposed a novel data injection cyber-attack against dynamic state estimation. In this method, unlike the previous injection attack perfectly escaping the BDD, one kind of the imperfect attack targeting state variables is firstly investigated, and these targeted state variables are then determined by a new search approach, i.e., a ε-feasible injection attack strategy. Simulation results confirm the feasibility of the proposed attack strategy. To study the detection of the data injection attack against dynamic state estimation can be considered as the further research work.

Acknowledgments. This work was supported in part by the National Science Foundation of China under Grant Nos. 61633016 and 61533010, and project of Science and Technology Commission of Shanghai Municipality under Grants No. 15220710400 and 15JC1401900.

References

1. Fang, X., Misra, S., Xue, G., et al.: Smart grid – the new and improved power grid: a survey. IEEE Comun. Surv. Tut. **14**(4), 944–980 (2012)
2. Ciavarella, S., Joo, J.Y., Silvestri, S.: Managing contingencies in smart grids via the internet of things. IEEE Trans. Smart Grid **7**(4), 2134–2141 (2016)
3. De Persis, C., Tesi, P.: Input-to-state stabilizing control under Denial-of-service. IEEE Trans. Autom. Control **60**(11), 2930–2944 (2015)
4. Liang, G., Zhao, J., Luo, F., et al.: A review of false data injection attacks against modern power systems. IEEE Trans. Smart Grid **8**(4), 1630–1638 (2016)
5. Zhang, X., Chi, K.T.: Assessment of robustness of power systems from a network perspective. IEEE J. Emerg. Sel. Topics Circuits Syst. **5**(3), 456–464 (2015)
6. Liu, Y., Ning, P., Reiter, M.K.: False data injection attacks against state estimation in electric power grids. ACM Trans. Comput. Syst. **14**(1), 21–32 (2011)
7. Ozay, M., Esnaola, I., Vural, F.T.Y., et al.: Sparse attack construction and state estimation in the smart grid: centralized and distributed models. IEEE J. Sel. Areas Commun. **31**(7), 1306–1318 (2013)
8. Rahman, M.A., Mohsenian-Rad, H.: False data injection attacks against nonlinear state estimation in smart power grids. In: Power and Energy Society General Meeting, pp. 1–5 (2013)

9. Hug, G., Giampapa, J.A.: Vulnerability assessment of AC state estimation with respect to false data injection cyber-attacks. IEEE Trans. Smart Grid **3**(3), 1362–1370 (2012)

10. Su, C.L., Lu, C.N.: Interconnected network state estimation using randomly delayed measurements. IEEE Trans. Power Syst. **6**(4), 870–878 (2001)

11. Monticelli, A.: Electric power system state estimation. Proc. IEEE **88**(2), 456–464 (2015)

12. Zimmerman, R.D., Murillo-Sanchez, C.E., Gan, D.: MATPOWER: a MATLAB power system simulation package (2009). http://pserc.cornell.edu/matpower

A Novel Combination of Forecasting Model Based on ACCQPSO-LSSVM and Its Application

Nan Xiong[1], Minrui Fei[1(✉)], Sizhou Sun[1,2], and Taicheng Yang[3]

[1] School of Mechatronics Engineering and Automation, Shanghai University,
Shanghai 200072, China
mrfei@staff.shu.edu.cn
[2] School of Electrical Engineering, Anhui Polytechnic University,
Wuhu 241000, China
[3] Department of Engineering and Design, University of Sussex,
Brighton BN1 9QT, UK

Abstract. This paper proposed a novel combination of prediction model based on Adaptive Cauchy and Chaos Quantum-behaved Particle Swarm Optimization (ACCQPSO) and Least Squares Support Vector Machine (LSSVM) to forecast the short-term output power more accurately. To improve the performance of QPSO, chaotic sequences are used to initialize the origin particles, and particle premature convergence criterion, Cauchy and Chaos algorithm are employed, which can effectively increase the diversity of population and avoid the premature convergence. The kernel parameters of LSSVM are optimized by ACCQPSO to obtain hybrid forecasting model. To verify the proposed method, the seven days actual data recorded in a wind farm located in Anhui of China are utilized for application validation. The results show that the proposed combinational model achieves higher prediction accuracy.

Keywords: LSSVM · QPSO · Cauchy mutation · Chaotic · Wind power

1 Introduction

Support Vector Machine (SVM) algorithm has been successfully used to forecast time-series with satisfactory forecasting results [1,2]. SVM algorithm trained as a convex optimization problem has many advantages in solving small sample, non-linearity and pattern recognition. However, SVM algorithm has defects of rather complicated high-dimensional computation. In order to solve this shortcoming, the Least Squares SVM (LSSVM) was presented. Wind power energy is clean and renewable resource, and the installed capacity of global wind power

This work was supported in part by the Natural Science Foundation of China under Grant 61633016 and 61533010, in part by the Key Project of Science and Technology Commission of Shanghai Municipality under Grant No. 15220710400.

© Springer Nature Singapore Pte Ltd. 2017
K. Li et al. (Eds.): LSMS/ICSEE 2017, Part III, CCIS 763, pp. 616–625, 2017.
DOI: 10.1007/978-981-10-6364-0_62

becomes the fastest growing renewable electricity generation around the world [3]. In China, the cumulative installed wind capacity reached 91412.89 MW in 2013 while the new installed worldwide wind capacity was about 23370.56 MW. However, due to the stochastic nature of wind [2], the rapid integration of large capacities of wind power into the electricity grid could affect power systems security, stability, scheduling. Accurate wind power forecasting can deal with wind power intermittency effectively, which can improve the system operational security, stability and the sustainable development of the wind power industry [4,5]. The focus of this paper is to establish a hybrid model used in prediction the output power of wind farm, namely a novel combination of prediction model based on Adaptive Cauchy and Chaos Quantum behaved Particle Swarm Optimization and LSSVM (ACCQPSO-LSSVM), and also verify the proposed model by using wind power datasets.

In the existing work, there are difficulties in selection of hyper-parameters of LSSVM model, which have great impact on the generalization and regression performance of the LSSVM directly. With the development of artificial intelligence algorithm, many artificial intellectual methods are employed to optimize the regularization parameter γ and the kernel parameter δ of the LSSVM model [6]. A body of studies that establish a hybrid different intelligent algorithm and LSSVM model for wind power prediction obtains higher accuracy. A hybrid bat algorithm (BA) and LSSVM model was proposed to improve wind speed prediction accuracy and the prediction results validated the learning ability and generalization of LSSVM [1]. A hybrid LSSVM and gravitational search algorithm (GSA) model was conducted to forecast the short-term wind power production of a farm, and the prediction accuracy of the model is higher than the back propagation (BP) neural network, PSO-LSSVM and HS (harmony search)-LSSVM model [7]. Then, a hybrid Seasonal Auto-Regression Integrated Moving Average (SARIMA) and LSSVM model was developed to forecast wind speed, which obtained more powerful forecasting capacity for monthly wind speed prediction compared with ARIMA, SARIMA, LSSVM and ARIMA-SVM models [8]. A hybrid empirical wavelet transform (EWT) coupled simulated annealing (CSA) and LSSVM model was proposed to improve short-term wind speed forecasting accuracy, the EWT was used to extract true information from wind speed series, and the LSSVM parameters were optimized by CSA algorithm [9].

2   Principle of LSSVM and QPSO Algorithm

2.1   Least Squares Support Vector Machine

The training process of LSSVM is equivalent to dealing with a quadratic programming with linear and equality constraints, which is an improvement of the standard SVM and promotes the applications of SVM successfully in system modeling and pattern recognition [10]. The input samples x_i are mapped to feature space by nonlinear mapping function $\varphi(x_i)$ (such as kernel function) according to the principle of SVM, which converts regress estimation from nonlinear original space to linear feature space. The structure of the LSSVM can be

equivalent to that as shown in Fig. 1, it is observed from Fig. 1 that there are
input vectors, intermediate nodes and outputs; input vectors are processed by
kernel function $k(x_i, x)$ and combined linearly as output.

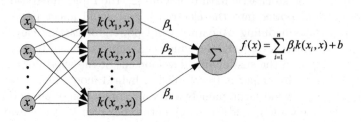

Fig. 1. Structure of the LSSVM

2.2   Particle Swarm Optimization Algorithm

Let the speed and the position of the ith particle in D-dimensional space are
expressed as $v_i = (v_{i1}, v_{i2}, \cdots, v_{iD})$ and $x_i = (x_{i1}, x_{i2}, \cdots, x_{iD})$, respectively.
The best previous position of the ith particle is recorded and expressed as
$(P_{i1}, P_{i2}, \cdots, P_{iD})$ or P_{ibest}, P_{gbest} represents the best global optimum solution
among all the particles. In each iteration, the particle location changes with its
velocity. The velocity v_i and the location x_i are updated towards its P_{ibest} and
P_{gbest} according to Eqs. (1) and (2).

$$v_{id}(t+1) = \omega v_{id}(t) + c_1 \gamma_1 [p_{idbest}(t) - x_{id}(t)] + c_2 \gamma_2 [P_{dgbest}(t) - x_{id}(t)], \quad (1)$$

$$x_{id}(t+1) = x_{id}(t) + v_{id}(t+1), \quad (2)$$

where $v_{id}(t+1)$ and $x_{id}(t+1)$ are the velocity vector and position vector of the
$(t+1)$th particle , respectively; P_{idbest} and P_{dgbest} represent personal best position of the ith particle and global best position of population until tth iteration. c_1 and c_2 are cognitive and social coefficients, respectively, which are
non-negative constants and not more than 2 generally; γ_1 and γ_2 are generated randomly in the interval $(0, 1)$; v_{id} must be in the interval $[-v_{max}, v_{max}]$,
v_{max} is a designated maximum velocity. ω represents the inertia coefficient
and is set to decrease linearly from 0.9 to 0.4, which is calculated as follow
$\omega = \omega_{\max} - \frac{\omega_{\max} - \omega_{\min}}{t_{\max}} t$, where $w_{max} = 0.9$, $w_{min} = 0.4$; t_{max} is the maximum
iteration number.

2.3   Quantum Behaved Particle Swarm Optimization Algorithm

Quantum-behaved particle swarm optimization(QPSO) is improved based on the
principle of PSO algorithm. QPSO algorithm adapts the quantum mechanics to
obtain global optimal solution which is superior to the PSO algorithm. The
particle renewing method is as follows

$$S(k) = (S_1(k), S_2(k), \cdots, S_m(k)) = \frac{1}{M} \sum_{i=1}^{M} P_{ibest}(k), \quad (3)$$

$$p_{ibest}(k+1) = u_{id}(k)P_{ibest}(k) + [1 - u_{id}(k)]P_{gbest}(k), \tag{4}$$

where $S(k)$ represents the mean of the p_{best} positions of all particles; α is contraction-expansion coefficient, u_{id} and ρ are generated randomly in the interval $(0, 1)$. Parameter α is known as contraction-expansion coefficient that is used to control the convergence speed of the QPSO algorithm, and it must be set in the interval $[0, 1.781]$ to guarantee convergence of QPSO. Generally, the value of the parameter α is fixing or it changes linearly in a certain range during the search process, the former is sensitive to maximum number of iterations and population size while the latter can surmount these problems effectively. QPSO algorithm can obtain good performance by linearly decreasing value of α as follow $\alpha = \alpha_{\min} + (\alpha_{\max} - \alpha_{\min})\frac{(k_{\max}-k)}{k_{\max}}$, where k and k_{max} are the current and maximum number of iteration, respectively; α_{min} and α_{max} are the initial and maximum values of α, respectively. In this paper, $\alpha_{min} = 0.5$ and $\alpha_{max} = 1$. P_{ibest} and P_{gbest} update according Eqs. (5) and (6) respectively.

$$P_{ibest}(k+1) = \begin{cases} x_m(k+1) \ f(x_m(k+1)) < f(P_{ibest}(k)), \\ P_{ibest}(k) \ f(x_m(k+1)) \geq f(P_{ibest}(k)), \end{cases} \tag{5}$$

$$P_{gbest}(k+1) = \arg\{\min_{1 \leq m \leq M}[f(P_{best}(k))]\}, \tag{6}$$

where $f(\cdot)$ represents fitness function.

3 Optimization Methodology

Chaos QPSO. Chaos is utilized instead of random to generate sequences which have dynamic characteristic and ergodic properties and obtain satisfactory results in many evolutionary algorithms. Logistic map proposed by Sir Robert May in 1976 has been validated good chaotic behavior. In this study, Logistic sequence is utilized to initialize the particles. Logistic map mathematical expression is defined as follows

$$\theta(k+1) = \mu\theta(k)[1 - \theta(k)], \ 0 < \theta(0) < 1, \ k = 0, 1, 2, \cdots, \tag{7}$$

where μ is the control parameter that reflects the chaotic state, Eq. 7 generates different sequences with different μ values, however, Logistic map sequence exhibits chaotic dynamics when $\mu = 4.0$; $\theta(k)$ is a variable and $\theta(1)$ must be in the interval $(0, 1)$ and $\theta(0) \neq 0.25, 0.5$ and 0.75.

Premature Criterion. When the global optimal position traps into a local optimal location, the current position of the particle will be attracted toward the point, which renders QPSO converging prematurity; the fitness of each particle updates with the position of particle and tends to be the same in that the particles are attracted towards the optima or sub-optima. The particles present aggregated distribution in the search space under the state of premature convergence that can be distinguished by $\chi^2 = \sum_{i=1}^{N}(\frac{f_i-f_{avg}}{f})^2$, where N is the population number, f_i and f_{avg} are the fitness of the ith particle and the average fitness of the current particle swarm respectively, f is normalized calibration factor and expressed as follow

$$f = \begin{cases} \max_{1<i<N} |f_i - f_{avg}|, \max |f_m - f_{avg}| > 1. \\ 1, others. \end{cases} \tag{8}$$

Cauchy QPSO. Although QPSO algorithm can solve the global optimization problems effectively, it also encounters premature convergence problems the same as PSO algorithm. If QPSO traps into premature convergence, the Cauchy mutation strategy is introduced into QPSO algorithm to disturb the mean best position $S(k)$; by comparison with the Gaussian distribution, the random number generated from Cauchy distribution has broader distribution scope. It's easy to know that the particle position of next iteration is affected by the distance between the particle's current position and the mean P_{ibset} position $S(k)$ directly. When the Cauchy mutation performs, the mean optimal position $S(k)$ will be pull away from its original position, which can increase population diversity and expand the search range of the particles in the search process. In this study, Cauchy mutation operator is adopted to make particles diverse in the larger search space and discard local optima, the Cauchy mutation operator is express as follow

$$\tau_k = (c_{\min} + (c_{\max} - c_{\min})\frac{k}{k_{\max}})C_k(0,1), \tag{9}$$

where $c_{max} = 2.0$ and $c_{min} = 0.05$, k and k_{max} are the current and and the maximum iteration number respectively, $C_k(0,1)$ is random number obeying Cauchy distribution. The mean optimal position $S(k)$ can be mutated as follow

$$\hat{S}(k) = S(k)(1 + \tau_k), \tag{10}$$

where $S(k)$ is obtained by Eq. (3), then the particle position is updated as Algorithm 1.

Algorithm 1

---

    if $\rho \geq 0.5$ **then**
        $x_{id}(k+1) = P_{ibest}(k) + \alpha|\hat{S}_{id}(k) - x_{id}|ln(\frac{1}{u_{id}})$
    else
        $x_{id}(k+1) = P_{ibest}(k) - \alpha|\hat{S}_{id}(k) - x_{id}|ln(\frac{1}{u_{id}})$
    end if

Implementation of ACCQPSO Algorithm

1. Initialize the QPSO parameter, such as the number of iteration, the dimension of particles, etc.;
2. Initialize the particle population to chaotic variables;

Algorithm 2

---
if $\rho \geq 0.5$ then
$$x_{id}(k+1) = p_{ibest}(k) + \alpha|\hat{S}_{id}(k) - x_{id}|ln(\frac{1}{\theta_{id}})$$
else
$$x_{id}(k+1) = p_{ibest}(k) - \alpha|\hat{S}_{id}(k) - x_{id}|ln(\frac{1}{\theta_{id}})$$
end if

---

3. Calculate the fitness value of each particle with the initial particle value, local optima P_{best} and P_{gbest};
4. Update the position of every particle;
5. Calculate the fitness value of each particle;
6. Confirm personal optima P_{best} and global optima P_{gbest} according to Eqs. (5) and (6), respectively; if current fitness value is better than previous P_{best}, replace the previous P_{best} with current particle position; if current P_{best} value is better than global optima P_{gbest} value then set current P_{best} value as the P_{gbest} value;
7. Calculate the aggregation χ^2;
8. If the aggregation χ^2 is less than a set value, it is considered that particles fall into state of premature convergence; if population drops in local optima, Cauchy mutation is introduced in the QPSO algorithm and the particle position updates according to Algorithm 3, else the particle position updates;
9. If termination criterions are not met, repeat step 5 to step 8, else output the P_{gbest} value.

Algorithm 3

---
if $\chi^2 \leq \nu$ then
  if $\rho \geq 0.5$ then
$$x_{id}(k+1) = p_{ibest}(k) + \alpha|\hat{S}_{id}(k) - x_{id}|ln(\frac{1}{\theta_{id}})$$
  else
$$x_{id}(k+1) = p_{ibest}(k) - \alpha|\hat{S}_{id}(k) - x_{id}|ln(\frac{1}{\theta_{id}})$$
  end if
else
  if $\rho \geq 0.5$ then
$$x_{id}(k+1) = p_{ibest}(k) + \alpha|S_{id}(k) - x_{id}|ln(\frac{1}{u_{id}})$$
  else
$$x_{id}(k+1) = p_{ibest}(k) - \alpha|S_{id}(k) - x_{id}|ln(\frac{1}{u_{id}})$$
  end if
end if

---

4   Application Studies

The proposed ACCQPSO-LSSVM model is established to predict wind power for a certain wind farm located in Anhui province of China. The historical data including wind power, temperature and humidity recorded in the wind farm is adopted as the input vector of LSSVM to forecast the future 24 h wind power.

4.1   Wind Farm Data Pretreatment

The total rated active power of the wind farm is 37.5 MW and data sampling interval time is 15 min. In order to meet the need of the prediction model, 4 wind power sampling data within one hour is averaged as one. 168 hourly (7 days) sampling data, including temperature, humidity and wind power, are utilized to predict wind power from 1 h to 24 h ahead.

The normalization of the original data can contribute to accelerate the training speed and reduce the prediction error of LSSVM model and the method is described as follow $y'(i) = \frac{y(i)-y_{min}}{y_{max}-y_{min}}$, where y_{max} and y_{min} are the maximum and minimum of the original data, respectively. The LSSVM prediction results $\hat{y}(i)$ are expressed as follow $\hat{y}(i) = \hat{y}'(i)[y_{max} - y_{min}] + y_{min}$, where $\hat{y}'(i)$ is the prediction of normalization value on the ith time point.

4.2   Choice of the Fitness Function and Evaluation Indexes

The mean absolute percentage error (MAPE) between the actual and the predicted output is selected as the fitness function to evaluate the performance of ACCQPSO-LSSVM prediction model. The fitness function is described as follow

$MAPE = fitness = \frac{1}{N} \sum_{i=1}^{N} \frac{|W_{Ai}-W_{Fi}|}{W_{Ai}} \times 100\%$, where W_{Ai} is the ith hour actual value of wind power; W_{Fi} is the ith hour forecasting value of wind power; N is the total number of wind power training sample.

Absolute percentage error (APE), root mean square error (RMSE), mean sum squared error (MSSE) and mean absolute error (MAE) are adopted as evaluation indexes to assess the prediction performance of the proposed ACCQPSO-LSSVM model quantitatively. These indexes are described as follow $APE = \frac{|W_{Ai}-W_{Fi}|}{W_{Ai}} \times$

100%, $RMSE = \sqrt{\frac{1}{N}\sum_{i=1}^{N}(\frac{W_{Ai}-W_{Fi}}{W_{Ai}})^2} \times 100\%$, $MSSE = \frac{1}{N}\sum_{i=1}^{N}(W_{Ai}-W_{Fi})^2$,

$MAE = \frac{1}{N}\sum_{i=1}^{N}|W_{Ai}-W_{Fi}|$. The APE and RMSE are unit-free measure prediction accuracy and reveal sensitivity to small changes in the wind power; MAE can reveal the similarity between the predicted and actual wind power; whereas the RMSE may indicate the overall deviation between the measured and forecasting wind power. These statistic indexes are used to evaluate the performance of LSSVM prediction model effectively although the wind speed has a character of the inherent uncertainty. So the above indexes are adopted to evaluate the proposed forecasting model in the same way as the existing literatures.

4.3   The Overall Implementation of the Proposed Model for the Application of Short-Term Wind Power Prediction

1. Normalize the training sample data in the interval [0, 1] and set up initial input parameter set of LSSVM model;

2. Compare different kernel function and select the type of kernel function to build LSSVM prediction model;
3. Initialize the ACCQPSO and establish chaotic particle;
4. Update particle position, calculate particle fitness with new particle position and find out the local optimal solution and global optimal solution, if the current fitness is better than the previous one, replace previous P_{best} with current particle position; if current P_{best} is superior to the global optima P_{gbest}, then replace previous P_{gbest} with current P_{best};
5. Calculate particle fitness using MAPE;
6. Calculate the aggregation χ^2;
7. If the aggregation χ^2 is less than a set value, it is said that particles trap in state of premature convergence; if population drops in local optima, Cauchy and Chaos algorithm is introduced in the QPSO algorithm and the particle position updates according to Algorithm 3, else the particle position updates;
8. Calculate particle fitness with new particle position and confirm the P_{best} and P_{gbest} according to Eqs. (5) and (6), respectively;
9. If termination criterions are not met, repeat step5 to step8, else output optimal parameter combination including the regularization parameter γ and the kernel parameter δ;
10. Establish LSSVM prediction model with optimal parameters combination;
11. Evaluate the predictable results of the LSSVM with the error indexes.

4.4 Comparison of the Validation Effect of Wind Power with Other Existing Models

Figure 2 displays prediction curves generated by the different hybrid models according to the corresponding parameter combination. From the Fig. 2, it can be seen that the prediction curve generated by the ACCQPSO- LSSVM model resembles the real wind power curve mostly. Figure 3 shows the absolute percentage errors of every hourly point wind power. Most of the APE value is not more than 10% while all the index APE value of the ACCQPSO-LSSVM prediction model is below 6%. The APE value in the interval [10h, 19h] is bigger since the output power of the wind farm fluctuates up and down as shown in the Fig. 3.

From Table 1, it can be seen that there are 12 predicted points of which the APE value of the ACCQPSO-LSSVM model is less than 1%, which takes 50% of the total amount; 17 predicted points less than 2%, which takes 70.83%; 19 predicted points less than 3%, which takes 79.16%; only 2 predicted points more than 5% but less than 6%.

The parameter combination and the corresponding model indexes MAPE, RMSE, MSSE and MAE are shown in Table 2. From Table 2, it can be seen that: (a) all proposed hybrid prediction models can forecast the wind power effectively; (b) Comparing ACCQPSO-LSSVM model with GA-LSSVM model, PSO-LSSVM model and QPSO-LSSVM model, the proposed hybrid model has improved the indexes MAPE, RMSE, MSSE and MAE of the latter ones obviously. The MAPE promoting percentages of ACCQPSO-LSSVM prediction model are 3.12%, 2.13% and 0.83%, respectively. The RMSE promoting percentages of ACCQPSO-LSSVM prediction model are 3.51%, 2.41% and

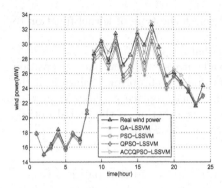
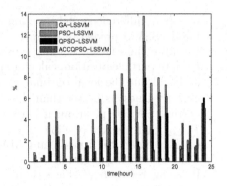

Fig. 2. Prediction curves of models optimized by different hybrid models.

Fig. 3. APE of different hybrid models.

1.03%, respectively. The MSSE promoting errors of ACCQPSO-LSSVM prediction model are 2.2694 MW, 1.3665 MW and 0.4886 MW, respectively. The MAE promoting errors of ACCQPSO-LSSVM prediction model are 0.8564 MW, 0.5955 MW and 0.2522 MW, respectively.

Table 1. APE distribution number of different models.

Hybrid model	<1%	1–2%	2–3%	3–4%	4–5%	>5%
GA-LSSVM	3	3	4	10	2	2
PSO-LSSVM	2	5	7	6	3	1
QPSO-LSSVM	8	5	6	2	2	1
ACCQPSO-LSSVM	12	5	2	3	0	1

Table 2. Performance indexes optimized by four algorithms.

Hybrid model	MAPE (%)	RMSE (%)	MSSE (MW)	MAE (MW)
GA-LSSVM	4.65	5.66	2.5749	1.2439
PSO-LSSVM	3.66	4.56	1.6720	0.9830
QPSO-LSSVM	2.36	3.18	0.7941	0.6397
ACCQPSO-LSSVM	1.53	2.15	0.3055	0.3875

Hence, it is obvious that the parameter combination optimized by ACCQPSO algorithm is of better choice to construct LSSVM wind power prediction model than the ones by GA, PSO and QPSO algorithms. The indexes MAPE, RMSE, MSSE and MAE provided by ACCQPSO-LSSVM are better than those of GA-LSSVM, PSO-LSSVM and QPSO-LSSVM, because ACCQPSO adds chaotic sequence initialization of particles and Cauchy mutation into the basis of QPSO.

5 Conclusion

This paper utilized hybrid ACCQPSO and LSSVM model to establish the wind power forecasting model. In the modeling process, we put forward a forecasting model by LSSVM to approach the nonlinear feature of wind power, and utilize ACCQPSO to train the kernel function parameter and the regularization parameter of LSSVM. The quantitative analyses of the forecasting validation showed that the proposed model had higher accuracy, and the global search ability of the combined optimal model is better than single optimal model. The results of the paper can clearly be of much help to increase the economic benefits of new energy into the prospective smart grids.

References

1. Wu, Q., Peng, C.Y.: Wind power grid connected capacity prediction using LSSVM optimized by the Bat algorithm. Energies 8(12), 14346–14360 (2015)
2. Foley, A.M., Leahy, P.G., Marvuglia, A., et al.: Current methods and advances in forecasting of wind power generation. Renew. Energy 37(1), 1–8 (2012)
3. Montoya, F.G., Manzano-Agugliaro, F., Lopez-Marquez, S., et al.: Wind turbine selection for wind farm layout using multi-objective evolutionary algorithms. Expert Syst. Appl. 41(15), 6585–6595 (2014)
4. Chitsaz, H., Amjady, N., Zareipour, H.: Wind power forecast using wavelet neural network trained by improved clonal selection algorithm. Energy Convers. Manag. 89, 588–598 (2015)
5. Jiang, P., Qin, S.S., Wu, J., et al.: Time series analysis and forecasting for wind speeds using support vector regression coupled with artificial intelligent algorithms. Math. Probl. Eng. 2015, 1–14 (2015)
6. Amjady, N., Keynia, F., Zareipour, H.: Wind power prediction by a new forecast engine composed of modified hybrid neural network and enhanced particle swarm optimization. IEEE Trans. Sustain. Energy 2(3), 265–276 (2011)
7. Yuan, X., Chen, C., Yuan, Y.B., et al.: Short-term wind power prediction based on LSSVM-GSA model. Energy Convers. Manag. 101, 393–401 (2015)
8. Guo, Z.H., Zhao, J., Zhang, W.Y., et al.: A corrected hybrid approach for wind speed prediction in Hexi Corridor of China. Energy 36(3), 1668–1679 (2011)
9. Hu, J.M., Wang, J.Z., Ma, K.L.: A hybrid technique for short-term wind speed prediction. Energy 81, 563–574 (2015)
10. Li, G., Shi, J.: On comparing three artificial neural networks for wind speed forecasting. Appl. Energy 87(7), 2313–2320 (2010)

Research on Power Terminal Access Control Technology Supporting Internet Interactive Service in Smart Grid

Song Deng$^{(\boxtimes)}$, Liping Zhang, and Dong Yue

Institute of Advanced Technology,
Nanjing University of Posts and Telecommunications, Nanjing 210003, China
ds16090311@163.com, 295345439@qq.com,
medongy@vip.163.com

Abstract. With the continuous development of smart grid interactive business applications, the existing terminal access mechanism is difficult to ensure that all types of power terminals access to power information network security. Based on the idea of trusted computing and trusted network connection, this paper proposes a power terminal security access architecture for power interactive business from terminal security access and access control, terminal encryption and content filtering, and establishes architecture of supporting for interaction business from Internet and terminal security access, enhances the external network interactive services and terminal access authentication and access control capabilities in order to improve the security of terminal access to protect the strength of the Internet to ensure interactive services and terminal access to the trustworthy.

Keywords: Interactive business · Power terminal · Access control · Smart grid

1 Introduction

In recent years, the related research of information network terminal, channel and border protection has made a series of results for smart grid. The security access system for internal power application is gradually formed. Two-way authentication and data encryption technology have been in the information network security protection equipment has been applied. As the power business is highly professional, the key goal of network access protection is to ensure that the terminal identity cannot be forged, the power control command cannot be tampered with, The core of the relevant research is also focused on solving the "mutual trust" problem between terminals and services, while the research on the different access control needs of different terminals is not much.

At present, the authentication and access control technologies used in different terminal accesses of different application systems are different, which is not conducive to the unified management of terminal identity and access rights. This improves the difficulty of collecting and processing information security data, so that the state of Internet interactive security cannot be analyzed and displayed in a unified form. Some of the business system to use the security technology strength and its own level of

© Springer Nature Singapore Pte Ltd. 2017
K. Li et al. (Eds.): LSMS/ICSEE 2017, Part III, CCIS 763, pp. 626–632, 2017.
DOI: 10.1007/978-981-10-6364-0_63

security requirements do not match, and the use of security technology is external general technology, autonomous control is not strong, and specific business needs cannot meet the situation. Due to the lack of security awareness or ease of development and other reasons, in the practical application of security technology, business system developers use a fixed key, fully open access control strategy, etc., has brought great security risks. Therefore, the Internet-oriented business system urgently needs to establish a self-controllable unified encryption, authentication and access control strategy to protect the security and trustworthy of the Internet interactive business and terminal access, and to meet the national grid company information network and the Internet trusted transmission requirements.

To solve these problems, the contributions of this paper are as follows:

(1) In order to solve the problem of access authentication and access control of power interactive service terminal, this paper constructs the terminal access system for power interactive business from terminal trusted access authentication, terminal encryption, access control and content filtering.
(2) The key technologies such as terminal trusted access authentication, terminal encryption, access control and content filtering are described in detail, and the key points and difficulties are analyzed.
(3) On the basis of the above, it describes the technical route for access authentication and access control of power interconnection service terminal.

The remainder of this paper is organized as follows. In Sect. 2, we briefly describe related works. In Sect. 3, we introduce a terminal trusted access architecture for power interactive business. In Sect. 4 we propose key technologies. And we conclude this paper in Sect. 5.

2 Related Work

Lamport [1] first proposed a password-based remote user authentication protocol in 1981 that allows users to perform authentication operations on a non-secure channel [2–4] with a password to achieve secure communication. In 2000, Hwang [5] pointed out that Lamport's agreement provides a security risk of centralized storage of the password table, that is, once the attacker once broke the server database, you can tamper with the user password, and completely block the legitimate user's communication services. To compensate for the security vulnerabilities of the Lamport protocol, Elgamal [6] designed and proposed a password authentication protocol based on the ELGamal public key cryptosystem. Sun [7] pointed out that Hwang and Li's protocol has a performance disadvantage. The protocol uses modulo arithmetic to create a large computational overhead and affect the computational efficiency of the system. But the two protocols cannot achieve mutual authentication between the user and the remote server, and cannot provide password replacement operation. In 2002, Chien et al. [8] proposed a smart card password authentication protocol that would enable mutual authentication. However, the agreement cannot resist internal attacks, guess attacks and reflection attacks. Ku and Chen [9] proposed a corresponding improvement agreement in 2004. However, Yoon et al. [10] found that the Ku–Chen protocol had a parallel session attack problem and could not provide a password replacement operation, thereby proposing further improvements. In 2007, Wang et al. [11] analyzed that the agreement proposed by Yoon and Ku-Chen et al. was susceptible to guessing attacks, counterfeiting attacks and denial of service attacks, and proposed a smart card password authentication protocol in a resource-constrained environment. In 2010, Chen et al. [12] pointed out that the Wang's protocol cannot resist counterfeiting attacks and parallel session attacks. To compensate for the above security vulnerabilities, a smart card-based remote user authentication protocol, the CHS protocol, is proposed. However, the CHS protocol cannot resist known outburst attacks and off-line guessing attacks. And in the same year, Song [13] proposed an enhanced smart card password authentication protocol, the Song protocol, and claimed that the protocol was resistant to multiple attacks, and the performance advantage was significant. However, the Song protocol has an offline password guessing attack problem, and does not have the system's self-recovery.

In the traditional authentication technology, the network authentication device only judges whether the access device is allowed to access the protected network only if the terminal has a password or a key, but ignores whether the access terminal itself is safe and reliable. If the access terminal has been attacked, the attacker is implanted with a virus or Trojan horse program, then the attacker can not only effortlessly monitor the authentication process, steal the user critical information, and can access the protected network to steal important information, or use the attack equipment to be attack as a springboard, attack the protected network, and cause serious security problems. The existing authentication method ignores the security status of the devices to be accessed, and prevents the network attacks in a passive way. With the continuous expansion of the scope of the Internet, hackers also expand the activities of space, attack means will be endless, the traditional certification technology is powerless.

Therefore, in the certification to introduce a trusted access system, only meet the requirements of the terminal was allowed to access the network, and access to the terminal access and control so that the behavior and process of the terminal are predictable. Trusted authentication technology can fundamentally solve the terminal due to most of the security risks to solve the traditional authentication technology is difficult to solve the security problem.

3 Terminal Trusted Access Architecture for Power Interactive Business

With the gradual deepening of power information and the construction of strong smart grid, the entire power information system will continue to be complicated, whether it is network architecture, application system scale, or coverage area, terminal type and scale have experienced rapid growth. Based on the idea of trusted computing and trusted network connection, this paper proposes a terminal access architecture for power interactive business, as shown in Fig. 1. A network access to the whole trusted, controllable active defense system is built from trusted access terminal identity, trusted access terminal security status, controllable access behavior, trusted network access channel, controllable network access content, comprehensive detection and audit and integrated security management. The system gives full play to the strategy-based, interactive coordination as the goal of the overall protection of thinking out of the traditional rely on a single security technology and product access control deficiencies, to protect the power business data carrying all kinds of terminals and network access full credible and controllable. The system not only gets rid of deficiencies which the traditional rely on a single security technology and product, but also protects the various types of terminals and network access credible and controllable.

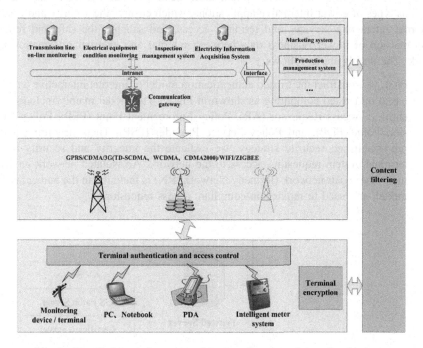

Fig. 1. Terminal trusted access architecture for power interactive business

(1) Terminal authentication and access control

This paper constructs the trustworthy authentication model of Internet interactive terminal based on trusted computing and the Internet access control and authorization model based on RBAC, and establishes the role-based security access control and management mechanism in Internet access environment.

(2) Terminal encryption

In view of the lack of SSL security transmission protocol, the key technologies such as terminal encryption transmission protocol, algorithm and dynamic key negotiation are focused on.

(3) Content filtering

This paper implements data filtering based on label and pattern matching.

4 Key Technologies

With the continuous development of power interactive business, all kinds of power access terminal certification and access control for the power of interactive business development is essential. This paper focuses on the construction of secure access architecture that supports Internet interactive services and terminals from the aspects of terminal authentication and access control, terminal encryption and content filtering. The system provides a unified encryption, authentication and access control strategy for various types of power interactive applications, which ensures the security and trust of Internet interactive services and terminal access, and satisfies the safe and reliable transmission of power business data between internal and external information network. The details are as follows.

(1) First, we construct the trusted authentication model of Internet interactive terminal based on trusted computing as shown in Fig. 2. The model mainly includes four entities: Access Requestor (AR), Policy Enforcement Point (PEP), Policy Decision Point (PDP), and Policy Service Provider (PSP). Then, according to the pre-established security strategy, we evaluate the integrity and security of the terminal system requesting access to the network. An isolated network environment, the repair network (Remedy Network, RN) is included in the authentication model, and used to repair non-compliant access requestor.

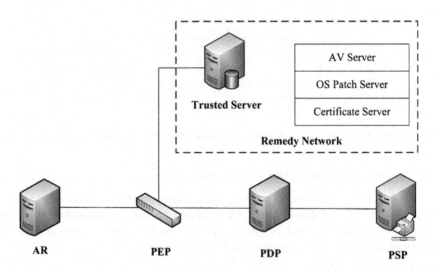

Fig. 2. Trusted authentication model of Internet interactive terminal based on trusted computing

(2) On the basis of the above model, this paper realizes the access authentication of the terminal based on the trusted network connection, and achieves the SSL terminal encryption and transmission protocol based on the SM1 and SM2 algorithms to implement the gateway to the gateway protocol.

(3) In order to better control the terminal, the RBAC model is introduced. By connecting the user and the authority, the access to the system resources is controlled by the authorization of the role, so that the access control management becomes simple and flexible.

(4) Finally, in order to correctly and quickly analyze the multi-type data of power business and ensure the controllability and stability of the data exchange between the internal and external networks of power information, this paper proposes a content for multi-type power business data Depth filtration method. Steps are as follows:

> Step 1: Analyzing user behavior data, and determining user safety behavior;
> Step 2: Designing security tags, and encapsulating user behavior data and multiple types of data based on label;
> Step 3: Designing a policy expression, and configuring user behavior and multiple types of data;
> Step 4: Designing matching algorithm, and achieving multi-type data content filtering based on the policy expression.

5 Conclusions

With the continuous development of the power interactive business, Information external network applications in the frequency of information interaction and interactive content greatly exceeded the traditional power business applications. The existing isolation technologies have been unable to effectively support the safety of interactive business development. Based on the existing power information security protection, This paper focuses on the terminal secure access authentication technology from the aspects of terminal trusted access, terminal encryption and access control, and deep content filtering, and elaborated its key technology and implementation ideas for the smart grid interactive business security and stability to provide a strong technical support.

Acknowledgments. This work was supported in part by the National Natural Science Foundation of China under Grant 51507084, in part by NUPTSF under Grant NY214203.

References

1. Lamport, L.: Password authentication with insecure communication. Commun. ACM **24**, 770–772 (1981)
2. Chang, C.-C., Lee, C.-Y., Chiu, Y.-C.: Enhanced authentication scheme with anonymity for roaming service in global mobility networks. Comput. Commun. **32**, 611–618 (2009)
3. Peyravian, M., Zunic, N.: Methods for protecting password transmission. Comput. Secur. **19**, 466–469 (2000)
4. Wu, T.C., Hung-Sung, S.: Authenticating passwords over an insecure channel. Comput. Secur. **15**, 431–439 (1996)

5. Hwang, M.S., Li, L.H.: A new remote user authentication scheme using smart cards. IEEE Trans. Consum. Electron. **46**, 28–30 (2000)
6. Elgamal, T.: A public key cryptosystem and a signature scheme based on discrete logarithms. IEEE Trans. Inf. Theory **31**, 469–472 (1985)
7. Sun, H.M.: An efficient remote use authentication scheme using smart cards. IEEE Trans. Consum. Electron. **46**, 958–961 (2000)
8. Chien, H.Y., Jan, J.K., Tseng, Y.M.: An efficient and practical solution to remote authentication: smart card. Comput. Secur. **21**, 372–375 (2002)
9. Ku, W.C., Chen, S.M.: Weaknesses and improvements of an efficient password based remote user authentication scheme using smart cards. IEEE Trans. Consum. Electron. **50**, 204–207 (2004)
10. Yoon, E.J., Ryu, E.K., Yoo, K.Y.: Further improvement of an efficient password based remote user authentication scheme using smart cards. IEEE Trans. Consum. Electron. **50**, 612–614 (2004)
11. Wang, X.M., Zhang, W.F., Zhang, J.S., Khan, M.K.: Cryptanalysis and improvement on two efficient remote user authentication scheme using smart cards. Comput. Stand. Interfaces **29**, 507–512 (2007)
12. Chen, T.H., Hsiang, H.C., Shih, W.K.: Security enhancement on an improvement on two remote user authentication schemes using smart cards. Fut. Gener. Comput. Syst. **27**, 377–380 (2011)
13. Song, R.: Advanced smart card based password authentication protocol. Comput. Stand. Interfaces **32**, 321–325 (2010)

Research on Model and Method of Maturity Evaluation of Smart Grid Industry

Yue He[1], Junyong Wu[1], Yi Ge[2(✉)], Dezhi Li[3], and Huaguang Yan[3]

[1] School of Electrical Engineering, Beijing Jiao Tong University,
Haidian District, Beijing 100044, China
1131931300@qq.com, wujy@bjtu.edu.cn
[2] State Grid Jiangsu Economic Research Institute, 251 Zhongshan Road,
Nanjing 210000, China
geyi0820@163.com
[3] China Electric Power Research Institute, 15 Xiaoying East Road,
Haidian District, Beijing 100192, China
ldz97@126.com, hgyan@epri.sgcc.com.cn

Abstract. Smart grid has become the inevitable development trend of the modern power grid. The vigorous development of the smart grid led to the rise and development of the smart grid industry, seize the smart grid industry development opportunities, also has become one of the important choice of regional planning and construction. Scientifically reflecting the effect of regional development of smart grid industry of the conditions and the development of the industries to the region will guide regional smart grid industry planning, and encourage regional investment in the development of smart grid industry. This paper established smart grid industry maturity comprehensive evaluation index system from five aspects, the technical performance, industrial facilities, market environment, policy environment and social influence, to put forward to smart grid industry maturity evaluation algorithm, thus smart grid industry maturity assessment model is established, in order to provide reference for regional planning and smart grid industry.

Keywords: Smart grid · Maturity modeling · Fuzzy comprehensive evaluation · Maturity assessment

1 Introduction

The contradiction between the development of social economy and the protection of ecological environment is increasingly intensified, so the conversion of clean energy into electricity and the enlargement of proportion of electricity consumption in energy end-user become one of the main means to tackle the above-mentioned contradiction. Smart grid has realized a highly-informative, automatic and interactive power grid, and also become an inevitable trend in the development of modern power grids. Therefore, seizing the opportunities of smart grid industry development has become an important choice for regional plan and construction. In the context of the current economic and social development, it forms a positive significance to guide planning of regional smart grid industry and to encourage regional investment in the development of smart grid

K. Li et al. (Eds.): LSMS/ICSEE 2017, Part III, CCIS 763, pp. 633–642, 2017.
DOI: 10.1007/978-981-10-6364-0_64

industry by scientific reflection of various aspects of regional development of smart grid industry and impact of industrial development on the region as well.

The existing comprehensive evaluation of the smart grid includes the following three aspects: assessment of enterprise operating maturity such as IBM's smart grid maturity model; assessment of project construction like U.S EPRI's cost/benefit assessment indicators of smart grid construction project, European smart grid revenue evaluation index system, China's smart grid pilot project evaluation index system method; assessment of power grid's construction and operation, such as US Department of Energy's smart grid development evaluation index system. As analyzed, it is observed that assessments related to the development of industrial planning are lack of the views from the perspective of development of smart grid industry. Therefore, the author hereby puts forward the model of smart grid industry maturity evaluation so as to perform the maturity assessment of various manufacturing industries at the initial stage of smart grid industry planning and development.

2   Maturity Evaluation Index System of Smart Grid Industry

Smart grid has become an inevitable choice in response to energy and environmental challenges, to achieve sustainable development of energy and power, and to promote economic development. In return, rapid development of smart grid has also driven the rise and development of smart grid industry, forming a huge industrial chain.

2.1   Characteristics of Smart Grid Industry Development

The development of smart grid industry mainly has the following characteristics:

(1) Large Industrial Scale: The traditional transmission network is underway to configure energy, industry, information and other resources, to strive for intelligent power grid changes of smart home, intelligent transportation, community, and city development. With reference to the 'views on accelerating the construction of strong smart grid' issued by State Grid Corporation, it pointed out that from 2009 to 2020, the total investment in smart grid construction scale will reach 4 trillion yuan. It is foreseeable that smart grid has large industry investment, big industrial scale, broad space for future development, with huge development opportunities.

(2) Technical Orientation: The smart grid is based on the physical power grid, supported by information platform and featured with information, automation, and interaction. With the deepening of pilot project of smart grid, many power grid equipment needs to have "intelligent" functions like the technology of integrated digit, sensor, information and communication as well as modern control, since that led to the rapid development of China's smart grid equipment. That is to say, development of smart grid industry is essentially the innovation of core technology of smart grid. Consequently, concerning the maturity evaluation of smart grid industry, technical performance should be acted as the core guide.

(3) Extension of Industrial Chain: In the construction of smart grid, extensional development of its related industries has been promoted with the widely-used technology of new energy access, energy storage, power electronics, control,

information and communication and others. In the planning and development of smart grid industry, industrial transformation and technological upgrading can be realized. Also, smart grid industry's development can effectively be promoted, depending on the original industrial base and use of existing research advantages.

(4) Policy Orientation: One major different feature of smart grid industry compared to other manufacturing industries is its strong oriented-policy. Therefore, smart equipment manufacturing is supposed to keep pace with the development of smart grid in order to enter the market for seeking development. From the industrial development point of view, all regions have been strengthening the planning of smart grid industry, and introduced the relevant policy measures to promote industrial development, which has a vital impact on development of smart grid industry.

2.2 Maturity Evaluation Index System of Smart Grid Industry

The maturity evaluation index system of smart grid industry is an important foundation and basis for carrying out the maturity evaluation of smart grid industry. Based on the aforesaid characteristics of smart grid industry, this paper establishes a comprehensive maturity evaluation index system of smart grid industry from five aspects to wit,

Fig. 1. The index of maturity evaluation of smart grid industry

technical performance, industrial matching, market environment, policy environment and social influence, as shown in Fig. 1. The index system consists of five primary indicators, 17 secondary indicators and 47 three indicators (Fig. 2).

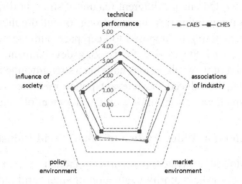

Fig. 2. The comparison of maturity evaluation between CAES and CHES

3 Evaluation Model of Industrial Maturity Based on Multi-level Comprehensive Evaluation

3.1 Model of Smart Grid Industry Maturity

Smart grid industry maturity refers to a smart grid industry in the regional planning and development in the development stage, but also the regional development of the smart grid industry potential and level. The smart grid industry maturity model is divided into five steps, and the maturity stage of the development of smart grid is given, and the development suggestions of different stages are given. Smart grid industry maturity is divided into five levels: germination period, starting period, growth period, maturity, mature optimization period.

Corresponding to the five smart grid industry maturity level, the smart grid industry maturity evaluation index system in the five criteria for the stage described in Table 1:

Table 1. Descriptions of smart grid industry grading system

Stage	Tech-performance	Supporting	Market environment	Policy environment	Social impact
1	Conceptual stage	None	None	None/rare	None
2	Industrialization on going	Partial	Some but replaceable	Trail in progress	Some
3	Can be industrialized	Basic industry supporting	Local needs and market space	Need to strengthen policy support and management	Create an employment environment, driving economic development
4	Good performance	Well-prepared supporting	Local needs and industry requires	Efficient policy support and management	Driving industrial chain
5	Strong formation	Powerful support	International market	Well-arranged policy	Regional backbone

The evaluation of the five criteria has given the maturity stage of the development of a specific industry in the region. The evaluation values of the five maturity levels are shown in Table 2. After giving the maturity level of the smart grid industry, the model also gives a description of the industrial development conditions at this stage, and gives the development suggestions of the industry in the maturity stage.

Table 2. Grading levels of maturity evaluation

Stage	Initial	Beginning	Growing	Mature	Optimization
Score	0–30	30–50	50–70	70–90	90–100
Conditions	Technology does not yet have the industrialization conditions or regional industrial development foundations	The technology is in industrialization process and related foundations are weak	The technology can be industrialized, and the development of the power grid needs a certain development foundation	The technical performance is good, and the development for the power grid requires a better foundation	The technical performance is perfect, and the development for the power grid requires a better foundation
Suggestions	It is not recommended to develop related industries	Introduce the relevant research institutes and focus on the development of technology	The basic conditions of development industry is met so it can be industrialized	The basic conditions of development industry is met and focus on the optimization of industry	Industry can be developed and focused on technical innovation to dominate the market

3.2   The Maturity Evaluation of Smart Grid Industry

The maturity evaluation of smart grid industry adopts the method of combination of analytic hierarchy process and fuzzy comprehensive evaluation method to carry out comprehensive evaluation. Specific steps are as follows:

(1) Analysis the problem, hierarchy and structure; (2) Determine the evaluation factors and evaluation level; first, determine the set of factors U to be judged and the evaluation set. (ie, evaluation index), to characterize each factor V in which the N kinds of decisions (that is, the evaluation level, that is, germination, starting, growth, maturity, optimization) (3) Determine the weight set;

(a) Constructing hierarchical structure

In each layer elements m between the two pairs of comparison, structure judgment matrix $C = (c_{ij})_{m \times m}$. Which c_{ij} indicates the importance of factor i and factor j relative to the target.

(b) Single hierarchical arrangement

The single order of the hierarchy is based on the judgment matrix to calculate the relative importance of a hierarchy factor relative to an element of the previous layer. Calculate the intermediate quantity according to the importance level:

$$\overline{W}_i = \sqrt[n]{\prod_{j=1}^{n} c_{ij}} \quad i = 1, 2, \ldots, n \quad W_i = \frac{\overline{W}_i}{\sum_{j=1}^{n} \overline{W}_j} \quad (1)$$

(c) Consistency verification

The consistency verification formula is:

$$CI = \frac{\lambda_{\max} - n}{n - 1} \quad CR = \frac{CI}{RI} \quad (2)$$

In the formula: $\lambda_{\max}$ is the maximum eigenvalue of the judgment matrix C; the number of elements; RI is the mean random consistency index. CI is called the compatibility index, and CR is called the random consistency ratio. At that time, it can be considered that the result of single order is satisfactory. When $CR = \frac{CI}{RI} < 0.1$.

(4) Determine the fuzzy matrix and introduce the weight assignment set

The general evaluation matrix R is constructed, in which the r_{ij} membership degree of the subject can be rated by the factor u_i, which makes it to meet the condition that $\sum r_{ij} = 1$ and the fuzzy relationship $\mathbf{R} = (r_{ij})_{m \times n}$ is determined from U (evaluation index) to V (evaluation grade).

In addition, we introduce a fuzzy subset $\mathbf{A}$ as a set of weights, which suits $a_i \gg 0$ and $\sum a_i = 1$.

(5) For fuzzy synthesis and Blurring two fuzzy sets $\mathbf{B} = \mathbf{A} * \mathbf{R}$

In the formula: * refers to the operator symbol, $\mathbf{A}$ refers to the distribution of weights, $\mathbf{R}$ for the fuzzy relationship.

The fuzzy subset $\mathbf{B}$ is introduced as a decision set, b_j indicating that the object B_k to be evaluated has a corresponding degree $v_j (0 \ll b_j < 1$ and $\sum b_j = 1)$, which reflects the distribution of the subject B_k in the judged feature, and so called the fuzzy evaluation.

(6) Calculate the synthesis parameters

When the parameter column vector F is introduced, the evaluation result of the grade parameter is obtained as follows: $\mathbf{B} * \mathbf{F} = p$

Using the above method, we can calculate the comprehensive parameters of each criterion layer to fully reflect the impact of various factors on the maturity of the smart grid industry, and then by fuzzy synthesis, intelligent grid industry maturity can be fuzzy evaluation, and ultimately get mature Degree level.

4 Case Analysis

Take city A as an example, city A is important port city in upstream of Yangtze River area, marine, land and airline transportations are available, it has a good heavy industry base and energy reserves. Currently city A is building the state standard high-tech zones, we planned taking the smart power grid industry as one of key construction projects, and related policies will be issued. In the early stage of the smart grid industry planning, a comprehensive evaluation of the maturity of various smart power grid industries in the development of city A was carried out to provide reference for the planning. City A takes energy reserves as one of main planning issues. The following are comprehensive evaluations of the maturity for Compressed-Air Energy Storage, CAES and Compressed-Hydrogen Energy Storage, CHES to verify the correctness and validity of the model method.

4.1   Maturity Evaluation Analysis of Compressed Air Storage Energy Industry of City A

The fuzzy-AHP composite model was used for the evaluation of maturity for smart power grid industry. Based on this composite model that each index is calculated as following. Take technical performance as an example, setting up a comprehensive technical performance evaluation structure, construct judgment matrix according to the statistic structures, through the judgment matrix that the technical performance level of single sequence is:

$$a = (0.243, \quad 0.255, \quad 0.079, \quad 0.119, \quad 0.093, \quad 0.212)$$
$$a_1 = (0.0904, \quad 0.0904, \quad 0.2713, \quad 0.2362, \quad 0.3116)$$
$$a_2 = (0.3874, \quad 0.4434, \quad 0.1692)$$
$$a_3 = (0.2, \quad 0.4, \quad 0.4)$$
$$a_4 = (0.3333, \quad 0.3333, \quad 0.3333)$$
$$a_5 = (0.4126, \quad 0.3275, \quad 0.2599)$$
$$a_6 = (0.5, \quad 0.5)$$

In the same way, we can determine the judgment matrix for the association of industries, market environment, policy environment and influences of society with principles respectively. And the result of single sequence can be calculated with each principles.

Under the principles of association of industries, we judge each bottom level principle and the fuzzy relations are determined. Thus, each sub-principle for fuzzy relations and ratings of parameters which are under principle of technical performance can be calculated, shown as Table 3.

Table 3. The sub-principle grading evaluation of CAES

Sub-principle	1	2	3	4	5	Score
Degree of advantage	0.033	0.157	0.451	0.270	0.087	3.22
Technological economy	0	0.035	0.071	0.639	0.253	4.11
Technical, environmental benefits	0	0.084	0.389	0.326	0.200	3.64
Standardization level	0	0.473	0.280	0.157	0.087	2.86
Levels of application	0.027	0.213	0.233	0.330	0.195	3.45

In the same way, the decision sets and the rating parameters can be determined for associations of industry, the market environment, the policy environment and the influence of society. Based on the evaluation results of each principle for the compressed air storage industry in city A.

With fuzzy transformation that the results of the comprehensive evaluation for the maturity of compressed air storage industry in city A are:

$$\mathbf{K} = (0.028, \quad 0.155, \quad 0.369, \quad 0.358, \quad 0.085) \quad p_K = 66.0$$

Generally speaking, the compressed air energy storage industry is still in growing process. The technology has industrialized foundation, the region has the developmental foundation, and the smart power grid development demands for the compressed air storage energy technology. If city A develops the compressed air energy storage industry, We can introduce relevant technologies and talents with special focus on the optimization and upgrading of the technologies, and pay close attention on the field of large volume compressed air storage.

4.2   Maturity Evaluation Analysis of CHES of City A

With the principle of association of industry and judge with each index of bottom level and the fuzzy relations can be determined. Thus, the fuzzy relation and ratings of parameter of each sub principles which are under technical performance principles can be calculated.

In the same way we can determine the decision sets and the rating parameters under each principle for associations of industry, the market environment, the policy environment and the influence of society. Based on the evaluation of principles of CHES in city A. With fuzzy transformation that the result of maturity evaluation of CHES for city A is

$$\mathbf{H} = (0.172, \quad 0.376, \quad 0.277, \quad 0.131, \quad 0.043) \quad p_H = 49.9$$

In general, the development of CHES in A city is in transition from the beginning to the growing stage. The technology is advanced and the potential for future development is vast, but it is still in the process of industrialization. The regional fundamental is relatively weak. If city A develops CHES industry, the relevant research institutes should be introduced, and the technology developed should be focused and then industrialized when time is right.

4.3   Comparison of Maturity Evaluation Between CAES and CHES of City A

Through the comparison of CAES industry technical performance and CHES industry technical performance we can find that the CHES is better in advanced technology and environmental benefits which are better than that of the CAES technology, so CHES has better prospects for development. CAES has been invested in commercial operations on a large scale, while CHES is still in the experimental stage of development, and its economy and applications are much lower than CAES. Overall, the performance of CAES is on average, and the technical performance is good. The CHES technology is advanced but it is immature, and it is in the experimental stage of development.

The contrast between CAES industry and CHES industry in association of industry and market environment shows that in city A, the development of CHES industry in the upstream manufacturers supply capacity and middle technology development capabilities is significantly lower than that of CAES industry. In addition, they both lack the international competitiveness for the products. In general, the development of CAES industry in the city A is better than the development of CHES industry in the area of associations of industry and market environment.

Through the contrast of policy environment and influence of society between CAES and CHES shows that they both have low score for policy environment, for city A that development for CAES industry is better than CHES industry due to imperfect of related policies.

It can be seen that the overall score of CAES industry is higher than that of CHES, especially in the associations of industry, and the scores for policy environment ratings are low. Due to the limitation of the technical difficulties of CHES, the associations of industry and the market environment are restricted too, but the long term development prospects are better. The development conditions of the CAES industry are more even, and the conditions for the industrialization of the city are preliminarily available. The maturity for City A development of CAES industry is higher than CHES industry, CAES industry for city A is at the growth stage, and CHES industry is in the initial stage at transition to period of growth. If city A develops CAES industry, then it can be introduced with mature production lines, specially focus on the optimization of technology and upgrade, pay attention on the large capacity of CAES technology. If city A develops CHES industry, the relevant research institutes will be introduced, and the technology will be developed and the industrialization will be carried out in right time.

5   Conclusion

Based on the analytic hierarchy process and the fuzzy theory, this paper puts forward the evaluation model and comprehensive evaluation method of the smart grid industry maturity. On the basis of analyzing the characteristics of smart grid industry, the author establishes the comprehensive evaluation index system of smart grid industry maturity from the aspects of technical performance, industry supporting, market environment, policy environment and social influence. The fuzzy theory is applied to the smart grid industry, and fuzzy-AHP combination evaluation method is adopted to evaluate the maturity of smart grid industry. The maturity stage of smart grid industry is expounded, and the development of the stage industry is given as well.

Smart grid has become an inevitable trend of modern power grid development, and development of smart grid led the rise and development of smart grid industry, which gives birth to the development of many new industries. Furthermore, seizing the opportunities of smart grid industry development becomes an important choice for regional planning and construction. By the means that scientific reflection on the regional development of smart grid industry in all aspects of the conditions and the impact of industrial development on the region, it shall have a positive significance in guiding the regional smart grid industry planning, encouraging regional investment in the development of smart grid industry.

Acknowledgements. This work is supported by the State Grid project: Research on technology system of global energy internet.

References

1. DOE: GRID 2030: a national vision for electricity's second 100 years (2003)
2. European Commission: European technology platform smart grids: vision and strategy for Europe's electricity networks of the future (2006)
3. Wang, M.: Smart grid and smart energy resource grid. Power Syst. Technol. **34**, 1–5 (2010)
4. Lin, Y., Zhong, J., Felix, W.: Discussion on smart grid supporting technologies. Power Syst. Technol. **33**, 8–14 (2009)
5. Zhang, J., Hu, J.: Design of smart grid oriented adaptive planning system. Autom. Electr. Power Syst. **35**, 1–7 (2011)
6. Sun, Q., Ge, X., Liu, L.: Review of smart grid comprehensive assessment systems. In: ICSGCE, pp. 219–229
7. Wang, Z., Li, H., Li, J., Han, F.: Assessment index system for smart grids. Power Syst. Technol. **33**, 14–18 (2009)
8. Tan, W., He, G., Liu, F.: A preliminary investigation on smart grid's low-carbon index system. Autom. Electr. Power Syst. **34**, 1–5 (2010)
9. Jia, W., Kang, C., Liu, C., Li, M.: Capability of smart grid to promote low-carbon development and its benefits evaluation model. Autom. Electr. Power Syst. **35**, 7–12 (2011)
10. Zhang, J., Pu, T., Wang, W.: A comprehensive assessment index system for smart grid demonstration projects. Power Syst. Technol. **35**, 5–9 (2011)
11. Duncan, S.J., Griendling, K., Mavris, D.N.: An assessment of ROSETTA for smart electricity grid system-of-systems design. In: Proceedings of 2011 6th International Conference on System of Systems Engineering, pp. 231–236 (2011)
12. Wang, B., He, G., Mei, S.: Construction method of smart grid's assessment index system. Autom. Electr. Power Syst. **35**, 1–5 (2011)

An Improved Multi-objective Differential Evolution Algorithm for Active Power Dispatch in Power System with Wind Farms

Shu Xia[1], Yingcheng Xu[1], and Xiaolin Ge[2(✉)]

[1] Shibei Electricity Supply Company of State Grid Shanghai Municipal Electric Power Company, Shanghai 200072, China
[2] College of Electrical Engineering, Shanghai University of Electric Power, Shanghai 200090, China
gexiaolin2005@126.com

Abstract. For the uncertainty of wind power and load, a reserve risk index is defined from minimum of load loss and maximum of utilizing wind power. Then, the index is introduced into optimizing for active power dispatch. Considering three indexes which consist of fuel cost, pollutant emission amount and the reserve risk index, a multi-objective optimization model for active power dispatch in power system with wind farms is established. For better solving model, an improved multi-objective differential evolution algorithm is proposed. This algorithm contains chaos initialization strategy, parameter adaptive strategy, dynamic non-dominated sorting strategy introduced to enhance the global searching ability. With the Pareto solution set, the entropy-based TOPSIS (technique for order performance by similarity to ideal solution) is adopted to sort the optimal solution set for the final scheme. The results and data analysis demonstrates the model is reasonable and the algorithm is valuable.

Keywords: Wind farm · Reserve risk · Optimizing for dispatching of active power · Multi-objective differential evolution algorithm · Improvement strategies

1 Introduction

To realize coordinated sustainable development of energy production and the environment, the focus of research on the active power dispatch shift from the traditional economic dispatch optimization with single target of minimum total energy consumption, to environmental economic dispatch optimization with comprehensive target of energy conservation [1, 2]. Meanwhile in recent years, with the development of wind power, the uncertainty of wind power brings new problems to establish the reserve capacity in active power dispatch. Many scholars have studied these issues. In [3], the impact of wind power prediction errors gets reduced by increasing the output of wind farm by 20% as the upper and lower spinning reserve. In [4], the reserve costs and penalty costs are brought in the objective function to punish wind power costs which are estimated to be too large and too small. In [5], based on wind power forecast confidence, the index of reserve costs is introduced into the objective function. In [6], based on risk reserve constraints, a dynamic economic dispatch model in wind power

© Springer Nature Singapore Pte Ltd. 2017
K. Li et al. (Eds.): LSMS/ICSEE 2017, Part III, CCIS 763, pp. 643–652, 2017.
DOI: 10.1007/978-981-10-6364-0_65

system is proposed. The above works have made contributions to the reverse constraints, but it is difficult to obtain the trade-off relationship between reserve and cost, and the losses of load and wind power cannot be directly reflected. Therefore, in this paper, the reserve index is studied as a separate objective. Due to taking into account of the cost, pollutant emissions amount and reserve index, optimal active power dispatch is a multi-objective optimization model. Genetic algorithm (GA) [7], particle swarm optimization (PSO) [8], nonlinear fractional programming approach [9] and other methods have been used to solve multi-objective optimization problem, however, in these methods, the weighting factors are introduced to convert multiply objectives into single objective. However, it is often difficult to determine weighting factors in reality. In recent years, with the development of multi-objective optimization algorithm (MOEA), such as, improved strength Pareto evolutionary algorithm (SPEA2) [10] and improved non-dominated sorting genetic algorithm (NSGA-II) [11] which provides a new approach for solving multi-objective problems. The MOEA does not need to set the weighting factors, and get a uniform distribution of the Pareto-optimal set. Decision-makers can select one or more optimal solutions according to demand. Therefore, in this paper, multi-objective optimization algorithm is applied to solve multi-objective optimization of active power dispatch.

Differential evolution (DE) algorithm is an intelligent optimization method, which is of ease to use, robustness, good global search capability, etc., so it have been widely applied in practical work [1, 12]. In [13], combining the advantages of DE and fast non-dominated sorting (FNS) of NSGA-II, multi-objective differential evolution algorithm (MODE) is proposed. This algorithm shows better performance than the NSGA-II in multiple test problems. However, in MODE algorithm randomly selection of initial population may cause "premature", and the crossover and mutation operations depend on the parameters strongly, in addition, fast non-dominated sorting could lead to inhomogeneous distribution of Pareto-optimal set. Therefore, the solving performance of MODE still needs further improvement.

Based on the above researches, considering the uncertainty of wind and load, combined with the risk theory, the reserve risk index is defined, and then a multi-objective optimal active power dispatch model with wind farm is proposed. To get better Pareto-optimal set, chaotic searching strategy, factors adjusting strategy and dynamic non-dominated sorting strategy are introduced, and an improved multi-objective differential evolution algorithm (IMODE) is proposed. Illustrated by the example of IEEE-118 bus system, from the aspects of outer solution, index C and index S, compared with IMODE and other multi-objective algorithms, an excellent candidate for optimal active power dispatch can be provided by IMODE. Then, sorting by the technique for order performance by similarity to ideal solution (TOPSIS), the final optimal dispatch scheme is acquired.

2 Mathematical Model

2.1 Reserve Risk Index

Power system operation risk is the comprehensive measurement of the loss [14]. Its mathematical expression given as follows:

$$I_R = \sum_j P_{\text{rob}}(X_j) V_{\text{ev}}(X_j) \tag{1}$$

where I_R is the operation risk assessment index, X_j and $P_{\text{rob}}(X_j)$ are the j-th uncertain disturbance and its occurrence probability respectively, $V_{\text{ev}}(X_j)$ is the j-th loss degree of uncertain disturbance. As shown in formulation (1), the reserve risk index reflects the uncertainty of wind power and load is defined as follows:

$$I_R = \sum_s P_{\text{rob}}^s \omega_1 \frac{V_D^s}{P_D} + \sum_s P_{\text{rob}}^s \omega_2 \frac{V_w^s}{P_{\text{WR}}} \tag{2}$$

where s is all the possible operational scenarios due to the prediction error of wind power and load, P_{rob}^s is the probability of the s-th scenario, V_D^s, V_W^S are the load loss and wind power loss in the s-th scenario respectively, P_D is the forecasted value of load, P_{WR} is the total rated capacity of wind farms, ω_1 and ω_2 are the weights. The greater the reserve level in dispatching scheme, the smaller the loss of load and wind power, and the lower the reserve risk index(and vice versa). The details of calculating method for the reverse risk index I_R is shown as follows.

Based on the method of [15], the joint probability density function of load-wind $f_Z(P_Z)$ is presented, and its diagram is shown in Fig. 1. P_{WA} is the total planned output of wind power; S^u and S^d are the upper and lower adjustment capacity. When the variable P_Z waves in the interval $[P_{\text{WA}} - S^u, P_{\text{WA}} + S^d]$, units have enough reserve to ensure security of the system without loss of load and wind; When P_Z is more than $P_{\text{WA}} + S^d$, part of the wind power needs to be cut; when P_Z is less than $P_{\text{WA}} - S^u$, part of load needs to be cut. Based on the above analysis, formulation (2) is converted into following form:

$$
\begin{aligned}
I_R = \omega_1 &\int_{-\infty}^{P_{\text{WA}}-S^u} \left[\frac{f_Z(P_Z) \times (P_{\text{WA}} - S^u - P_Z)}{P_D} \right] dP_Z \\
+ \omega_2 &\int_{P_{\text{WA}}+S^d}^{+\infty} \left[\frac{f_Z(P_Z) \times (P_Z - P_{\text{WA}} - S^d)}{P_{\text{WR}}} \right] dP_Z
\end{aligned}
\tag{3}
$$

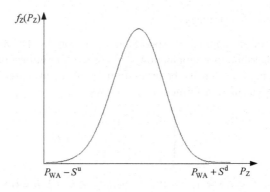

Fig. 1. The probability density function of load and wind

2.2 Objective Function and Constraints

Considering the units cost of fuel I_C, the amount of pollutant emission I_E, and the reserve risk I_R, multi-objective optimal active power dispatch model with wind farm can be expressed as:

$$\min y = [I_C, I_E, I_R] \tag{4}$$

$$I_C = \sum_{i=1}^{N_I} \left[a_i + b_i P_{i,t} + c_i P_{i,t}^2 + \left| d_i \sin\left\{ e_i \left(P_i^{\min} - P_{i,t} \right) \right\} \right| \right] \tag{5}$$

$$I_E = \sum_{i=1}^{N_I} \left[\alpha_i + \beta_i P_{i,t} + \gamma_i P_{i,t}^2 + \eta_i \exp\left(\delta_i P_{i,t} \right) \right] \tag{6}$$

where N_I is the number of conventional power generation units, $P_{i,t}$ is the output of unit i in period t, $P_i^{\min}$ is the minimum output of unit i, a_i, b_i, c_i, d_i, e_i is the fuel cost coefficients of unit i, α_i, β_i, γ_i, η_i, δ_i is the pollutant gas emissions coefficients of unit i.

For there is a conflict among these objectives, not all target values can converge to a minimum at the same time. Therefore, it only got a set of Pareto optimal solution set.

The constraints include conventional power generation units constraint, load balance constraint, reserve constraint, line security constraint, which can be referred to [16].

3 IDEMO Algorithm

3.1 DEMO Algorithm

The process of mutation and crossover operations in DEMO are the same as in DE. In selection operation, Pareto non-dominated level and crowding distance of NSGA-II are employed, which make DEMO can be applied to multi-objective problem, leading to a group of Pareto optimal solution sets. The detailed calculation is given in [13].

3.2 IDEMO

(1) Chaos Initialization Strategy

Random method is applied in species initialization by DEMO, which causes inhomogeneous distribution in population, and reduces the utilization of initial population. Therefore, in order to make better uniformity, the chaos initialization strategy is adopted [17]. The calculation formula is as follows:

$$r_{k+1} = \begin{cases} r_k^p, r_k \in (0, a) \\ \mu r_k (1 - r_k), r_k \in [a, b] \\ r_k^q, r_k \in (b, 1) \end{cases} \tag{7}$$

where $0 < a < b < 1$, $0 < p < 1$, $q > 0.1$. Generally, $a = 0.2$, $b = 0.8$, $p = 0.5$, $q = 15$.

(2) Parameter Adaptive Strategy

In DEMO, the mutation factor F and the crossover factor C_R have great influence on its convergence performance and searching efficiency. There are various parameter control strategies in DEMO. In [18], based on random method and evolutional generation, the adjustment strategy is proposed. The above strategies have achieved good results. The method involved in [18] is so flexible that it can be applied to DEMO, however this method induces F values between [0.5,1], which causes unreasonableness (F values by practical problem). Therefore, the improved method of control parameter adjustment is proposed as follows:

$$F_i^k = F_{\min} + rand \cdot (F_{\max} - F_{\min}) \tag{8}$$

$$C_{Ri}^k = (C_{R\max} - C_{R\min}) \cdot \frac{(K_{\max} - k)}{K_{\max}} \tag{9}$$

where F_i^k and C_{Ri}^k are mutation factor and crossover factor of individual i in the k-th generation; $F_{\min}$ and $F_{\max}$ is the upper and lower limits of crossover factor respectively. $C_{R\max}$ and $C_{R\min}$ are the upper and lower limits of mutation factor respectively; $K_{\max}$ is maximum iteration number.

(3) Dynamic Non-dominated Sorting Strategy

In DEMO, the fast non-dominated sorting (FNS) is adopted. The FNS only needs one sorting operation for all individuals, and takes a short time. However, this method does not provide good result. For example, in Fig. 2, six individuals need to be selected from eleven members. The selected results by the fast non-dominated sorting are shown in Fig. 2(a). As it can be seen, all individuals between A and B are not selected, which losses some good solutions. To get better distribution of solutions, dynamic non-dominated sorting (DNS) is adopted [19]. When n_d individuals are chosen from a non-dominated set H including N_d individuals at the same non-dominated rank, the operation of the DNS algorithm is:

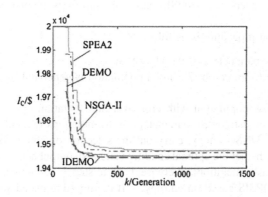

Fig. 2. Convergence curves of outer solutions for cost

Step 1: Calculate a crowding distance for each individual in the non-dominated set H.

Step 2: Remove an individual with a minimum crowding distance value from H and update H.

Step 3: If the number of individuals in set H equals to n_d, output H, otherwise go to Step4.

Step 4: Calculate new crowding distance for each individual in the set H.

Step 5: Remove an individual with a minimum crowding distance value from H and update H. Go to Step 3.

3.3   Performance Evaluation Index

There are many evaluation indexes for the performance of multi-objective algorithm, and the outer solution, index C and index S are introduced in this paper. The outer solution is the solutions that one certain objective component is optimal. Generally, by comparison between the evolutionary process of relevant objective component for outer solutions and its value in last generation, the calculation speed and robustness are clearly shown [20]. Index C is relative coverage ratio between two solution sets that can judge the quality of them [11]. Index S measures the distribution of the points on solution sets: the lower the index S reaches, the more uniform the distribution of solution set is [20].

4   Calculation Progress

After Pareto-optimal set is got by solving multi-objective optimal problem, the selection of optimal solution on demand is a multi-attribute decision making problem. Technique for order preference by similarity to an ideal solution (TOPSIS) is one of the commonly used methods [21], its target is to minimum the gap between the selected scheme and the ideal solution, and maximum the gap from negative ideal one. However, this scheme has to give out the weights of every attributes, which is hard to realize. Therefore, in [22], the selection method based on entropy weight is proposed, namely establishing the weights by contrast information of dataset. For example, in view of one certain attribute, this attribute makes a small contribution when all schemes have same results. Therefore, TOPSIS based on entropy is adopted to get the optimal solution.

The calculation procedure is as follows:

Step 1: Set the parameters of the algorithm.

Step 2: By the formulation (7), initial population is generated by chaotic searching strategy.

Step 3: Get new population with crossover and mutation operations, in this progress, adjust control parameters by formulations (8) and (9).

Step 4: By the DNS strategy, carry out the selection operation for populations.

Step 5: Judge whether the optimal condition attained: output non-dominated solutions in the last generation; otherwise, back to step 3, continue the calculation.

Step 6: By TOPSIS based on entropy, sort the non-dominated solution sets, make decisions and output the optimal scheme.

5 Numerical Examples

5.1 Testing System

Based on the IEEE-118 bus systems, generator type is modified. The systems include 118 nodes, 184 lines, 14 conventional generators and 2 wind farms. The parameters of conventional generators is referred to [16], conventional generator C1, C3, C5, C10 and C12 participate in real-time control, and regulating capacity is 22 MW. Wind farm 1 is connected with node 54 and wind farm 2 is connected with node 103. Every wind farm has 100 wind generators. Forecasted wind speed of wind farm 1 is 10 m/s, and the forecasted speed of wind farm 2 is 12 m/s. The nominal capacity of wind generator is 2 MW, cut-in speed is 4 m/s, rated speed is 12.5 m/s, and cut-out speed is 20 m/s. Forecasted data of system load is 4242 MW, standard deviation of load forecast error is 0.8% of the predictive value, and standard deviation of wind speed forecast error is 0.8% of the predictive value. ω_1 and ω_2 are 0.9 and 0.1 respectively.

5.2 Analysis of IDEMO

In order to compare with the performance of multi-objective optimal algorithms, SPEA2, NSGA-II, DEMO and IDEMO are applied to solving multi-objective dispatch optimization. The parameters of IDEMO are set as follows: population size $N_P = 100$, maximum iterations $K_{max} = 1500$, upper and lower limits of mutation factor $F_{max} = 0.8$ and $F_{min} = 0.3$, upper and lower limits of crossover factor $C_{Rmax} = 0.8$ and $C_{Rmin} = 0.3$. The population size and maximum iterations of the others is the same as IDEMO.

To analyze the outer solution, every algorithms run 30 times. For every objective functions, 30 relevant objective components can be obtained in every algorithms in k_{th} generation, namely fuel cost $x_C^{(k)}$, pollutant emissions amount $x_E^{(k)}$, reserve risk index $x_R^{(k)}$. Take census of these objective components $x_C^{(1500)}$, $x_E^{(1500)}$, $x_R^{(1500)}$ in 1500th generation, as shown in Table 1. In Table 1, it is can be seen that the approximate optimal solution and robustness by IDEMO and DEMO are better than SPEA2 and NSGA-II, and IDEMO is optimal. Take census of these average objective components, and evolutionary curve is drawn as in Fig. 2. In Fig. 2, it is can be seen that the calculation speed of IDEMO is fast than the other algorithms.

Table 1. The outer solutions at the final generation by different algorithms

Algorithm	I_C ($x_C^{(1500)}$)			I_E ($x_E^{(1500)}$)			I_R ($x_R^{(1500)}$)		
	Best value/$	Worst value/$	Deviation/ %	Best value/$	Worst value/$	Deviation/ %	Best value/$	Worst value/$	Deviation/ %
SPEA2	19461.62	19479.37	0.091	16964.06	17020.25	0.331	0.00333	0.00333	0
NSGA-II	19450.76	19470.81	0.103	16920.17	16991.33	0.421	0.00333	0.00333	0
DEMO	19437.21	19442.26	0.026	16916.62	16951.25	0.205	0.00333	0.00333	0
IDEMO	19436.43	19439.48	0.013	16910.82	16937.79	0.160	0.00333	0.00333	0

Average index C between any two algorithms (30 operations) is shown in Table 2. From this table, there is 78% and 49% of solutions that dominant solution of SPEA2 and NSGA-II, which shows its high solution quality. To compare the uniform distribution, the index S of Pareto frontier in last generation (30 operations) is shown in Table 3. This table clearly shows that index S is between 0.0323 and 0.0378, which is less and better than others.

Table 2. Average of metric C at the final generation (%)

Algorithm	SPEA2	NSGA-II	DEMO	IDEMO
SPEA2	–	6	0	0
NSGA-II	39	–	2	0
DEMO	71	42	–	3
IDEMO	78	49	11	–

Table 3. Metric S at the final generation

Algorithm	Best value/$	Worst value/$	Deviation/%
SPEA2	0.0504	0.0697	0.0193
NSGA-II	0.0462	0.0589	0.0127
DEMO	0.0404	0.0469	0.0065
IDEMO	0.0323	0.0378	0.0055

The above analysis shows that robustness, calculations speed and solution precision of IDEMO is much better than SPEA2 and NSGA-II, and it is less superior to DEMO; its uniformity is best. Therefore, IDEMO can provide an excellent candidate for optimal active power dispatch with wind farm.

5.3   Analysis of Different Optimal Scheme

To research the effect among different dispatch schemes, 4 schemes are selected as follows:

Scheme 1: The solution with the minimum fuel cost.
Scheme 2: The solution with the minimum pollutant emissions.
Scheme 3: The solution with the optimal reserve risk index.
Scheme 4: The optimal solution by TOPSIS based on entropy.

For 4 schemes, the results are shown in Table 4. Through the comparison of scheme 1 and scheme 2, only with consideration of fuel cost, when $I_C = 19436.43\$$, pollutant emissions amount in scheme 1 is 3500.91 lb more than scheme 2, moreover, the reserve risk index for both schemes is relatively large. For scheme 3, the obtained optimal risk index is 0.00333 with minimum reserve risk index. According to the solution in scheme 4, fuel cost is 19783.64$, pollutant emission amount is 17423.03 lb and reserve risk index is 0.00333, in which the reserve risk index is minimal. Therefore, TOPSIS based on entropy can coordinate each objective well, and provide the comprehensive decision-making scheme.

Table 4. Results in different scheduling modes

Scheme	Cost/$	Emission/lb	Risk
Scheme 1	19436.43	20411.73	0.00767
Scheme 2	20138.99	16910.82	0.00883
Scheme 3	19676.85	17813.84	0.00333
Scheme 4	19783.64	17423.03	0.00333

6 Conclusions

Considering three indexes which consist of fuel cost, pollutant emission amount and the reserve risk index, a multi-objective optimization model for active power dispatch in power system with wind farms is established. And IMODE and TOPSIS based on entropy are introduced to solve the optimal active power dispatch problem. An example is analyzed, and the conclusions are as follows:

(1) The defined reserve risk index can directly reflect the extent of the losses of load and wind power in the cases of different reserve capacity.
(2) Combining with chaotic searching strategy, parameter adaptive strategy, dynamic non-dominated sorting, IMODE is verified to has the ability of providing an excellent candidate for optimal active power dispatch.
(3) TOPSIS based on entropy can coordinate well each objective, and provide the comprehensive decision-making scheme.

Acknowledgements. This work was sponsored in part by National Natural Science Foundation of China (No. 51507100), and in part by Shanghai Sailing Program (No. 15YF1404600), and in part by the "Chen Guang" project supported by the Shanghai Municipal Education Commission and Shanghai Education Development Foundation (No. 14CG55).

References

1. Soumitra, M., Aniruddha, B., Sunita, H.D.: Multi-objective economic emission load dispatch solution using gravitational search algorithm and considering wind power penetration. Int. J. Electr. Power Energy Syst. **44**(1), 282–292 (2013)
2. Lamadrid, A.J., Shawhan, D.L., Murillo, S.C., et al.: Stochastically optimized, carbon-reducing dispatch of storage, generation, and loads. IEEE Trans. Power Syst. **30**(2), 1064–1075 (2015)
3. Lee, T.Y.: Optimal spinning reserve for a wind-thermal power system using EIPSO. IEEE Trans. Power Syst. **22**(4), 1612–1621 (2007)

4. Hetzer, J., Yu, D.C., Bhattarai, K.: An economic dispatch model incorporating wind power. IEEE Trans. Energy Convers. **23**(2), 603–611 (2008)

5. Junli, W., Buhan, Z., Weisi, D.: Application of cost-CVaR model in determining optimal spinning reserve for wind power penetrated system. Int. J. Electr. Power Energy Syst. **66**(1), 110–115 (2015)

6. Wei, Z., Yu, P., Hui, S.: Optimal wind-thermal coordination dispatch based on risk reserve constraints. Eur. Trans. Electr. Power **21**(1), 740–756 (2013)

7. Yasar, C., Ozyon, S.: Solution to scalarized environmental economic power dispatch problem by using genetic algorithm. Int. J. Electr. Power Energy Syst. **38**(1), 54–62 (2012)

8. Guo, C.X., Bai, Y.H., Zheng, X., et al.: Optimal generation dispatch with renewable energy embedded using multiple objectives. Int. J. Electr. Power Energy Syst. **42**(1), 440–447 (2012)

9. Chen, F.H., Huang, G.H., Fan, Y.R., et al.: A nonlinear fractional programming approach for environmental-economic power dispatch. Int. J. Electr. Power Energy Syst. **78**(1), 463–469 (2016)

10. Gang, Y., Tianyou, C., Xiaochuan, L.: Multiobjective production planning optimization using hybrid evolutionary algorithms for mineral processing. IEEE Trans. Evol. Comput. **15**(4), 487–514 (2011)

11. Deb, K., Pratap, A., Agareal, S., et al.: A fast and elitist multiobjective genetic algorithm: NSGA-II. IEEE Trans. Evol. Comput. **6**(2), 182–197 (2002)

12. Elsayed, S.M., Sarker, R.A., Essam, D.L.: An improved self-adaptive differential evolution algorithm for optimization problems. IEEE Trans. Industr. Inf. **9**(1), 89–99 (2013)

13. Robic, T., Filipic, B.: DEMO: differential evolution for multiobjective optimization. Lect. Notes Comput. Sci. **3410**(1), 520–533 (2005)

14. Wei, Q., Jianhua, Z., Nian, L., et al.: Multi-objective optimal generation dispatch with consideration of operation risk. Proc. CSEE **32**(22), 64–72 (2012). (in Chinese)

15. Shu, X., Ming, Z., Gengyin, L., et al.: On spinning reserve determination and power generation dispatch optimization for wind power integration systems. In: 2012 IEEE Power and Energy Society General Meeting, 22 July 2012, San Diego, CA, United states, pp. 1–8 (2012)

16. Xia, S., Zhou, M., Li, Y.: A coordinated active power and reserve dispatch approach for wind power integrated power systems considering line security verification. Proc. CSEE **33**(13), 18–26 (2013). (in Chinese)

17. Zifa, L., Jianhua, Z.: An improved differential evolution algorithm for economic dispatch of power systems. Proc. CSEE **28**(10), 100–105 (2008)

18. Das, S., Abraham, A., Konar, A.: Automatic clustering using an improved differential evolution algorithm. IEEE Trans. Syst. Hum. **38**(1), 218–237 (2008)

19. Wang, L., Wang, T.G., Luo, Y.: Improved non-dominated sorting genetic algorithm (NSGA)-II in multi-objective optimization studies of wind turbine blades. Appl. Math. Mech. **32**(6), 739–748 (2011)

20. Zhihuan, L., Xianzhong, D.: Comparison and analysis of multiobjective evolutionary algorithm for reactive power optimization. Proc. CSEE **4**(5), 57–65 (2010)

21. Olson, D.L.: Comparison of weights in TOPSIS models. Math. Comput. Model. **40**(1), 721–727 (2004)

22. Jie, W., Jiasen, S., Liang, L., et al.: Determination of weights for ultimate cross efficiency using Shannon entropy. Expert Syst. Appl. **38**(5), 5162–5165 (2011)

Integration of the Demand Side Management with Active and Reactive Power Economic Dispatch of Microgrids

Mohammed K. Al-Saadi[1,2], Patrick C.K. Luk[1(✉)], and John Economou[3]

[1] Cranfield University, Bedford, UK
p.c.k.luk@cranfield.ac.uk
[2] University of Technology, Baghdad, Iraq
[3] Cranfield University, Swindon, UK

Abstract. This paper presents a fully developed integration of the demand side management (DSM) into multi-period unified active and reactive power dynamic economic dispatch of the microgrids (MGs) combined with unit commitment (UC) to reduce the total operating cost or maximizes the profit with higher security. In the proposed optimization approach all consumers, such as residential, industrial, and commercial one can involve simultaneously in the DSM techniques. The shifting technique is applied to the residential load, while demand bidding programme (DBP) is applied to the industrial and commercial loads. The proposed optimal approach is tested on a low voltage (LV) hybrid connected MG including different types of loads and distributed generators (DGs). The results reveal that the proposed optimization approaches reduce the operating cost of the MG, while there are no impacts of the DSM on the profit.

Keywords: Demand response · Microgrid optimization · UC

1 Introduction

The DSM has several potential benefits not only for the utilities but also for customers as well. Therefore, there are many approaches and models to formulate the optimization of MGs with the integration of the DSM techniques efficiently. Authors in [1, 2] incorporated the DSM as load cutting with an optimization problem of the MG. These papers found that trade-off between the peak reduction, users comfort and cost can be achieved by tuning the penalty factor, while author in [3] incorporated the DSM as shifting algorithm with optimization problem and they claimed that the cost decreases when the load is shifted to the period when the renewable generation are available. The impacts of load cutting on the operating cost and the profit are addressed in [4, 5]. They stated that the load cutting reduces the operating cost, while it is not encouraged for maximizing profit. In [6, 7] authors proposed a shifting algorithm which is mathematically formulated to minimize the difference between the objective load curve and the actual load curve. This DSM technique was applied to the MG to study the impacts of DSM on the loads and on the operating cost. They claimed that the considering the

© Springer Nature Singapore Pte Ltd. 2017
K. Li et al. (Eds.): LSMS/ICSEE 2017, Part III, CCIS 763, pp. 653–664, 2017.
DOI: 10.1007/978-981-10-6364-0_66

DSM reduces the operating cost. In [8] authors proposed the integration of the DSM with the profit of the MG; however, they considered the DSM algorithms as input to the optimization algorithm and not as decision variable. They claimed that the DSM as shifting technique lead to reduce the operating cost. The impacts of the DSM as load shifting on the minimizing both operating cost and emission level of greenhouse gases are proposed in [9], where it is found that the management of the LV load can result in significant reduction in the operating cost.

In this paper, an optimal integration of the active and reactive DSM with security-constrained optimization algorithm of the MG is proposed. Both proposed DSM techniques are applied to the active and reactive loads, where these techniques are mathematically modelled and incorporated with optimization algorithms as decision variables. Besides, the reactive power production cost, battery degradation cost, and environmental cost are considered in the optimization problem. In addition, a comprehensive set of constraints including active and reactive steady state security and limitation of greenhouse gases constraints with constraints relevant to the active and reactive DSM techniques are taken into consideration in the formulation of the optimization problem. Further, both active and reactive powers are considered when formulating and solving the UC.

2   System Modelling

The components of the MG are modelled in this section as follows

2.1   Model of DGs Fuel Cost

The ith DG fuel cost can be calculated by the following equation [10]:

$$CP_{DG_i}(P_{DG_i}(t)) = \left(a_i + b_i.P_{DGi}(t) + c_i.P_{DGi}^2(t)\right) \tag{1}$$

where a_i (€/h), b_i (€/kWh), and c_i (€/kW^{2}h) are the coefficients of the fuel cost function, and $P_{DG_i}(t)$ is the output active power of ith.

2.2   Reactive Power Production Cost of the DGs

The production cost of the reactive power can be determined as follows [11]:

$$CQ_{DG_i}(Q_{DG_i}(t)) = \left(ar_i + br_i.Q_{DGi}(t) + cr_i.Q_{DGi}^2(t)\right) \tag{2}$$

where ar_i (€/h), br_i (€/kVArh), and cr_i (€/kVAr^{2}h) are the coefficients of the cost function of the reactive power, and $Q_{DG_i}(t)$ is the output reactive power of ith DG.

2.3   Model of the Cost of the DGs Maintenance

This cost can be calculated by the following equation:

$$COM_{DG_i}(P_{DG_i}(t)) = KOM_{DG_i}.P_{DG_i}(t) \tag{3}$$

where KOM_{DG_i} (€/kWh) is the coefficient of the maintenance cost of the ith DG.

2.4   Model of Operation Cost of the Batteries

The operating cost of the battery can be calculated by the using this equation

$$C_b(t) = C_d \times P_b(t) \times \Delta T \tag{4}$$

where C_d is the battery degradation cost (€/kWh), $P_b(t)$ is either charging or discharging power of the storage battery. Details about calculation of the C_d can be found in [12].

2.5   Model of the Exchanging Active and Reactive Power with the Main Grid

The cost of trading power with the main grid can be determined as:

$$C_{gP}(t) = c_{gP}(t).P_g(t) \tag{5}$$

$$C_{gQ}(t) = c_{gQ}(t).Q_g(t) \tag{6}$$

where $c_{gP}(t)$ in (€/kWh) and $c_{gQ}(t)$ in (€/kVArh) are the open market prices (OMPs), $P_g(t)$ and $Q_g(t)$ are the active and reactive trading power with utility grid.

2.6   Model of Greenhouse Gases Emission Cost

The cost of the emission of carbon dioxides CO_2, sulfur dioxides SO_2, nitrogen oxides NO_x, and particle matter PM can be calculated by:

$$C_e(t) = \sum_{j=1}^{M} \sum_{i=1}^{N} E_{j,i}.C_j.P_{DG_i}(t) \tag{7}$$

where C_j (€/kg) is a price of emission of jth greenhouse gas, and $E_{j,i}$ (kg/kWh) is the emission rate of greenhouse gas from the ith DG. M and N are the total number of greenhouse gases and DGs respectively.

2.7   Model of Star-up and Shutdown Cost of the DGs

The start-up and shutdown costs of the DGs can be formulated as follows:

$$SU_{DG_i}(t) = Sc_i.(\delta_{DG_i}(t) - \delta_{DG_i}(t).\delta_{DG_i}(t-1)) \tag{8}$$

$$SD_{DG_i}(t) = Sd_i.(\delta_{DG_i}(t-1) - \delta_{DG_i}(t).\delta_{DG_i}(t-1)) \tag{9}$$

where $\delta_{DG_i}(t)$ is the state of ith DG. Sc_i and Sd_i are the price of the start-up and shutdown cost of the ith DG.

3 Proposed Models of the DSM Techniques

In this paper, three types of loads are considered and different types of the DSM techniques applied to these loads namely residential (R), commercial (C) and industrial (I), where the DSM as shifting technique is applied to the residential load, while DBP technique is applied to the industrial and commercial loads.

3.1 Proposed Shifting Strategy

The shifting strategy is implemented to shift the connection time of the household smart appliances such as washing clothes machine (WM) and dishwasher (DW). The estimated number of the smart appliances can be obtained from this equation [13].

$$C_t = \sum_{w=1}^{d} D_{t-(w-1)}.p_w \tag{10}$$

where D_t represent the number of devices should be determined that starting time of consumption at time t. C_t is the consumption power at time t read from the diversified consumption, where diversified consumption profile for WMs and DWs can be obtained from households survey data presented in reports [14]. d is the duration of device consumption cycle and p_w is the device consumption at each time interval, $(w = 1, 2 \ldots d)$.

3.2 Proposed Mathematical Models of the Shifting Technique

By applying the shifting DSM techniques, the load will be changed according to the demand shifted and recovered as follows

$$P_{Dres}^{DSM}(t) = P_{Dres}(t) - P_{Dres}^{shft}(t) + P_{Dres}^{reco}(t) \tag{11}$$

where $P_{D}^{DSM}(t)$ is the total load after applying the DSM.

$$P_{Dres}^{reco}(t) = \sum_{i \in T} \sum_{k=1}^{n_1} X_{kit}.P_{1k} + \sum_{l=1}^{d_k-1} \sum_{i \in T} \sum_{k=1}^{n_1} X_{ki(t-1)}.P_{(1+l)k} \tag{12}$$

where X_{kit} is the number of smart appliances of type k that are shifted from time step i to t, P_{1k} and $P_{(1+l)k}$ are the power consumption at time steps 1 and $(1+l)$, n_1 is the number of device types, T is the optimization horizon.

Mathematical formulation of the shifting demand can be formulated in a similar manner as above

$$P_{Dres}^{shft}(t) = \sum\nolimits_{j\in T}\sum\nolimits_{k=1}^{n_1} X_{ktj}.P_{1k} + \sum\nolimits_{l=1}^{d_k-1}\sum\nolimits_{j\in T}\sum\nolimits_{k=1}^{n_1} X_{k(t-1)j}.P_{(1+l)k} \qquad (13)$$

where X_{ktj} is the number of the appliances of type k which are delayed from time step t to j.

3.3 Proposed DBP

The cost of the load shedding of the commercial and industrial areas can be formulated as follows:

$$C_{Pindshd}(t) = \sum\nolimits_{t=1}^{T}\delta_{ind}(t).P_{Dindshd}(t).c_{Pind}(t) \qquad (14)$$

$$C_{Pcomshed}(t) = \sum\nolimits_{t=1}^{T}\delta_{com}(t).P_{Dcomshd}(t).c_{Pcom}(t) \qquad (15)$$

where $\delta_{ind}(t)$ and $\delta_{com}(t)$ are the binary variables are employed to accept or reject the load curtailments for industrial and commercial sectors respectively. $P_{Dindshd}(t)$ and $P_{Dcomshd}(t)$ are the load cutting that they are offered by industrial and commercial consumers. $c_{Pind}(t)$ and $c_{Pcom}(t)$ are the prices that the industrial and commercial consumers would be willing to cut.

4 Formulation of the Proposed Optimization Problems

Both the shifting and DB techniques are considered simultaneously when formulating the optimization problem.

4.1 Minimizing the Total Operating Cost

The objective function of minimizing the total operating cost involves the fuel cost of the DGs, reactive power production cost, maintenance cost, start-up and shut down cost, environmental cost, battery degradation cost, cost of purchasing active and reactive from the utility grid, and the cost of the active and reactive industrial and commercial loads shed. The objective function also contains the revenue from selling power to the utility grid.

$$\begin{aligned}
F = Min \sum\nolimits_{t=1}^{T} \Big\{ &\sum\nolimits_{i=1}^{N} [[CP_{DG_i}(P_{DG_i}(t)) + CQ_{DG_i}(Q_{DG_i}(t)) \\
&+ COM_{DGi}(P_{DG_i}(t))]\delta_{DG_i}(t) + SU_{DG_i}(t) + SD_{DG_i}(t)] \\
&+ \sum\nolimits_{j=1}^{M}\sum\nolimits_{i=1}^{N} E_{j,i}.C_j.\delta_{DG_i}(t).P_{DG_i}(t) + C_d.P_b(t).\Delta T + c_{gP}(t).P_g(t) \\
&+ c_{gQ}(t).Q_g(t) + C_{Pindshd}(t) + C_{Qindshd}(t) + C_{Pcomshd}(t) + C_{Qcomshd}(t) \Big\}
\end{aligned} \qquad (16)$$

4.2 Maximizing the MG Profit

The goal of this objective function is to maximize the MG profit, where the MG sells the electricity to the consumers and to the utility grid in the (OMPs). The profit can be expressed as follows:

$$Maximaze(Revenue - Expense) = maximize(profit) \tag{17}$$

The objective function is as follows:

$$
\begin{aligned}
F = \text{Max} \sum_{t=1}^{T} & \left\{ \sum_{i=1}^{N} \left[c_{gP}(t).P_{DG_i}(t) + c_{gQ}(t).Q_{DG_i}(t) \right] \delta_{DG_i}(t) \right. \\
& + c_{gP}(t).P_{bdis}(t).\Delta t + c_{gP}(t).P_{PV}(t) + c_{gP}(t).P_W(t) \} \\
& - \sum_{t=1}^{T} \sum_{i=1}^{N} \left[[CP_{DG_i}(P_{DG_i}(t)) + CQ_{DG_i}(Q_{DG_i}(t)) \right. \\
& + COM_{DG_i}(P_{DG_i}(t)) \right] \delta_{DG_i}(t) + SU_{DG_i}(t) + SD_{DG_i}(t) \\
& + \sum_{j=1}^{M} \sum_{i=1}^{N} E_{j,i}.C_j.\delta_{DG_i}(t).P_{DG_i}(t) + C_d.P_b(t).\Delta T + c_{gP}.P_{bch}(t).\Delta T \\
& + b_W.P_W(t) + b_{PV}.P_{PV}(t) + C_{Pindshd}(t) + C_{Qindshd}(t) + C_{Pcomshd}(t) \\
& \left. + C_{Qcomshd}(t) \right\}
\end{aligned}
\tag{18}
$$

5 Models of Constraints

The following constraints are considered when solving the optimization problem

5.1 Power Balance Constraint

These constraints for active and reactive power are expressed as follows.

$$
\begin{aligned}
\sum_{t=1}^{T} & \left\{ \sum_{i=1}^{N} \delta_{DG_i}(t).P_{DG_i}(t) + P_w(t) + P_{PV}(t) + P_b(t) + P_g(t) \right. \\
& = (P_{Dres}(t) - P_{Dres}^{shft}(t) + P_{Dres}^{reco}(t)) + (P_{Dcom}(t) - P_{Dcomshd}(t)) \\
& \left. + (P_{Dind}(t) - P_{Dindshd}(t)) \right\}
\end{aligned}
\tag{19}
$$

$$
\sum_{t=1}^{T} \left\{ \sum_{i=1}^{N} \delta_{DG_i}(t).Q_{DG_i}(t) + Q_g(t) = Q_{Dres}^{DSM}(t) + (Q_{Dcom}(t) - Q_{Dcomshd}(t)) + (Q_{Dind}(t) - Q_{Dindshd}(t)) \right.
\tag{20}
$$

5.2 Ramp Rate Limits

This constraints is formulating as follows [15]:

$$-DR_i.\Delta T \le P_{DG_i}(t+1) - P_{DG_i}(t) \le UR_i.\Delta T \tag{21}$$

5.3 Emission Limitation Constraints

These constraints can be expressed as follows:

$$\sum_{i=1}^{N} E_{j,i}.P_{DG_i}(t) \le L_j \tag{22}$$

where L_j (kg/h) is the emission limitation of the jth greenhouse gas.

5.4   Steady Sate Security Constraints (SSSCs)

The SSSCs are essential for the reliable and secure operation of MGs and it is noteworthy that the constraints have to be satisfied at each time interval, and they are formulated as:

$$\sum_{i=1}^{T} \left\{ \sum_{i=1}^{N} \delta_{DG_i}(t).P_{DG_imax}(t) \geq P_{Dres}^{DSM}(t) + (P_{Dcom}(t) - P_{Dcomshd}(t)) + \atop (P_{Dind}(t) - P_{Dindshd}(t)) \right\} \tag{23}$$

$$\sum_{i=1}^{T} \left\{ \sum_{i=1}^{N} \delta_{DG_i}(t).P_{DG_imax}(t) \geq Q_{Dres}^{DSM}(t) + (Q_{Dcom}(t) - Q_{Dcomshd}(t)) + \atop (Q_{Dind}(t) - Q_{Dindshd}(t)) \right\} \tag{24}$$

5.5   Constraints Relevant to the Shifting DSM

The following constraints should be considered when formulating and solving the optimization problem with considering DSM.

A. The number of shifting appliances could not be negative

$$X_{kit} \geq 0 \qquad\qquad \forall i, t \in T, k \in n_1 \tag{25}$$

B. The number of the appliances that are shifted at a time step could not be greater than the appliances available for manage at the time step i.

$$D_{kit} \geq \sum_{t \in T} X_{kit} \qquad\qquad \forall i, t \in T \tag{26}$$

The expected numbers of smart appliances D_{kit}, which start their operation at time step i are calculated by using Eq. 10.

C. The appliances could not be moved back in the past

$$X_{kit} = 0 \qquad\qquad if\ t < i \tag{27}$$

D. The shifted appliances should be recovered within the scheduling day

$$X_{kit} = 0 \qquad\qquad if\ t > T - 1 \tag{28}$$

6   MG System Under Case Study

The proposed optimization algorithms are tested on the LV multi-feeders hybrid MG as shown in Fig. 1. The corresponding DGs technical parameters and the emission rates of the DGs are based on credible sources [2, 8, 16–19]. Moreover, the system includes a storage battery with a capacity of 50 kWh and the maximum energy at 22.5 kWh with charging and discharging efficiency assumed to be 0.9 both and DoD at 50%. The hourly profile for a typical day for wind and solar generation, open market prices (OMPs) [5, 20], total active and reactive load, and number of WMs and DWs for 192 UK household are shown in Table 1.

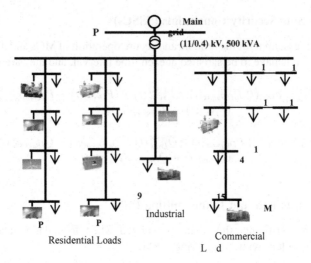

Fig. 1. The proposed multi-feeders MG

Table 1. Hourly profiles of the WTs and PV generation, OMPs, total active and reactive load, and number of WMs and DWs

Time (h)	Wind power (kW)	Solar power (kW)	Active price (€/kWh)	Reactive price (€/kWh)	Active load (kW)	Reactive load (kW)	No. WMs	No. DWs
1	8	3.916	0.050	0.005	186.800	90.467	9	6
2	8.444	6.721	0.070	0.007	221.060	107.059	8	6
3	6.667	9.014	0.080	0.008	250.400	121.269	8	6
4	5.111	10.760	0.090	0.009	266.500	129.066	7	6
5	5.556	11.589	0.120	0.012	265.080	128.378	7	5
6	6.222	11.431	0.225	0.023	286.500	138.752	8	6
7	7.333	10.408	0.100	0.010	297.400	144.031	8	5
8	10	8.414	0.085	0.009	283.380	137.241	8	7
9	11.111	5.962	0.150	0.015	270.100	130.809	8	5
10	14.667	3.352	0.450	0.045	258.000	124.949	8	10
11	15.556	1.411	0.150	0.015	292.600	141.706	9	6
12	13.333	0.3	0.180	0.018	300.200	145.387	9	13
13	12.222	0	0.160	0.016	313.000	151.586	9	7
14	10.667	0	0.310	0.031	277.800	134.539	8	13
15	8.889	0	0.050	0.005	233.000	112.842	7	5
16	8.444	0	0.040	0.004	186.000	90.080	4	6
17	9.111	0	0.025	0.003	104.600	50.658	3	1
18	10.667	0	0.035	0.004	100.240	48.546	3	3
19	13.333	0	0.032	0.003	92.000	44.556	3	3
20	14.444	0	0.030	0.003	90.260	43.713	3	3
21	16.667	0	0.030	0.003	84.600	40.972	3	3
22	14.889	0	0.033	0.003	90.820	43.984	4	4
23	12.444	0.335	0.050	0.005	100.040	48.449	6	5
24	8.889	1.693	0.047	0.005	136.100	65.913	9	6

7   Result and Discussion

Software tool ILOG CPLEX version 12.6 with interfacing with Microsoft Excel is used to solve the optimization problem. The shifted loads are moved from peak to off peak hours and the industrial and commercial consumers offer active and reactive load cutting for each hour from 11 to 13.

7.1   Minimizing the Total Operating Cost

It can be observed from Fig. 2 that the peak of the active total load is reduced from 313 kW to 267.93 kW. Similarly, the peak of the reactive load is reduced from 151.586 kVAr to 129.756 kVAr. However, the total peak load is still at hour 13. The reduction of the peak total load is resulted from both the shifting and shedding loads. Besides, the proposed DSM strategies reduce the peak of the residential and commercial loads because their peaks coincide with the peak of the total load. In contrast, the peak of the industrial load is not changed because its peak occurs in different hours of the peak of the total load, where the DSM is designed to reduce the peak of the total load. Furthermore, the proposed DSM increases the grid security by reducing the total load.

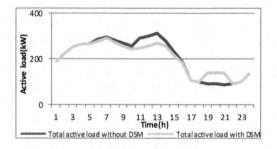

Fig. 2. Impacts of the DSM on the total load

Figures 3 and 4 illustrate that the active and reactive power are sold to the utility grid when the OMPs have the highest values and exceed the generation cost, in order to reduce the overall cost and the MG purchases active and reactive power from the utility grid when the OMPs reach low values. Besides, the battery is discharged the highest discharging power at hour 10 when the OMP reaches the highest value, while the battery is charged at low OMP. Therefore, the battery operations are managed to reduce the total operating cost. Moreover, the loads are recovered at hours 19 to 22 because the OMPs have the lowest values at these hours. It is found that the total operating cost is 288.5 € per scheduling day, where the results in the Table 2 reveals that the considering the DSM reduces the total operating cost by 8.9%.

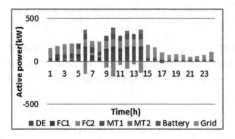

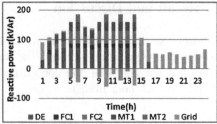

Fig. 3. Optimal active power scheduling   **Fig. 4.** Optimal reactive power scheduling

Table 2. Summarize the results with and without DSM

Cost without DSM (€)	Cost with DSM (€)	Percentage cost reduction %	Peak load reduction with DSM (kW)	Percentage load reduction %	No. shifting WMs	No. shifting DWs
316.7	288.5	8.9	45.1	14.4	82	66

7.2   Maximizing the MG Profit

Figure 5 reveals that the peak of the total load is not changed because the MG sells the electricity to the consumers and to the utility grid by the OMPs; therefore there are no economic incentives for shifting the peak load. Besides, the cutting of the industrial and commercial loads are equal to zero because the load cutting reduces the profit; therefore, the load cutting in the case of the maximizing profit is not encouraged. Figures 6 and 7 show that the MG sells power to the upstream grid when the OMPs have the highest values and purchases power when the prices have the lowest values. In addition, the active load at hours 17 to 24 are supplied from the utility grid, the DE, and the renewable generation, where the DE is operated with minimum output to satisfy the SSSCs. In contrast, the reactive load at the same hours is met from the utility grid solely because the minimum output reactive power of the DE is equal to zero. It is found that the profit is 247.2 € per scheduling day, where the results in the Table 3 demonstrates that the profit is slightly increased.

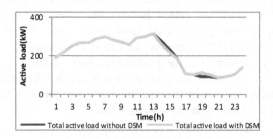

Fig. 5. Impacts of the DSM on the total load

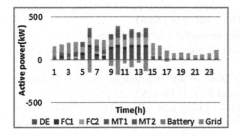

Fig. 6. Optimal active power scheduling

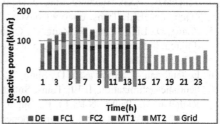

Fig. 7. Optimal reactive power scheduling

Table 3. Summarize the results with and without DSM

Profit without DSM (€)	Profit with DSM (€)	Profit increasing %	Peak load reduction with DSM (kW)	Percentage reduction %	No. WMs	No. DWs
246.8	247.2	0.2	0	0	15	18

8 Conclusions

A novel UC energy management system with integration of the DSM is presented. Different strategies of the DSM are applied to the different types of loads simultaneously. The results of minimizing the total operating cost reveal that the proposed DSM reduces the total operating cost. Besides, the proposed DSM also reduces the peak of the total load and reduces the peak of residential and commercial loads. In addition, the grid security is improved. In case of maximizing the profit, the obtained results demonstrate that the proposed DSM has insignificant impact on the profit and the peak load is not changed. The number of the shifted appliances also is quite less in comparison with the case of the minimizing the operating cost. Particularly, the OMPs values have significant impacts on the time and amount of the recovered loads for minimizing the operating cost. Furthermore, the operations of the battery typically reduce the total operating cost and increase the MG profit in spite of considering the battery degradation cost.

References

1. Parisio, A., Rikos, E., Tzamalis, G., Glielmo, L.: Use of model predictive control for experimental microgrid optimization. Appl. Energy **115**, 37–46 (2014)
2. Parisio, A., Rikos, E., Glielmo, L.: A model predictive control approach to microgrid operation optimization. IEEE Trans. Control Syst. Technol. **22**(5), 1813–1827 (2014)
3. Palma-Behnke, R., Benavides, C., Lanas, F., Severino, B., Reyes, L., Llanos, J., Saez, D.: A microgrid energy management system based on the rolling horizon strategy. IEEE Trans. Smart Grid **4**(2), 996–1006 (2013)

4. Tsikalakis, A.G., Hatziargyriou, N.D.: Centralized Control for Optimizing Microgrids Operation. IEEE Trans. Energy Convers. **23**(1), 241–248 (2008)
5. Zhang, D., Li, S., Zeng, P., Zang, C.: Optimal microgrid control and power-flow study with different bidding policies by using powerworld simulator. IEEE Trans. Sustain. Energy **5**(1), 282–292 (2014)
6. Logenthiran, T., Srinivasan, D., Shun, T.Z.: Demand side management in smart grid using heuristic optimization. IEEE Trans. Smart Grid **3**(3), 1244–1252 (2012)
7. Kinhekar, N., Padhy, N.P., Li, F., Gupta, H.O.: Utility oriented demand side management using smart AC and micro DC grid cooperative. IEEE Trans. Power Syst. **31**(2), 1151–1160 (2016)
8. Logenthiran, T., Srinivasan, D., Khambadkone, A.M., Aung, H.N.: Multiagent system for real-time operation of a microgrid in real-time digital simulator. IEEE Trans. Smart Grid **3** (2), 925–933 (2012)
9. Tsagarakis, G., Camilla Thomson, R., Collin, A.J., Harrison, G.P., Kiprakis, A.E., McLaughlin, S.: Assessment of the cost and environmental impact of residential demand-side management. IEEE Trans. Ind. Appl. **52**(3), 2486–2495 (2016)
10. Fossati, J.P., Galarza, A., Martín-villate, A., Font, L.: A method for optimal sizing energy storage systems for microgrids. Renew. Energy **77**, 539–549 (2015)
11. Xie, K., Song, Y.H., Zhang, D., Nakanishi, Y., Nakazawa, C.: Calculation and decomposition of spot price using interior point nonlinear optimisation methods. Int. J. Electr. Power Energy Syst. **26**(5), 349–356 (2004)
12. Tomić, J., Kempton, W.: Using fleets of electric-drive vehicles for grid support. J. Power Sources **168**(2), 459–468 (2007)
13. Cobelo, I.: Active control of distribution networks, Ph.D. thesis, Manchester University (2005)
14. Mark Bilton, G.S., Aunedi, M., Woolf, M.: Smart appliances for residential demand response. In: Low Carbon London Learning Lab. Imperial College London (2014)
15. Santos, J.R., Lora, A.T., Expósito, A.G., Ramos, J.L.M.: Finding improved local minima of power system optimization problems by interior-point methods. IEEE Trans. Power Syst. **18**(1), 238–244 (2003)
16. MICROGRIDS large scale integration of micro-generation to low voltage grids. In: Eu contract ENK5-CT-2002-00610, Tech. Annex (2004)
17. Box, P.O., -Hut, F.: Online management of MicroGrid with battery storage using multiobjective optimization. In: Power Engineering, Energy and Electrical Drives, pp. 231–236 (2007)
18. di Valdalbero, D.R.: External costs and their integration in energy costs. In: Europen Sustain Energy Policy Seminar (2006)
19. García-Gusano, D., Cabal, H., Lechón, Y.: Evolution of NOx and SO2 emissions in Spain: Ceilings versus taxes. Clean Technol. Environ. Policy **17**(7), 1997–2011 (2015)
20. Zakariazadeh, A., Jadid, S., Siano, P.: Smart microgrid energy and reserve scheduling with demand response using stochastic optimization. Electr. Power Energy Syst. **63**, 523–533 (2014)

Unit Commitment Dynamic Unified Active and Reactive Power Dispatch of Microgrids with Integration of Electric Vehicles

Mohammed K. Al-Saadi[1,2], Patrick C.K. Luk[1(✉)], and John Economou[3]

[1] Cranfield University, Bedfordshire, UK
p.c.k.luk@cranfield.ac.uk
[2] University of Technology, Baghdad, Iraq
[3] Cranfield University, Swindon, UK

Abstract. Electric vehicles (EVs) play a vital role in the reduction of emission of the greenhouse gases by reducing the consumption of fossil fuel. This paper presents a fully developed integration of the EVs with a security-constrained unified active and reactive power dynamic economic dispatch of microgrids (MGs) to minimize the total operating cost or maximizes the profit. The formulation of the overall optimization problem considers the reactive power production cost and relevant constraints, the environmental costs, and the battery degradation cost. A comprehensive set of constraints including active and reactive security constraints, limitation of the greenhouse gases constraints, and constraints relevant to the integration of the EVs with the MG are considered as well. The bi-directional penetration of the EVs with the MG is modelled and incorporated with unit commitment (UC) optimization problem. The results show that the proposed approach of the integration of the EVs with the MG reduces the total operating cost and increases the profit.

Keywords: Optimal charging and discharging of EVs · Energy storage · UC

1 Introduction

Many researchers proposed the optimization problem of the MG with the integration of the EVs. Some of the researchers considered the integration of the EVs either as electricity suppliers (V2G) or as additional load (G2V) and some of them considered the EVs as bi-directional integration. Authors in [1] proposed the UC optimization problem to reduce the cost-emission problem of the conventional system with integration of EVs as V2G. They found that the UC with EVs reduce operational cost and emission, whereas in [2] authors presented optimization problem with bi-directional integration of the EVs to minimize the cost and emission of the system including renewable energy resources. The results showed that the proposed integration of the EVs reduced the operating cost and emission level of greenhouse gases. Authors in [3] addressed the optimization problem of distribution network for minimizing the operating cost of the hybrid MG with considering EVs as bi-directional integration. It is found that the integration of the EVs reduced the operating cost. In [4] authors

K. Li et al. (Eds.): LSMS/ICSEE 2017, Part III, CCIS 763, pp. 665–676, 2017.
DOI: 10.1007/978-981-10-6364-0_67

presented a day-ahead energy management system for hybrid LV connected residential MG with considering the V2G and the aim of optimization was to minimize the operating cost. The results obtained demonstrated that the managing strategy saved cost by 10%, while authors in [5] proposed energy management system for low voltage MG which includes photovoltaic (PV), energy storage and V2G and G2V. The paper concluded that the proposed approach reduced the operation cost and improved the reliability of the system. A quiet few researchers addressed the participation of the EVs in the deregulate market, where in [6] authors demonstrated day-ahead probabilistic optimal operation of the connected MG. The aim of the objective function was to maximize the total profit of the MG and to investigate the impacts of the integration of the EVs on the economic operation of the MG.

In this paper, a novel methodology to integrate the EVs with unified active and reactive power dynamic economic dispatch of the MG is presented. The UC is extended and developed to take into consideration both the active and reactive power. The overall formulation of the optimization problem includes emission of greenhouse gases cost, battery degradation cost, reactive power production cost, and the cost of buying power from the EVs and utility grid. Besides, the constraints such as active and reactive security, emission level limitation, the constraints relevant to the reactive power management, and the constraints of the operating of the EVs are considered when formulating the optimization problem. In addition, both the system operators and owners of the EVs requirements are satisfied in the proposed optimization approach.

2   System Modelling

The components of the MG are modelled in this section as follows:

2.1   Fuel Cost of the DGs

The fuel cost of the ith DG can be modelled as follows, [7]:

$$CP_{DG_i}(P_{DG_i}(t)) = \left(a_i + b_i \cdot P_{DGi}(t) + c_i \cdot P_{DGi}^2(t)\right) \tag{1}$$

where a_i (€/h), b_i (€/kWh), and c_i (€/kW^{2}h) are the respective coefficients of the fuel cost funct i^{th} ion, and $P_{DG_i}(t)$ is the output active power of ith DGs.

2.2   Reactive Power Production Cost of the DGs

The reactive power production cost of ith DG can be calculated as [8]:

$$CQ_{DG_i}(Q_{DG_i}(t)) = \left(ar_i + br_i \cdot Q_{DGi}(t) + cr_i \cdot Q_{DGi}^2(t)\right) \tag{2}$$

where ar_i (€/h), br_i (€/kVArh), and cr_i (€/kVAr^{2}h) are the respective coefficients of the reactive power cost function, and $Q_{DG_i}(t)$ is the output reactive power of ith DG.

2.3   Model the Operating and Maintenance Cost of DGs

Operating and maintenance cost can be calculated by the following equation:

$$COM_{DG_i}(P_{DG_i}(t)) = KOM_{DG_i} \cdot P_{DG_i}(t) \tag{3}$$

where KOM_{DG_i} (€/kWh) is the coefficient of the maintenance cost of the ith DG.

2.4   Model of Fixed Battery Cost

To allow for the cost of battery degradation, the following equation suffices to factor into the effect in the total MG cost, and is expressed as:

$$C_b(t) = C_d \times P_b(t) \times \Delta T \tag{4}$$

where C_d is the battery degradation cost (€/kWh), $P_b(t)$ is either charging or discharging power of the storage battery. Details about calculation of the C_d can be found in [9].

2.5   Model of the Exchanging Active and Reactive Power with the Utility Grid

The cost of trading active and reactive power with the main grid can be expressed as:

$$C_{gP}(t) = c_{gP}(t) \cdot P_g(t) \tag{5}$$

$$C_{gQ}(t) = c_{gQ}(t) \cdot Q_g(t) \tag{6}$$

where $c_{gP}(t)$ (€/kWh) and $c_{gQ}(t)$ (€/kVArh) are the active and reactive open market prices(OMPs). $P_g(t)$ and $Q_g(t)$ are the active and reactive trading power.

2.6   Model of Greenhouse Gases Emission Cost

The emission costs of carbon dioxides CO_2, sulfur dioxides SO_2, nitrogen oxides NO_x, and particle matter PM can be calculated by:

$$C_e(t) = \sum_{j=i}^{M} \sum_{i=1}^{N} E_{j,i} \cdot C_j \cdot P_{DG_i}(t) \tag{7}$$

where C_j (€/kg) is a price of emission of jth greenhouse gas, and $E_{j,i}$ (kg/kWh) is the emission rate of greenhouse gas from the ith DG. M and N are the total number of Greenhouse gases and DGs respectively.

2.7　Star-Up and Shutdown Cost of the DGs

The start-up and shutdown costs of the DGs can be formulated as follows [10]:

$$SU_{DG_i}(t) = Sc_i \cdot (\delta_{DG_i}(t) - \delta_{DG_i}(t) \cdot \delta_{DG_i}(t-1)) \tag{8}$$

$$SD_{DG_i}(t) = Sd_i \cdot (\delta_{DG_i}(t-1) - \delta_{DG_i}(t) \cdot \delta_{DG_i}(t-1)) \tag{9}$$

where $\delta_{DG_i}(t)$ is the state of ith DG. Sc_i and Sd_i are the price of the start-up and shutdown cost of the ith DG.

2.8　Proposed Model of the EVs

The EVs are modelled as a storage battery in the economic operation of the power system. The modelling of the EV is as follows:

$$E_{EV}(t) = E_{EV}(t-1) + \left(P_{EVch}(t) \cdot \eta_{EVch} - \left(\frac{P_{EVdis}(t)}{\eta_{EVdis}}\right)\right) \cdot \Delta t \tag{10}$$

When the EV is driven at hour t

$$E_{EV}(t) = E_{EV}(t-1) - E_{EV}^{Trip}(t) \tag{11}$$

$$E_{EV}^{Trip}(t) = C \cdot D(t) \tag{12}$$

where $E_{EV}(t)$, $E_{EV}(t-1)$ are the state charge of the battery at current and previous state respectively, $P_{EVch}(t)$, $P_{EVdis}(t)$ are the battery charging and discharging power respectively. η_{EVch}, η_{EVdis} are the corresponding charging and discharging efficiencies, Δt is the sampling time. $E_{EV}^{Trip}(t)$ is the energy consumption during the trip of the EV at period t, C is the driving energy consumption per km and $D(t)$ is the driving distance of the EV at hour t.

2.9　Model the Cost of the Integration of the EVs with the MG

The EVs can be connected to grid at different areas and the cost of the bi-directional integration of the EVs with optimization problem as follows:

A.The cost of the EVs that are connected to the residential area

$$C_{EV}^{Res}(t) = \sum_{t=1}^{T} N_{EV}^{Res}(t) \{ c_{EVdis}(t) \cdot P_{EVdis}^{Res}(t) - c_{EVch}(t) \cdot P_{EVch}^{Res}(t) \} \Delta T \tag{13}$$

B.The cost of the EVs that are connected to the industrial area

$$C_{EV}^{Ind}(t) = \sum_{t=1}^{T} N_{EV}^{Ind}(t) \{ c_{EVdis}(t) \cdot P_{EVdis}^{Ind}(t) - c_{EVch}(t) \cdot P_{EVch}^{Ind}(t) \} \Delta T \tag{14}$$

C. The EVs that are connected to the commercial area

$$C_{EV}^{Com}(t) = \sum_{t=1}^{T} N_{EV}^{Com}(t)\{c_{EVdis}(t) \cdot P_{EVdis}^{Com}(t) - c_{EVch}(t) \cdot P_{EVch}^{Com}(t)\}\Delta T \qquad (15)$$

where $c_{EVch}(t)$, $c_{EVdis}(t)$ are the price of charging and discharging respectively (€/kWh). $N_{EV}^{Res}(t)$, $N_{EV}^{Ind}(t)$, and $N_{EV}^{Com}(t)$ are the number of the EVs that are connected at each time interval to the residential, industrial, and commercial areas respectively. P_{EVdis}^{Res}, P_{EVdis}^{Ind}, and P_{EVdis}^{Com} are the discharging power of the EVs that are connected to the residential, industrial and commercial sectors respectively. P_{EVch}^{Res}, P_{EVch}^{Ind}, and P_{EVch}^{Com} are the charging power of the EVs that are connected to the residential, industrial and commercial sectors respectively.

3   Formulation of the Proposed Optimization Problems

3.1   Minimizing the Total Operating Cost

The goal of this objective function is to minimize the total operating cost, where the cost function includes the fuel cost of the DGs, reactive power production cost, maintenance cost, start-up and shut down cost, environmental cost, battery degradation cost, cost of purchasing active and reactive from the main grid, cost of the buying power from the EVs, and cost of the unserved EVs in all areas.

$$\begin{aligned} F = Min \sum_{t=1}^{T} \sum_{i=1}^{N} & \{[[CP_{DG_i}(P_{DG_i}(t)) + CQ_{DG_i}(Q_{DG_i}(t)) + COM_{DG_i}(P_{DG_i}(t))]\delta_{DG_i}(t) \\ & + SU_{DG_i}(t) + SD_{DG_i}(t)] + C_d \cdot P_b(t) \cdot \Delta T + \sum_{j=1}^{M} \sum_{i=1}^{N} E_{j,i} \cdot C_j \cdot \delta_{DG_i}(t) \cdot P_{DG_i}(t) \\ & + c_{gP}(t) \cdot P_g(t) \cdot \Delta T + c_{gQ}(t) \cdot Q_g(t) + C_{EV}^{Res}(t) + C_{EV}^{Ind}(t) + C_{EV}^{Com}(t) \\ & + c_{EVcut} \cdot P_{REVcut}(t) \cdot \Delta T + c_{EVcut} \cdot P_{IEVcut}(t) \cdot \Delta T + c_{EVcut} \cdot P_{CEVcut}(t) \cdot \Delta T\} \end{aligned}$$

$$(16)$$

where c_{EVcut} in (€/kWh) is the price of unserved the EVs charging, $P_{REVcut}(t)$, $P_{IEVcut}(t)$, and $P_{CEVcut}(t)$ are unserved power to the EVs in the residential, industrial and commercial areas.

3.2   Maximizing the MG Profit

The aim of this objective function is to maximize the profit of the MG, where the MG sells the electricity to the consumers and to the utility grid in the OMPs. The profit can be expressed as follows:

$$Maximaze(Revenue - Expense) = maximize(profit) \qquad (17)$$

$$F = \text{Max} \sum_{t=1}^{T} \left\{ \sum_{i=1}^{N} [c_{gP}(t) \cdot P_{DG_i}(t) + c_{gQ}(t) \cdot Q_{DG_i}(t)] \delta_{DG_i}(t) \right.$$

$$+ c_{gP}(t) \cdot P_{bdis}(t) \cdot \Delta t + c_{gP}(t) \cdot P_{PV}(t) + c_{gP}(t) \cdot P_W(t)$$

$$\left. + c_{gP}(t) \cdot [N_{EV}^{Res}(t) \cdot P_{EVdis}^{Res}(t) + N_{EV}^{Ind}(t) \cdot P_{EVdis}^{Ind}(t) + N_{EV}^{Com}(t) \cdot P_{EVdis}^{Com}(t)] \right\}$$

$$- \sum_{t=1}^{T} \left\{ \sum_{i=1}^{N} [[CP_{DG_i}(P_{DG_i}(t)) + CQ_{DG_i}(Q_{DG_i}(t)) + COM_{DG_i}(P_{DG_i}(t)) \delta_{DG_i}(t) \right.$$

$$+ SU_{DG_i}(t) + SD_{DG_i}(t)] + \sum_{j=1}^{M} \sum_{i=1}^{N} E_{j,i} \cdot C_j \cdot \delta_{DG_i}(t) \cdot P_{DG_i}(t) + C_d \cdot P_b(t) \cdot \Delta T$$

$$+ c_{gP} \cdot P_{bch}(t) \cdot \Delta T + b_W \cdot P_W(t) + b_{PV} \cdot P_{PV}(t) + c_{EVcut} \cdot P_{REVcut}(t) \cdot \Delta T$$

$$+ c_{EVcut} \cdot P_{IEVcut}(t) \cdot \Delta T + c_{EVcut} \cdot P_{CEVcut}(t) \cdot \Delta T + c_{EVdis} \cdot [N_{EV}^{Res}(t) \cdot P_{EVdis}^{Res}(t)$$

$$+ N_{EV}^{Ind}(t) \cdot P_{EVdis}^{Ind}(t) + N_{EV}^{Com}(t) \cdot P_{EVdis}^{Com}(t)] + (c_{gP}(t) - c_{EVch}) \cdot [N_{EV}^{Res}(t) \cdot P_{EVch}^{Res}(t)$$

$$\left. + N_{EV}^{Ind}(t) \cdot P_{EVch}^{Ind}(t) + N_{EV}^{Com}(t) \cdot P_{EVch}^{Com}(t)] \right\} \tag{18}$$

where b_{PV} and b_W are respectively yearly depreciation of generation per kWh for wind and photovoltaic panels.

4 Proposed Constraints

These constraints should be satisfied at each time interval.

4.1 Power Balance Constraint

These constraints for active and reactive power balance are expressed as follows.

$$\sum_{t=1}^{T} \left\{ \sum_{i=1}^{N} \delta_{DG_i}(t) \cdot P_{DG_i}(t) + P_w(t) + P_{PV}(t) + P_b(t) + P_g(t) = (P_{Dres}(t) \right.$$

$$- N_{EV}^{Res}(t) \cdot P_{EVdis}^{Res}(t) + N_{EV}^{Res}(t) \cdot P_{EVch}^{Res}(t)) + (P_{Dind}(t) - N_{EV}^{Ind}(t) \cdot P_{EVdis}^{Ind}(t)$$

$$\left. + N_{EV}^{Ind}(t) \cdot P_{EVch}^{Ind}(t)) + (P_{Dcom}(t) - N_{EV}^{Com}(t) \cdot P_{EVdis}^{Com}(t) + N_{EV}^{Com}(t) \cdot P_{EVch}^{Com}(t)) \right\} \tag{19}$$

$$\sum_{t=1}^{T} \left\{ \sum_{i=1}^{N} \delta_{DG_i}(t) \cdot Q_{DG_i}(t) + Q_g(t) = Q_{Dres}(t) + Q_{Dcom}(t) + Q_{Dind}(t) \right\} \tag{20}$$

4.2 Emission Constraints

These constraints can be expressed as follows:

$$\sum_{i=1}^{N} E_{j,i} \cdot P_{DG_i}(t) \leq L_j \tag{21}$$

where L_j is the emission limitation of the jth Greenhouse gas in the area of the MG.

4.3 Steady Sate Security Constraints (SSSCs)

These constraints are formulated as:

$$\sum_{i=1}^{T} \left\{ \sum_{i=1}^{N} \delta_{DG_i}(t) \cdot P_{DG_imax}(t) \geq P_{Dres}(t) + P_{Dcom}(t) + P_{Dind}(t) \right\} \tag{22}$$

$$\sum_{i=1}^{T} \left\{ \sum_{i=1}^{N} \delta_{DG_i}(t) \cdot P_{DG_imax}(t) \geq Q_{Dres}(t) + Q_{Dcom}(t) + Q_{Dind}(t) \right\} \tag{23}$$

4.4 Proposed Electric Vehicles Operation Constraints

These constraints should be satisfied at each time interval and the same constraints can be used for fixed storage battery.

A. State of Charge Constraints

This constraint is expressed in this equation

$$E_{EVmin} \leq E_{EV}(t) \leq E_{EVmax} \tag{24}$$

where E_{EVmax} and E_{EVmin} are the determined maximum and minimum of the battery state of charge.

B. Charging and Discharging Power Constraints

To prevent the simultaneous charging and discharging operations of batteries of the EVs at each time interval two binary variables, $\delta_{EVch}(t) \in [0, 1]$ and $\delta_{EVdis}(t) \in [0, 1]$, are assigned to formulate the status of battery operation and $\delta_{EVch}(t) + \delta_{EVdis}(t) \leq 1$ is set to prevent the battery of EV charging and discharging simultaneously during the optimization. The charging and discharging power is performed at the maximal power available that the charger provided. These constraints for EVs battery can be accordingly formulated as

$$\delta_{EVch}(t) \cdot P_{EVchmin} \leq P_{EVch}(t) \leq \delta_{EVch} \cdot P_{EVchmax} \tag{25}$$

$$\delta_{EVdis}(t) \cdot P_{EVdismin} \leq P_{EVdis}(t) \leq \delta_{EVdis} \cdot P_{EVcdismax} \tag{26}$$

C. The Owner of the EV Requirements

This constraint is formulated as

$$E_{EV}(t_{last}) \geq E_{EV}^{Trip,q}(t) \tag{27}$$

where t_{last} is the last connecting time of EV with grid before start q trip, $E_{EV}^{Trip,q}(t)$ is the required energy for the EV q trip which starts at hour $t_{last} + 1$.

4.5 Maximum System Capacity Constraint

This constraint is formulated as follows

$$\sum\nolimits_{t=1}^{T}\{P_{bch}(t) + (P_{Dres}(t) + N_{EV}^{Res}(t) \cdot P_{EVch}^{Res}(t)) + (P_{Dind}(t) + N_{EV}^{Ind}(t) \cdot P_{EVch}^{Ind}(t))$$
$$+ (P_{Dcom}(t) + N_{EV}^{Com}(t) \cdot P_{EVch}^{Com}(t)) \leq S_{sys} \cdot cos\theta\} \tag{28}$$

where S_{sys} is the KVA rating of the MG, $cos\theta$ is the power factor.

5 EVs Related Parameters

The percentage number of the drivers arriving their homes and works from the final trip for residential and commercial sectors are shown in Table 1. The batteries of all EVs are lithium ion and their capacities are 29.02 kWh for REVs and CEVs, while for IEVs are 15 kWh and the cycle life of the batteries and DoD are 2200 and 95% respectively [11]. It is supposed that the 33% of households have EVs [12] and 3 EVs in industrial area with 12 EVs in commercial area. The charging and discharging prices are assumed fixed over the entire scheduling horizon and they equal to 0.08 and 0.16 (€/kWh) respectively.

Table 1. Hourly profile of the wind and PV generation, OMP, total loads, and number of the EVs in the residential and commercial areas

T (h)	Wind power (kW)	Solar power (kW)	Active power price (€/kWh)	Reactive power price (€/kWh)	Total active load (kW)	EVs in the residential area	EVs in the commercial area
1	8	3.916	0.050	0.005	186.800	0	2
2	8.444	6.721	0.070	0.007	221.060	0	6
3	6.667	9.014	0.080	0.008	250.400	1	3
4	5.111	10.760	0.090	0.009	266.500	1	1
5	5.556	11.589	0.120	0.012	265.080	2	0
6	6.222	11.431	0.225	0.023	286.500	3	0
7	7.333	10.408	0.100	0.010	297.400	3	0
8	10	8.414	0.085	0.009	283.380	4	0
9	11.111	5.962	0.150	0.015	270.100	7	0
10	14.667	3.352	0.450	0.045	258.000	10	0
11	15.556	1.411	0.150	0.015	292.600	11	0
12	13.333	0.3	0.180	0.018	300.200	6	0
13	12.222	0	0.160	0.016	313.000	4	0
14	10.667	0	0.310	0.031	277.800	4	0
15	8.889	0	0.050	0.005	233.000	4	0
16	8.444	0	0.040	0.004	186.000	2	0
17	9.111	0	0.025	0.003	104.600	1	0
18	10.667	0	0.035	0.004	100.240	0	0
19	13.333	0	0.032	0.003	92.000	0	0
20	14.444	0	0.030	0.003	90.260	0	0
21	16.667	0	0.030	0.003	84.600	0	0
22	14.889	0	0.033	0.003	90.820	0	0
23	12.444	0.335	0.050	0.005	100.040	0	0
24	8.889	1.693	0.047	0.005	136.100	0	0

6 System Under Case Study

The proposed optimization approaches are validated by applying these approaches to the low voltage (LV) multi-feeders hybrid MG as shown in Fig. 1. The power factor is assumed to be typically 0.9 for the whole system. The corresponding DG technical parameters and the emission rates of the DGs are based on credible sources [13–18]. Moreover, the system includes a fixed storage battery with a capacity of 50 kWh and maximum charging and discharging energy at 22.5 kWh, and with charging and discharging efficiency assumed to be 0.9 both and DoD is 50%. The hourly profile for a typical day of wind and solar generation, OMPs, total load, and number of the EVs that are connected at each time interval to the residential and commercial areas are shown in Table 1 [19, 20].

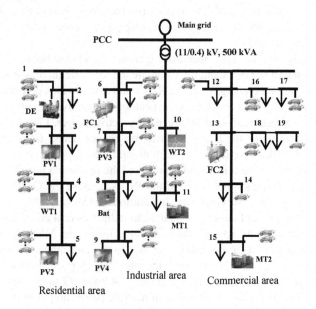

Fig. 1. Proposed multi-feeders MG in this paper

7 Result and Discussion

Software tool ILOG CPLEX version 12.6 which is interfaced with Microsoft Excel is used to solve optimization problem. The discharging operations are allocated on peak hours (6–14), while the charging operations are allocated at off peak hours (17–24).

7.1 Minimizing the Operating Cost and Maximizing the Profit Without EVs

In order to quantify the impacts of the EVs on the optimal operation of the MG, the comparisons with the case without EVs (base case) are conducted. The total operating cost and the profit for this case is 316.7 € and 246.8 € respectively.

7.2   Minimizing the Operating Cost and Maximizing the Profit with the EVs

Figure 2 depicts both the charging and discharging operations of the all EVs. It can be observed that the IEVs and CEVs are charged when the active OMP has low values and discharged when the price has the highest value, where the ICEVs, CEVs, and REVs are charged at hour 6 because the OMP has high value and higher than the discharging price. Besides, the highest charging power of IEVs and CEVs occurs at hour 8 because the OMP has low value and the owners should fully charge their vehicles before leaving the grid. In addition, the REVs are discharged when the load has high values and the OMP has high values at hours 6, 10, 11, 12, 13, and 14, where the highest discharging power occurs at hour 14 because the price has high value and higher than hours 12 and 13. Further, the REVs are charged over the entire charging period at hours 17 to 24 to prevent the base and EVs loads from increasing higher than the maximum capacity of the system, where the highest charging power occurs at hour 21 because the OMP has the lowest value.

Figure 3 shows that the EVs operations increase the load at hours 1, 2, 3, 4, and 8 due to the charging of the IEVs and CEVs, while the loads are increased significantly at hours 17 to 24 due to the charging of the REVs and it is obvious that peak of the total active load is shifted to hour 8, whereas the peak of the reactive power does not change because the EVs are charged or discharged only active power. Figures 4 and 5 show the optimal scheduling of the active and reactive power. These figures reveal that at hours 17 to 24 the DE is committed with minimum output power to satisfy the active and reactive SSSCs and the utility grid is provided the maximum possible active power to supply the base and charging loads of the REVs because the OMP has the lowest values and lower than the charging price. In addition, the MG sells the maximum active power to the utility grid at hours 6, 10, and 14 to reduce its cost and increases its profit because the OMP has the highest values. At these hours, the MG buys active power from the EVs and sells it to the utility grid and to the consumers because the active OMP has higher values and higher than the discharging price. Further, the reactive power load is provided from the utility grid solely at hours 17 to 24, although the DGs are committed. This is because the minimum reactive power output of the DGs is equal to zero.

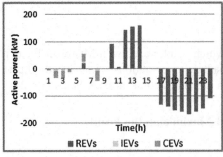

Fig. 2. Charging and discharging of the EVs

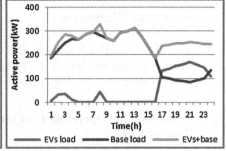

Fig. 3. Modified load with the EVs

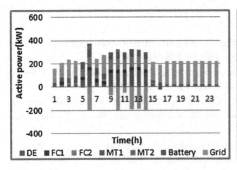

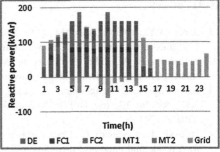

Fig. 4. Optimal active power scheduling

Fig. 5. Optimal reactive power scheduling

Table 2. Optimal hourly total cost

Time (h)	Cost (€/h)	Time (h)	Cost (€/h)	Time (h)	Cost (€/h)	Time (h)	Cost (€/h)
1	10.9	7	25.6	13	33.6	19	−3.0
2	15.1	8	22.7	14	−2.0	20	−3.8
3	19.4	9	23.8	15	15.6	21	−4.6
4	21.6	10	−37.9	16	12.4	22	−3.2
5	22.8	11	28.9	17	−2.7	23	1.3
6	9.1	12	26.8	18	−1.3	24	3.8

Table 2 shows the optimal hourly cost, where the highest cost is at hour 13 because the load has the highest value. In some hours, the hourly cost has negative values that mean revenue to the MG. It is found that the total cost is 235.2 € per scheduling day, where the proposed economic integration of the EVs reduces the total cost by 81.5 € or by 25.7%. The profit is 328.3 €, where the profit is increased by 81.5 € or by 33%. The charging and discharging costs are 104.2 and 98.1 € per scheduling day respectively, where the consumers will pay only 6.1 € to charge their vehicles. Therefore, the consumers gain 37.6 € from their participation in the optimal scheduling of the MG because the charging cost of the all EVs in all areas is 43.7 €.

8 Conclusions

A novel security-constraint UC unified active and reactive optimization problem with bi-directional integration of the EVs for minimizing the total operating cost or maximizes the MG profit is presented. The results demonstrate that the proposed economic integration of the EVs with grid reduces the total operating cost and increases the profit. The EVs also are discharged when the OMP has high values and charged when the OMP reaches low values to minimize the total operating cost or maximizes the profit. Besides, the EVs owners also gain revenue from participating in the optimal scheduling of the MG. In addition, the active OMP values determine the charging and discharging operations of the EVs.

676   M.K. Al-Saadi et al.

References

1. Saber, A.Y., Venayagamoorthy, G.K.: Intelligent unit commitment with vehicle-to-grid—a cost-emission optimization. J. Power Sources **195**(3), 898–911 (2010)
2. Zakariazadeh, A., Jadid, S., Siano, P.: Multi-objective scheduling of electric vehicles in smart distribution system. Energy Convers. Manag. **79**, 43–53 (2014)
3. Lan-xiang, H.Z., Change-nian, L., Yu-kai, L.: Optimal operation of complicated distribution networks. In: Electric Utility Deregulation and Restructuring and Power Technologies (DRPT) (2015)
4. Igualada, L., Corchero, C., Cruz-Zambrano, M., Heredia, F.J.: Optimal energy management for a residential microgrid including a vehicle-to-grid system. IEEE Trans. Smart Grid **5**(4), 2163–2172 (2014)
5. Wang, W., Jiang, X., Su, S., Kong, J., Geng, J., Cui, W.: Energy management strategy for microgrids considering photovoltaic-energy storage system and electric vehicles. In: IEEE Transportation Electrification Conference and Expo, ITEC Asia-Pacific 2014—Conference Proceedings, pp. 1–6 (2014)
6. Masoud, S., Tafreshi, M., Ranjbarzadeh, H., Jafari, M.: A probabilistic unit commitment model for optimal operation of plug-in electric vehicles in microgrid. Renew. Sustain. Energy Rev. **66**, 934–947 (2016)
7. Fossati, J.P., Galarza, A., Martín-villate, A., Font, L.: A method for optimal sizing energy storage systems for microgrids. Renew. Energy **77**, 539–549 (2015)
8. Xie, K., Song, Y.H., Zhang, D., Nakanishi, Y., Nakazawa, C.: Calculation and decomposition of spot price using interior point nonlinear optimisation methods. Int. J. Electr. Power Energy Syst. **26**(5), 349–356 (2004)
9. Tomić, J., Kempton, W.: Using fleets of electric-drive vehicles for grid support. J. Power Sources **168**(2), 459–468 (2007)
10. Al-saadi, M.K., Luk, P.C.K.: Impact of Unit Commitment on the Optimal Operation of Hybrid Microgrids, no. 2 (2016)
11. Chen, C., Duan, S.: Optimal integration of plug-in hybrid electric vehicles in microgrids. IEEE Trans. Industr. Inf. **3203**, 1 (2014)
12. Papadopoulos, P.: Integration of Electric Vehicles into Distribution Networks. Ph.D. thesis. Cardiff University (2012)
13. Parisio, A., Rikos, E., Glielmo, L.: A model predictive control approach to microgrid operation optimization. IEEE Trans. Control Syst. Technol. **22**(5), 1813–1827 (2014)
14. Logenthiran, T., Srinivasan, D., Khambadkone, A.M., Aung, H.N.: Multiagent system for real-time operation of a microgrid in real-time digital simulator. IEEE Trans. Smart Grid **3**(2), 925–933 (2012)
15. MICROGRIDS large scale integration of micro-generation to low voltage grids. In: Eu contract ENK5-CT-2002–00610, Technologies Annex (2004)
16. di Valdalbero, D.R.: External costs and their integration in energy costs. In: European Sustain Energy Policy Seminar (2006)
17. ATSE: The hidden costs of electricity: externalities of power generation in Australia. In: Australian Academy of Technology Science Engineering, Australia (2009)
18. García-Gusano, D., Cabal, H., Lechón, Y.: Evolution of NOx and SO2 emissions in Spain: ceilings versus taxes. Clean Technol. Environ. Policy **17**(7), 1997–2011 (2015)
19. Zhang, D., Li, S., Zeng, P., Zang, C.: Optimal microgrid control and power-flow study with different bidding policies by using powerworld simulator. IEEE Trans. Sustain. Energy **5**(1), 282–292 (2014)
20. Zakariazadeh, A., Jadid, S., Siano, P.: Smart microgrid energy and reserve scheduling with demand response using stochastic optimization. Electr. Power Energy Syst. **63**, 523–533 (2014)

Optimal Design and Planning of Electric Vehicles Within Microgrid

Mohammed Alkhafaji[1], Patrick Luk[1(✉)], and John Economou[2]

[1] Cranfield University, Bedford MK43 0SX, UK
{m.h.alkhafaji,p.c.k.luk}@cranfield.ac.uk
[2] Cranfield University, Shrivenham, Swindon SN6 8LA, UK
j.t.economou@cranfield.ac.uk

Abstract. Optimal allocation and economic dispatch of the distributed generators (DGs) and electric vehicles (EVs) are very important to achieve resilience operating of future microgrids. This paper presents a new energy management concept of interfacing EV charging stations with the microgrids. Optimal scheduling operation of DGs and the EVs is used to minimize the total combined operating and emission costs of a hybrid microgrid. The problem was solved using a mixed integer quadratic programming (MIQP) approach. Different kinds of distributed generators with realistic constraints and charging stations for various EVs with the view to optimizing the overall microgrid performance are investigated. The results have convincingly revealed that discharging EVs could reduce the total cost of the microgrid operation.

Keywords: Microgrid optimisation · Charging station operator · Unit commitment · Mixed integer quadratic programming (MIQP)

1 Introduction

Nowadays, electric power is mostly generated centrally in bulk quantity by large generation plants linked with long transmission lines that bring electric power to the end users via the utility grid. Smart grid is an electrical system that aims to distribute electric power from the producers to the consumers efficiently [1]. Since today's producers and consumers are very sophisticated actors in terms of their behaviour in the supply- demand dynamics, smart grid is a very complex system that deploys different communication protocols to deal with nonlinearity, security and bidirectional power flow [2]. Despite recent advances in the modern technology of communication protocols and monitoring devices, supervision of a large complex system remains very difficult [1, 3–8]. In general, smart grid is best facilitated within a microgrid, which is a relatively small scale localised energy network with the ability of connecting to or isolating from the main grid. A microgrid has the following characteristics [9–12]: Managing sources and demand locally; Reducing the expenditure of the whole system by offering economic dispatch and optimally scheduling of demand and micro-sources.

High penetration of intermittent renewable sources in the distribution network will deteriorate the immunity of network stability during various network contingencies [13, 14]. As a consumer of electricity when hooked on a charging station, EVs are also

© Springer Nature Singapore Pte Ltd. 2017
K. Li et al. (Eds.): LSMS/ICSEE 2017, Part III, CCIS 763, pp. 677–690, 2017.
DOI: 10.1007/978-981-10-6364-0_68

classified as a mobile energy storage system that distributes within a microgrid, and as such, produce significant uncertainty to the network [15]. Fixed and mobile storage devices, power exchange with the utility grid, lowest pollution treatment, and losses reduction of the network [16].

This paper focuses on primarily on the implementation of the 'microgrid operator'.

2   A Modular Power Management of the Microgrid Structure

The hierarchical structure of classical management methodology has been chosen as it fits well with the description of problems in this work. In Fig. 1. the modular structure involves several processes ranging from long-term period segment decisions of the MGO to very high-speed control of the power electronic of the PES to facilitate the power management by addressing each major part of the process independently. Then, the modular structure consolidates the multidisciplinary process to form a complete system. The structured format describes hierarchical tiers corresponding to the level of management mission, and is seen as the upside-down structure on the left-hand side of Fig. 2.

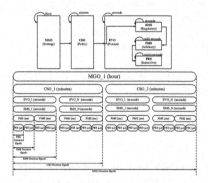
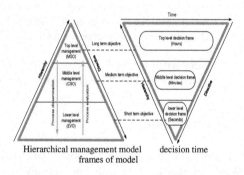

Hierarchical management model   decision time
frames of model

Fig. 1. Concept of a modular manage-   **Fig. 2.** Hierarchical and decision timeframes of
ment structure   management model

The complete modular structure can be described as a hierarchical decision epoch as presented in lower part of 错误!未找到引用源。 MGO-strategy is a plan of action designed to achieve a long-term or overall aim of the power generation from DGs of the MG network to balance the electrical demand.

3   Implementation of the Algorithm

The output of the optimisation model is the optimal configuration of the microgrid. The procedure to achieve optimal operation is listed below and explained in Fig. 3.

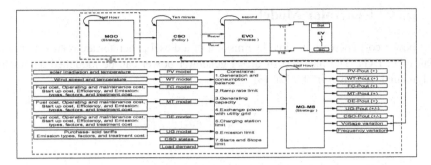

Fig. 3. Interrelationship of operational control of MGO

The decision of power amount from the DGs of an optimisation algorithm makes based on the fuel price, the maintenance cost, the startup cost, and pollutants treatment cost. The wind turbine and the photovoltaic are assumed to generate free emission. Because of the wind speed is very low in the Baghdad environment. Therefore, the wind turbine is not efficient. The output power of the wind turbine is very small or cut off at a wind speed below 3 m/s. Therefore, the wind turbine is replaced with the photovoltaic cell in this study.

4 Electric Vehicle Model

Energy storage system is the powertrain of the EVs which have bi-directional power flow characteristic. Therefore, the EVs provide a great opportunity to discharge some power of energy storage system to synergize the DGs of the MG in balancing the net demand. Discharging the EVs helps the microgrid to maintain the stability and reliability of the system.

Prediction the discharging power of the EVs depends mainly on the spatial characteristics of the EVs, the range of discharging state of charge of the energy sources, the required energy for next journey of the EVs after unplugged from the microgrid. The CSO is responsible for the centralised smart charger. The objective function of the CSO is either achieving minimum charging the cost of the EVs as applied in Eq. (1) or achieving a maximum discharging cost of the EVs as applied in Eq. (2). Further information about the operation of CSO to optimise charging and discharging power of the EVs is provided in CSO optimization paper.

$$Min - Cost = \sum_{i=t_a}^{t_d} \sum_{j=1}^{N} \alpha.s_1.\varepsilon_1.P_{ch,ij}.\Delta t.F.\rho.C_{rate,p,ch} + \alpha.s_1.\varepsilon_2.Q_{ch,ij}.\Delta t.F.\rho.C_{rate,q,ch}$$

$$(1)$$

$$Max - Cost = \sum_{i=t_a}^{t_d} \sum_{j=1}^{N} D.\beta.s_2.P_{dis,ij}.\Delta t.F.C_{rate,p,dis} + D.\beta.s_2.Q_{dis,ij}.\Delta t.F.C_{rate,q,dis} \quad (2)$$

5 Diesel Generator Model

The power cost and carbon dioxide emission of a diesel generator are relatively higher than other DGs. The fuel cost of diesel generator power can be modelled as quadric polynomial as shown in the Eq. (3).

$$C_{dg,i} = \sum_{i=1}^{N} \alpha_i + \beta_i P_{dg,i} + \gamma_i P_{dg,i}^2 \tag{3}$$

6 Cost Formulation

The operating cost of the Microturbine and fuel cell usually includes fuel cost, maintenance cost, and startup/shutdown cost in ($/h) as shown in Eq. 4 [18]. The fuel cost of the Microturbine and fuel cells are calculated as shown in equation 错误!未找到引用源。 [19].

The maintenance cost of the Microturbine, the fuel cell, and the diesel generator are proportion to the supplied power based on forecasting with minimal real life situation of microsources. Thus, the maintenance cost of unit i in time interval t is shown in equation 错误!未找到引用源。 The DGs startup cost depends mainly on time the unit which has been off before it is startup again. The startup cost of the generators is calculated from the equation 错误!未找到引用源。

$$CF(P) = \sum_{i=1}^{N} (C_i F_i + OM_i + SC_i) \tag{4}$$

$$C_{MT,i} = C_{nl} \sum_{J} \frac{P_J}{\eta_{lJ}} \tag{5}$$

$$OM_i = K_{om} \sum_{i=1}^{N} P_i \tag{6}$$

$$SC_{i,t} = KSC_i \times (\delta_{dgi,t} - \delta_{dgi,t-1}) \tag{7}$$

Where

$$SC_{i,t} \geq 0 \qquad \delta_{dgi} = \begin{cases} 1 & if\ DG\ ON \\ 0 & otherwise \end{cases}$$

The pollutant treatment cost of the microgrid including The cost of the emission of carbon dioxide CO_2 and particulate matter such as sulphur dioxide SO_2, nitrogen oxide NO_x can be described as in Eq. 8.

$$E(P) = E1(P) + E2(P) \tag{8}$$

$$E1(P) = \sum_{i=1}^{N}\sum_{k=1}^{N}(C_k\gamma_{ik})P_i + \sum_{k=1}^{N}(C_k\gamma_{grid,k})P$$

$$E2(P) = \sum_{i=1}^{N}\sum_{k=1}^{N}(C_{co2}\gamma_{ico2})P_i + \sum_{k=1}^{N}(C_{co2}\gamma_{grid,co2})P_{grid}$$

7 Multiobjective Functions

Multiobjective optimisation is a technique to find the optimum solution between different objectives. Usually, the objectives are conflict and possibly contradict. That means there is no decision variables which can set all the objectives function simultaneously. The multiobjective function is used to search the efficient decision variables of optimisation function. A general mathematical formulation of multi-objective optimisation which has n-dimensional decision variables is expressed as shown in the Eq. 9. There is a different method to solve multi-optimization problems. One of them is the weighting method. The general idea of the weighting method is to associate each objective function with a weighting coefficient and minimise the weighted sum of the objectives. The multiobjective roles in this approach are transformed into a single objective function.

$$\min\{f_1(x), f_2(x), \ldots, f_k(x)\} \tag{9}$$

$$\text{subject to } \begin{array}{l} g_i(x) \leq 0, i = 1, 2, \ldots, q \\ h_j(x) = 0, j = 1, 2, \ldots, p \end{array}$$

Where $x = (x_1, x_2, \ldots, x_n)$ is n dimensional decision variables

$f_k(x)$ is the k-th objective function
$g_i(x) \leq 0$ is inequality constraints
$h_j(x) = 0$ is equality constraints

Mathematically, the economic dispatch/environmental pollutant of microgrid problem are considered in this paper. The mathematical equation is formulated to find the lowest operation cost, the lowest environment pollutions (sulphur dioxide SO2 and nitrogen oxide NOx), and the lowest carbon dioxide (CO2) of the generation unit as shown in Eq. 10.

$$\text{objective function} = \min \sum_t \{w_1(CF(P(t)) + \delta_g P_{grid}(t)) + w_2 E1(P(t)) + w_3 E2(P(t))\}$$

$$\tag{10}$$

Where w_1, w_2, w_3 are the weight coefficient of the fuel cost, pollutant treatment cost, and carbon dioxide treatment cost objective functions respectively.

The criteria to calculate the judgment matrix of the objective function is based on an analytic hierarchy process (AHP) method which was analytically proposed on [20] and shown in Table 1.

Table 1. The criteria of the judgment matrix

Judgment value	Explanation
1	Two activities contribute equally to the objective
2	Two activities contribute slightly equally to the objective
5	Experience and judgment strongly favour one activity over another
Reciprocals	If activity x has one of the above values assigned to it when compared with activity y, then y has reciprocal value when compared with x

The objective function is classified into three levels; operating cost as the first level, pollutants treatment cost as the second level, and carbon dioxide treatment cost as the third level. The rank of the criteria subjective is explained below: The operating cost is five times as important as the pollutants treatment cost. The operating cost is three time as important as the carbon dioxide treatment cost. The pollutants treatment cost is two time as important as the carbon dioxide treatment cost. The judgment values of the criteria subjective matrix (11) are $w_1 = 0.64833$, $w_2 = 0.22965$, and $w_3 = 0.12202$.

$$J = \begin{bmatrix} 1 & 3 & 5 \\ 1/3 & 1 & 2 \\ 1/5 & 1/2 & 1 \end{bmatrix} \tag{11}$$

The objective function of the optimisation problem is subject to: Generation and consumption balance: The microgrid operator should balance between the total load demand and the total power generation as represented by the Eq. (12). Ramp rate limit: The ramp rate should be meet at each sampling time as expressed in equation 错误!未找到引用源。

$$\sum_t \left(CF(P(t)) + \delta_g P_{grid}(t) \right) = P_d(t) \tag{12}$$

$$-DR_i \Delta T \le P_i(t+1) - P_i(t) \le UR_i \Delta T \tag{13}$$

Generating capacity: For stable operation, each DG should limit output power according to its capacity as shown in equation 错误!未找到引用源。 Charging station limit: The battery state at each sampling time can be represented based on the state of charge into either charging, discharging or not act. The decision of battery state is made by CSO depending on how many numbers of EVs connected to the network at instance time, the resources capacity of each EV, and the state of charge of the resources.

The objective function of discharging mode is formulated to get the maximum discharging power cost as shown in equation 错误!未找到引用源。

$$P_{i,min}(t) \leq P_i(t) \leq P_{i,max}(t) \tag{14}$$

$$P_{EVs,min}(t) \leq P_{cs,i}(t) \leq P_{EVs,max}(t) \tag{15}$$

- Exchange power with utility grid:
  The power exchange between microgrid and utility grid at connected mode bidirectional where each direction either purchasing or selling has its tariff and limits. It is not possible to purchase and sell at the same time. Therefore, binary number $\delta_{g,b}(t) \in [0\ 1]$ and $\delta_{g,s}(t) \in [0\ 1]$ is introduced selling and purchasing modes where $\delta_{g,b}(t) + \delta_{g,s}(t) = 1$. The exchange power between the microgrid and utility grid can be represented as in Eq. (16).

$$\begin{cases} P_{g,smin}(t) \leq P_{g,s}(t) \leq P_{g,smax}(t) \\ P_{g,bmin}(t) \leq P_{g,b}(t) \leq P_{g,bmax}(t) \end{cases} \tag{16}$$

- Emission limit
  The total environment pollutant should not exceed certain limit for each DG as shown in equation 错误!未找到引用源。
- Starts and stops limit
  For stable operation of the DGs, some start-up of each generator should be limit to a certain number based on generator type as shown in equation 错误!未找到引用源。

$$E(P(t)) \leq L_j \tag{17}$$

$$\sum_i N_{start-stop,i} \leq N_{start-stop,max} \tag{18}$$

8 The Mathematical Model for the Microgrid Optimization Problem

The optimisation algorithm of the DGs deals with non-linear, non-convex, and highly dimension problems. It is also dealing with the unit commitment problem. The achievement should be balance the supply of the thirteen DGs, CSO operation, the exchange power with the utility grid, and the consumer electrical demand with the lowest cost of operation and pollutants treatment. The cost function of the Microturbine and fuel cell are linear while the cost function of the diesel generator and utility grid are non-linear. The main utility grid (UG) is treating as another source connected to microgrid with its operation prices and pollutants treatment in the optimisation problem.

It also balances the difference between the consumer demand and the microsources output power, whenever the DGs could not cover the consumer demand. On the other hand, the microgrid DGs could operate as stability balance of the utility grid whenever the voltage and frequency stability reach to collapse point to enhance the stability margin.

The states between the microgrid and the utility grid could be either exchange power from the microgrid to the utility grid (buying power) or exchange power from the utility grid to the microgrid (selling power). If no exchange power between the microgrid or utility grid means that either the point of common coupling isolates the systems from each other or the microgrid covers itself with cheaper operating cost than utility grid without penalty of power to sell. Therefore, the condition term of exchange power between the microgrid and the utility grid are defined as in equation 错误!未找到引用源。 and 错误!未找到引用源。

$$F_{ug,s} = C_{ug,s}.P_{ug,s} \tag{19}$$

$$F_{ug,b} = C_{ug,b}.P_{ug,b} \tag{20}$$

Where $\delta_{ug,b}(t) + \delta_{ug,s}(t) = 1, \delta_{ug} \in [0\ 1]$

$$\begin{cases} \delta_{ug,b}(t).P_{ugmin,b}(t) \le P_{ug,b}(t) \le \delta_{ug,b}(t).P_{ugmax,b}(t) \\ \delta_{ug,s}(t).P_{ugmin,s}(t) \le P_{ug,s}(t) \le \delta_{ug,s}(t).P_{ugmax,s}(t) \end{cases}$$

The EVs have dual functions from the view of the optimisation problem, which are either behaves as electrical demand or resources. The charging station operator sends the states of the connected EVs such as the total electrical demand required to charge them as well as the available power could be discharged from them. The MGO collects the information from all CSO within the microgrid. The discharging decision takes based on the voltage variation, frequency variation, or both. After activating the discharging option, the MGO treats the CSO as a resource of the power. Therefore the power available on the EVs connected at CSO compete with the other DGs according to the objective function. The CSO formula includes the operation cost where the emission cost equal to zero as shown in Eq. (21). Further detail about CSO is available in CSO optimization paper.

$$F_{cso} = \left(C_{ug,s}.\delta_{cso,d} + C_{ug,b}.\delta_{cso,c} \right).P_{cso,i} \tag{21}$$

Where

$$\delta_{cso,d} = \begin{cases} 1 & \text{if discharge mode is applied} \\ 0 & \text{otherwise} \end{cases}$$

$$\delta_{cso,c} = \begin{cases} 1 & \text{if charge mode is applied} \\ 0 & \text{otherwise} \end{cases}$$

The unit commitment application is applied to schedule the DG operation at every period of executed optimisation algorithm. It is introduced a binary variable u_a to express logical statement of the unit commitment implementation.

Where

$$u_a = \begin{cases} 1 & \text{if unit commetment is applied} \\ 0 & \text{otherwise} \end{cases}$$

$$u_n = \begin{cases} 1 & \text{if unit commetment is not applied} \\ 0 & \text{otherwise} \end{cases}$$

The proposed cost function of microgrid in connected mode with DGs, CSO, UG, and unit commitment consideration is formulated as in equation 错误!未找到引用源。

$$
\begin{aligned}
CF(P) = {} & w_1 \left(\sum_{t=1}^{24} \sum_{i=1}^{N} (u_a \delta_{dgi} + u_n)(C_i F_i(t) + OM_i + SC_{i,t}) + F_{cso,i}(t) + F_{ug}(t)\delta_{ug} \right) \\
& + \left(\sum_{i=1}^{N} \sum_{k=1}^{N} (u_a \delta_{dgi} + u_n)((w_2 C_k \gamma_{ik} + w_3 C_{co2}\gamma_{ico2})P_i) \right) \\
& + \sum_{k=1}^{N} (w_2 C_k \gamma_{grid,k} + w_3 C_{co2}\gamma_{grid,co2})P_{grid}\delta_{u,g})
\end{aligned}
\tag{22}
$$

9 Case Study

A case study can offer a useful insight to simplify a very complex problem by focusing on the snapshot operation of a typical network.

Figure 4 shows the distribution network under study. It consists of 7 feeders and 49 busbars to supply the electricity loads of a local community in the City of Baghdad.

- Interfacing power electronics and communication electronics allowing bi-directional flow of power at the selected nodes connected with DGs.
- A pseudo-isolated operating mode such that the entire local loads are met by thirteen DGs located at selected nodes as shown in the case study. The DGs compromised as one diesel generator, three fuel cell, four microturbine, five photovoltaic cells, and two electric vehicle charging stations. The wind turbine excluding from this study because the analysis shows that the output power of wind turbine is very small due to low wind speed in Baghdad environment.

The typical aggregated daily load curve pattern of the microgrid network under study based on appliances ownership is shown in Fig. 5 [21]. There is some fluctuation in demand during the day time. The DGs to meet the daily load have chosen according to type and efficiency of DG in [22], the solid oxide fuel cell has much better efficiency than micro gas turbine for range 10–100 kW. The power range location of DGs have chosen based on the voltage stability distributed location algorithm as shown in Table 2. The parameters of the DGs are shown in Table 3.

After midday, the tariff is started to reduce gradually until reaching to minimum at midnight; then it is kept constant until early morning as shown in Fig. 6

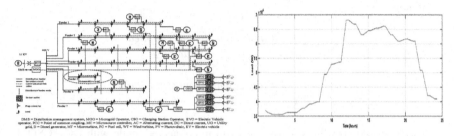

Fig. 4. Typical microgrid structure   **Fig. 5.** Daily load curve

Table 2. Distributed generators type, range, and location

Type	Power range (kW)	No. of busbars
PV	$0 \leq P_{PV} \leq 50$	17, 20, 24, 34, 32
FC	$0 \leq P_{FC} \leq 50$	16, 23, 45
MT	$0 \leq P_{MT} \leq 75$	10, 15, 25, 39
DE	$25 \leq P_{DE} \leq 250$	18
Utility grid	$-200 \leq P_{UG} \leq 200$	1

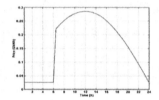

Fig. 6. Utility grid Tariff

Table 3. Parameter content of the grid equipment

Emission type	Pollutant disposal coefficient $/kg	Factors of FC	Factors of MT	Factors of DE	Factors of UG	Factors of PV
NO_x	9.34	0.01	0.2	9.89	13.6	0
SO_2	2.21	0.003	0.0036	0.206	1.8	0
CO_2	0.032	489	724	649	889	0
Other parameters						
Fuel cost $/kWh	/	0.0294	0.0457	$\alpha = 0.4$ $\beta = 0.0185$ $\gamma = 0.0042$	See Fig. 6	0.024
Kom $/kWh	/	0.00419	0.00587	0.01258	/	0.001
KSC_i $	/	0.96	1.65	1.75	/	0

10 Results

The main objective function minimises the operation cost and the treatment pollutant cost of the microgrid with different weight; the operation cost has a higher weight than the treatment cost. Therefore, the results classify into two scenarios as described below: Scenario A: Minimum operational and pollutant treatment policy with unit commitment consideration at Isolated mode. Scenario B: Minimum operational and pollutant treatment policy of microgrid with unit commitment consideration at connected mode.

A. Scenario: Minimum operation and pollutant treatment policy with unit commitment consideration at isolated mode

The microgrid in isolated mode should cover all sensitive demand without any exchange power with the utility grid where the point of common coupling has isolated the microgrid from the utility grid. The isolated mode cost function with unit commitment consideration is formulated as in equation 错误!未找到引用源。

$$
\begin{aligned}
CF(P) = {} & w_1\left(\sum_{t=1}^{24}\sum_{i=1}^{N}\left(u_a\delta_{dgi}\right)\left(C_iF_i(t)+OM_i+SC_{i,t}\right)+F_{cso,i}(t)\right) \\
& + \left(\sum_{i=1}^{N}\sum_{k=1}^{N}\left(u_a\delta_{dg,i}\right)\left((w_2C_k\gamma_{ik}+w_3C_{co2}\gamma_{ico2})P_i\right)\right)
\end{aligned}
\tag{23}
$$

Figure 7 shows the Hourly optimal power schedule of DGs at isolated mode with unit commitment consideration. Figure 8 reveals the hourly operating cost of the DGs, pollutants treatment cost, and the carbon dioxide treatment cost of the microgrid at isolated mode with unit commitment consideration. The total operation and emission cost is presented in Table 4.

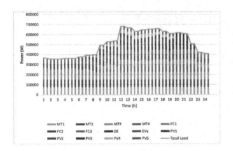

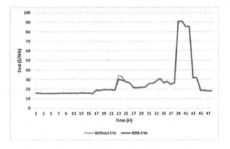

Fig. 7. Hourly optimal power schedule of DGs

Fig. 8. Hourly total DGs operating cost

There are 74 electric vehicles at each CSOs could deliver power to the network at voltage deviation time. Figure 8 showed the difference of using the diesel generator instead of the CSOs to compensate the voltage stability where using the diesel generator recorded higher price than using the CSOs.

Table 4. daily cost of total DGs operating cost

	Daily operation cost ($)	Daily emission treatment cost ($)	Daily overall cost ($)
Without Evs	1000.9424	279.146	1280.088
With EVs	999.9678	272.850	1272.818

B. Scenario: Minimum operation and pollutant treatment policy of microgrid with grid connected mode operation and unit commitment consideration at connected mode

The operation of the microgrid at connected mode is expressed as the island mode connected to infinity bus. Therefore, there is a penalty of power could move bidirectionally between the microgrid and utility grid. The direction of power decides according to the cooperation between the DMS and MGO according to availability, required, and total cost. The proposed cost function of the microgrid in connected mode with unit commitment consideration is formulated as in 错误!未找到引用源。Figure 9 shows the hourly optimal schedule power of the DGs and exchange power with the utility grid.

The objective function of the optimization problem decides the sufficient power amount to buy or sell from the microgrid. For example after 6:0. Figure 10 depicted the whole hourly operation and pollutant treatment cost of the islanded microgrid with considering unit commitment operation. The operation, emission, and total cost of microgrid are shown in Table 5. The total cost of the EVs is slightly better than replacing them with DE. At high penetration of the EVs, the cost difference between using EVs and DE become clear as the DE cost curve quadratic which increases nonlinearly when power increased.

$$F(P) = w_1 \left(\sum_{t=1}^{24} \sum_{i=1}^{N} \left(u_a \delta_{dgi} \right) \left(C_i F_i(t) + OM_i + SC_{i,t} \right) + F_{cso,i}(t) + F_{ug}(t) \delta_{ug} \right)$$
$$+ \left(\sum_{i=1}^{N} \sum_{k=1}^{N} \left(u_a \delta_{dg,i} \right) \left((w_2 C_k \gamma_{ik} + .w_3 C_{co2} \gamma_{ico2}) P_i \right) \right)$$
$$+ \sum_{k=1}^{N} \left(w_2 C_k \gamma_{grid,k} + w_3 C_{co2} \gamma_{grid,co2} \right) P_{grid} \delta_{ug} \right) \tag{24}$$

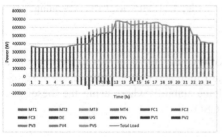

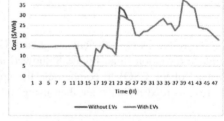

Fig. 9. Hourly optimal power schedule of DGs   **Fig. 10.** Hourly total DGs operating cost

Table 5. Daily cost of total DGs operating cost

	Daily operation cost ($)	Daily emission treatment cost ($)	Daily overall cost ($)
Without Evs	667.46276	291.488	958.9503
With Evs	666.4882	285.192	951.6803

11 Conclusion

This paper presented a holistic approach to the managing electric vehicles charging/discharging within microgrid in three major processes of hierarchical arrangement. The paper focused on the first part of hierarchical management which is microgrid operator. MGO is proposed an optimal management system to minimise the total operation and emission pollutants cost of the hybrid microgrid. The results show that considering of unit commitment in optimisation problem reduced the total cost of the microgrid operation and increased the flexibility of optimal microgrid optimisation. Furthermore, it is not necessary that the utility grid operation cost less than the other DGs. Finally, it can be seen that discharging electric vehicles reduced the total operation cost of the microgrid.

References

1. Fang, X., Misra, S., Xue, G., Yang, D.: Smart grid — the new and improved power grid: a survey. IEEE Commun. Surv. Tutorials **14**(4), 944–980 (2012)
2. Güngör, V.C., Sahin, D., Kocak, T., Ergüt, S., Buccella, C., Cecati, C., Hancke, G.P.: Smart grid technologies: communication technologies and standards. IEEE Trans. Industr. Inf. **7**(4), 529–539 (2011)
3. Coll-Mayor, D., Paget, M., Lightner, E.: Future intelligent power grids: analysis of the vision in the European Union and the United States. Energy Policy **35**(4), 2453–2465 (2007)
4. Fan, J., Borlase, S.: The evolution of distribution. IEEE Power Energy Mag. **7**(2), 63–68 (2009)
5. Liang, H., Tamang, A.K., Zhuang, W., Shen, X.S.: Stochastic information management in smart grid. IEEE Commun. Surv. Tutorials **PP**(99), 1–25 (2014)
6. El-hawary, M.E.: The Smart grid—State-of-the-art and Future Trends. Electr. Power Compon. Syst. **42**(3–4), 239–250 (2014)
7. Galli, S., Scaglione, A., Wang, Z.: For the grid and through the grid: the role of power line communications in the smart grid. In: Proceedings of IEEE, vol. 99, no. 6, pp. 998–1027 (2011)
8. Wang, W., Xu, Y., Khanna, M.: A survey on the communication architectures in smart grid. Comput. Netw. **55**(15), 3604–3629 (2011)
9. Lasseter, R.H.: MicroGrids. In: IEEE Power Engineering Society Winter Meeting, pp. 305–308 (2002)
10. Olivares, D.E., Mehrizi-Sani, A., Etemadi, A.H., Cañizares, C.A., Iravani, R., Kazerani, M., Hajimiragha, A.H., Gomis-Bellmunt, O., Saeedifard, M., Palma-Behnke, R., Jiménez-Estévez, G.A., Hatziargyriou, N.D.: Trends in microgrid control. IEEE Trans. Smart Grid **5**(4), 1905–1919 (2014)
11. Parhizi, S., Lotfi, H., Khodaei, A., Bahramirad, S.: State of the art in research on microgrids: a review. IEEE Access **3**, 890–925 (2015)

12. Abusharkh, S., Arnold, R., Kohler, J., Li, R., Markvart, T., Ross, J., Steemers, K., Wilson, P., Yao, R.: Can microgrids make a major contribution to UK energy supply? Renew. Sustain. Energy Rev. **10**(2), 78–127 (2006)
13. Soshinskaya, M., Crijns-Graus, W.H.J., Guerrero, J.M., Vasquez, J.C.: Microgrids: experiences, barriers and success factors. Renew. Sustain. Energy Rev. **40**, 659–672 (2014)
14. Majumder, R.: Some Aspects of Stability in Microgrids. IEEE Trans. Power Syst. **28**(3), 3243–3252 (2013)
15. Tie, S.F., Tan, C.W.: A review of energy sources and energy management system in electric vehicles. Renew. Sustain. Energy Rev. **20**, 82–102 (2013)
16. Gaur, P., Singh, S.: Investigations on Issues in Microgrids. J. Clean Energy Technol. **5**(1), 47–51 (2017)
17. Weather Undeground. https://www.wunderground.com/history. Accessed 25 Dec 2016
18. Zhao, B., Shi, Y., Dong, X., Luan, W., Bornemann, J.: Short-term operation scheduling in renewable-powered microgrids: a duality-based approach. IEEE Trans. Sustain. Energy **5**(1), 209–217 (2014)
19. Mohamed, F.A., Koivo, H.N.: Online management of MicroGrid with battery storage using multiobjective optimization. In: International Conference on Power Engineering Energy Electrical Drives, pp. 231–236 (2007)
20. Saaty, T.L.: Decision making with the analytic hierarchy process. Int. J. Serv. Sci. **1**(1), 83 (2008)
21. Hague, T.: Iraq electricity masterplan (2010)
22. Energy Transition Group: http://www.energytransitiongroup.com/vision/localenergy.html. Accessed 25 Dec 2016

Security-Constrained Two-Stage Stochastic Unified Active and Reactive Power Management System of the Microgrids

Mohammed K. Al-Saadi[1,2] and Patrick C.K. Luk[1(✉)]

[1] Cranfield University, Cranfield, Bedfordshire, UK
p.c.k.luk@cranfield.ac.uk
[2] University of Technology, Baghdad, Iraq

Abstract. This paper presents a developed robust two-stage scenario-based stochastic unified active and reactive power economic management system of microgrids (MGs) based on the unit commitment (UC) to minimize the total operating cost. The security constraints, the environmental costs, and the storage battery operating cost are considered in the proposed optimization approach. The mathematical stochastic models of the generation fluctuation of wind turbines (WTs) and photovoltaic panels (PV), and open market prices (OMPs) are developed and incorporated with UC optimization problem of the MG. The proposed stochastic approach is a two-stage optimization, where the first stage is the day-ahead scheduling based on the forecasted data, whereas the second stage mimics the real-time by considering the WT, PV, and OMP variability, where the UC is not changed in the second stage. The proposed optimization algorithm is tested on the low voltage connected MG. The results reveal that the feasible solution can be obtained for all scenarios.

Keywords: Stochastic optimization · Energy storage · Hybrid microgrid

1 Introduction

New methodologies and approaches have been proposed to incorporate the uncertainties with optimization problems and study their impacts on the optimal operation of the MG to increase the robustness of the optimization approaches.

In [1] stochastic optimization was formulated as a probabilistic constrained approach. In this approach, the hard constraint on exact balance power is relaxed by introducing a probabilistic constraint, which contains renewable power, and load demands as random variables. The power balance constraint is considered a high probability, while penalty is added to the cost function for violation of the constraint. On the other hand, a two-stage scenario-based stochastic programming was presented. In the first stage the decision variables of the UC and the day-ahead energy scheduling of the DGs and other generation resources are taken, while in the second stage the decision variables of the first stage have been changed according to the realization of generation fluctuation of the renewable energy resources [2]. Similarly, in [3, 4] authors formulated the stochastic optimization problem as a two-stage, where the UC decisions

© Springer Nature Singapore Pte Ltd. 2017
K. Li et al. (Eds.): LSMS/ICSEE 2017, Part III, CCIS 763, pp. 691–703, 2017.
DOI: 10.1007/978-981-10-6364-0_69

were taken in the first stage were not change in the second stage, while authors in [5] addressed that the decisions of the UC and battery charging and discharging operations were taken in the first stage are not change in the second stage. In [6, 7] authors presented a two-stage stochastic approach with the first stage UC and the day-ahead energy scheduling of DGs and other generation resources are taken and the second stage the reserve and security cost of the each scenarios are considered, wherein the variables of the first stage are not changed. Authors in [8] proposed the stochastic optimization as single stage based multi-scenarios stochastic programming. The two-stage is more accurate than single stage [7], where the UC is defined at the first stage and the UC is not changed at second stage.

In this paper, a two-stage scenario-based stochastic optimization is introduced. This approach includes scheduling active and reactive power with considering the emission cost and battery degradation cost. Besides, the UC is developed to take into consideration the active and reactive power. In addition, the security constraints and emission limitation constraint with other constraints relevant to the reactive power are considered in the proposed optimization approach. The models of the uncertainties that come from the renewable generation fluctuation and forecast error of the OMPs are developed and incorporated with optimization approach to explore the impacts of these uncertainties on the economic dispatch of the MG.

2 System Modelling

The components of the MG are modelled as follows:

2.1 Fuel Cost of the DGs

This cost can be calculated from the following equation [9]:

$$CP_{DG_i}(P_{DG_i}(t)) = \left(a_i + b_i.P_{DGi}(t) + c_i.P_{DGi}^2(t)\right) \tag{1}$$

where a_i (€/h), b_i (€/kWh), and c_i (€/kW^{2}h) are the respective coefficients of the fuel cost function, and $P_{DG_i}(t)$ is the output active power of ith DGs.

2.2 Reactive Power Production Cost of the DGs

This cost of ith DG can be calculated as [10]:

$$CQ_{DG_i}(Q_{DG_i}(t)) = \left(ar_i + br_i.Q_{DGi}(t) + cr_i.Q_{DGi}^2(t)\right) \tag{2}$$

where ar_i (€/h), br_i (€/kVArh), and cr_i (€/kVAr^{2}h) are the respective coefficients of the reactive power cost function, and $Q_{DG_i}(t)$ is the output reactive power of ith DG.

2.3   Model the Operating and Maintenance Cost of DGs

The operating and maintenance cost can be calculated by the following equation:

$$COM_{DG_i}(P_{DG_i}(t)) = KOM_{DG_i}.P_{DG_i}(t) \tag{3}$$

where KOM_{DG_i} (€/kWh) is the coefficient of the maintenance cost of the ith DG.

2.4   Model of Operating Cost of Storage Battery

The battery operating cost can be expressed as:

$$C_b(t) = C_d \times P_b(t) \times \Delta T \tag{4}$$

where C_d is the battery degradation cost (€/kWh), $P_b(t)$ is either charging or discharging power of the storage battery. Details about calculation of the C_d can be found in [11].

2.5   Model of the Exchanging Active and Reactive Power with the Main Grid

The cost of trading power with the main grid can be calculated as:

$$C_{gP}(t) = c_{gP}(t).P_g(t) \tag{5}$$

$$C_{gQ}(t) = c_{gQ}(t).Q_g(t) \tag{6}$$

where $c_{gP}(t)$ (€/kWh) and $c_{gQ}(t)$ (€/kVArh) are the active and reactive cost of the exchanging power with the utility grid. $P_g(t)$ and $Q_g(t)$ are the active and reactive exchanging power.

2.6   Model of Greenhouse Gases Emission Cost

The emission cost of jth Greenhouse gas from i^{th} DG can be calculated by:

$$C_e(t) = \sum_{j=1}^{M} \sum_{i=1}^{N} E_{j,i}.C_j.P_{DG_i}(t) \tag{7}$$

where C_j (€/kg) is a price of emission of jth greenhouse gas, and $E_{j,i}$ (kg/kWh) is the emission rate of greenhouse gas from the ith DG. M and N are the total number of greenhouse gases and DGs respectively.

2.7   Start-Up and Shutdown Cost of the DGs

The start-up and shutdown costs of the DGs can be formulated as follows [12]:

$$SU_{DG_i}(t) = Sc_i.(\delta_{DG_i}(t) - \delta_{DG_i}(t).\delta_{DG_i}(t-1)) \tag{8}$$

$$SD_{DG_i}(t) = Sd_i.(\delta_{DG_i}(t-1) - \delta_{DG_i}(t).\delta_{DG_i}(t-1)) \tag{9}$$

where Sc_i and Sd_i are the price of the start-up and shutdown cost of the ith DG.

2.8   Models of the Stochastic Variables

The stochastic model of wind, solar irradiance, and OMP are as follows:

A.   Stochastic Model of the Wind Generation

Rayleigh distribution can be used to formulated the stochastic nature of the wind speed [7] and the following equation expressed the Rayleigh distribution.

$$f(v) = \left(\frac{2v}{c^2}\right)^{k_1-1} e^{-(v/c)^2} \tag{10}$$

$$c \cong 1.128 v_{mean} \tag{11}$$

where v_{mean} is the hourly forecasted wind speed, $k_1 = 2$. The power of the WTs can be calculated by the following equation [7].

$$P_W = \begin{bmatrix} 0 & v \leq v_{ci} \text{ or } v \geq v_{co} \\ P_{W-r} * \frac{v-v_{ci}}{v_r-v_{ci}} & v_{ci} \leq v \leq v_r \\ P_{W-r} & v_r \leq v \leq v_{co} \end{bmatrix} \tag{12}$$

where P_{W-r} is the rated power of wind turbine. v_{ci}, v_{co}, v_r are cut in, cut out and rated wind speeds in (m/s).

B.   Stochastic Model of the PV Generation

The output power of the PV depends on the solar irradiance. It is assumed that the fluctuation of the irradiance follows the normal distribution, where the Monte Carlo simulation is used to obtain the stochastic solar irradiance. The proposed stochastic solar irradiance is as:

$$G(t) = G(t)^{forec} + \mu(t)^{solar}.\sigma(t)^{solar} \tag{13}$$

where $G(t)^{forec}$ and $\sigma(t)^{solar}$ are the forecasted solar irradiance at hour t and its standard deviation of the irradiance. $\mu(t)^{solar}$ is the random variable generated for the solar irradiance at time t by using the normal distribution with a mean of zero and a standard deviation is one. $G(t)^{forec}$ is from Table 1. The PV output power can be calculated from the following equation [6].

$$P_{PV}(G(t)) = G(t) \times A_{pv} * \eta_{pv} \tag{14}$$

where A_{pv} is the area of the PV panel, η_{pv} is the efficiency of PV panel.

Table 1. Hourly profile of the wind speed, solar irradiance, active and reactive OMPs, and total active and reactive loads

T(h)	Wind speed (m/s)	Solar irradiance (kW/m^2)	Active power price (€/kWh)	Reactive power price (€/kWh)	Total active load (kW)	Total reactive load (kW)
1	7.8	0	0.035	0.0035	100.240	48.546
2	9	0	0.032	0.0032	92.000	44.556
3	9.5	0	0.030	0.003	90.260	43.713
4	10.5	0	0.030	0.003	84.600	40.972
5	9.7	0	0.033	0.0033	90.820	43.984
6	8.6	0.019	0.050	0.005	100.040	48.449
7	7	0.096	0.047	0.0047	136.100	65.913
8	6.6	0.222	0.050	0.005	186.800	90.467
9	6.8	0.381	0.070	0.007	221.060	107.059
10	6	0.551	0.080	0.008	250.400	121.269
11	5.3	0.61	0.090	0.009	266.500	129.066
12	5.5	0.657	0.120	0.012	265.080	128.378
13	5.8	0.648	0.225	0.0225	286.500	138.752
14	6.3	0.59	0.100	0.010	297.400	144.031
15	7.5	0.477	0.085	0.0085	283.380	137.241
16	8	0.338	0.150	0.015	270.100	130.809
17	9.6	0.19	0.450	0.045	258.000	124.949
18	10	0.08	0.150	0.015	292.600	141.706
19	9	0.017	0.180	0.018	300.200	145.387
20	8.5	0	0.160	0.016	313.000	151.586
21	7.8	0	0.310	0.031	277.800	134.539
22	7	0	0.050	0.005	233.000	112.842
23	6.8	0	0.040	0.004	186.000	90.080
24	7.1	0	0.025	0.0025	104.600	50.658

C. Stochastic Model of the OMP

The forecast error of the OMP follows a normal distribution and Monte Carlo simulation is used to obtain the stochastic OMP. The proposed stochastic of the OMP as follows.

$$c_g(t) = c_g(t)^{forec} + \mu(t)^{price} . \sigma(t)^{price} \tag{15}$$

where $c_g(t)^{forec}$ and $\sigma(t)^{price}$ are the forecasted OMP at hour t and its standard deviation of the OMP. $\mu(t)^{price}$ is a random variable generated for the OMP at time t by using normal distribution with the mean of zero and a standard deviation is one. $c_g(t)^{forec}$ is from Table 1.

3 Formulation of the Proposed Stochastic Optimization Problems

A number of scenarios is generated to represent the uncertainty of each variable then a clustering technique is considered as a scenario reduction technique to reduce the number of the generated scenarios [13]. The first stage of the cost function includes the start-up and shut down cost of the DGs, whereas the second stage involves the fuel cost of the DGs, reactive power production cost, maintenance cost, emission cost, battery degradation cost, the cost of purchasing power from the utility grid, production power cost of WT and PV, and cost of the unserved load in different sectors. The cost function can be calculated by the following equation

$$
\begin{aligned}
F = Min \sum_{t=1}^{T} \sum_{i=1}^{N} & [SU_{DG_i}(t) + SD_{DG_i}(t)] + \sum_{s=1}^{S} \lambda_s \sum_{t=1}^{T} \left\{ \sum_{i=1}^{N} \left[CP_{DG_i}\left(P_{DG_i}^s(t)\right) \right. \right. \\
& + CQ_{DG_i}\left(Q_{DG_i}^s(t)\right) + COM_{DG_i}\left(P_{DG_i}^s(t)\right) \right] \delta_{DG_i}(t) + \sum_{j=1}^{M} \sum_{i=1}^{N} E_{j,i}.C_j.\delta_{DG_i}(t).P_{DG_i}^s(t) \\
& + C_d.P_b^s(t).\Delta T + c_{gP}^s(t).P_g^s(t) + c_{gQ}^s(t).Q_g^s(t) + \sum_{i1=1}^{N1} b_{W_{i1}}.P_{W_{i1}}^s(t) \\
& + \sum_{i2=1}^{N2} b_{PV_{i2}}.P_{PV_{i2}}^s(t) + c_{Pres}.P_{Drescut}^s(t) + c_{Qres}.Q_{Drescut}^s(t) + c_{Pind}.P_{Dindcut}^s(t) \\
& + c_{Qind}.Q_{Dindcut}^s(t) + c_{Pcom}.P_{Dcomcut}^s(t) + c_{Qcom}.Q_{Dcomcut}^s(t) \right\}
\end{aligned}
$$

$$(16)$$

where c_{Pres}, c_{Pind} and c_{Pcom} (€/kWh) are the cost of the active involuntary cut of residential, industrial, and commercial loads respectively, c_{Qres}, c_{Qind} and c_{Qcom} (€/kVAr) are the cost of the reactive involuntary cut of residential, industrial, and commercial loads respectively. $P_{Drescut}^s(t)$, $P_{Dindcut}^s(t)$ and $P_{Dcomcut}^s(t)$ are the amount of curtailing active power from residential, industrial and commercial loads, while $Q_{Drescut}^s(t)$, $Q_{Dindcut}^s(t)$ and $Q_{Dcomcut}^s(t)$ are the amount of curtailing reactive power from residential, industrial and commercial loads for scenario (s), λ_s is the probability of the joint scenario (s), $b_{W_{i1}}$ and $b_{PV_{i2}}$ are the production costs of the WTs and PV panels.

4 Proposed Constraints

The first stage constraints are developed as follows.

4.1 Power Balance Constraint

These constraints are expressed as follows.

$$
\begin{aligned}
\sum_{t=1}^{T} & \left\{ \sum_{i=1}^{N} \delta_{DG_i}(t).P_{DG_i}(t) + \sum_{i1=1}^{N1} P_{W_{i1}}(t) + \sum_{i2=1}^{N2} P_{PV_{i2}}(t) + P_b(t) + P_g(t) \right. \\
& = P_{Dres}(t) + P_{Dind}(t) + P_{Dcom}(t) \right\}
\end{aligned}
$$

$$(17)$$

$$
\sum_{t=1}^{T} \left\{ \sum_{i=1}^{N} \delta_{DG_i}(t).Q_{DG_i}(t) + Q_g(t) = Q_{Dres}(t) + Q_{Dcom}(t) + Q_{Dind}(t) \right\} \quad (18)
$$

4.2 Ramp Rate Limits

This constraint can be expressed as follows [14]:

$$-DR_i.\Delta T \leq P_{DG_i}(t+1) - P_{DG_i}(t) \leq UR_i.\Delta T \tag{19}$$

4.3 Emission Constraints

These constraints can be expressed:

$$\sum_{i=1}^{N} E_{j,i}.P_{DG_i}(t) \leq L_j \tag{20}$$

where L_j is the emission limitation of the jth Greenhouse gas in the area of the MG.

4.4 Steady State Security Constraints (SSSCs)

The SSSCs are essential for the reliable and secure operation of MGs and it is noteworthy that the constraints have to be satisfied at each time interval, and they are formulated as:

$$\sum_{t=1}^{T} \left\{ \sum_{i=1}^{N} \delta_{DG_i}(t).P_{DG_imax}(t) \geq P_{Dres}(t) + P_{Dcom}(t) + P_{Dind}(t) \right\} \tag{21}$$

$$\sum_{t=1}^{T} \left\{ \sum_{i=1}^{N} \delta_{DG_i}(t).P_{DG_imax}(t) \geq Q_{Dres}(t) + Q_{Dcom}(t) + Q_{Dind}(t) \right\} \tag{22}$$

The second stage of the objective function are subjected to the same constraints of the first stage, however some of these constraints are modified as follows to accommodate the uncertainties.

A. Active and reactive power balance constraints

$$\sum_{t=1}^{T} \left\{ \sum_{i=1}^{N} \delta_{DG_i}(t).P_{DG_i}^s(t) + \sum_{i1=1}^{N1} P_{W_{i1}}^s(t) + \sum_{i2=1}^{N2} P_{PV_{i2}}^s(t) + P_b^s(t) + P_g^s(t) \right.$$
$$= (P_{Dres}(t) - P_{Drescut}^s(t)) + (P_{Dind}(t) - P_{Dindcut}^s(t)) + (P_{Dcom}(t) - P_{Dcomcut}^s(t)) \right\} \tag{23}$$

$$\sum_{t=1}^{T} \left\{ \sum_{i=1}^{N} \delta_{DG_i}(t).Q_{DG_i}^s(t) + Q_g^s(t) = (Q_{Dres}(t) - Q_{Drescut}^s(t)) + (Q_{Dind}(t) \right.$$
$$- Q_{Dindcut}^s(t)) + (Q_{Dcom}(t) - Q_{Dcomcut}^s(t)) \right\} \tag{24}$$

B. SSSCs

$$\sum_{t=1}^{T} \left\{ \sum_{i=1}^{N} \delta_{DG_i}(t).P_{DG_imax}(t) \geq (P_{Dres}(t) - P_{Drescur}^s(t)) + (P_{Dind}(t) - P_{Dindcut}^s(t)) \right.$$
$$+ (P_{Dcom}(t) - P_{Dcomcut}^s(t)) \right\} \tag{25}$$

$$\sum_{i=1}^{T}\left\{\sum_{i=1}^{N}\delta_{DG_i}(t).Q_{DG_imax}(t) \geq (Q_{Dres}(t) - Q_{Drescut}^{s}(t)) + (Q_{Dind}(t)\right.$$
$$\left. - Q_{Dindcut}^{s}(t)) + (Q_{Dcom}(t) - Q_{Dcomcut}^{s}(t))\right\} \tag{26}$$

5   System Under Case Study

The proposed optimization algorithms are validated by applying the proposed optimization approach on the low voltage multi-feeders hybrid MG as shown in Fig. 1. The power factor is assumed to be typically 0.9 for the whole system. The corresponding DG technical parameters and the emission rates of the DGs are based on credible sources [15–20]. Moreover, the system includes a storage battery with a capacity of 50 kWh and maximum charging and discharging energy at 22.5 kWh, with charging and discharging efficiency assumed to be 0.9 both and DoD at 50%. The input data are shown in Table 1 [6, 21]. Software tool ILOG CPLEX version 12.6 is used to solve the optimization problem.

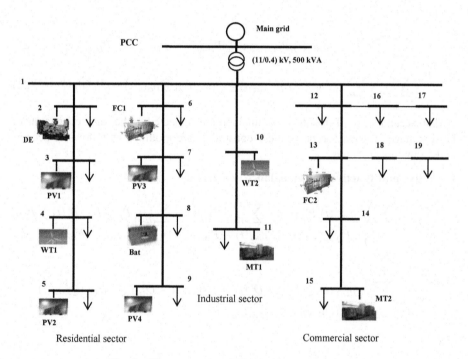

Fig. 1. The proposed multi-feeders MG in this paper

6 Results and Discussion

6.1 Minimizing the Total Operating Cost Under Deterministic Case

Figures 2 and 3 show the optimal scheduling of active and reactive. These figures show that the MG sells active and reactive power to the main grid at hours 12, 13 and 16 to 21 when the OMPs have high values and exceed the cost of generation of the DGs to minimize the total operating cost. Exactly for the same purpose, the MG purchases active and reactive power from the main grid at hours 1 to 11, 14, 15 and 22 to 24 when the OMPs reach low values. In addition, Fig. 3 shows that at hours 1 to 7 the reactive loads are met by buying power from the main grid solely because the minimum output reactive power of the DE is equal to zero. Further, the storage battery is discharged maximum discharging power at hour 17 when the OMP reaches the highest value, while it is charged at the low OMP as shown in Fig. 2. It is found that the total cost per scheduling day is 408.1 €.

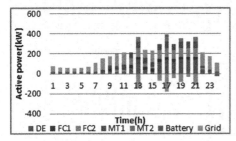   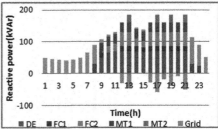

Fig. 2. Optimal active power scheduling   **Fig. 3.** Optimal reactive power scheduling

6.2 Minimizing the Total Operating Cost Under Stochastic Environment

Figures 4 and 5 reveal that the storage battery in scenarios 1, 3 and 4 is discharged twice at hours 13 and 17 because the OMP has high values at these hours. Whereas, in the scenarios 2 and 5, the battery is discharged once at hour 17 when the price has the highest value because at hour 13 the OMP is lower than other scenarios; therefore, there are no economic incentives for operating the battery. Further, the MG at hours 4, 5, and 24 in the scenarios 3 and 5 purchases more active power from the utility grid than other scenarios because the wind and solar generation are equal to zero at hours 4 and 5 and quite low at hour 24. Furthermore, the MG at hour 21 in the scenarios 3 and 5 sells higher power to the utility grid than other scenarios because the renewable generation significantly low in other scenarios. Figure 5 reveals that in the scenarios 2 and 5 at hours 1 to 6 the reactive load is supplied by buying power from the utility grid, while in the scenarios 1, 3, and 4 at hour 6 the DE with the utility grid supply the loads because the DE is already committed. This is because the reactive OMP at hour 6 of the scenarios 1, 3 and 4 higher than the other two scenarios. Besides, the MG at hours 22 and 23 in the scenarios 2 and 5 purchases more reactive power from the utility grid than other scenarios because the reactive OMP is lower. Table 2 illustrates that the Det.

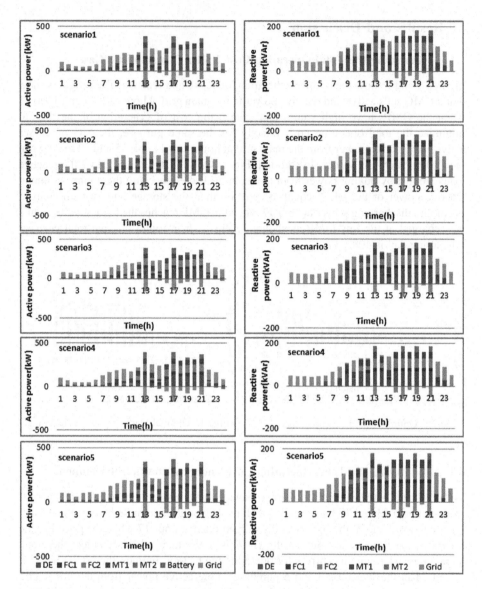

Fig. 4. Optimal active power scheduling   **Fig. 5.** Optimal reactive power scheduling

Case has the lowest operating cost, although it has the lowest renewable generation per scheduling day. This is because the cost of the renewable power production is included in the cost function. This table reveals that at hour 17 the cost has negative values because the OMPs have the highest values at this hours; therefore the MG sells high power to the utility grid to reduce its cost.

Table 2. Optimal hourly cost of the five highest probability scenarios and Det. Case

T(h)	Sc1	Sc2	Sc3	Sc4	Sc5	Det. Case
1	6.8	6.4	7.8	6.8	7.5	7.9
2	6.8	6.4	6.5	6.8	6.0	7.2
3	8.0	8.0	8.0	8.0	8.0	7.2
4	7.6	7.7	4.5	7.6	4.6	7.4
5	8.3	8.2	5.4	8.3	5.2	7.5
6	9.0	8.1	9.1	9.0	8.3	8.5
7	7.7	9.5	10.5	7.6	11.7	10.0
8	13.8	13.6	14.7	13.7	14.6	14.1
9	20.8	18.4	21.1	20.7	18.9	20.2
10	22.3	25.7	22.4	22.5	25.8	24.8
11	27.7	27.4	27.4	27.9	27.1	27.2
12	29.1	28.8	29.1	29.1	28.8	29.1
13	8.2	17.3	11.3	7.7	19.9	21.4
14	32.5	31.7	32.7	32.5	31.9	31.8
15	28.8	29.1	28.8	28.6	29.1	28.3
16	28.9	25.7	28.9	28.9	25.7	28.8
17	-23.9	-21.0	-16.6	-25.0	-14.0	-28.5
18	32.6	32.4	32.4	32.5	32.2	32.8
19	34.1	31.1	35.8	34.1	34.0	32.5
20	35.8	35.9	35.6	35.8	35.7	36.3
21	21.3	25.5	19.0	21.3	23.8	13.5
22	20.8	20.3	20.3	20.8	19.8	19.1
23	13.5	12.0	14.3	13.5	12.9	12.5
24	10.4	9.9	7.8	10.4	7.1	8.5
Total	410.9	418.1	416.8	409.1	424.6	408.1

7   Conclusions

A novel multi-period active and reactive two-stage scenario-based stochastic optimization approach combined with the UC is proposed. The results obtained through cases study demonstrate that the proposed optimization approach can accommodate the uncertainties. In addition, it can be obtained feasible solution for all scenarios. Besides, the uncertainties that come from the OMPs affect the operations of the battery during the scheduling horizon. Further, the battery operations are minimized the operating cost for all scenarios. Particularly, it is not necessary that the scenario that has the highest renewable generation per scheduling day has the lowest total cost per scheduling day.

References

1. Nguyen, T.A., Crow, M.L.: Stochastic optimization of renewable-based microgrid operation incorporating battery operating cost. IEEE Trans. Power Syst. **31**(3), 2289–2296 (2016)
2. Parisio, A., Rikos, E., Glielmo, L.: Stochastic model predictive control for economic/ environmental operation management of microgrids: an experimental case study. J. Process Control **43**, 24–37 (2016)
3. Daniel, A., Olivares, E., Lara, J.D., Cañizares, C.A., Kazerani, M.: Stochastic-predictive energy management system for isolated microgrids. IEEE Trans. Smart Grid **6**(6), 2681–2692 (2015)
4. Liu, C., Wang, J., Botterud, A., Zhou, Y., Vyas, A.: Assessment of impacts of PHEV charging patterns on wind-thermal scheduling by stochastic unit commitment. IEEE Trans. Smart Grid **3**(2), 675–683 (2012)
5. Li, Z., Zang, C., Zeng, P., Yu, H.: Combined two-stage stochastic programming and receding horizon control strategy for microgrid energy management considering uncertainty. Energies **9**(7), 499 (2016)
6. Zakariazadeh, A., Jadid, S., Siano, P.: Smart microgrid energy and reserve scheduling with demand response using stochastic optimization. Electr. Power Energy Syst. **63**, 523–533 (2014)
7. Talari, S., Haghifam, M.-R., Yazdaninejad, M.: Stochastic-based scheduling of the microgrid operation including wind turbines, photovoltaic cells, energy storages and responsive loads. IET Gener. Transm. Distrib. **9**(12), 1498–1509 (2015)
8. Mohammadi, S., Soleymani, S., Mozafari, B.: Scenario-based stochastic operation management of microgrid including wind, photovoltaic, micro-turbine, fuel cell and energy storage devices. Int. J. Electr. Power Energy Syst. **54**, 525–535 (2014)
9. Fossati, J.P., Galarza, A., Martín-villate, A., Font, L.: A method for optimal sizing energy storage systems for microgrids. Renew. Energy **77**, 539–549 (2015)
10. Xie, K., Song, Y.H., Zhang, D., Nakanishi, Y., Nakazawa, C.: Calculation and decomposition of spot price using interior point nonlinear optimisation methods. Int. J. Electr. Power Energy Syst. **26**(5), 349–356 (2004)
11. Tomić, J., Kempton, W.: Using fleets of electric-drive vehicles for grid support. J. Power Sources **168**(2), 459–468 (2007)
12. Al-saadi, M.K., Luk, P.C.K.: Impact of unit commitment on the optimal operation of hybrid microgrids, no. 2 (2016)
13. Sumaili, J., Keko, H., Miranda, V., Zhou, Z., Botterud, A., Wang, J.: Finding representative wind power scenarios and their probabilities for stochastic models. In: Conference on Intelligent System Applications to Power Systems, ISAP 2011 (2011)
14. Santos, J.R., Lora, A.T., Expósito, A.G., Ramos, J.L.M.: Finding improved local minima of power system optimization problems by interior-point methods. IEEE Trans. Power Syst. **18**(1), 238–244 (2003)
15. Parisio, A., Rikos, E., Glielmo, L.: A model predictive control approach to microgrid operation optimization. IEEE Trans. Control Syst. Technol. **22**(5), 1813–1827 (2014)
16. Logenthiran, T., Srinivasan, D., Khambadkone, A.M., Aung, H.N.: Multiagent system for real-time operation of a microgrid in real-time digital simulator. IEEE Trans. Smart Grid **3**(2), 925–933 (2012)
17. MICROGRIDS Large Scale Integration of Micro-Generation to Low Voltage Grids. In: Eu contract ENK5-CT-2002-00610, Technical Annex (2004)

18. Mohamed, F.A., Koivo, H.N.: Online management of microgrid with battery storage using multiobjective optimization. In: Power Engineering, Energy and Electrical Drives, pp. 231–236 (2007)
19. di Valdalbero, D.R.: External costs and their integration in energy costs. In: Europen Sustain Energy Policy Seminar (2006)
20. García-Gusano, D., Cabal, H., Lechón, Y.: Evolution of NOx and SO2 emissions in Spain: ceilings versus taxes. Clean Technol. Environ. Policy **17**(7), 1997–2011 (2015)
21. Zhang, D., Li, S., Zeng, P., Zang, C.: Optimal microgrid control and power-flow study with different bidding policies by using powerworld simulator. IEEE Trans. Sustain. Energy **5**(1), 282–292 (2014)

Charging and Discharging Strategy of Electric Vehicles Within a Hierarchical Energy Management Framework

Mohammed Alkhafaji[1], Patrick Luk[1(✉)], and John Economou[2]

[1] Cranfield University, Bedford MK43 0AL, UK
{m.h.alkhafaji,p.c.k.luk}@cranfield.ac.uk
[2] Cranfield University, Shrivenham, Swindon SN6 8LA, UK
j.t.economou@cranfield.ac.uk

Abstract. As the number of EVs is increasing modern methods are required to understand their impact to the power grid (operators and users). In order to reduce/manage fluctuations on voltage stability and angle stability there is a need for a management control strategy. This paper presents an energy management concept of Charging Station System (CSS) to charge or discharge power of EVs in different situations while retaining system integrity. A suitable objective function is formulated of frequency deviation and voltage deviation on the optimal operation of the charging station are evaluated by formulating and solving the optimisation problem using mixed integer linear programming. The results show that EVs act as a regulator of the microgrid which can control their participation role by discharging active or reactive power in mitigating frequency deviation and/or voltage deviation. The optimisation algorithm is evaluated by formulating and solving the optimisation problem using mixed integer linear programming. Case studies are used to show the viability of the proposed energy management concept.

Keywords: Electric vehicles · Charging station operator · Optimization · Mixed integer linear programming (MILP)

CSA	Charging station agency
CSO	Charging station operator
CSS	Charging station systema
EMS	Energy management shell
EVA	Electric vehicle agency
EVO	Electric vehicle operator
EVs	Electric vehicles
FAN	Frequency above nominal
FBN	Frequency below nominal
FN	Nominal frequency
FOA	Frequency over above nominal
FOB	Frequency over below nominal
MGO	Microgrid operator
PES	Power electronic shell
PMS	Power management shell

© Springer Nature Singapore Pte Ltd. 2017
K. Li et al. (Eds.): LSMS/ICSEE 2017, Part III, CCIS 763, pp. 704–716, 2017.
DOI: 10.1007/978-981-10-6364-0_70

RS   Recharging socket
RSA   Recharging socket agency
VAN   Voltage above nominal
VBN   Voltage below nominal
VN   Nominal voltage
VOA   Voltage over above nominal
VOB   Voltage over below nominal

1 Introduction

The uptake of the EV market is set to create a considerable amount of loads for the electrical power distribution networks in the near future. In general, the EVs will be connected to various locations of a microgrid at different times of a day for recharging of the batteries. The charging demands of the EVs may tend to coincide with the peak demands of the microgrid power. Therefore, EVs are adding extra burdens to the generation capacity. On the other hand, the incorporation of microgrids to the existing distribution networks not only enables the penetration of renewable energy resources but also allows the EVs to be connected as mobile energy storage system.

The electric vehicles are classified according to using their fuel into "wholly electrically fueled" such as battery EVs (BEVs) or "partly electrically fueled" such as hybrid EVs (HEVs) and Plug-In Hybrid EVs (PHEVs). The PHEVs and BEVs are the two main technologies which are considered suitable for grid connection mode.

An EV from the power distribution point of view are a simple load that draws a continuous current independently from discrete network nodes; flexible load from an aggregation EVs with coordinated charging; generation units where EVs are using their storage devices to inject power into the grid according to available resources [1–4].

Multiple challenges could be created for generation units of a microgrid due to integration the EVs. Intelligent controller and communication link between the microgrid utilities are necessary to control and schedule the power flow between them efficiently [5]. To achieve optimal bi-directional power control between the microgrid and the EVs, the ability to predict spatiotemporal connection of EVs and the state of charge of resources of each EV is required. There are many methods proposed for charging EVs either as single charging or aggregator charging which are classified in the following: the electric vehicle starts charge immediately after plugged-in to ensure fully charging the battery. The electricity tariff usually kept fixed for the whole day [6]. The dual tariff charging strategy is that the day tariff is more expensive than the night tariff. The EV starts charging overnight when the demand and the power tariff are lower than the rest of the day. Scheduling charging of the EVs to achieve specific objective function which is either single objective such as achieving minimum charging cost or maximum operator profit, or multiple objectives such as achieving both previous objectives. The latter method is further classified into: **Centralised smart charger (CSC)**: the electric vehicle agent collaborates with the plug-in socket outlet to decide to charge or to discharge the EVs [5, 7–13]. Secondly a **decentralised smart charger (DSC)**: the supplier disperses the EVs demand in time and space according to the electricity tariff and the EVs states. [6, 14–17].

This paper aims to provide further understanding of the impact of management and control of charging and discharging the EVs in a microgrid. In particular, the paper focuses primarily on the implementation of the 'charging station operator' using an innovative 3-tier strategy hierarchical management architecture.

2   A Modular Power Management of the Microgrid Structure

The hierarchical structure of classical management methodology has been chosen as it fits well with the description of problems in this work. The system implementation has been designed as a unified systematic framework where the power management decomposes into modular blocks in chronological execution. The execution of the framework proceeds to demonstrate a reconstruction of the power management problem. Framework management divides the problem into smaller groups to control specific functions which are MGO, CSO, and EVO as shown in 错误!未找到引用源。. Microgrid Management focuses on the entire organisation of all available resources and electrical demand from short-term to long-term perspective to achieve resilience operation with maximum efficiency. It includes all equipment of the microgrid network such as supplies, loads, electric vehicles, conductors, capacitors and reactors, power electronic devices, protective devices, and communication devices. Management consists of the interlocking functions of creating a corporate strategy, policy, and process tiers for organising, planning, controlling, and directing the microgrid resources to achieve resilience operation of the microgrid (Fig. 1).

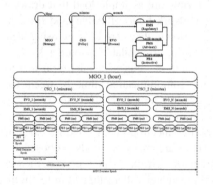

Fig. 1. Concept of a modular management structure

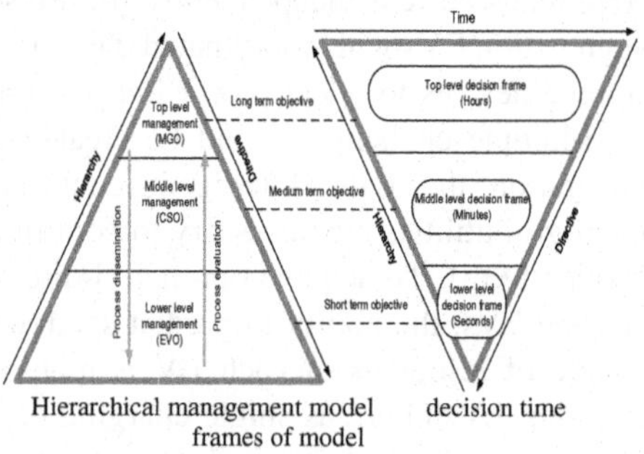

Fig. 2. Hierarchical and decision timeframes of management model

The concept of the modular power management into hierarchical process levels presents an approach to define functionality and interface multidisciplinary objects in a systematic in-depth manner. This is further illustrated in 错误!未找到引用源。. The modular structure involves several processes ranging from long-term period segment decisions of the MGO to very high-speed control of the power electronic of the PES to facilitate the power management by addressing each major part of the process independently. Then, the modular structure consolidates the multidisciplinary process to form a complete system. The structured format describes hierarchical tiers corresponding to the level of management mission, and is seen as the upside-down structure on the left-hand side of 错误!未找到引用源。. The breadth of the tier organisation management structure increases lower down the management hierarchy in relation to time execution. The top tier of the modular management structure handles the strategic planning or long-term strategy of the balancing the generation and demand power. The middle tier decides appropriate actions of the policy planning or medium-term policy of the centralised charging/discharging aggregated EV. The decision which has been taken by middle tier feeds to the lower tier of the execution. The lower tier, which handles the lower-term planning of the EV power management, processes the decision of the middle tier in three stages which are Energy Management Shell (EMS), Power Management Shell (PMS), and Power Electronic Shell (PES). The EMS involves long-term periodic time segment decisions of the energy expenditure for process tier. The decisions of the energy management are accordingly termed as the regulatory of the system which provides a regulated periodic manipulation of a set of control parameters to the next stage of the processing tier. The PMS involves medium-term of the periodic time segment decisions. The PMS is termed as an advisory of the processing tier which is responsible for determining the power split decision for the multiple energy storage systems of the electric vehicle. The two objectives of managing energy and power cannot be completely decoupled due to the different objective response of maintaining SoC level as well as maximise the usable energy of both resources and determining the instantaneous power split between the resources. The iteration rate of the PMS is several magnitudes higher than the EMS. The final stage in the processing tier is the PES which is termed as instructive of the process. The PES is responsible for the actual power blending of energy storage systems of EV using power split ratio which determined by the PMS. The PES decomposes the reference power trajectories into appropriate control switching function of power electronic devices.

The complete modular structure can be described as a hierarchical decision epoch as presented in lower part of Fig. 2. MGO-strategy is a plan of action designed to achieve a long-term or overall aim of the power generation from distributed generator of the MG network to balance the electrical demand. The CSO-policy is a mid-term level of the power management framework. CSO-policy and EVO-procedure dictate how charging/discharging occur in each area of the MG. The CSO-policy could be driven in either a top-down as a centralised charger or a bottom-up as decentralised charger method of the organisation. The EVO-process is the lowest level of power management framework that must adhere to CSO guidelines. It addresses the CSO guides and implements its philosophy to the energy storage systems of the EV. The EVO-process implements by linking three sub-process involved MES-regulatory for energy management system, PMS-advisory for power management system, and PES-instructive for power electronic interface. Therefore, the holistic power management of the microgrid network is addressed in the proposed methodology. The market operator in territory control level is responsible for identifying the different rate of a tariff to power direction mode during a day through all operators of the microgrid.

3 Problem Formulation

The charging station operator may work in two modes:

Mode 1: if $P_{net} > 0$, the EVs work in charging mode (G2V).
Mode 2: if $P_{net} < 0$, the EVs work in discharging mode (V2G). Where $P_{net} = P_s - P_d$

In addition to these two modes, it is possible to discharge the energy of the EVs at $P_{net} > 0$ during a sudden change either in voltage or frequency. According to these two modes, the objective function arranges based onto the microgrid operator requirement into charging EVs mode or discharging EVs mode.

The CSO knows the charging and discharging power tariff at each time, the frequency deviation, and the voltage deviation by the MGO. The CSO labels all the connected EVs and solves the objective function problem periodically to book slots for charging or discharging each EV. The CSO sends the required power to charge all the connected EVs and the available power that could be discharged from the EVs to the MGO. According to the information provided from the microgrid components, the MGO decides to cover all demand of the charging station or part of it and sends the power reference to the CSO. The power reference provides from the MGO either cover the full demand of the charging station or part demand of it; then the charging station builds the charging strategy.

For the charging mode, the CSO sets the objective function of the optimisation problem to achieve minimum power charging cost as shown in Eq. (1). There are two states in charging mode either the MGO could cover all demand or part demand of the EVs. The CSO books charging slots for each EV by solving Eq. (1) based on the information provided from the EVA at covering full demand state. For part demand, the strategy of charging depends on the voltage and frequency level either schedule charging applies when the voltage level at VAN or VOA and the frequency level at FAN or FOA, or curtail some EVs applies when the voltage level at VBN or VOB and the frequency level at FBN or FOB as shown in Table 1. The schedule or curtail charging EVs strategies based mainly on the state of charge of battery and supercapacitor of EVs. The CSO predicts the optimum power charging slots of each EV based on the level of power demand and time required to reach the desired to leave state of charge whereas the low state of charge resources has high charging priority by applying a k scale factor to the Eq. (1) as shown in the Eq. (3). If the EV does not reach the desired to leave state of charge, the CSO applies the priority charging based on giving priority to the earliest EV departure that has low state of charge level compared with the desired leaving state of charge. Furthermore, if the CSO predicts that the EV barely reaches to the desired leaving state of charge or does not reach to it, The CSO informs the owner that the CSO could not charge the EV to selected leaving state of charge and suggests a new level of the desired leaving state of charge to the owner.

In the discharging mode, the MGO sends the information to the CSO to discharge the qualified EVs due to frequency deviation or voltage deviation. Hence no EV could discharge power lower than the predefined state of charge limit. The objective function of discharging mode would get the maximum discharging power cost as shown in

Table 1. Frequency and voltage range of operation

Term	Range	
FN	$-0.001f \le f \le 0.001f$	
VN	$-0.03V \le V \le 0.03V$	
FAN	$0.001f \le f \le 0.01f$	Charging reference
FOA	$0.01 < f$	
VAN	$0.03V \le V \le 0.1V$	
VOA	$0.1V < V$	
FBN	$-0.001f \le f \le -0.01f$	Discharging reference
FOB	$-0.01 < f$	
VBN	$-0.03V \le V \le -0.1V$	
VOB	$-0.1V < V$	

The procedure of these references operation is described in Table 2

Table 2. Map range of frequency and voltage deviation

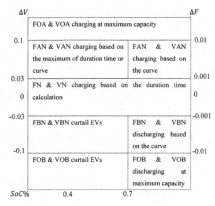

Eq. (2). The priority of discharging classified according to the voltage deviations at VBN or VOB or frequency deviations at FBN or FOB into either discharging the resources of the EV at schedule discharging for VBN and FBN or at maximum discharging for VOB and FOB as explained in Table 2.

The health battery starts discharging to compensate the voltage divisions in case the state of charge of the supercapacitor goes under control limit of 40%. Moreover, the same procedures applied to discharge the healthy supercapacitor to compensate the frequency divisions in case the state of charge of battery to under the limit of 70%. According to the lower limit, the schedule discharging factor of the battery is shown in the Eq. (4) whereas the schedule discharging factor of the supercapacitor is shown in the Eq. (5).The detail of calculation required and available energy and power in battery and supercapacitor are shown in (6 to 错误!未找到引用源。 whereas the control parameters of charging current and discharging current are shown in (11) to (14).

$$Min - Cost = \sum_{i=t_a}^{t_d} \sum_{j=1}^{N} \alpha.s_1.\varepsilon_1.P_{ch,ij}.\Delta t.\rho.T_{P,ch} + \alpha.s_1.\varepsilon_2.Q_{ch,ij}.\Delta t.\rho.T_{Q,ch} \qquad (1)$$

$$Max - Cost = \sum_{i=t_a}^{t_d} \sum_{j=1}^{N} D.\beta.s_2.\vartheta_1.P_{dis,ij}.\Delta t.F.T_{P,ch} + D.\beta.s_2.\vartheta_2.Q_{dis,ij}.\Delta t.F.T_{Q,ch} \qquad (2)$$

$$k_{b/sc,ch} = 15 \times \log_2(SoC - 101) \qquad (3)$$

$$k_{b,dis} = 20.18 \times \log_2(SoC - 69) \qquad (4)$$

$$k_{sc,dis} = 16.86 \times \log_2(SoC - 39) \qquad (5)$$

$$E_{j,b/sc}^{required} = V_{t,b/sc}Q_{j,b/sc}^{rated}\left(\left(SoC_{j,b\backslash sc}^{desired} - SoC_{j,b\backslash sc}^{initial}\right)/100\right) \qquad (6)$$

$$E_{j,b/sc}^{available} = V_{t,b/sc}Q_{j,b/sc}^{rated}\left(\left(SoC_{j,b\backslash sc}^{initial} - SoC_{j,b\backslash sc}^{min}\right)/100\right) \qquad (7)$$

$$P_{j,b}^{required} = \frac{E_{j,b}^{required}}{\left(t_j^{departure} - t_j^{connected}\right).\eta_{ch}}; \quad Q_{j,sc}^{required} = \frac{E_{j,sc}^{required}}{t_j^{departure} - t_j^{connected}}.\eta_{ch} \qquad (8)$$

$$P_{j,b}^{available} = \frac{E_{j,b}^{available}}{\left(t_j^{departure} - t_j^{connected}\right).\eta_{ch}}; \quad Q_{j,sc}^{available} = \frac{E_{j,sc}^{available}}{\left(t_j^{departure} - t_j^{connected}\right).\eta_{ch}} \qquad (9)$$

$$C_{rate}^{required} = \frac{\varepsilon_1 P_{j,b}^{required} + \varepsilon_2 Q_{j,sc}^{required}}{V_{tb/sc}Q_{j,b,sc}^{rated}}; \quad C_{rate}^{actual} = \frac{\varepsilon_1 P_{j,b}^{available} + \varepsilon_2 Q_{j,sc}^{available}}{V_{tb/sc}Q_{j,b/sc}^{rated}} \qquad (10)$$

At charging mode
$$\begin{aligned} I_{j,b}^{ch} &= Q_{j,b}^{rated}C_{rate}^{required} \text{ if } s_1 = 1, \varepsilon_1 = 1 \\ I_{j,sc}^{ch} &= Q_{j,sc}^{rated}C_{rate}^{required} \text{ if } s_1 = 1, \varepsilon_2 = 1 \end{aligned} \qquad (11)$$

At schedule charging mode
$$\begin{aligned} I_{j,b}^{sch} &= Q_{j,b}^{rated}C_{rate}^{required}.K_{f,ch} \text{ if } s_1 = 1, \varepsilon_1 = 1 \\ I_{j,sc}^{sch} &= Q_{j,sc}^{rated}C_{rate}^{required}.K_{f,ch} \text{ if } s_1 = 1, \varepsilon_2 = 1 \end{aligned} \qquad (12)$$

At discharging mode
$$\begin{aligned} I_{j,b}^{dis} &= Q_{j,b}^{rated}C_{rate}^{actual} \text{ if } s_2 = 1, \vartheta_1 = 1, D = 1 \\ I_{j,sc}^{dis} &= Q_{j,sc}^{rated}C_{rate}^{actual} \text{ if } s_2 = 1, \vartheta_2 = 1, D = 1 \end{aligned} \qquad (13)$$

At schedule discharging mode
$$\begin{aligned} I_{j,b}^{sdis} &= Q_{j,b}^{rated}C_{rate}^{actual}.K_{f,dis}; \text{ if } s_2, \vartheta_1, D = 1 \\ I_{j,sc}^{sdis} &= Q_{j,sc}^{rated}C_{rate}^{actual}.K_{f.dis}; \text{ if } s_2, \vartheta_2, D = 1 \end{aligned} \qquad (14)$$

Where $I_{j,b/sc,min}^{ch/dis} \leq I_{j,b/sc}^{ch/dis} \leq I_{j,b/sc,max}^{ch/dis}; \; I_{j,b/sc,min}^{sch/sdis} \leq I_{j,b/sc}^{sch/sdis} \leq I_{j,b/sc,max}^{sch/sdis}$

$$|s_1| + |s_2| = 1; s_1 = \begin{cases} 1 & \text{if } \sum_{j=1}^{N} P_{ch,ij} \geq P_{net}(t) \text{charging mode apply} \\ 0 & \text{otherwise} \end{cases}$$

$$s_2 = \begin{cases} 1 & \text{if } \sum_{j=1}^{N} P_{dis,ij} \geq P_{v2g}(t) \text{discharging mode apply} \\ 0 & \text{otherwise} \end{cases}$$

ε is the binary number either 1 or 0, ε_1 refers to active power charging mode and ε_2 refers to reactive power charging mode, where $|\varepsilon_1| + |\varepsilon_2| = 1$.

$$\varepsilon_1 = \begin{cases} 1 & \textit{if SoC}_B^d - \textit{SoC}_B^a \leq 0 \\ 0 & \textit{otherwise} \end{cases}; \; \varepsilon_2 = \begin{cases} 1 & \textit{if SoC}_B^d - \textit{SoC}_B^a > 0 \, \& \textit{SoC}_{sc}^d - \textit{SoC}_{sc}^a \leq 0 \\ 0 & \textit{otherwise} \end{cases}$$

$$\rho(\textit{factor of charging}) = \begin{cases} 1 & \textbf{\textit{Apply optimization algorithem}} \\ 1.5 & \textbf{\textit{Charge EV at rated power with}} \; 50\% \, \textbf{\textit{extra price}} \end{cases}$$

$$D(\textit{owner choice}) = \begin{cases} 1 & \textit{if bidirection} \\ 0 & \textit{if unidirection} \end{cases}$$

ϑ is the binary number either 1 or 0, ϑ_1 refers to active power discharging mode at frequency deviation and ϑ_2 refers to reactive power discharging mode at voltage deviation, where $|\vartheta_1| + |\vartheta_2| = 1$.

$$\vartheta_1 = \begin{cases} 0 & \textit{if } \Delta\omega > 0.001 \textit{ and } \textbf{SoC}_b \geq 70\% \textit{ OR } \Delta V > 0.3 \textit{ and } \textbf{SoC}_{sc} \leq 40\% \\ 0 & \textit{otherwise} \end{cases}$$

$$\vartheta_2 = \begin{cases} 1 & \textit{if } \Delta V > 0.3 \textit{ and } \textbf{SoC}_{sc} \geq 40\% \textit{ OR } \Delta\omega > 0.001 \textit{ and } \textbf{SoC}_b \leq 70\% \\ 0 & \textit{otherwise} \end{cases}$$

The degradation cost (C_d) is set to be 6.5 cents / kWh as adopted from [22].

α and β are weighting dynamic (delay) coefficients for the charging and discharging power. It represents the rate of convergence for charging or discharging to an equilibrium point. It is required at high disturbance to reduce the oscillation in frequency or voltage variation by adjust it to a high value. It could be increased those values or choose it minimum at normal state.

4 Constraints

There are several constraints to be considered.

- Power constraint
  In charging mode, the MGO should cover all demand for the EVs as shown in Eq. (15). In case the MGO provides less than the net power of the EVs then schedule charging should apply based on the SoC of resources and period of parking. In discharging mode, the electric vehicles should supply the required power to maintain the network stability as represented by Eq. (16). In case the total power of qualified EVs is less than the required power of the MGO then the MGO should use the available power from EVs and search for extra power from the rest of distributed generators to balance the load. It may be that the MGO could not find the surplus power to cover the load then the MGO applies the curtail strategy starting with the non-sensitive load.

$$\sum_{n=1}^{N} P_{EV,n} \leq |P_{net}| \tag{15}$$

$$P_{min} \leq \sum_{i \in T} \sum_{j \in N} P_{ch \backslash dis,ij} \leq P_{max}, \forall i \in T, \forall j \in N \tag{16}$$

- Power of battery and supercapacitor constraint
  For safe operation of battery and supercapacitor, the current flow through the energy storage system should be limited. The maximum current flow from the battery depends mainly on the chemistry reaction and ambient temperature. Exceed the maximum current of the energy storage system causes heat flame smoke leading to destroy the device. Operate the energy storage system beyond rated value causes overuse of the device leading to degrading their cycle life that adversely affects the lifetime of the energy storage system. The limited charging and discharging current are shown in Eqs. (17) and (18) respectively.

$$I_{j,b/sc,min}^{ch/dis} \leq I_{j,b/sc}^{ch/dis} \leq I_{j,b/sc,max}^{ch/dis} \tag{17}$$

$$I_{j,b/sc,min}^{sch/sdis} \leq I_{j,b/sc}^{sch/sdis} \leq I_{j,b/sc,max}^{sch/sdis} \tag{18}$$

- State of charge constraint
  The state of charge of the battery and the supercapacitor represent the amount of stored energy inside them. To maintain long life of the energy storage system at charging mode, the state of charge should be limited to prevent degrading their cycle life. At discharging mode, the discharging power from the energy storage system should be limited to keep the energy for next journey and increase the lifetime of the device. The Eq. (19) shows the limitation of the state of charge.

$$SoC_{min} \leq \sum_{i \in T} \sum_{j \in N} SoC_{b \backslash sc,ij} \leq SoC_{max}, \forall i \in T, \forall j \in N \tag{19}$$

5   Case Study

To explain the implementation of optimisation algorithm strategy under various circumstances, battery and supercapacitor operation of numerical simulation are explored in this paper:

The case study has 15 EVs from different brands with different capacity of battery and supercapacitor connected at time 9–15 of a day as shown in Table 3. The desired leaving state of charge of the battery and the supercapacitor are assumed to be 100% whereas the initial state of charge of the supercapacitor is assumed 45%, and the initial state of charge of the battery is shown in Table 3. The capacity of the supercapacitor has chosen lower than the capacity of the battery. Thus, the first minutes of processing time are booked for the charging supercapacitor. The power tariff is chosen variable based on the time of a day which starts at a lower price during night. The tariff steps at early morning and increases gradually until reaching a maximum at mid-day. After that, it starts to reduce gradually until reaching to minimum tariff at midnight; then it is kept constant until early morning as shown in Fig. 3.

Table 3. Type of electric vehicles and resources capacity

	EV make	Energy_B (kWh)	Q_{sc} (kWh)	Initial SoC_b
1.	Toyota prius PHEV	4.4	0.44	45%
2.	Chevy volt PHEV	16	1.6	45%
3.	Mitsubishi iMiEV	16	1.6	45%
4.	Smart fortwo ED	16.5	1.65	45%
5.	Honda fit	20	2	45%
6.	GM spark	21	2.1	45%
7.	BMW i3	22	21	45%
8.	Ford focus	23	2.3	45%
9.	Fiat 500e	24	2.4	45%
10.	Nissan leaf	25	2.5	45%
11.	Mercedes B	28	2.8	45%
12.	Nissan leaf	30	3.0	45%
13.	Tesla S 60	60	6	50%
14.	Tesla S70	70	7	55%
15.	Tesla S85	90	9	70%

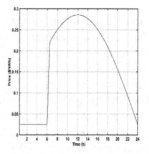

Fig. 3. Microgrid utility tariff

Battery operation:

The EVs can represent as a resource for microgrid in a congestion electricity market such as generator outage or sudden increasing in demand. Moreover, the reference of discharging the EVs is VBN, VOB, FBN or FOB. The CSO receives the discharging references to start discharging all qualified EVs either at scheduled discharging by applying (2) and (4) when the reference is VBN or FBN, or at maximum converter rating power by applying (2) only when the reference is VOB or FOB. The CSA applies the selling tariff of discharging EV. The performance of each EV regarding the state of charge and power consumed depicted in Figs. 4 and 5 respectively. The cost of power charging and discharging for each EV is shown in Fig. 6.

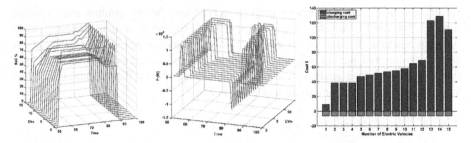

Fig. 4. SoC of each vehicle   **Fig. 5.** Power of each vehicle   **Fig. 6.** The cost of power charging for each vehicle

Supercapacitor operation:

The supercapacitor can charge and discharge a virtually unlimited number of times. Charging supercapacitor has faster response times than a battery, typically at about 1–10 s. However, the rated of the converter limits charging supercapacitor. In this paper, the first minutes of connection are reserved to charge the supercapacitor of the EV fully by applying (1), whereas the discharging of the supercapacitor strategy is the same as discharging battery strategy as in case four. Fuzzy interference programmatic solution is employed as a solution to make a logical decision of discharging battery and supercapacitor within EVO level. The fuzzy rule antecedents are the frequency of microgrid, the voltage of microgrid, the state of charge of the battery, and the state of charge of supercapacitors whereas the fuzzy rule consequent is the battery reference trajectory and the supercapacitor reference trajectory. The fuzzy power management for the battery and supercapacitor is the subject of another paper.

The performance of each EV regarding the state of charge and consumed power depicted in Figs. 7 and 8 respectively. The results clearly show that all EVs reached to the desired state of charge before leaving the charging station. The cost of power charging for each vehicle is shown in Fig. 9.

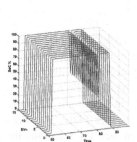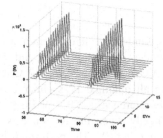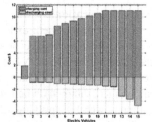

Fig. 7. SoC of each vehicle   **Fig. 8.** The power of each vehicle   **Fig. 9.** The cost of power charging for each vehicle

6   Conclusions

The proposed power management strategy presented a new and holistic approach in managing the operation of EVs (charging/discharging) within a microgrid in three major processes of hierarchical arrangement. The paper focuses on the second tier of the hierarchical management system (CSO) in minimising the total charging power cost or maximising the discharging power cost of an optimal management system of EVs. By using representative case study, a number of benefits emerge from the proposed strategy. It has been demonstrated that discharging EVs have high impact in frequency deviation or voltage deviation of the microgrid. The impact of the optimal operation of CSO is quantified and evaluated using mixed integer linear programming. The results showed that applying the proposed management of CSO to charge the EVs is very effective in preventing overload in the microgrid and achieving optimum charging or discharging power costs. Scheduling slots for each EV to meet full demand from MGO are the best charging operation, which could achieve the desired leaving state of charge at minimum charging cost. In discharging mode, the owner of the EV benefits from the unit discharging price which is normally higher than the unit charging price.

References

1. Battistelli, C., Baringo, L., Conejo, A.J.: Optimal energy management of small electric energy systems including V2G facilities and renewable energy sources. Electr. Power Syst. Res. 92, 50–59 (2012)
2. Khorramdel, B., Khorramdel, H., Aghaei, J., Heidari, A., Agelidis, V.G.: Voltage security considerations in optimal operation of BEVs/PHEVs integrated microgrids. IEEE Trans. Smart Grid 6(4), 1575–1587 (2015)
3. Beer, S., Gómez, T., Dallinger, D., Marnay, C., Stadler, M., Lai, J.: An economic analysis of used electric vehicle batteries integrated into commercial building microgrids. IEEE Trans. Smart Grid 3(1), 517–525 (2012)
4. Rahbari-Asr, N., Chow, M.Y.: Cooperative distributed demand management for community charging of PHEV/PEVs based on KKT conditions and consensus networks. IEEE Trans. Ind. Inform. 10(3), 1907–1916 (2014)
5. Hu, J., Morais, H., Sousa, T., Lind, M.: Electric vehicle fleet management in smart grids: a review of services, optimization and control aspects. Renew. Sustain. Energy Rev. 56, 1207–1226 (2016)
6. Waraich, R.A., Galus, M.D., Dobler, C., Balmer, M., Andersson, G., Axhausen, K.W.: Plug-in hybrid electric vehicles and smart grids: investigations based on a microsimulation. Transp. Res. Part C Emerg. Technol. 28, 74–86 (2013)
7. Turker, H., Radu, A., Bacha, S., Frey, D., Richer, J., Lebrusq, P.: Optimal charge control of electric vehicles in parking stations for cost minimization in V2G concept. In: 2014 3rd International Conference Renewable Energy Research Application ICRERA, pp. 945–951 (2014)
8. Acha, S., Member, S., Green, T.C., Member, S., Shah, N.: IEEE Xplore - effects of optimised plug-in hybrid vehicle charging strategies on electric distribution network losses. In: IEEE, pp. 1–6 (2010)

9. Galus, M.D., Waraich, R.A., Noembrini, F., Steurs, K., Georges, G., Boulouchos, K., Axhausen, K.W., Andersson, G.: Integrating power systems, transport systems and vehicle technology for electric mobility impact analysis and efficient control. IEEE Trans. Smart Grid 3(2), 934–949 (2012)
10. Clement, K., Haesen, E., Driesen, J.: Coordinated charging of multiple plug-in hybrid electric vehicles in residential distribution grids. In: 2009 IEEE/PES Power Systems Conference and Exposition PSCE 2009, pp. 1–7 (2009)
11. Liu, H., Hu, Z., Song, Y., Wang, J., Member, S., Xie, X.: Vehicle-to-grid control for supplementary. 30(6), 1–10 (2014)
12. Yao, W., Zhao, J., Wen, F., Xue, Y., Ledwich, G.: A hierarchical decomposition approach for coordinated dispatch of plug-in electric vehicles. IEEE Trans. Power Syst. 28(3), 2768–2778 (2013)
13. Xu, Z., Su, W., Hu, Z., Song, Y., Zhang, H.: A hierarchical framework for coordinated charging of plug-in electric vehicles in China. IEEE Trans. Smart Grid 7(1), 428–438 (2016)
14. Peças Lopes, J.A., Soares, F.J., Rocha Almeida, P.M.: Identifying management procedures to deal with connection of electric vehicles in the grid. In: 2009 IEEE Bucharest PowerTech Innovative Ideas Towar. Electr. Grid Futur., pp. 1–8 (2009)
15. Ahn, C., Li, C.T., Peng, H.: Optimal decentralized charging control algorithm for electrified vehicles connected to smart grid. J. Power Sources 196(23), 10369–10379 (2011)
16. Liu, H., Hu, Z., Song, Y., Lin, J.: Decentralized vehicle-to-grid control for primary frequency regulation considering charging demands. IEEE Trans. Power Syst. 28(3), 3480–3489 (2013)
17. Beaude, O., He, Y., Hennebel, M.: Introducing decentralized EV charging coordination for the voltage regulation. In: 2013 4th IEEE/PES Innovative Smart Grid Technologies Europe ISGT Europe 2013, pp. 1–5 (2013)
18. Tennessee Valley Authority: Types of Electric Vehicles (2015). http://www.tva.com/environment/technology/car_vehicles.htm
19. Williamson, S.S.: Electric drive train efficiency analysis based on varied energy storage system usage for plug-in hybrid electric vehicle applications. In: 2007 IEEE Power Electronics Specialists Conference, pp. 1515–1520 (2007)
20. OECD/IEA: Global EV Outlook 2016 Electric Vehicles Initiative (2016). www.iea.org/t&c/
21. Anker, P., Ch, G.: (19) United States (12) Reissued Patent (2013)
22. White, C.D., Zhang, K.M.: Using vehicle-to-grid technology for frequency regulation and peak-load reduction. J. Power Sources 196(8), 3972–3980 (2011)

Optimization Methods

Optimization Allocation of Aerospace Ground Support Vehicles for Multiple Types of Military Aircraft

Fuqin Yang[1]([✉]), Jinhua Li[2,3], and Mingzhu Zhu[4]

[1] Department of Military Logistics, Military Transportation University,
Tianjin 300161, China
yfqfjwfjl@163.com
[2] Department of Automation, TNList, Tsinghua University,
Beijing 100084, People's Republic of China
[3] The Logistics Information Center of PLA,
Beijing 100842, People's Republic of China
[4] Innovation and Enterpreneurship Development Center,
Tianjin Sino-German University of Applied Sciences, Tianjin 300350, China

Abstract. As an important class of support resources, the allocation of aerospace ground support vehicles (AGSV) has important impact on the sortie generation rate of multiple types of military aircraft (MTMA). In this paper, a general queueing model of AGSV has been built to describe the features of the support process of MTMA using the multi-class closed queueing networks. To satisfy constraints on each sortie generation rate (SGR) of MTMA and get good economic benefits, an optimization allocation model of AGSV has been developed to minimize the total value of AGSV. Based on mean value analysis and marginal analysis, a solving algorithm determines the numbers of AGSV at each station. The results of a case study show applicability of the optimization model.

Keywords: Optimization allocation · Ground support vehicles · Multiple types of military aircraft · Closed queueing network

1 Introduction

The success of a modern air combat often depends on the support level of logistics of military aircraft [1]. As one of the most important parameters of integrated support, sortie generation rate (SGR) of military aircraft refers to the times of mission flying by an individual aircraft within one day according to maintenance support plan. In the sortie generation process, numerous support activities are required for a mission-ready aircraft to fly a specified sortie [2]. To increase the SGR, most studies concentrate on how to optimize the support resources allocation and improve the level of aircraft availability besides enhancing performance of military aircraft [3–5].

As an important class of support resources, aerospace ground support vehicles (AGSV), such as tractor, oxygen-filling vehicle, nitrogen-filling vehicle, refuelling vehicle and power vehicle, are used to supply power, gas, or oil respectively in the

© Springer Nature Singapore Pte Ltd. 2017
K. Li et al. (Eds.): LSMS/ICSEE 2017, Part III, CCIS 763, pp. 719–728, 2017.
DOI: 10.1007/978-981-10-6364-0_71

ground support process of military aircraft. With coordination among multiple types of military aircraft (MTMA) becoming a basic style of aviation air operations in modern warfare, the configuration of various ground support vehicles has turned into an important problem in the support resources allocation at airport of air force. In order to gain insight into configuration and allocation issues, one of the most common methods for quantitative research is simulation. Some simulation models of logistics support such as autonomic logistics system (ALS) [6], logistics composite model (LCOM) [7] have been developed and widely used in the air force logistical support bases, greatly improving logistical support effectiveness and promoting logistical plan adaptability. Fitting the minimizing constraints on support time of one aircraft, the optimal configuration of support vehicles has been discussed based on the simulation software of Arena [8]. The support sequences of aircraft have been researched with particle swarm optimization (PSO) to satisfy the demand of dynamic flight support vehicles scheduling [9]. The simulation model of support resources configuration of multi-aircraft formation has been developed in executing assignment based on the Anylogic and Arena software by Ma et al. [10] and Luo et al. [11] respectively.

Using simulation to solve the allocation problem of support resources at airport has some advantages, such as rather precise and reliable calculating results and being capable of solving highly complicated problems completely. But it should be noted that it is also very tedious and time-consuming. As alternatives to simulation, analytical approaches of closed queueing network have been offered to address the allocation problem. For example, Dietz and Jenkins [12], Xia et al. [13] proposed an optimization model for support resources allocation of one kind of military aircraft at airport.

If MTMA such as fighters, bombers and airborne early warning aircraft land and take off at one airport at the same time, the optimization allocation of support resources becomes a very complex problem because different types of aircraft may need different kinds of ground support vehicles at the same support activity service center. Meanwhile, it is unlikely to provide too many of various types of vehicles due to support area and cost constraint, and collisions among support vehicles and aircraft may take place when there are too many vehicles moving outfield of airport. However, few papers have yet proposed a method to address the configuration problem of multi-servers with multiple constant unit rates for MTMA. In this paper, an analytical approach of allocation of AGSV has been presented that obtains excellent results for a support resources allocation system modelled using a multi-class closed network of multi-server with multiple constant unit rates and marginal analysis. This approach extends the recent research of Xia et al. to multi-type aircraft and multi-kind vehicles in the same support activity service center.

The structure of the paper is as follows. The complex system model of MTMA support process is presented in Sect. 2. In Sect. 3, the assumptions, notations and optimization allocation model are described. To achieve pre-defined objectives of sortie generation rates, corresponding optimization algorithm steps of support resources allocation of the MTMA are given based on mean value analysis (MVA) and marginal analysis. A case study of the real work situation is conducted in Sect. 4. Finally, the results and future works are given in Sect. 5.

2   Model Description

Although the support system of MTMA can be quite complex, a simple model suffices to illustrate the fundamental behaviour of aircraft sortie generation process. To demonstrate this concept, it is assumed that the MTMA operate from a single base in a multi-class closed-loop system. From the time of arriving at the airport to the time of taking off, the MTMA flow through a network of support activity services which include towing, oxygen-filling, refueling, power, air condition and munitions upload. The aircraft usually return to the same airport after flying a sortie, so the sortie generation process of MTMA can be considered as a multi-class closed network in which R classes of aircraft circulate among M multiple server stations. Finishing the support activity at any station i, each aircraft moves to station j with time invariant probability p_{ij}. Figure 1 shows the movement of MTMA among stations in a typical sortie generation cycle through the network.

The support activity of each class-r aircraft starts with tractor towing to parking apron, during which any of the tractors can be selected because various vehicles are assumed belonging to general purpose machinery in this model. The number of each type of AGSV is usually limited. If a service center is free when one aircraft of class-r arrives, then the aircraft enters support activity immediately. Otherwise, it waits in the queue for service. In the support process, the class-r aircraft awaits oxygen-filling, refuelling and perfect checks in turn as shown in Fig. 1. If at least one malfunction is indicated in the course of prefight checks, the class-r aircraft waits for repairs with probability p_{45} and any one of air conditioner vehicles can be chosen. If no malfunction is detected, the class-r aircraft waits for munitions upload with probability p_{46}. Finally, the class-r aircraft is ready to take off and the cycle is repeated.

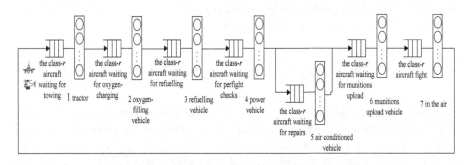

Fig. 1.   Aerospace ground support process of MTMA

3   Optimization Allocation Model of AGSV of MTMA

3.1   Model Assumptions

For purposes of analytical tractability, each type of aircraft, service ability of each service center, and the queueing discipline must fulfill the following assumptions in the model of AGSV of MTMA:

(1) The multi-class closed network has M stations and there are mi or infinite independent servers at station i.

(2) The multi-class closed network has R classes of aircraft, the population vector k = (k1, k2, ..., kR) expresses the network containing the number of the class-r aircraft. Each aircraft cannot change its class, that is, the total number of aircraft in each class is constant.

(3) All support service time at each service center should be exponentially distributed including the time of aircraft taxiing from the previous service center to present one.

(4) The queueing discipline is first come first service (FCFS) at the finite server station and infinite servers (IS) at the infinite server station.

3.2   MVA for Multi-class Multi-server Queueing Networks

If the MTMA flow through the closed network as shown in Fig. 1, performance measures can be computed using the MVA presented by Reiser and Lavenberg [14]. According to the queueing theory, the mean response time $R_{ir}(k)$ of class-r aircraft at any station i is equal to service time S_{ir} and waiting time. For the queueing networks with product-form solution, the waiting time is equal to the average service time S_{ir} multiplied by the mean queue size of the system with one aircraft less based on the arrival theorem, which is the critical foundation for MVA. Therefore,

$$R_{ir}(\vec{k}) = S_{ir} \cdot [1 + Q_{ir}(\overrightarrow{k - 1_r})], \qquad (1)$$

Where $Q_{ir}(k - 1_r)$ is the mean queue size of the system with one aircraft less including the aircraft waiting for support activity.

If station i has m_i ($m_i \geq 1$) servers, then:

$$R_{ir}(\vec{k}) = \frac{S_{ir}}{m_i} \left[1 + \sum_{s=1}^{R} K_{is}(\overrightarrow{k - 1_r}) + \sum_{j=0}^{m_i-2} (m_i - j - 1) P_i(j|(\overrightarrow{k - 1_r}) \right], \qquad (2)$$

Where i is the ith station, $i = 1, ..., M$; The r is the class-r aircraft, $r = 1, ..., R$; The S_{ir} is the service time of a class-r aircraft at station i; The m_i is the number of servers at station i; The k is a vector of aircraft denoted by $(k_1, k_2,..., k_R)$; The $(k - 1_r)$ is the aircraft vector k with one aircraft removed from class-r and $P_i(j/k)$ is the marginal probability of aircraft j at station i given population k.

$$P_i(j|\vec{k}) = \frac{1}{j} \left[\sum_{r=1}^{R} e_{ir} \cdot s_{ir} \cdot \lambda_r(\vec{k}) \cdot P_i(j - 1|\overrightarrow{k - 1_r}) \right], \qquad (3)$$

If $j = 0$, then:

$$P_i(0|\vec{k}) = 1 - \frac{1}{m_i}\left[\sum_{r=1}^{R} e_{ir} \cdot s_{ir} \cdot \lambda_r(\vec{k}) + \sum_{j=1}^{m_i-1}(m_i - j) \cdot P_i(j|\vec{k})\right], \qquad (4)$$

Where $e_{ir} = \sum_{j=1}^{N}\sum_{s=1}^{R} e_{js} \cdot p_{js,ir}$ is the mean number of visits of class-r aircraft to station i.

If the number of servers is infinite, then:

$$R_{ir}(\vec{k}) = S_{ir}, \qquad (5)$$

for all k.

Throughput of class-r aircraft can be computed as:

$$\lambda_r(\vec{k}) = \frac{k_r}{\sum_{i=1}^{N} e_{ir} \cdot T_{ir}(\vec{k})}, \qquad (6)$$

Throughput of class-r aircraft at station i can be computed as:

$$\lambda_{ir}(\vec{k}) = \lambda_r(\vec{k}) \cdot e_{ir}, \qquad (7)$$

By applying Little's law, the mean queue length and utilization of class-r aircraft at station i can be determined as:

$$Q_{ir}(\vec{k}) = \lambda_{ir}(\vec{k}) \cdot R_{ir}(\vec{k}), \qquad (8)$$

$$\rho_{ir}(\vec{k}) = \frac{\lambda_{ir}}{m_i} \cdot S_{ir}, \qquad (9)$$

Based on MVA, the mean response time, throughput, mean queue length and utilization of class-r aircraft at each station can be obtained. It is noted that the throughput at fight station is the SGR of aircraft in the sortie generation process. By analyzing the relation between numbers and SGR of MTMA, the bottleneck factors restricting further increase in the SGR can be identified and viewed in the original configuration of AGSV. The following sections show how to optimize the allocation of AGSV on the basis of attaining the objectives of SGR.

3.3   Optimization Model and Marginal Analysis

Supporting decision-making based on marginal or incremental changes to support resources instead of one based on totals or averages, the process of marginal analysis identifies the benefits and costs of different alternatives by examining the incremental effect on total revenue and total cost caused by a very small (just one unit) change in the output or input of each alternative. The marginal analysis is used to optimize the spare parts configuration in recent years [15, 16]. In this paper, marginal analysis is used to solve the model of optimization allocation of AGSV.

The allocation of AGSV at each station has to fit a cost constraint and to reach a target level of SGR of the whole complex system. The optimization model AGSV allocation can be expressed by:

$$\begin{cases} \min C = \sum_{i=1,2,\dots,6} m_i c_{ir} \\ s.t. \lambda_{7r}(\vec{k}) \geq \lambda_{tr} \end{cases}, \tag{10}$$

Where C is the total value of vehicles at all stations, and C_{ir} is the value coefficient of vehicles for class-r aircraft at station i.

The allocation of AGSV can be tested on the marginal ability gene $\delta(m_i)$ that can be defined as follows:

$$\delta(m_{ir}) = \frac{\lambda_{7r}(m_{ir}+1) - \lambda_{7r}(m_{ir})}{c_{ir}}, \tag{11}$$

Where $\lambda_{7r}(m_{ir} + 1) - \lambda_{7r}(m_{ir})$ is the SGR increment of class-r aircraft due to the increase of one vehicle at station i.

Implemented in MATLAB, the optimization model can be solved by the MVA and marginal analysis. Beginning with an initial configuration of vehicle at each station, an iteration consists in finding the best value of C_{min} and $\lambda_{7r}(k)$ considering all the possible allocations of a new vehicle at a station. If the SGR objectives constraints of the class-r aircraft are not attained, the allocation is rejected and the second best is tested. When it is attained, the selected vehicle is allocated at station i to check the total value of vehicles. The algorithm continues to execute a new iteration until the SGR objectives constraints of the class-r aircraft are all satisfied.

3.4 Steps of Optimization Algorithm

Based on the above analysis, the iteration calculation steps of the optimization algorithms can be obtained and described as follows.

Step 1. Initialization. For all stations $i = 1,\dots, N$, and all classes $r = 1, \dots, R$, let the number of vehicle at each finite server station $m_i = 1$ and calculate SGR value of the class-r aircraft based on the MVA.

Step 2. Add one vehicle at every finite server station and calculate the corresponding SGR value of the class-r aircraft using the MVA.

Step 3. Calculate the marginal ability gene $\delta(m_i)$ of every vehicle and order them from largest to smallest, then add one vehicle to the station which corresponds to the maximum marginal ability gene.

Step 4. Determine whether the SGR objectives constraints of the class-r aircraft are satisfied. If the SGR target constraints are satisfied, then stop iterating at this point. If not, then go to step 2.

4   A Case Study

To test the optimization method, a series of historical data were collected from an airport of Chinese Air Force. A support process model of MTMA was selected, considering their fight plan and various ground support vehicles, identifying the most critical ground support process to implement a case. The model includes two types of aircraft and six kinds of supply vehicles. Aircraft have two types: the fighters and attackers, which take off from or land at the same airport. The supply vehicles have six types: tractor, oxygen-filling vehicle, refuelling vehicle, power vehicle, air condition vehicle and munitions upload vehicle, which are considered as servers at the support service stations, respectively. According to a time series analysis, the service time of support service stations are in accord with negative exponential distribution at each station. The queuing disciplines are on the basis of FCFS at the finite server station and on the basis of IS at the infinite server station. The values of each variable at each service station in the support process of MTMA are shown in Table 1.

Table 1. Values of each variable at each service station.

Service stations i	Service time S_{ir} (h)	Objectives of SGR		Optimization allocation of servers number	Value coefficient	Total value of optimization allocation
		Fighters	Attackers			
1. Tractor	$S_{11} = 0.10,$ $S_{12} = 0.15$			3	5	
2. Oxygen-filling vehicle	$S_{21} = 0.08,$ $S_{22} = 0.15$			3	4	
3. Refuelling vehicle	$S_{31} = 0.30,$ $S_{32} = 0.40$			8	2	
4. Power vehicle	$S_{41} = 0.15,$ $S_{42} = 0.20$	12	9	4	6	136
5. Air condition vehicle	$S_{51} = 0.20,$ $S_{52} = 0.25$			2	3	
6. Munitions upload vehicle	$S_{61} = 0.34,$ $S_{62} = 0.45$			9 $/$	7 $/$	
7. Fight	$S_{71} = 2.00,$ $S_{72} = 2.50$					

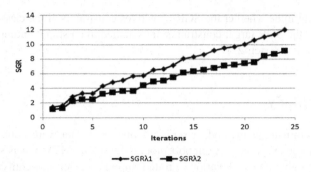

Fig. 2. Evolution of SGR

According to the data analysis, the state transfer matrices are as follow:

$$P = \left[p_{js,ir}\right] = \begin{bmatrix} 0 & 1 & 0 & 0 & 0 & 0 & 0 \\ 0 & 0 & 1 & 0 & 0 & 0 & 0 \\ 0 & 0 & 0 & 1 & 0 & 0 & 0 \\ 0 & 0 & 0 & 0 & 0.3 & 0.7 & 0 \\ 0 & 0 & 0 & 0 & 0 & 1 & 0 \\ 0 & 0 & 0 & 0 & 0 & 0 & 1 \\ 1 & 0 & 0 & 0 & 0 & 0 & 0 \end{bmatrix}$$

The movement time of each aircraft from one service station to the next one is included in the service time of the next service station. Each vehicle at the same station has the same service level. All the malfunctions can be repaired, that is, no aircraft exit fight.

To show the evolution process of the total value of allocation, the initial configuration of vehicle is set to one for each service station during the analysis in the design stage. The numbers of fighters and attackers are both 48. The SGR constraints of fighters and attackers are more than 12 and 9 respectively. The results of optimization allocation have been obtained after 24 iterations and shown in Table 1.

As shown in Figs. 2 and 3, each SGR and vehicles at each station both increase with the number of iterations. The system achieves its goals of each SGR in the 24 iterations. In the evolution, the two fastest growing numbers in each station are munitions upload vehicles and refuelling vehicles, which demonstrates that the two types of vehicles have a significant impact on each SGR of MTMA. Meanwhile, the numbers of tractors, power vehicles and oxygen-filling vehicles increase more quickly than that of air condition vehicles.

Figure 4 presents the correlation between each SGR of two types of military aircraft and total value of various vehicles. This is realized by cumulating the total value of various vehicles results given in each iteration of the model, where the total value equals the numbers of various vehicles times their value coefficient. As the total value of various vehicles increases, each SGR of two types of military aircraft increases almost linearly, which shows this method can achieve good economic benefits. On the basis of fitting each SGR constraint of fighters and attackers more than 12 and 9, the optimization allocation of AGSV has been obtained, as presented in Table 1.

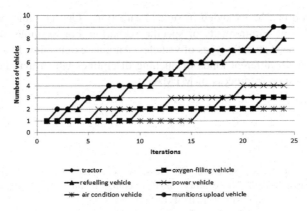

Fig. 3. Evolution of vehicles allocation

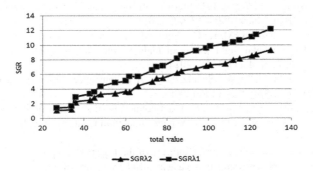

Fig. 4. SGR-cost plot

5 Conclusions

In this paper, a general queueing model of AGSV was set up in the support process of MTMA using the multi-class closed queueing networks to describe the features of a complex system. To satisfy constraints on the each SGR of MTMA and get good economic benefits, an optimization allocation model of AGSV was developed to minimize the total value of AGSV. Based on MVA and marginal analysis, a solving algorithm determines the numbers of AGSV at each service station. The results of a case study show applicability of the optimization model and present the evolution of each SGR of the system according to the increase of the total value due to vehicles allocation.

The research motive and objective of this paper are strictly related to the support requirements of military aircraft in not only Air Force but also the Army, where there are many types of vehicles and the SGR cannot be decreased. Logistics department or fight command at airport can utilize the presented optimization model to obtain the optimization allocation and good economic benefits at the moment of making the logistic support plans.

The research can be extended in several directions. One possible extension is to model the different requirements of specific scenarios. For example, some types of military aircraft need specific types of vehicles. Another extension is to analyse system characteristics with different kinds of queuing disciplines which may be more applicable to special support process of MTMA.

References

1. Henley, S., Currer, R., Scheuren, B., Hess, A., Goodman G.: Autonomic logistics-the support concept for the 21st century. In: Proceedings of the 2000 Aerospace Conference Proceedings, vol. 6, pp. 417–421. IEEE (2000)
2. Mackenzie, A., Miller, J.O., Hill, R.R., Chambal, S.P.: Application of agent based modelling to aircraft maintenance manning and sortie generation. Simul. Model. Pract. Theory 20(1), 89–98 (2012)
3. Liu, Q.H., Wang, N.J., Wang, J.H.: Study on the optimizing scheduling model of ship-based aircrafts. Key Eng. Mater. 450, 417–420 (2011)
4. Sabatini, R., Richardson, M.A., Cantiello, M., Toscano, M., Fiorini, P., Zammit, M.D., Gardi, A.: Experimental flight testing of night vision imaging systems in military fighter aircraft. J. Test. Eval. 42(1), 1–17 (2014)
5. Guarnieri, J., Johnson, A.W., Swartz, S.M.: A maintenance resources capacity estimator. J. Oper. Res. Soc. 57, 1188–1196 (2006)
6. Fass, P.D., Miller, J.O.: Impact of an autonomic logistics system on the sortie generation process. In: Proceedings of the 2003 Winter Simulation Conference, vol. 1, pp. 1021–1029. IEEE (2003)
7. Pettinggill, K.B.: An analysis of the efficacy of the logistics composite model in estimating maintenance manpower productive capacity. BiblioScholar (2012)
8. Fang, S.Q., Wei, K., Chen, W.P., Zhao, S.H., Xu, J.: Modelling and simulation of flight support process of airfield station based on ARENA. J. Syst. Simul. 20(2), 746–749 (2008)
9. Wang, S.H., Yong, Q.D., Xia, Z.Y., Li, Y.: Quantum–behaved dispersive PSO algorithm approach for vehicles scheduling optimization problem to solve disperse warplane refueling. Logist. Technol. 32(4), 265–268 (2013)
10. Ma, L., Wu, Q., Li, L., Lian, C.: Study on the support resources configuration of multi-aircraft formation. In: Prognostics and System Health Management, vol. 1, pp. 1–6. IEEE (2012)
11. Luo, X.M., Zhang, Z.Y., Xie, L.J., Yang, H.J.: Simulation of four station equipment requirements among multi-aircraft support. Ordnance Ind. Autom. 28(9), 45–48 (2009)
12. Dietz, D.C., Jenkins, R.C.: Analysis of aircraft sortie generation with the use of a fork-join queueing network model. Naval Res. Logist. 44(2), 153–164 (1997)
13. Xia, G.Q., Chen, H.Z., Wang, Y.H.: Analysis of aircraft sortie generation rate based on closed queueing network model. J. Syst. Eng. 26(5), 686–693 (2011)
14. Reiser, M., Lavenberg, S.S.: Mean-value analysis of closed multi-chain queueing networks. J. ACM 27(2), 313–322 (1980)
15. Costantino, F., Di Gravio, G., Tronci, M.: Multi-echelon, multi-indenture spare parts inventory control subject to system availability and budget constraints. Reliab. Eng. Syst. Saf. 119, 95–101 (2013)
16. Mao, D.J., Li, Q.M., Ruan, M.Z., Huang, A.L.: System availability evaluation and support project optimization for anti–aircraft system of ship formation. Syst. Eng. Theory Pract. 31(7), 1394–1402 (2011)

Multi-level Maintenance Economic Optimization Model of Electric Multiple Unit Component Based on Shock Damage Interaction

Hong Wang[1(✉)], Yong He[1], Lv Xiong[1], and Zuhua Jiang[2]

[1] School of Mechatronic Engineering, Lanzhou Jiao Tong University,
Lanzhou 730070, People's Republic of China
wh@mail.lzjtu.cn
[2] School of Mechanical Engineering, Shanghai Jiao Tong University,
Shanghai 200240, People's Republic of China

Abstract. In order to simulate the reliability evolution process of Electric Multiple Unit (EMU) components under external shock and improve maintenance economy. The multi-level preventive maintenance method is established and the influence of maintenance period and allocation of multi-level imperfect maintenance on the maintenance economy are discussed respectively. Numerical experiments show that the multi-phase preventive maintenance model can reduce the maintenance cost rate. The analysis of bi-level imperfect maintenance capacity indicates that two-level preventive maintenance can extend the mileage of four-level preventive maintenance and three-level preventive maintenance can reduce the maintenance cost rate. Finally, some recommendations for the allocation of maintenance efforts are provided according to the different railway route features.

Keywords: Shock damage · Multi-phase · Preventive maintenance (PM) · Bi-level imperfect PM capacity · Equivalent failure rate increase factor

1 Introduction

Reliability and economy is the two most important indicators of EMU maintenance plan. EMU which runs in special environment, its PM is same as others which run in the normal environment, can easily lead to inadequate reserves of the reliability. Therefore, the establishment of the model that can describe the evolution of such components' reliability is essential.

Anderson [1] establish the external shock model, then two-unit system model with failure rate interaction is defined [2], but the determine of the maintenance period is too conservative. Ruey [3] proposed a two-phase preventive maintenance model for lease equipment and set the length of two period, in order to avoid the lack of periodic PM and flexible period PM. Qu and Wu [4] uses the sequential preventive maintenance strategy to optimize the number of period and imperfect maintenance cost in each period. Although the above research methods can be optimized to obtain the length of

© Springer Nature Singapore Pte Ltd. 2017
K. Li et al. (Eds.): LSMS/ICSEE 2017, Part III, CCIS 763, pp. 729–739, 2017.
DOI: 10.1007/978-981-10-6364-0_72

each period, but each phase of the model only use single imperfect PM, it can't meet the multi-level imperfect PM model's requirement.

Motivated by the above discussion, this article further investigates multi-level imperfect PM economy. The contributions of this work are listed as follows: (i) An external shock damage interaction model based on multi-level imperfect PM to simulate the reliability evolution process of EMU components put into operation in special environment. (ii) Assuming that the different PM methods are linear with the maintenance cost. Obtain the influence of the different maintenance period and allocation of different PM capacity on the multi-level maintenance economy.

The remainder of the paper is organized as follows. Section 2 introduces maintenance period model and formulates the evolution process of reliability and failure rate. Section 3 introduces the process of solution. Numerical examples are illustrated in Sect. 4. Conclusion remarks are outlined in Sect. 5.

2 Model Derivation

Based on the bi-level imperfect PM model, two kinds of maintenance period of EMU component are proposed to explore the influence of maintenance period on the economy. The model 1 is used to periodic PM. The model 2 is used to multi-phase PM. The periodic preventive maintenance model requires that the maintenance intervals Δl_i within the entire operating range is consistent ($\Delta l_1 = \Delta l_2 = \cdots = \Delta l_i = \Delta l$).

In the multi-phase preventive maintenance model, the four-level PM mileage of the component is divided into multiple phases (p is the serial number of phase in the whole mileage, q is the serial number of maintenance interval in each phase). In the same phase, maintenance interval Δl_q^p is consistent. In different phases, the maintenance interval Δl_q^p and number of maintenance n^p are allowed to be different as Fig. 1.

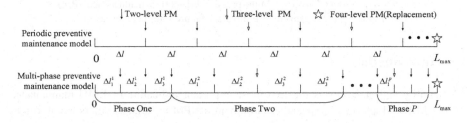

Fig. 1. Two maintenance model under different maintenance period

2.1 Model of External Shock

The model can be impacted of the external shock is established, the process of external shock is shown in Fig. 2. Component is subjected to external shocks during the kth PM. The number of component fails by external shock is defined as H_w^k.

The probability which component failed after external shocks is related to the frequency and the damage of the external shock [5]. The occurrence of external shocks

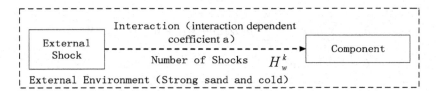

Fig. 2. The process of external shock

is subject to random distribution, and the Poisson distribution is used to describe the shock behavior. The distribution function is:

$$\Pr(N_k = d) = \left[(\lambda \Delta l_k)^d / d! \right] \exp(-\lambda \Delta l_k),$$
$$d = 0, 1, 2, \cdots, \infty \tag{1}$$

where: Δl_k is maintenance interval of $(k\text{-}1)^{\text{th}}$ and k^{th} preventive maintenance; N_k is number of shock in maintenance interval Δl_k. λ is the parameter of the Poisson distribution, indicating the average incidence of random shocks per unit of mileage.

Each shock causes damage to the component, the damage caused by external shocks obey the distribution function is [2]:

$$P\big(W_{kf} \leq x\big) = 1 - \exp(-\mu x) = G(x)$$
$$f = 0, 1, 2, \cdots, \infty \tag{2}$$

where: μ is the parameter of the exponential distribution, W_{kf} indicates damage caused to the component by the fth shock. According to the theory of shock, the damage caused by the shock of parts is cumulative, the probability distribution of the sum of the damage caused by the f times in the maintenance interval Δl_k can be expressed as:

$$\Pr\big(W_{k1} + W_{k2} + \cdots + W_{kf} \leq x\big) = G^f(x) \tag{3}$$

where: $G^f(x)$ is the f-fold Stieltjes convolution of $G(x)$.

Assuming that Z_k represents the cumulative damage value of the component from the new state to the kth PM, since the component has been repaired at the $(k\text{-}1)^{\text{th}}$ PM, the cumulative damage will be reduced $\Delta Z_{k-1} = (1 - \alpha)Z_{k-1}$, where α is recovery factor of the cumulative damage. Then the cumulative damage occurred in the mileage of l_k:

$$Z_k = \alpha Z_{k-1} + W_{k1} + W_{k2} + \cdots + W_{kf} \tag{4}$$

where: $Z_0 = 0$.

The probability of failure is related to the current cumulative damage value, the probability of failure is subjected to exponential distribution $p(x) = 1 - \exp(-\theta x)$, θ is an exponential distribution parameter. The probability of the failure to be caused by external shock is as follows:

$$h_w^k = \sum_{f=1}^{N_k} p(Z_k) = \sum_{f=1}^{N_k} p(\alpha Z_{k-1} + W_{k1} + W_{k2} + \cdots + W_{kf}), \tag{5}$$

$$k = 1, 2, \cdots, m$$

The number of failure within Δl_k:

$$H_w^k = E(h_w^k)$$
$$= \lambda \Delta l_k - \frac{G^*(\theta)}{1 - G^*(\theta)} \times \exp\left\{ -\sum_{f=1}^{k-1} \lambda \Delta l_f \left[1 - G^* \left(\theta \prod_{r=f}^{k-1} \alpha_r \right) \right] \right\}$$
$$\times \{1 - \exp(-\lambda[1 - G^*(\theta)])\Delta l_k\} \tag{6}$$

where: $G^*(\theta) = \int_0^\infty e^{-\theta x} dG(x) = \mu / (\mu + \theta)$.

2.2 Interaction Modeling

If the component is not affected by external shock, the failure rate can be expressed as [6]:

$$\begin{cases} h_k(l) = h_k^e + h_k^s(l), l \in \Delta l_k \\ h_k^s(l) = \frac{\phi_k^1 \beta}{\eta} \left(\frac{1}{\eta}\right)^{\beta-1} \end{cases} \tag{7}$$

where: h_k^e is the initial failure rate at the beginning of the kth PM stage, $h_k^s(l)$ is the failure rate of component which is using on this stage with no external shock, ϕ^1 is equivalent failure rate increase factor under imperfect PM. In the imperfect PM, the failure rate increase factor is first proposed by Nakagawa [7] to simulate the process of imperfect PM with the deterioration of the working condition of the equipment. Xia et al. [8] had further research, but did not give a solution for multi-level imperfect PM. In this section, the equivalent failure rate increase factor is defined as:

$$\phi_k^1 = \delta_k^{b + (2 - x_k)c} \tag{8}$$

where: δ is recession factor under imperfect maintenance, b is the cumulative number of the three-level PM on the kth stage, c is the cumulative number of two-level PM after the last time three-level PM, x_k means different imperfect PM method, $x_k = 1$ means to use the two-level PM at the time of the kth PM; $x_k = 2$ means to use the three-level PM at the time of the kth PM. The improvement process of failure rate is shown in Fig. 3 and it is explained as follows: when the component is only used with two-level PM, the equivalent failure rate increase factor increases with the cumulative number of two-level PM. After the components using three-level PM, the failure rate of the component will be restored to a lower level. The bi-level imperfect PM method is better than the traditional single imperfect PM to reduce failure rate.

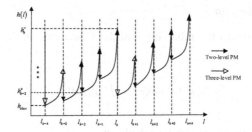

Fig. 3. The change of failure rate under different PM methods

The failure rate $h_k^s(l)$ is affected by the external shocks. $h_k^s(l)_j$ is converted to $h_k^s -^c (l)$, $h_k^s -^c (l)$ is the conditional failure rate $h_k^s -^c (l| H_w^k = j) = (1+a)^j h_k^s(l)$, where: a is the interaction dependent coefficient, indicates the degree which is affected by external shocks. Therefore, $h_k^s -^c (l)$ is increasing with the increase of its own mileage l and the number of external shocks j. References to research results by Lai and Chen [2] and Gao [9] on the failure rate interaction and the external shock, $h_k^s -^c (l)$ is defined as follows:

$$h_k^s -^c (l) = \sum_{j=0}^{+\infty} h_k^s(l)^j P[H_w^k] = \sum_{j=0}^{+\infty} h_k^s(l) \frac{\left[(1+a)H_w^k\right]^j e^{-H(l_k)}}{j!} = h_k^s(l)e^{aH_w^k} \quad (9)$$

Due to the presence of external shocks, $h_k^s -^c (l)$ increases by $e^{aH_w^k}$ times than $h_k^s(l)$. The failure rate of the component $h_k(l)$ under external shock is:

$$h_k(l) = h_k^s -^c (l) \quad (10)$$

In addition, the evolution rule of initial failure rate h_k^c after PM is:

$$\begin{cases} 0, k = 1 \\ h_k^- - \phi_c^2(h_k^- - h_{k-1}^+), x_k = 1, k = 2, 3, \cdots m \\ h_k^- - \phi_g^2(h_k^- - h_{slast}), x_k = 2, k = 2, 3, \cdots m \end{cases} \quad (11)$$

where: h_k^- is the failure rate of the component before k^{th} PM; h_k^+ is the failure rate of the component after k^{th} PM; h_{k-1}^+ is the failure rate of the component after $(k-1)^{\text{th}}$ PM; h_{slast} is the failure rate after the last time three-level PM; ϕ_c^2 is the reliability improvement factor of two-level PM; ϕ_g^2 is the reliability improvement factor of three-level PM; m is the total number of PM. In the case of $(k-1)^{\text{th}}$ PM, if the implementation of two-level PM, the component failure rate is higher than the last two-level PM h_{k-2}^+. In the case of kth PM, if the implementation of three-level PM, the component failure rate is higher than the last three-level PM h_{slast}, as shown in Fig. 3.

2.3   Reliability Modeling

The relationship between the reliability and the failure rate in the operating mileage $[A, B]$ is:

$$R(l) = \exp\left[-\int_A^B h(l)\mathrm{d}l\right] \tag{12}$$

The component reliability can be calculated:

$$R_k(l) = R_k^c \exp\left[-\phi^1 e^{aH_k^w}\left(\frac{1}{\eta}\right)^\beta\right] \tag{13}$$

where: R_k^c is the initial reliability of the component after preventive maintenance, and the evolution rules as follow:

$$R_k^c = \begin{cases} 1, k = 1 \\ R_k^- + \phi_c^2(R_{k-1}^+ - R_k^-), x_k = 1, k = 2, 3, \cdots m \\ R_k^- + \phi_g^2(R_{slast} - R_k^-), x_k = 2, k = 2, 3, \cdots m \end{cases} \tag{14}$$

where: R_k^- is the reliability of the component before k^{th} PM; R_k^+ is the reliability of the component after k^{th} PM; R_{k-1}^+ is the reliability of the component after $(k\text{-}1)^{\text{th}}$ PM; R_{slast} is the reliability after the last three-level PM.

2.4   Objective Function

EMU component life cycle costs are divided into five parts, respectively: component acquisition costs C_{pr}, performance inspection costs C_{jc}, preventive maintenance costs C_1, wasting at the time of replacement costs C_2, and minor repair costs C_3. The maintenance cost rate C_z is:

$$c_z = \frac{C_{pr} + C_{jc} + C_1 + C_2 + C_3}{L_{\max}} \tag{15}$$

The bi-level imperfect PM cost for the component C_1 is:

$$C_1 = \sum_{k=1}^m \left\{[2 - x_k]C_{pc} + [x_k - 1]C_{pg}\right\} \tag{16}$$

where: C_{pc} is a single two-level PM costs, C_{pg} is a single three-level PM costs. Due to the high reliability requirements of the EMU, the component often need to limit their reliability. C_2 is the wasting costs when replacing, to avoid implementing PM near the replacement node. The closer the reliability of the component to the minimum reliability threshold, the smaller C_2, the specific definition is:

$$C_2 = \frac{R_m^- - R_{\min}}{1 - R_{\min}} C_{pr} \tag{17}$$

where: R_m^- is the reliability at the time of the component replacement and $R_{\min}$ is the minimum reliability threshold for the component. In addition, the component occurs unexpected failure in the operating mileage, the minor repair costs C_3 is:

$$C_3 = C_x \sum_{k=1}^{m} [\int_{l_{k-1}}^{l_k} \lambda(l)\mathrm{d}l] \tag{18}$$

where: C_x is a single minor repair cost of a unexpected failure.

In order to explore the economic maintenance, the objective function of the two maintenance models are as follows:

$$\min\{c_z\} \tag{19}$$

$$s.t. \forall l, \exists R(l) \geq R_{\min}, l \in [0, L_{\max}] \tag{20}$$

$$\Delta l \in [l_{\min}, l_{\max}] \tag{21}$$

$$\Delta l_q^p \in [l_{\min}, l_{\max}] \tag{22}$$

Equation (20) indicates that the component must meet the reliability threshold requirement within the four-level maintenance range; Eq. (21) indicates maintenance interval of the component is restricted to avoid excessive maintenance and poor maintenance; Eq. (22) indicates that the maintenance intervals within each phase are limited.

3   Solution for Maintenance Model

Because the maintenance interval, number of maintenance and maintenance method of each phase need to be optimized at the same time, the solution by iterative method is not efficient, so the model is solved by genetic algorithm (GA). Although the calculation scale of this paper is not large, in order to avoid the premature problem, the adaptive method is used to control the crossover and mutation parameters [10]. The steps of GA are described below:

(1) Encoding and initial population X_0

The real-valued encoding method is applied. Each chromosome is represented using a numerical string.

(2) Crossover

The operation of crossover is separated into two parts. Firstly, generate cross point of phase, crossover point y_{c_p} belongs to $[1, p]$. Secondly, generate cross point of maintenance method, crossover point y_{c_m} belongs to $[1, n^p]$. The crossover probability is P_c.

(3) Mutation

This section uses a single point of mutation, the mutation point belongs to $[1, p]$, the mutation point is corresponding to the phase will be re-generated. The mutation probability is P_m.

(4) Parameter adaptive adjustment

When the number of cycles is not reached, all the populations of the two models are solved according to Eq. (19), the fitness is determined in the range of $[0, 1]$. Then the fitness value of X_b is as follows:

$$f_b = \frac{X_{max} - X_b}{X_{max} - X_{min}} \tag{23}$$

where: X_{max} and X_{min} represent the maximum and minimum objective function values in the contemporary population respectively. P_c and P_m are adjusted during the operation of the algorithm, so that it decreases with the increase of individual fitness value, which increases with the decrease of individual fitness value. The formula is [10]:

$$P_c = \begin{cases} e1 \dfrac{X_{max} - X'}{X_{max} - X_{avg}}, X' \geq X_{avg} \\ e2, X' < X_{avg} \end{cases} \qquad P_m = \begin{cases} e3 \dfrac{X_{max} - X''}{X_{max} - X_{avg}}, X' \geq X_{avg} \\ e4, X' < X_{avg} \end{cases} \tag{24}$$

where: X_{avg} represents the average objective function values in the contemporary population, X' represents crossover individual fitness, X'' represents mutation individual fitness.

(5) Selection

According to the individual fitness value obtained by Eq. (23), the probability that individuals remain to the next generation is proportional to their fitness, like roulette wheel selection. In order to avoid the "precocious" phenomenon, increase the overall optimization ability, the population set up gap, 1/4 lower fitness individuals of each generation will be eliminated, the remaining location by the algorithm to re-generate.

4 Numerical Examples

Taking the mechanical system of the EMU as an example, according to the maintenance parameters of EMU in the literature [11] and the external shock model in the literature [12], interaction dependent coefficient $a = 0.05$, recovery factor of the cumulative damage $\alpha = 0.5$, parameter of the Poisson distribution $\lambda = 0.1$, $\mu = 5$, $\theta = 0.03$, mileage range of four-level PM is $L_{max} = 2.4 \times 10^6$ km, performance inspection cost $C_{jc} = 500$ yuan, $R_{min} = 0.75$, other parameters shown in Table 1. The genetic algorithm of the model is implemented, population size = 150, $e1 = 0.6$, $e2 = 0.8$, $e3 = 0.05$, $e4 = 0.08$, the number of iteration is 100. In this section, model 1 is optimized firstly and the optimal PM interval is 30×10^4 km (plan 1), the specific results are shown in Table 2.

Table 1. Factors of maintenance

η	β	C_{pc} / Yuan	C_x / Yuan	C_{pg} / Yuan	C_{pr} / Yuan	ϕ^1	ϕ_c^2	ϕ_g^2
100	2.2	525	780	1275	1500	1.03	0.35	0.85

In order to verify the impact of the shock damage interaction to reliability, the sensitivity analysis of the external shock parameter λ and the interaction dependent coefficient a are carried out with the plan 1 as the basic maintenance model. The reliability changes under different a and λ are shown in Figs. 5 and 6, respectively. It can be seen that the reliability of the component is accelerated with the increase in the external shock frequency and the degree of interaction, and is more pronounced in the second half of the service mileage. When $\lambda = 0.3$, the minimum reliability of components has been unable to meet the threshold requirements. The two parameters increase by the same order of magnitude, the shock parameter λ has a stronger impact on reliability than the interaction dependent coefficient a.

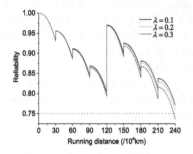

Fig. 5. Reliability change of component under different value of λ

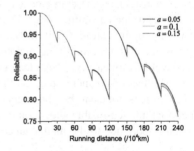

Fig. 6. Reliability change of component under different value of a

Table 2. Optimization results of two models

Factors	Plan 1	Plan 2	Plan 3
PM interval	30	38-34	36-32-34
Number of PM	7	3-3	2-2-2
PM method	1-1-1-2-1-1-1	1-1-2-1-1-1	1-1-1-2-1-1
c_z (yuan/10^4 km)	36.553	34.516	34.366

4.1 Comparison of Maintenance Period

The maintenance cost rate of two-phase and three-phase PM models are 5.6% and 6% lower than plan 1 respectively. The optimization results show that the economy of multi-phase PM model is better than the periodic preventive maintenance model. Then, the maintenance cost rate is optimized by periodic and three-phase PM model under different $L_{\max}$, respectively. The results are shown in Table 3. Compared with the

Table 3. Optimization results of two models under different L_{max}

Factor	Model one	Plan 3
Four-level maintenance mileage	190	220
PM interval	38	40-34-38
Number of PM	4	2-2-1
PM method	1-1-2-1	1-1-2-1-1
C_z (Yuan/10^4 km)	35.039	33.764

periodic PM model, the maintenance cost rate of the three-phase PM model is dropped by 3.6% and replacement range of the component is a little longer.

4.2   Analysis of the Impact of Maintenance Effort

As the PM capacity is often associated with the cost of PM, assuming that the maintenance capacity of the component is linearly related to the cost of maintenance; and the impact of the capability of two-level and three-level PM is carried out respectively. From the optimization result in Tables 5 and 6, with the cost of two-level PM increasing, although improving maintenance capacity of two-level PM can extend the four-level PM mileage, but also lead to maintenance cost rate increase. Then it can be seen the effect of the increase of the three-level PM cost on the length of the four-level PM mileage is not clear, but the effect of reducing the maintenance cost rate is obvious. Therefore, the reasonable allocation of the maintenance resources of the three-level PM is vital importance to the maintenance economy.

Table 5. Optimization results of different two-level PM capacity

ϕ_c^2	C_{pc}/Yuan	ϕ_g^2	C_{pg}/Yuan	Mileage of four-level PM/10^4 km	C_z (Yuan/10^4 km)
0.35	525	0.85	1275	190	35.093
0.75	1125			280	36.901
0.85	1275			320	38.423

Table 6. Optimization results of different three-level PM capacity

ϕ_c^2	C_{pc}/Yuan	ϕ_g^2	C_{pg}/Yuan	Mileage of four-level PM/10^4 km	C_z (Yuan/10^4 km)
0.35	525	0.75	1125	220	35.726
		0.85	1275	190	35.093
		0.95	1425	250	34.053

5 Conclusions

This study combines the special modeling requirements of the EMU components run in special environment, establishes multi-level imperfect PM. Numerical experiments verifies the necessary of considering the influence of external shock on reliability and multi-phase PM model can significantly reduce the maintenance cost rate. Based on assumption between maintenance capacity and bi-level imperfect PM maintenance cost. The length of route is short, it is recommended to improve investment of three-level PM to reduce maintenance cost rate. The length of route is long, the situation which the EMU doesn't repair on time is easily happen. Therefore, it is recommended to improve investment of two-level PM to extend the service mileage of components, to improve the efficiency of EMU operations.

Acknowledgement. This research is support in part by a grant (NO. 71361019) from the National Natural Science Foundation, P.R. China

References

1. Anderson, K.K.: A note on cumulative shock models. J. Appl. Probab. **25**, 220–223 (1988)
2. Lai, M.T., Chen, Y.C.: Optimal replacement period of a two-unit system with failure rate interaction and external shocks. IEEE Trans. Reliab. **39**(1), 71–79 (2008)
3. Yeh, R.H., Wen, L., Lo, H.C.: Optimal threshold values of age and two-phase maintenance policy for leased equipment using age reduction method. Ann. Oper. Res. **181**(1), 171–183 (2010)
4. Qu, Y., Wu, S.: Phasic sequential preventive maintenance policy based on imperfect maintenance for deteriorating systems. J. Mech. Eng. **47**(10), 164–170 (2011). [in Chinese]
5. Allan, G., Jurg, H.: Shock Models. In: Nikulin, M., Limnios, N., Balakrishnan, N., Kahle, W., Huber-Carol, C. (eds.) Advances in Degradation Modeling: Reliability, Survival Analysis, and Finance, pp. 59–76. Springer, Heidelberg (2010). doi:10.1007/978-0-8176-4924-1_5
6. Tsai, Y.-T., Wang, K.-S., Tsai, L.-C.: A study of availability-centered preventive maintenance for multi-component systems. Reliab. Eng. Syst. Saf. **84**, 261–270 (2004)
7. Nakagawa, T.: Sequential imperfect preventive maintenance policies. IEEE Trans. Reliab. **37**(3), 295–298 (1988)
8. Xia, T., Xi, L., Zhou, X., Du, S.: Modeling and optimizing maintenance schedule for energy systems subject to degradation. Comput. Ind. Eng. **63**(3), 607–614 (2012)
9. Gao, W.: Reliability Modeling and Dynamic Replacement Policy for Two-Unit Parallel System with Failure Interaction. Comput. Integr. Manufact. Syst. **21**(2), 511–515 (2015). [in Chinese]
10. Grefenstette, J.J.: Optimization of control parameters for genetic algorithms. IEEE Trans. Syst. Man Cybern. **16**(1), 122–128 (1986)
11. Wang, L.: Research on reliability-centered maintenance decision and support system for high-speed train equipments. Beijing, Beijing Jiaotong University (2011). [in Chinese]
12. Cha, J.H., Lee, E.Y.: An extended stochastic failure model for a system subject to random shocks. Oper. Res. Lett. **38**(5), 468–473 (2010)

A Composite Controller for Piezoelectric Actuators with Model Predictive Control and Hysteresis Compensation

Ang Wang and Long Cheng$^{(\boxtimes)}$

State Key Laboratory of Management and Control for Complex Systems,
Institute of Automation, Chinese Academy of Sciences, Beijing 100190, China
long.cheng@ia.ac.cn

Abstract. Piezoelectric actuators (PEAs) are ubiquitous in nanopositioning applications due to their high precision, rapid response and large mechanical force. However, precise control of PEAs is a challenging task because of the existence of hysteresis, an inherent strong nonlinear property. To minimize its influence, various control methods have been proposed in the literature, which can be roughly classified into three categories: feedforward control, feedback control and feedforward-feedback control. Feedforward-feedback control combines the advantages of feedforward control and feedback control and turns into a better control scheme. Inspired by this strategy, a composite controller is proposed for the tracking control of PEAs in this paper. Specifically, the model of PEAs is constructed by a multilayer feedforward neural network (MFNN). This model is then instantaneously linearized, which leads to an explicit model predictive control law. Then, an inverse Duhem hysteresis model is adopted as a feedforward compensator to mitigate the hysteresis nonlinearity. Experiments are designed to validate the effectiveness of the proposed method on a piezoelectric nanopositioning stage (P-753.1CD, Physik Instrumente). Comparative experiments are also conducted between the proposed method and some existing control methods. Experimental results demonstrate that the root mean square tracking error of the proposed method is reduced to 16% of that under the previously proposed model predictive controller [16].

Keywords: Feedforward-feedback control · Hysteresis compensation · Model predictive control (MPC) · Instantaneous linearization · Piezoelectric actuators (PEAs)

1 Introduction

Recent decades have witnessed the explosive development of nanotechnology which has been recognized as the fundamental requirement of modern manufacturing and process industry. To fulfill this requirement, piezoelectric actuators (PEAs) have been widely applied in nanopositioning applications such as

© Springer Nature Singapore Pte Ltd. 2017
K. Li et al. (Eds.): LSMS/ICSEE 2017, Part III, CCIS 763, pp. 740–750, 2017.
DOI: 10.1007/978-981-10-6364-0_73

micromanipulators [1], nanopositioning stages [2] and atomic force microscopes [3] because of their predominant capabilities of rapid response, high resolution, large mechanical force and wide operating bandwidth. However, the intrinsic nonlinear hysteresis of PEAs has the potential of degrading the control accuracy and system stability. Hysteresis is a memory effect that the current output of PEAs depends on the historical operations and the current control input. Moreover, it is also rate-dependent, which means the dynamic behavior of PEAs changes with the frequency of the control input. Therefore, how to deal with these difficulties has been drawing considerable attention.

Various control methods have been proposed for accurate tracking control of PEAs. These approaches are generally divided into feedforward control, feedback control and feedforward-feedback control. Feedforward control methods are naturally exploited to compensate the hysteresis nonlinearity with the inverse model of hysteresis. Before constructing the inversion, the feedforward hysteresis model needs to be obtained, which is usually described by Duhem model [4], Preisach model [5], Prandtl-Ishlinskii model [6] and Bouc-Wen model [7]. Then, a feedforward controller is constructed by cascading the inversion of these models for canceling the hysteresis effect [8]. However, the dependency on the accuracy of the inverse hysteresis model and the vulnerability to external disturbances may degrade the control performance of feedforward control methods. Feedback control methods treat the hysteresis nonlinearity as bounded nonlinear disturbances, and a feedback controller is designed based on control strategies like proportional-integral-derivative (PID) control, adaptive control [9] and sliding-mode control [10] to suppress the disturbances. However, the main drawbacks of feedback control are the difficulty of obtaining robust stable results and the performance maintenance when operating at high-frequency situation [2]. Furthermore, feedforward-feedback control methods have been developed with the idea of combining feedforward and feedback control schemes, where feedforward control mitigates the hysteresis nonlinearity and feedback control compensates the inaccuracy of the hysteresis model and unknown disturbances. It is noted that adding feedforward terms to feedback control systems can improve the tracking performance even in the presence of modelling error [11]. Inspired by this strategy, extensive studies have been conducted by combining different inverse hysteresis models and different feedback control approaches [12,13]. It is noted that the tracking error of these control methods is more than 20 nm when tracking low-frequency references. However, extreme high precision is required in some nanopositioning and nanomanipulation applications. For instance, the required accuracy of atom manipulation is less than 10 nm [14]. Therefore, more advanced control methods should be exploited for the precise control of PEAs in order to satisfy the extreme demand.

Recently, model predictive control (MPC) is developed for the tracking problem of PEAs, which has demonstrated a great performance in industrial applications due to its robustness and disturbance rejection properties. In [15], a nonlinear model predictive control (NMPC) method is proposed for tracking control of PEAs based on an MFNN model, where the control law is obtained by solving a complicated nonlinear optimization problem. The dynamic linearization is carried out to the MFNN model of PEAs in [16], then an explicit predictive control law can be obtained, which leads to a faster computational rate. In order to solve the off-line training accuracy problem, a predictive controller based on an adaptive Takagi-Sugeno (T-S) fuzzy model is designed for PEAs, where the model parameters can be on-line adjusted to achieve a better control performance [17]. All these control approaches belong to the feedback control method. It is known that the tracking performance can be improved by adding feedforward terms to feedback control systems. It is natural to ask whether the control performance can be improved if the feedforward compensator is added to the model predictive control.

In this paper, a composite controller with model predictive control and hysteresis compensation is proposed to achieve high-precision tracking control of PEAs. First, the complicated nonlinear mapping between the driving voltage and the output displacement of PEAs is modeled by the MFNN because of its strong approximation capability. In order to accelerate the on-line calculation, the MFNN model is linearized in each sample interval, and a predictive controller is designed based on the instantaneously linearized MFNN model, which yields an explicit control law and enables PEAs to track high-frequency references. Then, the inverse Duhem hysteresis model is adopted as a feedforward compensator to mitigate the hysteresis nonlinearity. To validate the effectiveness of the proposed method, extensive experiments are performed on a piezoelectric nanopositioning stage (P-753.1CD, Physik Instrumente). Comparisons are conducted between the proposed method and some existing control methods. Experimental results and comparisons demonstrate that the control accuracy of the proposed control method is superior to the majority of control approaches mentioned in the literature.

2   Composite Controller for PEAs with MPC and Hysteresis Compensation

First, the model of PEAs is approximated by an MFNN, which is further linearized instantaneously to mitigate the computational burden. A model predictive controller is designed based on this instantaneously linearized MFNN model. Then, the hysteresis model is constructed by the Duhem model, which is used to derive the inverse hysteresis model as a feedforward compensator. Synthesizing the predictive controller and the hysteresis compensator yields the proposed composite controller.

2.1   Instantaneously Linearized MFNN Model of PEAs

According to [16], a three-layer MFNN is utilized to approximate the complex nonlinear mapping between the driving voltage and the output displacement of PEAs. The input vector of the MFNN is defined as $X(k) = [y(k-1), \cdots, y(k-n), u(k), \cdots, u(k-m)]$, where $y(k)$ is the output displacement of PEAs, $u(k)$ is the driving voltage, and nonnegative integers n and m are the maximum time delays for $y(k)$ and $u(k)$, respectively. The nonlinear relation between the input and output of the MFNN is described as

$$y(k) = \sum_{h=1}^{q} w_h^2 \mathscr{F}(\sum_{i=1}^{p} w_{ji}^1 x_i(k) + w_{j0}^1) + w_0^2, \qquad (1)$$

where $p = n + m + 1$ is the length of the input vector, q is the number of the hidden-layer neurons, and the output layer only has one neuron. w_h^2 is the weight between the output neuron and the hth hidden-layer neuron, and w_{ji}^1 denotes the weight between the jth hidden-layer neuron and the ith input-layer neuron, which can be obtained by the off-line training. $x_i(k)$ is the ith element of input vector $X(k)$. The input-layer and output-layer neurons possess linear unit activation function, while the tangent sigmoid function $\mathscr{F}(x) = (e^x - e^{-x})/(e^x + e^{-x})$ is chosen for the hidden-layer neurons.

Then, the MFNN model is instantaneously linearized [16]. The instantaneously linearized model is able to express the behaviors of PEAs around the current operation point with a reasonable error. Taylor expansion is carried out for instantaneously linearizing the MFNN model,

$$\begin{aligned} y(k) - y(l) &= a_1(l)(y(k-1) - y(l-1)) + \cdots + a_n(l)(y(k-n) - y(l-n)) \\ &\quad + b_0(l)(u(k) - u(l)) + \cdots + b_m(l)(u(k-m) - u(l-m)), \end{aligned} \qquad (2)$$

where l is the current operating time, $a_i(l)$ $(i = 1, \cdots, n)$ and $b_i(l)$ $(i = n+1, j = 0, \cdots, m)$ are the partial derivative terms. Rewrite (2) as

$$\begin{aligned} y(k) &= a_1(l)y(k-1) + \cdots + a_n(l)y(k-n) \\ &\quad + b_0(l)u(k) + \cdots + b_m(l)u(k-m) + \varepsilon(l), \end{aligned} \qquad (3)$$

where the bias term

$$\begin{aligned} \varepsilon(l) &= y(l) - a_1(l)y(l-1) - \cdots - a_n(l)y(l-n) \\ &\quad - b_0(l)u(l) - \cdots - b_m(l)u(l-m) \end{aligned} \qquad (4)$$

can be regarded as the comprehensive effect of the hysteresis nonlinearity and external disturbances.

2.2   Inversion of the Duhem Hysteresis Model

The Duhem model is used to describe the hysteresis nonlinearity of PEAs, which can be expressed as [18]

$$\dot{f} = |\dot{u}| (\alpha u + \gamma f) + \beta \dot{u}, \qquad (5)$$

where u and f are the input and output of the Duhen model, respectively. α, β and γ are model parameters which need to be identified in practice. For experimental implementation, the discrete-time form of (5) is required.

If $\dot{u} > 0$ (the input voltage is monotonically increasing), the discrete-time hysteresis model can be derived with the trapezoid estimation [18], which is given by

$$f(k+1) = \alpha\frac{\lambda(k+1)}{2-\gamma\phi(k+1)} + \frac{2+\gamma\phi(k+1)}{2-\gamma\phi(k+1)}f(k) + \beta\frac{2\phi(k+1)}{2-\gamma\phi(k+1)}, \qquad (6)$$

where $\lambda(k+1) = u^2(k+1) - u^2(k)$, $\phi(k+1) = u(k+1) - u(k)$. Rewrite (6) as a quadratic function of $u(k+1)$, and the discrete-time inverse Duhem model can be obtained by solving the equation

$$\alpha u^2(k+1) + \delta_1 u(k+1) - \tau_1 = 0, \qquad (7)$$

where $\delta_1 = \gamma f(k+1) + \gamma f(k) + 2\beta$, $\tau_1 = \alpha u^2(k) + \delta_1 u(k) + 2\left[f(k+1) - f(k)\right]$. The solution of (7) is (see [19] for details)

$$u(k+1) = \frac{-\delta_1 + \sqrt{\delta_1^2 + 4\alpha\tau_1}}{2\alpha}. \qquad (8)$$

If $\dot{u} < 0$ (the input voltage is monotonically decreasing), similarly, the discrete-time inverse Duhem model is derived as

$$u(k+1) = \frac{\delta_2 - \sqrt{\delta_2^2 - 4\alpha\tau_2}}{2\alpha}, \qquad (9)$$

where $\delta_2 = -\gamma f(k+1) - \gamma f(k) + 2\beta$, $\tau_2 = -\alpha u^2(k) + \delta_2 u(k) + 2\left[f(k+1) - f(k)\right]$. Therefore, the inversion of the Duhem hysteresis model is described by (8) and (9).

2.3 MPC with Hysteresis Compensation

The principle of the proposed composite controller is shown in Fig. 1, which consists of a dynamic linearized MFNN model predictive controller and a hysteresis compensator. The feedforward compensator is directly constructed by the inversion of the Duhem hysteresis model, which has been derived in Sect. 2.2. Next, the predictive controller based on the instantaneously linearized MFNN model is designed to suppress the model uncertainties and other unknown disturbances.

In the predictive controller of PEAs, the control signal is generated by minimizing the differences between reference signal and predicted displacement. The instantaneously linearized MFNN model is adopted to predict the output displacement of PEAs, which is described by (3). The bias term $\varepsilon(l)$ only depends on the current operating point and can be modelled as integrated white noise. Therefore, it can be eliminated by transforming this equation to an adjacent difference form [20], and the displacement of the ith step is predicted as

$$\begin{aligned} y_p(k+i) =&(1+a_1(l))y_p(k+i-1) + (a_2(l) - a_1(l))y(k+i-2) + \cdots \\ &- a_n(l)y_p(k+i-n-1) + b_0(l)\Delta u(k+i) + \cdots \\ &+ b_m(l)\Delta u(k+i-m). \end{aligned} \qquad (10)$$

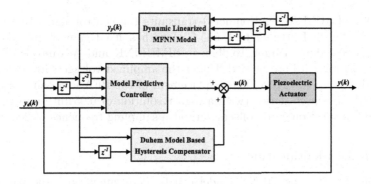

Fig. 1. Schematic diagram of the composite controller with MPC and hysteresis compensation.

Define P as the prediction horizon and the control horizon (for accuracy, P shouldn't be too large), the predicted displacements up to the Pth step are expressed as

$$Y_p(k) = G\Delta U(k) + H\Delta W(k) + SZ(k),\tag{11}$$

where $\Delta U(k) = [\Delta u(k+1), \Delta u(k+2), \cdots, \Delta u(k+P)]^T$, $\Delta W(k) = [\Delta u(k), \Delta u(k-1), \cdots, \Delta u(k-m+1)]^T$, $Z(k) = [y(k), y(k-1), \cdots, y(k-n)]^T$, and the definitions of G, H and S can be found in [16], which are matrices consisting of the coefficients of (3).

When the predicted displacement is available, a performance index is required to derive the control law of the predictive controller. Taking the error minimization and input voltage changing rate into account, the performance index is selected as [16]

$$V = [Y_d(k) - Y_p(k)]^T[Y_d(k) - Y_p(k)] + \rho \Delta U^T(k)\Delta U(k),\tag{12}$$

where $Y_d(k) = [y_d(k), \cdots, y_d(k+P)]^T$ is the reference signal, and penalty parameter $\rho > 0$ is used to limit $\Delta U(k)$. The predictive control law is obtained by solving the convex quadratic programming problem $\partial V/\partial \Delta U(k) = 0$, and the solution is

$$\Delta U(k) = (G^T G + \rho I)^{-1} G^T (Y_p(k) - H\Delta W(k) - SZ(k)).\tag{13}$$

Then, the control signal is

$$u(k+1) = u(k) + \Delta u(k+1),\tag{14}$$

where $\Delta u(k+1)$ is the increment of control signal for the next sampling interval.

3   Experiments and Comparisons

To validate the effectiveness of the proposed composite controller, extensive experiments have been conducted on a commercial piezoelectric nanopositioning stage (Physik Instrumente P-753.1CD). In this setup, a host computer (with

MATLAB/SIMULINK environment) transmits the control signal to a voltage amplifier (Physik Instrumente E-665.CR) with a fixed gain of 10, where the Real-Time Windows Target Toolbox in SIMULINK and Advantech PCI-1716 data acquisition card are needed. Under the amplified voltage, the travel range of the PEA is up to $12\,\mu$m. The displacement of the PEA is obtained from the integrated capacitive sensor (with a high resolution of 0.05 nm), and the PCI-1716 data acquisition card collects data at a sampling frequence of 200 kHz.

3.1 Model Identification

According to [16], the MFNN model can be determined with the parameters $n = 2$, $m = 1$, and $q = 5$. Before constructing the inversion of the Duhem hysteresis model as a feedforward compensator, parameters of Duhem model are required. Since the hysteresis nonlinearity is the dominant factor of the performance of PEAs under low-frequency activated voltage, a 80 V sinusoidal voltage of 1 Hz is used to activate the PEA. The forgetting factor recursive least squares algorithm is utilized to identify the parameters α, β, and γ with the input-output data, and the results are $\alpha = 0.7016$, $\beta = 1.0346$, and $\gamma = -0.4821$. Then, the hysteresis compensator is obtained by (8) and (9).

3.2 Verification of the Predictive Controller with Hysteresis Compensation

After the model parameters are determined, a set of experiments are conducted to test the proposed controller with the penalty parameter $\rho = 30$ and prediction horizon $P = 7$. The sinusoidal references $(4\sin(2\pi ft - \pi/2) + 5\,(\mu\mathrm{m}))$ with different frequencies are adopted as fixed-frequency references, and the tracking performances are provided in Figs. 2(a) and (b). For the 10 Hz reference signal, the steady-state tracking error is within the range of $[-0.0075, 0.0069]$ μm, and slightly increases to $[-0.0200, 0.0217]$ μm under 50 Hz reference signal.

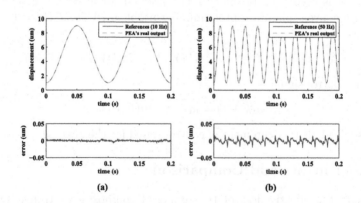

(a)   (b)

Fig. 2. Tracking performance of the PEA under different references: (a) 10 Hz; (b) 50 Hz.

Table 1. Comparison between the proposed method and the inversion-based MPC in [19]: the RMSE.

Reference frequency	The inversion-based MPC (RMSE, μm)	The proposed method (RMSE, μm)
$f = 1$ Hz	0.0083 μm	0.0014 μm
$f = 10$ Hz	0.0201 μm	0.0018 μm
$f = 50$ Hz	0.1669 μm	0.0060 μm

When tracking high-frequency trajectories, the proposed controller also has an acceptable performance. For instance, even for the reference of 200 Hz, the steady-state tracking error is only between -0.0549 and 0.1113 μm, which is still suitable for some nanopositioning applications like the DNA manipulation [14].

3.3 Comparisons with Other Methods

To further illustrate the effectiveness and superiority of the proposed method, comparisons are also conducted between the proposed control method and several other control methods mentioned in the literature.

Comparison with Inversion-Based MPC. In [19], an inversion-based MPC method with an integral-of-error state variable was developed, where an inverse Duhem hysteresis model was applied as a feedforward term to compensate the hysteresis nonlinearity. However, the MPC approach is only used for a global linear model of PEAs. This control method is considered for the performance comparison because the model predictive control and the inversion of the hysteresis model are utilized in both this method and the control method proposed in this paper. Comparison of the tracking performance is listed in Table 1, where the reference signal in [19] is $4\sin(2\pi ft - \pi/2) + 5$ (μm). It can be found that the RMSE of the proposed method is reduced to 4% of that of the inversion-based MPC in [19] under 50 Hz reference.

Comparison with Inversion-Free MPC. The inversion-free model predictive control proposed in [16] is a kind of feedback control schemes without using the feedforward compensator. It is noted that the use of feedforward terms can improve the tracking performance compared to the one using the feedback control alone [11]. It is likely that the proposed method has a better tracking performance than the inversion-free model predictive control proposed in [16]. This point has been validated by experiments under a sinusoidal reference $4\sin(2\pi ft - \pi/2) + 5$ (μm). The tracking performances of both methods are summarized in Table 2, which demonstrates that the RMSE of the proposed method is less than one sixth of that of the inversion-free MPC proposed in [16].

Comparison with Adaptive Fuzzy Internal Model Control [21]. The last comparison is conducted with the adaptive fuzzy internal model control [21],

Table 2. Comparison between the proposed method and the inversion-free MPC in [16]: the RMSE and the MAXE.

References frequency	The method in [16] (RMSE/MAXE, μm)	The proposed method (RMSE/MAXE, μm)
$f = 1\,\text{Hz}$	0.0022/0.0094	0.0014/0.0058
$f = 5\,\text{Hz}$	0.0042/0.0125	0.0015/0.0064
$f = 10\,\text{Hz}$	0.0080/0.0184	0.0018/0.0075
$f = 50\,\text{Hz}$	0.0395/0.0618	0.0060/0.0217
$f = 100\,\text{Hz}$	0.0794/0.1189	0.0122/0.0460
$f = 150\,\text{Hz}$	0.1182/0.1771	0.0194/0.0741

which belongs to the feedforward-feedback control scheme. A fixed-frequency signal $y_{d3}(t) = 0.8sin(100\pi t) + 1$ and a mixed-frequency signal $y_{d4}(t) = 0.5sin(100\pi t) + 0.35sin(50\pi t) + 1.1$ are set as references. Table 3 gives the tracking performances of the adaptive fuzzy internal model control and the proposed method. The control accuracy is quite close when tracking the fixed-frequency signal, however, for the mixed-frequency reference, the RMSE and MAXE of the proposed method are significantly lower than those of the approach proposed in [21].

Table 3. Comparison between the proposed method and the adaptive fuzzy internal model control in [21]: the RMSE and the MAXE.

References	The method in [21] (RMSE/MAXE, μm)	The proposed method (RMSE/MAXE, μm)
$y_{d3}(t)$	0.0033/0.0058	0.0018/0.0069
$y_{d4}(t)$	0.0085/0.0290	0.0016/0.0064

4 Conclusion

In this paper, a composite controller with the model predictive control and hysteresis compensation is proposed to achieve high-precision tracking control of PEAs. The overall feedforward-feedback control scheme consists of a feedforward compensator and a feedback predictive controller. First, the inverse Duhem model of PEAs is used to construct the feedforward compensator, aiming at mitigating the hysteresis nonlinearity of PEAs. Then, an MFNN model is adopted to approximate the dynamic behavior of PEAs, and the instantaneous linearization is implemented to this model. Based on the instantaneously linearized MFNN model, an explicit model predictive controller is designed to suppress the model inaccuracy and other unknown disturbances. Extensive experiments are conducted to verify the effectiveness of the proposed method on the P-753.1CD nanopositioning stage. Comparisons are also made with some existing control approaches in the literature. The experimental results demonstrate the superior of the proposed method for tracking control of PEAs.

Acknowledgments. This work was supported in part by the National Natural Science Foundation of China (Grants 61422310, 61633016, 61370032) and Beijing Natural Science Foundation (Grant 41620667).

References

1. Xu, Q.: Robust impedance control of a compliant microgripper for high-speed position/force regulation. IEEE Trans. Ind. Electron. **62**(2), 1201–1209 (2015)
2. Gu, G.-Y., Zhu, L.-M., Su, C.-Y., Ding, H., Fatikow, S.: Modeling and control of Piezo-actuated nanopositioning stages: a survey. IEEE Trans. Autom. Sci. Eng. **13**(1), 313–332 (2016)
3. Wu, J., Lin, Y., Lo, Y., Liu, W., Chang, K., Liu, D., Fu, L.: Effective tilting angles for a dual probes AFM system to achieve high-precision scanning. IEEE/ASME Trans. Mechatron. **21**(5), 2512–2521 (2016)
4. Ruiyue, O., Jayawardhana, B.: Absolute stability analysis of linear systems with duhem hysteresis operator. Automatica **50**(7), 1860–1866 (2014)
5. Li, Z., Zhang, X., Su, C.-Y., Chai, T.: Nonlinear control of systems preceded by preisach hysteresis description: a prescribed adaptive control approach. IEEE Trans. Control Syst. Technol. **24**(2), 451–460 (2016)
6. Gu, G.-Y., Zhu, L.-M., Su, C.-Y.: Modeling and compensation of asymmetric hysteresis nonlinearity for Piezoceramic actuators with a modified Prandtl-Ishlinskii model. IEEE Trans. Ind. Electron. **61**(3), 1583–1595 (2014)
7. Habineza, D., Rakotondrabe, M., Gorrec, Y.L.: Bouc-Wen modeling and feedforward control of multivariable hysteresis in piezoelectric systems: application to a 3-DoF Piezotube scanner. IEEE Trans. Control Syst. Technol. **23**(5), 1797–1806 (2015)
8. Wang, D., Yu, P., Wang, F.F., Chan, H.Y., Zhou, L., Dong, Z.L., Liu, L.Q., Li, W.J.: Improving atomic force microscopy imaging by a direct inverse asymmetric PI hysteresis model. Sensors **15**(2), 3409–3425 (2015)
9. Chen, X., Su, C.-Y., Li, Z., Yang, F.: Design of implementable adaptive control for micro/nano positioning system driven by Piezoelectric actuator. IEEE Trans. Ind. Electron. **63**(10), 6471–6481 (2016)
10. Ma, H., Wu, J., Xiong, Z.: Discrete-time sliding-mode control with improved quasi-sliding-mode domain. IEEE Trans. Ind. Electron. **63**(10), 6292–6304 (2016)
11. Devasia, S., Eleftheriou, E., Moheimani, S.O.R.: A survey of control issues in nanopositioning. IEEE Trans. Control Syst. Technol. **15**(5), 802–823 (2007)
12. Liu, L., Tan, K.K., Lee, T.H.: Multirate-based composite controller design of piezoelectric actuators for high-bandwidth and precision tracking. IEEE Trans. Control Syst. Technol. **22**(2), 816–821 (2014)
13. Janaideh, M.A., Rakotondrabe, M., Aljanaideh, O.: Further results on hysteresis compensation of smart micropositioning systems with the inverse Prandtl-Ishlinskii compensator. IEEE Trans. Control Syst. Technol. **24**(2), 428–439 (2016)
14. Fukuda, T., Nakajima, M., Pou, L., Ahmad, M.: Bringing the nanolaboratory inside electron microscopes. IEEE Nanotechnol. Mag. **2**(2), 18–31 (2008)

15. Cheng, L., Liu, W., Hou, Z.-G., Yu, J., Tan, M.: Neural Network based nonlinear model predictive control for Piezoelectric actuators. IEEE Trans. Ind. Electron. **62**(12), 7717–7727 (2015)
16. Liu, W., Cheng, L., Yu, J., Hou, Z.-G., Tan, M.: An inversion-free predictive controller for Piezoelectric actuators based on a dynamic linearized neural network model. IEEE/ASME Trans. Mechatron. **21**(1), 214–226 (2016)
17. Liu, W., Cheng, L., Hou, Z.-G., Huang, T.W., Yu, J., Tan, M.: An adaptive Takagi-Sugeno fuzzy model based predictive controller for Piezoelectric actuators. IEEE Trans. Ind. Electron. (2016, in press). doi:10.1109/TIE.2016.2644603
18. Cao, Y., Chen, X.B.: A novel discrete ARMA-based model for Piezoelectric actuator hysteresis. IEEE/ASME Trans. Mechatron. **17**(4), 737–744 (2012)
19. Cao, Y., Cheng, L., Chen, X.B., Peng, J.Y.: An inversion-based model predictive control with an integral-of-error state variable for Piezoelectric actuators. IEEE/ASME Trans. Mechatron. **18**(3), 895–904 (2013)
20. Norgaard, M., Ravn, O., Poulsen, N.K., Hansen, L.K.: Neural Networks for Modelling and Control of Dynamic Systems. Springer, New York (2000)
21. Li, P., Li, P., Sui, Y.: Adaptive fuzzy hysteresis internal model tracking control of Piezoelectric actuators with nanoscale application. IEEE Trans. Fuzzy Syst. **24**(5), 1246–1254 (2016)

Computational Methods for Sustainable Environment

Numerical Investigation of the Environment Capacity of COD, Inorganic Nitrogen and Phosphate in the Bohai Bay

Hao Liu[(✉)] and Zhi-kang Zhang

Shanghai Ocean University, Shanghai 201306, China
haoliu@shou.edu.cn

Abstract. An ocean dynamic model is used to simulate the tides and currents in the Bohai Bay. Model results are validated by comparing with observations. Furthermore, the conservative tracer is used to estimate the water exchange rate of the Bohai Bay, and it is found that about 62% of the seawater is transported out of the bay annually. At last, the grade 2 quality of the seawater is taken as the criteria to investigate the environment capacity of three major pollutants. It is found that the static capacity of COD, inorganic nitrogen and phosphate is about 3.999×10^5, 3.999×10^4 and 3.999×10^3 t/a, respectively, if the water exchange is not considered. furthermore, the process-controlled environment capacity for three pollutants can be 6.478×10^5, 6.478×10^4 and 6.478 t/a, respectively, and the consequence-controlled environment capacity may be as high as 1.052×10^6, 1.052×10^5 and 1.052×10^4 t/a, respectively.

Keywords: Bohai Bay · Environment capacity · Ocean Model · Water exchange

1 Introduction

Generally, the environment capacity indicates the maximum pollutant load that cannot be sustained by an environment until the environment has been destroyed. The environment capacity is scaled by the annual value, which means the maximum pollutant budget accumulated in a year [1]. Therefore, it can be deduced that an environment can support more pollutants if its environment capacity is large; Vise versa. Furthermore, the pollutant emission must adapt to the environment capacity. If the load of the discharged pollutant exceeds the environment capacity, some measures must be taken to reduce the pollutant emission and improve the environment protection. The method of the total quantity control is generally used to improve the environment management. Therefore, the total load of the discharged pollutant must be estimated firstly, and then the target can be further distributed to the pollutant sources no matter how many there are. So understanding the environment capacity is essential to take the method of the total quantity control.

This study focuses on the environment capacity of the Bohai Bay, which is located to the west of the Bohai Sea, and surrounded by Hebei, Tianjin and Shandong provinces (see Fig. 1). With the rapid development of the regional economy, the artificial activities

© Springer Nature Singapore Pte Ltd. 2017
K. Li et al. (Eds.): LSMS/ICSEE 2017, Part III, CCIS 763, pp. 753–761, 2017.
DOI: 10.1007/978-981-10-6364-0_74

have the deep influence on the sea environment. The direct consequence is the continuous deterioration of the environment quality which is characterized by the seawater eutrophication, the onset of the harmful algae bloom and the decrease of biological diversity. Therefore, better understanding the environment capacity of major pollutants is undoubtedly helpful to the implement of the total quantity control in the Bohai bay.

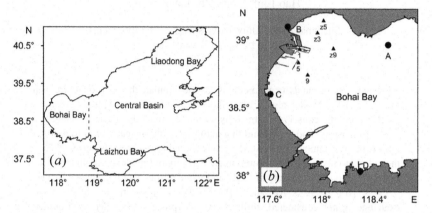

Fig. 1. (a) Location and scope of the Bohai Bay, (b) A, B, C, D are four tidal gages, and 1, 5, 9, z3, z5, z9 indicate six ship moorings used to measure the currents.

2 Research Approach

Unlike the environment capacity in an enclosed water mass, the maximum pollutant load in the Bohai Bay should comprise two parts, namely the static content dependent on the water quality standard and the dynamic pollutant content that is carried out of the bay through the process of the water exchange. The static pollutant content depends on the volume and the water quality standard of the Bohai Bay. The dynamic pollutant content is dependent on the rate of the water exchange between the Bohai Bay and the adjacent sea.

There are many methods to study the water exchange, such as tracers, ADCP measurement, numerical simulations, and so on [2–4]. In recent years, the numerical method is widely used due to its distinct merits. By means of the numerical modeling, the long-term water exchange can be reproduced, which is almost impossible for the in situ investigation. Besides, the numerical method is cheap to carry on and won't produce the second pollution. In this study, the Princeton Ocean Model (POM) is modified to simulate the water exchange of the Bohai Bay. The governing equation of the model can be found in "Appendix A", and its modification and application are given in previous studies [5, 6]. Based on the validated simulations, a numerical experiment is designed, in which the temporal variation of the conservative pollutant mass is used to understand the water exchange of the Bohai Bay. The choice of the conservative pollutant as the tracer is to avoid the matter reduction due to the bio-chemical process happening in the sea.

Just as illustrated in Fig. 1a, the dashed line is set to be the open lateral boundary linking the Bohai Bay and the adjacent sea. The initial concentration of the conservative tracer is set to be 1 gm^{-3} in the Bohai Bay, corresponding to the total tracer mass about 1.333×10^5 t. At the meanwhile, it is assumed that there are no such tracers out of the bay initially. Based on the above conditions, the ocean model is run for one year under the genuine external forces like the wind stress, the tidal force, the Coriolis force and the frictional force. Accordingly, the tracers begin to advect and diffuse with the water motion. The time series of the tracer mass in the bay are recorded to examine the water exchange of the bay.

Firstly, the water exchange is numerically simulated in the Bohai Sea, in which the grid resolution is set as 2 min by 2 min in the horizontal and 10 sigma layers are divided in the vertical. Eight major astronomic tides are used to drive the model along the open lateral boundary. Meanwhile, the monthly-mean wind fields are used to force the model on the sea surface. The model is run for one year, and then the time series of the variation in the tracer concentration can be reproduced. Secondly, the tidal feature is reproduced in the Bohai Bay, in which the fine grid resolution of 0.5 min by 0.5 min is adopted in the horizontal. The harmonic constants of the major tides on the open lateral boundary are obtained by interpolation of the model results of the Bohai Sea. The model is run for one month, and the model results are validated and analyzed by comparing with the observations.

3   Simulation Validation and Analysis

The following Table 1 gives a comparison of the simulated and observed harmonic constants of K_1 and M_2 tides on four stations. It is found that the root mean square error of the amplitude and phase lag is 2.18 cm and 2.98° for K_1 tide and 3.64 cm and 3.16° for M_2 tide, respectively. Therefore, the simulations agree with observations rather well. Moreover, Table 1 also shows that the model results avoid the system error, since the error comprises both the positive and the negative values.

Table 1. Comparison of computed and observed harmonic constants of M2 and K1 tides.

Tidal station	Name	Longitude (°E)	Latitude (°N)	K_1 tide		M_2 tide	
				ΔH (cm)	Δg (°)	ΔH (cm)	Δg (°)
A	Caofeidian	118°29′	38°58′	1.6	0.9	1.1	4.4
B	Tanggu	117°43′	39°06′	2.4	−2.6	−4.2	−0.2
C	Qihekou	117°35′	38°36′	−2.8	−5.0	−5.7	−0.5
D	Wawagou	118°16′	38°02′	1.7	1.7	−1.3	4.5

The computed co-amplitude and co-phase lag of M_2 and K_1 tides are given in Fig. 2. It can be seen that M_2 tide is stronger than K_1 tide in the Bohai Bay, since its amplitude exceeds 1 m near the west coast of the Bohai Bay, meaning that the semi-diurnal tide dominates in the sea region. It needs to be pointed out that the amphidromic point of M_2 tide appears off the old mouth of the Yellow River, and the diurnal tide dominates there since the amplitude of M_2 tide tends to be 0. Therefore, the open lateral boundary is just in the transition area from the diurnal tide domination to the semi-diurnal tide domination, and it is characterized by the mixed tides. The model results shown in Fig. 2 agree with observations [7] very well.

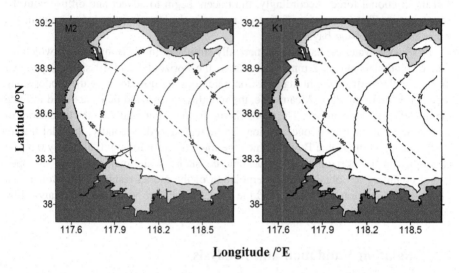

Fig. 2. Co-amplitude (solid lines) and co-phase lag (dashed lines) of M2 and K1 tides in the Bohai Bay.

In order to examine the simulations further, the tidal current observed during the spring and neap tide is also used to compare with the model results, respectively. The spring tide observation is from August 2 to 3, 2007, and the neap tide observation is from July 27 to 28, 2006.

Figure 3 shows the comparisons between simulated and observed tidal currents on six stations (see Fig. 1b). It can be seen that the simulations are reasonably consistent with the observations during the neap tide. Both simulated and observed tidal currents show the semi-diurnal type. By comparisons, it is also found that the current speed on the No. 9 station is apparently stronger than that on the No. 1 station, which means the bottom friction plays the important role in shaping the tidal current due to the more and more shallow water near the shore. Additionally, according to the time series of the current direction, it can been that the rotary current mainly appears in the open sea since the track streaked by the current vector is like an ellipse, whereas the rectilinear current appears close to coast since the track of the current vector is more like an line.

Comparing to the simulations on the neap tide, the model results show the more inconsistency with observations on the spring tide, though the basic characters of the tide and current are not very different. Except the influences of the weather disturbance, the equipment trouble and the data treatment problem, the artificial project like the sea reclamation may play the considerable role to the change of the local dynamics. Based

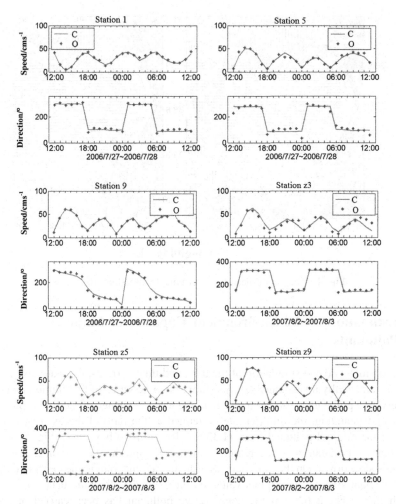

Fig. 3. Comparisons of computed and observed currents: C and O indicate the computed and observed values, respectively.

on the simulated and observed results, all three stations are characterized by the semi-diurnal tidal current on the spring tide just like that on the neap tide. The major type of the tidal current belongs to the rectilinear current along the coast due to the restriction of the coastline. The observation on No. z5 station shows some rotary current characters, whereas the simulation does not. This significant error needs to be further examined and analyzed.

Figure 4 gives the tidal current field on the flooding and ebbing tide, respectively, in order to understand the current character of the whole bay. It can be seen that the maximum current speed appear on the open lateral boundary, which may exceed 50 cms^{-1}. The tidal current shows the slower speed as it is closer to the coast, since the bottom friction effectively reduces the current speed near the shore.

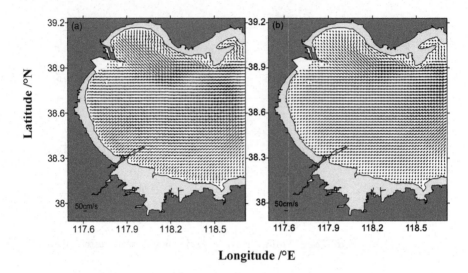

Fig. 4. The depth-mean tidal currents: (a) flooding, (b) ebbing

4 Estimation of the Environment Capacity of Major Pollutants

Based on the validated simulation of the tides and tidal currents, the water exchange is further numerically investigated. In this study, the time series of the dilution rate of the conservative pollutant is used to understand the water exchange of the Bohai Bay. For a better understanding, the dilution rate is defined as the ratio of the existing pollutant mass to the original pollutant mass (1.333×10^5 t) in the bay. From Fig. 5 it can be seen that the pollutant mass decreases continuously, and the declination is fast at first and then slow down. On the 280th day after the model integration, the existing pollutant mass is about 40% of the original value; on the 365th day, the ratio decreases to 38%, which means about 62% of the conservative pollutant has been carried out of the bay through the water exchange process. This value is consistency with the previous study [4].

The grade 2 quality of the seawater is taken as the standard to estimate the static capacity of COD, inorganic nitrogen and phosphate in the Bohai Bay. The concentration of three pollutants is 3, 0.3 and 0.03 gm^{-3}, respectively according to the grade 2 quality of the seawater. The volume of the research region is about 1.333×10^{11} m^3 which can be obtained by computing each grid volume. Hence, the static environmental capacity for three pollutants is about 3.999×10^5, 3.999×10^4 and 3999 t/a, respectively. These values do not include the pollutant loss due to the water exchange. If the water exchange is considered, then two situations may determine the dynamic pollutant content. One situation is the process control, namely the pollutant concentration is never beyond the seawater the grade 2 quality at any time in an annual cycle, which means that the load of the discharged pollutant would not exceed the pollutant loss correlated with the water exchange. Under this condition, the maximum pollutant replenish should equal to the pollutant loss, and the total environmental capacity of COD can be estimated as $3.999 \times 10^5 \times (1 + 62\%) = 6.478 \times 10^5$ t for 1 year. Based on the same calculation, the environmental capacity of the inorganic nitrogen and phosphate may be 6.478×10^4 and 6478 t/a, respectively. Another situation is the consequence control, which means that the pollutant concentration in the end of the year should not exceed the seawater quality of grade 2. Under this condition, the environmental capacity of COD can be calculated as $3.999 \times 10^5/(1 - 62\%) = 1.052 \times 10^6$ t for one year. Based on the same calculation, the environmental capacity of the inorganic nitrogen and phosphate may be 1.052×10^5 and 1.052×10^4 t/a. Obviously, the process control seems safer compared to the consequence control in terms of the management of the pollutant emission. In fact, the genuine environment capacity may be between the above two values.

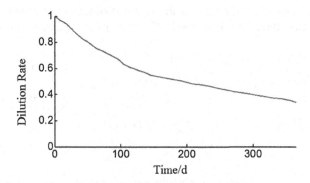

Fig. 5. Time series of the dilution rate of conservative tracers

5 Discussion and Conclusion

The Princeton Ocean Model is used to simulate the tides and tidal currents in the Bohai Bay, and the model results agree with observations rather well. Based on the validated simulations, a numerical experiment is conducted to investigate the water exchange of the Bohai Bay, in which the conservative pollutant is used as the tracer. The choice of the conservative pollutant as the tracer seems inconsistent with the character of COD, nitrogen and phosphate. In fact, the budget of the biogeochemical elements seems stable in an annual cycle, simply they exist in the water in some time and enrich in the body of the creatures in other time [8, 9]. The enriched part of elements can be released into the water again by means of the bio-chemical processes. Therefore, considering the period of one year, it is reasonable to take the conservative pollutant as the tracer. Based on the simulations, the annual water exchange is about 62% for the Bohai Bay, and the environmental capacity for COD, inorganic nitrogen and phosphate are about 6.478×10^5–1.052×10^6, 6.478×10^4–1.052×10^5, and 6.478×10^3–1.052×10^4 t/a, respectively.

It needs to be pointed out that the above values are based on the grade 2 quality of the seawater in the whole bay. In fact, the seawater quality is different for each sea region depending on its concrete function division. Therefore, understanding the regional function of each sea region and its water exchange is necessary to precisely investigate the water exchange of the Bohai Bay, which means the fine grids need to be adopted to simulate the small sea regions.

Acknowledgement. The work is financially supported by the Special Project of Marine Strategy Guide, Chinese Academy of Sciences (XDA11020305.2) and the Innovation Project of Shanghai Education Committee (12ZZ165).

Appendix A: Ocean Dynamic Model

A primitive equation ocean circulation model is used in this study, in which the orthogonal curvilinear coordinate and the sigma coordinate are adopted in the horizontal and vertical direction, respectively. The governing equations can be written as:

$$
\begin{cases}
\frac{\partial DU}{\partial x} + \frac{\partial DV}{\partial y} + \frac{\partial w}{\partial \sigma} + \frac{\partial \eta}{\partial t} = 0 \\
\frac{\partial UD}{\partial t} + \frac{\partial U^2 D}{\partial x} + \frac{\partial UVD}{\partial y} + \frac{\partial Uw}{\partial \sigma} - fVD + gD\frac{\partial \eta}{\partial x} \\
\quad + \frac{gD^2}{\rho_o} \int_\sigma^0 \left[\frac{\partial \rho'}{\partial x} - \frac{\sigma'}{D}\frac{\partial D}{\partial x}\frac{\partial \rho'}{\partial \sigma'} \right] d\sigma' = \frac{\partial}{\partial \sigma}\left[\frac{K_M}{D}\frac{\partial U}{\partial \sigma} \right] + F_x \\
\frac{\partial VD}{\partial t} + \frac{\partial UVD}{\partial x} + \frac{\partial V^2 D}{\partial y} + \frac{\partial Vw}{\partial \sigma} + fUD + gD\frac{\partial \eta}{\partial y} \\
\quad + \frac{gD^2}{\rho_o} \int_\sigma^0 \left[\frac{\partial \rho'}{\partial y} - \frac{\sigma'}{D}\frac{\partial D}{\partial y}\frac{\partial \rho'}{\partial \sigma'} \right] d\sigma' = \frac{\partial}{\partial \sigma}\left[\frac{K_M}{D}\frac{\partial V}{\partial \sigma} \right] + F_y
\end{cases} \tag{1}
$$

where U and V are the horizontal components of current speed and w indicates the current speed in the sigma direction; K_M is the vertical eddy viscosity; D is the genuine water depth and can be written as $D = H + \eta$, in which H and η indicate the mean water depth and the sea surface elevation, respectively.

In the above equation, F_x and F_y are the horizontal diffusion terms in the momentum equation and can be written as:

$$
\begin{cases}
F_x = \frac{\partial}{\partial x}(H\tau_{xx}) + \frac{\partial}{\partial y}(H\tau_{xy}) \\
F_y = \frac{\partial}{\partial x}(H\tau_{xy}) + \frac{\partial}{\partial y}(H\tau_{yy})
\end{cases} \tag{2}
$$

where τ_{xx}, τ_{yy} and τ_{xx} are the horizontal shear stresses and can be further written as:

$$
\begin{cases}
\tau_{xx} = 2A_M\frac{\partial U}{\partial x} \\
\tau_{yy} = 2A_M\frac{\partial V}{\partial y} \\
\tau_{xy} = \tau_{yx} = A_M\left(\frac{\partial U}{\partial y} + \frac{\partial V}{\partial x}\right)
\end{cases} \tag{3}
$$

where A_M is the horizontal eddy viscosity.

The wind stress and the bottom friction play the significant role on the sea surface $(\sigma \rightarrow 0)$ and the sea floor $(\sigma \rightarrow -1)$, respectively. Therefore, the vertical boundary conditions for the momentum equations are expressed as follows:

$$\begin{cases} \frac{K_M}{D}\left[\frac{\partial U}{\partial \sigma}, \frac{\partial V}{\partial \sigma}\right] = -\rho_a C_D \left[U_{10}^2 + V_{10}^2\right]^{1/2}(U_{10}, V_{10}) & \sigma \to 0 \\ \frac{K_M}{D}\left[\frac{\partial U}{\partial \sigma}, \frac{\partial V}{\partial \sigma}\right] = C_z \left[U^2 + V^2\right]^{1/2}(U, V) & \sigma \to -1 \end{cases} \tag{4}$$

where ρ_a is the air density; U_{10} and V_{10} are the wind speed 10 m above the sea surface in the x and y direction, respectively; C_D is the drag coefficient on the sea surface, and C_Z is the bottom friction coefficient on the sea floor.

The so-called Flather radiation condition is adopted at the open lateral boundary for the normal component of the external velocity field, as follows:

$$u_n = u_T + \sqrt{g/H}(\eta - \eta_T) \tag{5}$$

where u_n and u_T are the depth-mean current speed and prescribed tidal current speed, respectively; n denotes the normal vector on the open boundary; η indicates the sea surface elevation on the nearest interior non-open-boundary, which is calculated from the continuity equation; η_T indicates the prescribed tidal height on the open lateral boundary, and is expressed as the superposition of 8 major tides—K_1, O_1, P_1, Q_1, M_2, S_2, N_2 and K_2.

$$\eta_T = \sum f_i H_i \cos\left(\omega_i \cdot t + (v + u)_i - g_i\right) \tag{6}$$

where f_i is the nodal factor; v_i and u_i are the Greenwich initial phase and nodal correction, respectively; ω_i is angular frequency; g_i and H_i are the tidal phase lag and amplitude, respectively; the subscript i indicates the tidal constituent.

References

1. Liu, H., Yin, B.S.: Numerical calculation on the content of nitrogen, phosphate and COD in the Liaodong Bay. Mar. Sci. Bull. **25**(2), 46–54 (2006). (in Chinese)
2. Liu, Z., Wei, H., Liu, G.S., et al.: Simulation of water exchange in Jiaozhou Bay by average residence time approach. Estuar. Coast. Shelf Sci. **61**, 25–35 (2004)
3. Liu, H., Pan, W.R., Luo, Z.B.: Study on water exchange characters in the Shenhu Bay. Mar. Environ. Sci. **27**(2), 157–160 (2008). (in Chinese)
4. Wei, H., Tian, T., Zhou, F., et al.: Numerical study on the water exchange of the Bohai Sea: simulation of the half-life time by dispersion model. J. Qingdao Ocean Univ. **34**(2), 519–525 (2002). (in Chinese)
5. Liu, H.: Annual cycle of stratification and tidal fronts in the Bohai Sea: a model study. J. Oceanogr. **63**(1), 67–75 (2007)
6. Liu, H., Yin, B.S., Xu, Y.Q., et al.: Numerical simulation of tides and tidal currents in Liaodong Bay with POM. Prog. Nat. Sci. **15**(1), 47–55 (2005)
7. Chen, G.Z., Niu, G.Y., Wen, S.C., et al.: Marine Atlas of Bohai Sea, Huanghai Sea, East China sea-Hydrology, p. 504. Ocean Press, Beijing (1992). (in Chinese)
8. Liu, H., Yin, B.S.: Annual cycle of carbon, nitrogen and phosphorus in the Bohai Sea: a model study. Cont. Shelf Res. **27**, 1399–1407 (2007)
9. Liu, H., Yin, B.S.: Numerical investigation of nutrient limitations in the Bohai Sea. Mar. Environ. Res. **70**, 308–317 (2010)

An Artificial Neural Network Model for Predicting Typhoon Intensity and Its Application

Ruyun Wang[1(✉)], Tian Wang[2], Xiaoyu Zhang[3], Qing Fang[4], Chumin Wu[1], and Bin Zhang[1]

[1] College of Oceanography, Hohai University, Nanjing, China
wangry@hhu.edu.cn
[2] College of Harbour, Coastal and Offshore Engineering, Hohai University, Nanjing, China
[3] Department of Mathematics, College of Science, Beijing Forestry University, Beijing, China
[4] Faculty of Science, Yamagata University, Yamagata 990-8560, Japan

Abstract. Considering that the typhoon intensity's statistical predictors have the characteristics of inaccuracy, incompleteness and uncertainty, and the optional factors are factors are usually lots in a practical application, but the predictive ability will decline if using too many factors in a model, and may also lost the important information by choosing the inappropriate factorsQuery. Latitude and longitude of storm center, minimum central pressure, maximum wind speed near the storm center were chosen to be predictors, and a neural network model for predicting typhoon intensity was established by using every 6 h of current and former 18 h of these information directly. In this study, 61-year data set from 1949 to 2009 was used to train the networks, and 5-year data set from 2010 to 2014 was used to test the trained network. Compared with other typhoon predicting models, and results showed that the model has obtained a good predicting accuracy.

Keywords: Typhoon intensity · Artificial neural network · Predicting model

1 Introduction

On a global scale, the disaster that is more frequent and more lethal than earthquakes, tsunamis, volcanic eruptions, floods and etc. is the tropical cyclone (typhoons, hurricanes, cyclonic storm). In the northwestern Pacific, when the sustained wind speed of a tropical cyclone center is up to level 12 (i.e., 32.7 m/s or more), this tropical cyclone can be called typhoon. Typhoon can cause the abnormal fluctuation of the seawater—the storm surge. On average, there are 7 typhoons land on the coast areas of China every year, and most of the areas are plains and harbors, thus China is a frequency-occurring district of storm surge disasters. Storm surges cause huge loss to national defense, industrial and agricultural production, and national economy in China every year. Thus timely and accurate prediction of storm surge is of important significance to the disaster prevention and mitigation. To build the numerical forecasting model of storm surge

© Springer Nature Singapore Pte Ltd. 2017
K. Li et al. (Eds.): LSMS/ICSEE 2017, Part III, CCIS 763, pp. 762–770, 2017.
DOI: 10.1007/978-981-10-6364-0_75

based on dynamic equation, relevant parameters, such as typhoon track (longitude, latitude), typhoon. Intensity (near center minimum atmospheric pressure, near center maximum wind speed), are needed in advance, which values directly affect the accuracy of storm surge forecast.

Cai et al. built a predicating model of tropical cyclone landing sites based on BP network [4]. Zhou et al. used the improved BP network for the predicating experiment of typhoon tracks, and pointed out that typhoon moving direction output by neural network covers 97% of the actual path [12]. Yu et al. did comparative predicating tests of tropical cyclone tracks by using different kind of neural networks respectively such as BP, LM, RBF [11]. Lv et al. used the BP network and least-squares regression method for the predicating test of tropical cyclone tracks respectively, and pointed out that the neural network method is better than the regression method [9]. Shao et al. compared the predicating result of typhoon tracks by using BP network with the result by using CLIPER model, and it shows that the predicating precision of the BP network is higher than that of the CLIPER [10]. Given that a large number of predicating model of the tracks of typhoon center have been built, and higher predicating precision has been achieved, this essay try to study the predicating of the typhoon intensity, in order to provide reliable parameters of wind field model for the predicating of storm surge.

The research work of the predicating of typhoon intensity has obtained some achievements. Especially the artificial neural network, which is rapidly developing, for its simple structure, stable working status, easy to implement, has already get the preliminary application in the predicating of typhoon intensity. Baik et al. used BP network method to study the predicating of typhoon tracks and intensity respectively [1–3]. Huang et al. built the predicating model of the tropical cyclone intensity by using the BP network method [6]. On the basis of artificial neural network model, people combine the fuzzy mathematics with and the genetic algorithm to get a predicating model of typhoon intensity of higher precision [7, 8]. The predicating models of typhoon parameters by using artificial neural network model has been established by now, are basically using the latitude and longitude of the typhoon center, minimum atmospheric pressure and maximum wind speed of the near center of this moment and the past moment and their combination to constitute candidate factors, then choose predicating factors with better relevance to the predict and according to statistic method, to constitute pattern pair to perform network training and predicating testing. Such as, Lv et al. used deviation method to construct sustainable climate factors, and use them for correlation analysis with the predicting object [9]. Huang et al. designed sustainable climate factors and weather dynamic factors, choose the factors of which its simple correlation coefficient is up to 0.3 as predicting factors of tropical cyclone [6]. Considering that the typhoon intensity's statistical predictors have the characteristics of inaccuracy, incompleteness and uncertainty, and the optional factors are usually lots in a practical application, but the predicting ability will decline if using too many factors in a model, and may also lost the important information by choosing the inappropriate factors. As the 24 h longitude predicting BP network model built in paper [6], it uses multivariate regression analysis method to select the 4 factors of the 24 h longitude predicting as the latitudinal moving speed, the longitudinal moving speed, and the latitude of typhoon center at this moment, and the square and the square root of the latitudinal displacement and longitudinal displacement from the previous 12 h to this moment, obviously a longitude factor, which is of the most important, of the typhoon center at this moment is missing. Assuming that, typhoon center is at two places of different longitude on the same latitude at two different moments, but the corresponding longitudinal moving speed and latitudinal moving speed, and the square and the square root of the longitudinal displacement and the latitudinal displacement from previous 12 h to this moment are the same, then the network will mapping the longitude of this two typhoons after 24 h to the same value, and it shouldn't. Besides, similar problems also exist in paper [9]. Considering the BP artificial neural network itself has the ability of self-learning, self-organization and feature extracting to simulate high nonlinear problems. Thus there is no necessary to combine and screen the existing data, so here 16 factors such as, the latitude and the longitude of the typhoon center and the minimum atmospheric pressure and maximum wind speed of the near typhoon center of the current moment and every 6 h in the previous 18 h were chosen directly as forecast factors, to build the BP artificial neural network forecasting model of the typhoon intensity every 6 h in the future 6 to 72 h, in this way can we guarantee the information of typhoon as complete as possible.

2   Introduction of the BP Artificial Neural Network

An input layer, a number of hidden layers and one output layer, constitutes the BP network, through researches, Homik et al. [5] show that the BP network which contains a hidden layer is enough to deal with most of the forecasting problems, so a three-layered BP network was adopted (e.g., "Fig. 1"). Each unit of the upper layer and the next layer is connected through weight. Set the number of the units of the network input layer as n; the number of the units of the hidden layers as p; the number of the units of the output layer as q; the network training pattern pair as (A^k, Y^k), and $A^k = (a_1^k, a_2^k, \ldots, a_n^k)$ is the input vector, $Y^k = (y_1^k, y_2^k, \ldots, y_q^k)$ is the expected output vector, $k = 1, 2, \ldots, m$, m is the total number of the pattern pairs. Connection weight of input layer and hidden layers is $w_{ij}(i = 1, 2, \ldots, n, j = 1, 2, \ldots, p)$; the threshold of each unit of the hidden layers is $\theta_j(j = 1, 2, \ldots, p)$; the output of each unit of the hidden layers is $b_j^k(j = 1, 2, \ldots, p, k = 1, 2, \ldots, m)$; the connection weight of the hidden layers and the output layer is $v_{jl}(j = 1, 2, \ldots, p, l = 1, 2, \ldots, q)$; the threshold of each unit of the output layer is $\gamma_l(l = 1, 2, \ldots, q)$; the actual output is $C^k = (c_1^k, c_2^k, \ldots, c_q^k)(k = 1, 2, \ldots, m)$.

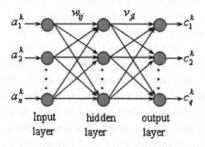

Fig. 1. Three-layer BP network model

The sigmoid function is adopted as the response function in the network training:

$$f(x) = \frac{1}{1 + e^{-x}} \tag{1}$$

Input the output vector into the network, the output results of the jth unit of the hidden layers and the lth unit of the output layer through the sigmoid function are

$$b_j^k = f(\sum_{i=1}^{n} \sum_{j=1}^{p} w_{ij} a_i^k - \theta_j) \tag{2}$$

$$c_l^k = f(\sum_{j=1}^{p} v_{jl}^k b_j^k - \gamma_l) \tag{3}$$

According to the actual network output c_l^k and the expected output y_l^k, the error d_l^k of the lth unit of the output layer can be calculated as

$$d_l^k = c_l^k(y_l^k - c_l^k)(1 - c_l^k) \tag{4}$$

According to the connection weight $\{v_{jl}\}$, the error $\{d_l^k\}$ of each unit of the output layer, and the output b_j^k of the hidden layers, the error e_j^k of the jth unit of the hidden layers can be calculated as

$$e_j^k = b_j^k(1 - b_j^k) \sum_{l=1}^{q} d_l^k v_{jl} \tag{5}$$

Set the global error of the network as E, then

$$E = \frac{1}{2} \sum_{k=1}^{m} \sum_{l=1}^{q} (y_l^k - c_l^k)^2 \tag{6}$$

Using the method of gradient descent to adjust the connection weigh and the threshold. The adjustment of w_{ij}, θ_j, v_{jl}, γ_l are:

$$\Delta w_{ij} = \alpha \sum_{k=1}^{m} e_j^k a_i^k, \quad \Delta\theta_j = -\alpha \sum_{k=1}^{m} e_j^k, \quad \Delta v_{jl} = \alpha \sum_{k=1}^{m} d_l^k b_j^k, \quad \Delta\gamma_l = \alpha \sum_{k=1}^{m} d_l^k.$$

The weight value and the threshold after adjustment are called w_{ij}, θ_j, v_{jl}, γ_l, then:

$$w_{ij} = w_{ij} + \Delta w_{ij}, \quad \theta_j = \theta_j + \Delta\theta_j, \quad v_{jl} = v_{jl} + \Delta v_{jl}, \quad \gamma_l = \gamma_l + \Delta\gamma_l.$$

The steps of the network training are as follow:

(1) First, set a minimal global error value $E_{\min}$ for the network; give Each connection weight $\left\{ w_{\overline{ijj}} \right\}$, $\{v_{jl}\}$ and threshold $\{\theta_j\}$, $\{\gamma_l\}$ a random value between the interval $(-1, +1)$;

(2) Provide pattern pair $(A^k, Y^k)(k = 1, 2, \ldots, m)$ to the network;

(3) Calculate $b_j^k, c_l^k, d_l^k, e_j^k$, E according to formulas (1)–(6);

(4) The value of α is 0.6;

(5) Substitute $b_j^k, c_l^k, d_l^k, e_j^k$ and α in the expressions of Δw_{ij}, $\Delta \theta_j$, Δv_{jl}, $\Delta \gamma_l$, to calculate Δw_{ij}, $\Delta \theta_j$, Δv_{jl}, $\Delta \gamma_l$;

(6) Substitute Δw_{ij}, $\Delta \theta_j$, Δv_{jl}, $\Delta \gamma_l$ in the expressions of w_{ij}, θ_j, v_{jl}, γ_l, to calculate the new weight value and threshold $\tilde{w}_{ij}$, $\tilde{\theta}_j$, $\tilde{v}_{jl}$, $\tilde{\gamma}_l$;

(7) Substitute $\tilde{w}_{ij}$, $\tilde{\theta}_j$, $\tilde{v}_{jl}$, $\tilde{\gamma}_l$ in formula (6) to calculate the new global error $\tilde{E}$;

(8) If $\tilde{E} < E$, turn to (10); or, when $\alpha > 0.0000001$, $\alpha = \alpha/2$, turn to (5), when α 0.0000001, then training attempt failed, program ends;

(9) $E = \tilde{E}$, $w_{ij} = \tilde{w}_{ij}$, $\theta_j = \tilde{\theta}_j$, $v_{jl} = \tilde{v}_{jl}$, $\gamma_l = \tilde{\gamma}_l$;

(10) If $E > E_{\min}$, turn to (3); or network training ends.

3 The Establishment of the Predicating Model of Typhoon Parameters

3.1 The Pretreatment of the Data of Typhoon Parameters

The activation function is sigmoid function; its range is (1,0), but the parameter values of typhoon are usually not in the range, so the selected typhoon data must be standardized, which is mapping it into (0,1). If a typhoon parameter is x, after standardize, it called $\hat{x}$, find the maximum $x_{\max}$ and minimum $x_{\min}$ of x from the selected typhoon historical data, then mapping all the x es into (0,1). But in the future forecasting of typhoon, typhoon parameter $x \notin [x_{\min}, x_{\max}]$ are likely to happen in the forecasting period, in order to left some scope for this situation, we mapping all the historical parameter data of typhoon into [0.01, 0.99], the standardized formula is

$$\hat{x} = \frac{0.01(x_{\max} - x) + 0.99(x - x_{\min})}{x_{\max} - x_{\min}} \tag{7}$$

the preparation of the training pattern pair, Set the matrix of the timing parameter of a typhoon with 6 J hours history as $A = (a_{ij})_{4 \times J}$, a_{1j}, a_{2j}, a_{3j}, a_{4j} are the latitude and longitude of the typhoon center and the minimum atmospheric pressure and maximum wind speed of the near center, respectively. The construction method of the artificial neural network model with the forecasting capability of 6 $N(N = 1, 2, \ldots, 12)$ hours is as follow:

$$\begin{pmatrix} a_{1,k} & a_{1,k+1} & a_{1,k+2} & a_{1,k+3} \\ a_{2,k} & a_{2,k+1} & a_{2,k+2} & a_{2,k+3} \\ a_{3,k} & a_{3,k+1} & a_{3,k+2} & a_{3,k+3} \\ a_{4,k} & a_{4,k+1} & a_{4,k+2} & a_{4,k+3} \end{pmatrix} \rightarrow \begin{pmatrix} a_{3,k+3+N} \\ a_{3,k+3+N} \end{pmatrix}$$

In here, $k = 1, 2, ..., J - 3 - N$, which means a typhoon with a life history of 6 J hours can construct $J - 3 - N$ pattern pairs.

3.2 Steps of Forecasting

After the network training, the actual forecasting work can go on, and has following two steps:

Get typhoon parameter data every 6 h from the previous 18 h to the current moment, standardize each parameter, to get the input matrix of the network;

Substitute input matrix into the corresponding BP network, then the network outputs are the atmospheric pressure and wind speed at the corresponding forecast time.

4 Test and Analysis of the Predicting

Selecting the northwestern pacific typhoon which affects the coastal areas of Jiangsu province of China as the research object, to build a ( ) hours artificial neural network predicting model. Data of typhoon tracks and parameters are acquired from "Datasets of the best tropical cyclone tracks by CMA-STI", which was edited by the Shanghai Typhoon Institute (STI) of China Meteorological Administration (CMA). Here typhoon parameters predicting is mainly required to build the storm surge numerical predicting model of the coastal areas of Jiangsu province, so the standard of choosing typhoon case is that, choose typhoon which enters 28°–38°N, 100°–150°E between 1949 to 2015. First, chose data of typhoon with more than 24 h life history from the datasets. Using the data of typhoon parameters from 1949 to 2009 into BP network training, from 2010 to 2015 into the forecast test of the network. The numbers of the pattern pairs of the network training of each time and pattern pairs of predicting test are as shown in Tables 1 and 2.

The mean absolute errors of the minimum atmospheric pressure and the maximum wind speed of typhoon near center predicted at different predicting time over 6 years (from 2010 to 2015) can be calculated, based on the measured data of typhoon, as shown in Table 3. The mean absolute errors of the maximum wind speed of the near center at 12th, 24th, 36th, 48th, 72th hours are 2.49, 3.95, 3.85, 4.65, 5.89 m/s, respectively, but if the calculation is based on "National Hurricane Center Forecast Verification Report (Verification_2010, 2011, 2012, 2013, 2014)" published by the National Hurricane Center (NHC) from 2010 to 2014, the mean absolute errors of the maximum wind speed of the near center at 12th, 24th, 36th, 48th, 72th hours are 3.05, 5.02, 6.41, 7.10, 7.74 m/s, showed a reduction of 0.56, 1.07, 2.56, 2.45, 1.85 m/s respectively than the American errors, that is decreased by 18.46%, 12.38%, 39.94%, 34.45%, 23.93% respectively. In the typhoon pressure data, which was using for

Table 1. The number of training pattern pairs

Time (h)	Number of pattern pairs
6	2819
12	2281
18	1838
24	1483
30	1207
36	989
42	808
48	662
54	547
60	447
66	362
72	291

Table 2. The number of forecast and test pattern pairs

Time (h)	Number of pattern pairs
6	259
12	203
18	153
24	117
30	91
36	71
42	57
48	46
54	37
60	29
66	23
72	17

training and predicting, the minimum atmospheric pressure and the maximum atmospheric pressure are 935, 1010 hPa respectively, and the minimum error and maximum error of all the predicting errors from the 6th hour to the 72th hour are 2.26, 5.97 hPa respectively.

Table 3. The mean predicting errors of minimum pressure and maximum wind speed near the typhoon center from 2010 to 2015

Forecast time (h)	6	12	18	24	30	36	42	48	54	60	66	72
Atmospheric pressure (hPa)	2.26	3.45	4.06	4.56	5.59	3.45	4.27	4.38	4.88	5.05	5.97	4.60
Wind speed (m/s)	1.58	2.49	3.41	3.95	5.06	3.85	4.19	4.65	4.67	4.42	5.17	5.89

Take the Nakri typhoon of 2014 and Molave typhoon of 2015 as examples, to show the mean absolute errors of a single typhoon forecasting, as shown in Tables 4 and 5.

Table 4. The mean forecast errors of minimum pressure and maximum wind speed near the typhoon center of Nakri

Forecast time (h)	6	12	18	24	30	36	42	48	54
Atmospheric pressure (hPa)	1.43	1.85	1.24	2.85	3.60	2.77	3.98	2.79	1.64
Wind speed (m/s)	1.65	2.31	2.56	3.37	0.93	2.24	2.48	1.38	2.26

Table 5. The mean forecast forecast errors of minimum pressure and maximum wind speed near the typhoon center of Molave

Forecast time (h)	6	12	18	24	30	36	42	48	54
Atmospheric pressure (hPa)	1.29	2.20	2.99	3.30	4.08	4.62	5.91	7.30	7.86
Wind speed (m/s)	0.82	1.29	1.97	2.33	2.84	3.35	3.97	4.83	5.03

The mean absolute errors of the minimum atmospheric pressure and the maximum wind speed of the near center, which were calculated based on data of the typhoon that land between 28°–38°N, 100°–150°E at 2015, and the predicting values of the National Meteorological Center (NMC), the Joint Typhoon Warning Center (JTWC), the Japan Meteorological Agency (JMA) and the Korea Meteorological Agency (KMA), were used to compare with the mean absolute errors calculated by the model in this essay, as shown in Tables 4 and 5. The forecast error at the 48th hour of the model in essay is much smaller than that at the 6th, 12th, 24th and 36th, and this model cannot predict the intensity of typhoon at the 60th hour and the 72th hour (Tables 6 and 7).

Table 6. The comparisons of 2015 mean predicting errors of minimum pressure near the typhoon center (hPa)

Forecast time (h)	6	12	24	36	48
This paper	3.19	3.33	3.16	2.63	0.75
NMC		4.77	5.17	8.93	12.39
JMA			10.12		15.18
KMA			5.98		13.09

Table 7. The comparisons of 2015 mean predicting errors of maximum wind speed near the typhoon center (m/s)

Forecast time (h)	6	12	24	36	48
This paper	1.92	2.18	2.36	1.57	0.42
NMC		2.58	2.75	4.88	6.70
JTWC		3.23	4.39	5.70	7.98
JMA			4.68		7.02
KMA			3.92		6.63

5 Conclusion

Results show that the predicting model of typhoon intensity of this paper has the advantage of high precision, and the reason might be that, it is appropriate to choose 16 factors, such as the latitude and the longitude of the typhoon center and the minimum atmospheric pressure and maximum wind speed of the near typhoon center of the current moment and every 6 h in the previous 18 h, directly as the forecast factors. Without the combination and selection of the predicting factors, the completeness of the typhoon information is guaranteed, so the predicting of the minimum atmospheric pressure and maximum wind speed of the near center is better, the further improvement of this model were about how to make the neural network algorithm more stable and superior.

References

1. Baik, J.J., Hwang, H.S.: Tropical cyclone intensity prediction using regression method and neural network. J. Meteorol. Soc. Jpn. **76**, 711–717 (1998)
2. Baik, J.J., Paek, J.S.: Performance test of back-propagation neural network in typhoon track and intensity prediction. Korean J. Atmos. Sci. **3**, 33–38 (2000)
3. Baik, J.J., Paek, J.S.: A neural network model for predicting typhoon intensity. J. Meteorol. Soc. Jpn. **78**, 857–869 (2000)
4. Cai, Y., Lu, W., Yao, L.: Artificial neural network method of forecasting tropical cyclone. J. Trop. Meteorol. **10**, 284–288 (1994)
5. Homik, K.M., Stinchcombe, M., White, H.: Multilayer feedforward networks are universal approximators. Neural Netw. **2**, 359–366 (1989)
6. Huang, X., Fei, J., Chen, P.: A neural network approach to predict tropical cyclone intensity. J. Appl. Meteorol. Sci. **20**, 699–705 (2009). (in Chinese)
7. Jin, L., Shi, X., Huang, X., et al.: A fuzzy neural network prediction model based on manifold learning to reduce dimensions for typhoon intensity. In: 2013 IEEE 8th Conference on Industrial Electronics and Applications (ICIEA), pp. 562–566 (2013)
8. Lin, K., Chen, B., Dong, Y.,et al.: A genetic neural network ensemble prediction model based on locally linear embedding for typhoon intensity. In: 2013 IEEE 8th Conference on Industrial Electronics and Applications (ICIEA), pp. 137–142 (2013)
9. Lv, Q., Luo, J., Zhu, K., Ren, J.: Experiments on predicting tracks of tropical cyclones base on artificial neural network. Guangdong Meteorol. **31**, 15–18 (2009). (in Chinese)
10. Shao, L., Fu, G., Cao, X.: Application of BP neural network to forecasting typhoon tracks. J. Nat. Disasters **18**, 104–111 (2009). (in Chinese)
11. Yu, S., Zhong, Y., Teng, W.: An experimental study on artificial neural network forecasting models of tropical cyclone path. J. Trop. Meteorol. **20**, 523–529 (2004). (in Chinese)
12. Zhou, Z., Han, G., Zhu, D.: A typhoon prediction system of artificial neural network. Meteorology **22**, 18–21 (1996). (in Chinese)
13. National Oceanic and Atmospheric Administration. http://www.Nhc.noaa.gov/verification/pdfs/verification_2010.pdf
14. National Oceanic and Atmospheric Administration. http://www.Nhc.noaa.gov/verification/pdfs/Verification_2011.pdf
15. National Oceanic and Atmospheric Administration. http://www.nhc.noaa.gov/verification/pdfs/Verification_2012.pdf
16. National Oceanic and Atmospheric Administration. http://www.nhc.noaa.gov/verification/pdfs/Verification_2013.pdf
17. National Oceanic and Atmospheric Administration. http://www.nhc.noaa.gov/verification/pdfs/Verification_2014.pdf

Analysis of Power Spectrum Feature Based on Slurry Noise in Electromagnetic Flowmeter

Jie Chen[✉], Qiong Fei, Bin Li, and Xiaojie Zheng

School of Mechatronic Engineering and Automation, Shanghai University,
Shanghai 200072, China
jane.chen@shu.edu.cn

Abstract. As an essential part of measuring technology, flowmeters have been widely used in industrial production. Accurate flow metering will not only improve the quality of products, promoting economic efficiency and **management** level, but lays foundation for the assessment of energy saving and environmental sewage discharging. Electromagnetic flowmeter is suitable for slurry flow measurement since it has high reliability, strong corrosion resistance, high measurement precision, and no stopped medium components in the measuring pipe. In order to improve the measurement of slurry noise problem, this paper will analyze the power spectrum of the slurry noise, and then find the relationship between slurry noise and excitation frequency as to provide theoretical basis for frequency switch of the variable frequency electromagnetic flowmeter.

Keywords: Electromagnetic flowmeter · Slurry noise · Power spectrum analysis · Frequency conversion

1 Introduction

Slurry which belongs to liquid-solid two-phase mixture, widely exist in nature and every field of industrial production and engineering. Compared with other kinds of flowmeter, the electromagnetic flowmeter has high reliability, strong corrosion resistance, high measurement precision, and no stopped medium components in the measuring pipe. It is suitable for the measurement of solid-liquid two-phase flow. While during the slurry flow measurement, the solid particles in liquid crash the electrods of sensor randomly and generate a kind of random noise. The amplitude of slurry noise is larger compared with the traffic signal. It is easily lead to sharp fluctuations in measurement.

Japanese Toshiba company had adopted high excitation frequency electromagnetic flowmeter to solve the problem of slurry noise [1]. The product FSM4000 of Germany ABB company used sine wave excitation with high frequency about 70 Hz. The electromotive force signal which generated by FSM4000 almost all involved in flow calculation. It can effectively measure the pulsating flow and suppress the slurry noise better [2]. Japanese Yokogawa company adopted the dual-band rectangular wave excitation, which can take into account both the zero stability of the signal and slurry noise problems [3]. Xu and Wang designed high frequency square wave excitation method. It realized the steady-state volatility of measurement is less than 4% [4]. Liang adopted the electromagnetic flowmeter signal processing method based on signal model, and can

© Springer Nature Singapore Pte Ltd. 2017
K. Li et al. (Eds.): LSMS/ICSEE 2017, Part III, CCIS 763, pp. 771–778, 2017.
DOI: 10.1007/978-981-10-6364-0_76

effectively weaken the slurry noise interference [5]. Pang et al. proposed a mixed digital filter method of signal disposal scheme. It can effectively reduce the noise [6]. Zhang put forward a kind of comb band-pass filter and amplitude demodulation method to process the electromagnetic flowmeter measurement signal in high frequency excitation. It can effectively inhibit the disturbance of the slurry noise to the measured signal [7].

The related research institutions develop solid-liquid two-phase flow slurry type electromagnetic flowmeter respectively from improving the excitation frequency, studying digital signal processing methods, and adopting new materials.

2   Theoretical Basis and Experimental Flatform

2.1   The Principle of Electromagnetic Flowmeter

The electromagnetic flowmeter which based on Faraday's law of electromagnetic induction is an instrument used to measure the flow of conductive liquid [8]. Experimental results show that on either side of the loop electrode will generate inductive electromotive force and inductive current when magnetic flux surrounded by the conductor loop changes. The induction electromotive force is proportional to the rate of change of magnetic flux, and its direction is determined by the Lenz's law [9].

$$E_i = -\frac{d\emptyset}{dt} \tag{1}$$

According to Faraday's law of electromagnetic induction, if only consider the size of the electromotive force, under the action of uniform magnetic field B (tesla), a conductor in length of D (meters) with velocity of V (m/s) along the direction perpendicular to the magnetic field can generate inductive electromotive force E (volt).

$$E_i = \frac{d\emptyset}{dt} = \frac{BdS}{dt} = \frac{BDdl}{dt} = BDV \tag{2}$$

If the conductive fluid with a uniform velocity V pass through the pipe with diameter of D composed of insulator, can also be induced the same amount of induction electromotive force.

As shown in Fig. 1, the volume flow through the cross section of the pipe must be equal to the product of the area of cross section and velocity. For roundness measuring tube, we can got the volume flow rate Q (m³/s).

$$Q = \frac{\pi}{4}D^2V \tag{3}$$

Simultaneous formula (2) and (3), Q can be obtained.

$$Q = \frac{\pi}{4}D\frac{E_i}{B} \tag{4}$$

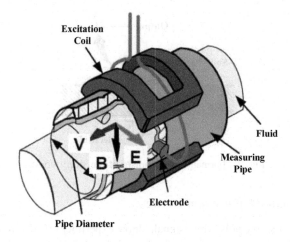

Fig. 1. Electromagnetic flow sensor measuring principle diagram.

Formula (4) illustrates that when pipe diameter D stays the same and the magnetic induction intensity B changes, flow rate is directly proportional to the induction electromotive force Ei and magnetic induction intensity B. We can find that the flow measurement of electromagnetic flowmeter has nothing to do with the change of other physical parameters, which is the biggest advantage of the electromagnetic flowmeter.

By formula (4), the induced electromotive force Ei is a very important parameter in flow measurement. Induction electromotive force is generated at the ends of the electrodes, and the electrodes direct contact with the liquid flows through the electromagnetic flowmeter. When measuring, the electrode can also be contact and impacted by the solid particles in the liquid, and this makes the slurry flow measurement process appears different problems with water measurement. The solid particles in liquid crash the electrods of sensor randomly and generate a kind of random noise. The amplitude of slurry noise is larger compared with the traffic signal. It is easily lead to sharp fluctuations in measurement.

2.2   The Formation of Slurry Noise

There is a very thin oxide film on the surface of electrodes, which makes the electrochemical reaction reaches equilibrium state. The formation mechanism of the slurry noise is shown in Fig. 2. The solid particles in the fluid impact the electrodes and the oxidation film on the surface is damaged [10]. While metal materials in contact with the fluid medium has re-generated oxidation film on the surface to keep the balance of the electrochemical equilibrium. During the period of reach the electrochemical equilibrium, free ions in the metal and fluid make electrochemical reaction under the action of the electric field. The passivation layer is destroyed due to the solid particles impact electrodes. The oxidation film is generated repeatedly under the electrochemical reaction and then the potential between the electrodes changes vary greatly. The change of the potential form noise in the flow signal. This is called the slurry noise in the electromagnetic flowmeter generally.

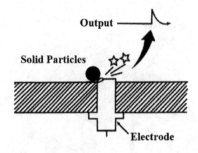

Fig. 2. Formation mechanism of the slurry noise.

2.3   Experimental Platform and Device

In order to collect slurry noise flow signal, slurry experiment platform is needed in the experiment. As it is shown in Fig. 3, the experimental devices mainly consist by the slurry pump, slurry tank, pipe, mixer and valves, etc. Two Yokogawa ADMAG AXF series slurry type electromagnetic flowmeter are in series on the pipe. The pipe size is 50 mm. Firstly, evenly mixing the slurry by the paddle mixer in the sink. Then open the water pumps and valves, make the slurry circulating in the closed loop pipe. Finally start the slurry measurement experiment when the panel of the electromagnetic flowmeter displays stability.

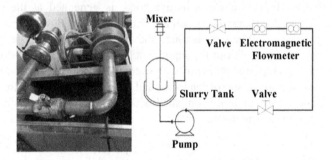

Fig. 3. Slurry experiment devices.

3   Signal Collection and Analysis of Slurry Noise

3.1   Signal Collection of Slurry Noise

The flow signal collection process of electromagnetic flowmeter is shown in Fig. 4 where Y is the collection point. Signal data are collected under the same slurry concentration and flow rate but different excitation frequency. The sampling frequency of the oscilloscope is 2500 Hz, the sampling time is 40 s and the sampling data length is 100000. Finally, the power spectrum analysis of these data is carried out on MATLAB.

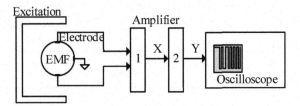

Fig. 4. Slurry flow signal collection process.

3.2   Power Spectrum Analysis of Slurry Noise

The analysis of slurry noise basically has the following steps.

(1) Remove the DC component of the sampling data. In order to avoid the influence of the DC component caused by power supply on the slurry noise signal analysis, each point value needs to subtract the average of all points to ensure the DC component has been removed from the sampling data.

(2) The power spectrum of the slurry noise at different excitation frequencies is obtained by the signal toolbox of MATLAB.

(3) Compare the power spectrum of the slurry noise at different frequencies under the same coordinates.

After 400 g of sand is added to 40 kg of water and flow rate is set as 2.0 m/s, the slurry noise data are collected and analyzed according to the steps described above. In this paper, the correlation function method (BT method) is used to estimate the power spectrum of the slurry noise signal. Its theoretical basis is the Wiener - Khintchine theorem, that is, the power spectral density of the stationary random signal is the Fourier transform of its autocorrelation function. The specific expression of the function is H = abs (fft (xcorr (y, 'unbiased'), N)), where y is the signal sequence.

Figure 5 shows the power spectrum of the slurry noise under the excitation frequency of 6.25 Hz, 12.5 Hz and 25 Hz at the flow rate of 2.0 m/s. It can be seen that the slurry noise presents a 1/f characteristic. Obvious frequency points also can be seen in the power spectra, which proves that the selected correlation function method can reflect the basic characteristics of slurry noise. Figure 6 shows the power spectra of three different frequency slurry noise signals under the same coordinates. The three curves from up to down represent the power spectrum curves at 6.25 Hz, 12.5 Hz and 25 Hz respectively. The ordinate represents the power spectral density in dB while the abscissa represents the frequency in Hz. It is obvious that the power spectrum curve of the slurry noise translates with the change of the excitation frequency and as the excitation frequency increases, the curve translates downwards. Based on the power spectrum curves of slurry noise, it can be concluded that increasing the excitation frequency can overcome the influence of slurry noise on slurry measurement.

To ensure the reliability of the experimental results, the flow rate is changed to 2.5 m/s and the power spectrum analysis of the slurry noise is carried out when the

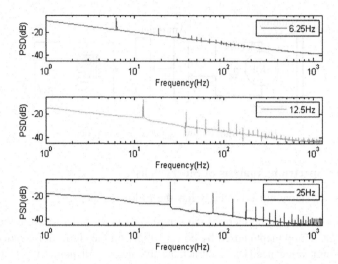

Fig. 5. Power spectrum of different frequency slurry noise at the flow rate of 2.0 m/s.

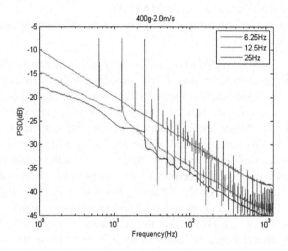

Fig. 6. Power spectrum comparison of different frequency slurry noise at the flow rate of 2.0 m/s

excitation frequency is 6.25 Hz, 12.5 Hz and 25 Hz respectively. The analysis results are shown in Figs. 7 and 8. The conclusion, the same as the previous one, is that with the increase of the excitation frequency, the power spectrum curve of the slurry noise translates downwards. It is proved again that the increasing the excitation frequency can improve the measurement results, so as to improve the measurement accuracy in slurry measurement.

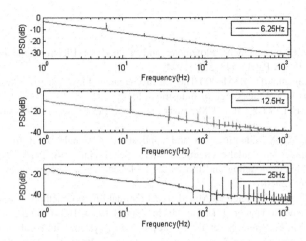

Fig. 7. Power spectrum of different frequency slurry noise at the flow rate of 2.5 m/s.

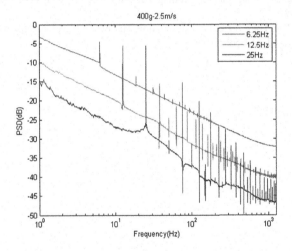

Fig. 8. Power spectrum comparison of different frequency slurry noise at the flow rate of 2.5 m/s.

4 Conclusion

In this paper, the power spectrum analysis of the slurry noise under different excitation frequencies by MATLAB is first put forward and the relationship between the excitation frequency of the electromagnetic flowmeter and the slurry noise is studied. In the frequency domain, it is proved that increasing excitation frequency of the electromagnetic flowmeter can overcome the influence of the slurry noise on flow measurement. This conclusion provides a theoretical basis for the frequency switching of electromagnetic flowmeter, which can effectively improve the slurry measurement.

References

1. Tomita, T.: Electromagnetic Flowmeter and Method for Electromagnetically Measuring Flow Rate. P. US Patent 5443552 (1994)
2. ABB Instrumentation: FSM4000-The Electromagnetic Flowmeter of Choice for Critical Applications in a Wide Range of Industries (2010)
3. Xiao, L.B., Hei, S.J.Y., Jun, T.M.: Electromagnetic Flowmeter. P. CN Patent 87101677A (1989)
4. Xu, K., Wang, X.: Identification and application of signal model for electromagnetic flowmeter under sinusoidal excitation. J. Meas. Sci. Technol. **18**, 1973–1978 (2007)
5. Liang, L.: Signal Modeling and Processing of Electromagnetic Flowmeter for Slurry-flow Measurement (2010)
6. Pang, B., Zhang, Z., Liang, Y.: Mixed-signal optimal filtering method of signal disposal for electromagnetic flowmeter. J. Electr. Mach. Control **19**(1), 102–106 (2015)
7. Zhang, R., Xu, K., Yang, S., et al.: Signal processing system with comb band-pass filter of electromagnetic flowmeter. J. Electron. Meas. Instrum. **26**(2), 177–183 (2012)
8. Su, Y.: Flow Measurement and Test. China Metrology Publisher, Beijing (1992)
9. Cai, W., Ma, Z., Zhai, G.: Electromagnetic Flowmeter. China Petrochemical Publisher, Beijing (2004)
10. Zhang, Z.: The High Frequency Excitation Hardware Development of Electromagnetic Flowmeter Based on DSP (2013)

A Two-Stage Agriculture Environmental Anomaly Detection Method

Lili Wang[1,2], Yue Yu[1,2], Li Deng[1,2(✉)], and Honglin Pang[1,2]

[1] School of Mechatronics Engineering and Automation, Shanghai University,
Shanghai 200072, China
dengli@shu.edu.cn
[2] Shanghai Key Laboratory of Power Station Automation Technology,
Shanghai 200072, China

Abstract. In order to process abnormal problems of massive distributed greenhouse environmental data, a novel anomaly detection algorithm based on the combination of Support Vector Machine (SVM) and Gaussian Mixture Model (GMM) is proposed and realized under the Spark framework, which is utilized to detect the environmental anomaly during crop growth. At the first stage, SVM is adopted to classify the data, Spark framework is utilized to solve optimization problem iteratively; at the second stage, GMM is used to do clustering on the classified data respectively. Spark framework is utilized to update the models internationally until stable, during every iteration. Map phase implements the distribution of the sample points to the models. Reduce phase renew the numbers of models and the parameters. Finally, the detection of environmental anomaly is completed by taking advantages of the clustering result. The results show that the proposed approach can be well applied to actual production.

Keywords: SVM · GMM · Agriculture environmental data · Anomaly detection · Spark

1 Introduction

With the development of information technology, cloud computing, agriculture is striding forward to the direction of digitalization, precision and intelligence. As a product of facility agriculture, intelligent greenhouse is capable to provide a favorable environment for the growth of crops. In order to ensure the stable growth of crops, special attention must be paid to the outlier data collected by sensor nodes in the greenhouse system. However, these abnormal data may not only be caused by the external environment, but also arise from the software or hardware failure of nodes. Therefore, it is of significance to analyze the abnormal data and its sources so as to exclude the anomalies in time.

As an important component of data mining, anomaly detection methods have been applied in agriculture. Yongwha et al. [1] have used the sound data to automatic detection of pig wasting diseases, a hierarchical two-level structure is proposed: the

© Springer Nature Singapore Pte Ltd. 2017
K. Li et al. (Eds.): LSMS/ICSEE 2017, Part III, CCIS 763, pp. 779–789, 2017.
DOI: 10.1007/978-981-10-6364-0_77

Support Vector Data Description (SVDD) as an early anomaly detector and the Sparse Representation Classifier (SRC) as a respiratory disease classifier.

Traditional anomaly detection methods are generally employ particular algorithm alone, classification or clustering method. In order to improve the accuracy and efficiency of detection, in past few years, approaches have been made by merging different algorithms together. Yasami et al. [2] have presented a combinatorial method based on k-Means clustering and ID3 decision tree learning algorithms for unsupervised classification of anomalous and normal activities in computer network ARP traffic. Chitrakar et al. [3] have proposed a hybrid approach of combining k-Medoids clustering technique with Support Vector Machine to detect anomalies in Intrusion Detection System, it is shown that the approach has outperformed the approach of combining Naïve Bayes with k-Means/k-Medoids. Varuna et al. [4] have introduced hybrid learning method, that integrates k-Means clustering and Naïve Bayes classification.

Considering the greenhouse agriculture data has the characteristics of differences in time, geographical, etc., these anomalies are conditional outliers. Unlike global anomaly, whether an object belongs to an anomaly is not only dependent on the behavior attribute, but also the context attribute. The traditional anomaly detection algorithms cannot detect anomalies effectively; also, the problem of traditional stand-alone processing mode gradually emerged due to that it can no longer meet the needs of a large number of data storage and computing.

Aiming at these problems, a two-stage anomaly detection algorithm based on the combination of Support Vector Machine (SVM) and Gaussian Mixture Model (GMM) is proposed and realized under the Spark framework in this paper. Clustering algorithm based on mixed probability model is widely used, and settled in the framework of Exception-Maximum (EM) algorithm [5]. However, traditional EM algorithm is quite sensitive to initial values, in which the number of components needs to be given a priori, such that it is easy to trap into local optimum. To resolve these drawbacks of EM algorithm, Figueiredo et al. [6] proposed an algorithm to deal with the number of clusters and also the estimates of parameters for mixture models simultaneously by using the particular form of a minimum message length (MML) criterion. Yang et al. [7] proposed a robust algorithm with a new objective function and model updating strategy, which can automatically determine the optimal number of clusters. Hence, combining SVM and GMM together, can ensure a higher detection accuracy.

In this paper, the Apache Spark is used to provide solutions for the storage and processing of a large number of greenhouse agricultural data. Apache Spark is an open source distributed computing framework based on memory computing, it outperforms MapReduce [8] when dealing with the iteration task in machine learning. Furthermore, it can meet the real-time requirements of data processing in large data environment, and has the characteristics of high fault-tolerance and high scalability. In recent years, with the rapid development of automatic control and cloud computing, many new technologies have increasingly applied to traditional agricultural production, for example, intelligent greenhouse monitoring system and agricultural expert knowledge database.

2   Greenhouse Data

About 45,000 data are used in this paper, which were collected from tomato greenhouse in ChongMing district, Shanghai between November 2014 and March 2015. The time intervals between samples are 5 min. Five features of the dataset are time, temperature (°C), relative humidity (%), concentration of carbon dioxide (ppm), photosynthetically active radiation ($\mu mol/s/m^2$), as shown in Table 1. Sampled data is stored in clusters deployed with Hadoop Distribute File System (HDFS).

Table 1. Data samples.

Time	TEMP	RH	CO2	PAR
...	...	...	...	...
2014-11-16 08:00:00	20.1	89	403.84	2.00
2014-11-16 08:05:00	20	89.98	424.58	2.00
...	...	...	...	...

2.1   Data Preprocessing

In the actual acquisition process, the dataset is very easy to be affected by noise, missing values, etc. For instance, there are 127 missing data between 3:40 and 14:10 on November 30, 2014. Taking into account the missing data may result from equipment problems or network failures and other factors, there are need for such anomalies in order to timely troubleshooting hardware and software failures. Refer to the common method to handle missing values, the data is deleted so as to reduce its impact on the subsequent data classification and clustering.

3   SVM-GMM Anomaly Detection Method

The tomatoes are in their growth stage when the dataset is collected. According to tomato growth model in greenhouse [9], effects of environmental factors on the growth of tomato [10] and other references during this phase, the appropriate temperature in daytime should be about 25 °C, ranges should not exceed ±5 °C. When the temperature is below 10 °C or more than 35 °C, the tomatoes will basically stop developing. The appropriate relative humidity should ranges from 80% to 90%. Carbon dioxide concentration should be more than 280 ppm, when the concentration is too low, it will affect the photosynthesis of tomatoes and will lead to growth retardation. In the nighttime, the appropriate temperature should between 15 °C and 20 °C, the relative humidity should not exceed 95%, the harm caused by low temperature and high humidity should be prevented.

Considering the abnormal of the greenhouse environment belongs to conditional abnormal, whether the data point is abnormal or not depends on the growth phase and time. In the proposed SVM-GMM method, the support vector machine is firstly used to classify diurnal and nocturnal data, the Gauss mixture model is secondly used to cluster these data respectively. After that, by screening the resulting clusters, abnormal data point will be detected.

3.1   Classification Phase Based on SVM

For one thing, regional, seasonal and other factors have an impact on the division of the day and night, for another thing, the differences are mainly reflected in carbon dioxide concentration and photosynthetic active radiation. It belongs to the problem of linear non separable, the classification strategy adopted in this paper is as follows.

Firstly, four features from cleaned dataset are selected to build the training set T, which are date, time, concentration of carbon dioxide (ppm), photosynthetically active radiation (μmol/s/m^2). $T = \{(x_1, y_1), (x_2, y_2), \ldots, (x_N, y_N)\}$ where x_i denote the four-dimensional training data, i.e. $x_i \in R^4$, y_i denote the label and $y_i \in \{-1, 1\}$, $y_i = -1$ and $y_i = 1$ represent nighttime data and daytime data respectively. Gaussian kernel function (Eq. 1) is used to map the samples in the original space to high dimension space, which can solve the problem of linear non separable in the original space.

$$K(x, z) = \phi(x)^T \phi(z) = exp\left(-\frac{||x - z||^2}{2\sigma^2}\right) \tag{1}$$

Secondly, in order to improve the robustness of the algorithm, slack variable ξ and l_1 regularization are introduced, and also the Lagrange multipliers $\alpha_i \geq 0, i = 1, 2, \ldots, m$ to get dual form of the problem (Eq. 2).

$$\max_\alpha \quad W(\alpha) = \sum_{i=1}^{m} \alpha_i - \frac{1}{2} \sum_{i,j=1}^{m} y^{(i)} y^{(j)} \alpha_i \alpha_j K\left(x^{(i)}, x^{(j)}\right)$$

$$s.t. \quad 0 \leq \alpha_i \leq C, \quad i = 1, \ldots, m \tag{2}$$

$$\sum_{i=1}^{m} \alpha_i y^{(i)} = 0$$

Thirdly, using stochastic gradient descent algorithm in Spark distributed computing framework to get the solution α_i^* of the optimization problem shown as Eq. (2). And then solve the parameters w^*, b^* of the separating hyperplane that determine the location of the hyperplane. The decision function is computed as follow (Eq. 3).

$$f(x) = sign\left(\sum_{i=1}^{N} \alpha_i^* y_i K(x^{(i)}, x) + b^*\right) \tag{3}$$

3.2   Clustering Phase Based on GMM

Both diurnal and nocturnal data obey an approximate Gaussian distribution, therefore Gaussian distribution model can be used as the basic model of each probability cluster, and clustering based on mixture models can be conducted. To avoid trapping into local optimum in the traditional EM algorithm, in which the number of clusters should be given a priori, in this paper, each data point will be viewed as a cluster at the beginning, and then the method iteratively optimize the model number c, the mixing proportions α_k and parameters θ_k of the k-th model until convergence.

To determine the optimal number of clusters, information entropy is introduced in the EM process. That is, regarding $-\ln \alpha_k$ as the information in the occurrence of one data point belongs to the k-th cluster, and $-\sum_{k=1}^{c} \alpha_k \ln \alpha_k$ is the entropy. Learning process can be used to estimate α_k by minimizing the entropy to get the most information for α_k. For this reason, $\sum_{k=1}^{c} \alpha_k \ln \alpha_k$ is used as a penalty term for the original objective function and the improved one (Eq. 4) is obtained.

$$J_{new}(\alpha, \theta) = \sum_{i=1}^{n} \sum_{k=1}^{c} \widehat{z}_{ki} \ln(\alpha_k) f(x; \theta_k) + \beta \sum_{i=1}^{n} \sum_{k=1}^{c} \alpha_k \ln \alpha_k \tag{4}$$

where β is the coefficient and $\beta \geq 0$.

By applying Lagrange Multiplier Approach and take the first derivative of the cost function with respect to α_k, and set it to be zero, we can derive the update equation for α_k (Eq. 5) and β (Eq. 6).

$$\alpha_k^{(new)} = \frac{\sum_{i=1}^{n} \widehat{z}_{ki}}{n} + \beta \alpha_k^{(old)} \left(\ln \alpha_k^{(old)} - \sum_{s=1}^{c} \alpha_s^{(old)} \ln \alpha_s^{(old)} \right) \tag{5}$$

$$\beta = \min \left\{ \frac{\sum_{k=1}^{c} \exp\left(-\eta n |\alpha_k^{(new)} - \alpha_k^{(old)}|\right)}{c}, \frac{1 - \max\limits_{1 \leq k \leq c} \left(\frac{\sum_{i=1}^{n} \widehat{z}_{ki}}{n} \right)}{\left(-\max\limits_{1 \leq k \leq c} \alpha_k^{(old)} \right) \sum_{k=1}^{c} \alpha_k^{(old)} \ln \alpha_k^{(old)}} \right\} \tag{6}$$

where $\eta = \min\left\{ 1, 0.5^{\lfloor \frac{d}{2} - 1 \rfloor} \right\}$.

The distributed Gaussian mixture model clustering algorithm used in this paper is stated as follows.

Step 1: Input the data set, initialize the mixture models, set the times of iterations $t = 0$ and other parameters $\beta^{(t)} = 1$, ε, view each data point as a model hence the number of models at the beginning is $c^{(t)} = n$, each mixing proportions $\alpha_k^{(t)} = \frac{1}{n}$, mean value $\mu_k^{(t)} = X$, and covariance matrix $\Sigma_k^{(t)} = d^2_{k(\lceil \sqrt{c_{initial}} \rceil)} I_d$. Where $\left\{ d^2_{k(1)}, d^2_{k(2)}, \ldots, d^2_{k(n')} \right\} = sort\left\{ d^2_{ki} = \|x_i - \mu_k\|^2 : d^2_{ki} > 0, i \neq k, 1 \leq i \leq n \right\}$ and I_d is an $d \times d$ identity matrix.

Step 2: Estimate the probability of each data point belongs to each model $z_{ki}^{(t)}$.

Step 3: $t = t + 1$, corresponding to the EM process, during which internationally updating and solving the model. And this can be divided into two stages: Map and Reduce. In Map stage, data in distributed file system will be allocated to each model, and parameters $(\mu_k^{(t)}, \Sigma_k^{(t)})$ of each Gaussian model are computed. In Reduce stage, by integrating calculation results in Map stage, $\alpha_k^{(t)}$ is computed by Eq. (5) and then we can get the numbers of data points belong to each model, discard the model with small proportion of mixture and empty model with no data point. After that, update $\beta^{(t)}$ by Eq. (6), the number of models and parameters $\widehat{z}_{k'i}$ and $\alpha_{k'}^{(t)}$.

Step 4: If the models are not stable, return to Step 3.

Step 5: Find the label of each sample and the model parameters according to $\widehat{z}_{(t)}_{ki}$.

3.3 Environmental Anomaly Detection

Usually, agricultural environment data points have similar characteristics, hence we can perform anomaly judgment on each class and find the abnormal classes. And because of the abnormal points are often small or remote cluster, the anomaly detection strategy can be taken, as shown in Fig. 1.

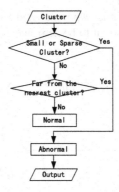

Fig. 1. Flow chat of anomaly judgment.

If the cluster is small or sparse, then it is determined to be abnormal directly. Else, determine whether the cluster is far from the nearest cluster according to the average distance between the clusters, if so, it is determined to be abnormal.

4 Experimental Results and Analysis

The experiment section is based on the 45,000 environmental data collected from tomato greenhouse stated above. Two kinds of anomaly detection methods are compared, only GMM and SVM-GMM. In the first experiment, three features are used in clustering stage and the same in the second experiment, which are temperature (°C), relative humidity (%), concentration of carbon dioxide (ppm). In the second experiment, four features are used in classification stage, which are date, time, and concentration of carbon dioxide (ppm), photosynthetically active radiation (μmol/s/m^2).

4.1 Experimental Environment

The experiments were carried out under the cluster environment built by two computers. One work as both the master node and the work node (Intel Core i3-6100 CPU

@3.70 GHz, RAM: 8 GB), the other one work as the work node only (Intel Core 2 Duo CPU E8200 @2.66 GHz, RAM: 4 GB).

In the production environment, Spark is mainly deployed in the Linux operating system and rely on JDK as well as Scala, so Ubuntu 14.04, jdk1.8 and Scala-2.11.8 were firstly installed on the two machines. Secondly, in order to facilitate the master node to send commands to the work node, Secure Shell (SSH) was configured to both node. Finally, Hadoop 1.2.1 and Spark 1.6.1 were installed respectively.

4.2 Anomaly Detection Based on GMM Clustering

After preprocessing, distributed GMM clustering approach is conducted. The dataset is grouped in 18 clusters eventually thereafter iterative calculation. Figure 2 is the visualization of clustering results, data points in the one class are represented in the same color. Three coordinates represent the temperature (°C), relative humidity (%), concentration of carbon dioxide (ppm) respectively.

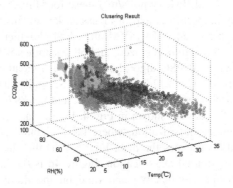

Fig. 2. Result of clustering.

According to the results of the above distributed Gauss mixture model, it is found that the abnormal clusters are small, and there are more data points in the normal clusters.

Based on clustering results, the small clusters, sparse clusters and remote clusters are regarded as outliers, and the visualization result of anomaly detection is obtained, shown as Fig. 3 and the dark blue dots represent abnormal data. Compare with Fig. 2, we can see that, outliers are distributed on the edge of the dataset, which consistent with the characteristics of the environmental anomaly in the crop growth procedure.

Single stage Gauss mixture model clustering method can only detect global anomaly and is incapable to correctly detect the anomaly if the feature apparently change over time, territory and other factors. For example, when the temperature value (°C) is 14.76 ± 0.075, which is judged to be normal in the method. However, in practical production environment, if the sample point is acquired in daytime, it should be viewed as the abnormal one, vice versa. To sum up, the single Gauss mixture model clustering can only detect global outliers, and the condition attribute set is empty. The average accuracy of detection is only 88.6%.

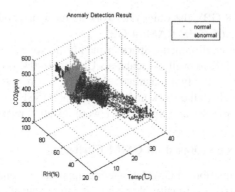

Fig. 3. Result of environmental anomaly detection. (Color figure online)

4.3   Anomaly Detection Based on SVM-GMM

Date, time, CO2 (ppm) and PAR ($\mu mol/s/m^2$) these four-dimension datasets are input to achieve classification. With the help of Gaussian kernel function and the machine learning library MLLib as well as LIBSVM, the SVM classifier is trained.

Firstly, the input data were normalized, and 60% samples were randomly selected for model training. We used PSO algorithm and cross validation dataset to optimize parameters, eventually get C = 212.0, $1/2\sigma^2$ = 2.0. Secondly, training support vector machine using stochastic gradient descent method, and get 897 support vectors. Parts of results in 2D visualization are shown in Fig. 4. Figure 4a, b are the relation of date and time with the diurnal classification, where the horizontal coordinates denote the time, and the vertical coordinates the date. Figure 4c is the effects of carbon dioxide (ppm) and time on the diurnal classification. Figure 4d is the influence of photosynthetically active radiation ($\mu mol/s/m^2$) and time on the diurnal classification. The horizontal coordinate is the time, and the vertical coordinate is the PAR. We can roughly find that during the daytime, the PAR rises and reaches its highest value at noon, while after noon, it decreases gradually and ultimately get about zero at night.

The remaining 40% samples were used to test the trained support vector machine classifier. Then we get the discrimination accuracy of day and night were 99.96% and 98.07%, respectively, and the average accuracy was 99.01%.

This distributed GMM clustering experiment was based on the classification of the dataset. There were 20855 diurnal data and 24768 nocturnal data. After clustering, 9 clusters were formed for daytime data while 13 clusters were formed for nighttime data. Figure 5 is the visualization of clustering results, data points in the one class are represented in the same color. Three coordinates represent the temperature (°C), relative humidity (%), concentration of carbon dioxide (ppm) respectively.

The suitable temperature for tomato in daytime during the growth stage is about 25 °C and should not fluctuate exceed ±5 °C, while in nighttime is about 15 °C. The appropriate relative humidity should ranges from 80% to 90% and should not exceed 95%. Carbon dioxide concentration should be more than 280 ppm. Combining with the parameters of the models (mean, variance, number of samples belong to the model), the information of each model was obtained. It is easy to find that the characteristics of

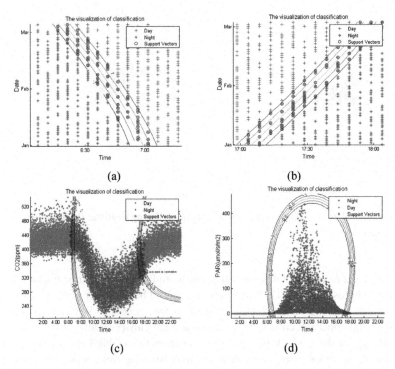

Fig. 4. Result of classification, (a) date-time; (b) time-PAR; **c** time-CO2.

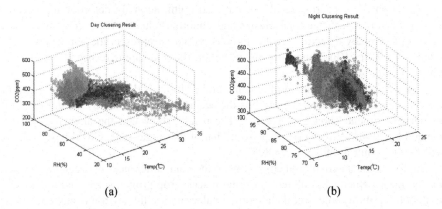

Fig. 5. Result of clustering, (a) during daytime; (b) during nighttime.

each cluster are relatively clear, which have practical significance to the analysis of agricultural environmental data.

If observe on the right side of Fig. 6a, we can see a few sparse samples gradually deviate from the main part of the dataset, towards the direction of high temperature and low humidity distribution, it indicated that the evaporation increases at higher temperature during daytime, in turn the relative humidity is lower. High temperature may

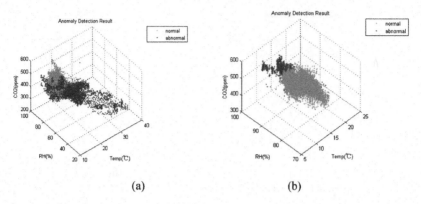

Fig. 6. Result of environmental anomaly detection, (a) during daytime; (b) during night. (Color figure online)

breed pests and cause plant diseases, and low humidity may lead to stomatal closure in turn affect photosynthesis. It is hazardous and should be avoid for crops to grow in this environment for long time which will lead to leaf necrosis. We can see that from Fig. 6b on the upper left corner, that, there is a small cluster significantly deviated from the main part of data, which presents the characteristics of high temperature and high humidity. If the crops grow in this environment for long time, it is prone to generate fog, breed bacteria and induce disease. It may also reduce crop transpiration, which is not conducive to crop growth.

Based on the clustering results of day and night, anomaly detection algorithm is used to distinguish the normal cluster and abnormal cluster. Figure 6 is the visualization of anomaly detection results, three coordinates represent the temperature (°C), relative humidity (%), concentration of carbon dioxide (ppm) respectively and dark blue dots represent abnormal data.

The actual number of abnormal points in the cluster is determined according to the environmental suitable value. And then the accuracy is calculated by the detect results of our experiment, shown as Table 1. The results show that the anomaly detection method based on SVM-GMM outperform that of the single-stage GMM based method and the detecting accuracy is 93.80%.

In the practical agricultural environment, there are differences between the environmental measurements in different time and places, crop growth stage, and this method has good feasibility to detect the abnormal value in specific environmental conditions. For instance, during daytime, the dataset labeled 1 and 2 which temperature (°C) is 16.05 ± 0.63 and 15.44 ± 1.45, should be treated as abnormal because of being low during the day. However, if these data are collected during nighttime, then should be viewed as normal ones. From the above, the two-stage anomaly detection method can effectively detect these abnormal data points and improve the detection accuracy.

5 Conclusions

This paper investigates the novel method that can be used to detect greenhouse environmental anomaly data accurately. Based on distributed classification and clustering a two-stage anomaly detection method is presented by using the support vector machine for classification and Gauss mixture model for clustering, with Spark distributed framework for iterative processing, Support vector machine is firstly used to classify diurnal and nocturnal data together with Spark stochastic gradient descent method to solve the optimization problem. Secondly, the Gauss mixture model is used to cluster the data separately by updating the model through the Spark iteration until the model is stable. Finally, the result of clustering is used to detect environment abnormal values. Compared the detect results of this method with single-stage clustering algorithm, the proposed detection method has higher accuracy, and is efficient to process large amounts of distributed agricultural data.

Acknowledgments. This work is supported by the Key Project of Science and Technology Commission of Shanghai Municipality under Grant No. 14DZ1206302. The authors would like to thank editors and anonymous reviewers for their valuable comments and suggestions to improve this paper.

References

1. Yongwha, C., Seunggeun, O., Jonguk, L., et al.: Automatic detection and recognition of pig wasting diseases using sound data in audio surveillance systems. Sensors **13**(10), 12929–12942 (2013)
2. Yasami, Y., Mozaffari, S.P.: A novel unsupervised classification approach for network anomaly detection by k-Means clustering and ID3 decision tree learning methods. J. Supercomput. **53**(1), 231–245 (2010)
3. Chitrakar, R., Huang, C.: Anomaly detection using support vector machine classification with k-Medoids clustering, pp. 1–5 (2012)
4. Varuna, S., Natesan, P.: An integration of k-means clustering and Naïve Bayes classifier for intrusion detection. In: International Conference on Signal Processing, Communication and Networking. IEEE (2015)
5. Bilmes, J.B.: A Gentle Tutorial of the EM Algorithm and Its Application to Parameter Estimation for Gaussian Mixture and Hidden Markov Models. International Computer Science Institute, Berkeley (1997)
6. Figueiredo, M.A.T., Jain, A.K.: Unsupervised learning of finite mixture models. IEEE Trans. Pattern Anal. Mach. Intell. **24**(3), 381–396 (2002)
7. Yang, M.S., Lai, C.Y., Lin, C.Y.: A robust EM clustering algorithm for Gaussian mixture models. Pattern Recogn. **45**(11), 3950–3961 (2012)
8. Zaharia, M., Chowdhury, M., Franklin. M.J., et al.: Spark: cluster computing with working sets. In: Usenix Conference on Hot Topics in Cloud Computing. USENIX Association, pp. 1765–1773 (2010)
9. Hou, J.L.: Study on Model to Greenhouse Tomato Growth and Development. China Agricultural University, Beijing (2005)
10. Liu, C.: Research on the Diagnosis of the Soil Volumetric Water Content During Tomato's Growth Period in Greenhouse. Graduate University of Chinese Academy of Sciences, Beijing (2012)

Building a Virtual Reality System for Intelligent Agriculture Greenhouse Based on Web3D

Qun Huang[1,2], Li Deng[1,2(✉)], Minrui Fei[1,2], and Huosheng Hu[3]

[1] School of Mechatronics Engineering and Automation, Shanghai University, Shanghai, China
dengli@shu.edu.cn
[2] Shanghai Key Laboratory of Power Station Automation Technology, Shanghai University, Shanghai, China
[3] School of Computer Science and Electronic Engineering, University of Essex, Colchester, UK

Abstract. The inevitable trend of agricultural development in China is to realize real-time monitoring, visualization and management of intelligent agriculture greenhouse. This paper presents a virtual reality system that is developed for an intelligent agricultural greenhouse using Web3D. Both virtual reality and image matching technology are integrated to construct such a VR system. The basic roaming interactive function of the system is realized using 3DS MAX modelling technology and virtual reality platform (VRP). An improved SIFT algorithm is proposed to complete panoramic image mosaic, which is on the basis of SIFT feature matching algorithm combined with Harris algorithm. Finally, the 3D visualization of agricultural greenhouse, the data management and real-time interactive control are realized and tested.

Keywords: Web3D · Virtual reality platform · Intelligent agriculture greenhouse · Panoramic image mosaic

1 Introduction

With the advance of science and technology in general and robotics technology in particular, the development of advanced agriculture facilities becomes the main trend of modern agricultural development worldwide. Agricultural greenhouses are one of advanced agriculture facilities, and their real-time monitoring and management is an urgent task for their successful deployment. This in turn has attracted many researchers worldwide to work on the challenge tasks.

To build a real-time monitoring and visualization for agriculture greenhouses, virtual reality technology could be used to generate realistic specific virtual scenes, and update the image information in the virtual scene in real time. In this way, users could achieve a combination of virtual and reality views through the scene roaming interaction and other functions. In the 1960s, researchers began to use computer simulation of plant growth process, such as a general framework for simulating the plant in 1968 [1].

K. Li et al. (Eds.): LSMS/ICSEE 2017, Part III, CCIS 763, pp. 790–799, 2017.
DOI: 10.1007/978-981-10-6364-0_78

After that, Honda realized the tree structure of the computer simulation of L system for the first time [2], which is a commonly used method of building virtual plant model.

Recently, Web3D based virtual reality technology gets rapid development. Many classic and famous Web3D virtual reality systems have been developed internationally. World Wind developed by NASA is a free and open source API for a virtual globe. It allows developers to quickly create interactive visualizations of 3D globe map and geographical information. Web3D refers to all interactive 3D content which are embedded into web pages. It allows users to watch any corner of the street and on the earth image, achieving a dream of travelling around the earth without leaving their homes. Google Street View [6] is produced by combining 3D map of the city Street View software systems, giving users the scene to see 360° View.

In general, 3D models not only require the storage space, but also impacts on its transmission and download over the network. It is of great significance to build efficiency, realistic and ease of use 3D virtual reality scene model. The image panoramic view of Web3D is currently the most widely used technology, and the fastest growing in the field of virtual reality. Panoramic view of the virtual scene building is to use the computer to build a realistic virtual environment by using the image splicing technology [7, 8], image registration, and the classic feature extraction algorithm with Harris corner, SUSAN corner point and SIFT feature point algorithm.

He et al. put forward a kind of high real time capability F-SIFT stitching algorithm [9], which is based on SIFT algorithm. It uses the phase correlation method in frequency domain image blocks, maintaining a fast operation characteristics and the accuracy of the SIFT algorithm. Ali et al. proposed a comprehensive SIFT and SURF algorithm for handling image scale and rotation changes and the illumination changes, which was highly robust and had the improved accuracy of image matching [10].

This paper integrates the virtual reality and image stitching technologies to build a virtual reality system for an intelligent agricultural greenhouse using Web3D. For achieving local strong interactivity and immersive virtual agricultural scenarios, we build lightweight 3D models. Image building is used to build the large-scale virtual scene in order to achieve small size of files and the fast download speed.

The rest of this paper is organized as follows. Section 2 describes the system VR design for an intelligent agriculture greenhouse based on Web3D. The static scene of agricultural virtual reality is investigated in Sect. 3, including a model for greenhouse environment, a model for Patrol robot, and a model for plant growth. Section 4 presents the dynamic function design of agricultural virtual reality. In Sect. 5, the panoramic image stitching technology is deployed based on Web3D. Section 6 presents the system test and analysis to show the feasibility and performance. Finally, a brief conclusion and future work are given in Sect. 7.

2 System Design

In our intelligent agricultural greenhouse, we consider the intelligent patrol robot as a core system, combined with fixed-point micro environmental monitoring sensors that monitor plants to collect data and image used in virtual reality system information. Greenhouse environmental condition is monitored and controlled in real time.

The collected data is stored in the database, which is used for updating virtual reality scene and model in real time. Therefore, users could view and monitor plant growth through 3D visualization of agricultural greenhouse, as well as for further analysis of data and interactive control.

The virtual agricultural greenhouse is developed using the Web3D virtual reality technique, which is mainly include virtual agricultural plants, scenes, virtual robots, as well as panoramic image mosaic technique. System structure is shown in Fig. 1. Agricultural virtual reality building part is mainly established through a static scene 3DS MAX scene model. Then the model was optimized and map, adding the UVW map coordinates. After that, animation work includes tomato plants growth and robot motion of rigid body, to make good models and animations for roasting. Therefore, the scenario models are imported into VRP, realizing the construction of virtual scene.

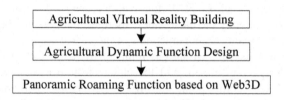

Fig. 1. System structure

Agricultural dynamic function design part of virtual reality is based on the realistic virtual scene design and interaction function by writing VRP script file. The system reads automatically database information, conducts real-time scene updating, and monitors the robot movement, to achieve real-time roaming system interaction.

Considering the agriculture building, there are two problems to be considered in the Web3D virtual reality system: (i) realistic scene and perfect function, (ii) fast Internet download speed. For a large-scale virtual panorama, it is necessary to adopt the Web3D virtual reality technology based on image. Improved SIFT algorithm is put forward, which is based on the SIFT feature matching algorithm combined with Harris algorithm. The advantages to adopt this algorithm include: improving the accuracy of image matching and quickness, completing the panoramic image mosaicking, realization of virtual reality system based on the Web3D panoramic roaming function.

3 Static Scene of Agricultural Virtual Reality

3.1 A Model of Greenhouse Environment

A greenhouse environment mainly includes the construction of a greenhouse building and the external environment rendering. First of all, according to the actual situation of aspect ratio, shape and structure, it is necessary to establish virtual reality scale relations, and then complete the layout of the environment in greenhouse in the AutoCAD floor plan. The floor plan will be imported to 3DS MAX, getting a 3DS model for layout, and a model for other single merger is added to the specified location. Finally, the texture images obtained from field are assigned to the corresponding material model. The designed greenhouse construction model is shown in Fig. 2.

The construction of external environment is mainly to increase the realistic effect of the scene, including the surrounding environment and the sky. The system uses the VRP box module simulation in the sky, importing the scenario model to the VRP, choosing the appropriate sky boxes with solar halo for setting, adding 3D shadows.

Fig. 2. Greenhouse construction model

3.2 Patrol Robot Model

The patrol robot system mainly consists of four parts: mobile robot platform, multi-axis manipulator, special actuators, and smart terminals. A multi-axis manipulator is fixed on the mobile robot, which has special actuators and smart terminals. The robot system has a camera and can complete image acquisition. It can travel along the guide rail to monitor the growth states of plants.

The designed mobile robot model is shown in Fig. 3, which is composed of mechanical, power and control systems. With them, it is possible to complete tour of the robot's motion control and information transmission and storage. According to practical requirement, a 6 DOF mechanical arm has fixed a smart sensor system to complete monitoring functions. The smart sensor system can collect soil PH data and humidity data. Smart terminals carry temperature sensors and CCD sensor to detect plant leaf conditions.

Fig. 3. Patrol robot model

3.3 Plant Growth Model

The current crop growth modelling method can be divided into empirical modelling and causal modelling. Empirical modelling of crop growth includes data statistics, simulation of crop growth trend; Causal modelling of crop growth includes the forecasting and the growth inherent law.

The tomato plant 3D model uses the empirical modelling of plant growth status related data and the collected image information. The 3DS MAX simulated building model is used to represent tomato plant growth process, the different stages of plant and the animation showing tomato plants to grow, as shown in Fig. 4.

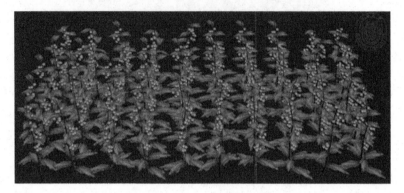

Fig. 4. Plant growth model

4 Dynamic Function Design of Agricultural Virtual Reality

4.1 Local Agricultural Roaming

Local agricultural roaming is to use real data to build a virtual scene, immersive users to experience a virtual intelligent agricultural greenhouse in real-time. This system has designed three buttons which use scripting language. Besides, three roaming ways are adopted, such as agricultural greenhouse tomato plants to grow scene roaming and patrol robot movement scene roaming. The overall scene roaming is realized by a flying camera. It is important to find the right point of view in the scene, and then create a flying camera, adjusting the camera parameters, writing VRP script file playback speed control tomato plants. Some methods are used to complete plant growth animation recording in a complete cycle such as branches and leaves and fruit of different colours and forms.

On the other hand, a fixed-point observation camera is used to watch the movement of the patrol robot. There are a number of actions to be taken, i.e. adjusting the camera parameters, writing VRP script file to control the robot motion, and collision detection. It is necessary to record the motion work scenarios of the animated patrol robot along the greenhouse, including dedicated actuators and mechanical arm operations, and the collected image data.

4.2   Interaction Design

To achieve real-time interactive performance, our virtual reality system should allow users to get feedback from the virtual reality scene and control the patrol robot by using the mouse and keyboard devices. In other words, the patrol robot could be controlled in real-time, and the sense of reality is realized via scene interaction function. Therefore, users can achieve immersion scene.

The first person perspective of virtual characters is under user controlling, including creating the role control camera, binding characters, designing role control button and writing VRP script file, adding the trigger function, opening the collision detection. Besides, it allows users to walk through the mouse and keyboard to control the role in the scene movement, walk around obstructions, and realize dynamic observation of agricultural environment and plant growth state in the scene.

Robot control includes the robot movement controlled by either user or the commands in its database. To control the robot movement, the user firstly selects the appropriate camera angle, sets the camera parameters, and then writes VRP script file to control the robot motion animation. Some settings include broadcast time frame and playback speed. The user could use a mouse to click robot control buttons in order to control the patrol moving to monitor the plant growth process.

In a VRP platform, the database access tools and the LUA script file enable us to access database information and control the robot movement. The system adopts the VRP database access tools to implement database access, Microsoft SQL Server as the background database system. By using a greenhouse sensor to monitor robot motion state in real time, the collected data is stored into the corresponding fields of database. In our virtual reality system, the VRP access tool is used to access the related field of robot motion state in the database. By writing good VRP script files, we could realize the real-time monitoring of the robot movement.

5   Panoramic Image Stitching Technology Based on Web3D

Agricultural greenhouse scene includes a lot of facility. In order to achieve the realistic effectiveness, we need to build a large number of models to represent the scene, which will greatly increase the computing cost and lead to the decrease of the scene running speed and the increased network download time.

Therefore, building a virtual reality system for an agriculture greenhouse can use the Web3D virtual reality technology which is based on the model and image. For interactive and immersive virtual scene, we use lightweight 3D modelling and the system function design to implement scenario updating in real time and interactive control. For a large-scale virtual view, we can use images to build a panoramic view. Harris corner detection algorithm of feature point detection is very effective, but it is sensitive to changes in scale. The SIFT algorithm (Scale Invariant Feature Transform) can solve the problem well. In order to improve the image registration and image matching speed and accuracy, improved SIFT algorithm is used here, which is based on Harris algorithm combining with the SIFT feature matching algorithm.

SIFT feature descriptor is based on gradient feature point neighbourhood pixels. Neighbourhood window pixels change of gradient, the greater the feature descriptor uniqueness, the better, the greater the probability of correct matching. Because Harris corner within the neighbourhood window of pixel gradient changes significantly, which can improve the SIFT feature extraction algorithm proposed. First, according to the traditional SIFT algorithm, extract multi-scale space of extreme value point for the reference image and stay registration respectively. Then, if there is an obvious angular point in the neighbourhood, reserve that point as characteristics point, or remove this point. The detection of corners on different scale spaces is as follows:

$$C(x, \delta_I, \delta_S) = G(\delta_I) \times \begin{bmatrix} I_x^2(x, \delta_S) & I_xI_y(x, \delta_S) \\ I_xI_y(x, \delta_S) & I_y^2(x, \delta_S) \end{bmatrix} \tag{1}$$

$$cornerness = \det(C(x, \delta_I, \delta_S)) - \alpha \times trace^2(C(x, \delta_I, \delta_S)) \tag{2}$$

where, δ_I means the scale of the integral, δ_S means the scale of the feature points.

By detecting the presence of Harris corner point in neighbourhood feature points, we further remove an unstable extremum points to reduce the number and the significance of feature points. This could improve the accuracy and speed of matching feature points matching effectively.

We use the RANSAC method for precise matching, and at the same time get the transformation of relationship between images. A threshold value T is firstly set for NN/SCN (nearest neighbour domain/second nearest neighbour domain). If the value of a point is less than the threshold value of NN/SCN, we reserve that point and then select randomly in M with false matching points. Three matching points are used to calculate a linear transform matrix H.

After image registration, image transformation relationship is uniquely identified. Due to different brightness between images, the stitching images have light and shade changes on both ends of the suture line. The weighted average of the fusion method can be used for the image smooth transition. There is apparent juncture getting fused image with visual consistency. Assuming that f_1 and f_2 are two stitching images, f is the fused image.

$$f(x, y) = \begin{cases} f_1(x, y) & (x, y) \in f_1 \\ d_1f_1(x, y) + d_2f_2(x, y) & (x, y) \in f_1 \cap f_2 \\ f_2(x, y) & (x, y) \in f_2 \end{cases} \tag{3}$$

where d_1 and d_2 are weight values, related to the width of the overlapping area.

Besides, d_1 and $d_2 = 1$, $0 < d_1, d_2 < 1$. In the overlapping area, d_1 gradient by 1 to 0, d_2 gradient from 0 to 1, so as to realize the smooth transition in the overlapping area from f_1 to f_2, seamless Mosaic.

6 System Test and Analysis

3D Studio MAX (3DS MAX) is used as the modelling software and MAYA as the rendering software to realize the construction of agricultural virtual reality scene and model animation. VR-Platform (VRP) is used to complete the virtual reality interactive design. Using C++ language to achieve based on Open CV development kit image mosaic algorithm to improve, stitching panoramic images, to achieve Web3D agricultural greenhouse panoramic roaming function.

A model in 3DS MAX and animation is still not roaming and interactive function, which need to import them into the VRP of software functional design. By writing a script file and increasing interactive system, agricultural greenhouse environment can be controlled and monitored successfully. Among them, two or more objects between the collision detection are very important part in the design of virtual reality system. Besides, objects collide is also an important part of the system roaming interaction.

(a) Browse animation   (b) Growth animation

(c) Robot animation   (d) Role roaming

(e) User control   (f) Automatic control

Fig. 5. System test.

Three buttons are designed with scripting language in the system, namely browse animation, growth animation and robot animation. As shown in Fig. 5(a–c), overall scene roaming, the growth scene roaming of tomato plants and patrol robot movement scene roaming are realized by clicking the buttons.

In addition, three buttons, namely role roaming, user control and automatic control, are designed to realize dynamic functions as shown in Fig. 5(d–f). By clicking the role roaming button, user can walk through the mouse and keyboard as the first person perspective in the scene movement. User control button and automatic control button are designed to control the robot movement by user and the database information.

The improved SIFT algorithm is based on Harris algorithm combining with the SIFT feature matching algorithm. Figure 6 shows the source images of panoramic image mosaic. The comparison between SIFT result and improved SIFT result is shown in Fig. 7. Compared with SIFT algorithm, the improved SIFT algorithm proved to be faster and more effective in the image registration and image matching. With the result of improved SIFT, agricultural panoramic roaming function is realized on Web 3D.

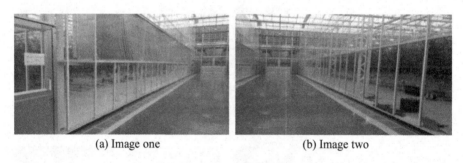

(a) Image one   (b) Image two

Fig. 6. Images to be matched.

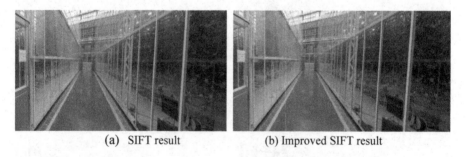

(a) SIFT result   (b) Improved SIFT result

Fig. 7. Comparison between SIFT result and improved SIFT result.

7   Conclusions

This paper developed a virtual reality system for an intelligent agricultural greenhouse using Web3D. Its scene model is firstly established using 3DS MAX, and then imported into VRP, realizing the greenhouse virtual scene roaming, the user control, the robot movement control, access database information, etc. As Web3D requires fast network download speed, image matching technology is used to realize agricultural panoramic roaming function. The experimental results show that the developed virtual reality system has realized real-time control, monitoring, visualization and management of an agricultural greenhouse.

Acknowledgments. This work is supported by Key Project of Science and Technology Commission of Shanghai Municipality under Grant Nos. 14DZ1206302 and 14JC1402200. The authors would like to thank editors and anonymous reviewers for their valuable comments and suggestions to improve this paper.

References

1. Ding, W., Jin, H., Cheng, Z., et al.: A visualization system for tomato plant modelling. In: 8th International Conference on Computer Graphics, Imaging and Visualization (CGIV), pp. 160–165 (2011)
2. Honda, H.: Description of the form of trees by the parameters of the tree-like body: effects of the branching angle and the branch length on the shape of the tree-like body. J. Theor. Biol. **31**(2), 331–338 (1971)
3. Wang, K.C., Ma, R.J.: Virtual reality technology and its application in agricultural machinery design. J. Syst. Simul. **18**(2), 500–503 (2006)
4. Takahashi, F., Abe, H., Koeduka, T., et al.: Development of design and operation supporting techniques for product inspection devices using virtual devices. In: The 14th IAPR International Conference on Machine Vision Applications (MVA), pp. 271–274 (2015)
5. McInerney, T., Tran, D.: Aperio: a system for visualizing 3D anatomy data using virtual mechanical tools. In: Bebis, G., et al. (eds.) ISVC 2015. LNCS, vol. 9474, pp. 797–808. Springer, Cham (2015). doi:10.1007/978-3-319-27857-5_71
6. Google, GoogleStreetView [EB/OL] (2013). http://www.google.com/intl/en_us/help/maps/streetview/
7. Yang, M.D., Chao, C.F., Huang, K.S., et al.: Image-based 3D scene reconstruction and exploration in augmented reality. Autom. Constr. **33**, 48–60 (2013)
8. Hauswiesner, S., Straka, M., Reitmayr, G.: Virtual try-on through image-based rendering. IEEE Trans. Visual Comput. Graph. **19**(9), 1552–1565 (2013)
9. He, B., Tao, D., Peng, B.: High real-time F-SIFT image mosaic algorithm. Infrared Laser Eng. **42**(52), 440–444 (2013)
10. Ali, N., Bajwa, K.B., Sablatnig, R., et al.: A novel image retrieval based on visual words integration of SIFT and SURF. PLoS ONE **11**(6), e0157428 (2016)

A Green Dispatch Model of Power System with Wind Energy Considering Energy-Environmental Efficiency

Daojun Chen[1](✉), Liqing Liang[1], Lei Zhang[2], Jian Zuo[1],
Keren Zhang[1], Chenkun Li[1], and Hu Guo[1]

[1] State Grid Hunan Electric Power Corporation Research Institute,
Changsha 410007, China
jihai77007@qq.com
[2] Hunan Xiangdian Test and Research Institute Company Limited,
Changsha 410007, China

Abstract. With rising pressure caused by decreasing natural resources, there emerges stronger call for environmental protection and sustainable development. As a kind of environmental-friendly energy, wind power has been increasing in capacity around the world in recent years. As more and more wind power gets connected with the grid, its impact on power dispatch should also be considered. Based on traditional power system optimized dispatch, this paper introduces the energy-environmental efficiency concept into construction of a green dispatch model for wind incorporated power systems. The strategy takes both minimum resource consumption and best energy-environmental efficiency as indexes to assess the optimization of wind power incorporated systems from an environmental-friendly point of view. Fuzzy technology is adopted along with the tabu search-based PSO algorithm to solve the problem. It is proven that the proposed model is reasonable and of good practicality.

Keywords: Wind power · Green dispatch model · Energy-environmental efficiency · Tabu search · Particle swarm algorithm

1 Introduction

With the global deterioration of energy crisis and environmental pollution, wind power has attracted more attentions as an inexhaustible energy source. Accordingly, optimal dispatch strategy for the wind power integrated systems is highly valued, as the integrated wind capacity increases [1–4].

Traditional power dispatch scheme aims to schedule the power generation properly in order to minimize generation cost within a certain period, while satisfying constraints like load balance, spinning reserves, and generation limits. A lot of research papers in this area have been published [5–10], but none of them involve the optimal dispatch problem after the wind power integrates into the system. In the dispatch model of the wind power integrated system, the impacts on the dispatch strategy caused by the stochastic and intermittent character of the wind should be considered from various aspects.

© Springer Nature Singapore Pte Ltd. 2017
K. Li et al. (Eds.): LSMS/ICSEE 2017, Part III, CCIS 763, pp. 800–812, 2017.
DOI: 10.1007/978-981-10-6364-0_79

The pollution caused by thermal power generation has become a bottleneck that hinders the global sustainable development. Correspondingly, many countries have issued relevant policies on thermal plant pollutant emission. Thus, the influence on the environment brought about by the power generation should be concerned in the process of system's optimal dispatch. Under the premise of secure and stable operation, the more important should be focused on how to utilize wind energy to optimize the power source structure, minimize resource consumption and environmental harm. In other words, the purpose of wind power development and utilization is the carbon emission reduction so as to gain environmental benefits. Therefore, it is significant how to assess those benefits.

This paper introduces the energy-environmental efficiency concept to assure the security of the systems incorporating large scale wind power and maximum wind utilization. A green dispatch model is developed, which includes minimum generating resource consumption and optimal energy-environmental efficiency, while comprehensively considering constraints of load balance, spinning reserves, the ramp rate requirement and start/shut time of generators, and wind power penetration limits. Due to the huge dimension of multi-objective optimization problems and complexity in solving the equation set, the paper adopts the fuzzy theory and a new particle swarm algorithm that combines tabu search to attain better solutions. Finally, to validate the rationality of the green dispatch model and the feasibility of the modified PSO, numerical simulations are reported based on the IEEE30 system including wind power plants.

2 Green Dispatch Model Incorporating the Energy-Environmental Efficiency

The paper introduces the energy-environmental efficiency concept and establishes the green dispatch model from the resource consumption and energy-environmental efficiency prospective.

2.1 Object Function

The traditional dispatch objective in a mixed system is to obtain an optimum allocation of power output among the available generators within given constraints. Minimum resource consumption still counts in power dispatch of the system with wind-powered generators. At any given time the objective function is as follows:

Minimize

$$\sum_{t=1}^{T}\sum_{i=1}^{G}(f_{it}(P_{it})+(1-I_{it-1})S_{it})I_{it} \tag{1}$$

where

$$S_{it} = \delta_i + \sigma_i \left(1 - e^{\left(-T_{it}^{off}/\tau_i\right)}\right) \tag{2}$$

where G is number of conventional power generators. T is scheduling period in hours; P_{it} is power from i th conventional generator at time t. I_{it} is status of generator i at time t, in this paper, it is ON('1')/OFF('0'). S_{it} is start-up cost of generator i at time t; δ_i, σ_i, τ_i is the start-up cost coefficients of generator i.

For the gradient-based optimization method, smooth quadratic function approximation of fuel cost function is needed to guarantee the optimal solution. However, it is not necessary for the proposed algorithm. The fuel cost function of each unit in this paper includes valve point loading effects [11, 12] and can be expressed as

$$f_{it}(P_{it}) = a_i + b_i p_{it} + c_i p_{it}^2 \tag{3}$$

where $f_{it}(P_{it})$ is individual generation production cost in terms of real power output P_{it} at time t. a_i, b_i, c_i are the constants of fuel cost.

The main problem traditional thermal plants face is the harmful effects of toxic gases emission. The paper takes CO_2, SO_2 and NO_x into consideration, based on the emission quantities per unit of useful energy produced. The higher emission limits of each gas are shown in Table 1.

Table 1. Maximum admissible concentration for some harmful gases in the atmosphere of the work place

Gas denomination	Average concentration (mg/m^3)	Maximum accepted concentration (mg/m^3)
Carbon dioxide	7000	10000
Sulphur dioxide	10	15
Nitrogen oxides	–	10

SO_2 and NO_x emission can be converted in equivalent CO_2 emission:

$$E_{CO_2} = C_{CO_2} + 700(C_{SO_2}) + 1000(C_{NO_x}) \tag{4}$$

In the above equation, E_{CO_2} is denoted as the "carbon dioxide equivalent", representing the emission per unit coal combustion. The unit of E_{CO_2} is kg/kg.

From the engineering thermodynamics perspective, the amount of heat emitted by unit fuel combustion is related to plant efficiency and power generation form. The former is defined as η_e, which equals the ratio of power output of generators to heat input per unit time.

The operation condition of the thermal generators affects η_e as well. To analyze the energy-environmental efficiency of the thermal units under different operation conditions, considering a functional relation between η_e and power generated by thermal generators denoted as P_{it}, the energy-environmental efficiency index is defined as follow [5]:

$$\varepsilon_{eve} = \frac{\eta_{ie}\eta_r(P_{it})\theta_\alpha}{\eta_{ie}\eta_r(P_{it})\theta_\alpha + kE_{CO_2}} \tag{5}$$

where ε_{eve} is energy-environmental efficiency index. η_{ie} is the efficiency of the ith thermal generator. θ_α is the net calorific value. k is heat loss coefficient of pollutant emission. $\eta_r(P_{it})$ is the functional relation between plant efficiency η_{ie} and P_{it}.

Literature [13] concludes that the best value of k is approximately 2 considering standard coal combustion. This conclusion is derived by comparison between the burning of hydrogen and coal.

$$\eta_r(P_{it}) = \gamma_i + \beta_i P_{it} + \alpha_i P_{it}^2 \tag{6}$$

where $\eta_r(P_{it})$ is the functional relation between plant efficiency η_{ie} and P_{it}. $\alpha_i, \beta_i, \gamma_i$ is coefficients of the efficient function.

The index ε_{eve} that refers to energy-environmental efficiency is dimensionless, whose value varies between the limits of unity and zero. The larger the value of ε_{eve}, the better the energy-environmental efficiency.

The specific energy-environmental efficiency model the paper proposed is as follows:

Maximize

$$F_2 = \max_{t \in T} \sum_{i=1}^{G} \frac{\eta_{ie}\eta_r(P_{it})\theta_\alpha^i}{\eta_{ie}\eta_r(P_{it})\theta_\alpha^i + 2E_{CO_2}^i} \tag{7}$$

where P_{jt} is power from jth wind-powered generator at time t.

The energy-environmental efficiency describes total energy-environmental efficiency of operating units in the th period. It refers to the pollution level of resource consumption brought by power generation per unit time. A larger value of the index refers to a better energy-environmental efficiency.

The green dispatch model for power system with wind power generator mentioned in this paper is constructed by Eqs. (1) and (7). By building an energy and environment beneficial function which breaks through the limitations of traditional dispatch model that concerns only about fuel consumption of thermal plants, the model deals with power dispatch problems from an environmental-friendly perspective. It focuses on the study of the influence on ecological environment.

2.2 Constraints

System power balance, ignore system losses power

$$\sum_{i=1}^{G} p_{it} + \sum_{j=1}^{M} p_{jt} = P_{Dt}, \quad t \in T \tag{8}$$

where P_{jt} is power from jth wind-powered generator at time t. M is number of wind-powered generators. P_{Dt} is system load demand at time t.

Spinning reserve requirement

$$\sum_{i=1}^{G} \left(P_{it}^{\max} - P_{it} \right) \geq \eta_1 P_{Dt}^{\mathrm{up}}, \quad t \in T \tag{9}$$

$$\sum_{i=1}^{G} \left(P_{it} - P_{it}^{\min} \right) \geq \eta_2 P_{Dt}^{\mathrm{down}}, \quad t \in T \tag{10}$$

where $P_i^{\max/\min}$ is maximum/minimum generation limit of i th conventional generator. η_1, η_2 is represent upward/downward regulating power respectively. $P_{Dt}^{\mathrm{up}}/P_{Dt}^{\mathrm{down}}$ is corresponding system load when upward/downward regulating power is used. $P_{it}^{\max/\min}$ is maximum/minimum real power on the transmission lines.

Real power operating limits

$$P_i^{\min} \leq P_{it} \leq P_i^{\max}, \quad i \in G \quad t \in T \tag{11}$$

Generating unit ramp rate limit

$$P_{it} - P_{i(t-1)} \leq r_i^{\mathrm{up}} \Delta T, \quad i \in G \quad t \in T \tag{12}$$

$$P_{i(t-1)} - P_{it} \leq r_i^{\mathrm{down}} \Delta T, \quad i \in G \quad t \in T \tag{13}$$

where $r_i^{\mathrm{down/up}}$ is ramp-down and ramp-up rate limits of ith conventional generator. ΔT is the operational period of a conventional power unit (in this paper $\Delta T = 1$ h)

Real power limits on transmission lines.

$$P^{L\min} \leq P_l \leq P^{L\max} \tag{14}$$

where P_l is real power limit on transmission lines. $P^{L\max/L\min}$ is maximum/minimum real power on the transmission lines.

Wind power penetration limit

$$P_{wt} \leq \delta_w P_{Dt}, \quad t \in T \tag{15}$$

where δ_w is wind power penetration coefficient. P_{wt} is real power output of wind farms (difference between generators are neglected).

3   Mathematical Model for Wind Farm Power Output

The stochastic variation of wind power output is mainly affected by the fluctuation of wind speed and wind direction. Different types of wind generators located in the same wind farm have nearly the same wind speed and direction. Each of them can be viewed as identical for wind turbines within a certain wind farm, so it can be assumed that all of the wind generators within a certain wind farm have the same parameters. Thus a wind farm can be simulated by a single wind unit, with a wake coefficient of 0.9.

The real power output P_w of wind generators is dependant mainly on wind velocity. The relationship between P_w and the specific wind speed on the hub height v_w can be expressed by the piecewise function as follows [14]:

$$P_{\mathrm{w}} = \begin{cases} 0, & v_{\mathrm{w}} \leq v_{\mathrm{CI}}, v_{\mathrm{w}} \geq v_{\mathrm{CO}} \\ \frac{P_{\mathrm{R}}}{v_{\mathrm{R}}^3 - v_{\mathrm{CI}}^3} v_{\mathrm{w}}^3 - \frac{v_{\mathrm{CI}}^3}{v_{\mathrm{R}}^3 - v_{\mathrm{CI}}^3} P_{\mathrm{R}}, & v_{\mathrm{CI}} \leq v_{\mathrm{w}} \leq v_{\mathrm{R}} \\ P_{\mathrm{R}}, & v_{\mathrm{w}} \geq v_{\mathrm{R}} \end{cases} \qquad (16)$$

where P_R is rated power output of wind generators. v_{CI} is cut-in wind speed. v_{CO} is cut-out wind speed. v_R is rated wind speed.

Now, stator resistance and power loss in iron core of common cage induction generator can be neglected compared to the real power output P_w. Assume the real power output equals the mechanical power input, the reactive power output of wind generators can be expressed as follows [15]:

$$Q_w = \frac{V^2}{X_m} - \frac{V^2 - \sqrt{V^4 - 4P_w^2(X_1 + X_2)^2}}{2(X_1 + X_2)} \qquad (17)$$

where Q_w is reactive power output of the wind farm. X_m is the excitation reactance. V is terminal voltage. X_1 is the stator reactance. X_2 is rotor reactance.

The solid line in Fig. 1 is the average power output prediction curve in different periods of certain wind farm. Errors between the prediction value and the field statistics of power output of this wind farm over the past two years are given to calculate the mean absolute percentage error (MAPE), which is 21.38%. Thus, an allowable

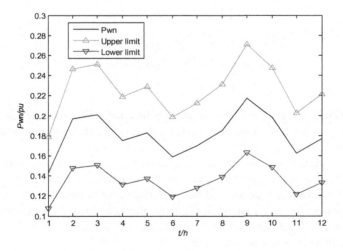

Fig. 1. Hourly average wind power and the rang of power fluctuating

deviation of ±25% is taken, which results in the upper and lower dashed in the figure. Assume that the average power output fluctuates between the two dashed lines, and then the range of wind power output is specified by them.

4   The Green Dispatch Model Processing and Solving

A. Fuzzy Modeling on the Dispatch Strategy

The approach to dealing with multi-objective optimal problem with the fuzzy optimization method is to convert the original problem into a single objective function. The key to the fuzzy modeling of the multi-objective function is to search for a proper membership function. The paper chooses the falling semi-liner style for the membership function for the resource consumption model of the generators and rising semi-liner style for the energy-environmental efficiency model [15]. The two models are represented by Eqs. (18) and (19) respectively. The functions are plotted in Fig. 2.

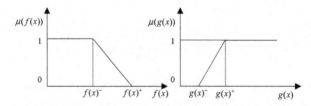

Fig. 2. Fuzzy membership function of smallest and largest

$$
\mu(f(x)) = \begin{cases} 1 & f(x) \leq f(x)^- \\ \frac{f(x)^+ - f(x)}{f(x)^+ - f(x)^-} & f(x)^- \leq f(x) \leq f(x)^+ \\ 0 & f(x) \geq f(x)^+ \end{cases} \tag{18}
$$

$$
\mu(g(x)) = \begin{cases} 1 & g(x) \geq g(x)^+ \\ \frac{g(x) - g(x)^+}{g(x)^- - g(x)^+} & g(x)^- \leq g(x) \leq g(x)^+ \\ 0 & g(x) \leq g(x)^- \end{cases} \tag{19}
$$

where $\mu(f(x))$ and $\mu(g(x))$ are the membership functions. $f(x)^+$ and $f(x)^-$ are the maximum and minimum of F_1 respectively. $g(x)^+$ and $g(x)^-$ are the maximum and minimum of F_2 respectively.

Set an expected reference membership degree μ_{rm} to each objective function and calculate the maximum absolute value of the difference between the membership degree of the objective function and the reference membership. Take the optimization rule of searching for the minimum of the above mentioned maximum absolute value to convert the non-linear multi-objective problem into a liner single-objective one to solve. The process is represented by Eq. (20):

$$\min\{\max|\mu_{rm} - \mu(F_m(x))|\}, m = 1, 2, 3$$
$$s.t. \begin{cases} \mu(F_m(x)) \geq \mu_{rm}, m = 1, 2, 3 \\ \text{Eq. (8)} \sim \text{Eq. (15)} \end{cases} \qquad (20)$$

B. Numerical Solution

(1) *Features of Particle Swarm Algorithm*

Kennedy and Eberhart first introduced the PSO method, which is an evolutionary computation technique [16, 17]. This method is derived from the social-psychological theory and has been found to be robust for solving problems featuring non-linearity and non-differentiability, and high dimensionality through adaptation [18, 19]. The fundamental idea is that the optimal solution can be found through cooperation and information sharing among individuals in the swarm.

(2) *Tabu Search Algorithm*

Tabu Search (TS) was introduced by Fred Glover as a high-level heuristic algorithm for solving combinatorial optimization problems [20–22]. The TS algorithm uses the past search to create and exploit better solutions. The main two components of the TS algorithm are the tabu list restrictions and the aspiration criterion.

(3) *Tabu Search-Based Refined PSO Algorithm*

Based on fuzzification of the model, the optimized dispatch discussed in this paper consist two issues of optimization: optimization of units combination and the loads allocation among given units. The standard PSO algorithm is easily trapped by the local optimum which frequently leads to prematurely. On the other hand, the tabu search algorithm distinguishes itself in local searching ability. It can skip from the local optimal solution and search for the global optimum. An obvious disadvantage of the tabu search algorithm is the rigid requirement on the initial values. The quality of those values directly affects the searching efficiency. On the contrary, The PSO algorithm does not emphasize on the initial search point. Therefore, the paper adopts a tabu search-based improved PSO to solve the green dispatch model, taking advantage of the complementary feature of the above mentioned two algorithms. It outweighs the standard PSO in both convergence rate and accuracy.

5 Case Study

The green dispatch model considering the energy-environmental efficiency and the tabu based PSO algorithm is tested on a typical IEEE-30 bus test system with 6-generators [23] for availability. The dispatch cycle is set 12 periods. The system contains a wind farm with 40 wind turbines. The rated power output is 0.45 pu. The cut-in wind speed, cut-out wind speed and rated wind speed are 3 m/s, 20 m/s, 14 m/s respectively. The excitation reactance X_m, stator reactance X_1, rotor reactance X_2 are set to 3.872, 0.012, 0.005 respectively, while the rotor resistance is set to 0.12, the rated is regulated to be $s_N = -0.005$. The wind farm is incorporated to the system in the nineteenth node, with a shunt capacitor of 10 Mvar to compensate for the reactive power. The spinning reserve is set to be 5% of the system load. The wind power penetration coefficient is represented by δ_w and regulated to be 7% of the system load. The control parameters of the algorithm are expressed as follows: the population size is set to 200; the maximum and minimum inertia weight factor ω_{max} and ω_{min} are set to be 0.9 and 0.3 respectively. Both the acceleration constant c_1 and c_2 are chosen to be 2, the maximum velocity V_{max} is regulated at 15 while the punishment factor λ is stipulated to 20, the maximum iteration K_{max} is set at 1000, the length of the tabu list is supposed to be 10. The initial reference membership value is chosen to be $\mu_{r1} = 0.7$, $\mu_{r2} = \mu_{r3} = 0.8$. During the process, the bus voltage is allowed to fluctuate with a secure range between 0.97 pu– 1.1 pu, while meeting the power balance constraint within the system. The base value of power is chosen to be 100 MVA. The system load statistics is in Table 2. Specifications of the fuels are listed in Table 3. Conventional generator parameters are shown in Table 4, where the serial numbers represent the node number that the generator is incorporated. The first generator serves as the balancing unit.

Table 2. Load demands

Hour	P_{Di}/pu	Hour	P_{Di}/pu	Hour	P_{Di}/pu
1	4.648	5	4.387	9	4.486
2	4.613	6	4.932	10	3.847
3	4.592	7	4.685	11	3.608
4	3.708	8	4.262	12	3.261

Table 3. Fuel characteristics of conventional generators

Unit no.	$E_{CO_2}^i$/(kg/kg)	θ_α^i/(MJ/kg)
1	23.9332	25.8627
2	38.9529	14.8726
5	34.2126	13.0714
8	40.5933	12.2880
11	42.0152	11.1917
13	44.1282	10.5303

Table 5 shows the power output of both conventional generators and wind turbines before and after the energy-environmental efficiency is considered. 1–6 corresponds to the conventional generators connected to the six buses. By comparison with the power output data in Table 5, it can be inferred that under the premise of load balance and secure and stable operation, the energy-environmental efficiency has significant influence on the dispatch strategy. For example, in the 10–12 periods with relatively low load, power output of those large capacity generators is reduced to support more spinning reserve capacity to the system. At the same time, small-capacity units are ensured to operate above the minimum stable power output condition. In this way, the energy-environmental efficiency of the whole system is improved, highlighting the advantages the green dispatch model has on environmental protection.

Table 6 shows the comparison of generation resource consumption between power dispatch strategies with and without the green dispatch model introduced. It can be

Table 4. Conventional generators' parameters

Unit no.	1	2	5	8	11	13
P_i^{max}/pu	4.00	1.30	1.30	0.80	0.55	0.55
P_i^{min}/pu	1.20	0.20	0.20	0.20	0.10	0.10
a_i/($/h)	663.3562	932.6582	876.7851	1235.2237	1332.3704	1658.1029
b_i/($/Mwh)	36.1938	45.6024	42.5818	38.2607	39.9277	35.2756
c_i/($/Mwh^2)	0.2048	0.1332	0.1298	0.0640	0.0254	0.0128
δ_i/$	4500	550	560	170	30	30
σ_i/$	4500	550	560	170	30	30
τ_i/h	4	2	2	2	1	1
r_i^{up}/(pu/h)	0.8	0.3	0.3	0.25	0.15	0.15
r_i^{down}/(pu/h)	0.8	0.3	0.3	0.25	0.15	0.15
η_{ie}	0.75	0.50	0.55	0.40	0.35	0.30
α_i/MW^{-2}	−0.0313	−0.2495	−0.1875	−0.1210	−0.7503	−0.6714
β_i/MW^{-1}	0.1375	0.4017	0.3775	0.3228	0.5301	0.4866
γ_i	0.8300	0.7466	0.7795	0.7496	0.7472	0.7520

Table 5. Comparison of generator outputs considering energy-environmental efficiency or not

Hour	Power output/pu energy-environmental efficiency not considered							Power output/pu energy-environmental efficiency considered							Energy-environmental efficiency
	1	2	3	4	5	6	P_{wt}	1	2	3	4	5	6	P_{wt}	
1	2.491	0.608	0.718	0.287	0.256	0.231	0.136	2.307	0.676	0.754	0.389	0.249	0.184	0.163	0.5704
2	2.473	0.695	0.697	0.338	0.154	0.183	0.151	2.418	0.591	0.764	0.339	0.227	0.161	0.188	0.5686
3	2.942	0.478	0.518	0.224	0.183	0.146	0.179	2.602	0.474	0.684	0.258	0.354	0.125	0.167	0.5649
4	2.303	0.332	0.481	0.221	0.178	0.118	0.140	2.062	0.440	0.546	0.220	0.214	0.115	0.144	0.5623
5	2.577	0.476	0.637	0.226	0.222	0.145	0.171	2.469	0.665	0.524	0.351	0.178	0.105	0.167	0.5653
6	2.953	0.433	0.651	0.321	0.289	0.179	0.185	2.561	0.654	0.728	0.493	0.172	0.208	0.189	0.5700
7	2.750	0.472	0.721	0.283	0.253	0.117	0.167	2.665	0.423	0.730	0.421	0.234	0.112	0.183	0.5658
8	2.357	0.563	0.686	0.221	0.195	0.132	0.168	2.267	0.651	0.609	0.318	0.187	0.141	0.154	0.5669
9	2.595	0.575	0.731	0.352	0.138	0	0.173	2.381	0.633	0.714	0.348	0.169	0.138	0.180	0.5679
10	2.272	0.481	0.631	0.369	0	0	0.153	2.051	0.456	0.650	0.467	0.136	0	0.152	0.5674
11	2.251	0.412	0.538	0.312	0	0	0.155	2.212	0.445	0.554	0.280	0	0	0.173	0.5585
12	1.742	0.424	0.547	0.354	0	0	0.154	1.809	0.526	0.540	0.279	0	0	0.161	0.5583

figured from Table 6 that considering the energy-environmental efficiency, the resource consumption of conventional generators changed from 49825.0508 $ to 49831.7544 $, which means consumption of resources doesn't change too much.

Table 6. Comparison of resource consumption considering energy-environmental efficiency or not

Green dispatch model	Resource consumption/$
Without energy-environmental efficiency	49825.0508
With energy-environmental efficiency	49831.7544

Figure 3 shows the comparison of the wind power output with and without the energy-environmental efficiency considered in the dispatch model. Curve 2 shows the situation when the energy-environmental efficiency is incorporated in the model while curve 1 corresponds to the one without energy-environmental efficiency considered. It can be figured from Fig. 3 that the power output specified by curve 2 outweighs that specified by curve 1 in all the 12 periods. This feature indicates that wind power is clean and environmental-friendly from the perspective of the energy-environmental efficiency. The value of the green dispatch model is also revealed. Additionally, the increase in wind power output improves the energy-environmental efficiency of conventional units and alleviates the impacts of generation pollutants on the environment. The energy-environmental efficiency included in the green dispatch model is represented more as the invisible capital, rather than the resource consumption of conventional units that can be estimated by specific values.

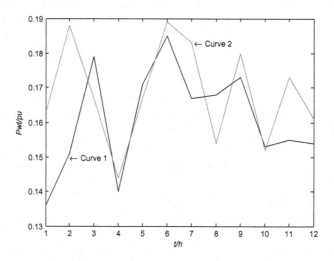

Fig. 3. Comparison of wind power output considering energy-environmental efficiency or not

By analysis of Tables 5 and 6, it can be inferred that when considering the energy-environmental efficiency, the green dispatch model will make little increase the resource consumption of conventional generator while at the same time improve the energy-environmental efficiency of the whole generating system. The increase this potential capital in turn, alleviates the harm to the environment by the power generation process. Thus, with the background of global development of low a carbon economy, energy-saving and emission-reduction, the green dispatch model displays a profound significance on optimal power dispatch.

6 Conclusion

The paper discusses the optimal dispatch strategy of wind power incorporated systems. In the modeling process, the energy-environmental efficiency concept is introduced and a multi-objective dispatch model for minimum resource consumption, the best energy-environmental efficiency and operation condition is established. From the biological and environmental perspective, the energy-environmental efficiency model focuses on the impact that power generation has on the ecological environment. The model accords with today's development strategy of promoting wind power to reduce environmental pollution and has positive influence on the dispatch of the existing power system. To search the optimal solution, fuzzy technology and tabu based PSO algorithm are adopted to improve the calculation precision. The test case verifies that the energy-environmental efficiency considered optimal multi-objective dispatch model for wind incorporated power systems the paper proposes has rationality and offers reference to traditional power generation and dispatch. Energy-environmental efficiency should become the heated issue in power system dispatch.

References

1. Damousis, I.G., Alexiadis, M.C., Theocharis, J.B., Dokopoulos, P.S.: A fuzzy model for wind speed prediction and power generation in wind parks using spatial correlation. IEEE Trans. Energy Convers. 19(2), 352–361 (2004)
2. Miranda, V., Hang, P.S.: Economic dispatch model with fuzzy wind constraints and attitudes of dispatchers. IEEE Trans. Power Syst. 20(4), 2143–2145 (2005)
3. Chen, C.L., Lee, T.Y., Jan, R.M.: Optimal wind-thermal coordination dispatch in isolated power systems with large integration of wind capacity. Energy Convers. Manage. 47(18–19), 3456–3472 (2006)
4. Eriksen, P.B., Ackermann, T., Abildgaard, H., Smith, P., Winter, W., Rodriquez Garcia, J. M.: System operation with high wind penetration. IEEE Power Energy Mag. 3(6), 65–74 (2005)
5. Marwali, M.K., Shahidehpour, S.M.: Coordination between long-term and short-term generation scheduling with net-work constraints. IEEE Trans. Power Syst. 15(3), 1161–1167 (2000)
6. Carrion, M., Arroyo, J.M.: A computationally efficient mixed-integer linear formulation for the thermal unit commitment problem. IEEE Trans. Power Syst. 21(3), 1371–1378 (2006)
7. Li, C.A., Jap, P.J., Streiffert, D.L.: Implementation of network flow programming to the hydrothermal coordination in an energy management system. IEEE Trans. Power Syst. 8(3), 1045–1093 (1993)
8. Kumar, S., Naresh, R.: Efficient real coded genetic algorithm to solve the non-convex hydrothermal scheduling problem. Electr. Power Energy Syst. 29(10), 738–747 (2007)
9. Yu, B.H., Yuan, X.H., Wang, J.W.: Short-term hydro-thermal scheduling using particle swarm optimization method. Energy Convers. Manag. 48(7), 1902–1908 (2007)
10. Damousis, I.G., Bakirtzis, A.G., Dokopoulos, P.S.: Network-constrained economic dispatch using real-coded genetic algorithm. IEEE Trans. Power Syst. 18(1), 198–205 (2003)
11. Yang, H.T., Yang, P.C., Huang, C.L.: Evolutionary programming based economic dispatch for units with nonsmooth fuel cost functions. IEEE Trans. Power Syst. 11, 112–118 (1996)

12. Walters, D., Sheble, C., Gerald, B.: Genetic algorithm solution of economic dispatch with valve point loading. IEEE Trans. Power Syst. **8**, 1325–1332 (1993)
13. Cardu, M., Baica, M.: Regarding a global methodology to estimate the energy-ecologic efficiency of thermopower plants. Energy Convers. Manag. **40**(1), 71–87 (1999)
14. Chen, H.Y., Chen, J.F., Duan, X.Z.: Study on power flow calculation of distribution system with DGs. Autom. Electr. Power Syst. **30**(1), 35–40 (2006)
15. Huang, C.M., Yang, H.T., Huang, C.L.: Bi-objective power dispatch using fuzzy satisfaction-maximizing decision approach. IEEE Trans. Power Syst. **12**(4), 1715–1721 (1997)
16. Kennedy, J., Eberhart, R.: Particle swarm optimization. In: IEEE Proceedings of the International Conference on Neural Networks, Perth, Australia, pp. 1942–1948 (1995)
17. Chen, D., Gong, Q., Zou, B., et al.: A low-carbon dispatch model in a wind power integrated system considering wind speed forecasting and energy-environmental efficiency. Energies **5** (4), 1245–1270 (2012)
18. Hirotaka, Y., Kenichi, K., Yoshikazu, F.: A particle swarm optimization for reactive power and voltage control considering voltage security assessment. IEEE Trans. Power Syst. **15**(4), 1232–1239 (2000)
19. Park, J.B., Jeong, Y.W., Shin, J.R., Lee, K.Y.: An improved particle swarm optimization for nonconvex economic dispatch problems. IEEE Trans. Power Syst. **25**(1), 156–166 (2010)
20. Saravuth, P., Issarachai, N., Waree, K.: Application of multiple tabu search algorithm to solve dynamic economic dispatch considering generator constraints. Energy Convers. Manage. **49**(4), 506–516 (2008)
21. Chen, D., Wang, X., Zuo, J.: A joint optimal dispatching method for pumped storage power station and wind power considering wind power uncertainties. Power Syst. Clean Energy **32** (8), 110–116 (2016)
22. Chen, D., Li, L., Yang, N.: Combination forecasting method of wind power based on optimal weight coefficient. Power Syst. Clean Energy **32**(4), 99–105 (2016)
23. University of Washington. Power systems test case archive [EB/OL]. http://www.ee. washington.edu/research/pstca/. 11 November 2010

Author Index